MECHANICS OF SOLIDS

MECHANICS OF SOLIDS

Pabitra Ranjan Maiti
Associate Professor
Department of Civil Engineering
Indian Institute of Technology (IIT)
Banaras Hindu University (BHU)
Varanasi (UP), India

Amitava Bhattacharya
Ex. Professor
Department of Mechanical Engineering
Indian Institute of Technology (IIT)
Banaras Hindu University (BHU)
Varanasi (UP), India

Vikash Khatri
Ex. Assistant Professor
THDC-IHET
Tehri (Uttarakhand), India

I.K. International Pvt. Ltd.
NEW DELHI

Published by
I.K. International Pvt. Ltd.
4435-36/7, Ansari Road, Daryaganj
New Delhi-110 002 (India)
E-mail: info@ikinternational.com
Website: www.techsarworld.com

ISBN 978-93-90620-69-2

Reprint 2024

Published/Printed by Krishan Makhijani for I.K. International Pvt. Ltd., 4435-36/7, Ansari Road, Daryaganj, New Delhi-110 002.

Preface

Mechanics of Solids is a basic engineering course that deals with the behaviour of solid bodies subjected to various types of loading. The basic objectives of this course are the determination of the stresses, strains and deformations produced by the loads. For a particular structure subjected to the given loading condition, what should be the suitable shape and size of the structure? What materials do we choose to make it safe, durable and economical? What should be the maximum load that one can apply on any structure? Mechanics of solids deals with the relations between externally applied loads and their internal effects on solid bodies. Moreover, the bodies are no longer assumed to be rigid; the deformations, however small, are of major interest in the subject matter of Mechanics of Solids. The study of mechanics of solids is aimed at predicting how geometry and physical properties of structure influence its behaviour under service condition. However, the term strength of materials most often refers to various methods of calculating stresses in structural members. To find the load carrying capacity of a material, the strength of the material has to be determined up to the failure of the structure. The basic interest is the complete mechanical behaviour of the material during loading process and up to its failure. It may be regarded as the statics of deformable or elastic bodies. The analysis is limited here up to elastic limit of the material.

The main objective of this book is to present the aspects of mechanics of materials in unified and integrated manner. The book has been divided in twelve chapters. In chapter 1, fundamentals of mechanics are discussed that emphasize on force equilibrium, drawing of free-body diagram and friction of simple problems. The mechanical behaviour of many materials is also presented graphically. The fundamental of stress and strain of materials body is discussed in details from chapter 2 to chapter 5. In chapter 2, the behaviour of material under axial loading is discussed. The stress-strain behaviour of one-dimensional structures, i.e., the structure shows only axial deformation subjected to tension, compression or thermal loads are also presented. Normal stress, axial strain, simple tension test, shear stress and strain, and strain energy under impact load is formulated and many solved problems are given. In chapter 3, analysis of two-dimensional stresses are focussed. Stresses at a point, Mohr's circle for plane stress and general stress system at a point is discussed. Analysis of strain, state of strain at a point, transformation of plane strain and strain rosset is dealt with in chapter 4. In chapter 5, stress-strain relation of three-dimensional body, plane stress and plane strain, relation between elastic constants and theories of elastic failure are discussed. A few applications of stress and strain in practical problems are discussed in chapters 6 to 8. Chapters 6 and 7 deal with the analysis of thin and thick pressure vessel. The behaviour of solid and hollow prismatic shaft under torsion is discussed in chapter 8. In chapter 9, bending of simple beams is discussed and bending moment and shear force diagram is presented for different types of loading and support conditions. The slope and deflection of simple beams are determined by different methods such as double integration method, moment area method, conjugate beam method, Castigliano's method in chapter 10. The bending normal and shear stresses of beams for different cross sections, unsymmetrical bending of beam are presented in chapter 11. Chapter 12 deals with the buckling of columns for simple columns for different end conditions.

Apart from this, a number of multiple-choice questions taken from different examination such as GATE, IES and Civil Services examinations are included in the appendix for the benefit of the students.

The authors will be grateful to his readers if any mistake in the book is brought to their notice or for any suggestions regarding the improvement of the book.

Pabitra Ranjan Maiti
Amitava Bhattacharya
Vikash Khatri

Contents

1
Introduction

1.1 INTRODUCTION

The primary objective of Mechanics of Solids is to develop a methodology about the response of material bodies to the applied forces. The applied forces may be external or internal to a body. Due to application of such forces, some quantities such as stress, strain, deformation, etc., develop in a body. These quantities can be determined in magnitude with the help of theoretical formulations and can be verified through experiments. For engineering the material bodies in the form of machine elements and structural components are of prime interest. The subject "Mechanics of Solids" finds its application in the design of structures and machine parts.

The strength of a material is the inherent property of a material to bear the load applied internally or externally on it up to a certain limit by virtue of its molecular arrangement. On application of load on deformable elastic bodies, the body changes its shape and size and deforms to resist the external force. This is the inherent property of the material to resist this external load by deforming itself. Deformation of the material is called *strain*, while the intensity of the internal forces is called *stress*. The strength of any material relies on three analytical methods: strength, stiffness and stability. Strength refers to the load carrying capacity, stiffness refers to the deformation or elongation, and stability refers to the ability to maintain its initial configuration.

The study of Mechanics of Solid is aimed at predicting how geometry and physical properties of structure influence its behaviour under service conditions. Knowing all the values of load up to failure load gives a complete picture of mechanical behaviour of the body. Beyond this limit the material will fail. The failure of the material does not mean the collapse of the structure. Elastic failure means yield stress is reached to all the points of the material and removing the forces the body will resume its original shape and configuration. The analysis of this book is limited up to the elastic limit of the material.

1.2 MECHANICAL PROPERTIES OF MATERIALS

The strength of a material is quantified by its elastic properties which can be found only experimentally. The relationship between load and the corresponding deformation for any material up to complete failure can be found by laboratory tests. The load deformation curve gives the complete picture of the strength behaviour of the material. Different materials show different load deformation curves. Stress-strain behaviour of any material can be predicted from this curve and different physical parameters have are derived to describe the mechanical behaviour of the material under applied loading.

The forces capable of deforming or changing the body temporarily (*elastic deformation*) or permanently (*plastic deformation* or *fracture*) are called deformable forces. Solid bodies which are acted upon by such forces are known as *deformable solids.* While studying mechanics of deformable solids we consider elastic behaviour of a material. So let us know what does elasticity means?

The property of a material to revert partially or completely to its initial shape after unloading is called *elasticity*. If the material completely recovers its initial length, the material is said to be perfectly elastic; if not, it is said to be only partially elastic.

The stress of a material is the maximum resistance which it can offer to the externally applied force.

When some external forces are applied to a body, then the body offers resistance to these forces. The magnitude of the internal forces is numerically equal to the applied forces. The internal resisting force per unit area is called *stress.*

Stress may be classified as *normal stress* (σ) and *shear stress* (τ). Shear stress acts parallel to the plane while normal stress is that component which acts perpendicular to the plane.

Stress, strain and stiffness are the three parameters related to the strength of any material. Stress is the internal forces developed due to external forces per unit cross-sectional area. Generally, stress is force per unit area. It can also be defined as resistance power of a body to resist external forces,

$$\sigma = \frac{F}{A}$$

Units of stress may be $\frac{\text{N}}{\text{m}^2}$ (or) pascal.

Elongation per unit length is defined as strain. Strain is a dimensionless quantity.

$$\varepsilon = \frac{\partial L}{L}$$

Stiffness decides the extent of deformation.

On the basis of stress-strain behaviour, materials are classified as *ductile* and *brittle.* Universal testing machine is used to find stress-strain of a bar. A device which measures the length elongated/extended is known as *extensometer.*

1.2.1 Stress-Strain Diagrams (Mild Steel) in Tension

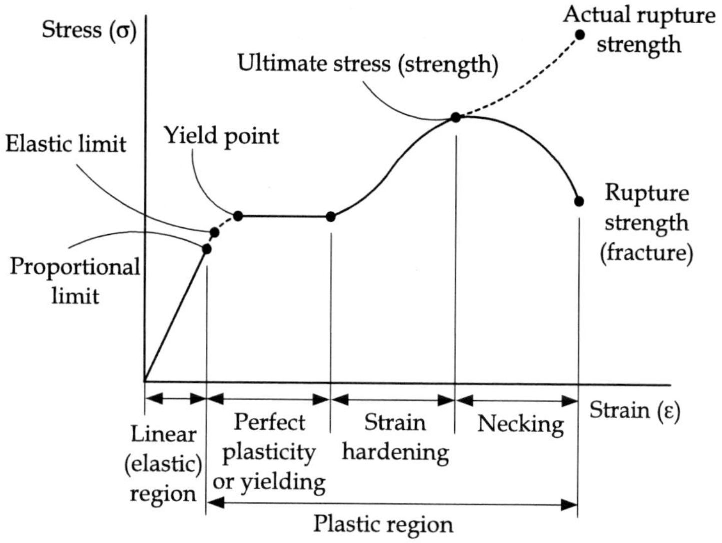

Fig. 1.1

1.2.2 Characteristics of a Stress-Strain Curve (Fig. 1.1)

- **Proportional limit:** An initial steep straight line portion (in this region σ is directly proportional to ε) is known as the proportional limit. The limit of proportionality is close to the point where the behaviour of the material ceases to be elastic (i.e., elastic limit). Below this, stress and strain have a linear relation. Above this, they have a nonlinear relation.
- **Elastic limit:** In the elastic range any displacement caused by the loading will be fully reversed on the removal of the load. Beyond elastic limit, deformation is permanent.
- **Yield point:** A turning point or knee of the data at the end of the elastic region is known as the yield point or stress (σ_y). At this point there is an appreciable elongation or yielding of the material without any corresponding increase of load. Note this phenomenon of yielding is peculiar to structural steel while other grades of steel and steel alloys do not possess it.
- **Plastic region:** A region of nonlinear extension accompanied by only a small increase in stress is known as the plastic region.
- **Strain hardening:** If a ductile material can be stressed considerably beyond the yield point without failure, it is said to be strain harden.
- **Ultimate stress:** It is the maximum stress known as the ultimate strength (σ_u) of the material (under the given test conditions).
- **Necking region:** It is the region where stress reduces while strain increases.

- Fracture strength: It is the failure stress known as the fracture (or breaking or rupture) strength (σ_f).

1.2.3 Stress-Strain Diagrams in Compression

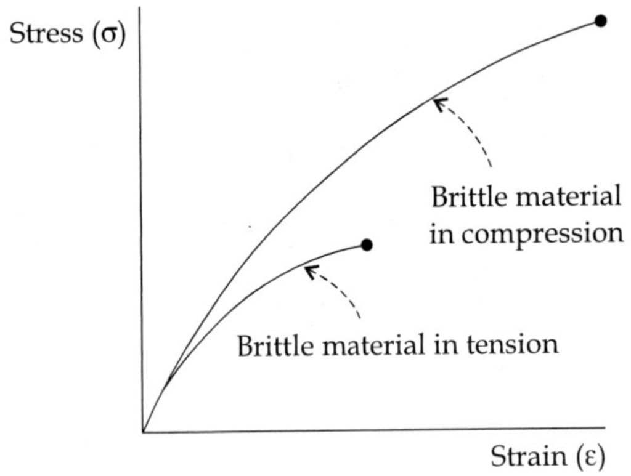

Fig. 1.2 (a)

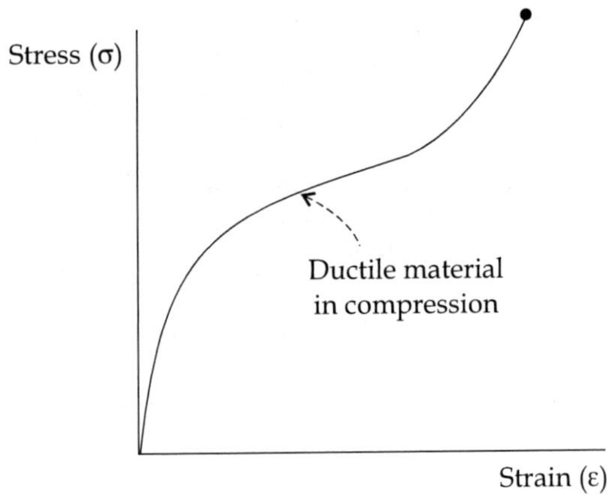

Fig. 1.2 (b)

- Ductile materials have proportional limits in compression very close to those in tension.
- Stress-strain curve of ductile a material in compression goes upward at the end because the specimen is flattened out due to compressive load which offers higher resistance to further shortening.
- Stress-strain curve of a brittle material in compression has similar shape as in tension with very high ultimate stress as compared to tension.
- Brittle materials fracture in compression, or break at the maximum load.

Ductile materials (Fig. 1.2 (b)) undergo large deformation prior to failure making the process somewhat gradual (e.g., steel and aluminium). They are strong in tension, moderate in compression and weak in shear. They possess high degree of plasticity and strength. Under tension they fail in shear and failure plane is at 45°. Under compression they fail due to bulging. Example: Mild steel (Fig. 1.2 (c))

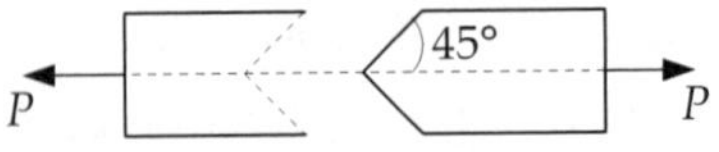

Shear failure in ductile metal

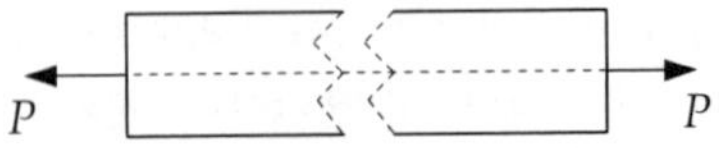

Tensile failure in brittle metal

Fig. 1.2 (c)

Brittle materials (Fig. 1.2 (a)) have a very small strain and hence fail suddenly. They are strong in compression, moderate in shear and weak in tension. Under tension they fail in tension and the failure plane is at 90°. Under compression they fail in shear and the failure plane is at 45°. Example: Concrete, cast iron, glass (Fig. 1.2 (d)).

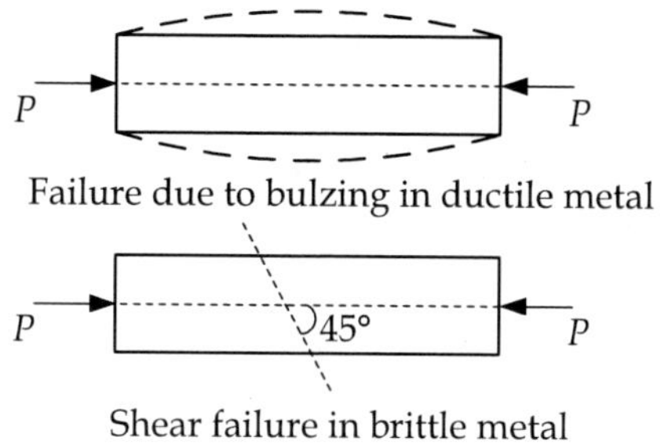

Fig. 1.2 (d)

Malleable materials can be made into thin sheets. (gold, silver). They are strong in tension, moderate in compression and weak in shear. They possess high degree of plasticity but may not necessary high strength.

Fatigue: In engineering, it is the manifestation of progressive fracture in a solid under cyclic loading as in the case of a metal strip that ruptures after repeated bending back and forth. Fatigue fracture begins with one or several cracks on the surface that spread inward in the course of repeated application of force until complete rupture occurs when the small unaffected portion is too weak to sustain the load. Structural and machine parts subject to vibrations and other cyclic loading must be designed to avoid fatigue fracture.

Toughness: The area under the complete stress-strain diagram is a measure of a material's ability to absorb the energy up to the failure and is called toughness.

Elastic Behaviour

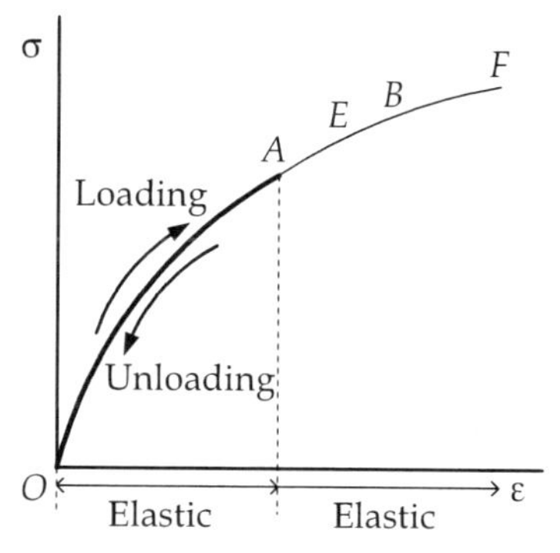

Fig. 1.3

- When the material is loaded and unloaded in the elastic region then the material follows the same stress-strain path and returns to the original shape (Fig. 1.3).

Partially Elastic Behaviour

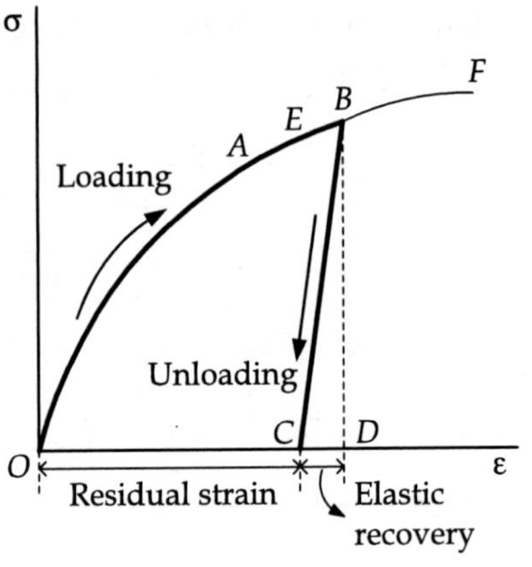

Fig. 1.4

- When a material is loaded and unloaded in the plastic (above elastic point) region then the material returns linearly (BC) parallel to a tangent at *O* at a certain permanent strain. It is called the residual strain (Fig. 1.4).

Carbon Content

- If carbon content is more strength, toughness and hardness increase but ductility decreases, i.e., chances of sudden failure increase (Fig. 1.5).

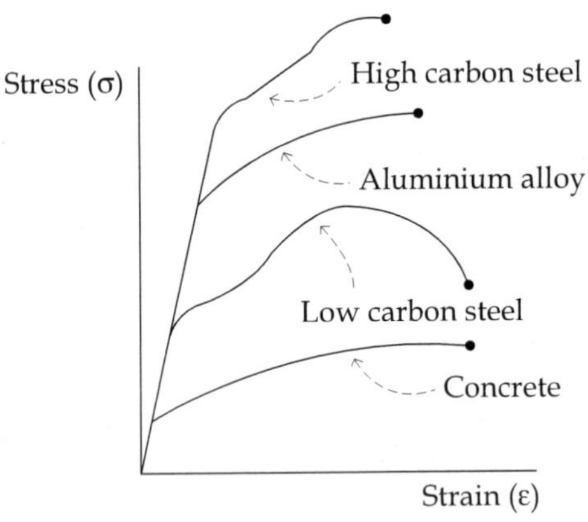

Fig. 1.5

Relaxation in a Cable

- When a cable or wire is stretched between two supports with initial constant stress then over a long period of time it develops additional strain called creep due to which the stress in the cable decreases. It is called relaxation of the material (Fig. 1.6).

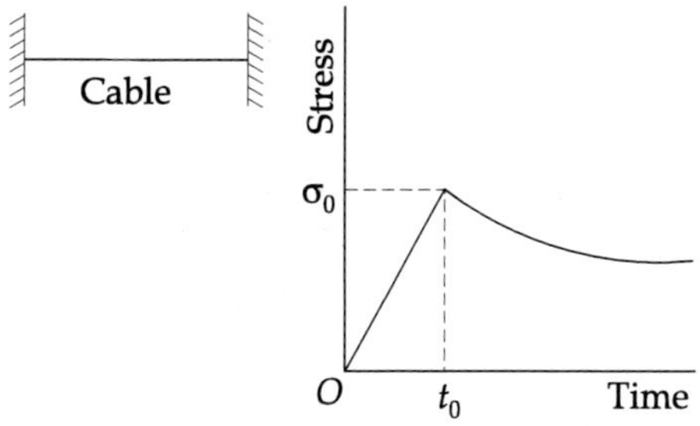

Fig. 1.6

Creep in a Bar Under Constant Load

- When a material is loaded over a long period of time it develops additional strain, that is, creep strain.
- The creep strain first increases with time and then remains constant (Fig. 1.7).

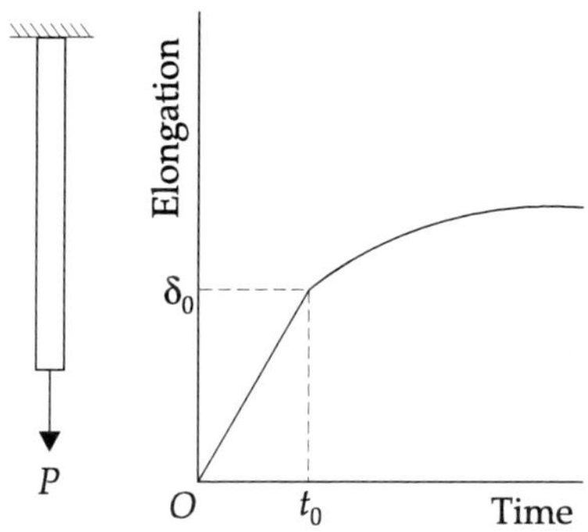

Fig. 1.7

1.2.4 Nonlinear Stress-Strain Diagram (Fig. 1.8)

Elastic Nonlinear Diagram

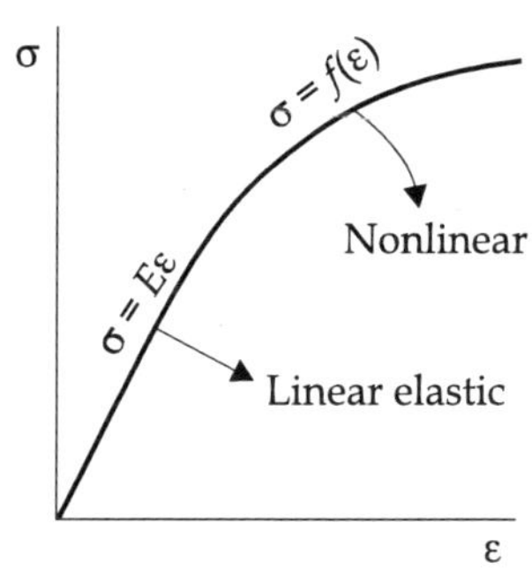

Fig. 1.8 (a)

Bilinear Elastic Diagram

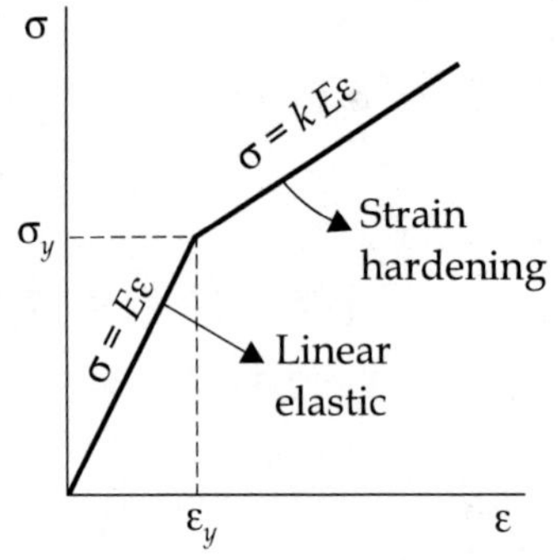

Fig. 1.8 (b)

Elastic-Plastic Diagram

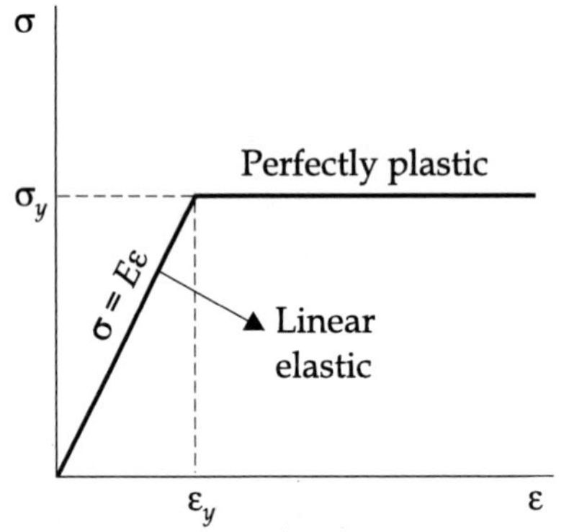

Fig. 1.8 (c)

Rigid Plastic Diagram

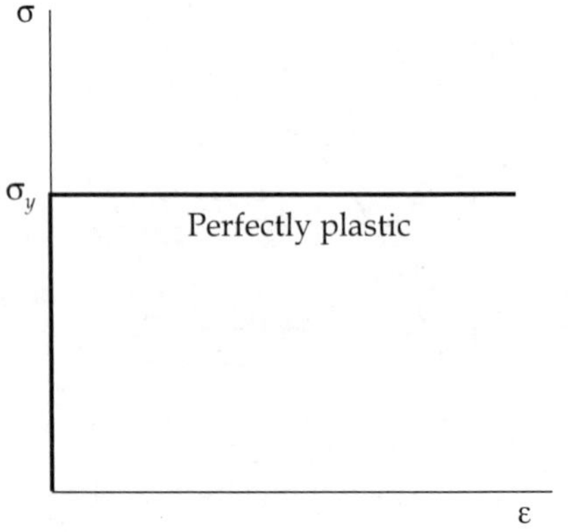

Fig. 1.8 (d)

Rigid Strain Hardening Diagram

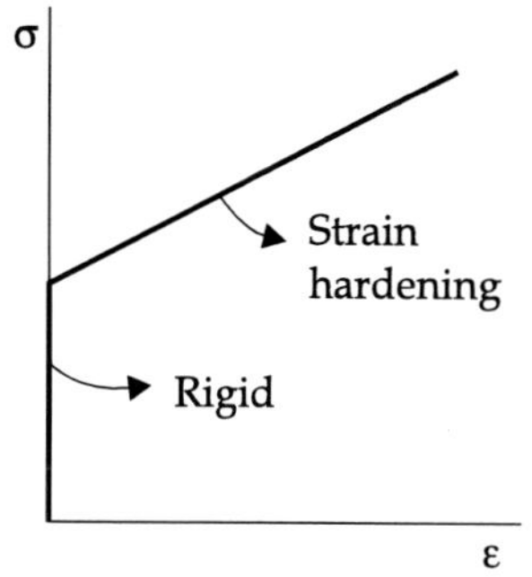

Fig. 1.8 (e)

Linear Elastic Strain Softening (Necking)

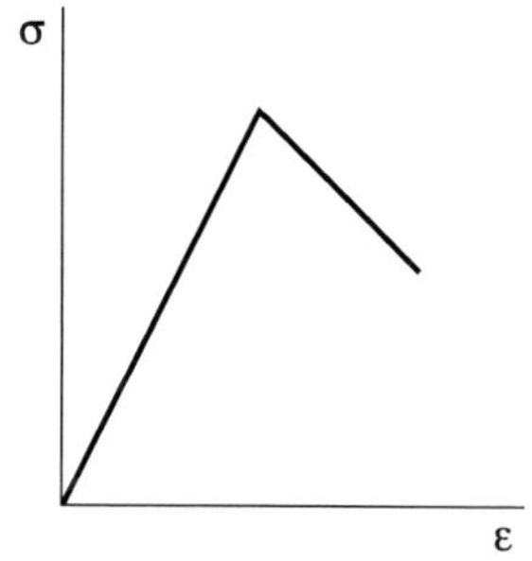

Fig. 1.8 (f)

Linear Elastic Yielding Diagram

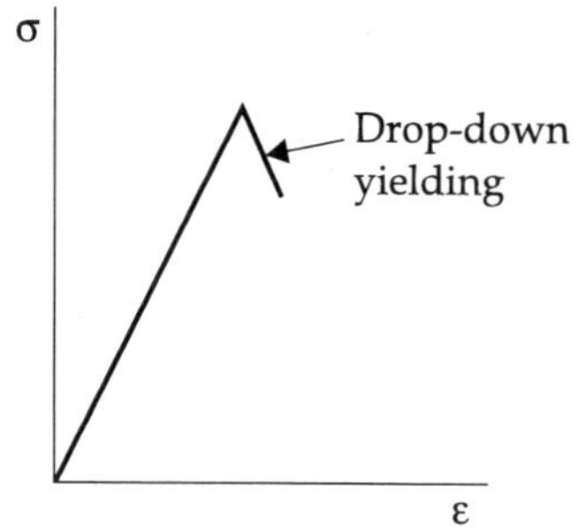

Fig. 1.8 (g)

Perfectly Rigid Diagram

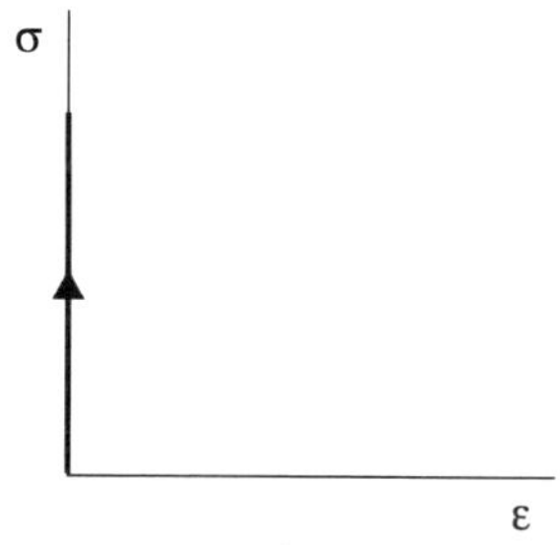

Fig. 1.8 (h)

Ideal Fluid Diagram (Continuous Deformation)

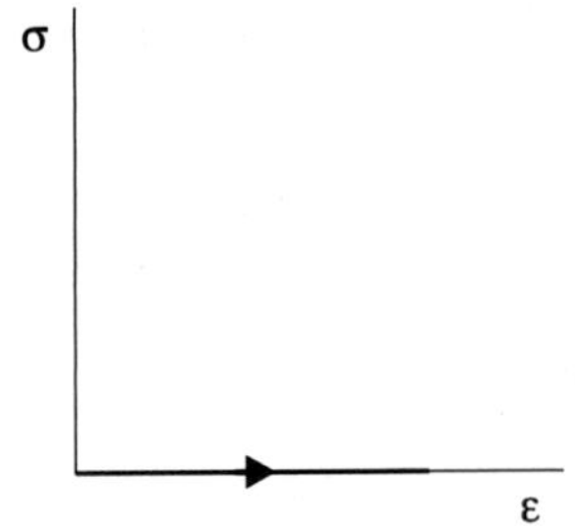

Fig. 1.8 (i)

Inelastic Diagram

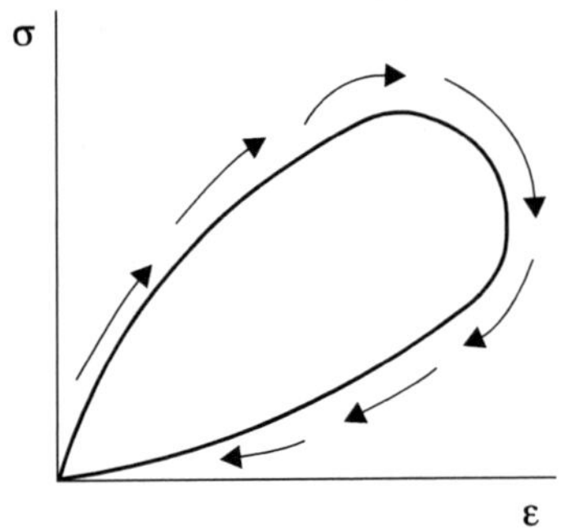

Fig. 1.8 (j)

Example 1.1: Consider the cable-beam system as shown in Fig. 1.9. Calculate the reactions at supports A and B.

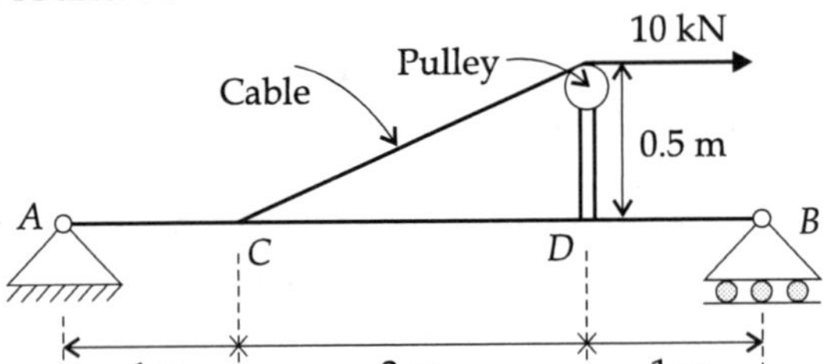

Fig. 1.9

Solution: Free-body diagram

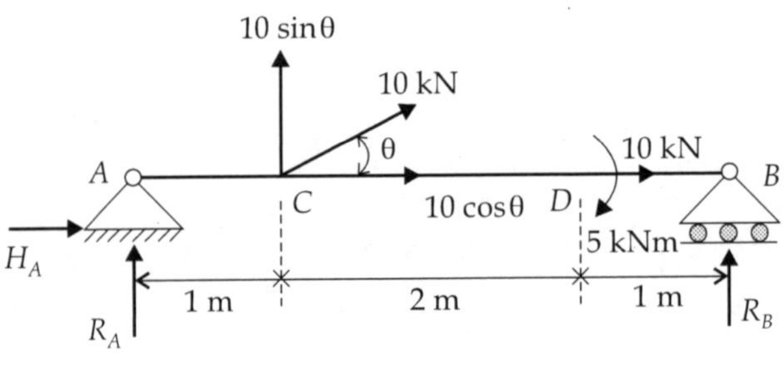

$\tan\theta = 0.25 \Rightarrow \sin\theta = 0.243$ and $\cos\theta = 0.971$

$$\sum M_B = 0 \Rightarrow R_A \times 4 + 2.43 \times 3 + 5 \Rightarrow R_A = -307$$

$$\sum F_y = 0 \Rightarrow R_A + R_B + 2.43 = 0 \Rightarrow R_B = 0.64$$

$$\sum F_x = 0 \Rightarrow H_A + 9.71 + 10 = 0 \Rightarrow H_A = -19.71$$

Example 1.2: Consider the cable-beam system as shown in Fig. 1.10. Calculate the reactions at supports A and B.

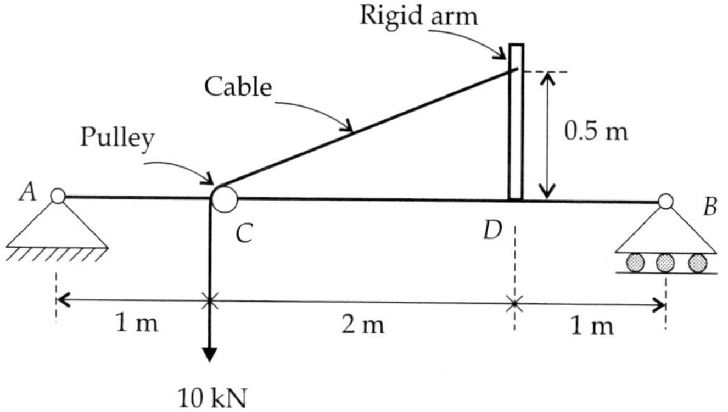

Fig. 1.10

Solution: Free-body diagram

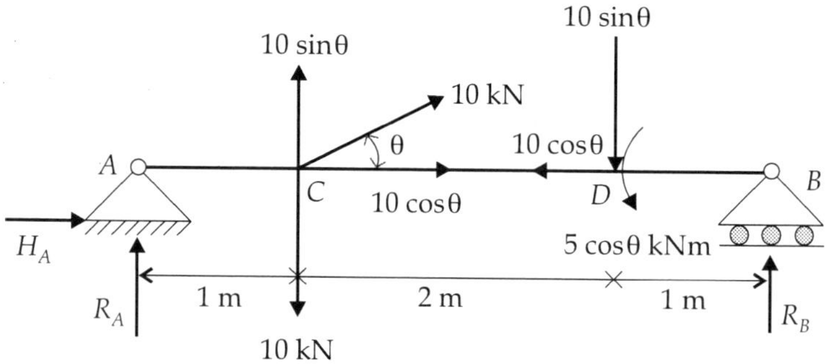

$\tan\theta = 0.25 \Rightarrow \sin\theta = 0.243$ and $\cos\theta = 0.971$

$$\sum M_B = 0 \Rightarrow R_A \times 4 + 2.43 \times 3 - 10 \times 3 - 2.43 \times 1 - 4.86 = 0 \Rightarrow R_A = 7.5$$

$$\sum F_y = 0 \Rightarrow R_A + R_B + 2.43 - 10 - 2.43 = 0 \Rightarrow R_B = 2.5$$

$$\sum F_x = 0 \Rightarrow H_A + 9.71 - 9.71 = 0 \Rightarrow H_A = 0$$

Example 1.3: Consider the truss as shown in Fig. 1.11. Calculate the value of force in members EF, HF and HI of the truss.

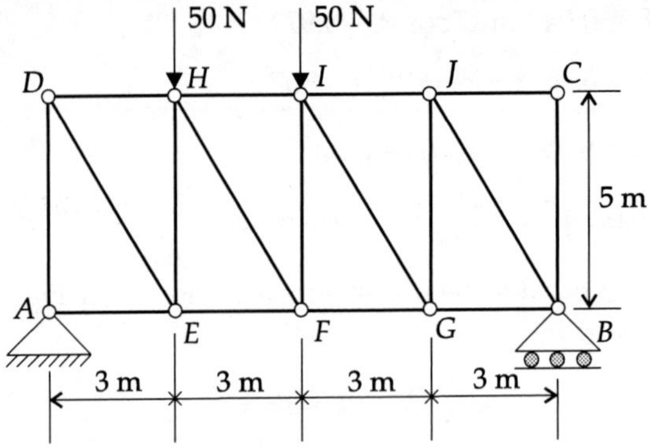

Fig. 1.11

Solution: FBD

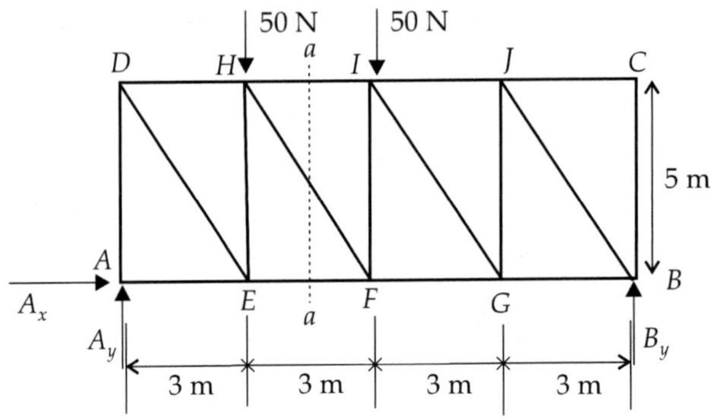

$$\sum F_x = 0; A_x = 0$$

$$\sum M_B = 0; \quad -A_y \times 12 + 50 \times 9 + 50 \times 6 = 0 \Rightarrow A_y = 62.5$$

$$\sum F_y = 0; \quad A_y + B_y - 100 = 0 \Rightarrow B_y = 37.5$$

The forces in members *EF*, *HF* and *HI* can be obtained by considering the section (*aa*) through the members *EF*, *HF* and *HI*. Assume all the members are in tension (*T*) and draw separate FBDs (free-body diagrams) of both the sections.

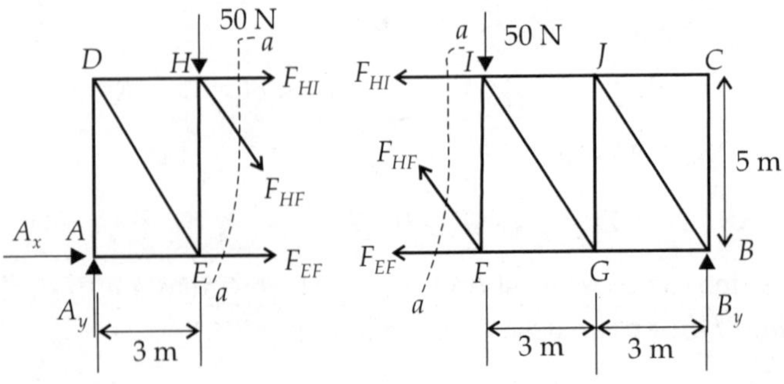

Equations of equilibrium for the left portion of truss

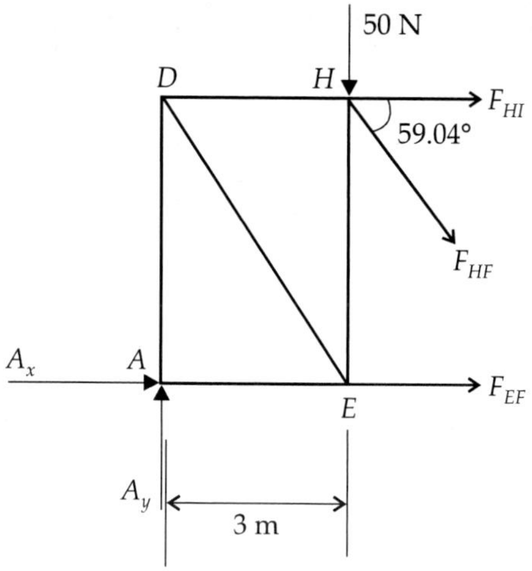

Consider the section *AEHD*, the force in member *EF* can be obtained by applying the moment about the joint *H* (since point *H* lies at the intersection of the lines of action of two unknown forces, F_{HF} and F_{HI}).

$$\sum M_H = 0; \quad A_x \times 5 - A_y \times 3 + F_{EF} \times 5 = 0 \Rightarrow F_{EF} = 37.5$$

$$\sum F_y = 0; \quad A_y - F_{HF} \sin 59.04° - 50 = 0 \Rightarrow F_{HF} = 14.58$$

$$\sum F_x = 0; \quad A_x + F_{EF} + F_{HF} \cos 59.04° + F_{HI} = 0 \Rightarrow F_{HI} = 45$$

Since values of forces F_{HF}, F_{HI} and F_{EF} are positive, it means our assumptions in these cases are correct, i.e., they are tensile forces.

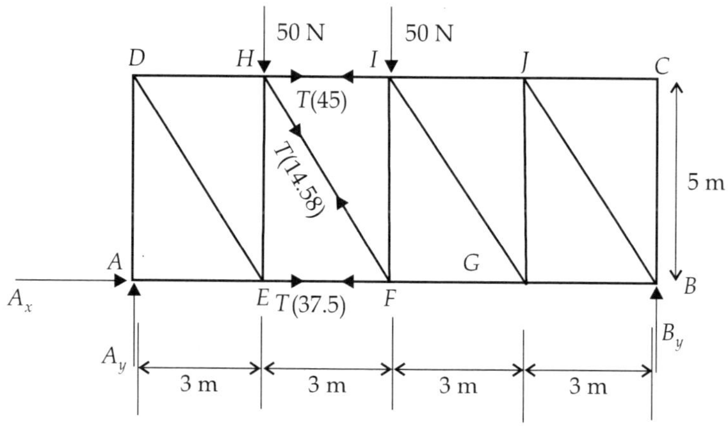

Example 1.4: A weight of 10 N is suspended at point *O* by two wires *OA* and *OB* as shown below. Calculate the tension in the wires *OA* and *OB*, if in equilibrium the sag in the wire at the suspended point is 2 m.

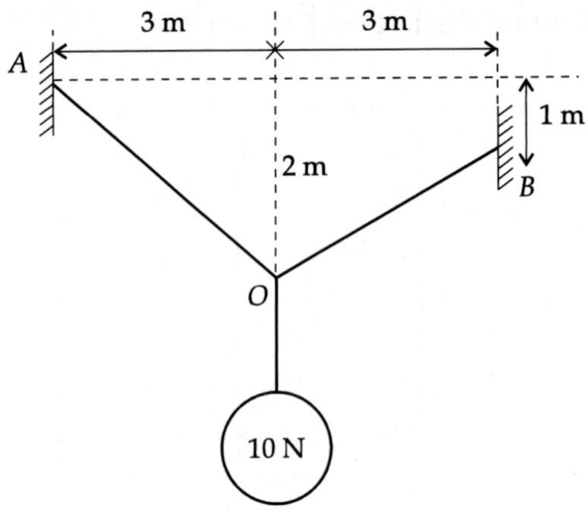

Fig. 1.12

Solution: Free-body diagram

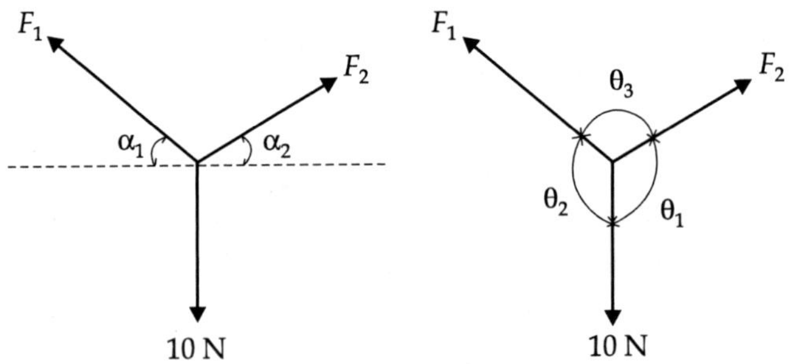

$\alpha_1 = \tan^{-1}(2/3) = 33.7°$ and $\alpha_2 = \tan^{-1}(1/3) = 18.4°$

$\theta_1 = 90 + \alpha_2 = 90 + 18.4 = 108.4°$

$\theta_2 = 90 + \alpha_1 = 90 + 33.7 = 123.7°$

$\theta_3 = 360 - \theta_1 - \theta_2 = 360 - 108.4 - 123.7 = 127.9°$

Using Lami's theorem

$$\frac{F_1}{\sin\theta_1} = \frac{F_2}{\sin\theta_2} = \frac{F_3}{\sin\theta_3} \Rightarrow F_1 = \frac{\sin 108.4°}{\sin 127.9°} \times 10 = 12 \text{ N}$$

and $F_2 = \dfrac{\sin 123.7°}{\sin 127.9°} \times 10 = 105 \text{ N}$

Example 1.5: Two smooth balls *A* and *B*, of weights 20 N and 10 N, and radii 8 cm and 4 cm, respectively are at rest in a smooth box of base 20 cm as shown in Fig. 1.13. Calculate the pressure exerted on the floor at the point of contact *P*.

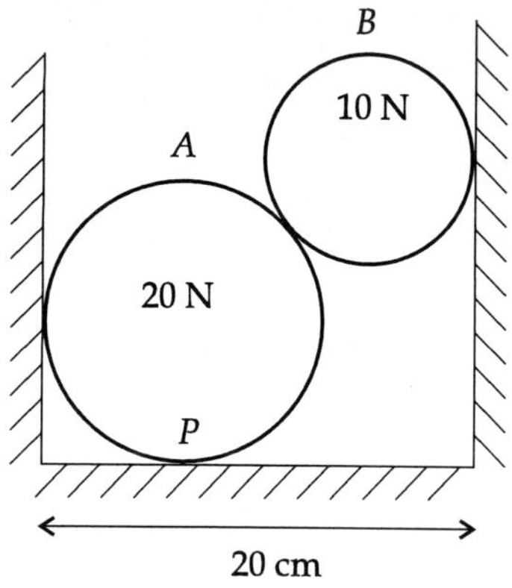

Fig. 1.13

Solution: Free-body diagram

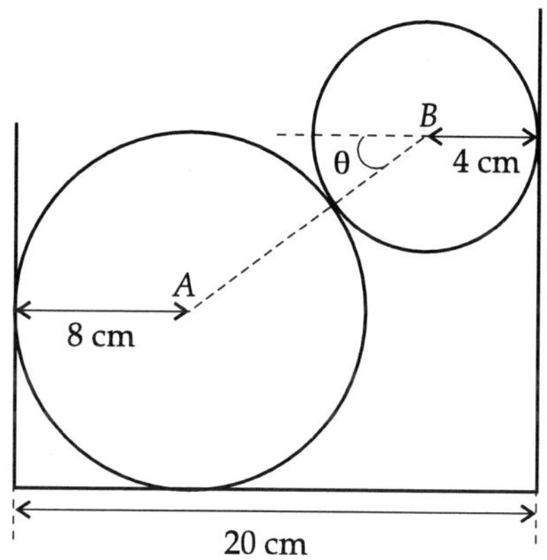

$$8 + AB\cos\theta + 4 = 20 \Rightarrow \theta = \cos^{-1}(8/12) = 48.2°$$

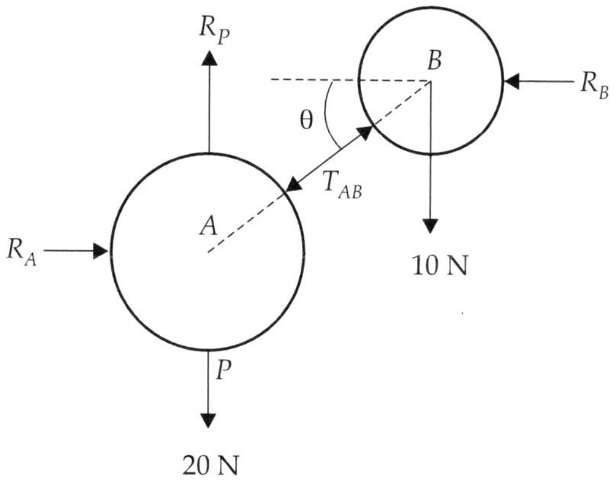

For equilibrium of ball A,

$$\sum F_x = 0 \Rightarrow R_A - T_{AB}\cos\theta = 0$$

$$\sum F_y = 0 \Rightarrow R_P - T_{AB}\sin\theta - 20 = 0$$

For equilibrium of ball B,

$\sum F_x = 0 \Rightarrow R_B - T_{AB} \cos \theta = 0$

$\sum F_y = 0 \Rightarrow T_{AB} \sin \theta - 10 = 0$

Solving the above four equations, we get $R_P = 30$ N

Example 1.6: A point load of 2 kN acts on a simply supported beam of 1 m span, as shown in Fig. 1.14. Find the reactions at the supports.

Solution: FBD of the beam AB

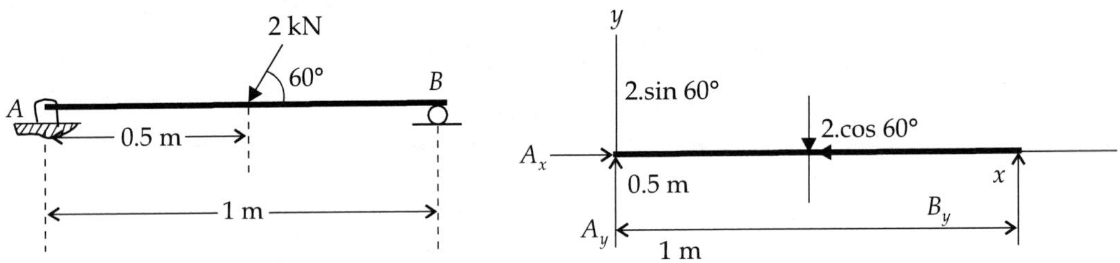

Fig. 1.14

It is found that the force-system acting on the beam is coplanar. Hence, the three equilibrium equations prescribed are $\sum F_x = 0$, $\sum F_y = 0$ and $\sum M = 0$.

Since 2 kN load has an axial component 2.cos 60°, a reaction A_x is also assumed at A.

From $\sum F_x = 0$, $A_x - 2.\cos 60° = 0$, or $A_x = 2.\cos 60°$. $\therefore A_x = 1$ kN

From $\sum F_y = 0$, $A_y + B_y - 2 \cdot \sin 60° = 0$, $\therefore A_y + B_y = 2 \cdot \sin 60° = \sqrt{3}$ kN ...(a)

From $\sum M_B = 0$, $A_y \times 1 - \sqrt{3} \times 0.5 = 0$, $\therefore A_y = \sqrt{3}/2$ kN

From $\sum M_A = 0$, $B_y \times 1 - \sqrt{3} \times 0.5 = 0$, $\therefore B_y = \sqrt{3}/2$ kN

Check: $A_y + B_y = (\sqrt{3}/2) + (\sqrt{3}/2) = \sqrt{3}$ kN

Example 1.7: Determine the reaction forces at A and B of the beam shown in Fig. 1.15.

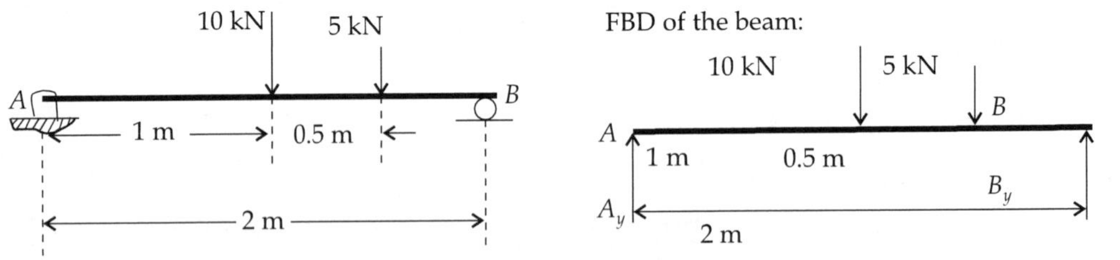

Fig. 1.15

From $\sum F_y = 0$, $A_y + B_y - 10 - 5 = 0$, or $A_y + B_y = 15$ kN ...(a)

From $\sum M_B = 0$, $A_y \times 2 - 10 \times 1 - 5 \times 0.5 = 0$. $\therefore$ $A_y = 6.25$ kN

From $\sum M_A = 0$, $B_y \times 2 - 10 \times 1 - 5 \times 1.5 = 0$. $B_y = 8.75$ kN

Check: $A_y + B_y = 6.25 + 8.75 = 15$ kN, which is satisfied according to Eqn. (a).

Example 1.8: A uniformly distributed load (u.d.l.) of 50 N/m is acting on a beam of 2 m span. Find the reactions at the supports (Fig. 1.16).

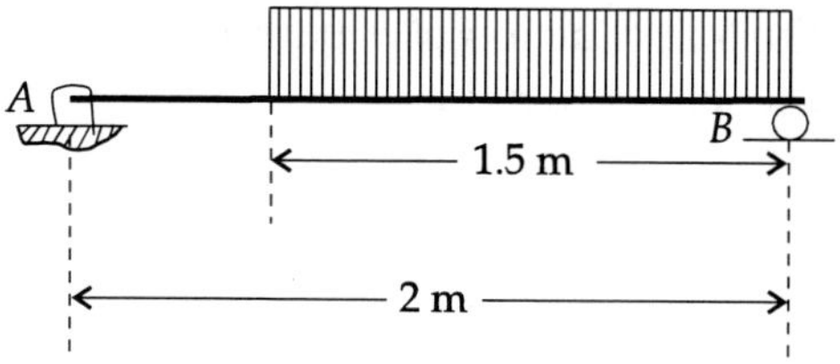

Fig. 1.16

Solution: For the given u.d.l., the equivalent point load, $W = 50 \times 1.5 = 75$ N

The point of application of W is at $(2 - 1.5) + 0.75 = 1.25$ m from support at A.

Hence, the *FBD* of the beam is drawn.

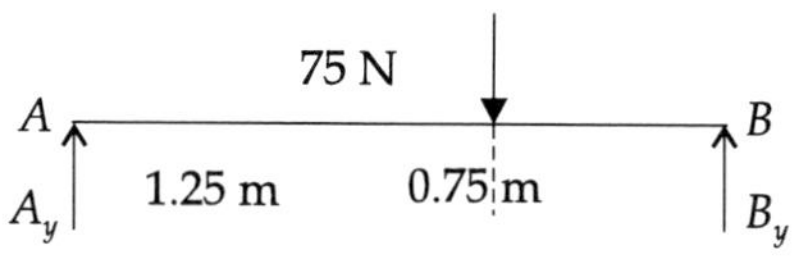

From $\sum F_y = 0,\ A_y + B_y = 75$ N

With $\sum M_B = 0,\ A_y \times 2 - 75 \times 0.75 = 0$

$\therefore\ A_y = 28.125$ N

With $\sum M_A = 0,\ B_y \times 2 - 75 \times 1.25 = 0$

$\therefore\ B_y = 46.875$ N

Check: $A_y + B_y = 28.125 + 46.875 = 75$ N.

Example 1.9: A cantilever beam of length 2 m is loaded with a uniformly varying load, as shown in Fig. 1.17. To Find the reactions at the root of the cantilever.

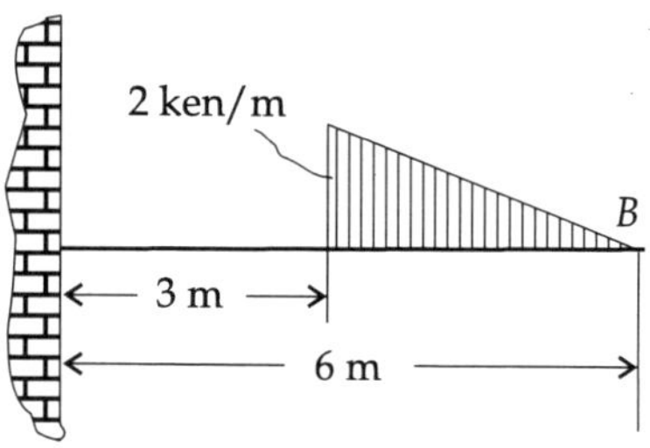

Fig. 1.17

Solution: For the given uniformly varying load, the equivalent point load, $W = (1/2) \times 3 \times 2 = 3$ kN and its point of application from the root of cantilever $= 3 + [3 - (2/3).3] = 4$ m.

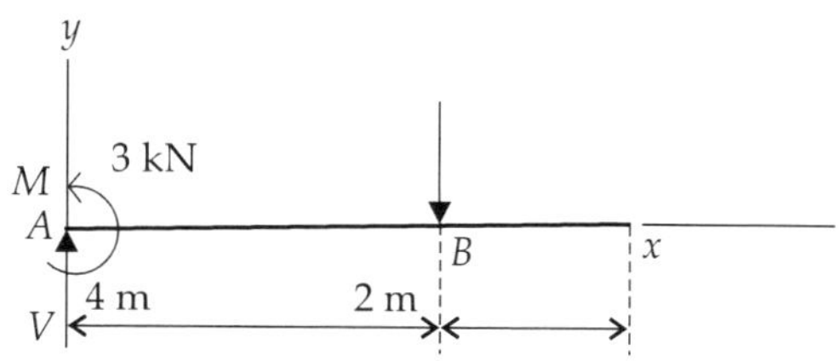

FBD of the cantilever beam

In the free-body diagram of cantilever, reactions at A comprise a vertically upward force (V) and a moment (M). Initially, both V and M are assumed in positive sense with respect to the chosen coordinates, $x - y$.

Next, applying equilibrium conditions, with $\sum F_y = 0$, $V - 3 = 0$ or $V = 3$ kN.

For $\sum M = 0$ at the point A, $M - (3 \times 4) = 0$ or $M = 12$ kNm

It should be noted that the line of action of V passes through the point A and hence there is no moment due to V about the point A.

Example 1.10: Determine the reactions at the supports A and B of the overhanging beam which is loaded as shown in Fig. 1.18.

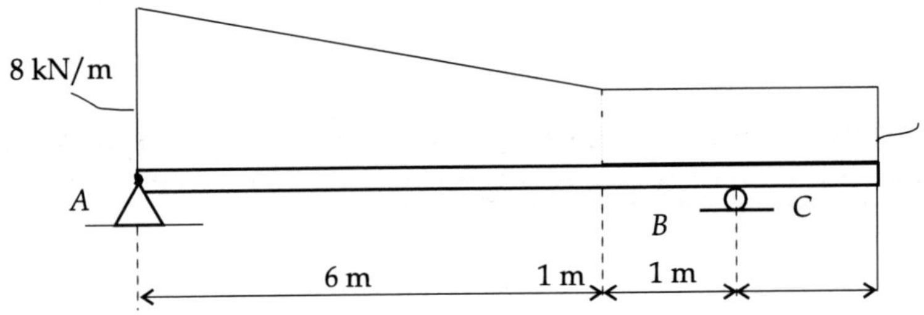

Fig. 1.18

Solution: Referring to Fig. 1.18, the distributed load can be divided into two parts. The left part is a u.v.l. With its base spread on 6 m length. The right part is a u.d.l. which is spread over 2 m length and distributed equally on both sides of support B. A point load replacing u.d.l. is of magnitude $4 \times 2 = 8$ N. The other point load replacing the trapezoidal load distribution = $(1/2) \times (8 + 4) \times 6 = 36$ kN To obtain the distance of the point of application of 36 kN load from support A, the trapezium can be divided into a triangle and a rectangle and executing further calculation.

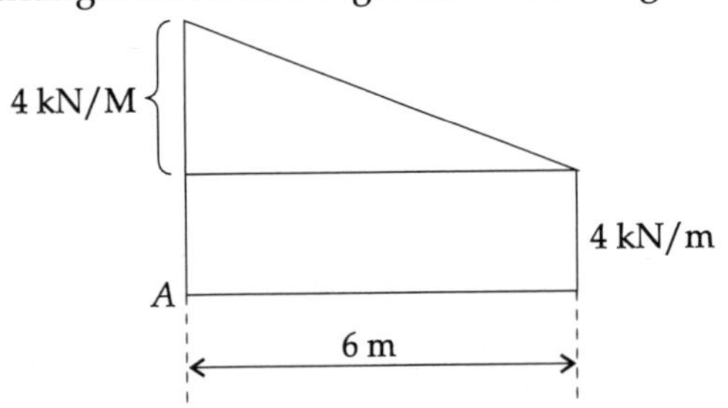

Load	Distance from A
$6 \times 4 = 24$ kN	3 m
$(1/2) \times 4 \times 6 = 12$ kN	$(1/3) \times 6 = 2$ m
Total 36 kN	

Assuming X_0 = the distance of 36 kN load from A, 36. $X_0 = (24 \times 3) + (12 \times 2)$, or $X_0 = 2.66$ m.

Therefore, the FBD of the entire beam is drawn as follows:

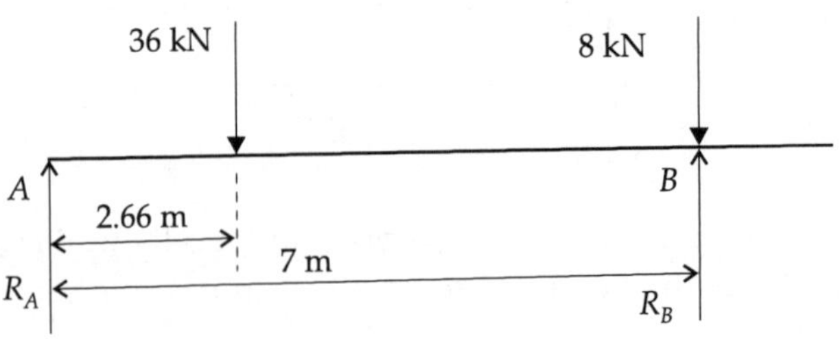

From $\sum F_y = 0$,

$R_A + R_B - 36 - 8 = 0$

or $R_A + R_B = 44$ kN ...(a)

From $\sum M_B = 0,\ R_A \times 7 - 36 \times (7 - 2.66) - 8 \times 0 = 0.$ $\therefore\ R_A = 22.3$ kN

From $\sum M_A = 0,\ R_B \times 7 - 36 \times 2.66 - 8 \times 7 = 0.$ $\therefore\ R_B = 21.7$ kN

Check : $R_A + R_B = 44$ kN, from (a).

1.3 MOMENT

Moment is a physical quantity which can turn a body about a certain axis which is called the axis of rotation of the body. While force is a polar vector (passing through a point), moment is an axial vector that acts along an axis. Moment can be produced by suitably applying a force on a body. In Fig. 1.19, a body of arbitrary shape is shown with the axis of rotation defined by a unit vector, **n**. The sectional plane shown is normal to the axis and intersects it at point O lying on it. A force vector (**F**) is lying on this sectional plane and cation of action of F which is of length d., then the moment due to **F** about the axis of rotation can be written in scalar form, as

$$M = I\,\mathbf{F}\,I \cdot d, \tag{1.1}$$

where, $I\,\mathbf{F}\,I$ is the magnitude of force F. The moment as a vector can be written as

$$\mathbf{M} = \mathbf{r} \times \mathbf{F} \tag{1.2}$$

where, **r** = displacement vector from the point O up to any point on the line of action of **F**.

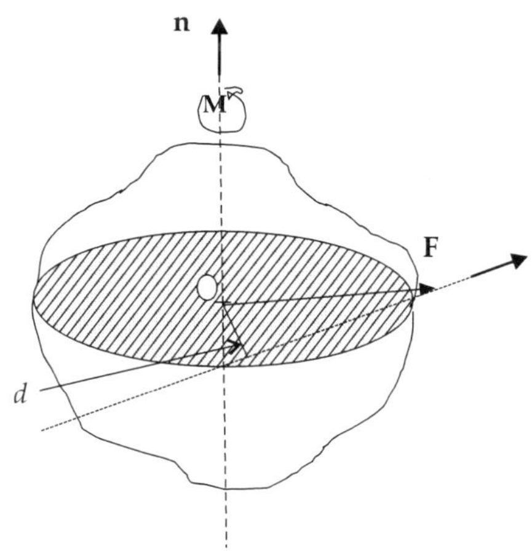

Fig. 1.19

According to the rules of cross-product and right-hand rule, the direction of M is along the unit vector, **n**. The moment (M) tends to rotate the body about the axis shown. When we apply a force to open or shut a door, the moment due to the force is responsible for swinging the door about an axis passing through its hinges.

Another type of moment is known as *couplemoment*. Two forces equal in magnitude but opposite in sense and having their lines of action parallel to each other result into a couple moment. A pair of such forces acting on a body are shown in Fig. 1.20. The perpendicular distance between the lines of action of forces is d. Hence, the couple moment is written as

$$C = F \cdot d, \tag{1.3}$$

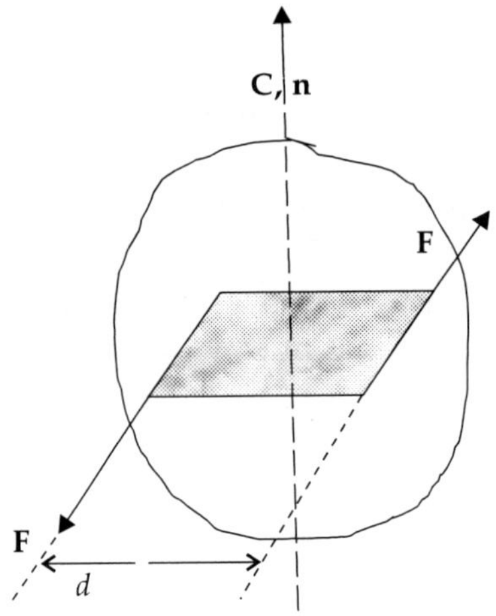

Fig. 1.20

where, F denotes the magnitude of each force. If $\mathbf{n}$ is a unit vector along the normal to the plane containing the two forces, then the vector representation of couple moment becomes

$$\mathbf{C} = C \cdot \mathbf{n} = (F \cdot d) \cdot \mathbf{n} \tag{1.4}$$

1.4 FORCE SYSTEMS AND EQUILIBRIUM

General force system: In the most general case a single body may be subjected to a force vector (**F**) and a moment vector (**M**), as shown in Fig. 1.21. The directions of F and M do not necessarily coincide. After arbitrarily selecting a coordinate system x-y-z the two vectors are expressed in terms of their components as

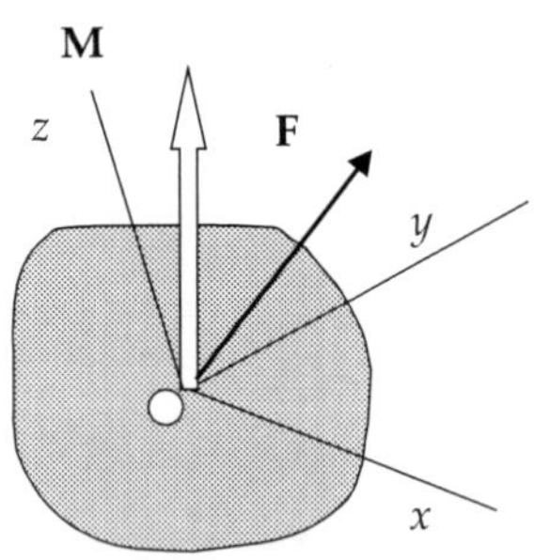

Fig. 1.21

$$\mathbf{F} = F_x \cdot i + F_y \cdot j + F_z \cdot k$$

and $$\mathbf{M} = M_x \cdot i + M_y \cdot j + M_z \cdot k, \quad \ldots 1.5$$

where F_x, F_y and F_z are orthogonal (mutually perpendicular) scalar components of **F** along x, y and z directions. Similarly, M_x, M_y and M_z are the orthogonal scalar components of **M**. Thus, there are six components of force and moment acting on the body and the body is said to be submitted to a general force system.

As it is common with a machine element or a structural part, several forces and couple moments are transmitted to it from the surrounding bodies. In such case, rules of mechanics can be used to translate all the forces and moments to pass through any conveniently chosen point so that there is a resultant force (F_R) and a resultant couple (C_R) acting about that point.

The process is explained in Fig. 1.22. A coordinate system *x*-*y*-*z* can be formed with *P* as origin and the components of $\mathbf{F_R}$ and $\mathbf{C_R}$ are written as

$$\mathbf{F_R} = \left(\sum F_{ix}\right)\cdot i + \left(\sum F_y\right)\cdot j + \left(\sum F_z\right)\cdot k,$$

$$\text{and } \mathbf{C_R} = \left(\sum C_x\right)\cdot i + \left(\sum C_y\right)\cdot j + \left(\sum C_z\right)\cdot k \qquad (1.6)$$

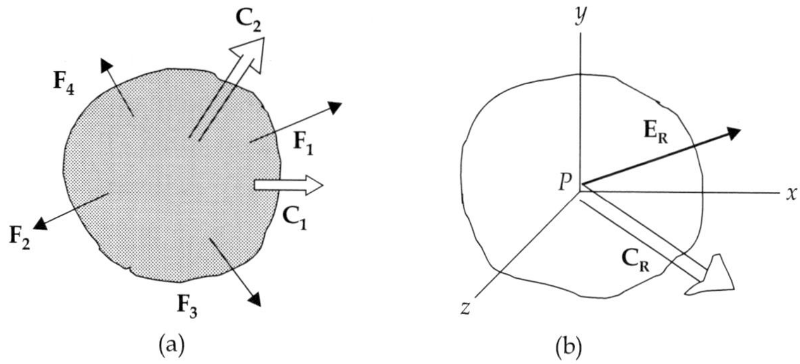

Fig. 1.22

In the above expression, $\sum F_x$ is the algebraic sum of scalar components of all the forces along *x*-direction. Similarly, $\sum C_z$ is the sum of scalar components of all the couple moments along *z*-direction.

Next, let us recall Newton's laws and the definition of equilibrium. If the body under consideration in Fig. 1.22 is in equilibrium, its translational and rotational accelerations are zero. It means that either the body is in a state of rest or it is under uniform translational and/or rotational velocity. Usually, each part of a structure and some parts of a machine are found to be at rest. Hence, equilibrium equations are needed to determine the unknown force or moment acting on a part. Therefore, considering equilibrium of the body in Fig. 1.22 we have the net force,

$\mathbf{F_R} = 0$, and net moment, $C_R = 0$ in Eqn. 1.6 to obtain

$$\sum F_x = 0,\ \sum F_y = 0.\ \sum F_z = 0,$$

$$\text{and } \sum C_x = 0,\ \sum C_y = 0,\ \sum C_z = 0 \qquad (1.7)$$

The six equations given above are known as the *fundamental scalar equations* of statics for a general force system. In a problem under this category if the unknown force or moment components are less than or equal to six, the problem can be solved for these unknown components. If the number of unknowns exceeds six, the problem is called a statically indeterminate problem and cannot be solved by applying the fundamental equations of statics.

Often the bodies in common applications are subjected to simpler force systems as discussed in the following sections.

Coplanar force system: If two or more force vectors acting on a body are coplanar and the couplemoment vectors, if any, act perpendicular to the plane of force vectors, then the forces and moments form a coplanar force system. In Fig. 1.23 (a), the forces F_1, F_2, etc., lie on a plane and the couplemoments $\mathbf{C}_1$, $\mathbf{C}_2$ etc., act perpendicular to this plane. Such a plane is not necessarily a horizontal or a vertical plane. The coordinates are set with x- and y-axes lying on this plane of force vectors, as is shown in Fig. 1.23 (b). All force vectors can be translated to pass through the origin, O. As such, some couple moments will also develop which are directed along z-axis. These can be added together and also with the already existing couple moment vectors C_1, C_2, etc., Translated at origin O. This exercise provides a resultant force FR and a resultant moment Cr, which are written as

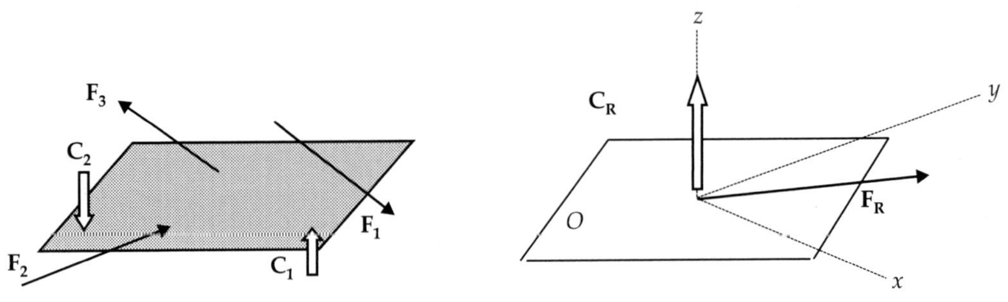

(a) (b)

Fig. 1.23

$$\mathbf{F_R} = \left(\sum F_x\right)\cdot i + \left(\sum F_y\right)\cdot j$$

and $$\mathbf{C_R} = \left(\sum C_z\right)\cdot k \tag{1.8}$$

From the above expression, only three scalar equations are available. To ascertain equilibrium the required condition will be

$$\sum F_x = 0,\ \sum F_y = 0,\ \sum C_z = 0, \tag{1.9}$$

which indicates that a coplanar system cannot be solved if the number of unknown components is more than three.

Coplanar and concurrent force system: In this case all force vectors lie on a plane and their lines of action pass through a single point. This is shown in Fig. 1.24. Selecting this point as origin with the x- and y-coordinates lying on the plane, a resultant force ($\mathbf{F_R}$) is obtained which lies on the same plane and passes through the same point (Fig. 1.24 (b)). Next, $\mathbf{F_R}$ can be expressed in the form

$$\mathbf{F_R} = \left(\sum F_x\right)\cdot i + \left(\sum F_y\right)\cdot j \tag{1.10}$$

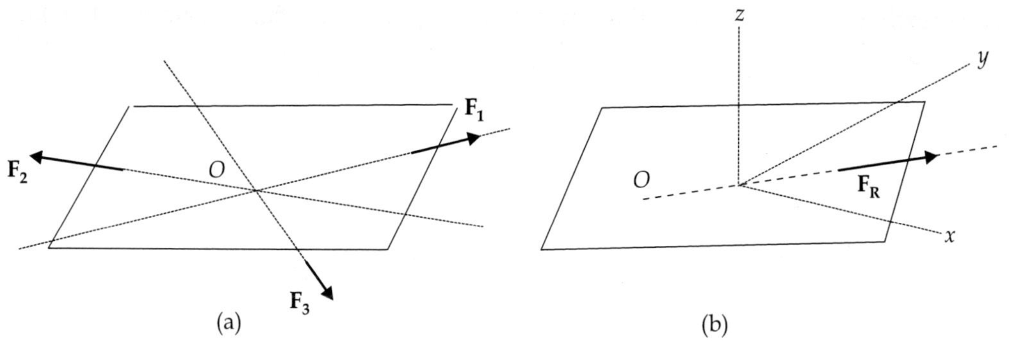

Fig. 1.24

From Eqn. 1.10, equilibrium will be attained when $\sum F_x = 0$ and $\sum F_y = 0$ which should decide whether a problem of this type is statically determinate or not.

Coplanar and parallel force system: A force system as this is formed when all force vectors are parallel to one another and lie on a plane, as shown in Fig. 1.25.

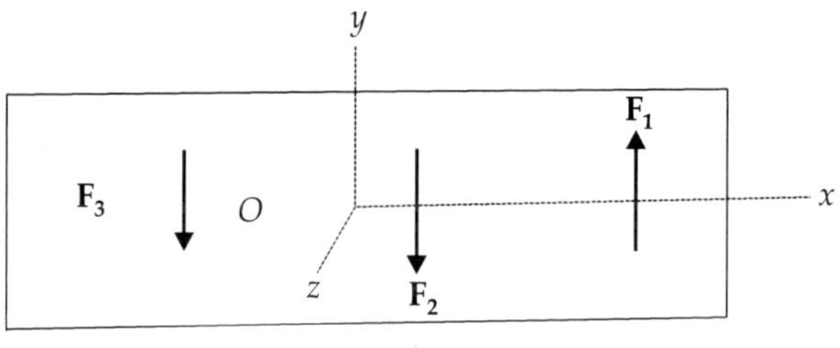

Fig. 1.25

After finding the resultant of these forces passing through the origin, we have

$$\mathbf{F}_R = \left(\sum F_y\right) \cdot j \text{ and } \mathbf{C}_R = \left(\sum C_z\right) \cdot k \tag{1.11}$$

Thus, only two scalar equations are available to solve a problem of this type.

1.5 FREE-BODY DIAGRAM

It is commonly observed that a body is in contact with some other bodies around it through which the forces are transmitted to this body. Also, the self-weight of the body cannot be ignored in some cases. Out of all these forces, some are known and some others may be unknown. It becomes necessary to determine the unknown forces acting on the body.

To accomplish this, a body can be mechanically isolated from other bodies in its surroundings and can be represented on paper in the form of a simple diagram. All the known forces in magnitude and sense are shown at their points of application. The position and sense of unknown forces are judiciously presented in this diagram. Such a figure drawn on paper is known as a free-body diagram (F.B.D.). In a free-body diagram the known and unknown forces comprise a force system. After identifying the force system, some appropriate equilibrium equations are used to determine unknown forces acting on the body.

A correctly drawn free-body diagram is very important for solving an equilibrium problem. Some examples of free-body diagrams are given below:

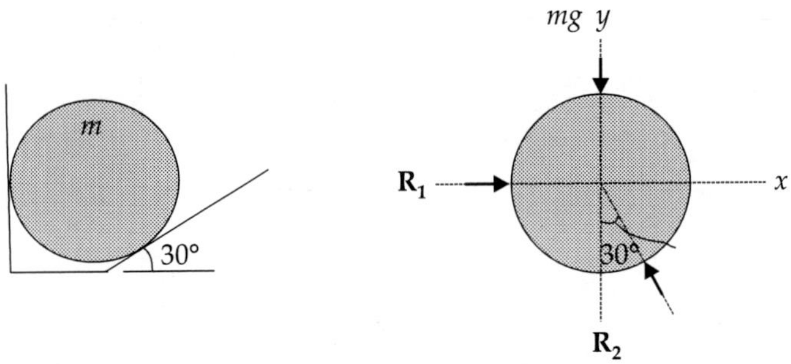

Fig. 1.26

In Fig. 1.26 (a), a sphere of mass (m) is supported against a vertical wall and also against a 30° incline. In its free-body diagram in Fig. 1.26 (b), the reaction forces (R_1 and R_2) are pushing forces on the sphere and mg is the self-weight of the sphere. It can be observed that the lines of action of the three forces pass through the centre of the sphere. While the self-weight is known beforehand, R_1 and R_2 are unknown forces. The force system is coplanar and concurrent. The coordinates, x and y, are set to coincide with the vertical plane. Therefore, R_1 and R_2 can be determined from

$$\sum F_x = 0, \text{ or } R_1 - R_2 \cdot \sin 30° = 0,$$

and $\sum F_y = 0$, or $R_2 \cdot \cos 30° - mg = 0$

Two unknowns (R_1 and R_2) are obtained from the two equations available.

Another example is that of a simply supported beam* which is subjected to a vertically downward point load (P) at a distance '(a) from the left support, as shown in Fig. 1.27 (a).

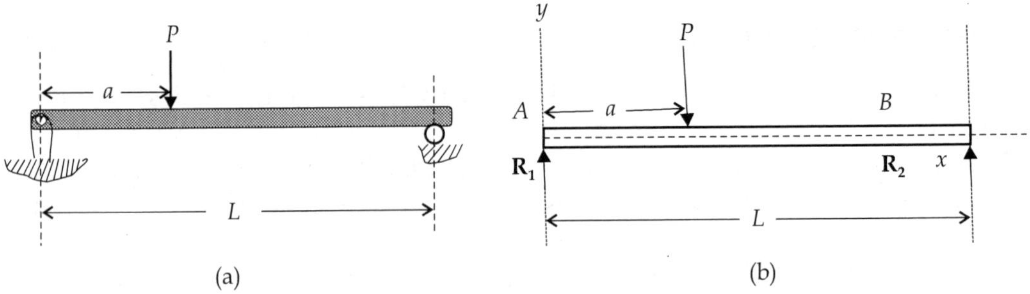

Fig. 1.27

The left support is a hinged support with a freely rotating pin, and the right support is on a roller. The distance between the supports is L. Free-body diagram of the beam is shown by Fig. 1.27 (b) in which the reactions R_1 and R_2 are shown as vertically upward pushing forces. It is assumed here that the self-weight of beam is negligibly small compared to the forces P, R_1 and R_2. As such it has practically little effect on the results and hence it is not included in the diagram. Since the beam is constrained to move horizontally at the hinged support (and not at roller support), a horizontal reaction (H) could be considered at the hinged support. However, equation ($\sum F_x = 0$) would have provided $H = 0$. This is the reason why H is discarded. A coordinate system, x-y, is chosen to coincide with the axis of beam AB with origin at the left support at A. The force system is found to be coplanar and parallel so that the two unknowns, R_1 and R_2, can be determined. Applying equilibrium condition, $\sum F_y = 0$, we have

$$R_1 - R_2 - P = 0, \text{ or } R_1 + R_2 = P$$

To ascertain that sum of moments is zero, either the support at A or that at B can be selected.

From $\sum M_A = 0$, $R_2 \cdot L - P.a = 0$, or $R_2 = P \cdot (L/a)$

and from $\sum M_B = 0$, $R_1 \cdot L - P.(L - a) = 0$. or $R_1 = P \cdot (L - a)/L$

When an external force is normal to the axis of a bar, the bar is called a "beam".

Thus, R_1 and R_2 are found and equation, $R_1 + R_2 = P$, can be used to check the calculations.

Instead of the above case, if the force (P) acts at an angle (θ) with the vertical (see Fig. 1.28), then a horizontal reaction (H) will also be present at the left support at A. and the three reactions, H, R_1 and R_2, become unknown.

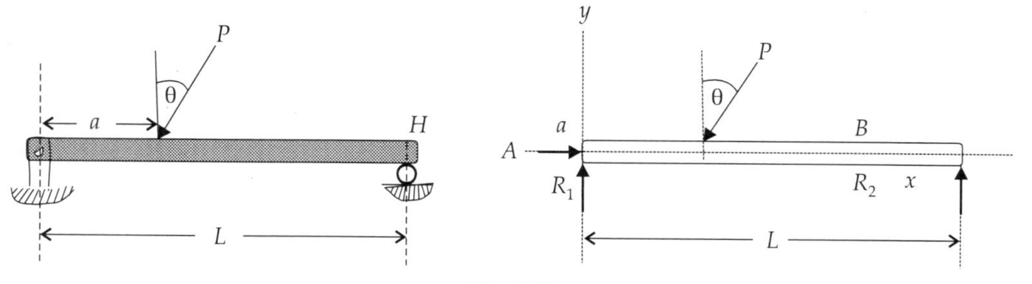

Fig. 1.28

Since it is a coplanar force system, the three unknowns can be determined.

From $\sum F_x = 0$, $H - P \cdot \sin\theta = 0$, or $H = P \cdot \sin\theta$

$\sum M_A = 0$, $R_2 \cdot L - P \cdot \cos\theta . a = 0$, or $R_2 = P \cdot \cos\theta . (a/L)$

and $\sum M_B = 0$. $R_1 \cdot L - P \cdot \cos\theta . (L - a) = 0$, or $R_1 = P \cdot \cos\theta \cdot [(L - a)/L]$

Also, from $\sum F_y = 0$, we have $R_1 + R_2 = P \cdot \cos\theta$ and the calculated values of R_1 and R_2 can be checked.

It should be noted that a hinged support and a roller support at the ends of a beam result into a statically determinate problem. However, the beam problems discussed above become statically indeterminate if both the supports are hinged supports. This is shown in Fig. 1.29.

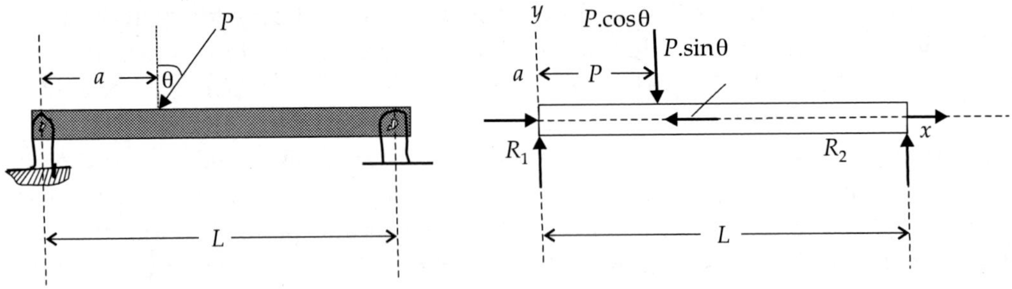

(a) (b)

Fig. 1.29

In this case there are four unknowns, such as H_1, H_2, R_1 and R_2. As it is a coplanar force system, these four unknowns cannot be obtained from only three scalar equations in hand.

Of course, a statically indeterminate problem with a beam can be solved by considering the displacements at specific points of the beam. As such, additional equations are obtained to find values of all the unknown forces. This will be explained later on in the coming chapters.

Example 1.11: A block of mass 500 kg is hanging through a cable to the pin-joint at *C* which is supported by another cable *AC* and a strut *BC*, as shown in Fig. 1.30. Determine the forces developed in the cable *AC* and the strut *BC*.

Solution: The forces in cable *AC* and strut *BC* must act along their axes so that the joint at *C* is in equilibrium under a tensile force *T* in from the cable *AC*. *A* compressive (pushing) force (*P*) from the strut (*BC*) and a tension of 500 g N due to the weight of the C block. Hence, the free-body diagram of the joint is drawn.

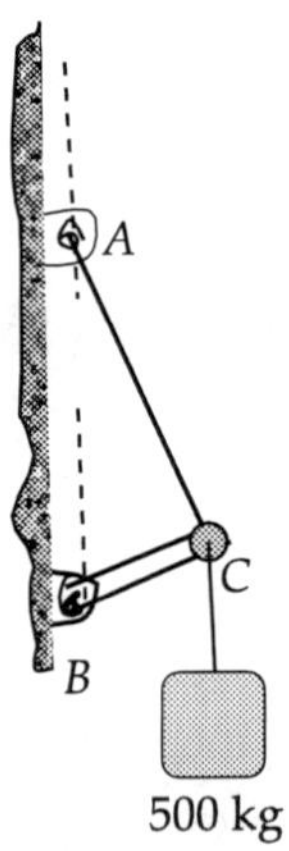

Fig. 1.30

For this coplanar and concurrent force system.

$$\sum F_x = 0$$

$\therefore$ $$P\cdot\cos 30° - T\cdot\sin 30° = 0$$

or $$T = P\cdot\cot 30° = 1.732\, P$$

Also, from $$\sum F_y = 0,$$

$$T\cdot\cos 30° + P\cdot\sin 30° - 500\text{ g} = 0 \quad (\text{where } g = 9.81 \text{ m/s}^2)$$

or $$1.732\, P\cdot\cos 30° + P\cdot\sin 30° = 500\text{ g}$$

$\therefore$ $P = 2452.5$ N, and $T = 1.732\, P = 4247.7$ N **Ans.**

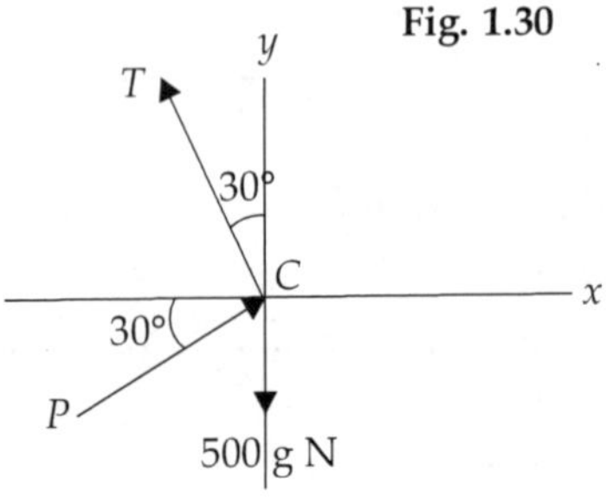

Since both *P* and *T* are positive, their initially assumed directions have been correct. As *T* is positive, it implies that the joint *C* is pulled by a force of 4247.7 N along the cable *AC*. Hence, the cable *AC* is also pulled by the same force in opposite direction so that equilibrium is maintained. Thus, there is a tension in the cable. Similarly, force P = 2452.5 N is pushing the joint and the joint pushes the strut with equal but opposite force. The strut is, therefore, under compression. This is shown by an figure below.

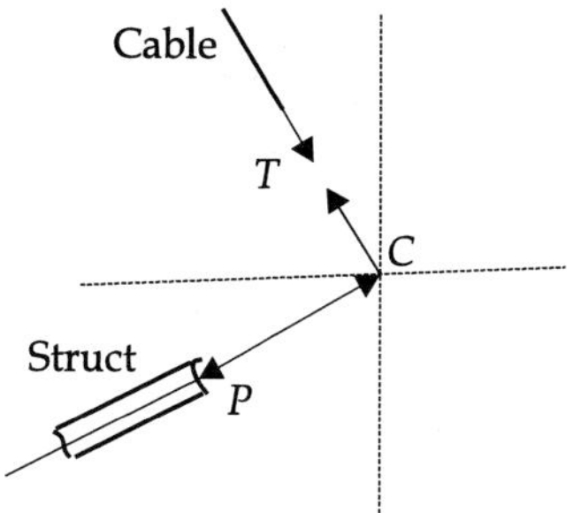

Example 1.12: The beam shown in Fig. 1.31 is hinged with a freely rotating pin at A and supported on roller at B. A uniformly distributed load of 1 kN/m, a point load of 5 kN, a moment of 9 kNm and a uniformly varying load acting on the beam are as shown. Determine the reaction at the supports A and B.

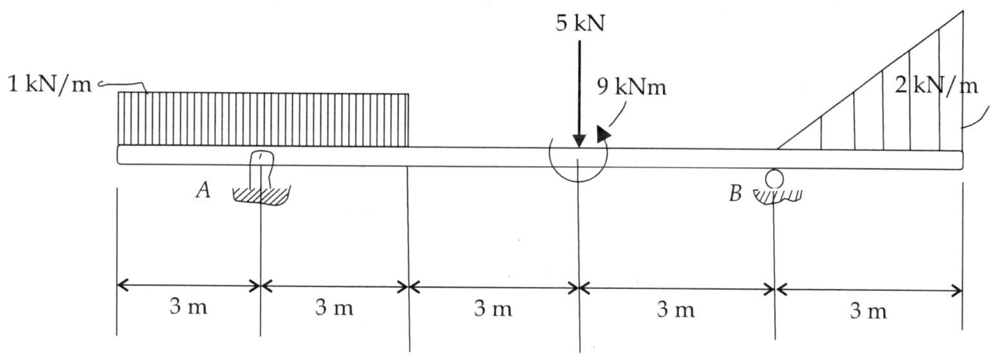

Fig. 1.31

Solution: In the F.B.D. of the beam, uniformly distributed load is shown by an equivalent point load having magnitude equal to the area of load diagram and its line of action passing through the centroid of load diagram. Similarly, the uniformly varying load is also replaced by an equivalent point load. Therefore, the complete free-body diagram follows.

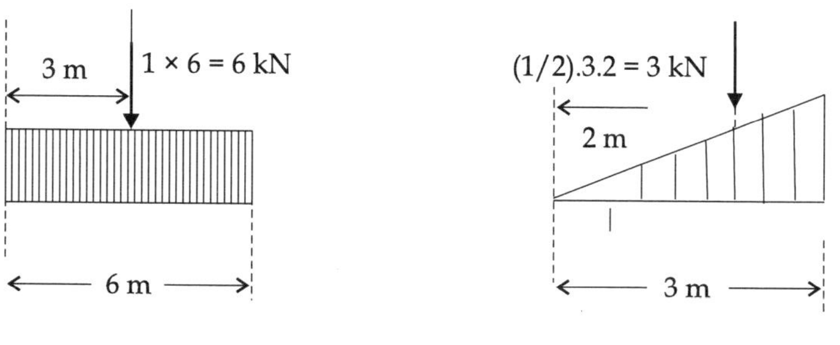

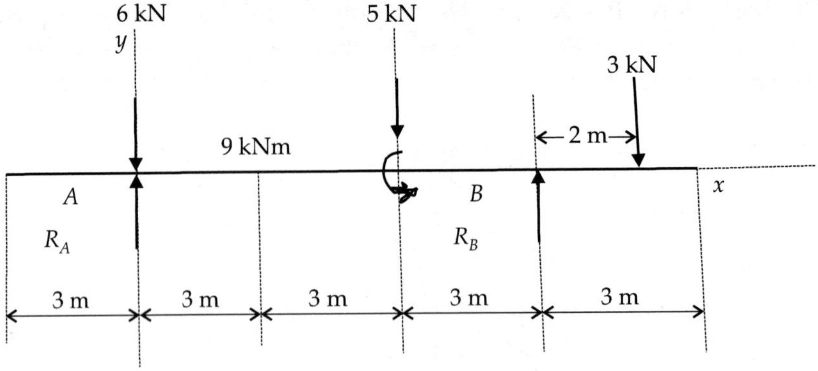

From $\sum F_y = 0$, $R_A + R_B - 6 - 5 - 3 = 0$, $R_A + R_B = 14$ kN ...(a)

From $\sum M_A = 0$, $R_B \cdot 9 - 5.6 + 9 - 3.11 = 0$, $\therefore R_B = 6$ kN **Ans.**

From $\sum M_B = 0$, $R_A \cdot 9 - 6.9 - 9 - 5.3 + 3.2 = 0$, $\therefore R_A = 8$ kN **Ans.**

It is verified from (a) that $R_A + R_B = 14$ kN

Example 1.13: *A bar AB of length 1 m and weighing 300 N is left to adjust itself in equilibrium over two inclined surfaces, as shown in Fig. 1.32. What will be the inclination (θ) of the bar?*

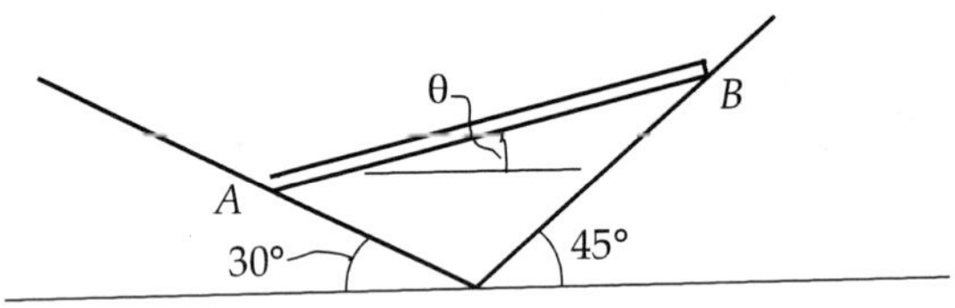

Fig. 1.32

Solution: The bar *AB* is in equilibrium under the action of three forces which are reaction R_1 normal to 30° incline, reaction R_2 normal to 45° incline, and self-weight (W) of the bar. The line of action of W passes through the mid-point of the bar.

The contact of bar at the two inclined surfaces can be assumed to be smooth, there will be no frictional force acting on the bar. As these three forces are coplanar on a vertical plane and are not parallel, they should be concurrent also. Hence, we can apply equilibrium condition at the point of concurrency. In the free-body diagram of the bar, all forces are shown to meet at the point *C*.

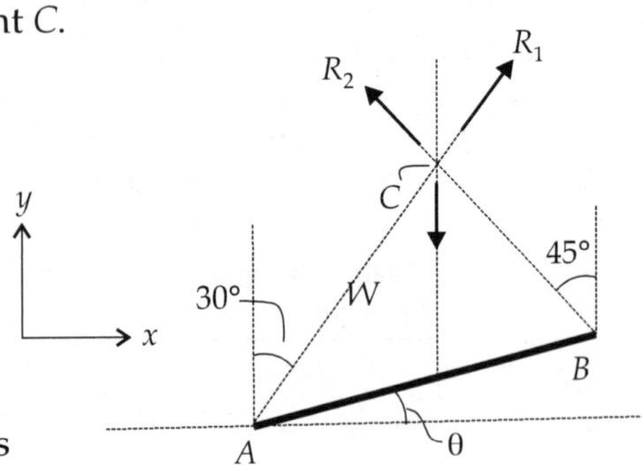

From $\sum F_x = 0$, $R_1 \cdot \sin 30° - R_2 \cdot \sin 45° = 0$

$\therefore$ $R_1 = R_2 . \sqrt{2}$.

Again from $\sum F_y = 0$,

$R \cdot \cos 30° + R_2 \cdot \cos 45° - W = 0$

or $R_2 \cdot \sqrt{2} \cdot \cos 30° + R_2 \cdot \cos 45° = 300$ N

$\therefore$ $R_2 = 155.3$ N

Next, sum of moments due to the forces about the point *A*, i.e. $\sum M_A = 0$ for equilibrium, or $\quad 155.3 \cdot \cos 45° \cdot \cos \theta + 155.3 \cdot \sin 45° \cdot \sin \theta = 300 \cdot (1/2) \cdot \cos \theta$

or $\quad 1 + \tan \theta = 300/(\sqrt{2}.155.3)$, $\quad \therefore \theta = 20.1°$ **Ans.**

Example 1.14: A simple truss is shown in Fig. 1.33. Each member is of length *a* and a load of 5 kN each is applied on the joints *E* and *B*. Determine the forces in members *AB*, *AE*, *EB* and *ED*.

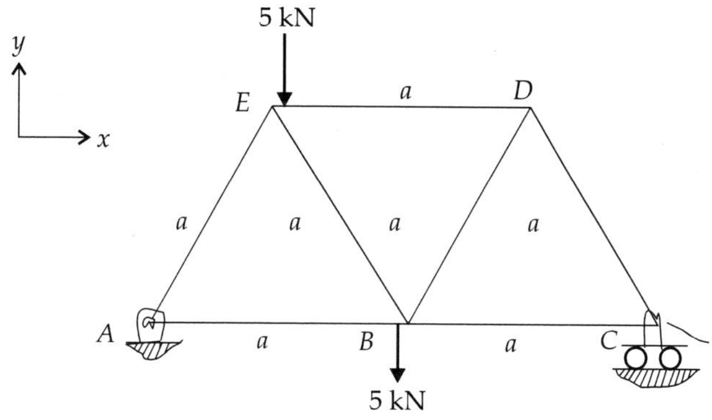

Fig. 1.33

Solution: In the F.B.D. of the entire structure, let R_A and R_C be the vertically upward reactions at *A* and *C*, respectively.

From $\quad \sum F_y = 0, R_A + R_C - 5 - 5 = 0$

or $\quad R_A + R_C = 19 \text{ kN}$

From $\quad \sum M_C = 0, R_A \cdot 2a - 5.1.5a - 5a = 0$

or $\quad R_A = 6.25 \text{ kN}$

From $\quad \sum M_A = 0, R_C \cdot 2a - 5 \cdot (a/2) - 5a = 0$

or $\quad R_C = 3.75 \text{ kN}$

Check : $R_A + R_C = 6.25 + 3.75 = 10 \text{ kN}$

To find the forces in members *AB* and *AE*, equilibrium of joint *A* is considered. The members *AB* and *AE* meet at this joint. Since these members are slender prismatic bars, the force developed in each member acts along its axis and also transmitted to joint *A*. Thus, the pin-joint at *A* is in equilibrium under the three coplanar and concurrent forces of which one is reaction force (R_A) and other two are transmitted by the two members. The unknown forces are denoted as F_{AB} and F_{AE} and initially both are assumed to act as tension forces on the joint.

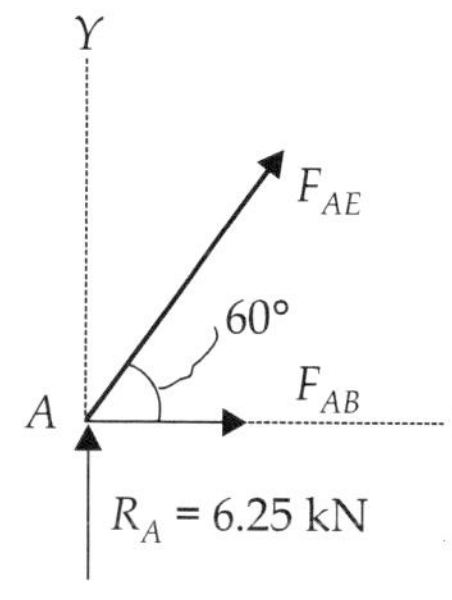

From $\quad \sum F_y = 0, F_{AE} \cdot \sin 60° + 6.25 = 0$

$\therefore \quad F_{AE} = -7.22 \text{ kN}$ **Ans.**

From $\quad \sum F_x = 0, F_{AB} + F_{AE} \cdot \cos 60° = 0$

or $\quad F_{AB} + (-7.22) \cdot \cos 60° = 0$

$\therefore \quad F_{AB} = 3.61 \text{ kN}$ **Ans.**

The negative sign of *FAE* indicates that the direction of this (F.B.D. of joint *A*) force should be just opposite to what has been initially assumed. It means that the member *AE* is pushing the joint *A* with a force F_{AE}. Therefore, a compressive force is transmitted through member *AE* to the joint *E* at the other end. This is shown in adjacent figure and we conclude that

$$F_{AE} = 7.22 \text{ kN (compressive)}$$

Similarly, force $F_{AB} = 3.61$ kN is positive and a pulling the joint *A*. As such, a tensile force develops in member *AB* which also pulls the Joint *B* at its other end. Therefore, $F_{AB} = 3.61$ kN (tension).

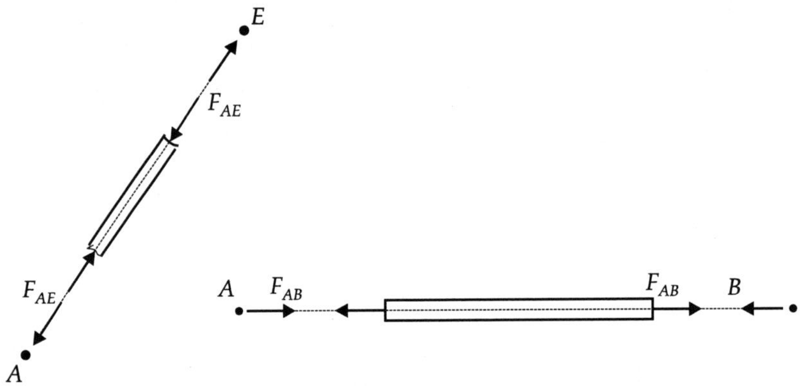

The method of analysis discussed here is known as the "method of joints". As the force system at each joint is coplanar and concurrent, only up to two forces can be determined at each joint. A joint should be judiciously selected for analysis so that the forces in only up to two members are unknown. The joint *E*, therefore can be considered for equilibrium to find the still-unknown forces in members *EB* and *ED*.

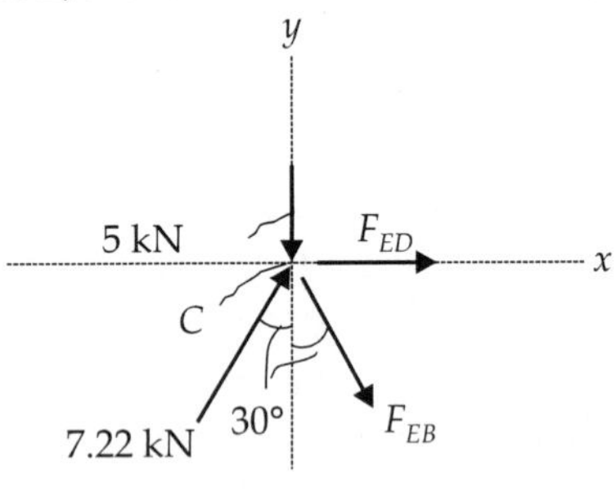

From $\sum F_y = 0, 7.22 \cdot \cos 30° - 5 - F_{EB} \cdot \cos 30° = 0$

$\therefore \quad F_{EB} = 1.44$ kN (tension) **Ans.**

From $\sum F_x = 0, F_{ED} + F_{EB} \cdot \sin 30° + 7.22 \cdot \sin 30° = 0$

$\therefore \quad F_{ED} + (1.44 + 7.22) \cdot \sin 30° = 0$

or $\quad F_{ED} = -4.33$ kN (compression) **Ans.**

Another approach of truss analysis is known as the "method of sections". Using this method forces in more than one member can be obtained at a time. The truss is cut by a section *x-x*, as shown, and equilibrium of the left part is considered. It can be seen that the sections of three members, *ED*, *EB* and *AB* are exposed. The forces in these members are initially assumed as positive or tensile in nature. Further, referring to F.B.D. of the left part as given below.

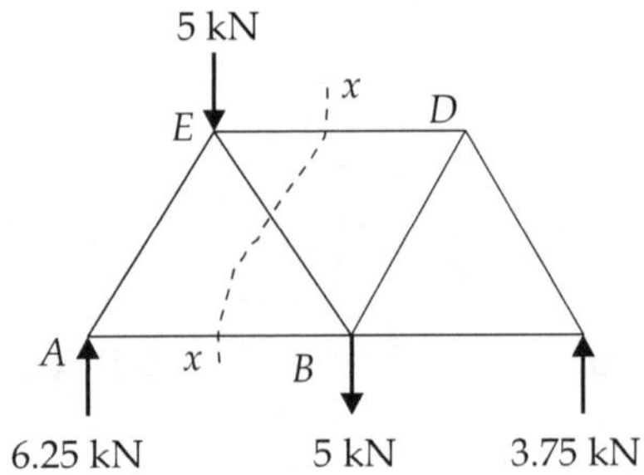

From $\sum F_y = 0$, $6.25 - 5 - F_{EB} \cdot \cos 30° = 0$

$\therefore \quad F_{EB} = 1.44$ kN (tension)

Next, from $\sum M_B = 0$,

$$F_{ED} \cdot a \cdot (\sqrt{3}/2) + 6.25 \cdot a - 5 \cdot (a/2) = 0$$

$\therefore \quad F_{ED} = -4.33$ kN (compression)

From $\sum F_x = 0$, it can be found that $F_{AB} = 3.61$ kN

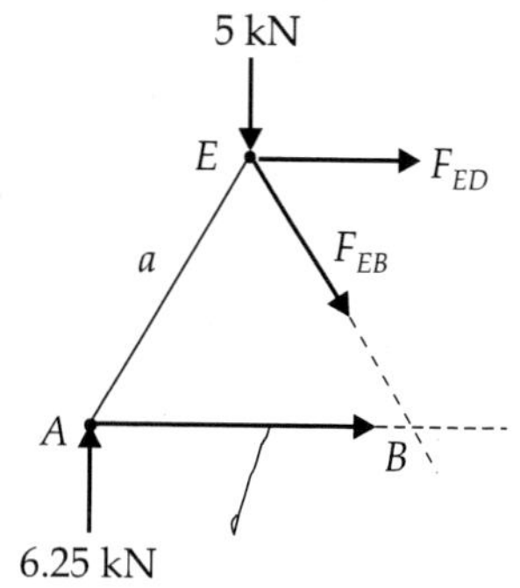

1.6 FRICTION

A friction force is a shear force that acts tangent to the surface of contact between two bodies. This force opposes the sliding motion between two bodies.

- For friction to exist, actual relative motion is not necessary. It is sufficient that there is a tendency to move.
- The direction of action of the frictional force on one of the bodies is such as to oppose the direction of the relative motion of one body with respect to other body.
- However, the friction forces are limited in magnitude and do not prevent motion if sufficiently large forces are applied.

Dry (or Coulomb friction) occurs when un-lubricated surfaces are in contact during sliding, hence friction force always opposes the sliding motion. Dry friction is of two types, i.e., if the two surfaces in contact do not move relative to each other, one has static friction or if the two contacting surfaces are moving relative to each other, then one has kinetic (dynamic or slipping) friction.

Let us consider a block assuming that initially it was at rest. A horizontal force (P) is then applied. The free-body diagram for the block is shown in Fig. 1.34 where F is the friction force.

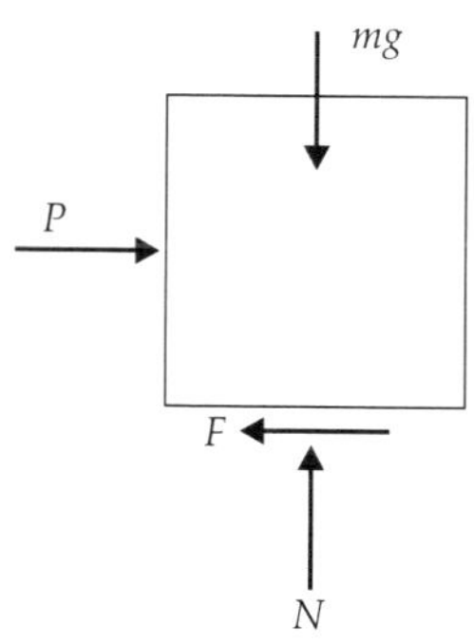

Fig. 1.34: Free-body diagram

Now increase the magnitude of the force (P) slowly from zero. As we do this we will get two stages as (Fig. 1.35):

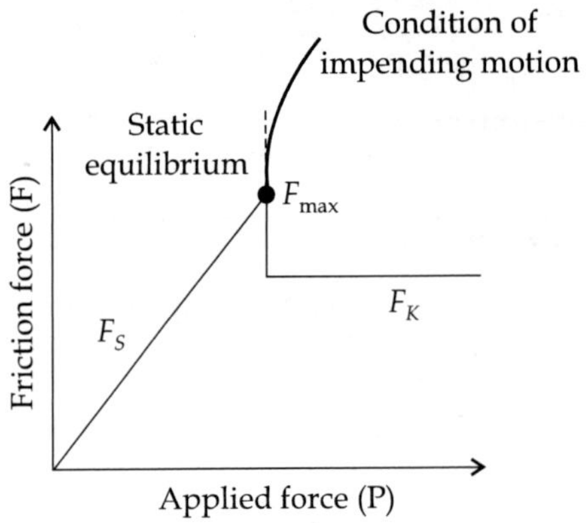

Fig. 1.35: Stage of friction

Static stage: In this stage there is no motion, that is, the body is in equilibrium $\left(\sum F_x = 0 \Rightarrow F_S = P\right)$

It means the friction force just matches the applied force. As P increases, the static-friction force (F_S) increases as well until it reach a maximum value, F_{max}.

Kinetic stage: After this upper limit of $P = F_{max}$ as P continues to increase the box will start sliding (motion is impending) and the magnitude of static friction force (F_S) drops to F_K (the kinetic friction force) and then remain constant, i.e., on further increase P, the magnitude of F_K, will remain constant. In other words there is an upper limit for the magnitude of the friction force.

Example 1.15: Draw the free-body diagram of a block of weight mg resting on an inclined plane and an external force (P) is applied on it (Fig. 1.36). Assume that the inclination of the plane α is greater than the angle of repose θ, so that the block will start moving downward.

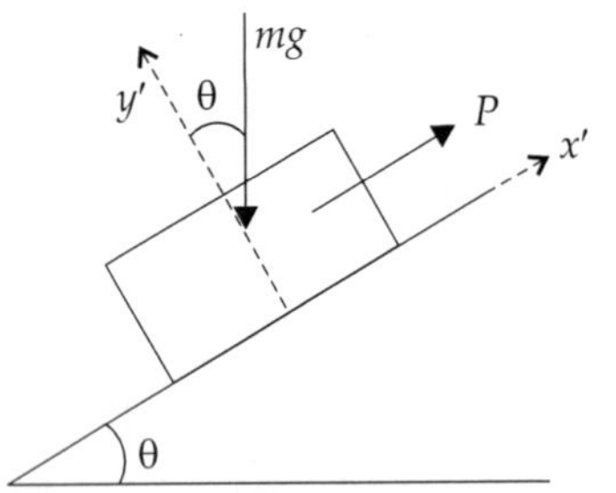

Fig. 1.36

Depending upon the value of external force (P) three types of cases will arise:

- As we increase the external force (P) gradually up to some force P_1, a stage is reached when the body is just on the point of moving downwards (Fig. 1.37), i.e., when $P \leq P_1$ block will not move downwards. At this stage, the frictional force is maximum and acts upwards.

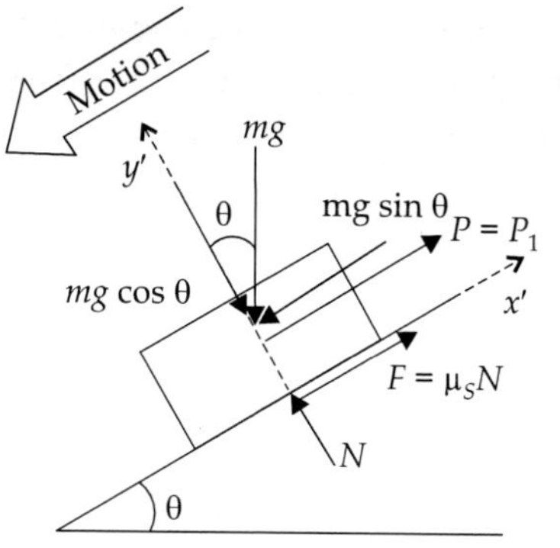

Fig. 1.37

- As the external force (P) further increased up to some force P_2, a stage is reached when the body is neither at the point of moving downwards nor upwards (Fig. 1.38), i.e., when $P \leq P_2$ block will not move upwards and when $P \geq P_2$ block will not move downwards. At this stage, the frictional force is zero.

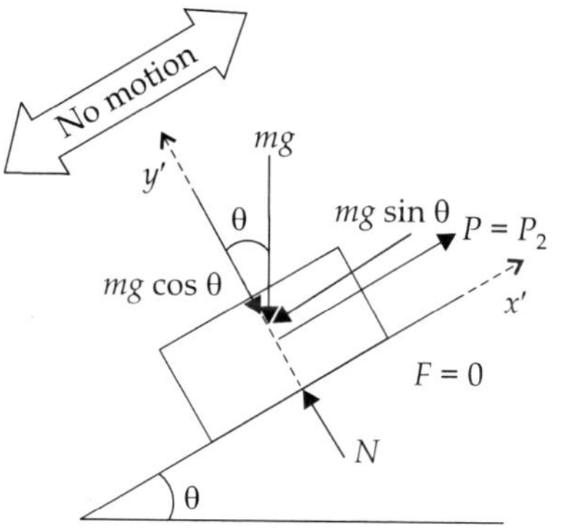

Fig. 1.38

- As the external force (P) further increased up to some force P_3, when the body is just on the point of moving upwards (Fig. 1.39), i.e., when $P > P_3$ the block will move upwards. At this stage, the frictional force is maximum and acts downwards.

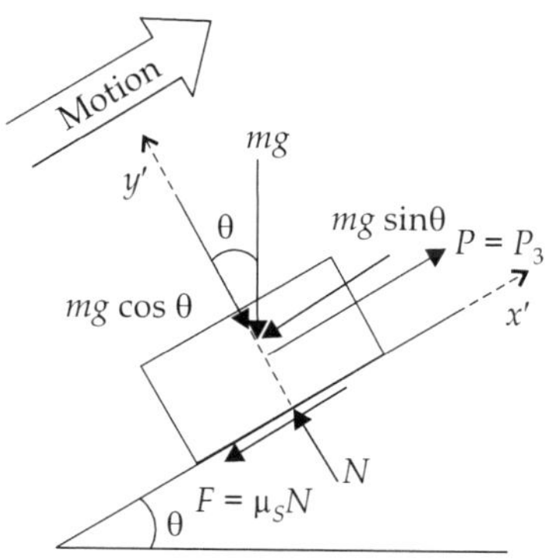

Fig. 1.39

Angle of repose (ϕ): If we continue to increase the angle of inclination, motion will soon become impending. At the time, the critical angle, θ, (Fig. 1.40) between R and the normal will have reached its maximum value, ϕ_s. The value of the angle of inclination (θ) corresponding to impending motion is called the **angle of repose.**

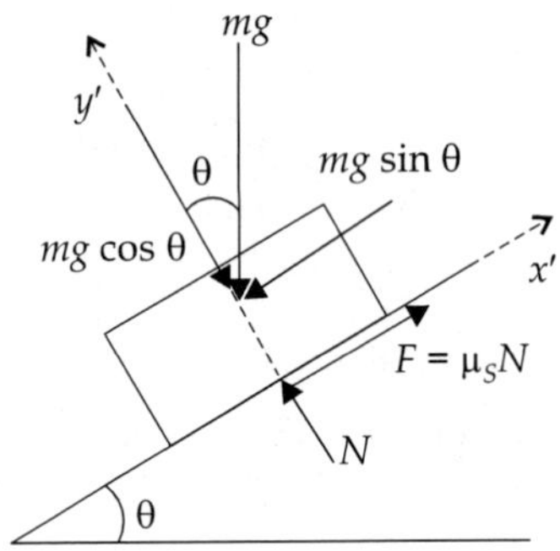

Fig. 1.40

1.6.1 Applications of Friction

Wedges

Wedges are the simple machines used to raise or move heavy loads. Force required for lifting or moving the block is significantly less than the block weight. Friction prevents the wedge from sliding by producing frictional force between all the moving surfaces.

Let the block of weight W (Fig. 1.41) be required to be lifted by the application of much smaller driving force, P.

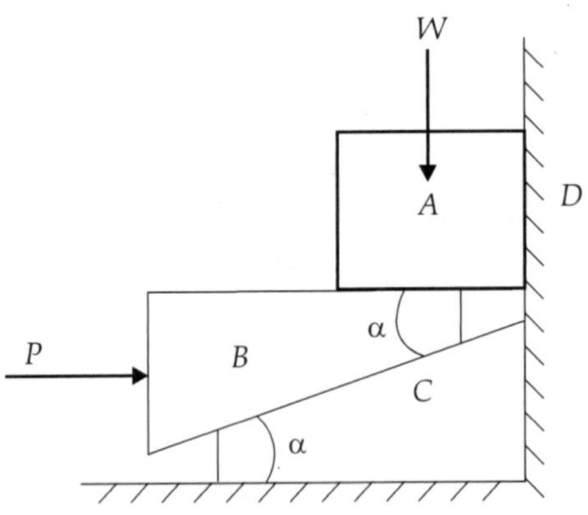

Fig. 1.41

Friction will occur between the surfaces A and B; B and C; and A and D.

To analyze the system, a free-body diagram of block is shown in Fig. 1.41 and a free-body diagram of wedge is shown in Fig. 1.42.

For Block A

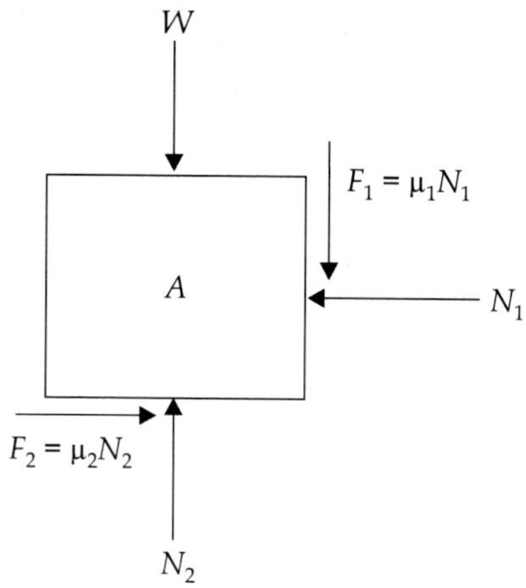

Fig. 1.42

$$\sum F_x = -N_1 + \mu_2 N_2 = 0$$

$$\Rightarrow \quad N_1 = \mu_2 N_2$$

$$\sum F_y = -W - \mu_1 N_1 + N_2 = 0$$

$$\Rightarrow \quad N_2 = W + \mu_1 N_1$$

For wedge B

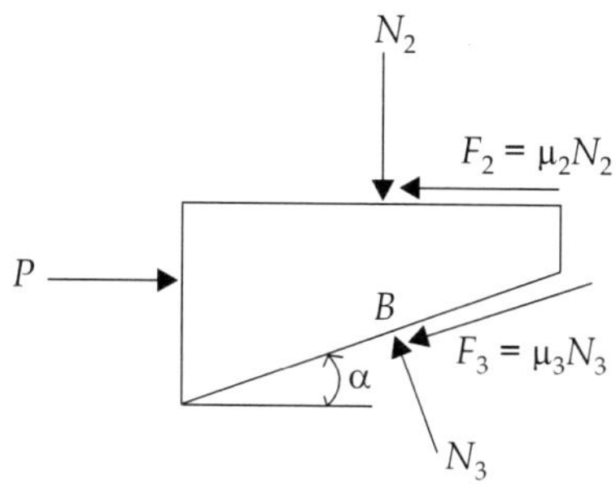

Fig. 1.43

$$\sum F_x = P - \mu_2 N_2 - \mu_3 N_3 \cos\alpha - N_3 \sin\alpha = 0 \quad \sum F_y = -N_2 - \mu_3 N_3 \sin\alpha + N_3 \cos\alpha = 0$$

$$\Rightarrow P = \mu_2 N_2 + \mu_3 N_3 \cos\alpha + N_3 \sin\alpha \qquad \Rightarrow N_2 = N_3(-\mu_3 \sin a + \cos\alpha)$$

By equating the above four equations, we get:

$$P = W\left[\frac{2\mu_2 \cos\alpha + \sin\alpha - \mu_2 \mu_3 \sin\alpha}{(1 - \mu_1\mu_2)(\cos\alpha - \mu_3 \sin\alpha)}\right]$$

PROBLEMS

1. A bent bar is fixed to the ground and a 2 kN force is applied at its free end, as shown in Fig. 1.44. Find the reactions at the support.

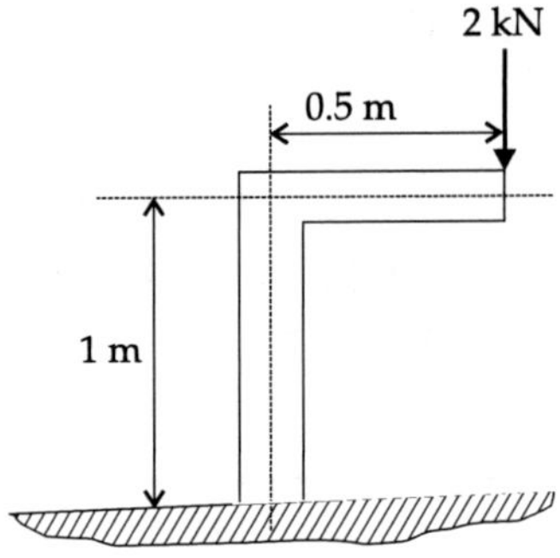

Fig. 1.44

Ans. 2 kN ( ), 1 kNm (Ↄ)

2. The force P is applied to a small roller which rolls on the cable ACB. If the tension in parts AC and BC are each 600 N at equilibrium, what is the magnitude of P and angle α?

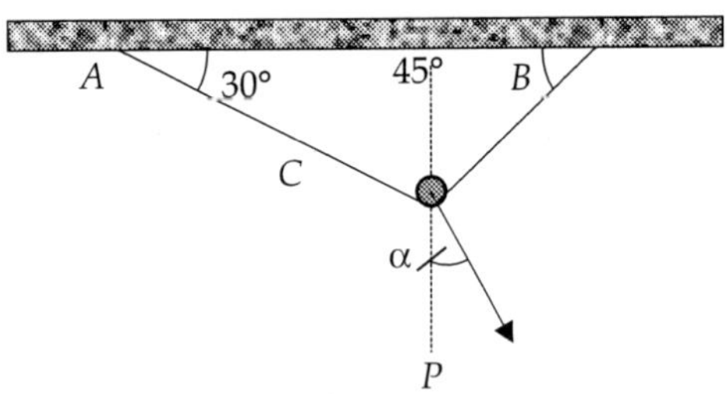

Fig. 1.45

Ans. P = 730.5 N, α = 7.5°

3. Determine the force in members BD, CD, DE, DF and DG of the truss shown (Fig. 1.46). Left support at A is a hinged support and the support at H is on rollers.

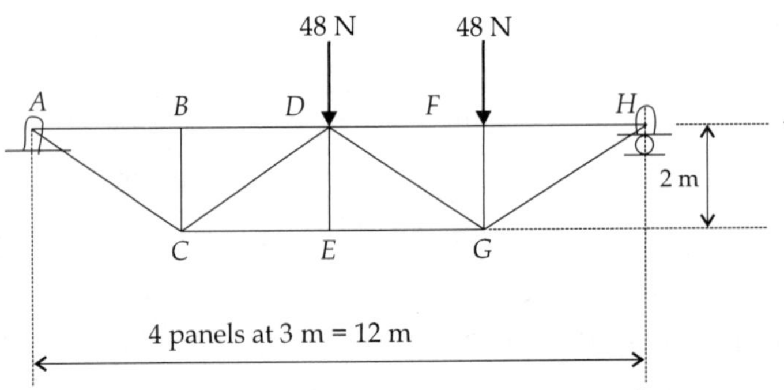

Fig. 1.46

Ans. *BD* : 54 N (C)
CD : 64.9 N (C)
DE : zero
DF : 90 N (C)
DG : 21.6 N (C)

4. Two solid spheres, each of weight W and diameter D, are made to touch each other and placed within an inverted cylindrical can. One side of the can is open, which is on the ground. If the weight of can is W_C, its inner diameter is 1.6 D and wall thickness is small. Find the ratio of weights W_C/W for which the can will tip over on the ground about the point T.

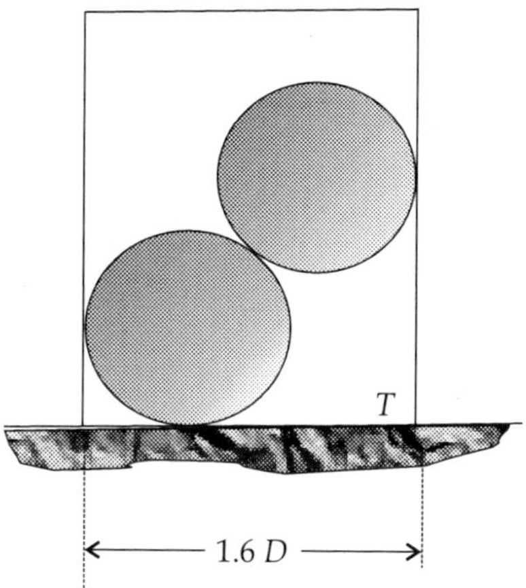

Fig. 1.47

Ans. $W_C/W = 0.75$

5. In the cantilever truss shown in Fig. 1.48, what are the forces in members *CE*, *CF*, *DF* and *EG* ?

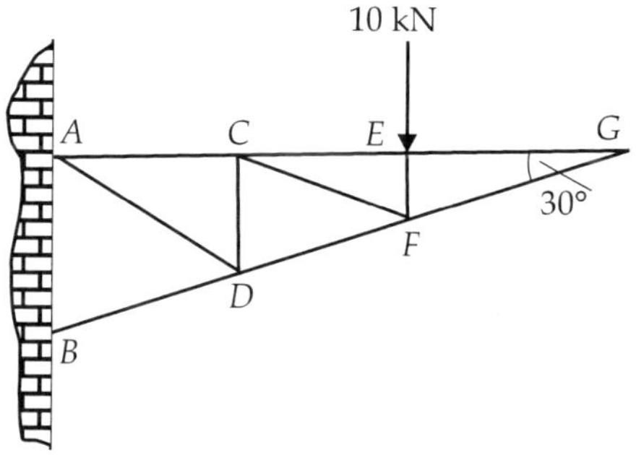

Fig. 1.48

Ans. CE : zero
CF : 10 kN (T)
DF : 10 kN (C)
EG : zero

6. A beam is supported at A and B. Apart from a uniformly distributed load of 6 kN/m and a point load of 5 kN a moment of 5 kNm is applied at the point C. Find the reactions at the supports. What will be the reactions if the moment at C is just reversed?

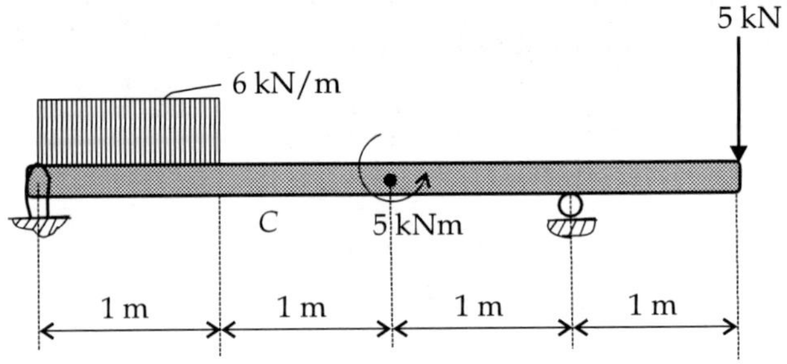

Fig. 1.49

Ans. $R_A = 5$ kN
$R_B = 6$ kN
(5/3 kN and 28/3 kN)

7. At the mid-point of beam shown in Fig. 1.50, there is a vertical extension 1 m long and a further horizontal extension of 0.5 m in length. What will be reaction at the supports A and B due to a 40 N force at the free end of extension?

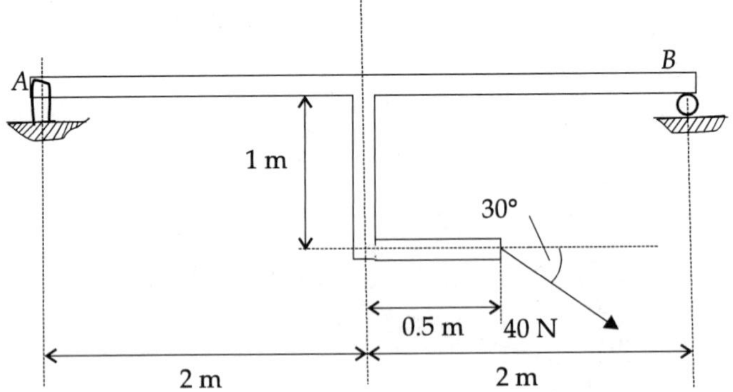

Fig. 1.50

Ans. $H_A = 34.6$ N (←)
$R_A = 16.2$ N (↑)
$R_B = 3.8$ N (↑)

8. Determine the reactions at the supports A and B of the beam shown in Fig. 1.51.

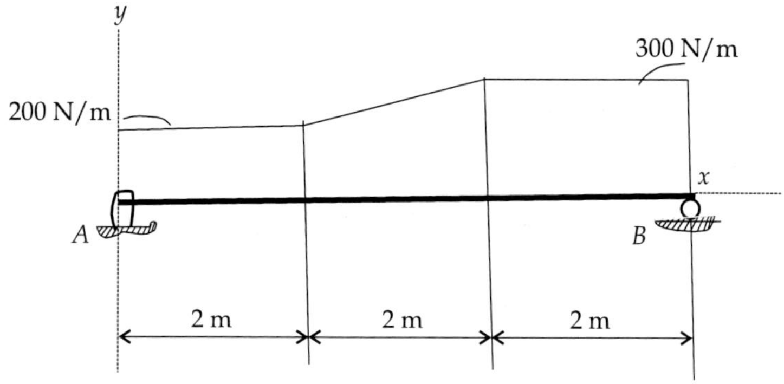

Fig. 1.51

Ans. $R_A = 677.5$ N (↑)
$R_B = 822.5$ N (↑)

9. Two bent bars ACD and EFB are hinged at B and supported at A and E, respectively with hinged supports. The hinges at B, A and E contain smooth pins and a load of 200 N is applied at the free end, D. Find the forces developed at A and E. (Note that both the supports at ground are hinged but they belong to two separate members and the problem is not statically indeterminate.)

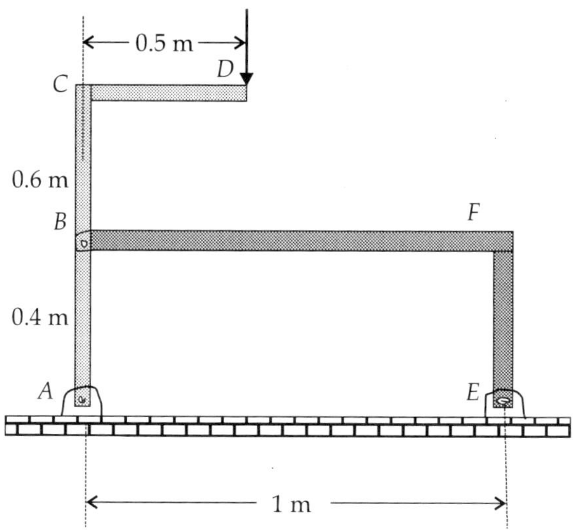

Fig. 1.52

Ans. At A, 269.3 N 21.8°
At E, 269.3 N 21.8°

10. A bracket is hinged to the ceiling at A and supported on roller at B against a vertical wall, as shown in Fig. 1.53. A 1 kN force is applied at C at an angle α to the vertical.

    (a) Determine the reactions at A and B when $\alpha = 0$.

    (b) Determine the value of α for which the reaction at A is vertical.

    (c) Reactions at A and B for the condition in part (b).

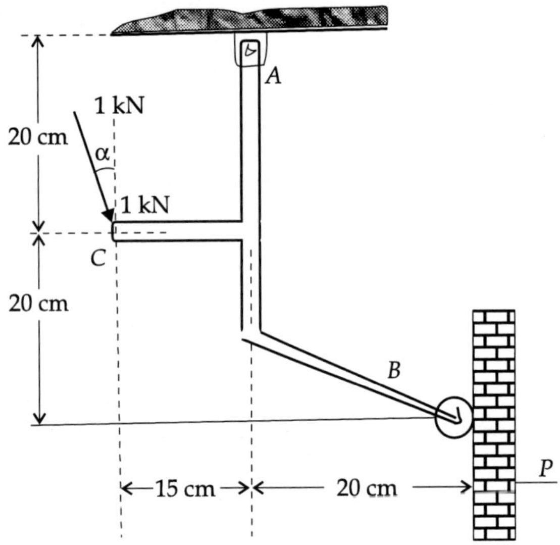

Fig. 1.53

Ans. (a) A : 1068 N at 69.4°

B : 375 N (←)

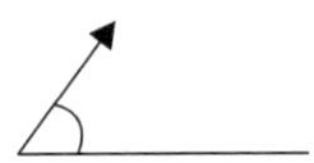

(b) $\alpha = 37°$

(c) A : 800 N (↑)

B : 600 N (←)

2

Axial Loading, Simple Stresses and Strains

2.1 INTRODUCTION

When a body is externally loaded, internal forces develop in the body and the intensity of internal force at any point within the body becomes important. Also, there may be some deformation of the body, i.e., change in its volume or shape associated with displacement at different points in the body. With the understanding of internal force intensity it is decided whether the body can sustain the known external load. As part of a machine or structure it will undergo a small permissible deformation or there will be rupture and/or failure due to excessive deformation. While the term "stress" is related to intensity of internal force, strain is a quantity which is related to deformation of the body. To understand stress and strain in their simplest form the axial loading of a prismatic bar is most appropriate.

2.2 AXIAL LOADING

A prismatic bar has uniform cross section throughout its length. Axis of the bar passes through the centroid of every cross section and is normal to it. A bar of circular cross section and another of square cross section are shown, respectively in Fig. 2.1 (a) and Fig. 2.1 (b).

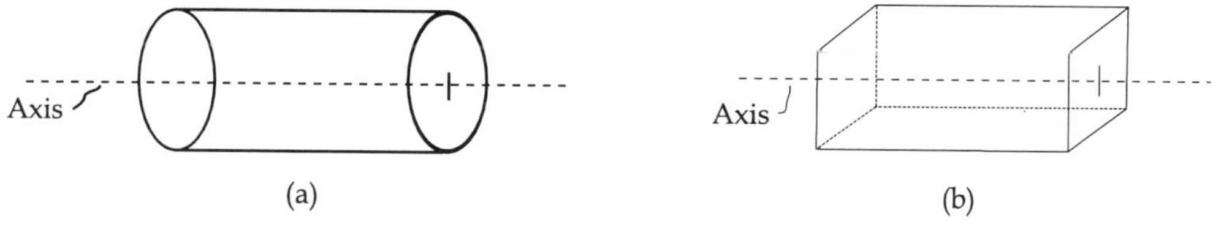

Fig. 2.1

When a prismatic bar is in equilibrium under the action of a pair of external forces which are equal in magnitude, opposite in direction and act along the axis of the bar, the loading is known as *axial loading*. A bar is under tension when the two forces as vectors are acting away from each other. Otherwise, there is compression if the force vectors are pointing towards each other.

2.3 NORMAL STRESS

In Fig. 2.2(a), a bar of uniform cross section is subjected to tension due to the forces P acting along its axis. Every cross section of the bar is normal to the plane of paper. Let us consider a sectional plane (m-n) perpendicular to the axis which contains a certain cross section and divides the bar in two parts, part-I to the left and part-II to the right of m-n. Free-body diagram of each part is shown in Fig. 2.2 (b). It is noticed that since the entire bar in Fig. 2.2 (a) is in equilibrium, each part obtained after sectioning should also be in equilibrium. Internal force on each of the exposed cross sections is denoted by F. For equilibrium according to $\sum F_x = 0$, we have $F - P = 0$, or $F = P$. If A = area of cross section and the force F is assumed to be uniformly distributed over the cross section, then the average normal stress is written as

$$\sigma = F/A, \text{ or } \sigma = P/A \tag{2.1}$$

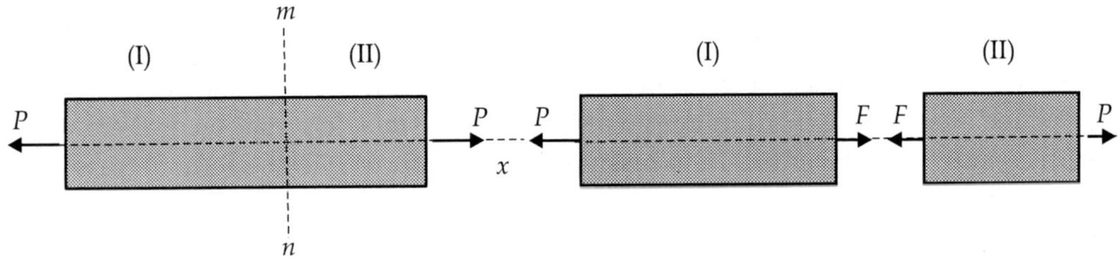

Fig. 2.2

As this stress (σ) has direction parallel to the axis and the bar is under tension, σ is also called the longitudinal stress which is tensile in nature. When the bar is under compression as shown in Fig. 2.3, a similar procedure is applied to write the normal stress as

$$\sigma = -F/A, \text{ or } \sigma = -P/A \tag{2.2}$$

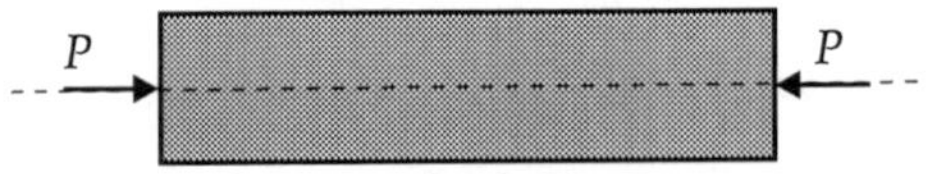

Fig. 2.3

The normal stress in Eqn. 2.2 is compressive and according to convention a negative sign is assigned to it. Tensile stress is always written with a positive sign.

If the internal force is not uniformly distributed over a cross section, as shown in Fig. 2.4(a), then the magnitude of distributed force may vary from point to point and the stress at a certain point may be different from average stress calculated for the entire cross-sectional area. In Fig. 2.4(b) there is a point C on the cross section of the bar. An infinitely small area dA surrounding the point C is considered over which a small force dF is acting. Hence, the normal stress at C is defined as

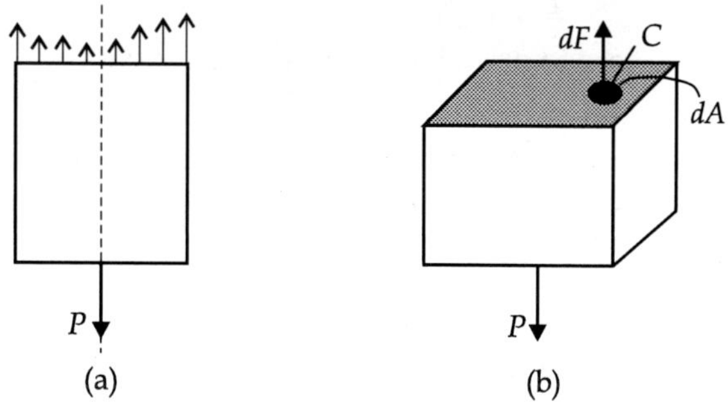

Fig. 2.4

$$\sigma_n = \underset{dA \to 0}{\text{Lim}} (dF/dA) \tag{2.3}$$

In axial loading, it is generally found that axial force is almost uniformly distributed the over cross section and the difference between stress at a particular point (σ_n) and the average stress (σ) can be ignored for all practical purposes. As such for solving problems the normal stress can be taken to be $\sigma = F/A$, where, F represents the total axial force and A is the area of cross section.

2.4 AXIAL STRAIN

When a rubber strip is pulled by equal forces applied to its ends an increase in its length is observed. Similarly, the bars made of common engineering materials like steel, copper, aluminium, etc., also increase in length when pulled by tensile forces. Changes in length cannot be observed with the naked eyes but can be detected with precision equipment. Strain is a term associated with this change in length.

A bar of some material having uniform cross section and length L is shown in Fig. 2.5. When tensile forces (F) are applied at the ends, the length of bar changes (increases) by an amount ΔL, as shown. Then the axial strain is defined as change in length divided by original length of the bar.

It is denoted by 'ε' and written as

$$\varepsilon = \Delta L/L \tag{2.4}$$

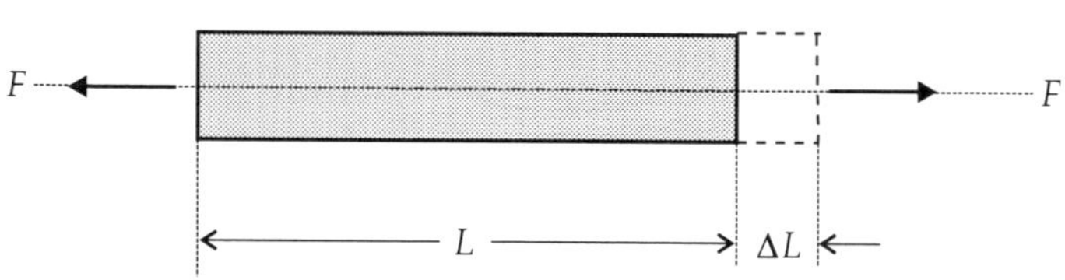

Fig. 2.5

If the forces acting on a bar are compressive in nature, there will be shortening of length and the axial strain will become

$$= - \Delta L/L \tag{2.5}$$

where, the negative sign refers to an axial strain in compression.

The normal stress and axial strain are related to one another by what is known as Hooke's law, which incorporates the property of material of the bar. To establish a relation between stress and strain, a graph of normal stress against axial strain should be obtained from simple tension test conducted in laboratory.

2.5 SIMPLE TENSION TEST

Mild steel is is extensively used in construction work, as well as for preparing machine parts used in industries. Hence, the tension test of mild steel is of prime interest though its test results differ from those of high carbon steel, alloy steels, cast iron, aluminium, copper, etc. In tension test, a slowly increasing (quasi-static) load is applied to a test piece and the resulting deformation is monitored at every step of the applied load. Many testing machines are equipped with devices to record continuous stress-strain data throughout the test.

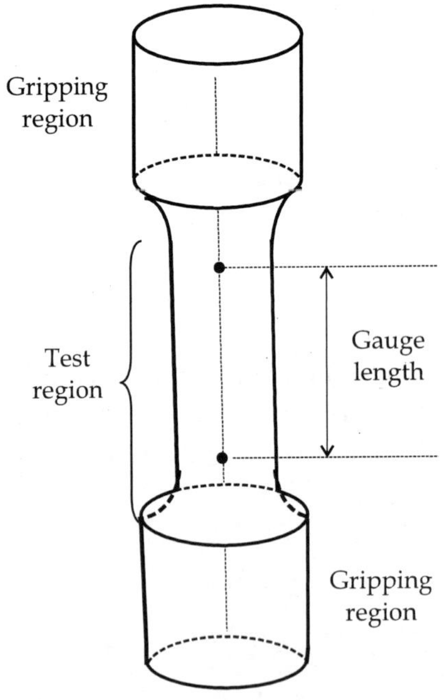

Fig. 2.6

Drawing of a typical test piece is given in Fig. 2.6. The gripping regions of test piece are made with larger cross-sectional area and these are accommodated at the arms of testing machine. A lower cross-sectional area is maintained at the test region so that rupture may always take place at this region. The test region is marked with an initial gauge length over which the deformation is measured. In a standard test piece the gauge length is usually between 2 and 8 inches and diameter of test region is in the range of ½ inch to 1 inch. The mild steel test piece is held firmly between the arms of the testing machine. Tensile load is increased gradually and extension in gauge length is observed. At any stage of loading, the load divided by original cross- sectional area is known as **engineering stress**. Also, a change in gauge length divided by the original gauge length is called **engineering strain**. A graph is plotted with engineering stress along the ordinate against the engineering strain along the abscissa, which is known as a **stress-strain diagram**.

A schematic drawing of stress-strain diagram is given in Fig. 2.7. Some important points located on this stress-strain graph are (1) proportionality limit, (2) elastic limit, (3) yield point, (4) ultimate strength, (5) rupture strength, and (6) true rupture strength.

(1) Proportionality limit: In a stress-strain diagram, there is a straight line up to a point which is known as proportional limit. Stress is proportional to strain ($\sigma \propto \varepsilon$) within the proportionality limit and Hooke's law is applicable. The ratio of stress and strain is defined as **modulus of elasticity** or **Young's modulus** which is a material property, i.e., its magnitude depends on the material of the test piece. Proportionality between stress and strain is no more valid beyond this point though the material (mild steel) still retains its elastic property. Modulus of elasticity is an important quantity because all subsequent theories involving the behaviour of elastic bodies are based on stress-strain proportionality.

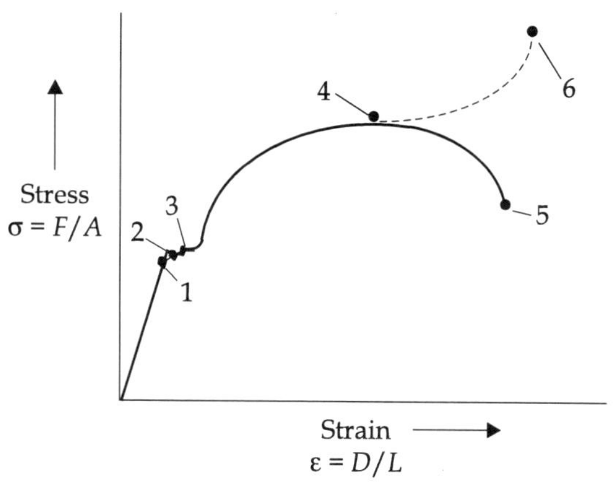

Fig. 2.7

The straight line up to proportionality limit in Fig. 2.7 is representative of linearly elastic behaviour of mild steel. However, there are some materials which show nonlinear elastic behaviour. Stress-strain graph of such materials has no straight line from the beginning of loading. A schematic drawing depicts stress-strain curves of some of these materials (Fig. 2.8). For such materials, therefore, the limit of proportionality is not clearly defined like that of mild steel. Since the design work with such materials requires some data equivalent to proportionality limit, a method of extrapolation is employed. For this, the proportionality between stress and strain is assumed to exist up to a stress at which the strain increases at a rate 50% greater than that given by initial tangent at the origin of stress-strain diagram. This is shown in Fig. 2.9.

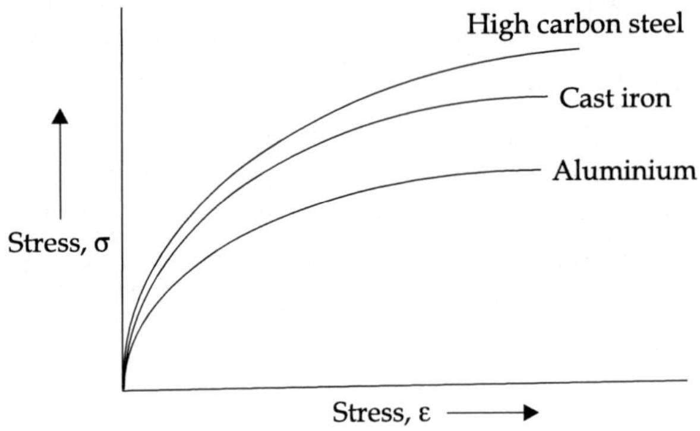

Fig. 2.8

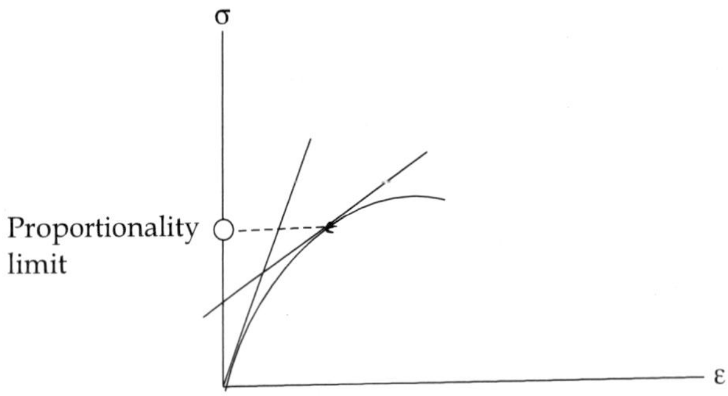

Fig. 2.9

(2) **Elastic limit:** Beyond proportionality limit and up to elastic limit, the proportionality between stress and strain ceases to exist though the material remains elastic, i.e., on removing the external load the strain becomes zero so that the original volume of test piece is retained.

(3) **Yield point:** This is a point just beyond elastic limit and very close to it. At yield point, there is a fairly detectable increase in strain and elongation without any corresponding increase of load or stress. Engineering stress at yield point is known as **yield-point stress** or simply **yield stress** of the material of test piece. As soon as yield point is reached, the material loses its elastic property and begins to flow plastically. If the load is removed at this stage, stress becomes zero but the strain attains a permanent value which is not zero. This is called **permanent set** as given in Fig. 2.10. A permanent set can be observed in the form of increased gauge length at test region. Yield strength is a material property, i.e., its value is fixed for a particular material.

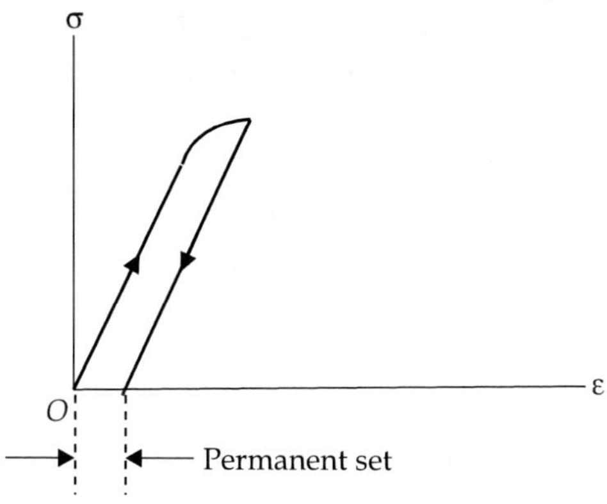

Fig. 2.10

For certain ductile, as well as brittle materials other than mild steel a nonlinear pattern of stress-strain curve results into the absence of any definite yield point. To arrive at an alternative to the yield strength of such materials, a stress level is chosen in the stress-strain diagram which corresponds to a specific strain lying in the range of 0.05 to 0.3 per cent. This stress was earlier called **proof stress** but has been accepted as **yield strength** by the present-day designers. This method is shown for a nonlinear stress- strain curve in Fig. 2.11. An amount of strain between 0.05% and 0.3% is shown by *OS*. This is known as **offset strain**. A tangent to the curve is drawn at the origin. From the point *S* a line parallel to this tangent is drawn to meet the curve at *P*. Then the stress level at *P* gives the yield strength and it is denoted by σ_y.

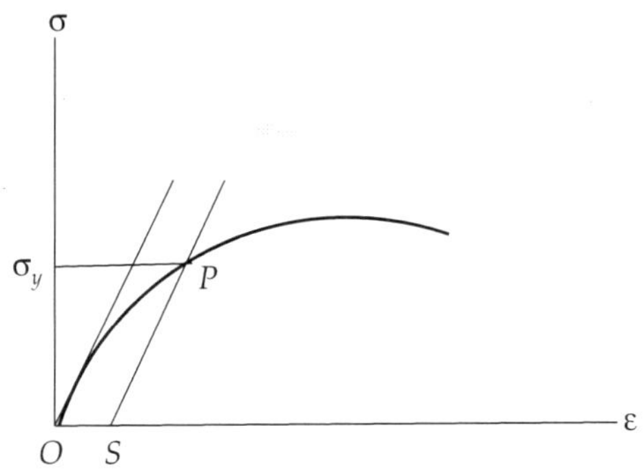

Fig. 2.11

(4) Ultimate strength: After a large strain appears at the yield point, the material begins to reveal strain hardening property. To increase strain any further, higher and higher load or stress is required so that the slope of stress-strain curve becomes positive. This can be seen in Fig. 2.7. In fact, once the plastic flow begins at the yield point the crystal structure within the material changes. Because of defects in crystals in the form of dislocations, displacements at the interior are constrained and higher stresses are required to bring about a specific amount of strain. This process continues further with higher loads till a point corresponding to highest stress is reached. This maximum stress is known as **ultimate tensile strength**(UTS). Like yield strength, ultimate strength is also a material property and both of these strength values are useful in the design of components made of mild steel.

(5) **Rupture strength:** Beyond the ultimate stress point, lower stresses are witnessed with further increments in strain. At this stage, the diameter at test region goes on reducing and a neck formation is visible. The minimum diameter at neck goes on decreasing in size till the rupture takes place, i.e., the test piece breaks into two parts. Load at rupture divided by original cross-sectional area of test region gives the rupture strength. Value of rupture strength is lower than the the ultimate strength which may be apparently paradoxical. However, true rupture strength is the load at rupture divided by the minimum cross-sectional area at the neck which is higher than the UTS. This is shown by point (6) in Fig. 2.7. After rupture of the test piece a cup and cone type formation is found at the neck, which is shown in Fig. 2.12. Failure at the point of rupture is due to a complex stress, and the shear stress also plays a dominant role. However, for consistency with other test data a tensile failure is assumed and rupture strength is calculated.

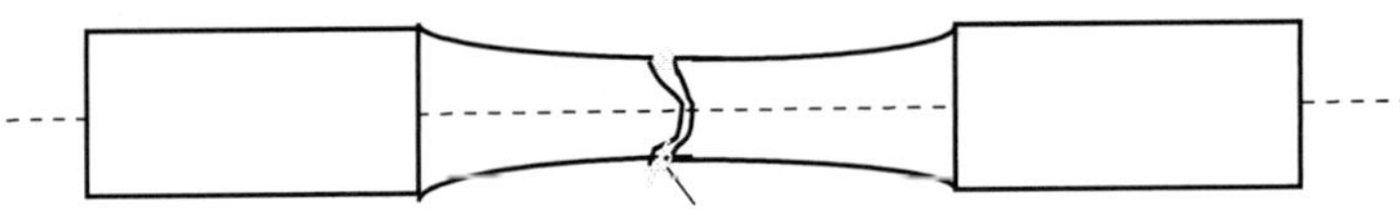

Cup and cone shape

Fig. 2.12

Some other results derived from tension test are **percentage elongation** and **percentage reduction of area**. These two are written as follows:

$$\text{Percentage elongation} = [(L - L_0)/L_0] \times 100 \qquad (2.6)$$

where, L = length in test area at rupture,

and L_0 = original gauge length.

Similarly, we have,

$$\text{Percentage reduction of area} = [(A_0 - A)/A] \times 100 \qquad (2.7)$$

where, A_0 = original cross-sectional area at test region,

and A = reduced cross-sectional area at rupture.

Percentage elongation is a measure of ductility of a material. A test piece made of more ductile material will undergo a larger extension before rupture and hence, a larger percentage elongation. A test piece of brittle material is hassmall amount of elongation before rupture.

(6) **True stress and natural strain:** It is found with a test piece of ductile material, like mild steel, that between the stages of yield point and ultimate stress the diameter at test region begins to reduce in size though a neck is formed only

after ultimate stress is reached. Hence, the engineering stress does not represent the actual value of stress. True stress is, therefore, defined as load divided by current cross-sectional area at test region (and not the original cross-sectional area). Since the volume remains practically unchanged during the stage of test considered,

$$LA = L_0 \cdot A_0 \tag{2.8}$$

where L = current length at test region,

L_0 = original gauge length,

A = current cross-sectional area at test region during the test,

A_0 = original cross-sectional area.

From Eqn. 2.8, $A = A_0 \cdot L_0/L = A_0/(L/L_0)$.

But, engineering strain, $\varepsilon = (L - L_0)/L_0$, or $L = L_0 \cdot (1 + \varepsilon)$.

Hence, $A = A_0/(1 + \varepsilon)$ and true stress, $\sigma = P/A = (P/A_0) \cdot (1 + \varepsilon)$

or

$$\sigma = \sigma_0 \cdot (1 + \varepsilon) \tag{2.9}$$

where, σ_0 = engineering stress.

Further, the incremental nature of strain can be defined as: $de = dL/L$.

Hence, the natural strain is written as

$$e = \int_{L_0}^{L} dL/L = \ln (L/L_0) = \ln (1 + \varepsilon) \tag{2.10}$$

For small deformations, natural strain (e) is very close to conventional strain (ε), whereas for large deformations their difference becomes more pronounced.

2.6 RELATION BETWEEN NORMAL STRESS AND AXIAL STRAIN

Referring back to axial loading in tension which is explained with the help of Fig. 2.2 and Fig. 2.5, it is a common pactice to apply a normal stress which is well below the proportionality limit of the material used. Therefore, normal stress and axial strain are assumed to be related according to $\sigma \propto \varepsilon$. Applying Hooke's law we have

$$\sigma = E \cdot \varepsilon \quad \text{or} \quad \varepsilon = \sigma/E \tag{2.11}$$

where, E = modulus of elasticity of material of the bar. Using Eqns. 2.1 and 2.4 we have

$$\Delta L/L = P/(A \cdot E), \text{ or } \Delta L = P/(AE/L) \tag{2.12}$$

The above expression gives a relation between force and deformation of the bar and indicates that the bar under the action of load (P) deforms in elastic manner in the same way as a helical spring. The quantity (AE/L) is stiffness of bar which depends on its material (E) and dimensions (A and L).

It should be noted that when the stress developed in a bar is below the yield stress, then the stress is independent of the type of material and depends only on the applied load and dimensions of the bar. It means that if bars made of mild steel, copper, aluminium, etc., are each of same cross section (A) and length (L), and subjected to same amount of force, then the same amount of stress will be induced in each of them. However, the strain (ε) or elongation (ΔL) will be different for each bar. This is due to the reason that the elasticity modulus € is different for the different materials and strain, $\varepsilon = \sigma/E$. The change in length, $\Delta L = P{\cdot}L/(AE)$, will vary from material to material. For example, stress-strain diagrams within the elastic range are shown together for the bars of three materials in Fig. 2.13. Against a common normal stress (σ_0), the strains developed in the bar are respectively: $\varepsilon_1 = \sigma_0/E_1$, $\varepsilon_2 = \sigma_0/E_2$, and $\varepsilon_3 = \sigma_0/E_3$, where, E_1, E_2 and E_3 are elastic moduli of the three types of materials. Similarly, the bars subjected to same force (P) will go through extensions given by $PL/(AE_1)$, $PL/(AE_2)$ and $PL/(AE_3)$, respectively.

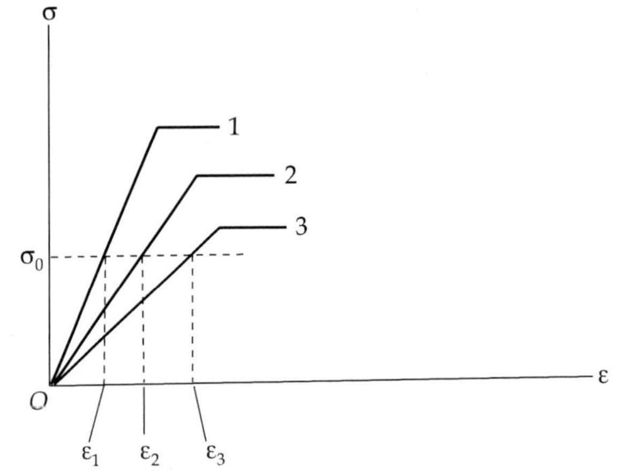

Fig. 2.13

Example 2.1: An axial tensile force of 20 kN is applied to a 300 mm long bar and its extension is 0.15 mm. If diameter of the bar is 16 mm, what is the elasticity modulus of the material of the bar?

Solution: $P = 20 \times 10^3$ N, $L = 300$ mm, $\delta = 0.15$ mm, $d = 16$ mm

$\therefore$ $A = (\pi/4){\cdot}d^2$ and $E = 4PL/(\pi{\cdot}d^2{\cdot}\delta)$

$\therefore$ $E = \dfrac{4 \times 20 \times 10^3 \times 300}{\pi \times 16^2 \times 10^{-6} \times 0.15}$ N/m² or $E = 199 \times 10^9$ N/m² = 199 MPa **Ans.**

Example 2.2: A 14 mm diameter aluminium bar is subjected to axial loads at the positions shown in Fig. 2.14. If $E = 70$ GPa, compute the total axial deformation of the bar.

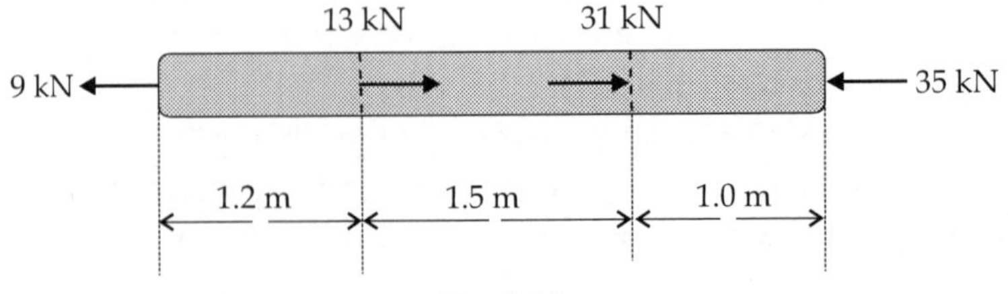

Fig. 2.14

Solution: Axial deformation in each part, $\delta = 4 \cdot P \cdot L / (\pi \cdot d^2 \cdot E)$.

For part I,

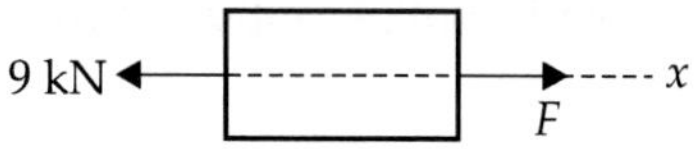

$$\sum F_x = 0, \therefore F - 9 = 0, F = 9 \text{ kN (tension)}$$

$\therefore$ $\delta_1 = 4 \times 9 \times 10^3 \times 1200/(\pi \times 14^2 \times 10^{-6} \times 70 \times 10^9)$ mm

or $\delta_1 = 1.00$ mm

For part II,

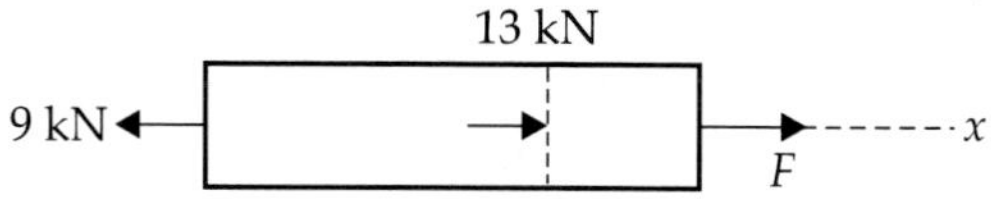

$F - 9 + 13 = 0, \therefore F = -4$ kN (compression)

$$\delta_2 = 4 \times (-4 \times 10^3) \times 1500/(\pi \times 14^2 \times 10^{-6} \times 70 \times 10^9)$$

or $\delta_2 = -0.56$ mm

For part III,

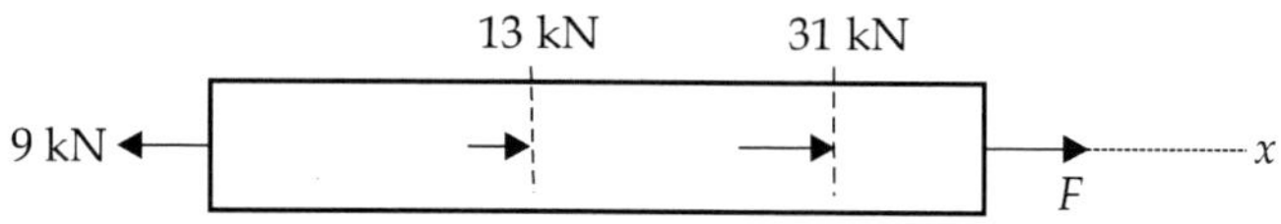

$F - 9 + 13 + 31 = 0,$

$\therefore$ $F = -35$ kN (compression)

$\therefore \delta_3 = 4 \times (-35 \times 10^3 \times 1000/(\pi \times 14^2 \times 10^{-6} \times 70 \times 10^9)$ mm

or $\delta_3 = -3.25$ mm

$\therefore$ Total axial deformation of bar $= \delta_1 + \delta_2 + \delta_3$

$= 1.00 + (-0.56) + (-3.25)$

$= -2.81$ mm (contraction) **Ans.**

Example 2.3: A uniform concrete slab shown (Fig. 2.15) is of weight, *W*. It is attached to a 1.5 m long aluminium rod and a 2.0 m long steel rod such that the lower ends of the two rods are at the same horizontal level. If the elasticity modulus of steel and aluminum are 210 GA and 70 GPa respectively, determine the ratio of cross-sectional areas of the two rods so that the concrete slab remains horizontal.

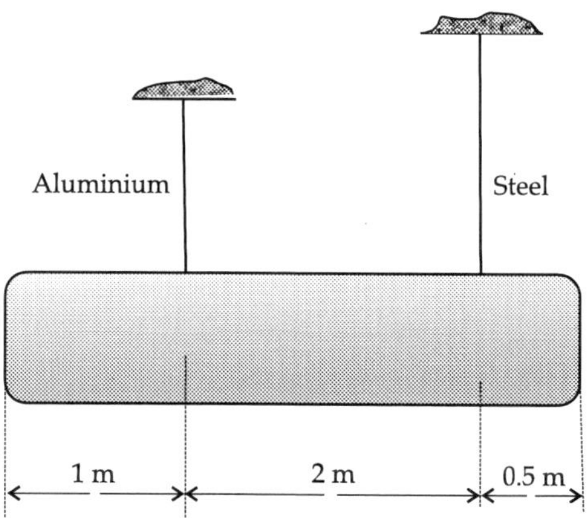

Fig. 2.15

Solution: On applying load, the extension of two rods must be same so that the slab remains horizontal. The forces developed in two rods can be obtained from the FBD of the slab.

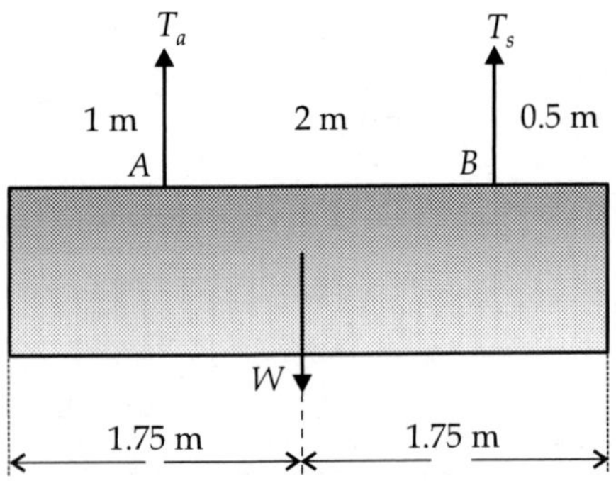

Let E_a = elasticity modulus of aluminium
E_s = elasticity modulus of steel
A_a = cross-sectional area of aluminium rod
A_s = cross-sectional area of steel rod
L_a = length of aluminium rod
L_s = length of steel rod
T_a = tensile force developed in aluminium rod
T_s = tensile force developed in steel rod

From $\sum M_B = 0$, $T_a \times 2 - W \times (1.75 - 0.5) = 0$, or $T_a = 0.625$ W

From $\sum M_A = 0$, $T_s \times 2 - W \times (1.75 - 1) = 0$, or $T_s = 0.375$ W

Check : $T_a + T_s = (0.625 + 0.375)$ W = W

$\therefore$ Extension of aluminium rod, $\delta_1 = [0.625 \text{ W}/(A_a \cdot E_a)] \times L_a$

Extension of steel rod, $\delta_2 = [0.375 \text{ W}/(A_s \cdot E_s)] \times L_s$

But, $\delta_1 = \delta_2$, or $0.625 \text{ W} \times 1.5/(A_a \times 70) = 0.375 \text{ W} \times 2.0/(A_s \times 210)$

$\therefore A_a/A_s = 3.75$ **Ans.**

Example 2.4: A two-step vertical bar supports a load of 7 kN acting at the top, as shown in Fig. 2.16. A second load (P) is uniformly distributed over the shoulder of bottom step as shown. The diameter at the upper and lower parts are 30 mm and 60 mm, respectively. Determine the normal stress induced in the upper part of the bar. If the lower part is to bear same compressive stress as the upper part, what should be the magnitude of load, P?

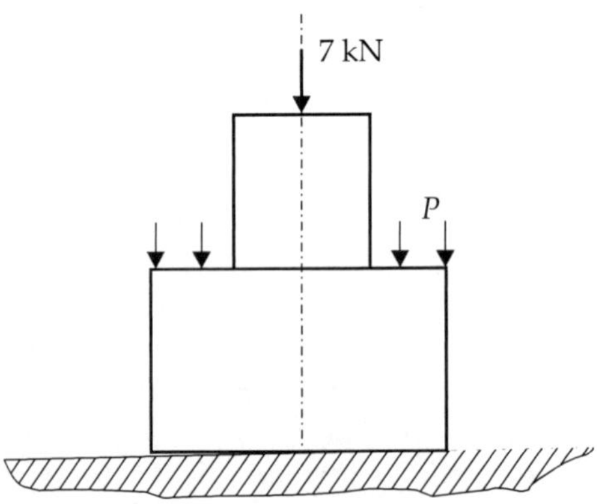

Fig. 2.16

Solution: Considering *FBD* of upper part up to an arbitrarily chosen cross section,

$$\sum F_y = 0$$

$$F - 7 = 0, \therefore F = 7 \text{ kN}$$

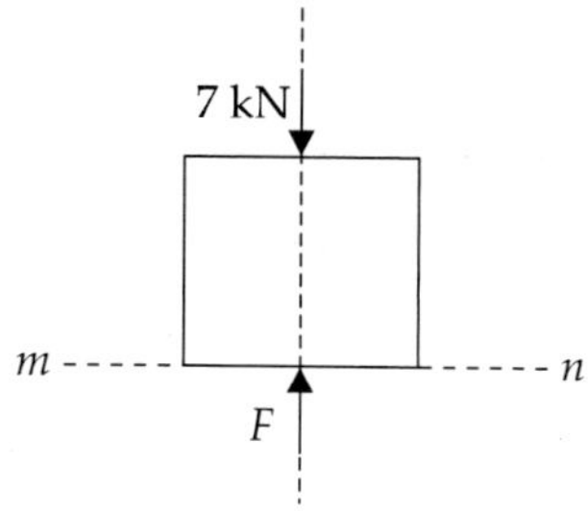

Thus, the upper part is subjected to a compressive force of 7 kN.

$\therefore$ Stress $= -(4 \times 7 \times 10^3)/(\pi \times 30^2 \times 10^{-6})$ N/m^2

$= -9.9 \times 10^6$ N/m^2 or 9.9 MPa (compressive) **Ans.**

Next, considering FBD after sectioning in the bottom part,

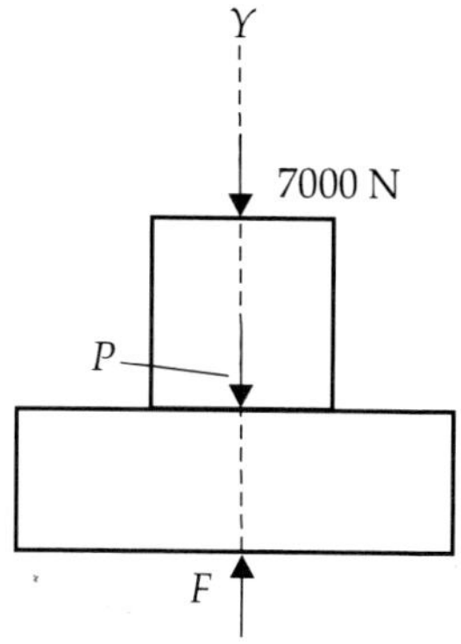

From $\sum F_y = 0, F - P - 7000 = 0$

$\therefore$ $F = P + 7000$ N (a compressive force)

To satisfy the required condition in the problem,

$4 \times (P + 7000)/(\pi \times 60^2 \times 10^{-6}) = 9.9 \times 10^6$

$\therefore P = 20991.5$ N, or $P = 21$ kN **Ans.**

Example 2.5: Two identical steel bars, *AB* and *BC*, are each 3 m in length and connected through pin-joint at *B* (see Fig. 2.17). A load of 445 kN is applied at the joint *B*. If the normal stress in each bar is not to exceed 200 MPa, determine the cross-sectional area of each bar. Also, find the vertical displacement of the joint *B*. Take $E = 200$ GPa.

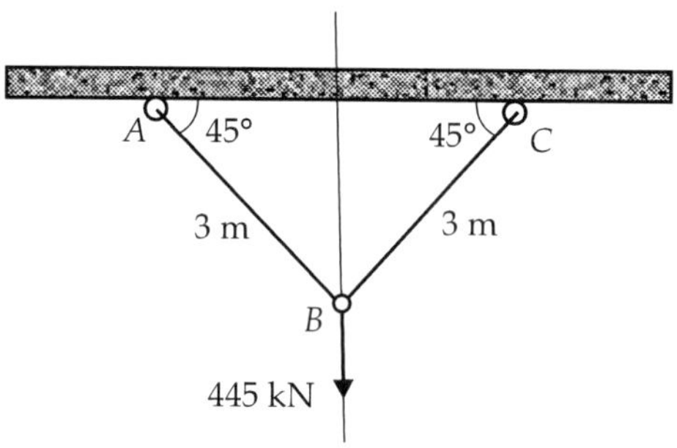

Fig. 2.17

Solution: Considering FBD of joint *B* in equilibrium, let T_1 and T_2 be the tensile forces in bar *AB* and *BC*.

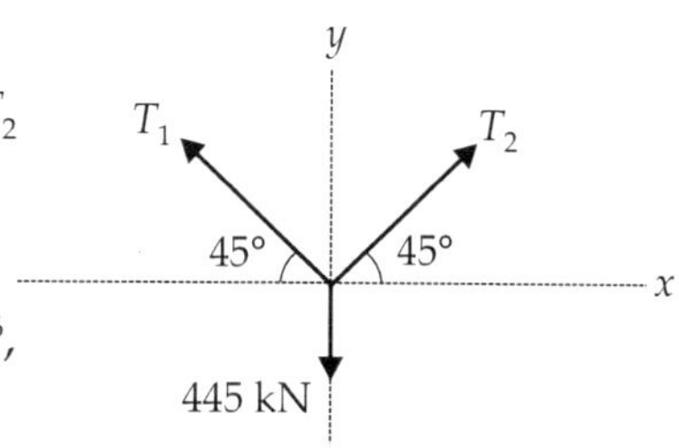

From $\sum F_x = 0, T_2 \cdot \cos 45° - T_1 \cdot \cos 45° = 0.$ or $T_1 = T_2$

From $\sum F_y = 0, (T_1 + T_2) \cdot \sin 45° - 445 = 0.$

$\therefore$ $T_1 = T_2 = 314.66$ kN or 315 kN.

The stress induced in each bar: $315 \times 10^3/A = 200 \times 10^6$,

where, A = area of cross section.

$\therefore$ $A = 1.575 \times 10^{-3}$ m^2 or $A = 15.75$ cm^2 **Ans.**

Elongation of each bar, $\delta = (\sigma/E)\cdot L = [200 \times 10^6/(200 \times 10^9)\cdot 3$, or $\delta = 0.003$ m

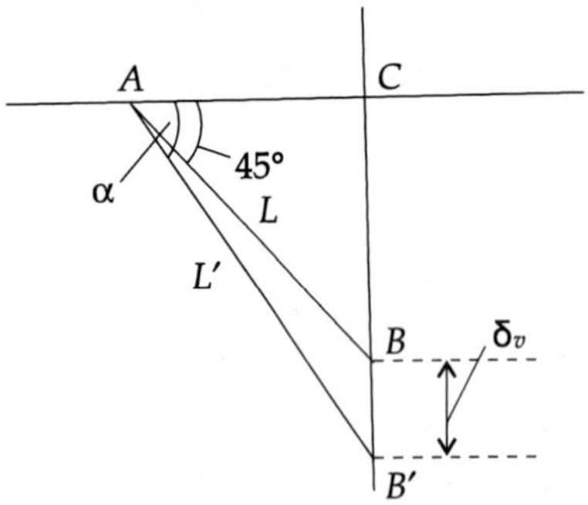

For $L = 3$ m, $L' = L + \delta = 3.003$ m

$$AC = L'\cdot\cos\alpha = L\cdot\cos 45°$$

or $3.003\cdot\cos\alpha = 3\cdot\cos 45°$

$\therefore \quad \alpha = 45.0572°$

Vertical deflection, $\delta_v = L'\cdot\sin\alpha - L\cdot\sin 45°$

or $\quad \delta_v = 3.003\cdot\sin 45.0572° - 3\cdot\sin 45°$

$\therefore \quad \delta_v = 4.24 \times 10^{-3}$ m or $\delta_v = 4.24$ mm **Ans.**

Example 2.6: A uniform bar of length L, cross-sectional area A, and weight W is suspended vertically from one end, as shown in Fig. 2.18. Elasticity modulus of the material of bar is denoted by E. Show that the total elongation of the bar is $\delta = WL/(2\cdot A\cdot E)$. If ρ = mass per unit volume, show that $\delta = \rho\cdot g\cdot L^2/(2E)$.

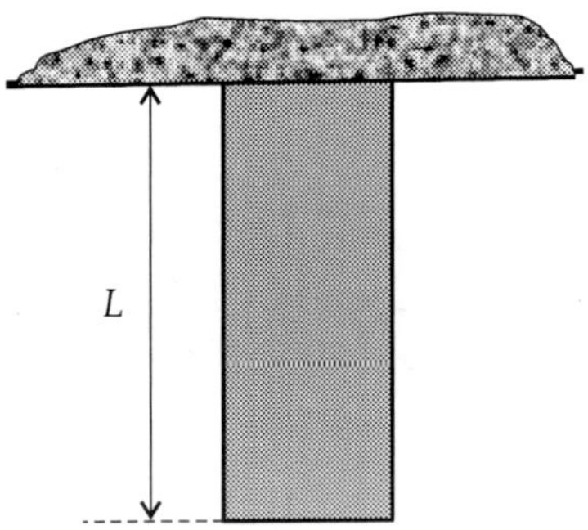

Fig. 2.18

Solution: Let F = vertically downward force normal to any cross section which varies along the length of the bar such that

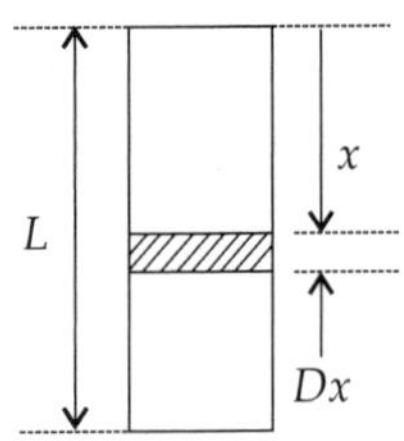

at $\quad x = 0, F = W$

$\quad x = L, F = 0$

Assuming F as a function of x measured along the axis of bar, let $F = m.x + C$.

Hence, applying the above conditions, we have

$$m = -W/L \text{ and } C = W, \text{ so that}$$

$$F = -(W/L)\cdot x + W \text{ or } F = W\cdot(L - x)/L$$

Next, considering an element of length dx at a distance x from the fixed end, the stress over the cross section

$= W\cdot(L - x)/(A\cdot L)$

Hence, the element's elongation, $d\delta = [W\cdot(L - x)/(A\cdot L\cdot E)]\cdot dx$

$\therefore$ Total elongation, $\delta = \int_0^L [W\cdot(L-x)/(A > L > E)]\cdot dx$, or $\delta = W\cdot L/(2\cdot A\cdot E)$ **Proved.**

Similarly, normal force over the cross section at a distance x is $F = A\cdot(L - x)\cdot\rho\cdot g$

$\therefore\ d\delta = [A\cdot(L - x)\cdot\rho\cdot g\ /(A\cdot E]\cdot dx$ or $d\delta = [\rho\cdot g\cdot(L - x)/E]\cdot dx$

$\therefore\ \delta = (\rho\cdot g\ /E)\cdot\int_0^L (L - x).dx$ or $\delta = \rho\cdot g\cdot L^2/(2.E)$ **Proved.**

Example 2.7: A bar between two built-in supports (Fig. 2.19) is acted upon by a force P. Determine reactions at the two ends of the bar.

Solution: This is a statically indeterminate problem and cannot be solved with $\sum F_x = 0$, only. Hence, the deformations of two parts of length a and b are to be used.

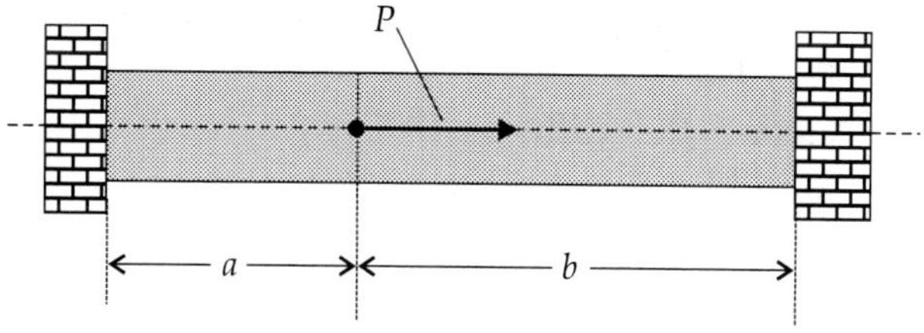

Fig. 2.19

Let R_1 and R_2 be reactions at the left end right end, respectively.

For equilibrium up to a cross section on the left part $(0 < x < a)$,

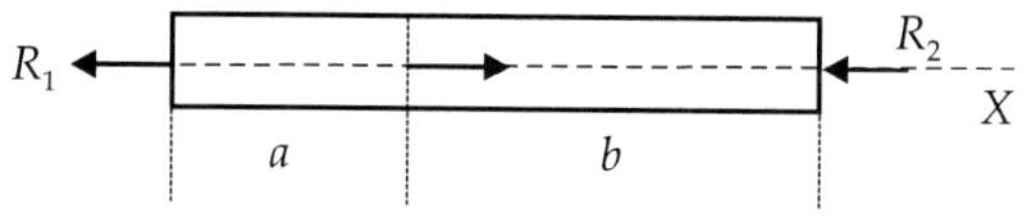

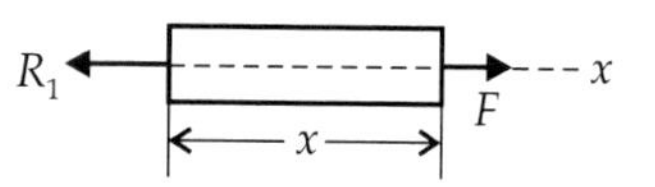

From $\sum F_x = 0$, $F - R_1 = 0$ or $F = R_1$, and F is a tensile force which is valid from $x = 0$ to $X = a$. Therefore, extension of the left part, $\delta_1 = R_1\cdot a/(A\cdot E)$.

Similarly, for equilibrium up to a cross section on the right part $(a < x < (a + b))$,

From $\sum F_x = 0$, $F + P - R_1 = 0$

$\therefore\quad F = -(P - R_1)$

$\therefore$ Contraction of right part,

$\delta_2 = -(P - R_1)\cdot b/(A\cdot E)$

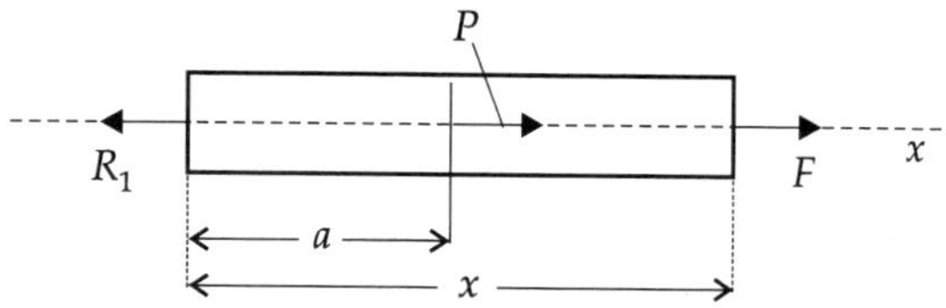

Since the linear space between supports is fixed, $|\delta_1| = |\delta_2|$

$\therefore\ R_1\cdot a/(A\cdot E) = (P - R_1)\cdot b/(a\cdot E)$, or $R_1 = P\cdot b/(a + b)$ **Ans.**

Applying equilibrium condition to the entire bar, $-R_1 + P - R_2 = 0$

$\therefore\ R_2 = P - R_1 = P - [(P\cdot b/(a + b)]$, or $R_2 = P\cdot a/(a + b)$ **Ans.**

Example 2.8: As shown in Fig. 2.20, three steel wires, each 32 mm^2 in area, are used to lift a load, W = 6.6 kN. The unstretched lengths of wires are 22.854 m, 22.857 m and 22.860 m, respectively.

(a) What will be the normal stress induced in the longest wire ?

(b) If W = 2.2 kN, determine the stress induced in the shortest wire.

Take E = 200 GPa.

Solution:

(a) Let us assume that the load (W) is slowly applied from zero to 6.6 kN. At first the shortest wire begins to stretch and bears the load till its length becomes equal to the second wire of the next higher length. After that the two wires simultaneously get stretched till their lengths are equal to that of the largest wire. Following this, all the three wires extend further till they bear 6.6 kN load together.

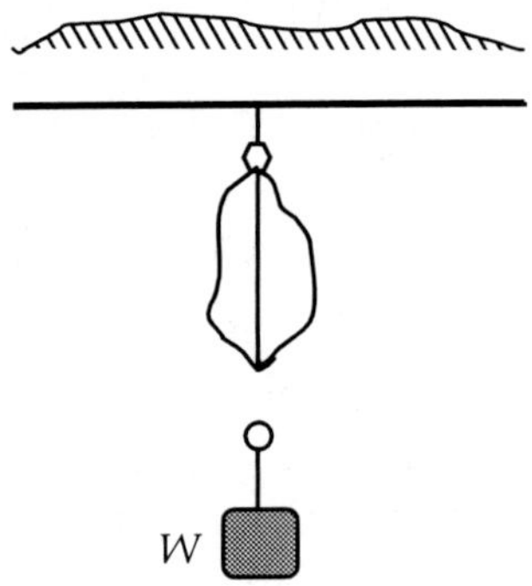

Fig. 2.20

Hence, the stiffness ($A{\cdot}E/L$) of each wire is first found because: stiffness × extension = load.

Stiffness of wires are as follows :

$k_1 = 32 \times 10^{-6} \times 200 \times 10^9/22.854$ N/m or $k_1 = 280.04 \times 10^3$ N/m

$k_2 = 32 \times 10^{-6} \times 200 \times 10^9/22.857$ N/m or $k_2 = 280.00 \times 10^3$ N/m

$k_3 = 32 \times 10^{-6} \times 200 \times 10^9/22.860$ N/m or $k_3 = 279.97 \times 10^3$ N/m

Force developed in the shortest wire when it attains the length equal to the wire of next higher length:

$F_1 = k_1{\cdot}(L_2 - L_1) = 280.04 \times 10^3{\cdot}(22.857 - 22.854)$ N = 840.12 N

Force developed further as the first two wires have attained length equal to that of of the longest wire:

$F_2 = (k_1 + k_2){\cdot}(L_3 - L_2) = (280.04 + 280.00) \times 10^3 \times (22.860 - 22.857)$ N = 1680.12 N

{Note that equivalent stiffness of two wires in parallel is $(k_1 + k_2)$}

Next, let the three wires extend further through x metres till the total force = 6.6 kN

$\therefore \quad F_1 + F_2 + (k_1 + k_2 + k_3){\cdot}x = W$

or $\quad 840.12 + 1680.12 + (280.04 + 280.00 + 279.97) \times 10^3{\cdot}x = 6600$ N

$\therefore \quad x = 4.857 \times 10^{-3}$ m

Normal stress induced in the longest wire, $\sigma = k_3{\cdot}x/A$

or $\sigma = 279.97 \times 10^3 \times 4.857 \times 10^{-3}/(32 \times 10^{-6})$ N/m² or $\sigma = 42.5$ MPa **Ans.**

(b) Extension of two wires together to lift the load of 2.2 kN (the longest wire still remains slack):

$F_1 + (k_1 + k_2){\cdot}x = W$, or $840.12 + (280.04 + 280.00) \times 10^3{\cdot}x = 2200$ N

$\therefore x = 2.43 \times 10^{-3}$ m

Stress developed in the shortest wire, $\sigma = (F_1 + k_1{\cdot}x)/A$

or $\sigma = (840.12 + 280.04 \times 10^3 \times 2.43 \times 10^{-3})/32 \times 10^{-6}$ N/m², or $\sigma = 47.5$ MPa

Ans.

Example 2.9: A copper tube between two rigid plates is to be pressed by a 16 mm diameter steel bolt shown in Fig. 2.21. The pitch of screw threads of bolt is 2 mm. Inner and outer diameteres of copper tube are 20 mm and 40 mm, respectively. If the nut is tightened through one quarter of revolution, what are the stresses produced in the bolt and the tube? Take E_s = 200 GPa for steel and E_c = 100 GPa for copper.

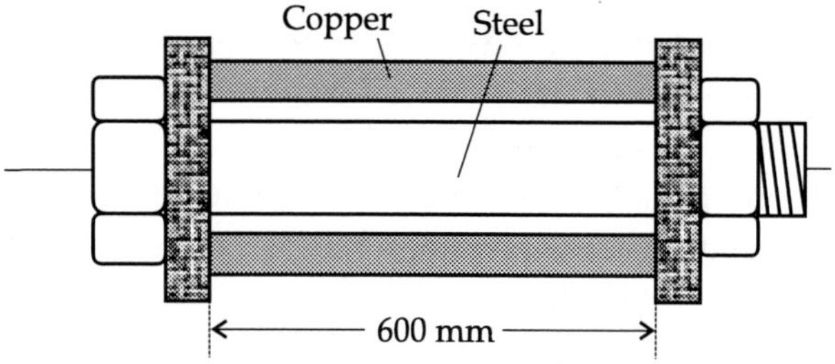

Fig. 2.21

Solution: When the nut is tightened, the steel bolt is in tension and copper tube is under compression.

Let σ_s = tensile stress produced in steel bolt

σ_c = compressive stress produced in copper tube

A_s = cross-sectional area of steel bolt

A_c = cross-sectional area of copper tube

A part of this assembly up to a cross-sectional plane can be considered to be in equilibrium so that there is a tensile force in the bolt which is equal and opposite to the compressive force in the tube.

From $\sum F_y = 0,$

$$\sigma_s \cdot A_s = \sigma_c \cdot A_c \quad \text{...(a)}$$

After tightening the nut through a distance along the axis of bolt by an amount $n{\cdot}p$, where, $n = 1/4$ and p = 2 mm, there will be an extension in bolt, given by $\delta_B = (\sigma_s / E_s)\cdot L$, and a contraction in tube, $\delta_T = (\sigma_c / E_c)\cdot L$ so that $\delta_B + \delta_T = n{\cdot}p$.

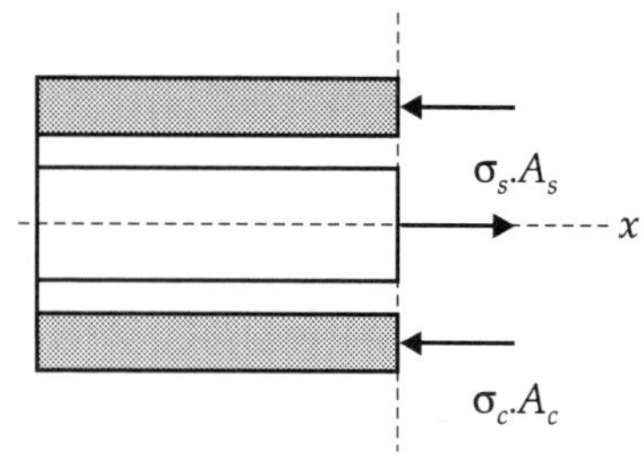

$$\therefore (\sigma_s / E_s) + (\sigma_c / E_c) = n{\cdot}p / L \quad \text{...(b)}$$

$$A_s = (\pi/4) \times 16^2 \times 10^{-6} = 64{\cdot}\pi \times 10^{-6} \text{ m}^2, \text{ and}$$

$$A_c = (\pi/4)\cdot(40^2 - 20^2) \times 10^{-6} = 300{\cdot}\pi \times 10^{-6} \text{ m}^2$$

From (a), $\sigma_s = \sigma_c \cdot (A_c / A_s) = \sigma_c \cdot (300/64)$, or $\sigma_s = 4.6875 \cdot \sigma_c$

In (b), $\sigma_c \cdot [(4.6875/200) + (1/100)] \times (1/10^9) = 2/(4 \times 600)$

$$\sigma_c = 24.9 \times 10^6 \text{ N/m}^2 = 24.9 \text{ MPa}$$

$$\sigma_s = 116.8 \times 10^6 \text{ N/m}^2 = 116.8 \text{ MPa}$$

Ans.

2.7 FACTOR OF SAFETY

Yield strength and ultimate strength discussed in section 2.4, get their practical application in design calculations with the use of a quantity known as 'factor of safety'.

When a structural part made from ductile material is loaded, it is expected that the magnitude of stress at any point should not exceed the yield strength of the material. Plastic yielding may cause deformation which results into malfunctioning of the part or unwanted additional load on the adjoining parts. A structural component or a machine element is designed with the purpose of retaining the stress at every point within the elastic limit. But, there are several factors which do not guarantee this. For example, there may be points of stress raisers at the surface due to manufacturing defect; there are occasional high loads producing higher stresses or because of size and shape of a part for which the tensile test with the same material may not be a true representative of the state of yielding anywhere within the part. To avoid such uncertainties, structural parts are designed with a limiting stress well below the yield strength which is called **working stress (σ_w), allowable stress** or **permissible stress,** and written as

$$\sigma_w = \sigma_y / n \tag{2.13}$$

where, n = factor of safety, and σ_y = yield strength.

Factor of safety is a number which is always greater than one. Higher the value of n, lower is the allowable stress and design calculation results into larger dimensions and higher weight of the component. On the other hand, with a lower factor of safety the permissible stress is closer to the yield strength and the designer faces a risk, especially with high strength materials. Selection of factor of safety is a complicated task for designers and it is accomplished on the basis of experience in workshops or experimentation with the prototypes of components.

For a material like high carbon steel, concrete, cast iron, etc., whose yield strength is not clearly defined, the permissible stress is selected against the ultimate strength. The philosophy supporting this is if the stress at any point reaches to a magnitude equal to ultimate strength, it will soon result into catastrophic rupture of the component. Therefore, the permissible stress is written as

$$\sigma_w = UTS / n \tag{2.14}$$

Factor of safety applied to ultimate strength is larger than that applied to yield strength so that the working stress value may remain also below the yield strength of the material.

2.8 POISSON'S RATIO

When a bar is under tension or compression, it is found that simultaneously with axial deformation there is also deformation in the transverse direction, i.e., in a direction perpendicular to the axis of bar. Poisson found a definite relation between these two types of deformations. A bar under tension is shown in Fig. 2.22(a) and another in compression is given in Fig. 2.22(b). It is found from experiments that an extension (δ_x) is always accompanied by a contraction ($-\delta_y$) in the transverse direction. Similarly, a contraction ($-\delta_x$) in axial direction occurs with an extension (δ_y) in transverse direction.

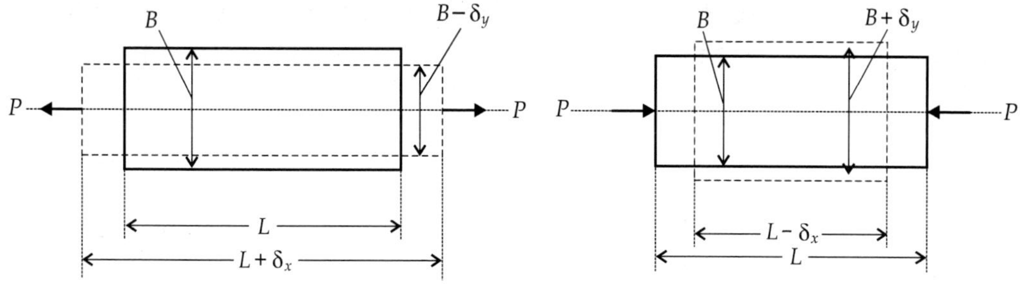

Fig. 2.22

If normal stress in the bar remains within the proportionality limit, it is found that the ratio of transverse strain (ε_y) to axial strain (ε_x) is always a constant for the material of the bar. This constant is known as **Poisson's ratio** and it is denote by ν.

Referring to Fig. 2.22(a), axial strain, $\varepsilon_x = \delta_x/L$, transverse strain, $\varepsilon_y = -\delta_y/B$.

Hence, the Poisson's ratio is written as: ν = –(transverse strain/axial strain)

$$\nu = -(\varepsilon_y/\varepsilon_x) \tag{2.15}$$

Similarly, in Fig. 2.22(b), $\varepsilon_x = -\delta_x/L$ and $\varepsilon_y = \delta_y/B$, and Poisson's ratio

$$\nu = -(\varepsilon_y/\varepsilon_x) \tag{2.16}$$

It should be noted that when ε_x is positive, ε_y is negative and vice versa. Poisson's ratio (ν) is, therefore, a positive quantity. Like elasticity modulus, ν is also a material constant.

The above proposition can be extended to the third dimension (*z*-direction) also. It is shown in Fig. 2.23 that with an extension in *x*-direction, there are contractions both in *y*- and *z*-directions. As such, Poisson's ratio, $\nu = -(\varepsilon_y/\varepsilon_x) = -(\varepsilon_z/\varepsilon_x)$. Thus, the ratio of lateral strain to longitudinal strain is same for both *y*- and *z*-directions due to an axial force (and normal stress) in *x*-direction.

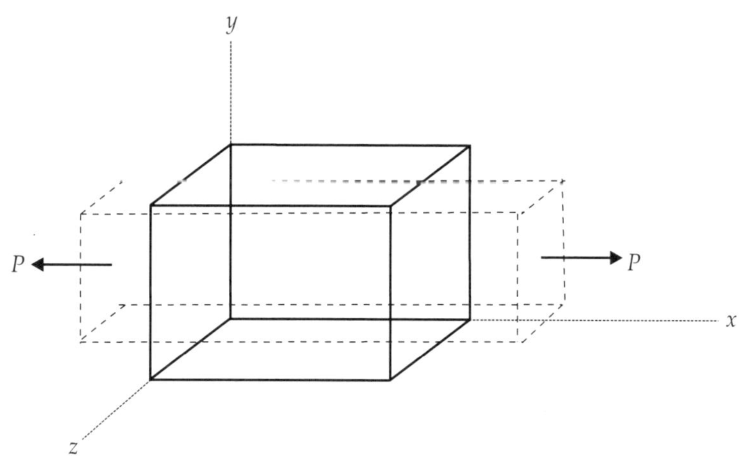

Fig. 2.23

Values of Poisson's ratio: Most of the common engineering materials are assumed to be homogeneous and isotropic. If the macroscopically observed physical properties such as density, thermal expansion, etc., are identical at all points in a material body, the material is said to be homogeneous. If the elastic property (stress-strain relation) at any point is same in every direction, the material is said to be isotropic. It can be shown that the maximum value of Poisson's ratio for a homogeneous and isotropic material is 0.5. Let us take a cubical element with L as length of each side and subjected to uniaxial tension (P), as shown in Fig. 2 .24. The volume of cube = L^3. .ε.

Let ε = the strain in x-direction so that after the force P is applied, the length of sides parallel to x-axis becomes $L_x = L + \varepsilon \cdot L$, or $L_x = L \cdot (1 + \varepsilon)$. With ν = Poisson's ratio and strains along y- and z-directions

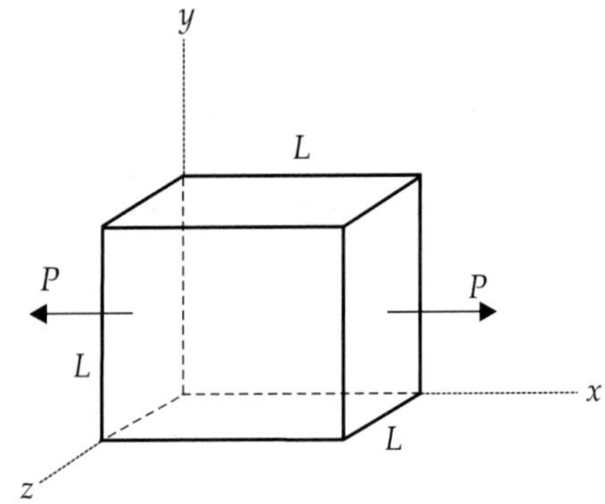

Fig. 2.24

As ε_y and ε_z, respectively,

$$\nu = -(\varepsilon_y/\varepsilon) \text{ or } \varepsilon_y = -\nu \cdot \varepsilon, \text{ and}$$

$$\nu = -(\varepsilon_z/\varepsilon) \text{ or } \varepsilon_z = -\nu \cdot \varepsilon.$$

∴ Length of sides parallel to y- and z-axes are

$$L_y = L_z = L - \nu \cdot \varepsilon \cdot L = L \cdot (1 - \nu \cdot \varepsilon)$$

Hence, the volume of cube after force (P) is applied

$$V' = L_x \cdot L_y \cdot L_z = L \cdot (1 + \varepsilon) \cdot L^2 \cdot (1 - \nu.\varepsilon)^2$$

Since ε is a small quantity, ε^2 and ε^3 are ignored, so that $V' = L^3 \cdot (1 + \varepsilon - 2 \cdot \nu \cdot \varepsilon)$

∴ Change in volume, $\Delta V = V' - V = L^3 \cdot \varepsilon \cdot (1 - 2\nu)$.

Because it is most unlikely that ΔV will be negative under a tensile load, $\Delta V > 0$,. or $(1 - 2\nu) > 0$.

∴ $\nu < 0.5$.

For the materials like rubber, paraffin, etc., Poisson's ratio is close to 0.5. Volume of these materials almost remains same on applying an axial force. On the other hand, for concrete ν is as small as 1/8 to ½, and for cork, ν = 0.

2.9 SHEAR STRESS AND SHEAR STRAIN

It was discussed earlier that a force acting perpendicular to a plane area gives rise to a normal stress. Unlike this, a force acting tangentially over a plane area results into a shear stress. In Fig. 2.25(a), a cube is shown to be fixed with the ground. Each side of the cube measures 'a'. A force (P) acts tangentially on its top surface. Then the shear stress can be written as

$$\tau = F/a^2 \tag{2.17}$$

Like normal stress the shear stress is also expressed in the unit of N/m². Just as an axial force is responsible for axial deformation, a shear force causes shear deformation. A side-view of shear deformation of the cube is shown in Fig. 2.25(b). For small deformation, shear stress only produces distortion, i.e., the change in shape of the body and no change in volume. The shear deformation can be visualized as sliding of different layers of cube (parallel to the ground) relative to one another. A layer just in contact with the ground has no displacement at all but a layer at a distance of '*a*' from the ground undergoes a displacement, δ_s. Therefore, the shearing strain is defined by tan (γ), where, γ is the change in 90° angle measures in radians. Since the deformation is small, we have tan (γ) $\simeq \gamma$ and the shear strain can be written as

$$\gamma = \delta_s / a \tag{2.18}$$

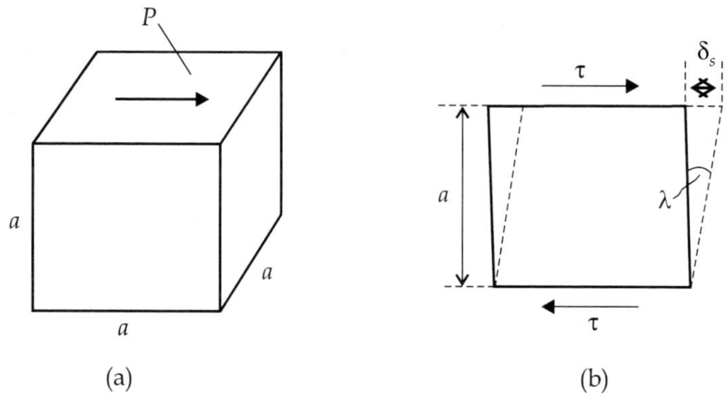

Fig. 2.25

Shear stress and shear strain are also related according to Hooke's law. Within the limit of proportionality, we have $\tau \propto \gamma$, or

$$\tau = G \cdot \gamma \tag{2.19}$$

where, G = shear modulus or modulus of rigidity of the material. Shear modulus (G) is related to the two fundamental elastic constants, E and ν, according to the relation

$$G = E / [2 \cdot (1 + \nu)] \tag{2.20}$$

Shear modulus (G) is also a constant for a given material.

Example 2.10: A bar of uniform cross section is axially loaded with a force (P) (see Fig. 2.26). What will be the normal and shear stresses on a section inclined at an angle θ with respect to the cross section ? If P = 200 N and cross-sectional area, A = 10 mm^2, what is the maximum shear stress and for what value of θ does it occur?

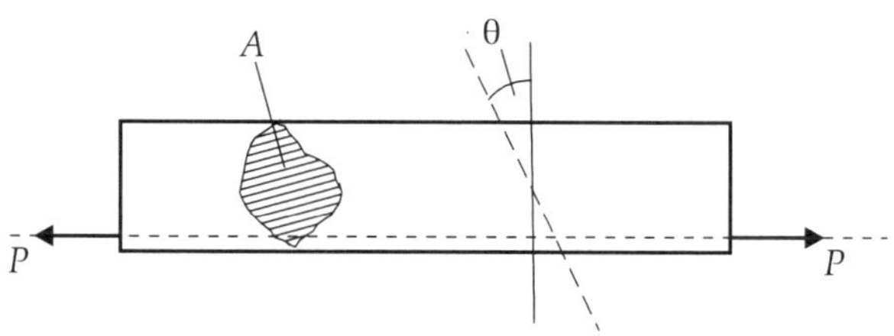

Fig. 2.26

Solution: Free-body diagram of the left part is drawn including the section.

Area of cross section at n-n is A. ∴ Area of section, n-p is $A/\cos\theta$.

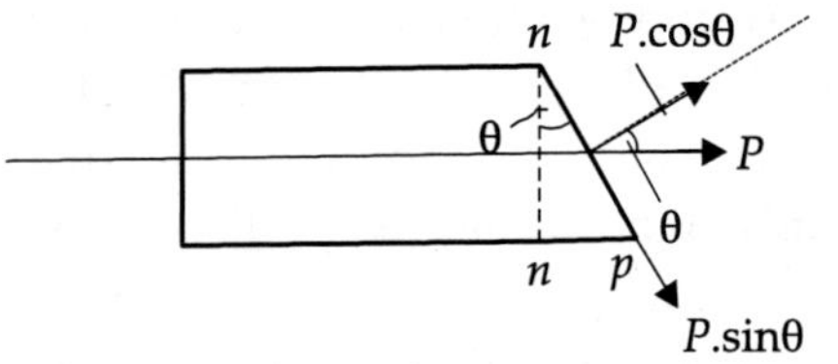

Force P has a component $P\cdot\cos\theta$ perpendicular to section n-p and another component $P\cdot\sin\theta$ tangential to this section.

∴ Normal stress, $\sigma_n = P\cdot\cos\theta/(A/\cos\theta)$

$= (P/A)\cdot\cos^2\theta$ **Ans.**

and shear stress, $\tau = P\cdot\sin\theta/(a/\cos\theta) = (P/A)\cdot\sin\theta\cdot\cos\theta$,

or $\tau = (P/2A)\cdot\sin 2\theta$ **Ans.**

For shear stress to be maximum, $d\tau/d\theta = 0$, or $(P/A)\cdot\cos 2\theta = 0 = \cos 90°$. ∴ $\theta = 45°$

At $\theta = 45°$, $\tau = \tau_{max} = P/2A$. ∴ $\tau_{max} = 200/(2\times10\times10^{-6})\ \text{N/m}^2 = 10$ MPa **Ans.**

Example 2.11: A plate of length L, width B and thickness t is subjected to uniform tensile stress σ at the two ends, as shown in Fig. 2.27. Elasticity modulus is E and Poisson's ratio is ν for the material of the plate. (a) Before σ is applied, the slope of diagonal OB is B/L. What will be the slope when σ is acting? (b) What is the increase in area of the front face $OABC$ of the plate after σ is acting? (c) What is the decrease in cross-sectional area after σ is acting?

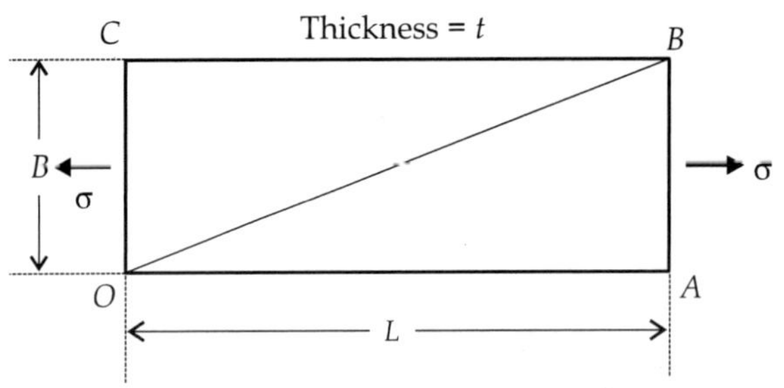

Fig. 2.27

Solution:

(a) $L_x = L + L\cdot\varepsilon = L + (\sigma/E)\cdot L$

or $L_x = L\cdot[1 + (\sigma/E)]$

$B_y = B - \nu\cdot\varepsilon\cdot B = B\cdot[1 - \nu\cdot(\sigma/E)]$

Slope $= B_y/L_x$

$= (B/L)\cdot(E - \nu\cdot\sigma)/(E + \sigma)$. **Ans.**

(b) $A' = B_y\cdot L_x = B[1 - \nu\cdot(\sigma/E)]\cdot L\cdot[1 + (\sigma/E)] = B\cdot L\cdot[1 - \nu\cdot(\sigma/E) + (\sigma/E)]$.

$A = B\cdot L$

∴ $\Delta A = A' - A = B\cdot L\cdot\sigma\cdot(1 - \nu)/E$ **Ans.**

(c) Initial area of cross section $= B\cdot t$

Final area of cross section $= B\cdot[1 - \nu\cdot(\sigma/E)]\cdot t\cdot[1 - \nu\cdot(\sigma/E)] = B\cdot t\cdot[1 - 2\nu\cdot(\sigma/E)]$

∴ Change in area $= -B\cdot t\cdot 2\cdot\nu\cdot\sigma/E$

Decrease in cross section $= 2Bt\nu\sigma/E$ **Ans.**

Example 2.12: A vertical shaft with a thrust collar is supported against a bearing plate shown in Fig. 2.28. If the average punching shear stress in the collar is 70 MPa and the average bearing stress between the collar and the plate is limited to 96 MPa, determine the maximum axial load that can be applied to the shaft.

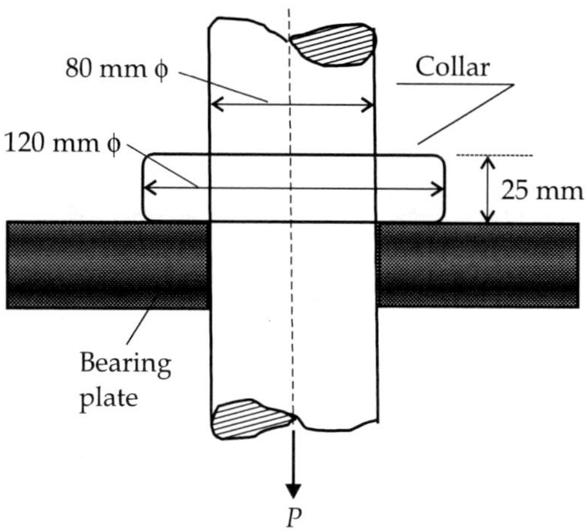

Fig. 2.28

Solution: If the collar fails by shearing, it should look like a lose ring such that the shear failure has occurred at the inner cylindrical surface of the ring.

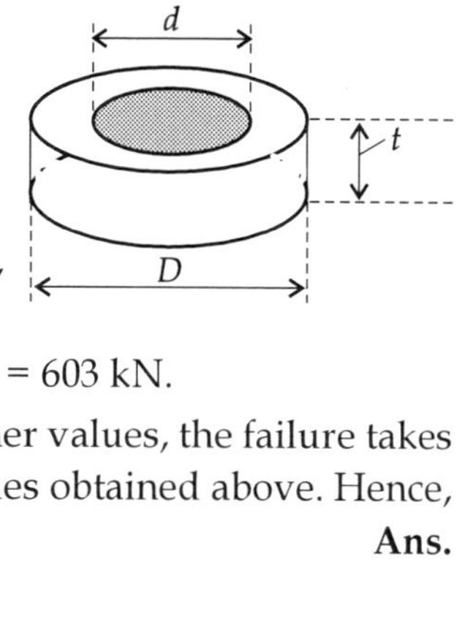

Let P = axial load for shear failure

$\therefore \quad P = \pi \cdot d \cdot t \cdot \tau_w$

or $\quad P = \pi \times 80 \times 25 \times 10^{-6} \times 70 \times 10^6$ N

$\therefore \quad P = 439.8$ kN

Otherwise, if the failure is due to the bearing pressure, i.e., the compressive stress between the collar and plate

$$P = (\pi/4)\cdot(D^2 - d^2) = (\pi/4)\cdot(120^2 - 80^2)\cdot 96 \text{ N}, \quad \text{or} \quad P = 603 \text{ kN}.$$

Assuming that P is increased gradually from zero to higher values, the failure takes place as soon as P = 439.8 kN which is lower of the two values obtained above. Hence, P = 439.8 kN **Ans.**

2.10 TEMPERATURE STRESSES

Every material body is known to have a natural volume at a certain temperature. The volume increases with the rise in temperature and decreases by lowering the temperature. Similarly, a bar made of some material has a natural length at a certain temperature. If there is a bar of length L at ambient temperature or room temperature (t_1) and temperature is then raised to t_2, an elongation of bar will be observed which can be written as

$$\delta = \alpha \cdot L \cdot (t_2 - t_1) = \alpha \cdot L \cdot \Delta t \tag{2.21}$$

where, α = coefficient of linear expansion (or contraction) of the material of the bar. Unit of α is length per unit length per degree change in temperature.

If the temperature is reduced from t_1 to t_3 ($t_1 > t_3$), then the contraction will be given by

$$\Delta = \alpha \cdot L \cdot (t_3 - t_1) = -\alpha \cdot L \cdot (t_1 - t_3) \tag{2.22}$$

Values of linear expansion coefficient (α) for some materials are given in Table 2.1

Table 2.1

Material	α (/°C)	Material	α (/°C)
Steel	12.0×10^{-6}	Cast iron	10.4×10^{-6}
Aluminium	22.5×10^{-6}	Zinc	35.4×10^{-6}
Copper	16.5×10^{-6}	Nickel	13.0×10^{-6}

When the deformation of a member is constrained at the supports or at the contact with the neighbouring members, temperature stresses may develop during heating or cooling of the member. Thus, unwanted stresses are likely to occur in a casting during its non-uniform cooling, which may result in the formation of cracks. Large amount of temperature stresses are present in a turbine disk because of difference in temperature at the centre and the periphery of disk. When a machine or structural part is formed by assembling two or more elements made from different materials, temperature stresses develop in the part due to unequal linear expansion coefficients of materials.

If a bar is prevented from expansion or contraction during a change in temperature, it develops normal stress which is either compressive or tensile in nature. A bar of length L and cross-sectional area A is shown in Fig. 2.29, which is constrained to deform due to the rigid supports at its ends. It is assumed that the ambient temperature is t_1 and the bar is just snugly fitted between supports. If the temperature is raised to t_2 and the bar is free to expand at the right end only, an extension in the bar would be $\delta = \alpha \cdot L \cdot (t_2 - t_1) = \alpha \cdot L \cdot \Delta t$. Since the length of the bar is to remain confined between the supports, the situation at temperature t_2 is equivalent to an axial compressive strain, $\varepsilon = -\delta / L = \alpha \cdot \Delta t$.

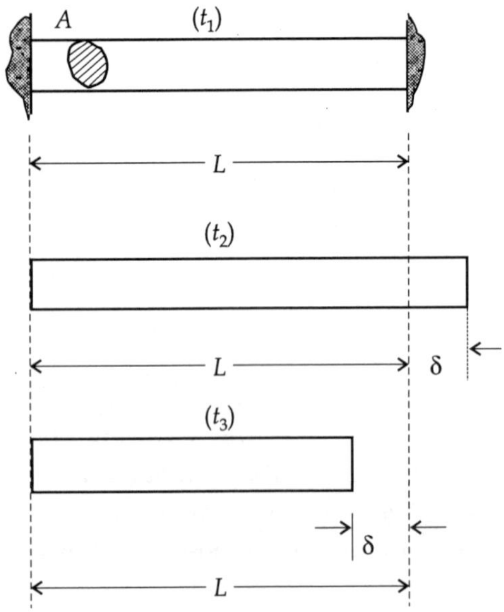

Fig. 2.29

Hence, a compressive stress is induced which is

$$\sigma = E \cdot \varepsilon = -E \cdot \alpha \cdot \Delta t \tag{2.23}$$

where, E = elasticity modulus of the material of the bar.

Also, the bar is then subjected to a compressive force given by

$$P = \sigma \cdot A = -A \cdot E \cdot \alpha \cdot \Delta t \tag{2.24}$$

where, A = area of cross section.

Otherwise, if temperature is reduced from t_1 to a lower amount t_3 and the bar is free at its right end, there will be a contraction $\delta = -\alpha \cdot L \cdot (t_1 - t_3)$. But the bar is fixed between supports and an equivalent axial strain will be $\varepsilon = \delta / L = \alpha \cdot \Delta t$. As such, the tensile stress will be

$$\sigma = E \cdot \varepsilon = E \cdot \alpha \cdot \Delta t \tag{2.25}$$

and there will be a tensile force

$$P = \sigma \cdot A = A \cdot E \cdot \alpha \cdot \Delta t \tag{2.26}$$

From Eqns. 2.23 and 2.25 it is found that the bar will be under compression at a higher temperature and in tension at a lower temperature.

Example 2.13: A steel rod with a cross-sectional area of 160 mm^2 is stretched between two fixed points such that a tensile force of 5.3 kN develops in the rod at 70° F. If the temperature is reduced to 0° F, what will be the normal stress in the rod? At what temperature will the stress be zero? Take $\alpha = 6.5 \times 10^{-6}/°\text{F}$ and E = 200 GPa.

Solution: At 70° F, normal stress in the rod $\sigma = P/A$

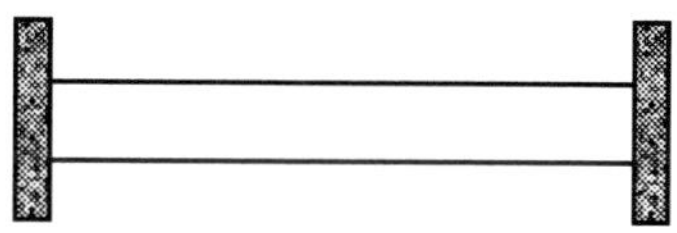

or $\sigma = 5.3 \times 10^3/(160 \times 10^{-6})$

$= 33.125 \times 10^6$ N/m^2

$\therefore$ strain, $\varepsilon = \sigma/E = 33.125 \times 10^6/(200 \times 10^9)$

$= 0.1656 \times 10^{-3}$

If the temperature is decreased to 0° F, there will be a further decrease in natural length provided that the rod is free to move. Hence, an additional strain develops which is $\varepsilon' = \alpha \cdot \Delta t = 6.5 \times 10^{-6} \times 70$

or $\varepsilon' = 0.455 \times 10^{-3}$.

At 0° F the total strain $= \varepsilon + \varepsilon' = (0.1656 + 0.455) \times 10^{-3} = 0.6206 \times 10^{-3}$.

$\therefore$ Stress $= E \cdot (\varepsilon + \varepsilon') = 200 \times 10^9 \times 0.6206 \times 10^{-3}$

$= 124.12 \times 10^6$ N/m^2 = 124.12 MPa **Ans.**

If the rod is not stretched initially, its natural length at 70° F is less than the length between fixed supports. On increasing temperature to a certain value the rod will increase in length to fill up the gap between the supports.

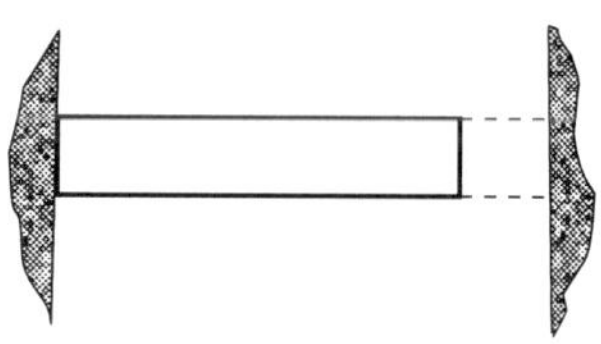

$\therefore \alpha \cdot \Delta t = \varepsilon$, or $6.5 \times 10^{-6} \cdot \Delta t = 0.1656 \times 10^{-3}$. $\therefore \Delta t = 25.5°$ F

Required temperature = 70 + 25.5 = 95.5° F **Ans.**

Example 2.14: A 0.5 m long bar of invar and a 1 m long bar of copper are placed end to end between two rigid supports (see Fig. 2.30) at 20° C. The diameter of invar bar is 50 mm and that of copper bar is 60 mm. Determine the force developed in each bar if the temperature is raised to 120°C. Take E = 144 GPa and $\alpha = 10^{-6}/°\text{C}$ for invar; E = 117 GPa and $\alpha = 16.5 \times 10^{-6}/°\text{C}$ for copper.

Solution: From $\sum F_x = 0$ along the common axis, it can be concluded that the same amount of compressive force (P) will develop in both invar and copper bars as the temp. is raised.

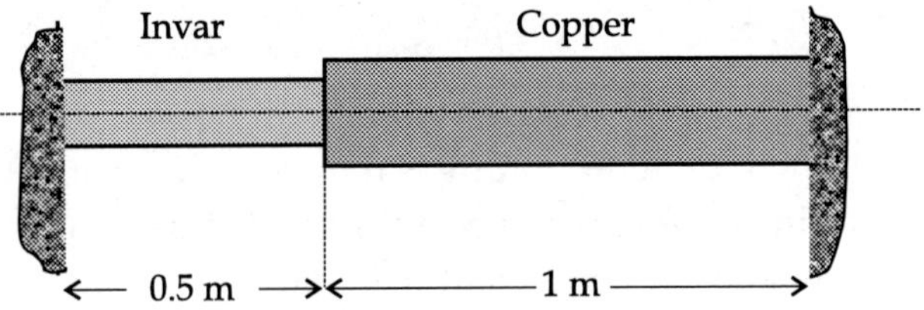

Fig. 2.30

Given that,

Invar : $d_1 = 50$ mm, $L_1 = 0.5$ m, $E_1 = 144$ GPa, $\alpha_1 = 10^{-6}/°C$

Copper: $d_2 = 60$ mm, $L_2 = 1.0$ m, $E_2 = 117$ GPa, $\alpha_2 = 16.5 \times 10^{-6}/°C$

$t_1 = 20°$ C, $t_2 = 120$, $\therefore \Delta t = t_2 - t_1 = 100°$ C

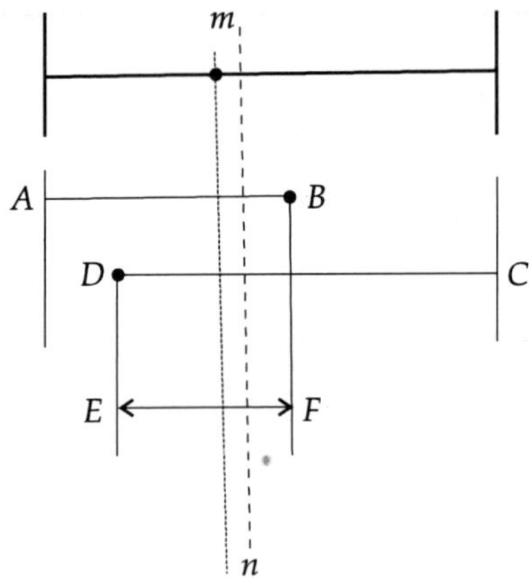

Because of temperature change (Δt) if the invar bar is free to move at the junction end, its new length will be AB. Similarly, the copper bar after free expansion will assume a length CD. The distance between the free ends of bars parallel to axis is shown by EF, so that

$EF = \alpha_1 \cdot L_1 \cdot \Delta t + \alpha_2 \cdot L_2 \cdot \Delta t = \Delta t \cdot (\alpha_1 \cdot L_1 + \alpha_2 \cdot L_2)$.

Since the bars will push each other to adjust at their new junction on the vertical plane, m-n, the sum of contractions of the two bars is

$EF = [P \cdot L_1 / (A_1 \cdot E_1)] + [P \cdot L_2 / (A_2 \cdot L_2)]$

$\therefore [P \cdot L_1 / (A_1 \cdot E_1)] + [P \cdot L_2 / (A_2 \cdot L_2)] = \Delta t \cdot (\alpha_1 \cdot L_1 + \alpha_2 \cdot L_2)$

With $A_1 = (\pi/4) \cdot 50^2 = 1963.5 \times 10^{-6}$ m^2, and $A_2 = (\pi/4) \cdot 60^2 = 2827.4 \times 10^{-6}$ m^2

$P \cdot [\{0.5/(1963.5 \times 144)\} + \{1.0/(2827.4 \times 117)\}] \cdot [1/(10^{-6} \times 10^9)] = 100 \cdot [(1 \times 0.5) + (16.5 \times 1.0)] \times 10^{-6}$

or $P = 354.8$ kN **Ans.**

Example 2.15: A steel tube of 2.5 cm outer diameter and 2.0 cm inner diameter is 75 cm long. It encloses a copper rod of same length and 1.5 cm in diameter, as shown in Fig. 2.31. The tube is firmly joined to the rod at both ends and the temperature of this assembly is raised by 120° C. Calculate the stresses in steel and copper. Find the increase in length of this composite system and external force required to prevent the increase in length. Take $E_s = 210$ GPa, $E_c = 100$ GPa,

$\alpha_s = 10 \times 10^{-6}/°C$ and $\alpha_c = 15 \times 10^{-6}/°C$.

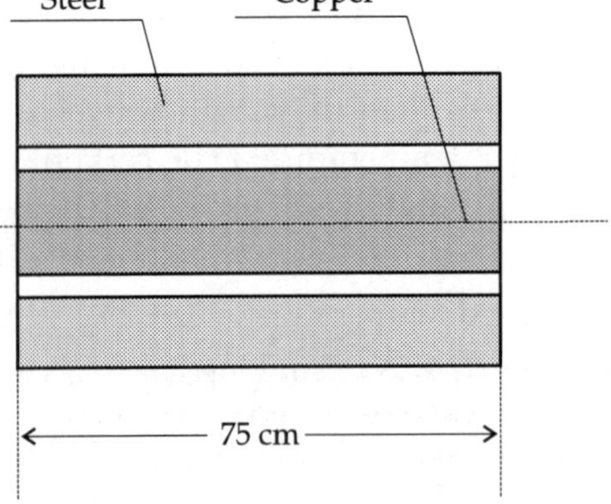

Fig. 2.31

Solution: Stresses: Suppose that the copper rod and steel tube of same length are free to move at the right end. Since $\alpha_c > \alpha_s$, extension of copper rod will be larger than that of the steel tube. As the right end is firmly joined, the copper rod will pull the steel tube and the steel tube will pull the copper rod so that at higher temperature both will assume a length up to a plane *m-n*.

Finally, the steel tube will be in a state of tension and the copper rod in compression.

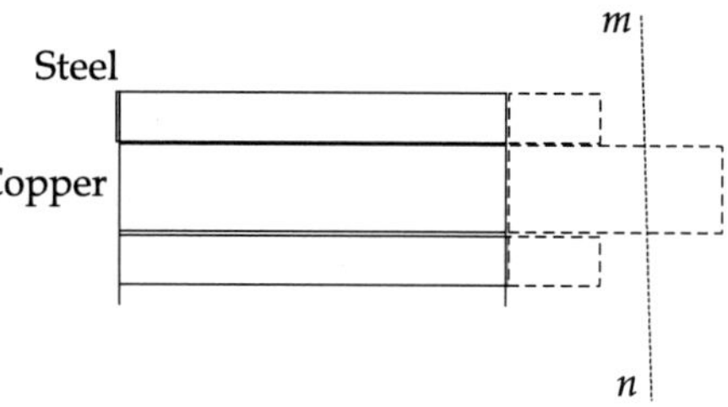

$\therefore\ \alpha_s \cdot \Delta t \cdot L + (\sigma_s / E_s) \cdot L = \alpha_c \cdot \Delta t \cdot L - (\sigma_c / E_c) \cdot L,$

or $\quad (\alpha_c - \alpha_s) \cdot \Delta t = (\sigma_s / E_s) + (\sigma_c / E_c)$ ... (a)

For equilibrium along the axial direction,

$\sigma_s \cdot A_s = \sigma_c \cdot A_c$ ... (b)

where $A_s = (\pi/4) \cdot (2.5^2 - 2^2) = 1.767\ \text{cm}^2$

and $A_c = (\pi/4) \cdot 1.5^2 = 1.767\ \text{cm}^2$. $\therefore\ A_c / A_s = 1$

From (a) and (b), $\sigma_c \cdot [\{A_c / (A_s \cdot E_s)\} + (1/E_c)] = (\alpha_c - \alpha_s) \cdot \Delta t$

or $\sigma_c \cdot [(1/210) + (1/100)] \cdot (1/10^9) = (15 - 10) \times 10^{-6} \times 120.$

$\therefore\ \sigma_c = \sigma_s = 40.65$ MPa **Ans.**

Increase in length

Increase in length $= \alpha_s \cdot \Delta t \cdot L + (\sigma_s / E_s) \cdot L$

$= 10 \times 10^{-6} \times 120 \times 75 + [40.65 \times 10^6 / (210 \times 10^9)] \times 75$

$= 0.1045$ cm or 1.045 mm **Ans.**

Force required

Force $= \alpha_s \cdot \Delta t \cdot E_s \cdot A_s + \alpha_c \cdot \Delta t \cdot E_c \cdot A_c$

$= [(10 \times 210) + (15 \times 100)] \times 10^{-6} \times 10^9 \times 1.767 \times 10^{-4} \times 120$ N

$= 76334$ N or 76.33 kN **Ans.**

Example 2.16: A vertical steel bar and two outer bars of brass are all stress free and joined by the pin at *O* (see Fig. 2.32) before a load of 20 kN is applied. The load is applied gradually and at the same time the temperature of all three bars is decreased by 15° F. The cross-sectional area of each of the outer brass bars is 250 mm² and that of the central steel bar is 200 mm². Calculate the normal stress developed in each bar. For steel, $E_s = 200$ GPa and $\alpha_s = 6.3 \times 10^{-6}$/°F. For brass, $E_b = 85$ GPa and $\alpha_b = 10.4 \times 10^{-6}$/°F.

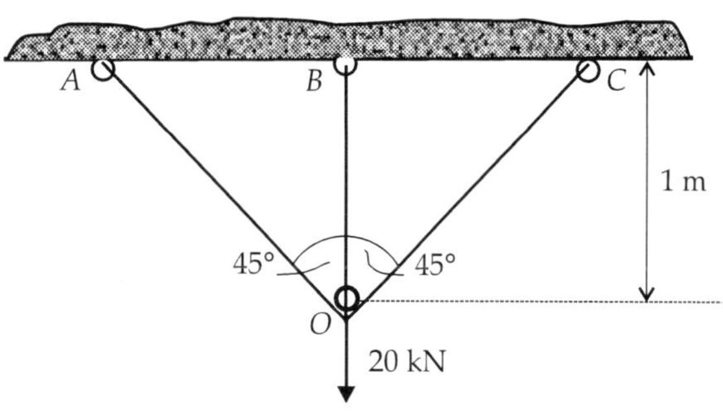

Fig. 2.32

Solution: Change of temperature, $\theta = 15° F$.

Considering free-body diagram of the joint O in equilibrium, let T_s and T_b be forces in steel bar and brass bar, respectively.

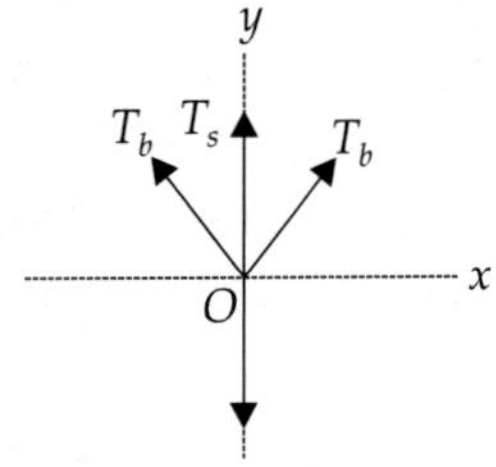

From $\sum F_x = 0$,

$T_s + 2 \cdot T_b \cdot \cos 45° = 20 \times 10^3$ kN ...(a)

Strain in steel bar due to load T_s and change of temperature θ,

$\varepsilon_s = \{T_s / (A_s \cdot E_s)\} - \alpha_s \cdot \theta$.

$\therefore$ Change in length of steel bar, $\delta_s = L \cdot \varepsilon_s = [\{T_s / (A_s \cdot E_s) - \alpha_s \cdot \theta\}] \cdot L$, where $A_s = 200$ mm² and $L = 1$ m.

Strain in brass bar, $\varepsilon_b = \{T_b / A_b \cdot E_b)\} - \alpha_b \cdot \theta$.

$\therefore$ Change in length of brass bar, $\delta_b = L\sqrt{2} \cdot \varepsilon_b = [\{T_b / (A_b \cdot E_b)\} - \alpha_b \cdot \theta] \cdot L\sqrt{2}$, where, $A_b = 250$ mm².

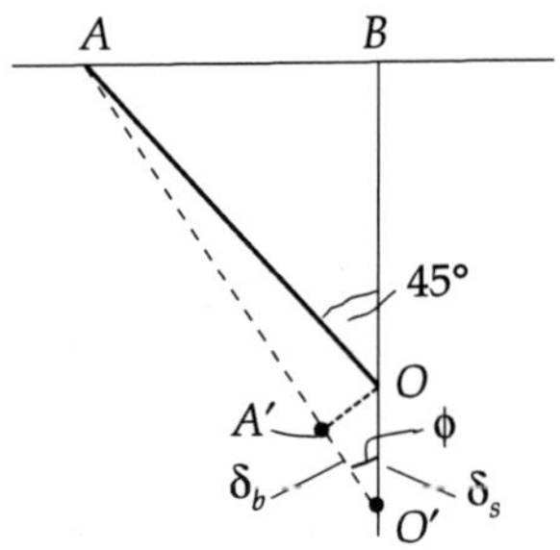

After strains occur, the joint O assumes a new position O' just vertically below the point O, so that $OO' = \delta_s$ and $AO' - AO = AO' - AA' = A'O' = \delta_b$.

In $\Delta A'O'O$, $A'O' = OO' \cdot \cos \phi$, or $\delta_b = \delta_s \cdot \cos \phi$.

As the deformation is small, $\phi \simeq 45°$.

$\therefore \delta_b = \delta_s \cdot \cos 45°$, or $\sqrt{2} \cdot \delta_b = \delta_s$.

$\therefore \quad [\{T_b / (A_b \cdot E_b)\} - \alpha_b \cdot \theta] \cdot L\sqrt{2} \cdot \sqrt{2} = [\{T_s / (A_s \cdot E_s) - \alpha_s \cdot \theta] \cdot L$

or $\quad \{2T_b / (A_b \cdot E_b)\} - 2\alpha_b \cdot \theta = \{T_s / (A_s \cdot E_s)\} - \alpha_s \cdot \theta$.

In the above equation, only T_b and T_s are unknowns. Hence, substituting for other data given in the problem, we have $T_s = 3.765 \cdot T_b - 8700$ N ... (b)

From (a) and (b), $T_b = 5541.4$ N and $T_s = 12163.3$ N

$\therefore$ Normal stress in brass bar, $\sigma_b = T_b / A_b$ or $\sigma_b = 5541.4 / (250 \times 10^{-6})$ N/m². $\sigma_b = 22.17$ MPa

Ans.

Normal stress in steel bar, $\sigma_s = T_s / A_s$ or $\sigma_s = 12163.3 / (200 \times 10^{-6})$ N/m², $\sigma_s = 60.82$ MPa.

Ans.

2.11 STRAIN ENERGY AND IMPACT LOAD

Strain energy is the energy absorption capacity of a material. Energy absorption capacity of a material within or up to elastic limit is known as **modulus of resilience** while energy absorption capacity of a material up to failure point is known as **modulus of toughness**.

In case of a ductile material, modulus of toughness is more than modulus of resilience. But in case of brittle or low ductile material they are more or less same or do not differ much.

Elastic strain energy for uniaxial stress

The internal work is stored in an elastic body as internal energy of deformation and $U = \dfrac{\sigma_x^2}{2E}$ for a particular material gives its ability to store or absorb energy without permanent deformation (Fig. 2.33).

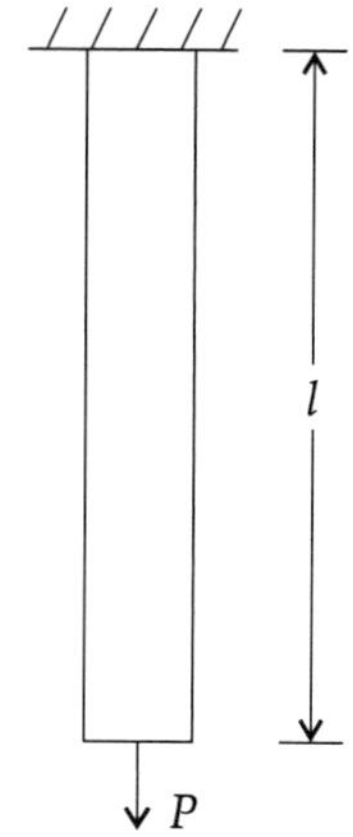

Fig. 2.33

Strain energy is the internal energy U.

$$U = \frac{1}{2}P\delta = \frac{1}{2}\sigma A\delta = \frac{1}{2}\frac{PL^2}{AE} = \frac{1}{2}\sigma A\frac{\sigma L}{E} = \frac{1}{2}\frac{\sigma^2 AL}{E}$$

$\Rightarrow \dfrac{U}{AL} = \dfrac{1}{2}\dfrac{\sigma^2}{E}$ = Energy stored per unit volume (or) Modulus of resilience.

Deflection due to impact (or sudden) load

Let us consider a bar of length (L) is subjected to sudden falling load (W) at its tip (Fig. 2.34). The bottom end of the bar is connected by a collar. The load (W) can freely fall on the collar and resist it. If the load is applied statically on the bar, then the maximum static deflection is $\delta_{st} = \dfrac{WL}{AE}$. It can be written by static deflection,

$\delta_{st} = \dfrac{W}{K}$, K = elastic constant, where, $K = \dfrac{AE}{L}$

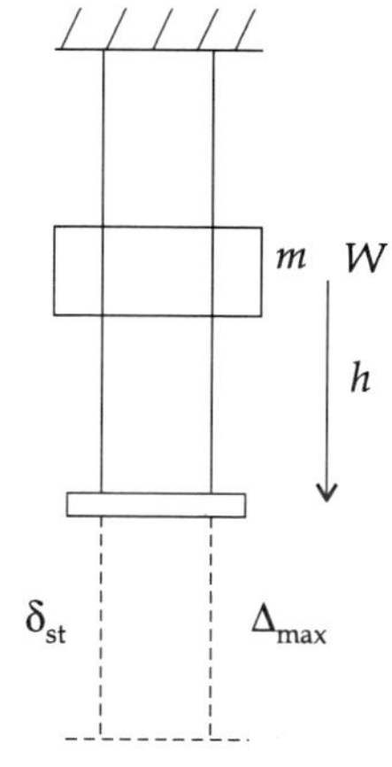

Fig. 2.34

If the same load (W) falls freely from a height (h) and maximum deflection occurs () Δ_{max}, then the dynamic load, $P_{dynamic}$, acts on the collar. For a linear elastic body, load deformation relationship is as follows and $K = \dfrac{W}{\delta_{st}}$

Force developed, $P_{dynamic} = K\,\Delta_{max} = \Delta_{max} \times \dfrac{W}{\delta_{st}}$

Work done by the load before rest on collar, work done $= \dfrac{1}{2}\cdot P_{dynamic} \times \Delta_{max}$

During falling of the body, the total potential and kinetic energy will be stored in the form strain energy of the bar.

From energy conservation law

Potential energy (strain energy) = $(h + \Delta_{max})\,W$

$$(h + \Delta_{max})\,W = \frac{1}{2}\cdot P_{dynamic} \times \Delta_{max}$$

$$\Delta_{max}^2 = 2\cdot\delta_{st}\,(h + \Delta_{max})$$

$$\Delta_{max}^2 - 2\delta_{st}\Delta_{max} - 2h\delta_{st} = 0$$

$$\Delta_{max} = \frac{2\delta_{st} \pm \sqrt{4\delta_{st}^2 + 4(2h\delta_{st})}}{2} = \delta_{st} \pm \sqrt{\delta_{st}^2 + 2h\delta_{st}}$$

$$\Delta_{max} = \delta_{st}\left[1 + \frac{2h}{\delta_{st}}\right]$$

The magnitude of a static force when dynamically applied is known as **impact factor**.

$$P_{dynamic} = W\left[1 + \sqrt{1 + \frac{2h}{\delta_{st}}}\right]$$

Putting $h = 0$ in the above expression, we get $\Delta_{max} = 2\delta_{st}$, this implies that if the load applied dynamically, the deformation will be double as compared to load applied statically.

PROBLEMS

1. A 10 mm thick flat steel plate is 600 mm long. The width of the plate at one end is 100 mm, which uniformly reduces to 40 mm at the other end. If an axial tensile force of 50 kN is acting, find the change in the length of the plate. Take E = 210 GPa.

Ans. 0.218 mm

2. A tie rod made of steel and a pipe strut of aluminium are used to support a load W, as shown in Fig. 2.35. The tie rod is 25 mm in diameter. The inner diameter of pipe is 50 mm and its wall thickness is 3 mm. If the allowable stresses in steel and aluminum are 510 MPa and 110 MPa, respectively, determine the maximum load (W) that can be supported by the structure.

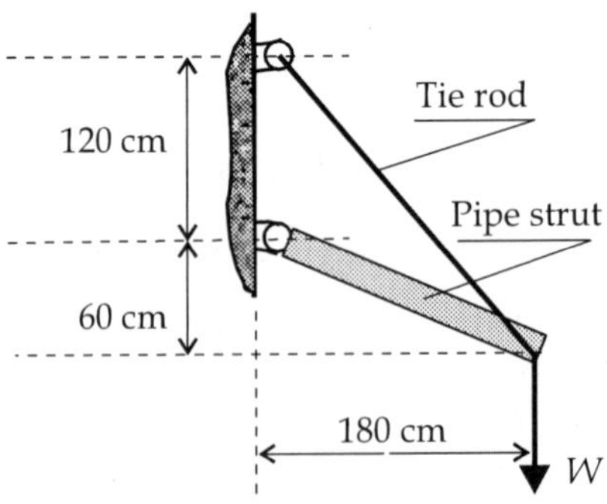

Fig. 2.35

Ans. 34.7 kN

3. Three co-axial steel bars are subjected to axial loads applied through rigid bearing plates, as shown in Fig. 2.36. The cross-sectional areas of bars *AB*, *BC* and *CD* are 100 cm², 25 cm² and 80 cm² respectively. Determine the normal stress induced over the cross section of each of the bars.

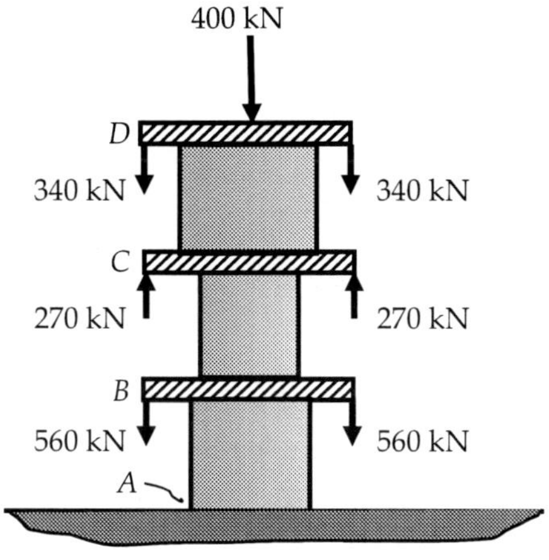

Fig. 2.36

Ans. $\sigma_{AB} = 166$ MPa (compressive)
$\sigma_{BC} = 216$ MPa (compressive)
$\sigma_{CD} = 135$ MPa (compressive)

4. An axial load of 25 kN is acting on a timber block with 100 × 100 mm cross section (see Fig. 2.37). Find the normal and shear stresses on the plane of wood-grain.

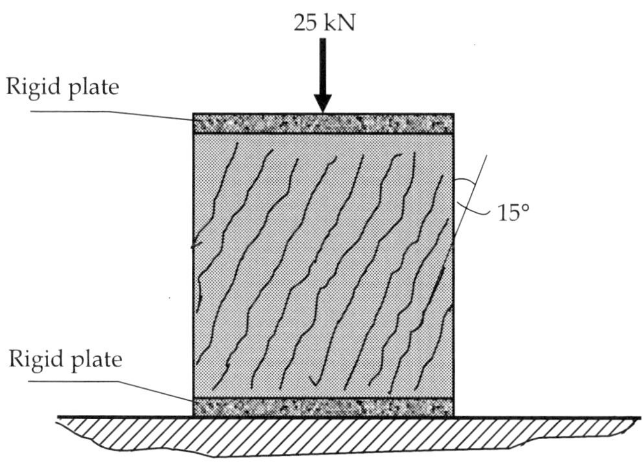

Fig. 2.37

Ans. $\sigma = 0.17$ MPa (compressive)
$\tau = 0.63$ MPa

5. A steel plate shown in Fig. 2.38 is to carry an axial tensile load of 500 kN. Find the minimum thickness (t) of the plate if the normal and shear stresses on plane AB are not to exceed 78 MPa and 55 MPa, respectively.

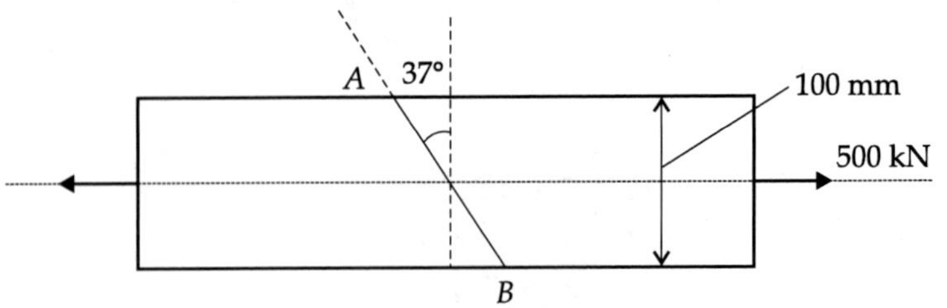

Fig. 2.38

Ans. 43.7 mm

6. A rigid bar is kept in horizontal position by a 4 m long steel rod at one end and a 3 m long brass rod at the other end (see Fig. 2.39). The diameter of steel rod is 30 mm and that of brass rod is 40 mm. Find the value of x so that the rigid bar remains horizontal after any load (W) is applied.

   Take E = 200 GPa for steel and 100 GPa for brass.

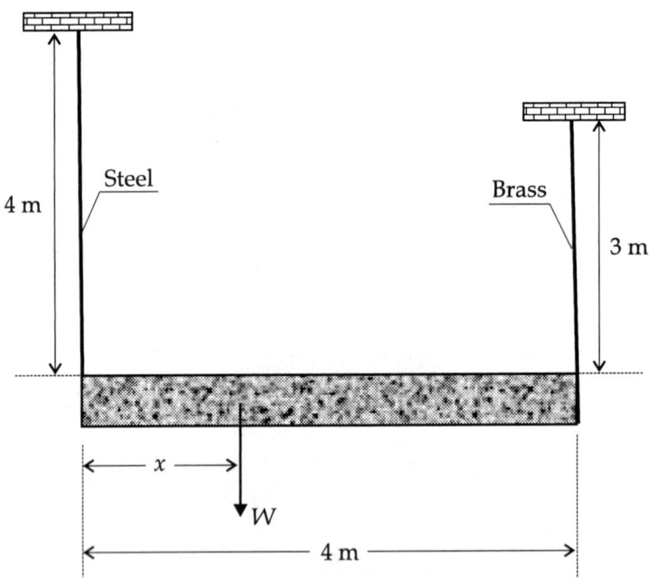

Fig. 2.39

Ans. X = 2.17 m

7. A lever is shown in Fig. 2.40. There are two plastic pins fixed with the lever and the system is used to operate a locking arrangement. The maximum force required at the lever arm is P = 110 N. If the average shearing stress in the plastic pin is 40 MPa, find the minimum diameter of each of the pins.

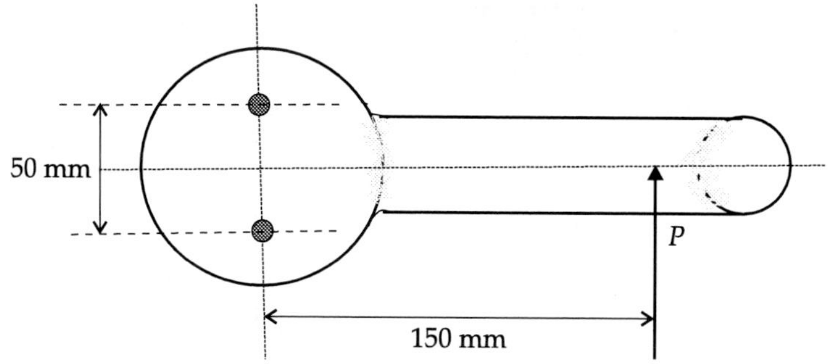

Fig. 2.40

Ans. 3.24 mm

8. A copper bar of length 1m is fixed to the ground as shown in Fig. 2.41. At room temperature, there is a gap of 0.02 cm between the top of the bar and rigid ceiling. If the temperature rises by 90°F, find the compressive stress induced in the bar. Take $E = 110$ GPa and $\alpha = 9.8 \times 10^{-6}$ /°F.

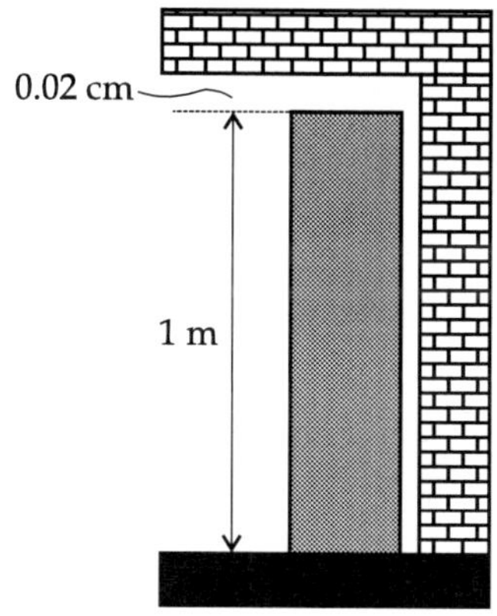

Fig. 2.41

Ans. 75 MPa

9. A rigid bar (*EH*) is hinged at *H* with a freely rotating pin and kept in horizontal position by a steel rod and a brass rod, as shown in Fig. 2.42. The cross-sectional area of steel rod is 13 cm^2 and that of the brass rod is 10 cm^2. The rods are unstressed at 65°F before the load *P* is applied. After applying the load, the temperature of rods increases to 95°F. Determine (a) the normal stresses in steel and brass rods, (b) maximum shearing stresses in both the rods, (c) shearing stress in the 20 mm diameter pin at *H*, which is in double shear ,and (d) the vertical deflection of free end € of the rigid bar. Take $E_s = 200$ GPa, $\alpha_s = 6.6 \times 10^{-6}$ /°F for steel, and $E_b = 100$ GPa, $\alpha_b = 9.8 \times 10^{-6}$ /°F for brass.

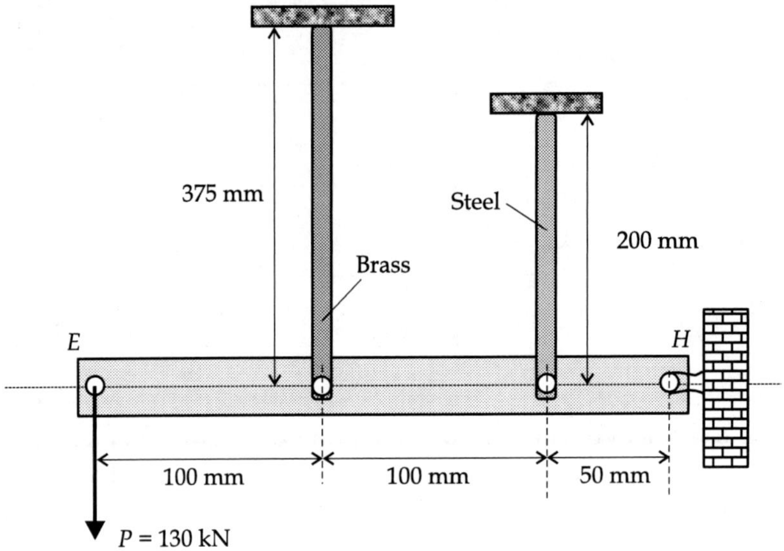

Fig. 2.42

Ans. (a) $\sigma_s = 174$ MPa

$\sigma_b = 141.3$ MPa

(b) $(\tau_s)_{max} = 87$ MPa

$(\tau_b)_{max} = 71.6$ MPa

(c) $\tau = 378$ MPa

(d) $\delta_E = 1.067$ mm

10. Bars of copper and aluminum, each 50 mm wide, are held together by pins at their ends (see Fig. 2.43). Thin spacers are used to separate the bars. Thickness of the copper bar is 10 mm and of aluminum bar is 20 mm. If the temperature of this assembly is raised by 40°C, determine the shearing stress in 12 mm diameter pins. Take $E_c = 120$ GPa, $\alpha_c = 16.5 \times 10^{-6}$ /°C for copper, and $E_a = 70$ GPa, $\alpha_a = 22.5 \times 10^{-6}$ /°C for aluminium.

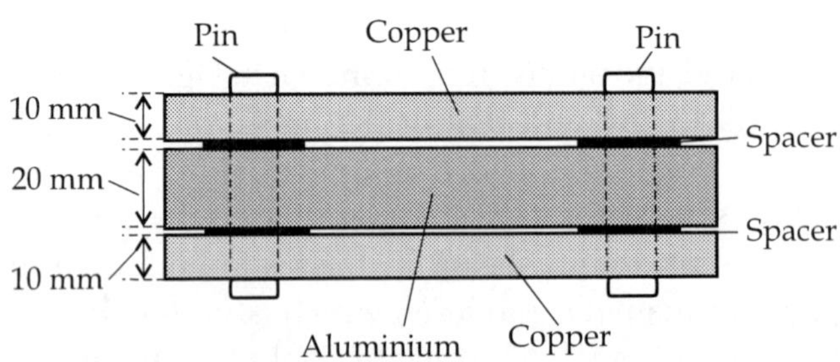

Fig. 2.43

Ans. 47 MPa

3
Analysis of Stress

3.1 STRESS AT A POINT INSIDE AN EXTERNALLY LOADED BODY

In the preceding chapter on axial loading, the applied load and resulting deformation were unidirectional, and so were the normal stress and the axial strain. However, a body, in general, may be subjected to external loads applied in different directions. In such case, the deformation of the body is not confined to a single direction. Also, at a point within the body the stresses in more than one direction are found to be present. Hence, a set of stresses can be obtained at a point within the externally loaded body which generally varies from point to point.

In the light of what has been stated above, it is necessary to investigate the significance of equilibrium principles applied to a point in a deformable body.

3.1.1 Stress at a Point

For a body in equilibrium under a large number of external forces, shown in Fig. 3.1(a), let us find the internal force intensity at a point P within the body. For this, a plane is assumed to pass through the point P which divides the body into two parts. A free-body diagram of the left part is given in Fig. 3.1(b) which is shown to be in equilibrium

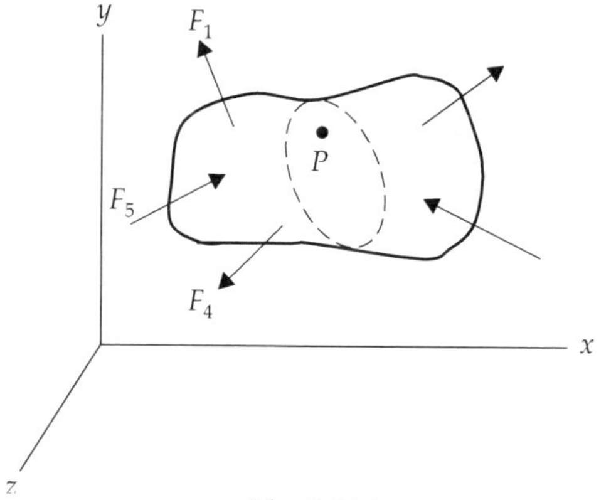

Fig. 3.1(a)

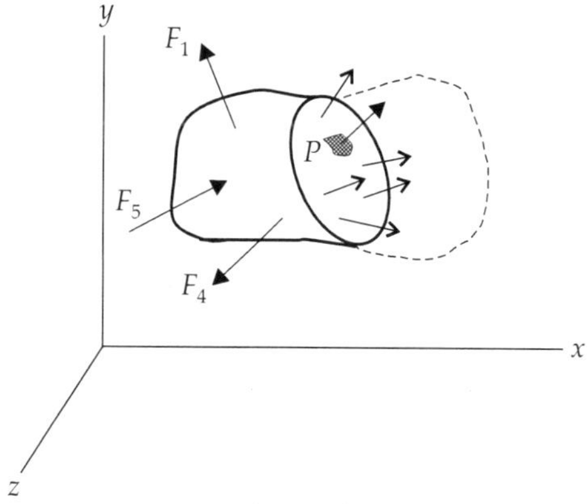

Fig. 3.1(b)

under the effects of external forces and the internal forces distributed over the exposed sectional plane passing through P. For the two parts of the body to be in equilibrium, the internal forces on the cutting plane must transmit across this plane such that the forces on the exposed section of the right part are equal in magnitude and opposite in sense. Next, we consider a very small area (ΔA) enclosing point P on the cutting plane and a force (ΔF) (acting on this area. This is shown separately in Fig. 3.2. If the exposed plane is defined by a unit vector (**n**) along its outward normal, the same vector (**n**) is normal to area, ΔA.

The force ($\Delta \mathbf{F}$) at P does not necessarily coincide with **n** but may act at some arbitrary direction.

For describing completely the action of force vector, $\Delta \mathbf{F}$, we have to specify:

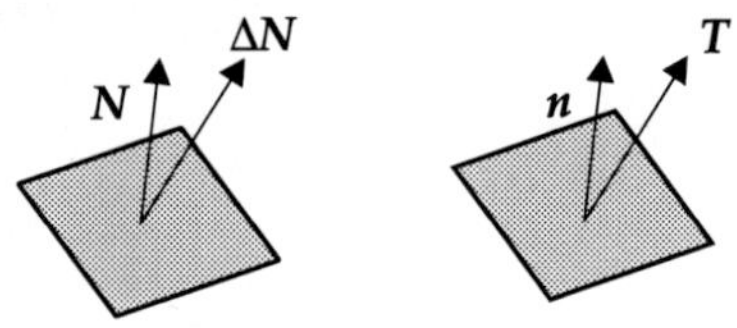

Fig. 3.2

(a) the orientation and size of the plane on which $\Delta \mathbf{F}$ acts,

(b) the magnitude of $\Delta \mathbf{F}$, and

(c) orientation of $\Delta \mathbf{F}$ with respect to the plane.

Let us introduce the stress vector ($\mathbf{T}^{(n)}$) (see Fig. 3.2) which is defined as

$$\mathbf{T}^{(n)} = \lim_{\Delta A \to 0} (\Delta \mathbf{F}/\Delta A) \tag{3.1}$$

The notation ($\mathbf{T}^{(n)}$) of stress vector indicates that it is a vector acting at P in a plane defined by the unit normal vector **n**. In general, the directions of $\mathbf{T}^{(n)}$ and **n** are different. It should be noted that defining the limit in the above equation is in accordance with the continuum hypothesis with respect to the distribution of internal forces.

There are four major characteristics of stress which are as follows:

(a) The physical dimension of stress is force per unit area.

(b) Stress is defined at a point on an imaginary plane dividing the material body into two parts.

(c) Stress at a point is a vector equivalent to the action of one part of body upon the other.

(d) The direction of stress vector is not restricted, i.e., it may change with the orientation of plane passing through the point.

The stress vector ($\mathbf{T}^{(n)}$) defined above can be written in terms of its components along the conveniently chosen coordinate directions as

$$\mathbf{T}^{(n)} = T_x \cdot \mathbf{i} + T_y \cdot \mathbf{j} + T_z \cdot \mathbf{k} \tag{3.2}$$

where, T_x, T_y and T_z are known as orthogonal scalar components of $\mathbf{T}^{(n)}$.

Thus, the stress vector acting on a plane defined by the unit normal vector (**n**) is expressed in terms of the three stress vectors in x-, y-, and z-directions. This is depicted by Fig. 3.3.

The process of finding stress vector at a point, as explained above, can be used more conveniently to obtain the stresses directly in x-y-z coordinate system. The first step to accomplish this is to consider a plane passing through the point P, which is parallel to y-z plane, shown in Fig. 3.4 (a). The body is in equilibrium, while the external forces are not shown. Let ΔA be a very small area enclosing the point P and $\mathbf{n} = \mathbf{I}$ along the outward normal towards positive x-direction. The force vector ($\Delta\mathbf{F}$) acting on this area can be expressed in terms of its components along x-, y- and z-directions as

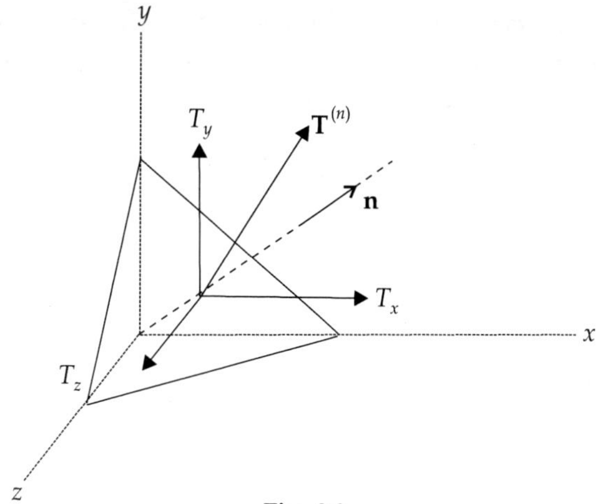

Fig. 3.3

$$\Delta\mathbf{F} = \Delta F_x\cdot\mathbf{i} + \Delta F_y\cdot\mathbf{j} + \Delta F_z\cdot\mathbf{k} \tag{3.3}$$

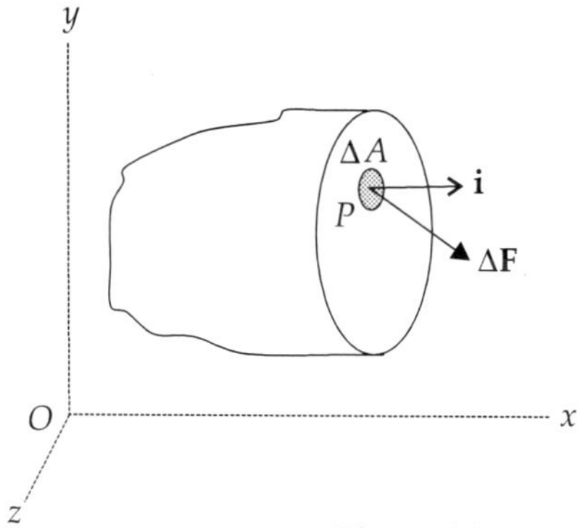

Fig. 3.4(a)

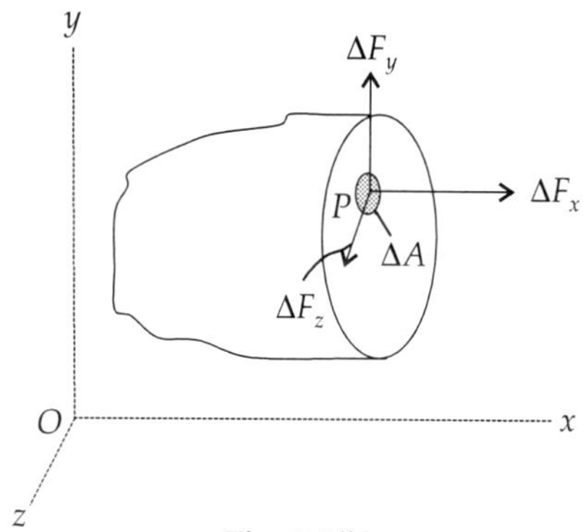

Fig. 3.4(b)

The force components acting at P are shown collectively in Fig. 3.4(b). It is observed that the force (ΔF_x) is perpendicular to area (ΔA), whereas the forces (ΔF_y) and (ΔF_z) are acting tangentially over the area. Also that the area vector ($\Delta A\cdot\mathbf{i}$) points towards the positive x-direction. Hence, we can define the three stresses in the form

$$\left.\begin{aligned} \sigma_{xx} &= \lim_{\Delta A \to 0}(\Delta F_x/\Delta A), \\ \tau_{xy} &= \lim_{\Delta A \to 0}(\Delta F_y/\Delta A), \\ \tau_{xz} &= \lim_{\Delta A \to 0}(\Delta F_z/\Delta A), \end{aligned}\right\} \tag{3.4}$$

where, σ_{xx} is a normal stress over the area ΔA and τ_{xy}, and τ_{xz} are shear stresses. In each of the notations for stresses, the first suffix indicates the position of plane over which the stress is acting, and the second suffix indicates the direction of stress. For example, the stress (τ_{xy}) is acting on a plane with its face pointing towards positive x-direction and according to the second suffix (y), the stress itself points towards positive y-direction. With such notations, all the stresses σ_x, τ_{xy} and τ_{xz} are positive in sense. The normal stress (σ_{xx}) is conventionally written as σ_x to avoid repeating of suffix x. The presence of these stresses at the point P are shown by Fig. 3.5. In the same way, an imaginary cutting plane parallel to x-z plane is assumed to pass through the point P. The stress vector over the area defined as $\Delta A \cdot \mathbf{j}$ on this plane can be used to define the stresses σ_y, τ_{yx} and τ_{yz}, all acting over the area ΔA. Similarly, on an area ΔA at P which is lying on a plane parallel to x-y plane, the three stresses σ_z, τ_{zx} and τ_{zy} are defined. Thus, the three normal stresses σ_x, σ_y and σ_z and six shear stresses τ_{xy}, τ_{xz}, τ_{yx}, τ_{yz}, τ_{zx}, τ_{zy} are obtained at the point P, which are collectively shown in Fig. 3.6.

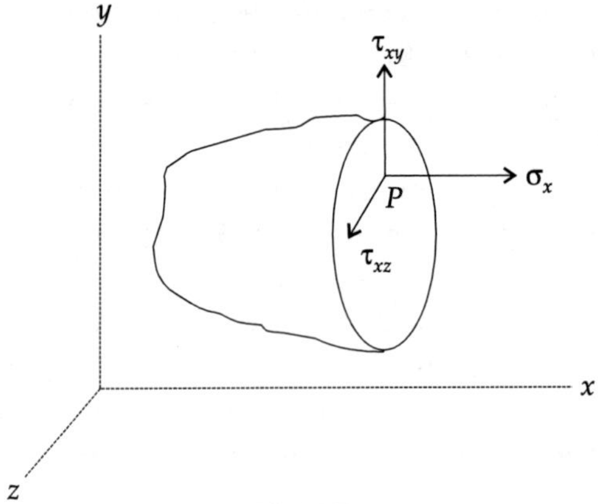

Fig. 3.5

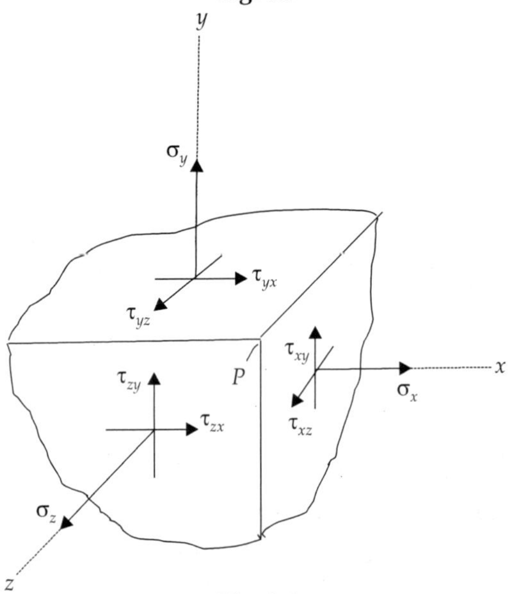

Fig. 3.6

Therefore, the state of stress at a point is defined by nine stress components which are written in the form of a matrix, as

$$\tau_{ij} = \begin{Bmatrix} \sigma_x & \tau_{xy} & \tau_{xz} \\ \tau_{xy} & \sigma_y & \tau_{yz} \\ \tau_{xz} & \tau_{yz} & \sigma_z \end{Bmatrix}, \tag{3.5}$$

where $\tau_{ij} = \sigma_i$, if $i = j$, and $i, j = x, y, z$

It can be seen that the stress components acting on a specific plane are arranged in each row of the above matrix. For example, τ_{yx}, σ_y and τ_{yz} are acting on a plane parallel to x-z plane whose outward normal points towards positive y-direction. All the nine stress components at P are required to determine the stress vector $\mathbf{T}^{(n)}$ acting on any arbitrarily chosen plane through this point.

The state of stress described above is found to vary from point to point in an externally loaded body. In order to establish equilibrium at a point, a very small parallelepiped of sides Δx, Δy and Δz surrounding the point can be chosen such that its faces are parallel to the coordinate planes, as shown in Fig. 3.7. The stress components are shown in positive sense which are acting on the faces having outward normal pointing to positive x-, y-, and z-directions. On the opposite faces with outward normal pointing to negative directions, all such stress components are presented in negative sense, so that we get an overall positive system of stress state. In Fig. 3.7 (a) and Fig. 3.7(b), the stress components are shown to be acting on the positive and negative faces of The parallelepiped, respectively.

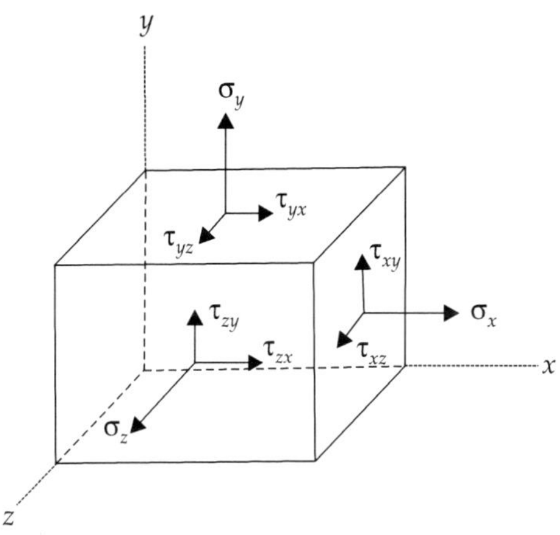

Fig. 3.7(a)

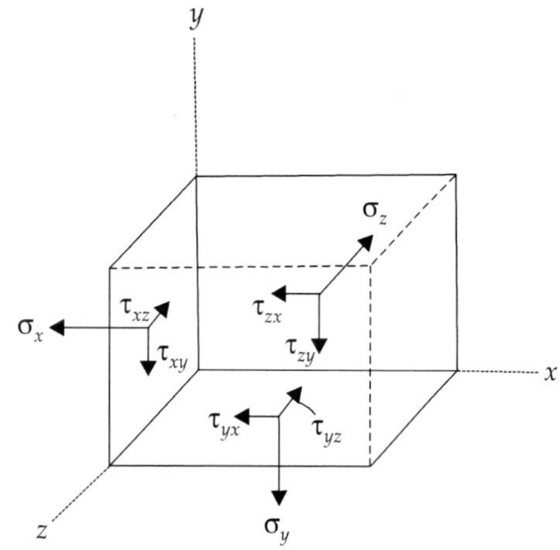

Fig. 3.7(b)

If all the force components acting on both positive and negative faces are used to satisfy the equilibrium condition such as $\sum M_z = 0$, it will be found that the moments other than $(\tau_{xy} \cdot \Delta y \cdot \Delta z)$ and $(\tau_{yx} \cdot \Delta x \cdot \Delta z)$ will cancel out, so that

$$(\tau_{xy} \cdot \Delta y \cdot \Delta z) \cdot \Delta x - (\tau_{yx} \cdot \Delta x \cdot \Delta z) \cdot \Delta y = 0.$$

After dividing by $(\Delta x \cdot \Delta y \cdot \Delta z)$ and simplifying, the above equation yields to

$$\tau_{xy} = \tau_{yx}\,. \tag{3.6}$$

Similarly, from $\sum M_x = 0$ and $\sum M_y = 0$, the following are obtained.

and
$$\left.\begin{aligned} \tau_{zy} &= \tau_{yz}, \\ \tau_{zx} &= \tau_{xz} \end{aligned}\right\} \tag{3.7}$$

From Eqns. 3.6 and 3.7 it can be inferred that the shear stresses on mutually perpendicular planes are equal in magnitude but opposite in sense (clockwise or counterclockwise). This is also verified with the help of Example 2.10 discussed in chapter 2. For a bar of cross-sectional area (A) and subjected to an axial force (P), shown in Fig. 3.8, the shear stress on two planes each inclined at 45° to the axis is equal to $P/2A$. Their directions are opposite in sense. Hence, using the relations from Eqns. 3.6 and 3.7, the stress matrix in Eqn. 5 can be written as

$$\begin{Bmatrix} \sigma_x & \tau_{xy} & \tau_{xz} \\ \tau_{xy} & \sigma_y & \tau_{yz} \\ \tau_{xz} & \tau_{yz} & \sigma_z \end{Bmatrix}, \tag{3.8}$$

which is a symmetric matrix.

In the state of stress given above, all the stress components have been assumed in positive sense. Also, if τ_{xy} is negative (clockwise) then τ_{yx} (= τ_{xy} in magnitude) is of opposite type, i.e., acting counterclockwise. The same is true for shear stresses, τ_{xz} and τ_{yz}.

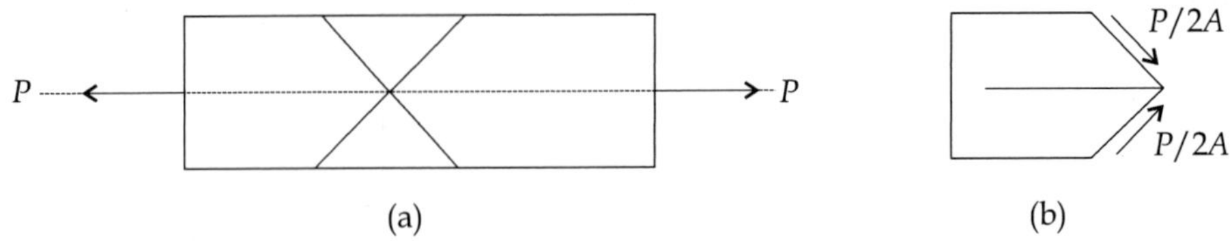

Fig. 3.8

3.2 TWO-DIMENSIONAL STRESS (PLANE STRESS)

In many instances, the state of stress is found to be simpler than the tri-axial state of stress discussed so far. In simpler cases, the stresses acting at a point are found to lie in a plane and there is absence of any stress perpendicular to that plane. This is known as a state of plane stress. A common example of plane stress is found in a thin plate which is subjected to external loads, as shown in Fig. 3.9. The plane of the plate is supposed to coincide with *x-y* plane. At any point within the plate, a small parallelepiped surrounding the point can be considered with its faces parallel to the coordinate planes. Then all the stresses acting on its faces will lie on *x-y* plane, as shown in Fig. 3.10. In fact, a thin plate is free to expand or contract in *z*-direction when loads are applied to the plate in *x-y* plane (let us recall Poisson's effect explained in chapter 2). In the absence of any constraint along *z*-axis, there is no stress related to *z*-direction. At the same time, the stresses shown by Fig. 3.10 are found to remain unchanged through the thickness of the plate.

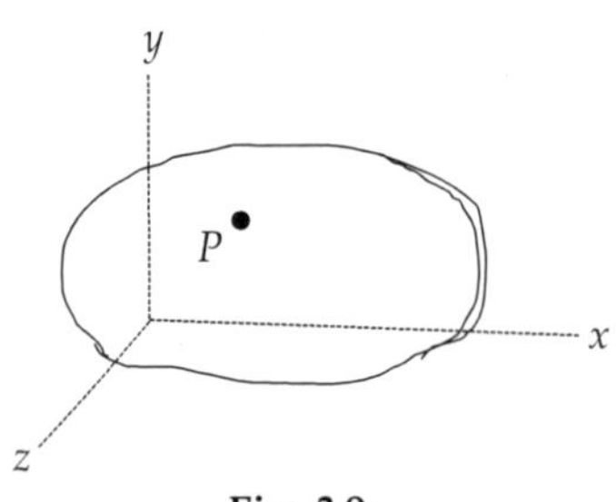

Fig. 3.9

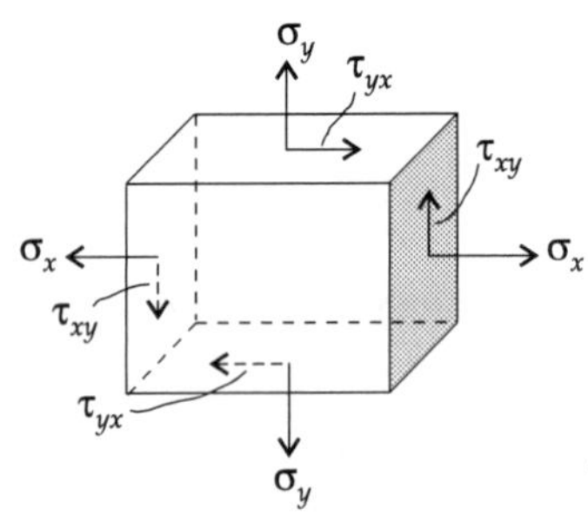

Fig. 3.10

The state of plane stress is conventionally shown with a square element, as in Fig. 3.11. The stress components are collectively arranged in the following matrix:

$$\begin{Bmatrix} \sigma_x & \tau_{xy} \\ \tau_{xy} & \sigma_y \end{Bmatrix} \tag{3.9}$$

The stress components in the above matrix are functions of x and y only, and these identify the state of plane stress in x-y plane.

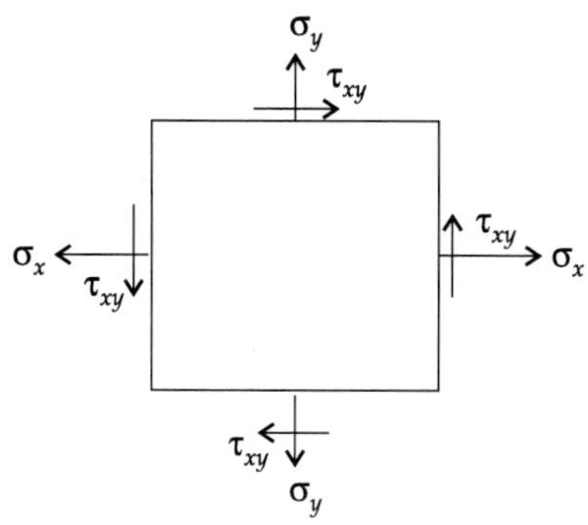

Fig. 3.11

There are many practical problems in which we come across the condition of plane stress and hence, it is necessary to investigate this special situation in some details.

3.3 STRESS TRANSFORMATION FOR PLANE STRESS

It was discussed earlier that individually a stress component can be represented as a vector. However, a set of stress components at a point such as that found in plane stress is a tensor of second order. As such, by rotating the orthogonal coordinates from x-y to x'-y' at the same point in x-y plane, a new set of stress components of plane stress are obtained. This is explained in Fig. 3.12.

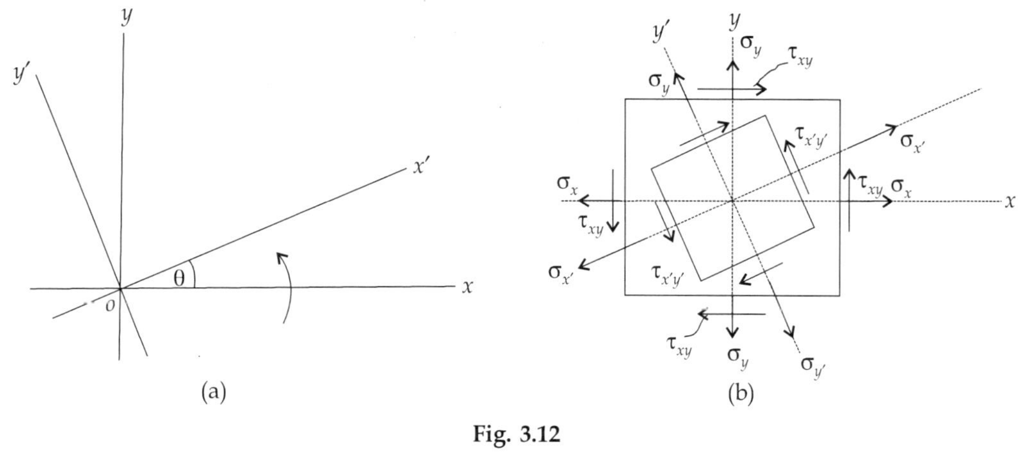

Fig. 3.12

It is shown in Fig. 3.12(a) that at a point O with already chosen coordinates x-y, a new set of coordinates x'-y' can be selected which is equivalent to the rotating x-y in counterclockwise manner through an angle θ to a new set of coordinates x'-y'. If the state of plane stress in x-y coordinates is defined with σ_x, σ_y and τ_{xy}, shown in Fig.

3.12(b), the plane stress condition in x'-y' coordinates provides a different set of stresses: $\sigma_{x'}$, $\sigma_{y'}$ and $\tau_{x'y'}$. A relation between the two sets of stresses can be obtained by including trigonometric functions of angle θ. To determine $\sigma_{x'}$, $\sigma_{y'}$ and $\tau_{x'y'}$ from the given stresses (σ_x, σ_y and τ_{xy}) is known as **stress transformation**, which is accomplished in the following paragraphs.

At a point *O* in an externally loaded thin plate as shown in Fig. 3.13, let us assume a plane stress condition given by Eqn. 3.9. A sectional plane (*m-n*) is assumed to pass through the point *O* such that the outward normal to this plane is defined by unit vector (**n**) at an angle (θ) with respect to *x*-direction. Our task is to determine the normal stress (σ_θ) and shear stress (τ_θ) on this sectional plane.

For this, we can consider a plane element *ABCD* of thickness (*t*) equal to the thickness of the plate surrounding the point *O* (see Fig. 3.14 (a)). A part of cutting plane (*m-n*) is shown by *DE*, which makes an angle θ with the face *DA*. Hence, a free-body diagram of wedge-shaped element (*DAE*) is taken separately in Fig. 3.14 (b). The stress vector ($\mathbf{T}^{(n)}$) on the plane *DE* has components T_x and T_y along *x*- and *y*-directions respectively, such that

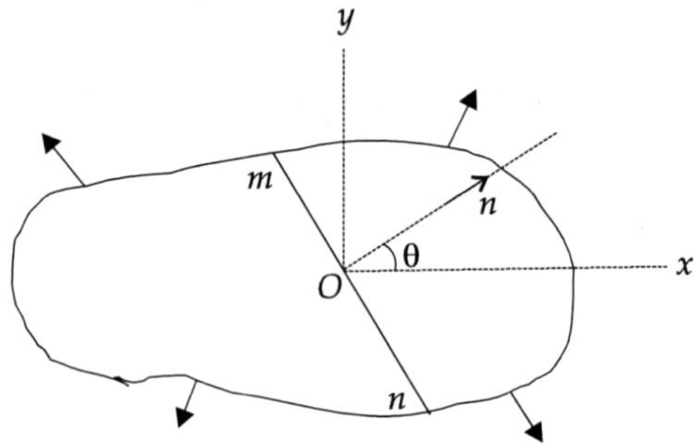

Fig. 3.13

$$\mathbf{T}^{(n)} = T_x \cdot \mathbf{i} + T_y \cdot \mathbf{j}.$$

Next, for the equilibrium of element *DAE* the condition $\sum F_x = 0$ leads to

$$T_x \cdot (DE \cdot t) - \sigma_x \cdot (DA.t) - \tau_{xy} \cdot (AE \cdot t) = 0$$

or $$T_x = \sigma_x \cdot (DA/DE) + \tau_{xy} \cdot (AE/DE)$$

∴ $$T_x = \sigma_x \cdot \cos\theta + \tau_{xy} \cdot \sin\theta. \quad (3.10)$$

Fig. 3.14

Also, from $\sum F_y = 0,\ T_y\cdot(DE\cdot t) - \sigma_y\cdot(AE\cdot t) - \tau_{xy}\cdot(DA\cdot t) = 0$

or $$T_y = \sigma_y\cdot(AE/DE) + \tau_{xy}\cdot(DA/DE)$$

$\therefore$ $$T_y = \sigma_y\cdot\sin\theta + \tau_{xy}\cdot\cos\theta \qquad (3.11)$$

Therefore, the normal stress on the plane DE is found as

$$\sigma_\theta = T\cdot\cos\theta + T_y\cdot\cos(\pi/2 - \theta), \quad \text{or} \quad \sigma_\theta = T_x\cdot\cos\theta + T_y\cdot\sin\theta,$$

or $$\sigma_\theta = \sigma_x\cdot\cos^2\theta + \sigma_y\cdot\sin^2\theta + 2\cdot\tau_{xy}\cdot\sin\theta\cdot\cos\theta. \qquad (3.12)$$

Also, $$\sigma_\theta = \{(\sigma_x + \sigma_y)/2\} + \{(\sigma_x - \sigma_y)/2\}\cdot\cos 2\theta + \tau_{xy}\cdot\sin 2\theta. \qquad (3.13)$$

Similarly, the shear stress on plane DE is written

$$\tau_\theta = T_y\cdot\cos\theta - T_x\cdot\sin\theta$$

or $$\tau_\theta = \{(\sigma_y - \sigma_x)/2\}\cdot\sin 2\theta + \tau_{xy}\cdot\cos 2\theta. \qquad (3.14)$$

In Eqns. 3.13 and 3.14, the normal and shear stresses are obtained in terms of the stress components in x-y plane. If the direction of the outward normal on the sectional plane at angle θ with respect to x-axis is chosen as x' coordinate, then the stresses (σ_θ and τ_θ) derived above can be replaced by notations $\sigma_\theta = \sigma_{x'}$ and $\tau_\theta = \tau_{x'y'}$, as given in Fig. 3.15. The coordinate y' is perpendicular to x', as shown. It implies that another plane perpendicular to the existing sectional plane can be assumed whose outward normal points to y' direction. The normal stress ($\sigma_{y'}$) and shear stress ($\tau_{y'x'}$) can be obtained on this plane. Therefore,

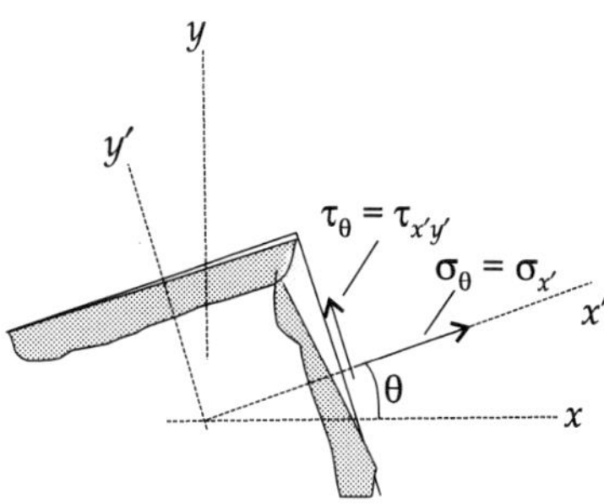

Fig. 3.15

with $$\left.\begin{aligned}\sigma_{x'} &= \{(\sigma_x + \sigma_y)/2\} + \{(\sigma_x - \sigma_y)/2\}\cdot\cos 2\theta + \tau_{xy}\cdot\sin 2\theta,\\ \tau_{x'y'} &= \{(\sigma_y - \sigma_x)/2\}\cdot\sin 2\theta + \tau_{xy}\cdot\cos 2\theta,\end{aligned}\right\} \qquad (3.15)$$

(the second line preceded by "and")

$\sigma_{y'}$ and $\tau_{y'x'}$ are found by replacing θ with $(\theta + \pi/2)$ in Eqns. 3.13 and 3.14.

Hence, $$\sigma_{y'} = \{(\sigma_x + \sigma_y)/2\} + \{(\sigma_x - \sigma_y)/2\}\cdot\cos(2\theta + \pi) + \tau_{xy}\cdot\sin(2\theta + \pi),$$

or $$\sigma_{y'} = \{(\sigma_x + \sigma_y)/2\} - \{(\sigma_x - \sigma_y)/2\}.\cos 2\theta - \tau_{xy}\cdot\sin 2\theta, \qquad (3.16)$$

and $$\tau_{y'x'} = \{(\sigma_y - \sigma_x)/2\}\cdot\sin(2\theta + \pi) + \tau_{xy}\cdot\cos(2\theta + \pi),$$

or $$\tau_{y'x'} = -[\{(\sigma_y - \sigma_x)/2\}\cdot\sin 2\theta + \tau_{xy}\cdot\cos 2\theta]. \qquad (3.17)$$

After comparing Eqns. 3.14 and 3.17 it is found that $\tau_{y'x'}$ and $\tau_{x'y'}$ are equal in magnitude but opposite in sense. Referring to Fig. 3.12(b), it is noted that τ_{xy} acts counterclockwise and τ_{yx} acts clockwise but they are equal in magnitude. Similarly, though equal in magnitude, $\tau_{x'y'}$ acts counterclockwise and $\tau_{y'x'}$ acts in clockwise manner.

The stress transformation discussed above can also be obtained by using tensor transformation rule. A vector is known as a tensor of the first order and vector transformation can be explained with the help of Fig. 3.16. For the point P shown in Fig. 3.16 the position vector in x-y coordinates can be written in the form

$$V = \begin{Bmatrix} x \\ y \end{Bmatrix}$$

In x' - y' coordinates which is rotated through angle θ, the position vector of point P becomes

$$V' = \begin{Bmatrix} x' \\ y' \end{Bmatrix}$$

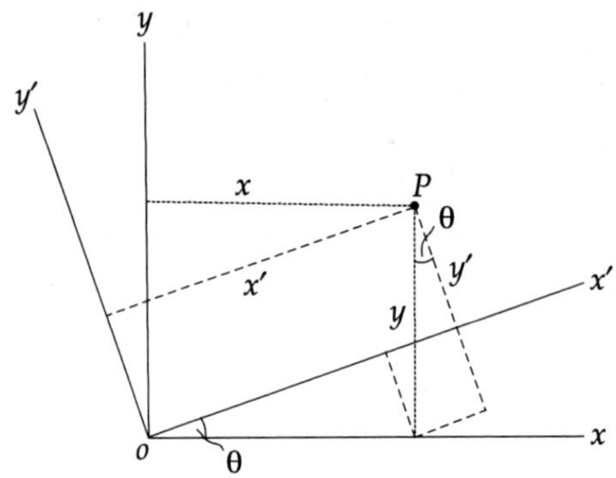

Fig. 3.16

From simple geometric relation in Fig. 3.16, it is found that

$$\left.\begin{aligned} x' &= x\cdot\cos\theta + y\cdot\sin\theta \\ \text{and} \quad y' &= -x\cdot\sin\theta + y\cdot\cos\theta \end{aligned}\right\} \tag{3.18}$$

The above equations can be written in the form

$$\begin{Bmatrix} x' \\ y' \end{Bmatrix} = \begin{Bmatrix} \cos\theta & \sin\theta \\ -\sin\theta & \cos\theta \end{Bmatrix} \begin{Bmatrix} x \\ y \end{Bmatrix} \tag{3.19}$$

As such, the above equation is written as $V' = R.V$,

where, $$R = \begin{Bmatrix} \cos\theta & \sin\theta \\ -\sin\theta & \cos\theta \end{Bmatrix},$$

and R is called **rotation matrix**.

Similarly, a tensor transformation is written as $T' = R\cdot T\cdot R^T$, where, R^T is transpose of R. It follows that with

$$T = \begin{Bmatrix} \sigma_x & \tau_{xy} \\ \tau_{xy} & \sigma_y \end{Bmatrix}, R = \begin{Bmatrix} \cos\theta & \sin\theta \\ -\sin\theta & \cos\theta \end{Bmatrix} \text{ and } T' = \begin{Bmatrix} \sigma_{x'} & \tau_{x'y'} \\ \tau_{x'y'} & \sigma_{y'} \end{Bmatrix},$$

we have

$$\begin{Bmatrix} \sigma_{x'} & \tau_{x'y'} \\ \tau_{x'y'} & \sigma_{y'} \end{Bmatrix} = \begin{Bmatrix} \cos\theta & \sin\theta \\ -\sin\theta & \cos\theta \end{Bmatrix} \cdot \begin{Bmatrix} \sigma_x & \tau_{xy} \\ \tau_{xy} & \sigma_y \end{Bmatrix} \cdot \begin{Bmatrix} \cos\theta & -\sin\theta \\ \sin\theta & \cos\theta \end{Bmatrix}.$$

The above matrix equation is solved and simplified further to get

$$\sigma_{x'} = \{(\sigma_x + \sigma_y)/2\} + \{(\sigma_x - \sigma_y)/2\}\cdot\cos 2\theta + \tau_{xy}\cdot\sin 2\theta,$$

$$\sigma_{y'} = \{(\sigma_x + \sigma_y)/2\} - \{(\sigma_x - \sigma_y)/2\}\cdot\cos 2\theta - \tau_{xy}\cdot\sin 2\theta,$$

and $$\tau_{x'y'} = \{(\sigma_y - \sigma_x)/2\}\cdot\sin 2\theta + \tau_{xy}\cdot\cos 2\theta,$$

which have been derived earlier through Eqns. 3.13-3.17.

Example 3.1: The stresses shown in Fig. 3.17 act at a point in a stressed body. Determine the normal and shear stresses at the point on the inclined plane *AB* shown in figure.

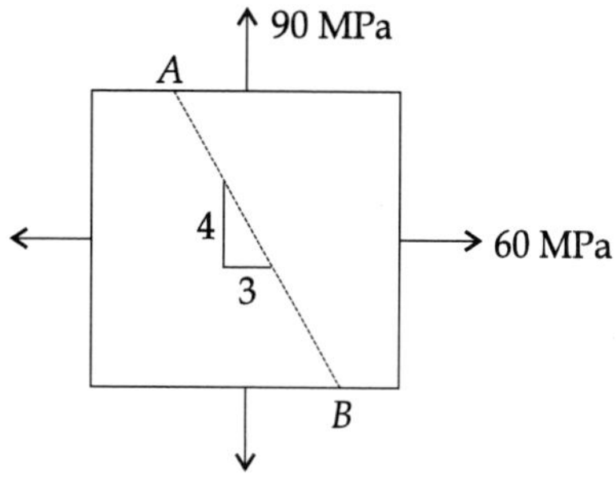

Fig. 3.17

Solution: For inclination of plane *AB*, $\sin\theta = 3/5$, $\cos\theta = 4/5$

Assuming *x*-*y* coordinates as per convention,

$\sigma_x = 60$ MPa, $\sigma_y = 90$ MPa, $\tau_{xy} = 0$.

From FBD of the left part and using Eqn. 3.13

$$\sigma_\theta = \{(\sigma_x + \sigma_y)/2\} + \{(\sigma_x - \sigma_y)/2\}\cdot(\cos^2\theta - \sin^2\theta)$$

$$= \{(60 + 90)/2\} + \{(60 - 90)/2\}\cdot\{(4/5)^2 - (3/5)^2\}\text{ MPa}$$

$$= 70.8 \text{ MPa (T)}$$ ... **Ans.**

Shear stress, $\tau_\theta = \{(\sigma_y - \sigma_x)/2\}\cdot 2\cdot\sin\theta\cdot\cos\theta$

$$= \{(90 - 60)/2\}\cdot 2\cdot(3/5)\cdot(4/5)\text{ MPa}$$

$$= 14.4 \text{ MPa}$$ ... **Ans.**

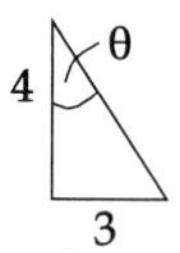

Example 3.2: At a point in a stressed thin plate, there are normal tensile stresses of 100 MPa on vertical plane and 140 MPa on horizontal plane, as shown in Fig. 3.18. Find the normal and shear stresses at this point on the inclined plane (*m*-*n*) shown in figure.

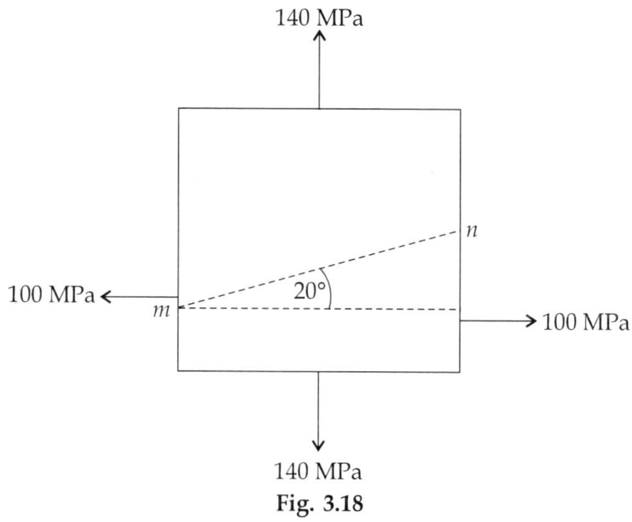

Fig. 3.18

Solution: Either upper or lower part of the element can be considered in equilibrium. Assuming equilibrium of the upper part, the components, T_x and T_y, of stress vector are introduced. In the related figure given here, a horizontal line (qn) is drawn to get the wedge shaped part (mnq), which is also in equilibrium.

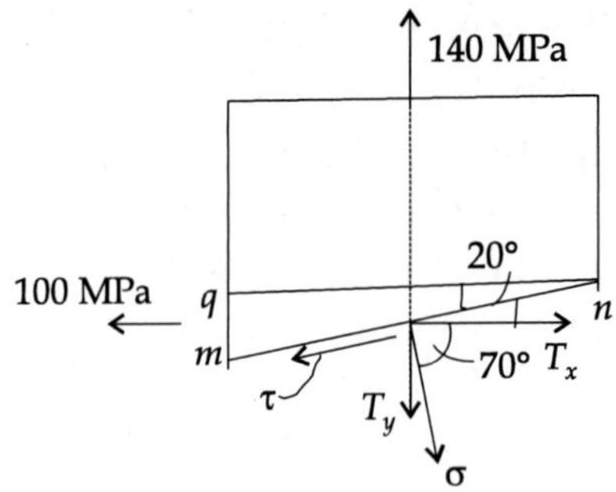

Let t = thickness of the plate. Hence, applying equilibrium conditions,

from $\sum F_x = 0$, $T_x \cdot (mn \cdot t) - 100 \cdot (qm \cdot t) = 0$, or $T_x = 100 \cdot (qm/mn) = 100 \cdot \sin 20°$

Similarly, from $\sum F_y = 0$, $T_y \cdot (mn \cdot t) - 140 \cdot (qn \cdot t) = 0$ (note that the same stress of 140 MPa transmits over area qn·t)

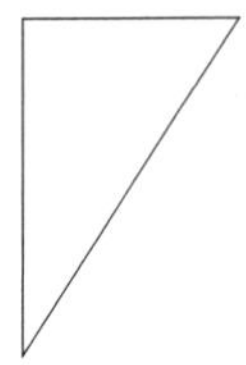

or $T_y = 140 \cdot (qn/mn) = 140 \cdot \cos 20°$

∴ Normal stress (σ) = $T_x \cdot \cos 70° + T_y \cdot \cos 20°$

$= 100 \cdot \sin 20° \cdot \cos 70° + 140 \cdot \cos 20° \cdot \cos 20°$ MPa

or $\sigma = 135.3$ MPa (T) **Ans.**

Shear stress (τ) = $T_x \cdot \cos 20° - T_y \cdot \sin 20°$ (expecting counterclockwise sense)

$= 100 \cdot \sin 20° \cdot \cos 20° - 140 \cdot \cos 20° \cdot \sin 20°$ MPa

or $\tau = -12.86$ MPa **Ans.** (calculation reveals clockwise sense)

The above results can also be obtained from a direct application of Eqns. 3.13 and 3.14. Thus, with $\sigma_x = 100$ MPa, $\sigma = 140$ MPa, $\tau_{xy} = 0$ and $\theta = -70°$,

$$\sigma = \{(100 + 140)/2\} + \{(100-140)/2\} \cdot \cos(-2 \times 70°) = 135.3 \text{ MPa(T)}$$

and $$\tau = \{(140 - 100)/2\} \cdot \sin(-2 \times 70°) = -12.86 \text{ MPa}$$

Example 3.3: For a shaft subjected to a twisting moment, the stresses at a point on the surface of shaft are shown in Fig. 3.19. What are the normal and shear stresses on a plane with the inclination shown in the figure?

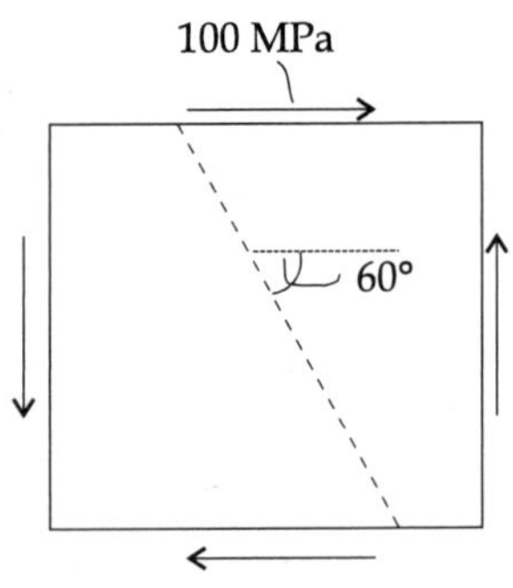

Fig. 3.19

Solution: For the left part of element in equilibrium,

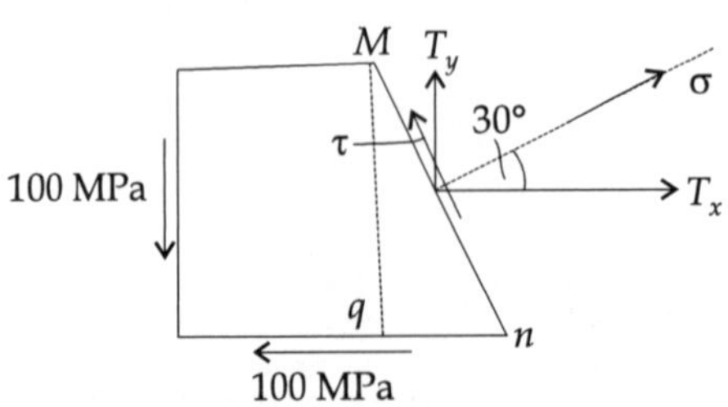

With $\sum F_x = 0$, $T_x \cdot (mn \cdot t) - 100 \cdot (qn \cdot t) = 0$

where t = plate thickness.

or $\quad T_x = 100\cdot(qn/mn) = 100\cdot\sin 30° = 50$ MPa

From $\sum F_y = 0$, $T_y\cdot(mn\cdot t) - 100\cdot(mq\cdot t) = 0$,

or $\quad T_y = 100\cdot(mq/mn) = 100\cdot\cos 30° = 86.6$ MPa

$\therefore$ Normal stress, $(\sigma) = 50\cdot\cos 30° + 86.6\cdot\cos 60°$ MPa, or $\sigma = 86.6$ MPa (T) **Ans.**

Shear stress $(\tau) = 86.6\cdot\sin 60° - 50\cdot\sin 30°$ MPa, or $\tau = 50$ MPa **Ans.**

Again, applying Eqns. 3.13 and 3.14, we have

$\sigma = 100\cdot\sin(2\times 30°) = 86.6$ MPa, and $\tau = 100\cdot\cos(2\times 30°) = 50$ MPa

Example 3.4: The stresses acting at a point on the surface of a beam are shown in Fig. 3.20. There is an inclined plane (*PQ*) through this point. Determine the normal and shear stresses on this sectional plane.

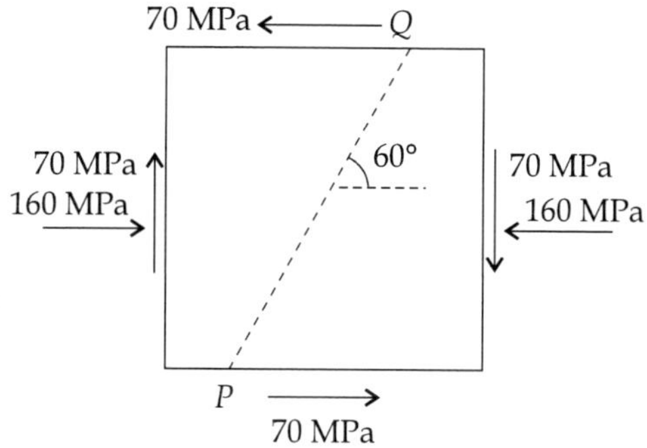

Fig. 3.20

Solution: Given that, $\sigma_x = -160$ MPa, $\sigma_y = 0$

$\tau_{xy} = -70$ MPa (because the direction is clockwise)

Using Eqn. 3.13, the normal stress

$\sigma = \{(-160)/2\}+\{(-160)/2\}\cdot\cos(-2\times 30°)+(-70)\cdot\sin(-2\times 30°)$

or $\sigma = -59.4$ MPa

$\therefore$ $\sigma = 59.4$ MPa (c) **Ans.**

Next, using Eqn. 3.14, the shear stress

$\tau = [\{0-(-160)\}/2]\cdot\sin(-60°) + (-70)\cdot\cos(-60°)$ MPa

$= -104.3$ MPa **Ans.** (acting clockwise)

The above results can be verified from FBD of wedge shaped element *PQR* given below.

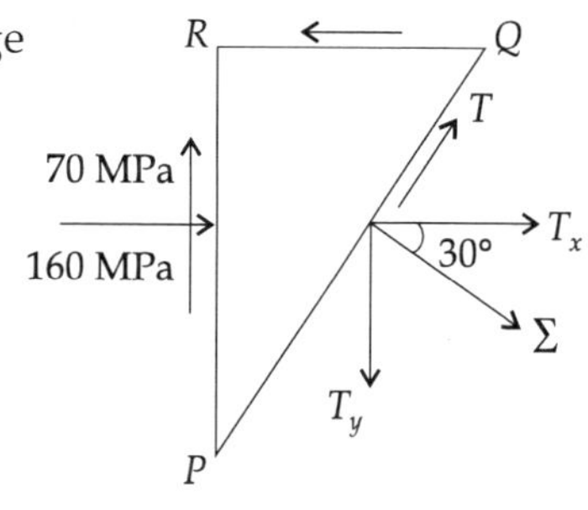

From $\sum F_x = 0$, $T_x\cdot(PQ\cdot t) + 160\cdot(PR\cdot t) - 70.(QR\cdot t) = 0$

$\therefore$ $\quad T_x = -160\cdot\cos 30° + 70\cdot\sin 30° = -103.56$

From $\sum F_y = 0$, $T_y\cdot(PQ\cdot t) - 70\cdot(PR\cdot t) = 0$

$\therefore$ $\quad T_y = 70\cdot\cos 30° = 60.62$

Hence, $\quad \sigma = (-103.56)\cdot\cos 30° + (60.62)\cdot\cos 60°$ MPa

$= -59.4$ MPa

Also, $\tau = (-103.56)\sin 30° - (60.62)\cdot\sin 60°$ MPa

$= -104.3$ MPa

Example 3.5: There is a plane stress condition at a point P of a machine part and the stresses at this point are $\sigma_x = -30$ MPa, $\sigma_y = 20$ MPa and $\tau_{xy} = 20$ MPa in x-y coordinates. An element at the same point is oriented counterclockwise at an angle θ with respect to x-axis and the stresses acting on this element are shown in Fig. 3.21. Determine the normal stress (σ_0), shear stress (τ_0) and angle (θ).

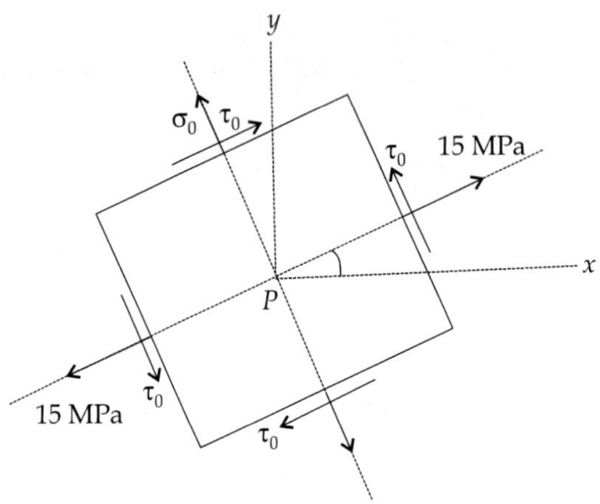

Fig. 3.21

Solution: Given that, $\sigma_x = -30$ MPa, $\sigma_y = 20$ MPa,

$\tau_{xy} = 20$ MPa,

$\sigma_{x'} = 15$ MPa, $\sigma_{y'} = \sigma_0$, $\tau_{x'y'} = \tau_0$

From Eqns. 3.15 and 3.16 derived earlier,

$$15 = \{(-30 + 20)/2\} + \{(-30 - 20)/2\}\cdot\cos 2\theta + 20\cdot\sin 2\theta$$

or $$4 = -5\cdot\cos 2\theta + 4\cdot\sin 2\theta \quad \text{...(a)}$$

$$\sigma_0 = \{(-30 + 20)/2\} - \{(-30 - 20)/2\}\cdot\cos 2\theta - 20\cdot\sin 2\theta$$

or $$\{(\sigma_0/5) + 1\} = 5\cdot\cos 2\theta - 4\cdot\sin 2\theta \quad \text{...(b)}$$

Also, $$\tau_0 = [\{20 - (-30)\}/2]\cdot\sin 2\theta + 20\cdot\cos 2\theta$$

or $$\tau_0 = 25\cdot\sin 2\theta + 20\cdot\cos 2\theta \quad \text{...(c)}$$

From (a) + (b), $4 + (\sigma_0/5 + 1) = 0$, or $\sigma_0 = -25$ MPa **Ans.**

In (b), $(-25/5 + 1) = 5\cdot\cos 2\theta - 4\cdot\sin 2\theta$

or $$-4 = 5\cdot\cos 2\theta - 4\cdot\sin 2\theta \quad \text{...(d)}$$

From (d) – (a), $-8 = 10\cdot\cos 2\theta - 8\cdot\sin 2\theta$.

Dividing throughout by $\sqrt{(10^2 + 8^2)} = 12.8062$, we have

$$0.7809\cdot\cos 2\theta - 0.6247\cdot\sin 2\theta = -0.6247 \quad \text{...(e)}$$

Let $\sin\phi = 0.7809$ and $\cos\phi = 0.6247$, so that $\phi = \tan^{-1}(0.7809/0.6247) = 51.34°$

Hence, the above expression (e) is written as $\sin(\phi - 2\theta) = -0.6247$

or $(51.34 - 2\theta) = \sin^{-1}(-0.6247)$

after solving, $\theta = 45°$ **Ans.**

Therefore, in (c), $\tau_0 = 25 \cdot \sin(2 \times 45°) + 20 \cdot \cos(2 \times 45°)$. $\therefore\ \tau_0 = 25$ MPa **Ans.**

Example 3.6: On an inclined plane passing through the point *A* of a stressed machine part, the normal and shear stresses are 35 MPa (T) and 10 MPa, respectively, as shown in Fig. 3.22. The normal stress on a vertical plane through the point is zero. Determine the shear stresses on horizontal and vertical planes passing through the point. Also, find the normal stress on horizontal plane at this point.

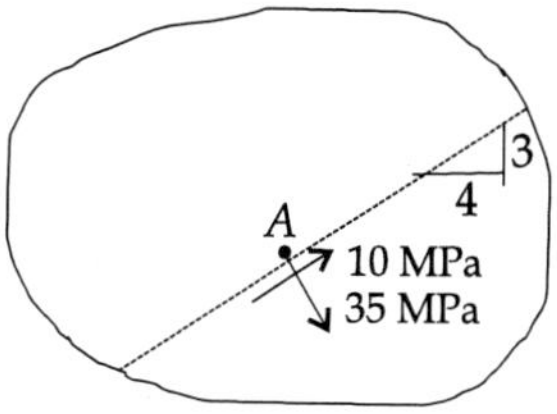

Fig. 3.22

Solution: $\sin\phi = 3/5$, $\cos\phi = 4/5$

Let us consider a wedge-shaped element enclosing the point *A*.

As $\theta = \pi/2 - \phi$, $2\theta = \pi - 2\phi$

$\therefore\ \cos 2\theta = -\cos 2\phi$

and $\sin 2\theta = -\sin 2\phi$

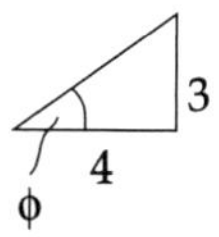

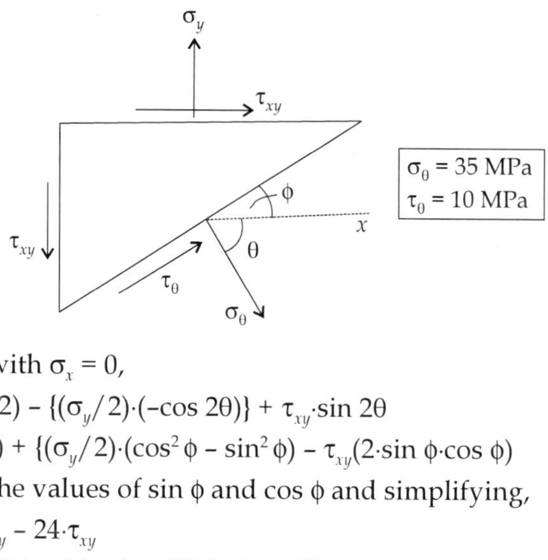

Using Eqn. 3.13 with $\sigma_x = 0$,

$$35 = (\sigma_y/2) - \{(\sigma_y/2) \cdot (-\cos 2\theta)\} + \tau_{xy} \cdot \sin 2\theta$$

or
$$35 = (\sigma_y 2) + \{(\sigma_y/2) \cdot (\cos^2\phi - \sin^2\phi) - \tau_{xy}(2 \cdot \sin\phi \cdot \cos\phi)$$

Substituting for the values of $\sin\phi$ and $\cos\phi$ and simplifying,

$$875 = 16.\sigma_y - 24 \cdot \tau_{xy} \quad \ldots \text{(a)}$$

Next, using Eqn. 3.14, $10 = \{(\sigma_y/2) \cdot (-\sin 2\phi)\} + \tau_{xy} \cdot (-\cos 2\phi)$,

or
$$250 = -12.\sigma_y - 7 \cdot \tau_{xy} \quad \ldots \text{(b)}$$

From $3 \times$ (a) $+ 4 \times$ (b), $3625 = -100.\ \tau_{xy}$, $\therefore\ \tau_{xy} = -36.25$ MPa **Ans.**

In (b), $250 = -12 \cdot \sigma_y - \{7.(-36.25)\}$ $\therefore\ \sigma_y = 0.31$ MPa **Ans.**

3.4 PRINCIPAL STRESSES AND MAXIMUM SHEAR STRESS

It was discussed earlier with the help of Eqn. 3.15 to Eqn. 3.17 that the normal and shear stresses ($\sigma_{x'}$, $\sigma_{y'}$, $\tau_{x'y'}$ and $\tau_{y'x'}$) on two mutually perpendicular planes defined by inclinations θ and $\theta + \pi/2$, are all functions of angle θ, i.e., the stresses change by changing the angle θ. It is found that at the two specific inclinations, $\theta = \theta_p$ and $\theta = \theta_p + (\pi/2)$, the shear stresses become zero ($\tau_{x'y'} = 0$ and $\tau_{y'x'} = 0$) and only the normal stresses ($\sigma_{x'} = \sigma_1$ and $\sigma_{y'} = \sigma_2$) are present on the two mutually perpendicular planes. Axial directions defined by θ_p and $\theta_p + (\pi/2)$ are known as **principal directions** and the corresponding planes are called **principal planes**. The normal stresses (σ_1 and σ_2) on these orthogonal planes are known as principal stresses. These are explained with the help of Fig. 3.23. Comparing normal stresses on planes of all possible orientations from $\theta = 0°$ to $\theta = 360°$, it is found that one of the principal stresses (σ_1) is the largest in magnitude and the other one (σ_2) is the smallest. Also, it is found that the maximum shear stress occurs on two mutually perpendicular planes which are at 45° angular separation from the principal planes.

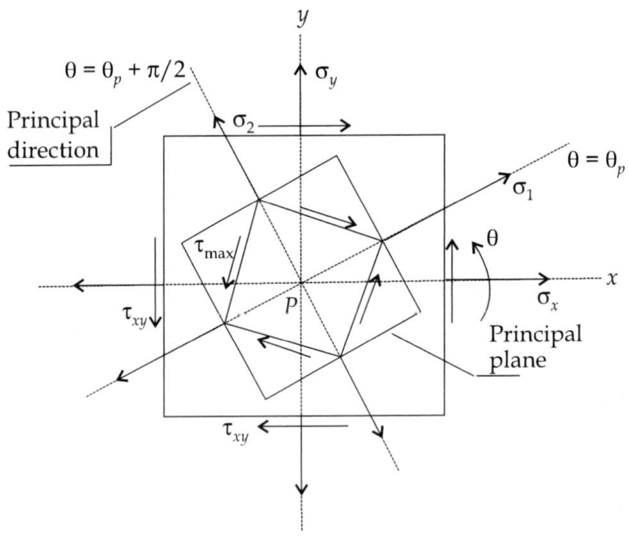

Fig. 3.23

Using Eqn. 3.13, the two principal stresses and maximum shear stress can be written in terms of the stresses, σ_x, σ_y and τ_{xy}.

With $\sigma_\theta = \{(\sigma_x + \sigma_y)/2\} + \{(\sigma_x - \sigma_y)/2\}\cdot\cos 2\theta + \tau_{xy}\cdot\sin 2\theta$, $d\sigma_\theta/d\theta = 0$ for maximum or minimum of σ_θ.

or $\{(\sigma_x - \sigma_y)/2\}\cdot(-2\sin 2\theta) + \tau_{xy}\cdot(2\cdot\cos 2\theta) = 0.$

$$\therefore \quad \tan 2\theta = (2\cdot\tau_{xy})/(\sigma_x - \sigma_y), \tag{3.20}$$

where, $\theta = \theta_p = (1/2)\cdot\tan^{-1}\{(2\cdot\tau_{xy})/(\sigma_x - \sigma_y)\}$, which gives a principal direction.

Also, $\tan\{2\cdot(\theta - \pi/2)\} = \tan 2\theta$, or $\tan(2\theta - \pi) = \tan 2\theta$.

$$\therefore \quad \tan(2\theta - \pi) = 2\cdot\tau_{xy}/(\sigma_x - \sigma_y). \tag{3.21}$$

Thus, $\theta = (\pi/2) + (12/2)\cdot\tan^{-1}\{(2\cdot\tau_{xy}/(\sigma_x - \sigma_y)\} = (\pi/2) + \theta_p$, the other principal direction.

From Eqn. 3.20 and using appropriate trigonometric relations,

$$\sin 2\theta = 2\cdot\tau_{xy}/\sqrt{(\sigma_x - \sigma_y)^2 + (4\cdot\tau_{xy}^2)}$$

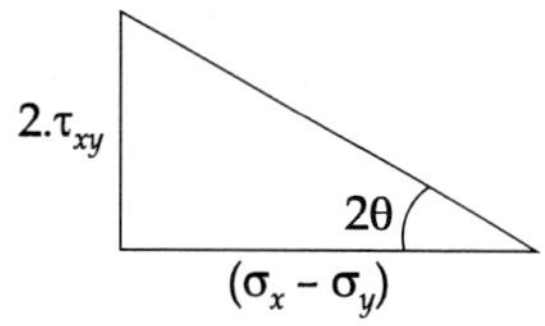

and $\cos 2\theta = (\sigma_x - \sigma_y)/\sqrt{(\sigma_x - \sigma_y)^2 + (4\cdot\tau_{xy}^2)}$

Hence, after substituting for sin 2θ and cos 2θ in Eqn. 3.13,

$$\sigma_\theta = \sigma_1 = \{(\sigma_x + \sigma_y)/2\} + (\sigma_x - \sigma_y)^2/\left\{2\cdot\sqrt{(\sigma_x - \sigma_y)^2 + (4\cdot\tau_{xy}^2)}\right\} + (2\cdot\tau_{xy}^2)/\sqrt{(\sigma_x - \sigma_y)^2 + (4\cdot\tau_{xy}^2)}$$

or $\sigma_1 = \{(\sigma_x + \sigma_y)/2\} + \sqrt{\{(\sigma_x - \sigma_y)/2\}^2 + \tau_{xy}^2}$ (3.22)

From Eqn. 3.21, $\sin(2\theta - \pi) = 2\cdot\tau_{xy}/\sqrt{(\sigma_x - \sigma_y)^2 + (4\cdot\tau_{xy}^2)}$

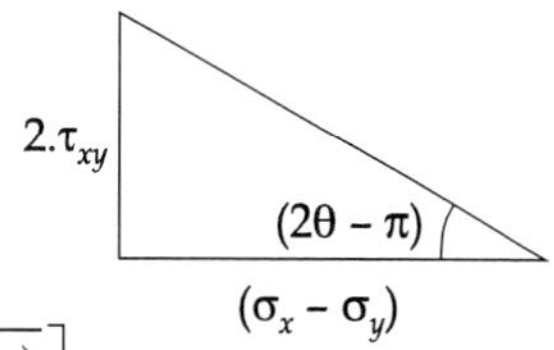

$\therefore$ $\sin 2\theta = -\left[2\cdot\tau_{xy}/\sqrt{(\sigma_x - \sigma_y)^2 + (4\cdot\tau_{xy}^2)}\right]$

Similarly, $\cos 2\theta = -\left[(\sigma_x - \sigma_y)/\sqrt{(\sigma_x - \sigma_y)^2 + (4\cdot\tau_{xy}^2)}\right]$

$$\therefore \quad \sigma_\theta = \sigma_2 = \{(\sigma_x + \sigma_y)\}/2 + \{(\sigma_x - \sigma_y)/2\}\cdot\left\{-(\sigma_x - \sigma_y)/\sqrt{(\sigma_x - \sigma_y)^2 + (4\cdot\tau_{xy}^2)}\right\} + 2\cdot\tau_{xy}\cdot\left\{-2\cdot\tau_{xy}/\sqrt{(\sigma_x - \sigma_y)^2 + (4\cdot\tau_{xy}^2)}\right\}$$

or $\sigma_2 = \{(\sigma_x + \sigma_y)\}/2 - \sqrt{\{(\sigma_x - \sigma_y)/2\}^2 + \tau_{xy}^2}$ (3.23)

Similarly, the shear stress (τ_θ) is maximum with $d\tau_\theta/d\theta = 0$, where, τ_θ is written according to Eqn. 3.13.

From $d\tau_\theta/d\theta = 0$, $\{(\sigma_y - \sigma_x)/2\}\cdot(2\cdot\cos 2\theta) - (\tau_{xy}\cdot\sin 2\theta) = 0$.

or $\tan 2\theta = (\sigma_y - \sigma_x)/(2\cdot\tau_{xy})$ (3.24)

From the above equation, we have $\sin 2\theta = (\sigma_y - \sigma_x)/\sqrt{(\sigma_y - \sigma_x)^2 + (4\cdot\tau_{xy}^2)}$,

and $\cos 2\theta = (2\cdot\tau_{xy})/\sqrt{(\sigma_y - \sigma_x)^2 + (4\cdot\tau_{xy}^2)}$,

so that
$$\tau_\theta = \tau_{max} = \{(\sigma_y - \sigma_x)/2\}\cdot\left\{(\sigma_y - \sigma_x)/\sqrt{(\sigma_y - \sigma_x)^2 + (4\cdot\tau_{xy}^2)}\right\}\left\{(\tau_{xy}\cdot 2\cdot\tau_{xy})/\sqrt{(\sigma_y - \sigma_x)^2 + (4\cdot\tau_{xy}^2)}\right\}$$

or $\tau_{max} = \sqrt{\{(\sigma_y - \sigma_x)/2\}^2 + \tau_{xy}^2}$ (3.25)

The normal stress on the plane of maximum shear stress is obtained as

$$\sigma_\theta = \sigma = \{(\sigma_x + \sigma_y)/2\} + \{(\sigma_x - \sigma_y)/2\}\cdot\left\{2\cdot\tau_{xy}/\sqrt{(\sigma_y - \sigma_x)^2 + (4\cdot\tau_{xy}^2)}\right\} + \left\{\tau_{xy}(\sigma_y - \sigma_x)/\sqrt{(\sigma_y - \sigma_x)^2 + (4\cdot\tau_{xy}^2)}\right\}$$

or $$\sigma = (\sigma_x + \sigma_y)/2 \qquad (3.26)$$

After adding the expressions of two principal stresses from Eqns. 3.22 and 3.23 we have $[(\sigma_1 + \sigma_2)/2] = [(\sigma_x + \sigma_y)/2]$, so that Eqn. 3.26 is also written as

$$\sigma = (\sigma_1 + \sigma_2)/2 \qquad (3.27)$$

The normal stress on the plane of maximum shear stress is the average of the two principal stresses.

If the plane of maximum shear stress is at an orientation $\theta = \theta_s$, Eqn. 3.24 can be rewritten as

$$\tan 2\theta_s = (\sigma_y - \sigma_x)/(2\cdot\tau_{xy})$$

and the product of the above equation and Eqn.3.20 gives

$$(\tan \theta_p)\cdot(\tan \theta_s) = \{(2\cdot\tau_{xy})/(\sigma_x - \sigma_y)\}\cdot\{(\sigma_y - \sigma_x)/(2\cdot\tau_{xy})\} = -1,$$

which implies that the angular separation between θ_p and θs is 45°, i.e., the plane of maximum shear stress is inclined at 45° relative to a principal plane (see Fig. 3.23).

Equations 3.22 to 3.27 are of immense importance in design analysis of structural parts based on strength of materials.

Example 3.7: The normal and shear stresses at a point in a structural component are shown to be acting on a plane stress element in Fig. 3.24. Determine the principal stresses and maximum shear stress at this point. Also, find inclinations of principal plane and the plane of maximum shear stress with respect to the horizontal direction. Show these results in a neat sketch.

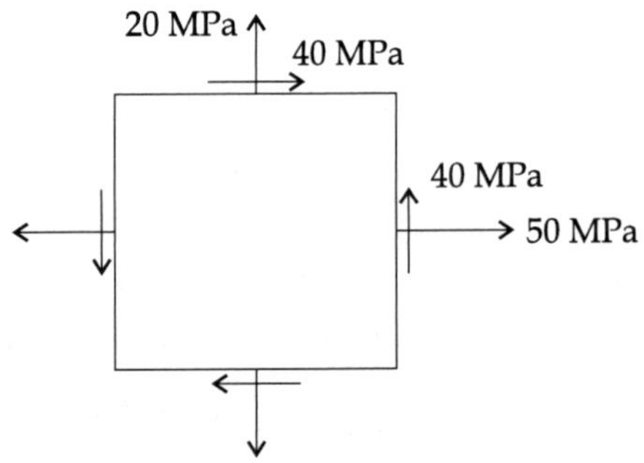

Fig. 3.24

Solution: Assuming the horizontal direction coinciding with x-axis,

$$\sigma_x = 50 \text{ MPa}, \sigma_y = 20 \text{ MPa and } \tau_{xy} = 40 \text{ MPa}.$$

The principal stresses are

$$\sigma_{1,2} = \{(50 + 20)/2\} \pm \sqrt{\{(50-20)/2\}^2 + (40^2)} = 35 \pm 42.72 \text{ MPa}$$

$\therefore\ \sigma_1 = 77.72$ MPa (T), $\sigma_2 = -7.72$ MPa

or $\sigma_2 = 7.72$ MPa (c) **Ans.**

Maximum shear stress,

$$\tau_{max} = \sqrt{\{(50-20)/2\}^2 + (40^2)}$$
$$= 42.7 \text{ MPa}$$ **Ans.**

Inclination of outward normal to the principal plane,

$\theta_p = (1/2)\cdot\tan^{-1}[(2 \times 40)/(50 - 20)]$,

or $\theta_p = 34.72°$ **Ans.**

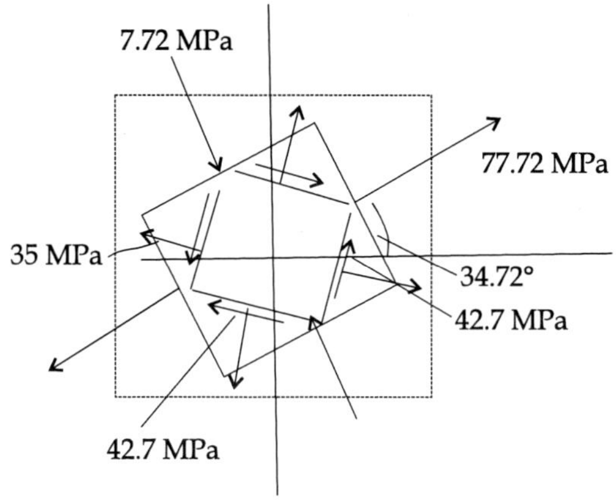

Orientation of the plane of maximum shear stress

$$\theta_s = (1/2)\cdot\tan^{-1}\{(20 - 50)/(2.40)\} = -10.28°$$ **Ans.**

$\therefore$ $\theta_s = 10.28°$ clockwise from x-axis.

Normal stress on the plane of maximum shear,

$$\sigma = (50 + 20)/2 = 35 \text{ MPa}$$ **Ans.**

Example 3.8: The plane stress element shown in Fig. 3.25 represents the state of stress at a point.

Determine: (a) the principal stresses, (b) maximum shear stress, (c) inclination of the two principal directions, and (d) normal stress on the plane of maximum shear stress.

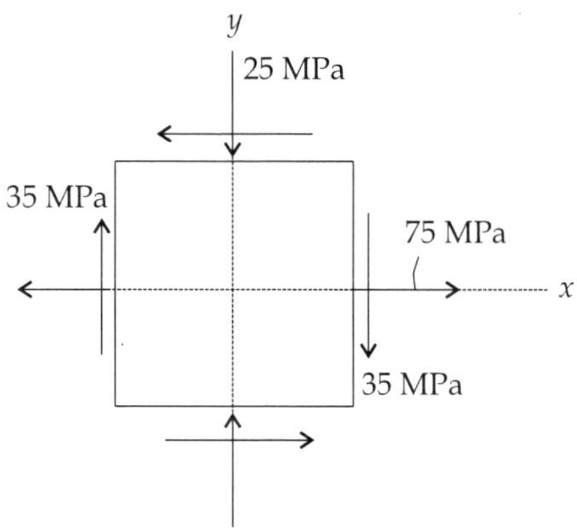

Fig. 3.25

Solution: Given that, $\sigma_x = 75$ MPa, $\sigma_y = -25$ MPa, $\tau_{xy} = -35$ MPa

(note the directions of σ_y and τ_{xy})

(a) $\sigma_{1,2} = \left[\{75+(-25)\}/2\right] \pm \sqrt{\left[\{75-(-25)\}/2\right]^2 + (-35)^2}$ MPa

$= 25 \pm 61$ MPa

∴ The principal stresses are

$\sigma_1 = 86$ MPa (t) and $\sigma_2 = -36$ MPa or 36 MPa (c). **Ans.**

(b) Maximum shear stress,

$$\tau_{max} = \sqrt{\left[\{75-(-25)\}/2\right]^2 + (-35)^2} \text{ MPa}$$

$= 61$ MPa **Ans.**

(c) $\theta_p = (1/2)\cdot\tan^{-1}[\{2\cdot(-35)\}/\{75-(-25)\}]$ or $\theta_p = -17.5°$ **Ans.**

(d) Normal stress on the plane of maximum shear, $\sigma = \{75 + (-25)\}/2 = 25$ MPa **Ans.**

Example 3.9: At a point on the surface of a machine element, the state of plane stress is shown in Fig. 3.26. The principal stresses at the point are 140 MPa (c) and 85 MPa (T). Determine the normal stresses, σ_y and σ_x, on the horizontal and vertical planes through the point and the angle between the x-axis and maximum principal stress at the point.

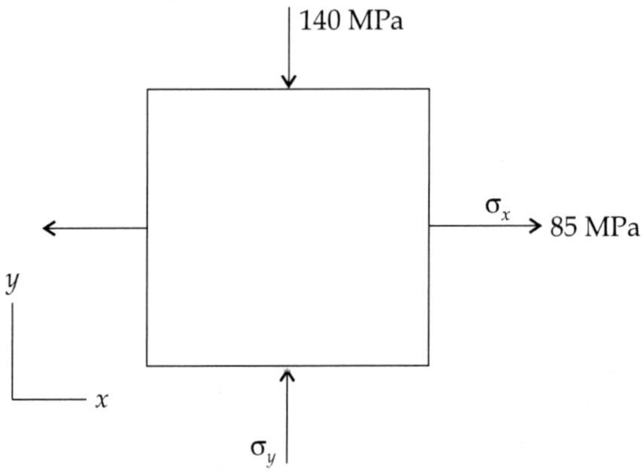

Fig. 3.26

Solution: Given that $\tau_{xy} = 60$ MPa, $\sigma_1 = 85$ MPa, $\sigma_2 = -140$ MPa .

It follows that $$85 = \{(\sigma_x + \sigma_y)/2\} + \sqrt{\{(\sigma_x - \sigma_y)/2\}^2 + (60^2)}, \quad \text{... (a)}$$

and $$-140 = \{(\sigma_x + \sigma_y)/2\} - \sqrt{\{(\sigma_x - \sigma_y)/2\}^2 + (60^2)}. \quad \text{... (b)}$$

In the above equations, σ and σ_y are unknown and initially both are assumed to be positive. The calculations reveal the magnitude and sense of the stresses.

From (a) + (b), $\sigma_x + \sigma_y = -55$ MPa. ... (c)

From (a) – (b) and further simplification,

$\sigma_x - \sigma_y = 190.3287$ MPa ... (d)

From (c) and (d), $\sigma_x = 67.66$ MPa (T) **Ans.**

and $\sigma_y = -122.66$ MPa

or $\sigma_y = 122.66$ MPa (c) **Ans.**

Also, angle $\theta_p = (1/2)\cdot\tan^{-1}\{(2 \times 60)/(67.66 - (-122.66)\}$

∴ $\theta_p = 16.1°$ **Ans.**

Example 3.10: A plane stress element is shown in Fig. 3.27. If one of the principal stresses is 26.5 MPa in tension, determine: (a) the stress σ_y, (b) the other principal stress and orientation of the principal planes. Show the principal stresses in a neat sketch with principal directions in *x*-*y* coordinates.

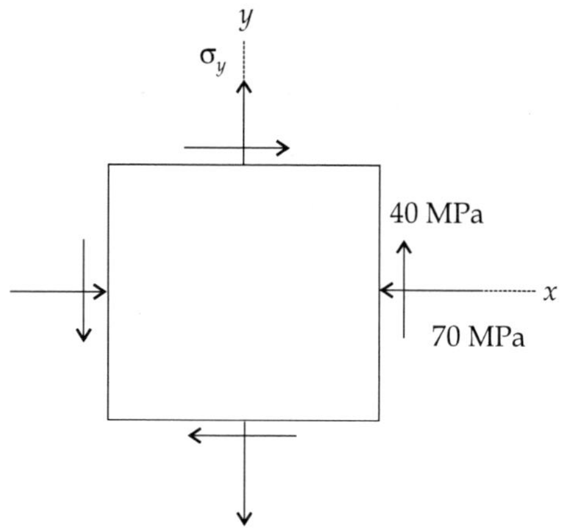

Fig. 3.27

Solution:

(a) Assuming σ_1 = 26.5 MPa,

$$26.5 = \{(-70+\sigma_y)/2\} + \sqrt{\{(-70-\sigma_y)/2\}^2 + (40^2)}$$

or $[26.5-\{(\sigma_y - 70)/2\}^2] = [\{(\sigma_y + 70)/2\}^2 + (40^2)]$

After simplifying, σ_y = 9.92 MPa (T) **Ans.**

(b) The other principal stress,

$$\sigma_2 = \{(-70+9.92)/2\} + \sqrt{\{(-70-9.92)/2\}^2 + (40^2)}$$

or σ_2 = –86.6 MPa. ∴ σ_2 = 86.6 MPa (T) **Ans.**

Also, $\theta_p = (1/2)\cdot\tan^{-1}\{(2\times 40)/(-70 - 9.92)\}$

θ_p = –22.5° **Ans.**

Principal stresses and principal direction are shown below.

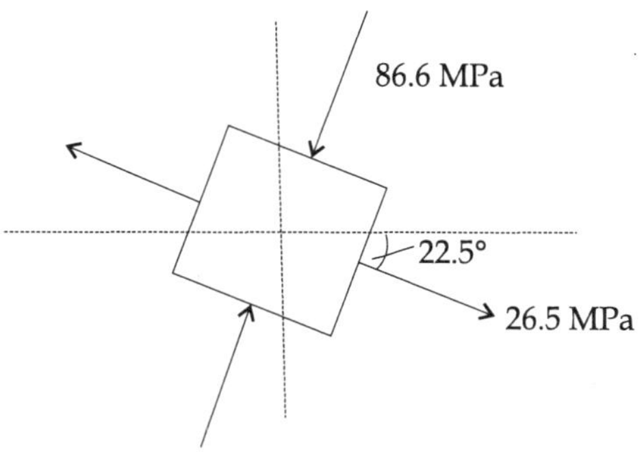

3.5 MOHR'S CIRCLE FOR PLANE STRESS

If the stresses are known at a point on two mutually perpendicular planes, as shown in Fig. 3.28, then the stresses on any other set of mutually perpendicular planes at the point can be obtained with the help of a simple geometric arrangement known as **Mohr's circle**. The concept of Mohr's circle was introduced by Otto Mohr (1886), which is applicable to physical quantities that are defined as the second order tensor. Let us first discuss the construction of Mohr's circle followed by its application.

In Fig. 3.28, a plane defined by unit vector (**n**) contains a normal stress (σ_n) with a shear stress (τ_n) in counterclockwise sense. There is another plane defined by **n′** which carries a normal stress ($\sigma_{n'}$) and also a sheear stress ($-\tau_n$) (in clockwise sense). It should be noted that the choice of *x*-*y* coordinates depends on the convenience and requirement of any problem. As such, the direction of **n** can be assumed as *x*-direction and that of **n′** as *y*-direction. We can further assume $\sigma_n = \sigma_x$, $\sigma_{n'} = \sigma_y$ and $\tau_n = \tau_{xy}$. On an inclined plane through the point having outward normal at an inclination θ relative to *x*-axis (see Fig. 3.28), the normal stress (σ) and shear stress (τ) are written using Eqns. 3.13 and 3.14 as

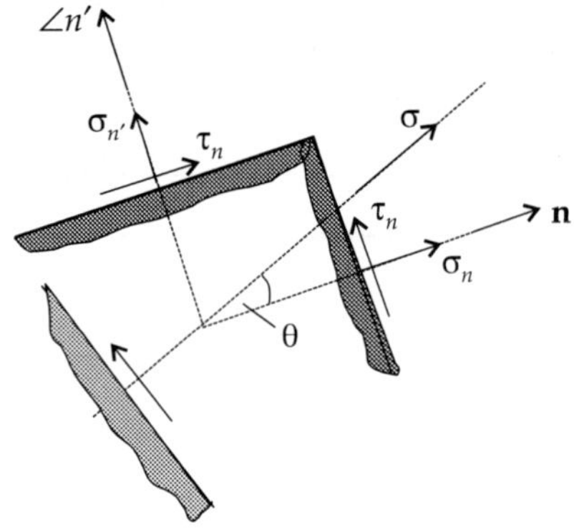

Fig. 3.28

$$\sigma = \{(\sigma_x + \sigma_y)/2\} + \{(\sigma_x - \sigma_y)/2\}\cdot\cos 2\theta + \tau_{xy}\cdot\sin 2\theta, \text{ and}$$

$$\tau = \{(\sigma_y - \sigma_x)/2\}\cdot\sin 2\theta + \tau_{xy}\cdot\cos 2\theta.$$

After rearranging the first expression and then sqaring and adding both, we have

$$[\sigma - \{(\sigma_x + \sigma_y)/2\}]^2 + (\tau - 0)^2 = \left[\sqrt{\{(\sigma_x - \sigma_y)/2\}^2 + \tau_{xy}^2}\right]^2 \qquad (3.28)$$

The above expression is in the form of equation of a circle in σ – τ coordinates, with centre at $[(\sigma_x + \sigma_y)/2, 0]$ and radius equal to $\sqrt{\{(\sigma_x - \sigma_y)/2\}^2 + \tau_{xy}^2}$. Hence, a Mohr's circle can be constructed as given in Fig. 3.29.

Construction of Mohr's circle: A σ – τ coordinate system is considered with its origin at O. A suitable scale (for example, 10 MPa = 1 cm) is chosen to fix σ_x, σ_y and τ_{xy}. Next, $OX = \sigma_x$ and $OY = \sigma_y$ are taken to scale to locate the points *X* and *Y* on σ-axis. A perpendicular ($XX = +\tau_{xy}$) is drawn at X which represents τ_{xy} acting on the plane facing *x*-axis (as in Fig. 3.28). Similarly, another perpendicular ($YY' = -\tau_{xy}$) is drawn at *Y*. The two points, *X′* and *Y′*, are joined to intersect σ-axis at the point *C*, which is the centre of Mohr's circle. With *C* as centre and *CX′* or *CY′* as radius, a circle is drawn to intersect σ-axis at *M* and *N*. It may be checked according to Eqn. 3.28 that the centre of the circle is at $OC = (OX + OY)/2 = (\sigma_x + \sigma_y)/2$, or $\sigma = (\sigma_x + \sigma_y)/2$ and $\tau = 0$. Also, the radius of the *X* circle, $CX' = \sqrt{(CX)^2 + (XX')^2} = \sqrt{\{(\sigma_x - \sigma_y)/2\}^2 + \tau_{xy}^2}$. Further, at the points *M* and N, $\tau = 0$, implies that these points correspond to the two principal planes with CM and CN representing the two principal directions. The angular separation between CM and CN is 180°, which is twice the angular separation actually observed (see Fig. 3.23). Similarly, actual angular separation (θ) between σ_x and principal stress (σ_1) is shown by $\angle X'CM = 2\theta$ in Mohr's circle, both measured counterclockwise.

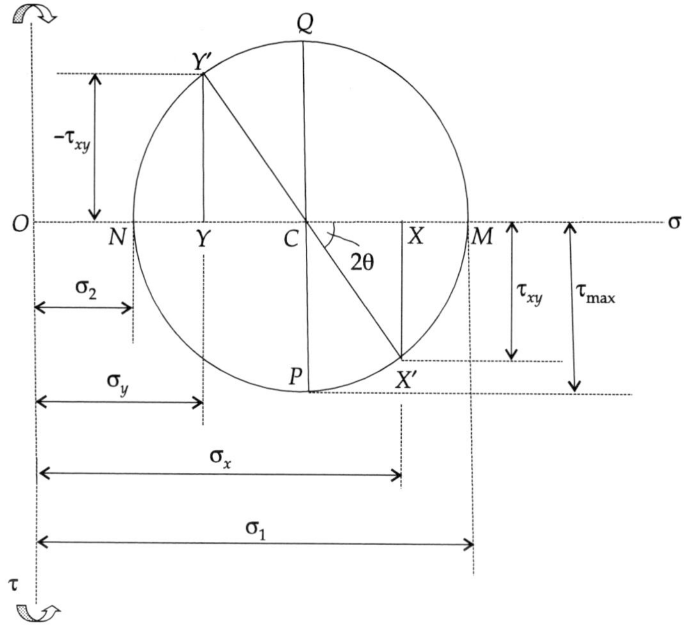

Fig. 3.29

Expressions of principal stresses (σ_1 and σ_2), which were derived in Eqns.3.22 and 3.23, are obtained from Mohr's circle as follows:

$$\sigma_1 = OM = OC + CM = OC + CX' = OC + \sqrt{(CX)^2 + (XX')^2}$$

$$= \{(\sigma_x + \sigma_y)/2\} + \sqrt{\{(\sigma_x - \sigma_y)/2\}^2 + \tau_{xy}^2}$$

Similarly, $\sigma_2 = ON = OC - CN = \{(\sigma_x + \sigma_y)/2\} - \sqrt{\{(\sigma_x - \sigma_y)/2\}^2 + \tau_{xy}^2}$

The stress transformation from x-y coordinates to x'-y' coordinates at angular separation (θ) can also be obtained from an appropriate Mohr's circle construction given in Fig. 3.30 as explained below.

The centre of Mohr's circle is at C. Also, $OX = \sigma_x$, $OY = \sigma_y$, $XA = \tau_{xy}$, $OX' = \sigma_{x'}$, $OY' = \sigma_{y'}$, $X'E = \tau_{x'y'}$, $\angle ACE = 2\theta$, and $\angle ECX'$ is assumed as α.

Let R = radius of Mohr's circle.

For $CX = CA \cdot \cos(\alpha + 2\theta)$,

$$R \cdot \cos(\alpha + 2\theta) = (\sigma_x + \sigma_y)/2, \quad \dots \text{(a)}$$

For $XA = CA \cdot \sin(\alpha + 2\theta)$

$R\cdot\sin(\alpha + 2\theta) = \tau_{xy'}$ ... (b)

Next, $CX' = CE\cdot\cos\alpha,$

or $R\cdot\cos\alpha = (\sigma_{x'} - \sigma_{y'})/2$ ... (c)

and $X'E = CE\cdot\sin\alpha$

or $R\cdot\sin\alpha = \tau_{x'y'}$ . ... (d)

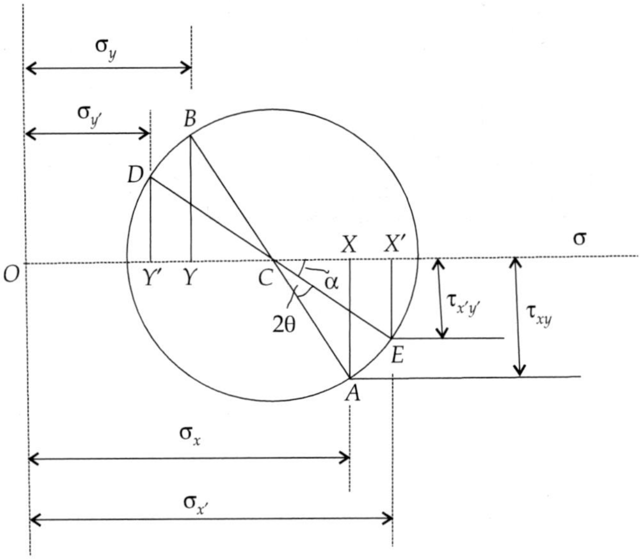

Fig. 3.30

From (a) and (b), $\left.\begin{aligned} R\cdot(\cos\alpha\cdot\cos 2\theta - \sin\alpha\cdot\sin 2\theta) &= (\sigma_x - \sigma_y)/2, \\ \text{and} \quad R.(\sin\alpha\cdot\cos 2\theta + \cos\alpha\cdot\sin 2\theta) &= \tau_{xy} \end{aligned}\right\}$

Next, from (c) and (d), we have

$$\left.\begin{aligned} \{(\sigma_{x'} - \sigma_{y'})/2\}\cdot\cos 2\theta - \tau_{x'y'}\cdot\sin 2\theta &= (\sigma_x - \sigma_y)/2 . \\ \text{and} \quad \tau_{x'y'}\cdot\cos 2\theta + \{(\sigma_{x'} - \sigma_{y'})/2\}\cdot\sin 2\theta &= \tau_{xy} . \end{aligned}\right\} \quad \text{... (e)}$$

After eliminating $\tau_{x'y'}$ from above expressions,

$$(\sigma_{x'} - \sigma_{y'})/2 = \{(\sigma_x - \sigma_y)/2\}\cdot\cos 2\theta + \tau_{xy}\cdot\sin 2\theta. \quad \text{... (f)}$$

Also from Fig. 3.30, $OC = (OX' + OY')/2 = (OX + OY)/2$

or $$(\sigma_{x'} + \sigma_{y'})/2 = (\sigma_x + \sigma_y)/2 \quad \text{... (g)}$$

Upon addition and subtraction between (g) and (f),

$$\sigma_{x'} = \{(\sigma_x + \sigma_y)/2\} + \{(\sigma_x - \sigma_y)/2\}\cdot\cos 2\theta + \tau_{xy}\cdot\sin 2\theta,$$

and $\sigma_{y'} = \{(\sigma_x + \sigma_y)/2\} - \{(\sigma_x - \sigma_y)/2\}\cdot\cos 2\theta - \tau_{xy}\cdot\sin 2\theta.$

From rearrangement of two equations in (e),

$$T_{x'y'} = \{(\sigma_y - \sigma_x)/2\}\cdot\sin 2\theta + \tau_{xy}\cdot\cos 2\theta.$$

The following are some important features of Mohr's circle:

1. Two diametrically opposite points lying on the circumference of the circle give normal stresses (along σ-axis) on two mutually perpendicular planes. Perpendicular distances of these points from σ-axis give shear stresses on two mutually perpendicular planes. Such shear stresses are equal in magnitude but opposite in sense (counterclockwise and clockwise).
2. Radius of the circle up to a point lying on the circumference represents the axis directed normal to that plane on which the normal and shear stresses are obtained from the position of the point in $\sigma - \tau$ coordinates.
3. The angle between radii of any two points lying on the circumference is two times the actual angle between the normal over two planes represented by the two radii in Mohr's circle.
4. It should be noted that the two principal stresses at a point are independent of the choice of coordinates. It may be observed in Mohr's circle that irrespective of the choice as x-y or x'-y' coordinates, the principal stresses (σ_1 and σ_2) are fixed. It is also found that $(\sigma_1 + \sigma_2)/2 = (\sigma_x + \sigma_y)/2 = (\sigma_{x'} + \sigma_{y'})/2$, which is known as **invariant of stresses**. It will be discussed later on that for three-dimensional stresses there are three invariants available for solving problems.

Example 3.11: A plane stress element is shown in Fig. 3.31 with only $\sigma_x = 100$ MPa acting. Using Mohr's circle, determine the stresses acting on the element oriented at a counterclockwise angle $\theta = 25°$ from x-axis. Also, find the maximum shear stress and associated normal stresses.

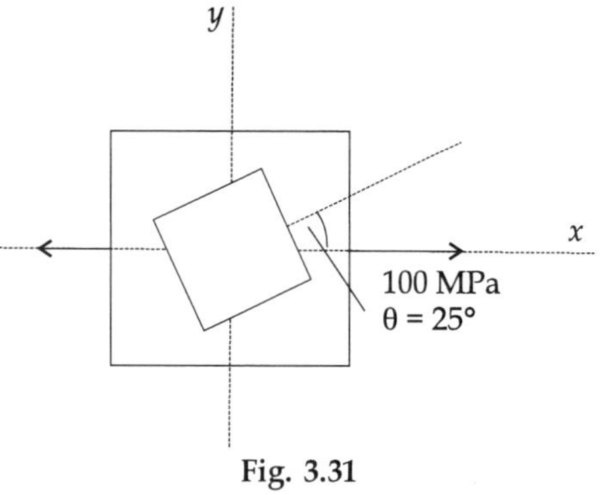

Fig. 3.31

Solution: Since the plane containing the normal stress of 100 MPa is free from any shear stress, the principal stress is $\sigma_1 = 100$ MPa in x-direction and $\sigma_2 = 0$ in y-direction. Hence, Mohr's circle is drawn showing these two principal stresses. At $2\theta = 2 \times 25 = 50°$ counterclockwise from σ-axis, the diameter AB is drawn.

The stresses are, $\sigma_{x'} = OE = OC + CE = OC + CA.\cos 2\theta$

$$= (100/2)+(100/2)\cdot\cos 50° \text{ MPa},$$

or $\sigma_{x'} = 82.14$ MPa (T) **Ans.**

$$\sigma_{y'} = OD = OC - CD = OC - CE = (100/2) - (100/2)\cdot\cos 50° \text{ MPa},$$

or $\sigma_{y'} = 17.86$ MPa (T) **Ans.**

$$\tau_{x'y'} = EA = DB = \pm CA\cdot\sin 50° = \pm(100/2)\cdot\sin 50° = 38.3 \text{ MPa}$$ **Ans.**

The maximum shear stress, $\tau_{max} = CP = (100/2) = 50$ MPa **Ans.**

Normal stress on the plane of maximum shear stress $= OC = (100/2) = 50$ MPa

Ans.

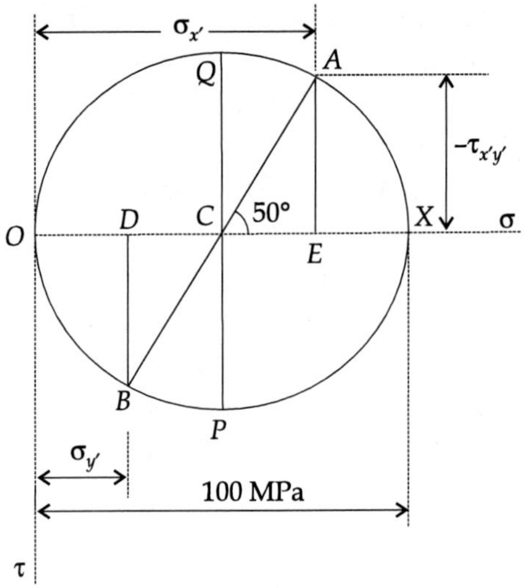

The stresses are shown over the element oriented at 25°.

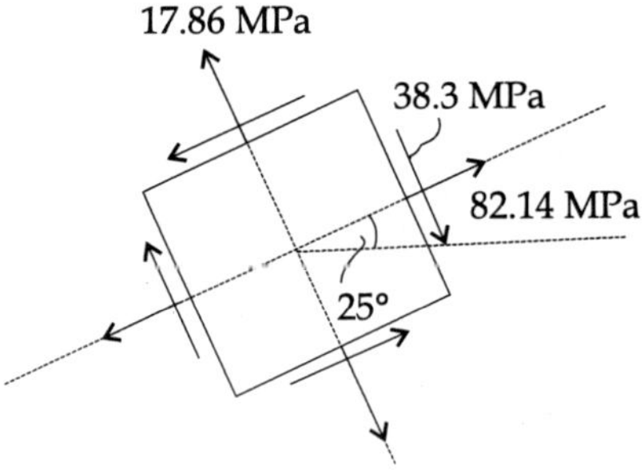

Example 3.12: A uniaxial compressive stress of 40 MPa is shown on a plane stress element in Fig. 3.32. Determine (a) the stresses acting on an element oriented at a slope 1 : 2, and (b) the maximum shear stress and associated normal stresses.

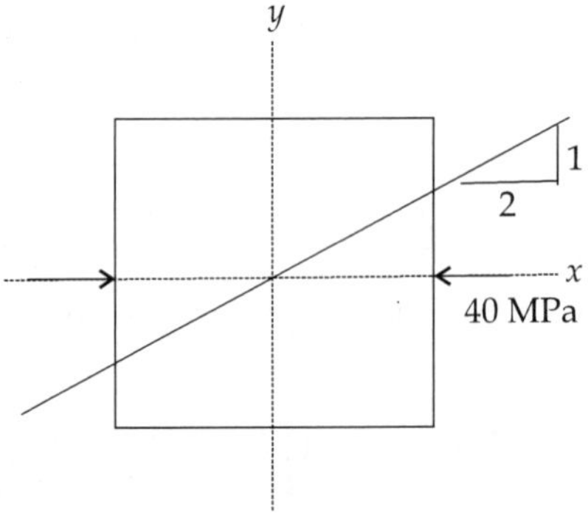

Fig. 3.32

Solution: The compressive stress of 40 MPa is a principal stress, written as $\sigma_1 = -40$ MPa. It is shown by *OM* in Mohr's circle. The line with slope 1:2 in Fig. 3.32 is at an angle counterclockwise relative to *x*-axis, given by $\theta = \tan^{-1}(1/2) = 26.565°$. Hence, $2\theta = 53.1°$. Through the centre (*C*) of Mohr's circle, a diameter *AB* is drawn counterclockwise such that $\angle ACM = 53.1°$. Two perpendiculars *AD* and *BE* are drawn on to σ-axis. Then $OD = \sigma_{x'}$, $OE = \sigma_{y'}$ and $DA = \tau_{x'y'}$.

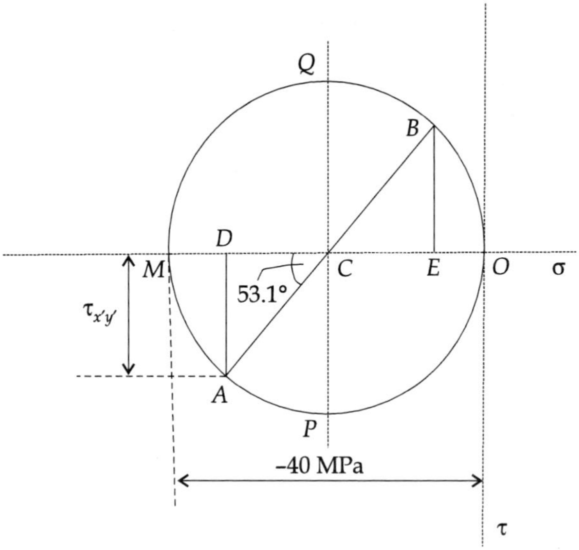

(a) $\sigma_{x'} = OD = OC + CD = (-40/2) + \{(-40/2)\cdot\cos 53.1°\}\cdot$ or $\sigma_{x'} = -32$ MPa (c) **Ans.**

$\sigma_{y'} = OE = OC - CE = (-40/2) - \{(-40/2)\cdot\cos 53.1°\}\cdot$ or $\sigma_{y'} = -8$ MPa (c) **Ans.**

$\tau_{x'y'} = DA = EB = \pm(40/2)\cdot\sin 53.1° = \pm 16$ MPa **Ans.**

The stresses on the element are as shown below:

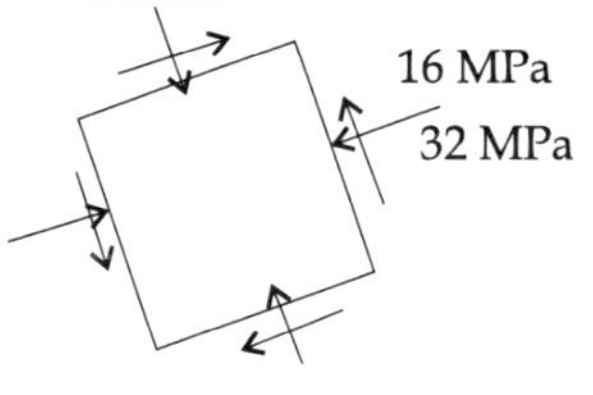

(b) $\tau_{max} = CP = CQ = \pm(40/2) = \pm 20$ MPa **Ans**

Associated normal stress, $\sigma = OC = -(40/2) = -20$ MPa **Ans.**

Example 3.13: The stresses, $\sigma_x = -5$ MPa and $\sigma_y = 65$ MPa, are acting at a point and shown with a plane stress element in Fig. 3.33. With Mohr's circle construction find the stresses on the element oriented at a slope of 1 on 2.5. What is the value of maximum shear stress?

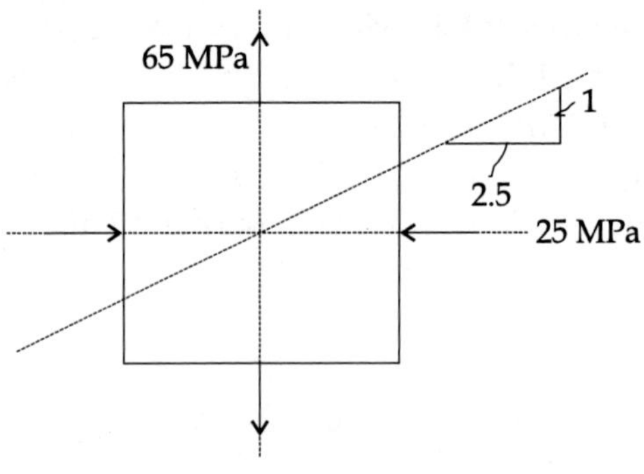

Fig. 3.33

Solution: Both σ_x and σ_y are principal stresses and these are shown on Mohr's circle by OX and OY, respectively.

Radius of Mohr's circle, $R = \{65 - (-25)\}/2 = 45$ MPa.

Orientation of element, $\theta = \tan^{-1}(1/2.5) = 21.8°$. $\therefore$ $2\theta = 43.6°$.

Therefore, the diameter AB is drawn such that $\angle DCA = 43.6°$.

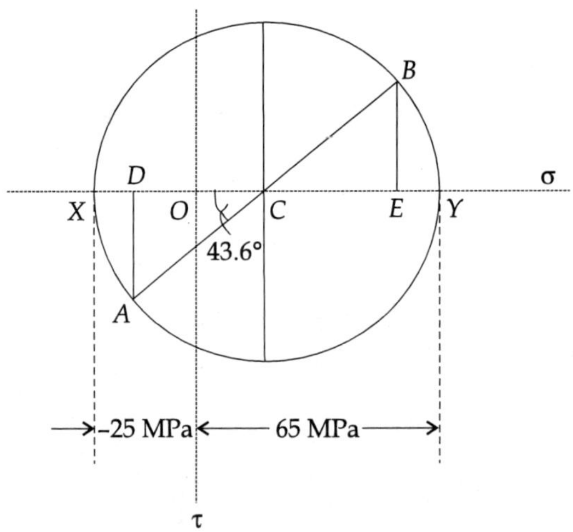

$$\sigma_{x'} = OD = CD - CO = -\{CA\cdot\cos 2\theta - CO\} = -\{CA\cdot\cos 2\theta - (CX - OX)\}$$

$$= -\{45\cdot\cos 43.6° - (45 - 25)\} \text{ MPa}$$

or $\sigma_{x'} = -12.6$ MPa (c) **Ans.**

$\sigma_{y'} = OE = CO + CE = CO + R\cdot\cos 2\theta = 20 + 45\cdot\cos 43.6°$, or $\sigma_{y'} = 52.6$ MPa (T) **Ans.**

$\tau_{x'y'} = \pm R\cdot\sin 2\theta = \pm 45\cdot\sin 43.6°$, or $\tau_{x'y'} = \pm 31$ MPa **Ans.**

Maximum shear stress, $\tau_{max} = R = 45$ MPa **Ans.**

Example 3.14: An element in pure shear with the stresses $\tau_{xy} = \pm 20$ MPa is shown in Fig. 3.34. Determine (a) the stresses on the element oriented counterclockwise at angle $\theta = 65°$ from x-axis, and (b) the principal stresses. Show the results on a sketch.

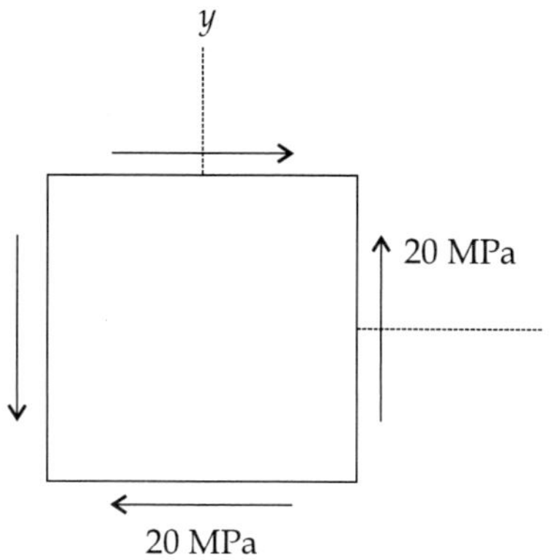

Fig. 3.34

Solution: When only shear stresses are acting on mutually perpendicular planes, it is known as a state of pure shear. In this, case the normal stresses are zero and shear stresses are shown by radii of Mohr's circle such as OX and OY. Also, the centre of Mohr's circle coincides with the origin of σ - τ coordinates.

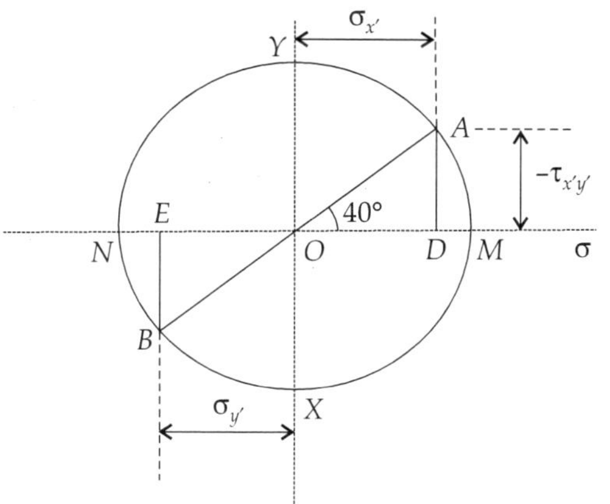

Referring to Mohr's circle, $OX = OY = 20$ MPa. Since $\theta = 65°$, $2\theta = 130°$. Hence, a diameter AB is drawn with $\angle XOA = 90° + 40° = 130°$.

$\therefore \quad \sigma_{x'} = OD = OA \cdot \cos 40° = 20 \cdot \cos 40°$ MPa, or $\sigma_{x'} = 15.3$ MPa (T) **Ans.**

$\sigma_{y'} = OE = OB \cdot \cos 40° = -20 \cdot \cos 40°$ MPa, or $\sigma_{y'} = -15.3$ MPa (c) **Ans.**

$\tau_{x'y'} = AD = EB = \pm 20 \cdot \sin 40° = \pm 12.86$ MPa **Ans.**

The principal stresses are, $\sigma_1 = OM = 20$ MPa (T), **Ans.**

and $\sigma_2 = ON = -20$ MPa (c) **Ans.**

The results are shown in the following figures:

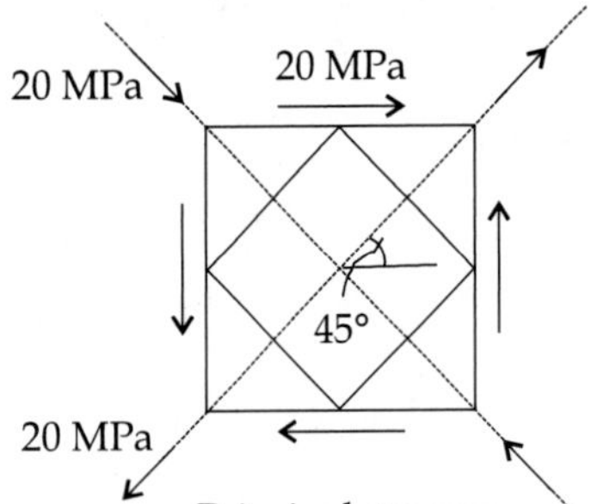

Principal stresses

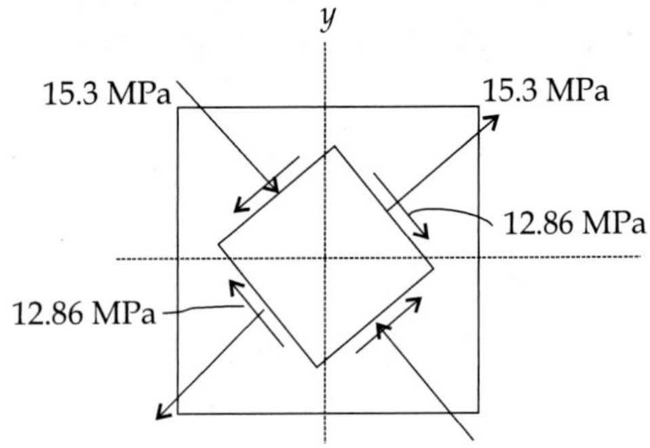

Example 3.15: The stresses, σ_x, σ_y and τ_{xy}, are acting at a point and shown by a plane stress element in Fig. 3.35. Determine with Mohr's circle construction the stresses acting on the element at $\theta = 50^{\circ}$ counterclockwise (ccw) from x-axis. Also find the principal stresses at the point.

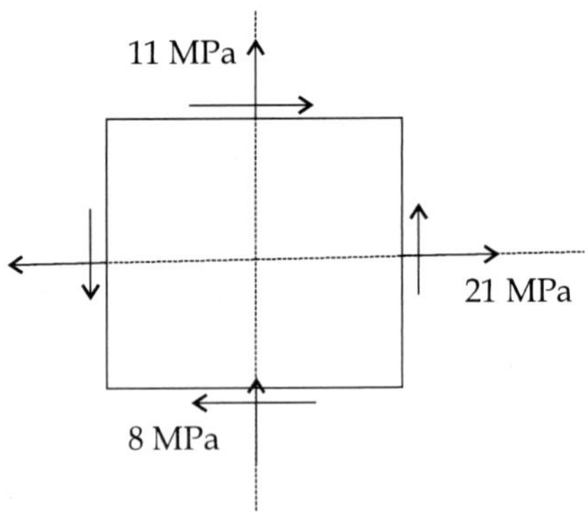

Fig. 3.35

Solution: Normal stresses on mutually perpendicular planes are both tensile (positive) and these are measured towards σ-axis. Therefore, in Mohr's circle,

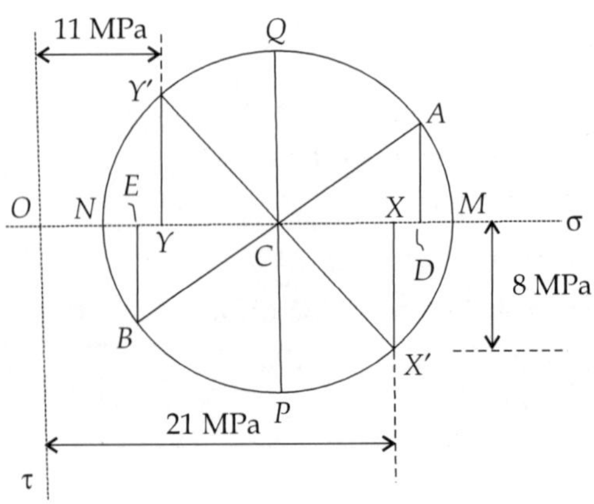

$OX = 21$, $OY = 11$, $XX' = 8$ and $YY' = -8$.

$\therefore \angle XCX' = \tan^{-1}(XX'/CX) = \tan^{-1}[8/\{(21 - 11)/2\}] = 58°$.

For an actual ccw orientation of 50° from x-axis, it will be $2 \times 50 = 100°$ in Mohr's circle. Hence a diameter AB is taken such that $\angle X'CA = 100°$. It follows that $\angle ACD = 100° - 58° = 42°$.

$\therefore \quad \sigma_{x'} = OD = OC + CD = \{(21 + 11)/2\} + \sqrt{8^2 + 5^2} \cdot \cos 42° = 23$ MPa **Ans.**

$\sigma_{y'} = OE = OC - CE = 16 - 7$ (since $CE = CD$), or $\sigma_{y'} = 9$ MPa **Ans.**

Also, $DA = CA \cdot \sin \angle ACD = \sqrt{8^2 + 5^2} \cdot \sin 42° = 6.3$.

$\therefore \quad \tau_{x'y'} = -6.3$ MPa, **Ans.**

and $EB = DA = 6.3$, or $\tau_{y'x'} = 6.3$ MPa **Ans.**

Principal stresses

$\sigma_1 = OM = OC + CM = OC + CX'$

$= \{(21 + 11)/2\} + \sqrt{8^2 + 5^2} = 25.4$ MPa (T) **Ans.**

$\sigma_2 = 16 - 9.43 = 6.57$ MPa (T) **Ans.**

$\tau_{max} = CP = CX' = 9.43$ MPa **Ans.**

Example 3.16: A structural component is made from a material for which the allowable stress in shear is 60 MPa. At a point on the surface of same component the state of plane stress is given by Fig. 3.36. Using Mohr's construction find the permissible range of the shear stress, τ_{xy}.

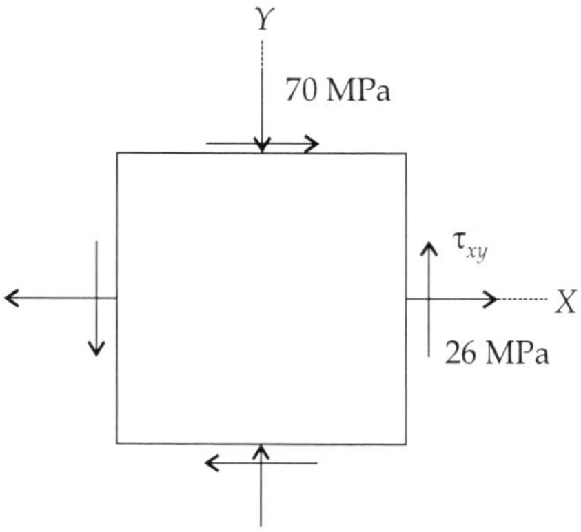

Fig. 3.36

Solution: Since the allowable shear stress is 60 MPa, the radius of Mohr's circle should be of this value to some suitable scale. Therefore, a $\sigma - \tau$ coordinate system with O as origin is drawn. Next, $OX = 26$ and $OY = -70$ is taken on σ-axis to some scale. Using average of the normal stresses (algebraically) the center of Mohr's circle is located at C. With C as centre and $R = 60$ as radius the circle is drawn. Then the perpendiculars are drawn at X and Y to meet the circle at X' and Y' respectively, so that $XX' = 36$ MPa and $YY' = -36$ MPa are obtained as two limiting values of permissible shear stress.

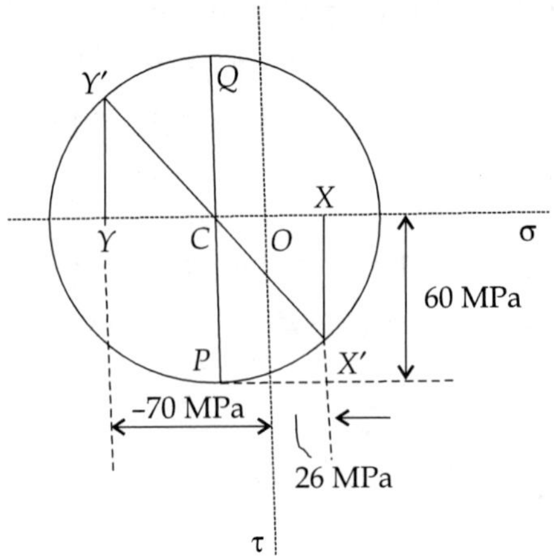

∴ the range of shear stress is $-36 \text{ MPa} \le \tau_{xy} \le +36 \text{ MPa}$ **Ans.**

3.6 GENERAL STRESS SYSTEM AT A POINT

It was discussed earlier that the general state of stress at a point is defined by the stresses σ_x, σ_y, σ_z, τ_{xy}, τ_{xz} and τ_{yz}, which can be arranged in a matrix as in Eqn. 3.8. All these stresses are shown over a very small cubical element surrounding a point in the body, as in Fig. 3.37. Also it is found that the shear stresses on mutually perpendicular planes are equal in magnitude and opposite in sense. With the known three-dimensional state of stress at a point, let us find the normal and shear stresses on an arbitrarily oriented plane passing through the point. A plane passing through the given point and defined by a unit vector (λ) towards its outward normal is shown in Fig. 3.37, in which the unit vector (λ) normal to plane *ABC* is written as

$$\lambda = l\cdot\mathbf{i} + m\cdot\mathbf{j} + n\cdot\mathbf{k}, \tag{3.29}$$

where, *l*, *m*, and n are direction cosines. For example, *m* is cosine of angle between *y*-axis and λ.

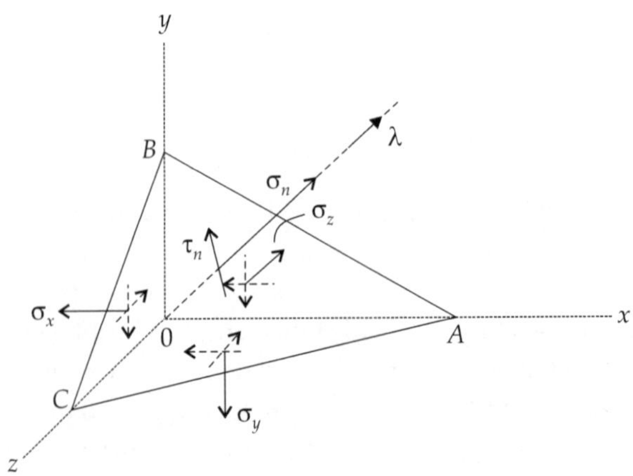

Fig. 3.37

Apart from σ_n and τ_n acting on plane ABC, other stresses are: σ_x, τ_{xy} and τ_{xz} on plane OBC, σ_y, τ_{yx} (= τ_{xy}) and τ_{yz} on plane OAC, σ_z, τ_{zx} (= τ_{xz}) and τ_{zy} (= τ_{yz}) on plane OAB. If $\mathbf{T}^{(n)}$ = the stress vector acting on plane ABC, and T_x, T_y and T_z its components, then

$$\mathbf{T}^{(n)} = T_x\cdot\mathbf{i} + T_y\cdot\mathbf{j} + T_z\cdot\mathbf{k}. \tag{3.30}$$

Next, considering equilibrium of the element in Fig. 3.37, $\sum F_x = 0$, $\sum F_y = 0$ and $\sum F_z = 0$ must be satisfied. To apply the first condition, all the forces parallel to x-axis are considered at a time in Fig. 3.38(a).

From $\sum F_x = 0$, $\quad T_x\cdot\Delta ABC - \sigma_x\cdot\Delta OBC - \tau_{xy}\cdot\Delta OAC - \tau_{xz}\cdot\Delta OAB = 0$,

or $\quad T_x = \sigma_x\cdot(\Delta OBC/\Delta ABC) + \tau_{xy}\cdot(\Delta OAC/\Delta ABC) + \tau_{xz}\cdot(\Delta OAB/\Delta ABC)$

$$\therefore \quad T_x = \sigma_x\cdot l + \tau_{xy}\cdot m + \tau_{xz}\cdot n \tag{3.31(a)}$$

Since the ratio of areas gives the cosine of angle between their normals.

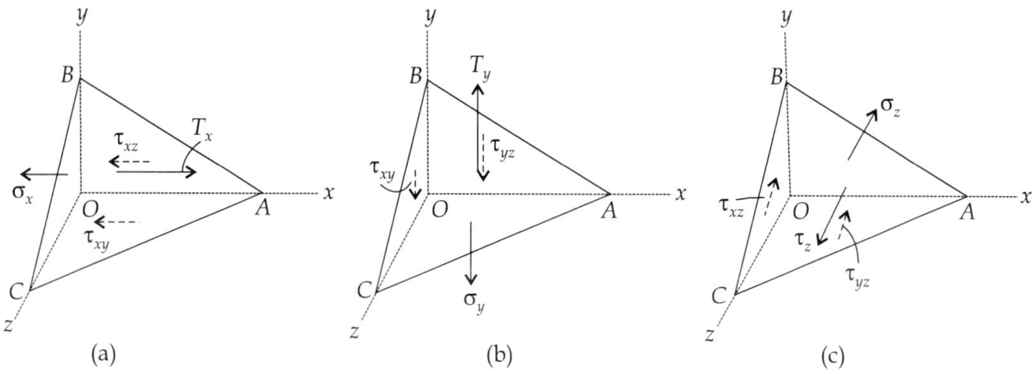

Fig. 3.38

Similarly, referring to Fig. 3.38(b) and from $\sum F_y = 0$,

$$T_y = \tau_{xy}\cdot l + \sigma_y\cdot m + \tau_{yz}\cdot n \tag{3.31(b)}$$

Also, from $\sum F_z = 0$ in Fig. 3.38(c),

$$T_z = \tau_{xz}\cdot l + \tau_{yz}\cdot m + \sigma_z\cdot n \tag{3.31(c)}$$

Therefore the normal stress on the plane ABC, $\sigma_n = T_x\cdot l + T_y\cdot m + T_z\cdot n$ and using Eqn. 3.38,

$$\sigma_n = \sigma_x\cdot l^2 + \sigma_y\cdot m^2 + \sigma_z\cdot n^2 + 2.\tau_{xy}\cdot lm + 2\cdot\tau_{xz}\cdot\text{ln} + 2\cdot\tau_{yz}\cdot mn \tag{3.32}$$

Next, according to Eqn. 3.30, $|\mathbf{T}^{(n)}|^2 = T_x^2 + T_y^2 + T_z^2$.

Since $\mathbf{T}^{(n)}$ is also a vector sum of the stresses σ_n and τ_n, the shear stress τ_n is written as

$$\tau_n = \sqrt{|\mathbf{T}^{(n)}|^2 - \sigma_n^2}. \tag{3.33}$$

Example 3.17: In a stressed body, the stresses at a point are given by $\sigma_x = \sigma_y = \sigma_z = 0$, τ_{xy} = +40 MPa, τ_{xz} = +55 MPa and τ_{yz} = +70 MPa. There is a plane passing through the point whose outward normal is equally inclined to x-, y- and z-axis. Determine the normal and shear stresses on this plane.

Solution: Given that, $\sigma_x = 0$, $\sigma_y = 0$, $\sigma_z = 0$, $\tau_{xy} = +40$ MPa, $\tau_{xz} = +55$ MPa, $\tau_{yz} = +70$ MPa.

The direction cosines of outward normal to the plane are $l = m = n$.

From $l^2 + m^2 + n^2 = 1$, we have $l = m = n = 1/\sqrt{3}$ (ignoring negative signs).

$$\therefore \quad T_x = 0\cdot(1/\sqrt{3}) + 40\cdot(1/\sqrt{3}) + 55\cdot(1/\sqrt{3}) = 95/\sqrt{3} \text{ MPa},$$

$$T_y = 40\cdot(1/\sqrt{3}) + 0\cdot(1/\sqrt{3}) + 70\cdot(1/\sqrt{3}) = 110/\sqrt{3} \text{ MPa},$$

$$T_z = 55\cdot(1/\sqrt{3}) + 70\cdot(1/\sqrt{3}) + 0\cdot(1/\sqrt{3}) = 125/\sqrt{3} \text{ MPa}.$$

$$\therefore \quad \sigma_n = (95/\sqrt{3})\cdot(1/\sqrt{3}) + (110/\sqrt{3})\cdot(1/\sqrt{3}) + (125/\sqrt{3})\cdot(1/\sqrt{3}),$$

or $\quad \sigma_n = 110$ MPa (T) **Ans.**

Also, $|\mathbf{T}^{(n)}|^2 = (95/\sqrt{3})^2 + (110/\sqrt{3})^2 + (125/\sqrt{3})^2 = 12250$

$$\therefore \quad \tau_n = \sqrt{12250 - (110)^2} = 12.25 \text{ MPa}$$ **Ans.**

3.6.1 Principal Stresses and Stress Invariants

Assuming general state of stress at a point, if a plane passing through the point is a principal plane then such a plane will be free from any shear stress ($\tau_n = 0$) and the stress vector ($\mathbf{T}^{(n)}$) will be perpendicular to this plane. As such, the normal stress $|(\mathbf{T}^{(n)}| = \sigma_n)$ will be the principal stress on this plane, as shown in Fig. 3.39. In Fig. 3.39, let λ be a unit vector along the outward normal to the principal plane ABC, which is defined by $\lambda = l\cdot\mathbf{i} + m\cdot\mathbf{j} + n\cdot\mathbf{k}$. If T_x, T_y and T_z are the scalar components of σ_n (as a vector) along x-, y- and z-directions, the following are obtained.

$$\left.\begin{aligned} T_x &= \sigma_n\cdot l, \\ T_y &= \sigma_n\cdot m, \\ \text{and} \quad T_z &= \sigma_n\cdot n. \end{aligned}\right\} \tag{3.34}$$

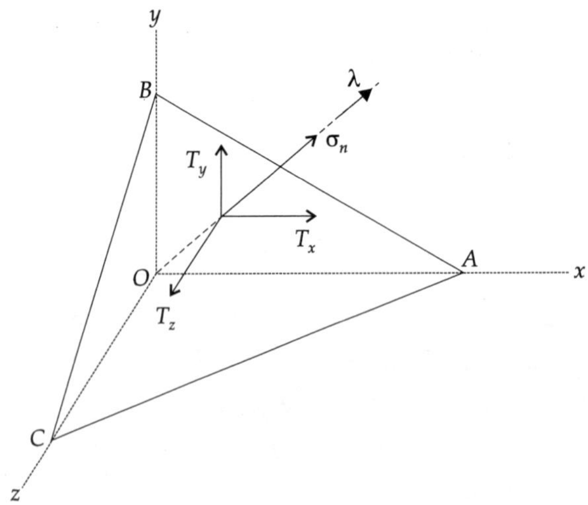

Fig. 3.39

Also, the stress components, T_x, T_y and T_z, can be written similar to Eqn. 3.31 in terms of the stresses acting on the three orthogonal planes through the point as

$$\left.\begin{aligned} T_x &= \sigma_x \cdot l + \tau_{xy} \cdot m + \tau_{xz} \cdot n, \\ T_y &= \tau_{xy} \cdot l + \sigma_y \cdot m + \tau_{yz} \cdot n, \\ T_z &= \tau_{xz} \cdot l + \tau_{yz} \cdot m + \sigma_z \cdot n. \end{aligned}\right\} \tag{3.35}$$

From Eqns. 3.34 and 3.35,

$$\left.\begin{aligned} (\sigma_x - \sigma_n) \cdot l + \tau_{xy} \cdot m + \tau_{xz} \cdot n = 0, \\ \tau_{xy} \cdot l + (\sigma_y - \sigma_n) \cdot m + \tau_{yz} \cdot n = 0, \\ \tau_{xz} \cdot l + \tau_{yz} \cdot m + (\sigma_z - \sigma_n) \cdot n = 0. \end{aligned}\right\} \tag{3.36}$$

Applying Crammer's rule the above equation can be solved for the direction cosines l, m, and n.

One such solution will be

$$l = \begin{vmatrix} 0 & \tau_{xy} & \tau_{xz} \\ 0 & (\sigma_y - \sigma_n) & \tau_{yz} \\ 0 & \tau_{yx} & (\sigma_z - \sigma_n) \end{vmatrix} \Bigg/ \begin{vmatrix} (\sigma_y - \sigma_n) & \tau_{xy} & \tau_{xz} \\ \tau_{xy} & (\sigma_y - \sigma_n) & \tau_{yz} \\ \tau_{xz} & \tau_{yx} & (\sigma_z - \sigma_n) \end{vmatrix}.$$

It is found in the above expression that the numerator is zero. It implies that the product of denominator and direction cosine l is zero. But, $l = 0$ gives a unique solution because it corresponds to a principal plane which is parallel to x-axis only. In general, therefore, for $l \neq 0$ the denominator must be zero. Hence, the above determinant in the denominator can be equated to zero to expand it in the form of the following cubic equation:

$$\sigma_n^3 - (\sigma_x + \sigma_y + \sigma_z) \cdot \sigma_n^2 + (\sigma_x \cdot \sigma_y + \sigma_x \cdot \sigma_z + \sigma_y \cdot \sigma_z - \tau_{xy}^2 - \tau_{xz}^2 - \tau_{yz}^2) \cdot \sigma_n$$
$$x - (\sigma_x \cdot \sigma_y \cdot \sigma_z + 2\tau_{xy} \cdot \tau_{xz} \cdot \tau_{yz} - \sigma_x \cdot \tau_{yz}^2 - \sigma_y \cdot \tau_{xz}^2 - \sigma_z \cdot \tau_{xy}^2) = 0 \tag{3.37}$$

The three roots of this cubic equation are denoted as σ_1, σ_2 and σ_3 which are the three principal stresses on three orthogonal planes at that point. For a known stress system in x-y-z coordinates the above equation provides the values of principal stresses: σ_1, σ_2 and σ_3. If these values of principal stresses are used in Eqn. 3.36, together with relation $l^2 + m^2 + n^2 = 1$, then the direction cosines can be obtained separately for three principal directions defined by unit vectors ($\mathbf{n}_1$, $\mathbf{n}_2$ and $\mathbf{n}_3$) shown in Fig. 3.40. The three orthogonal planes are the principal planes which are free from shear stresses.

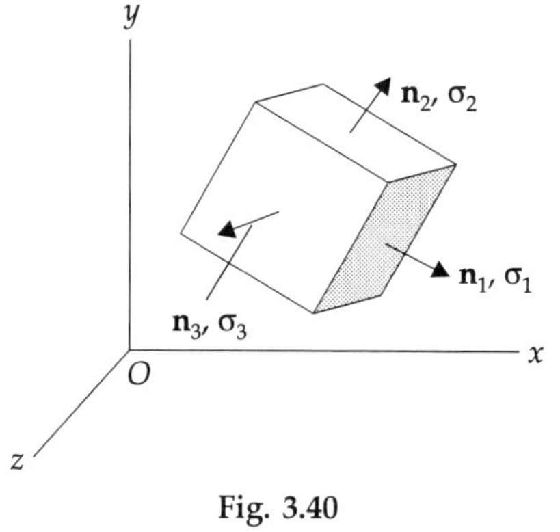

Fig. 3.40

Further, Eqn. 3.37 is written in the form, $\sigma_n^3 - I_1\cdot\sigma_n^2 + I_2\cdot\sigma_n - I_3 = 0$ (3.38)
where, I_1, I_2 and I_3 can be obtained from the stresses in x-y-z coordinates. The principal stresses at a point are known to be independent of the choice of coordinates. It follows that once the principal stresses are determined with the help of I_1, I_2 and I_3, derived from σ_x, σ_y, σ_z, τ_{xy}, τ_{xz} and τ_{yz}, these should not change with any other set of stresses $\sigma_{x'}$, $\sigma_{y'}$, $\sigma_{z'}$, $\tau_{x'y'}$ ...., etc. in x'-y'-z' coordinates. Therefore, I_1, I_2 and I_3 remain unchanged and these are known as invariants of stress. The three invariants are written as

$$I_1 = \sigma_x + \sigma_y + \sigma_z = \sigma_{x'} + \sigma_{y'} + \sigma_{z'} = \sigma_1 + \sigma_2 + \sigma_3,$$

$$I_2 = \sigma_x\cdot\sigma_y + \sigma_x\cdot\sigma_z + \sigma_y\cdot\sigma_z - \tau_{xy}^2 - \tau_{xz}\ \tau_{yz}^2$$

$$= \sigma_{x'}\cdot\sigma_{y'} + \sigma_{x'}\cdot\sigma_{z'} + \sigma_{y'}\cdot\sigma_{z'} - \tau_{x'y'}^2 - \tau_{x'z'}^2 - \tau_{y'z'}^2,$$

$$= \sigma_1\cdot\sigma_2 + \sigma_1\cdot\sigma_3 + \sigma_2\cdot\sigma_3.$$

and $$I_3 = \sigma_x\cdot\sigma_y\cdot\sigma_z + 2\cdot\tau_{xy}\cdot\tau_{xz}\cdot\tau_{yz} - \sigma_x\cdot\tau_{yz}^2 - \sigma_y\cdot\tau_{xz}^2 - \sigma_z\cdot\tau_{xy}^2,$$

$$= \sigma_1\cdot\sigma_2\cdot\sigma_3.$$

Maximum shear stress: From analysis of plane stress and discussions on Mohr's circle, the maximum shear stress was found from the difference of algebraically maximum and minimum principal stresses. Similarly, in a three-dimensional stress system the maximum shear stress is written as

$$\tau_{max} = (\sigma_1 - \sigma_3)/2 \tag{3.39}$$

where, σ_1 and σ_3 are maximum and minimum principal stresses obtained from the roots of Eqn. 3.37 or Eqn. 3.38.

Example 3.18: At a point in a stressed body, the known stresses are σ_x = 12 unit, σ_y = 4 unit, σ_z = 10 unit, τ_{xy} = 3 unit, τ_{xz} = 0 and τ_{yz} = 0, where, 1 unit = 10^3 kPa. Determine the principal stresses at the point and also their directions in x-y-z- coordinates.

Solution: The stress matrix can be written as

$$\begin{pmatrix} 12 & 3 & 0 \\ 3 & 4 & 0 \\ 0 & 0 & 10 \end{pmatrix} \times 1 \text{ MPa.}$$

Therefore, the determinant that leads to cubic equation (Eqn. 3.37) with roots as principal stresses is equated to zero, as

$$\begin{vmatrix} (12-\alpha) & 3 & 0 \\ 3 & (4-\sigma) & 0 \\ 0 & 0 & (10-\sigma) \end{vmatrix} = 0$$

or $$(12 - \sigma)\cdot(4 - \sigma)\cdot(10 - \sigma) - 3\cdot(3)\cdot(10 - \sigma) = 0$$

or $$(10 - \sigma)\cdot[(12 - \sigma)\cdot(4 - \sigma) - 9] = 0$$

or $$(10 - \sigma)\cdot[\sigma^2 - 16\cdot\sigma + 39] = 0$$

With $(10 - \sigma) = 0$, σ = 10 MPa is one of the principal stresses.

Next, with the quadratic equation, $\sigma^2 - 16\cdot\sigma + 39 = 0$,

$\sigma = \left\{16 \pm \sqrt{(16)^2 - (4 \times 39)}\right\}/2$, or $\sigma = 8 \pm 5$

∴ The three principal stresses are σ_1 = 13 MPa, σ_2 = 10 MPa and σ_3 = 3 MPa. **Ans.**

Direction cosines of principal stresses

For σ_1 = 13 MPa, Eqn. 3.36 is used to obtain

$$(12 - 13)\cdot l + 3\cdot m = 0, \quad \text{or} \quad -l + 3m = 0 \tag{a}$$

$$3.l + (4 - 13)\cdot m = 0, \quad \text{or} \quad 3l - 9m = 0 \tag{b}$$

and

$$(10 - 13)\cdot n = 0, \quad \text{or} \quad -3n = 0 \tag{c}$$

From (a) or (b), $l = 3\,m$, and from (c), $n = 0$

$$l^2 + m^2 + n^2 = 1 \tag{d}$$

With $l^2 + m^2 + n^2 = 1$, $9\,m^2 + m^2 = 1$, ∴ $m = \pm 1/\sqrt{10}$

Further, with $m = 1/\sqrt{10}$, $l = 3/\sqrt{10}$

∴ The direction of σ_1 = 13 MPa is defined by $(3/\sqrt{10}, 1/\sqrt{10}, 0)$. **Ans.**

For σ_2 = 10 MPa, $(12 - 10)\cdot l + 3\cdot m = 0$, or $2l + 3m = 0$ (e)

$$3.l + (4 - 10)\cdot m = 0, \quad \text{or} \quad 3l - 6m = 0 \tag{f}$$

$$(10 - 10)\cdot n = 0 \tag{g}$$

From (f), $l = 2\,m$, and in (e), $4m + 3m = 0$ or $m = 0$. Also, $l = 0$ in (e).

As n cannot be found from (g), from $l^2 + m^2 + n^2 = 1$, $0 + 0 + n^2 = 1$, ∴ $n = \pm 1$.

∴ Direction of σ_2 = 10 MPa is given by (0, 0, 1) **Ans.**

Therefore, σ_2 is perpendicular to x-y plane.

For σ_3 = 3 MPa, $(12 - 3)\cdot l + 3.m = 0$, or $9l - 3m = 0$ (h)

$$3\cdot l + (4 - 3)\cdot m = 0, \quad \text{or} \quad 3l + m = 0 \tag{I}$$

and

$$(10 - 3)\cdot n = 0 \tag{j}$$

From (j), $n = 0$.

From (i), $m = -3l$, ∴ $l^2 + (-3l)^2 + 0 = 1$, $l = \pm 1/\sqrt{10}$ and $m = \pm 3/\sqrt{10}$

With $l = 1/\sqrt{10}$, $m = -3/\sqrt{10}$, $n = 0$, the direction cosines of σ_3 = 3 MPa are $(1/\sqrt{10}, -3/\sqrt{10}, 0)$. **Ans.**

Example 3.19: For the state of stress at a point characterized by the components,

σ_x = 85 MPa (T), σ_y = 70 MPa (C), σ_z = 55 MPa (T),

τ_{xy} = 55 MPa, τ_{xz} = 85 MPa, and τ_{yz} = –70 MPa,

find the principal stresses and the maximum shearing stress at the point. Also, find the direction of maximum principal stress.

Solution: In order to make the calculations easier, let 1 unit = 10 MPa, so that the stresses in unit are written as $\sigma_x = 8.5$, $\sigma_y = -7$, $\sigma_z = 5.5$, $\tau_{xy} = 5.5$, $\tau_{xz} = 8.5$, $\tau_{yz} = -7$.

∴ The invariants are : $I_1 = (8.5) + (-7) + (5.5)$. or $I_1 = 7$.

$$I_2 = (8.5)\cdot(-7) + (8.5)\cdot(5.5) + (-7)\cdot(5.5) - (5.5)^2 - (8.5)^2 - (-7)^2.$$

or

$$I_2 = -202.75,$$

and $$I_3 = (8.5)\cdot(-7)\cdot(5.5) + 2\cdot(5.5)\cdot(8.5)\cdot(-7) - (8.5)\cdot(-7)^2 - (-7)\cdot(8.5)^2 - (5.5)\cdot(5.5)^2$$

or $$I_3 = -1058.875.$$

With the values of I_1, I_2 and I_3, a cubic equation (as in Eqn. 3.38) is formed which is satisfied only for the three particular values of principal stress, σ. Other than these three values the equation is not satisfied. Hence, a function is formed with the known invariants as

$$f(\sigma) = \sigma^3 - 7\cdot\sigma^2 + (-202.75)\cdot\sigma - (-1058.875)$$

or $$f(\sigma) = \sigma^3 - 7\sigma^2 - 203\sigma + 1059.$$

Three roots of σ can be obtained by applying a numerical method known as Newton's iterative method.

Thus, $f(\sigma) = \sigma^3 - 7\cdot\sigma^2 + (-202.75)\cdot\sigma - (-1058.875)$,

and $df(\sigma)/d\sigma = 3\sigma^2 - 14\sigma - 202.75$.

Assuming a suitable value, $\sigma = 7$, $f(\sigma) = -362$, $df(\sigma)/d\sigma = -154$.

A more correct value of $\sigma = \sigma - [f(\sigma)/\{df(\sigma)/d\sigma\}]$, or $\sigma = 7 - \{(-362)/(-154)\} = 4.65$.

Again, for $\sigma = 4.65$, $f(\sigma) = 64.23$ and $df(\sigma)/d\sigma = -203.23$, so that $\sigma = 4.65 - \{64.23/(-203.23)\}$

or $\sigma = 4.97$.

With $\sigma = 4.97$, $f(\sigma) = -0.05$, $df(\sigma)/d\sigma = -198.48$, it is found that $\sigma = 4.97$.

As the final value does not change appreciably, $\sigma = 4.97$, or one of the principal stresses is 49.7 MPa.

Therefore, the cubic equation is written as $(\sigma - 4.97)\cdot(\sigma^2 - 2.03\sigma - 213.09) = 0$ in which the second factor is in quadratic form. After solving this quadratic equation, two other values of σ are found to be 15.648 and –13.618. Hence, the three principal stresses are $\sigma_1 = 156.5$ MPa, $\sigma_2 = 49.7$ MPa and $\sigma_3 = -136.2$ MPa. While σ_1 and σ_2 are tensile in nature, σ_3 is a compressive stress.

Maximum shearing stress, $\tau_{max} = (\sigma_1 - \sigma_3)/2 = \{156.5 - (-136.2)\}/2 = 146.4$ MPa

Ans.

The maximum principal stress = 156.5 MPa. Therefore, from Eqn. 3.36,

$$(85 - 156.5)\cdot l + 55\cdot m + 85\cdot n = 0 \quad \text{(a)}$$

$$55\cdot l + (-70 - 156.5)\cdot m - 70n = 0 \quad \text{(b)}$$

$$85\cdot l - 70\cdot m + (55 - 156.5)\cdot n = 0 \quad \text{(c)}$$

Eliminating n from (a) and (b), $m = -0.021.l$, and eliminating m from (b) and (c), $n = 0.85.l$

Next, substituting for m and n in the relation $l^2 + m^2 + n^2 = 1$ and solving, we have $l = 0.76$, $m = -0.016$, $n = 0.65$, which are D'cosines for the direction of σ_1. **Ans.**

Example 3.20: The state of stress at a point is defined by the matrix

$$\begin{Bmatrix} \sigma_x & 2 & 1 \\ 2 & 0 & 2 \\ 1 & 2 & 0 \end{Bmatrix} \text{units}$$

Determine σ_x such that there is at least one principal plane passing through the point which is free from any stress. Find the direction cosines of the normal to that plane.

Solution: The invariants are:

$I_1 = \sigma_x + 0 + 0 = \sigma_x$, $I_2 = -2^2 - 1 - 2^2 = -9$, $I_3 = 2.2.1.2 - \sigma_x{\cdot}2^2 = 8 - 4\sigma_x$

$\therefore$ Cubic equation for principal stresses, $\sigma^3 - \sigma_x{\cdot}\sigma^2 - 9{\cdot}\sigma - (8 - 4\sigma_x) = 0$.

For a principal plane free from any stress, $\sigma = 0$. $(8 - 4\sigma_x) = 0$, or $\sigma_x = 2$ unit **Ans.**

With $\sigma = 0$, $$2{\cdot}l + 2{\cdot}m + n = 0 \quad \text{(a)}$$

$$2{\cdot}l + 2{\cdot}n = 0 \quad \text{(b)}$$

$$1{\cdot}l + 2{\cdot}m = 0 \quad \text{(c)}$$

From (b), $n = -l$, and from (c), $m = -l/2$. $\therefore\ l^2 + (-l/2)^2 + (-l)^2 = 1$, or $l = \pm\, 2/3$

Taking $l = 2/3$, $m = -1/3$, $n = -2/3$ **Ans.**

3.6.2 Octahedral Stresses

If the three principal stresses at a point are known, the principal directions can be assumed as x-, y- and z-directions. Then the planes which are equally inclined to principal directions are called octahedral planes because eight such planes are available that form an octahedron with eight equal faces. This is shown in Fig. 3.41.

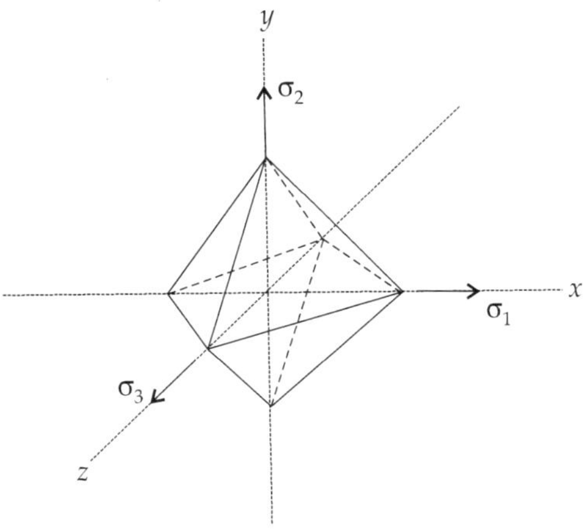

Fig. 3.41

For the octahedral planes in Fig. 3.41, outward normal to a plane is equally inclined to the principal axes x, y and z, so that the direction cosines of the normal are $l = m = n$.

From $l^2 + m^2 + n^2 = 1$, $l = m = n = 1/\sqrt{3}$. Hence, considering a plane with $l = m = n = 1/\sqrt{3}$, the normal stress $\sigma = \sigma_1 \cdot l^2 + \sigma_2 \cdot m^2 + \sigma_3 \cdot n^2$.

or $$\sigma = (1/3)\cdot(\sigma_1 + \sigma_2 + \sigma_3) = I_1/3, \tag{3.40}$$

where, I_1 = the first invariant.

If τ_{oct} = shear stress on the octahedral plane, $\sigma^2 + \tau_{oct}^2 = |T^{(n)}|^2 = \sigma_1^2 \cdot l^2 + \sigma_2^2 \cdot m^2 + \sigma_3^2 \cdot n^2$

or $$\tau_{oct}^2 = (1/3)\cdot(\sigma_1^2 + \sigma_2^2 + \sigma_3^2) - (1/9)\cdot(\sigma_1 + \sigma_2 + \sigma_3)^2$$

This can be simplified in the form $\tau_{oct}^2 = (1/9)\cdot[(\sigma_1 - \sigma_2)^2 + (\sigma_1 - \sigma_3)^2 + (\sigma_2 - \sigma_3)^2]$

or $$\tau_{oct}^2 = (2/9)\cdot[(\sigma_1 + \sigma_2 + \sigma_3)^2 - 3\cdot(\sigma_1\sigma_2 + \sigma_1\sigma_3 + \sigma_2\sigma_3)]$$

$\therefore$ $$\tau_{oct} = (\sqrt{2}/3)\cdot[I_1^2 - 3I_2]^{1/2} \tag{3.41}$$

where, I_1 and I_2 are stress invariants.

If the general state of stress is known in x-y-z-coordinates, the octahedral shear stress and normal stress can be found with the help of invariants, I_1 and I_2.

Thus, $$\sigma_{oct} = I_1/3 = (1/3)\cdot(\sigma_x + \sigma_y + \sigma_z). \tag{3.42}$$

Also, $$\tau_{oct} = (\sqrt{2}/3)\cdot[I_1^2 - 3I_2]^{1/2}$$
$$= (\sqrt{2}/3)\cdot[(\sigma_x + \sigma_y + \sigma_z)^2 - 3(\sigma_x\sigma_y + \sigma_x\sigma_z + \sigma_y\sigma_z - \tau_{xy}^2 - \tau_{xz}^2 - \tau_{yz}^2)]^{1/2}$$

or $$9\tau_{oct}^2 = (\sigma_x - \sigma_y)^2 + (\sigma_x - \sigma_z)^2 + (\sigma_y - \sigma_z)^2 + 6(\tau_{xy}^2 + \tau_{xz}^2 + \tau_{yz}^2). \tag{3.43}$$

The octahedral normal stress ($I_1/3$) can be interpreted as mean normal stress at a point. If in a certain state of stress the first invariant, $I_1 = \sigma_x + \sigma_y + \sigma_z = 0$, then the normal stress on the octahedral plane will be zero and only the shear stress will be present. This is important from the standpoint of strength and failure of some materials.

3.6.3 Hydrostatic and Deviatoric Stresses

A given state of stress can be resolved into a hydrostatic stress and a state of stress in pure shear. Let the state of stress be written in the form

$$\sigma_{ij} = \begin{Bmatrix} \sigma_x & \tau_{xy} & \tau_{xz} \\ \tau_{xy} & \sigma_y & \tau_{yz} \\ \tau_{xz} & \tau_{yz} & \sigma_z \end{Bmatrix}.$$

Then, the mean stress is written as $p = (1/3)\cdot(\sigma_x + \sigma_y + \sigma_z) = (1/3)\cdot I_1$.

Therefore, the above stress matrix is written as

$$\begin{Bmatrix} \sigma_x & \tau_{xy} & \tau_{xz} \\ \tau_{xy} & \sigma_y & \tau_{yz} \\ \tau_{xz} & \tau_{yz} & \sigma_z \end{Bmatrix} = \begin{Bmatrix} p & 0 & 0 \\ 0 & p & 0 \\ 0 & 0 & p \end{Bmatrix} + \begin{Bmatrix} (\sigma_x - p) & \tau_{xy} & \tau_{xz} \\ \tau_{xy} & (\sigma_y - p) & \tau_{yz} \\ \tau_{xz} & \tau_{yz} & (\sigma_z - p) \end{Bmatrix}$$

The first matrix on the right side represents a hydrostatic stress and the second matrix a state of pure shear. From the second matrix it can be checked that $I_1' = (\sigma_x - p) + (\sigma_y - p) + (\sigma_z - p) = 0$, which is a proof that the second matrix represents a state of pure shear. The pure shear state is also known as deviatoric stress or simply the stress deviator. While the hydrostatic stress at a point is responsible for change in volume, deviatoric stress causes the change in shape.

If a given state of stress is referred to principal axes,

$$\begin{Bmatrix} \sigma_1 & 0 & 0 \\ 0 & \sigma_2 & 0 \\ 0 & 0 & \sigma_3 \end{Bmatrix} = \begin{Bmatrix} p & 0 & 0 \\ 0 & p & 0 \\ 0 & 0 & p \end{Bmatrix} + \begin{Bmatrix} \sigma_1 - p & 0 & 0 \\ 0 & \sigma_2 - p & 0 \\ 0 & 0 & \sigma_3 - p \end{Bmatrix}$$

As before, the first matrix on the right side represents hydrostatic stress and the second matrix gives the deviatoric stress.

PROBLEMS

1. Determine the normal and shear stresses on the indicated planes in Fig. 3.42. Use (i) wedge method of analysis, (ii) stress transformation rules, and (iii) Mohr's circle construction to verify the results.

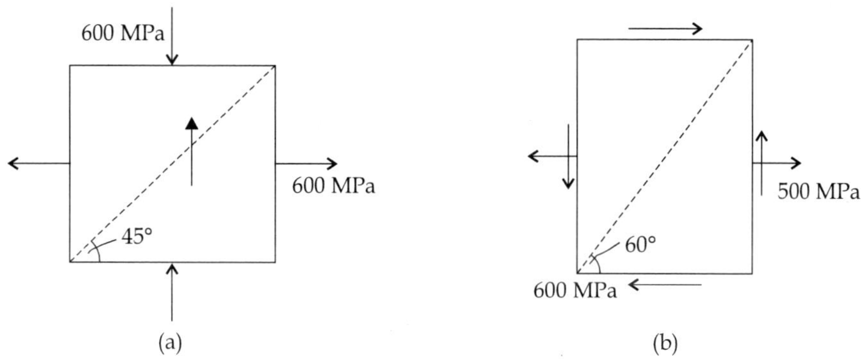

Fig. 3.42

Ans. (a) $\sigma = 0$, $\tau = 600$ MPa. (b) $\sigma = 145$ MPa (C), $\tau = 517$ MPa.

2. The state of stress at a point on the surface of a timber beam is shown in Fig. 3.43. The timber will split if the shear stress parallel to the grain exceeds 4×10^6 N/m^2. Is the beam safe ?

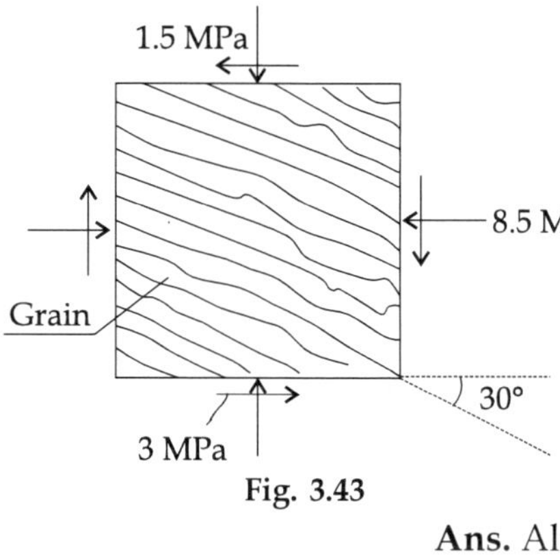

Fig. 3.43

Ans. Along the grain, $\tau = 4.53$ MPa

Hence, unsafe.

3. For the state of plane stress in Fig. 3.44, plot Mohr's circle to find the principal stresses and maximum shear stress in terms of τ. Also, find the inclination of larger principal stress with respect to x-axis.

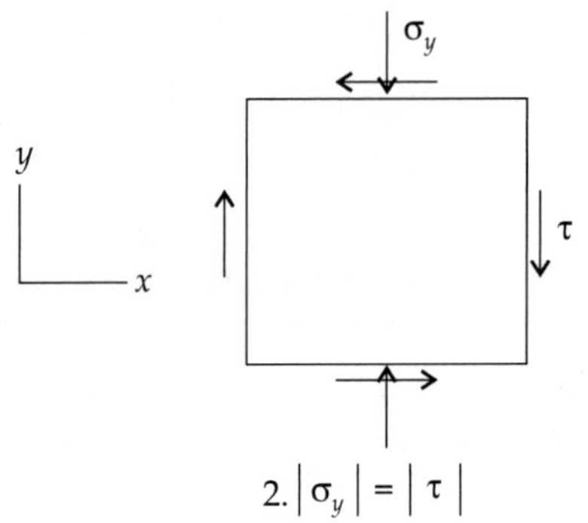

Fig. 3.44

Ans. $\sigma_1 = 0.781\tau$
$\sigma_2 = -1.281\tau$
$\tau_{max} = 1.03\tau$
$\theta = -38°$

4. Stresses acting on vertical and horizontal planes passing through a point are shown by Fig. 3.45. Determine the principal stresses and maximum shear stress at the point. Find the orientation of larger principal stress relative to x-axis.

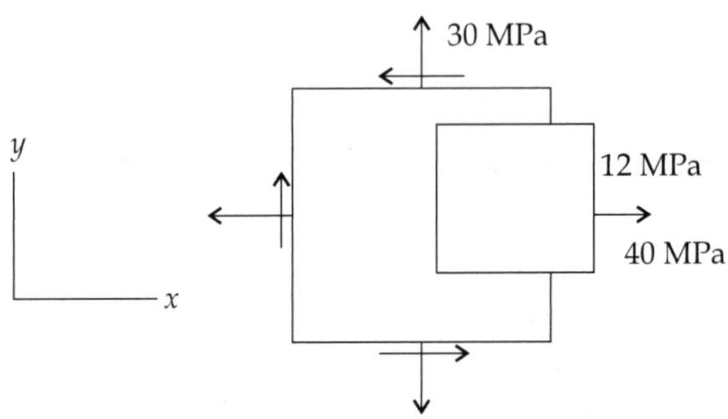

Fig. 3.45

Ans. $\sigma_1 = 48$ MPa
$\sigma_2 = 22$ MPa
$\tau_{max} = 13$ MPa
$\theta = -33.7°$

5. Some adhesive compound is used to join two wedges to form a cube (Fig. 3.46). The wedges will split apart if the normal stress on the joint is 15 MPa (T). What should be the maximum value of σ_y, if $\sigma_x = 8$ MPa and $\tau_{xy} = 5$ MPa?

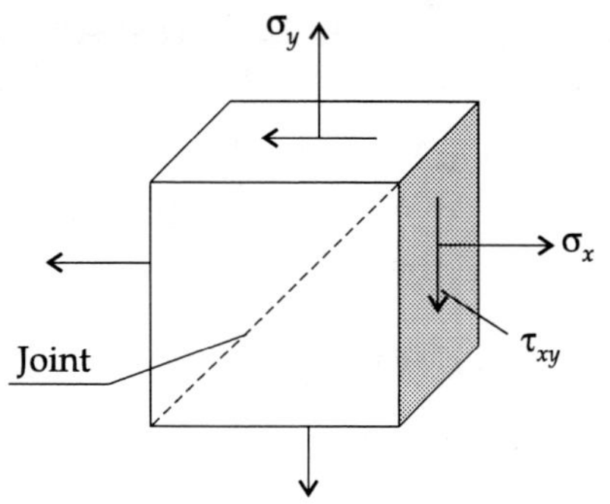

Fig. 3.46

Ans. $\sigma_y \leq 12$ MPa.

6. The state of stress at a point is obtained by superimposing two separate states of pure shear shown in Fig. 3.47. Hence, find the stresses σ_x, σ_y and τ_{xy} at the point. Also, determine the principal stresses and their directions with respect to x-axis.

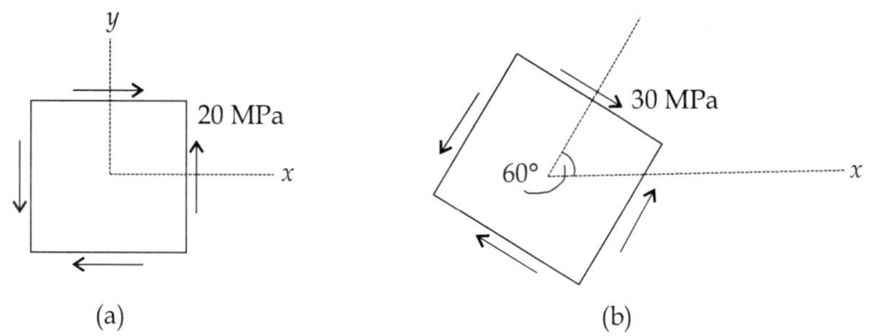

Fig. 3.47

Ans. σ_x = 26 MPa, σ_y = –26 MPa, τ_{xy} = 35 MPa,
σ_1 = 43.6 MPa, σ_2 = –43.6 MPa,
θ = 27° and 117°.

7. The state of stress at a point is found from superposition of three tensile stresses shown in Fig. 3.48. Determine the principal stresses and position of principal planes at the point.

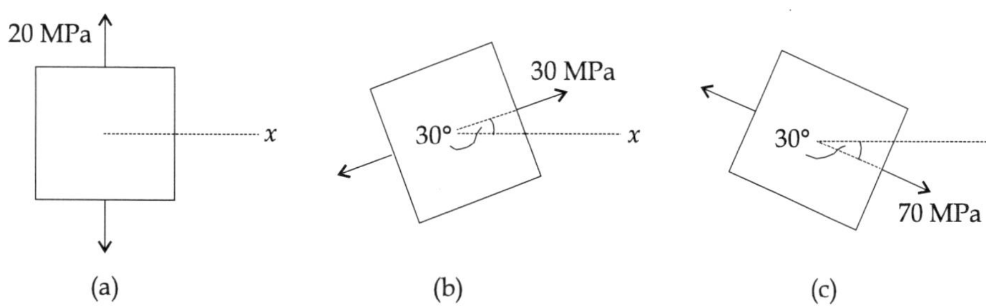

Fig. 3.48

Ans. $-\sigma_1 = 82.9$ MPa, $\sigma_2 = 37.1$ MPa
$\theta_1 = -24.5°$, $\theta_2 = 65.5°$

8. For the state of stress shown in Fig. 3.49, the stress (σ_x) is unknown. If the lower principal stress at the point is $\sigma_2 = 7$ MPa (C), find σ_x and the angle of inclination of larger principal stress with the x-direction.

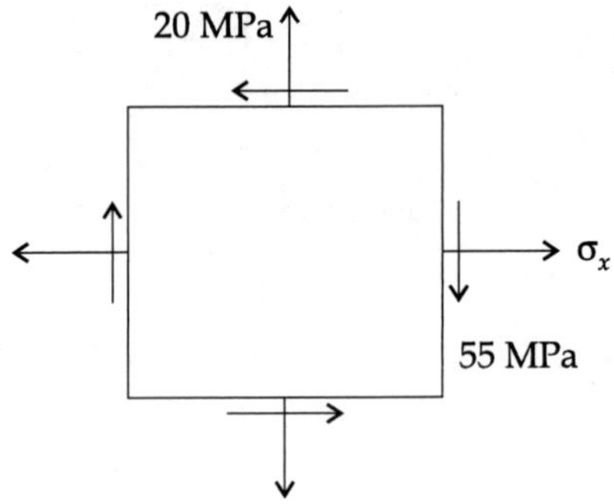

Fig. 3.49

Ans. $\sigma_x = 105$ MPa, $\theta = -26.2°$

9. If the stresses at a point are: $\sigma_x = 0$, $\sigma_y = 0$, $\sigma_z = 0$, $\tau_{xy} = 42$ MPa, $\tau_{xz} = 56$ MPa and $\tau_{yz} = 70$ MPa, determine the principal stresses and the maximum shear stress at the point. What is the orientation of plane containing the minimum principal stress ?

Ans. $\sigma_1 = 112.8$ MPa (T), $\sigma_2 = 40.3$ MPa (C)
$\sigma_3 = 72.5$ MPa (C), $\tau_{max} = 93$ MPa.
(0.24, 0.59, –0.77)

10. The stress components at a point in an externally loaded body are :
$\sigma_x = 18$ MPa (C), $\sigma_y = 15$ MPa (C), $\sigma_z = 12$ MPa (C),
$\tau_{xy} = -15$ MPa, $\tau_{xz} = -9$ MPa, $\tau_{yz} = +12$ MPa.
Determine the three principal stresses and the maximum shear stress at the point.

Ans. $\sigma_1 = 9.15$ MPa (T), $\sigma_2 = 22.4$ MPa (C), $\sigma_3 = 31.7$ MPa (C),
$\tau_{max} = 20.4$ MPa

4
Analysis of Strain

4.1 INTRODUCTION

The concepts of stress and strain were discussed in chapter 2 for axial loading of prismatic bars. It was explained that the axial strain or normal strain is associated with elongation/contraction of a bar and shear strain is a measure of angular distortion of a body. It was discussed in chapter 3 that the stress at a point in an externally loaded body is derived from the condition of equilibrium, which is valid for both rigid and deformable bodies. However, almost all the construction materials and the materials used in machine parts are practically elastic in nature and hence deformable. In general, the result of application of forces on a body appears in the form of rigid body motion and/or deformation of the body. Rigid body motion includes translation or rotation of the body as a whole, which is not our present subject of interest. The deformation of a body means the change in shape and volume that the body undergoes due to the forces applied on it.

When there is deformation, any point in the body shifts to a new position which is called displacement of the point. Also, the distance between two points in an undeformed body changes in magnitude and inclination when the body deforms under applied forces. An example for this can be drawn from deformation of a rubber balloon. In a partly inflated balloon, any two points close to each other can be marked with ink. On further increasing the pressure of gas, the balloon gets deformed and the relative displacement between the points will be observed. It should be noted that the displacement is a physical quantity from which the strain at a point can be derived.

At a point, the displacement as a vector is written as

$$\rho = u\cdot\mathbf{i} + v\cdot\mathbf{j} + w\cdot\mathbf{k}, \tag{4.1}$$

where, u, v and w are the displacement components in x-, y- and z-directions, respectively.

The displacements of points in a body are shown in Fig. 4.1. There are two points P and Q of un-deformed body (shown by dotted line). After deformation of the body, the points P and Q assume new positions at P' and Q'. As such, the displacement vectors are written as

$$\left.\begin{aligned} \overline{PP'} &= \rho_p = u_p\cdot\mathbf{i} + v_p\cdot\mathbf{j} + w_p\cdot\mathbf{k}, \\ \text{and } \overline{QQ'} &= \rho_Q = u_Q\cdot\mathbf{i} + v_Q\cdot\mathbf{j} + w_Q\cdot\mathbf{k}. \end{aligned}\right\} \quad (4.2)$$

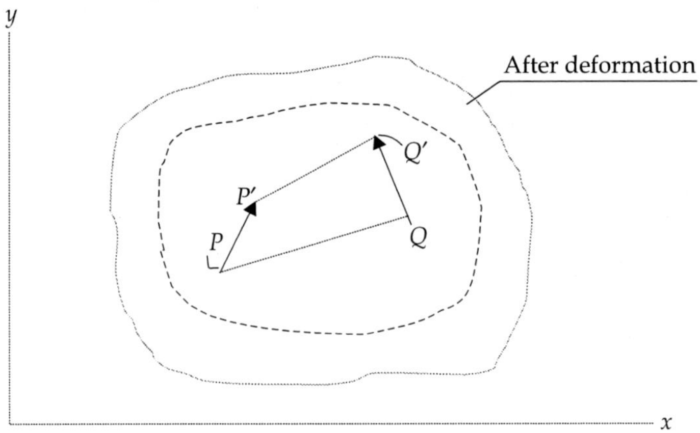

Fig. 4.1

It is apparent that the original distance PQ changes to $P'Q'$ indicating a relative displacement between the points. It will be shown later how strains at a point are derived from displacements.

4.2 STATE OF STRAIN AT A POINT

The state of strain at a point is obtained from deformation of a very small element at the point. This is accomplished with two basic assumptions. The first assumption is that the deformation of the element takes place in a state of uniform strain, i.e., any straight line remains straight after deformation and does not become curved though the orientation of the line may change with deformation. The second assumption is that the deformation should be geometrically compatible, which means that no void or hole should develop during deformation of the body.

In the light of the above discussions, a small element $OABC$ is chosen at a point O of a two-dimensional body lying in x-y plane, as shown in Fig. 4.2. The point O having coordinates $(x.y)$ is shown to be at a corner of the element $OABC$ with side $OA = \Delta x$ parallel to x-axis, and the side $OC = \Delta y$ parallel to y-axis. After deformation, the element $OABC$ transforms to $O'A'B'C'$ due to the displacement of four corner points. Then, the changes in length of OA and OC along x- and y-directions respectively, are measured per unit length and these are defined as normal strain components along x- and y-directions at the point O.

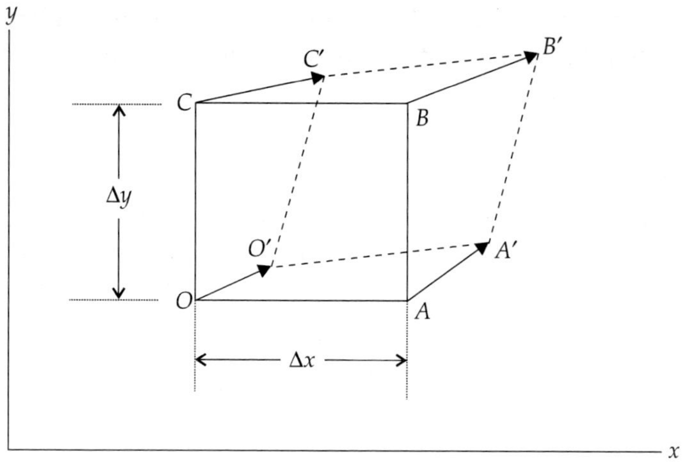

Fig. 4.2

Therefore, the normal strains, ε_x and ε_y along x- and y-directions, are written as

$$\left.\begin{aligned}\varepsilon_x &= \lim_{\Delta x \to 0} \{(O'A')_P - OA\}/OA,\\ \varepsilon_y &= \lim_{\Delta y \to 0} \{(O'C')_P - OC\}/OC,\end{aligned}\right\} \tag{4.3}$$

where, $(O'A')_P$ is projection of $O'A'$ along x-axis and $(O'C')_P$ is projection of $O'C'$ along y-axis.

From the above definition of normal strain it implies that the normal strain is positive if a line segment elongates, and negative if the line segment contracts.

The shear strain at a point is specified with respect to the two mutually perpendicular line segments in the un-deformed condition. It is denoted by symbol γ with two subscripts corresponding to the direction of two lines. Referring to Fig. 4.2, shear strain at the point O is defined as the tangent of change in $\angle AOC$, and written as

$$\gamma_{xy} = \tan(\angle AOC - \angle A'O'C')$$

It should be noted that small shear strains used in engineering applications are generally less than 0.01. Hence, it is sufficient to define shear strain as the change in angle only (in radians) since $\tan\theta \simeq \theta$ for a very small value of θ. As such, the above expression of shear strain is more precisely written as

$$\gamma_{xy} = \lim_{\substack{\Delta x \to 0 \\ \Delta y \to 0}} (\angle AOC - \angle A'O'C')$$

or

$$\gamma_{xy} = \lim_{\substack{\Delta x \to 0 \\ \Delta y \to 0}} (\pi/2 - \angle A'O'C') \tag{4.4}$$

For γ_{xy} to be positive, $\angle A'O'C'$ should be less than $\pi/2$. In other words, if the two mutually perpendicular lines, OA and OC, rotate relative to each other to form an acute angle $\angle A'O'C'$ after deformation of the element, the observed shear strain is positive. Otherwise, if this angle becomes obtuse, the shear strain is negative. In the present discussions, the deformation of element is considered in a limiting stage as the element shrinks to zero size and the strain components are defined at the point *O*.

4.3 RELATION BETWEEN STRAINS AND DISPLACEMENTS IN PLANE STRAIN

The three strain components (ε_x, ε_y and γ_{xy}) at a point in a plane body can be expressed in terms of two displacement components at the point, which are displacement (u) along x-direction and (v) along y-direction. For this, let us consider a plane element *OABC* of dimensions (Δx, Δy) shown in Fig. 4.3. The deformed state of element is given as *O′A′B′C′*. The displacement vector at the point *O* is taken as $\rho = OO' = u.\mathbf{i} + v.\mathbf{j}$. The displacement components (u and v) must be continuous functions of x and y in order to ensure geometrically compatible deformation. Using partial derivatives, the components of displacements at *A* and *C* are obtained along x- and y-directions.

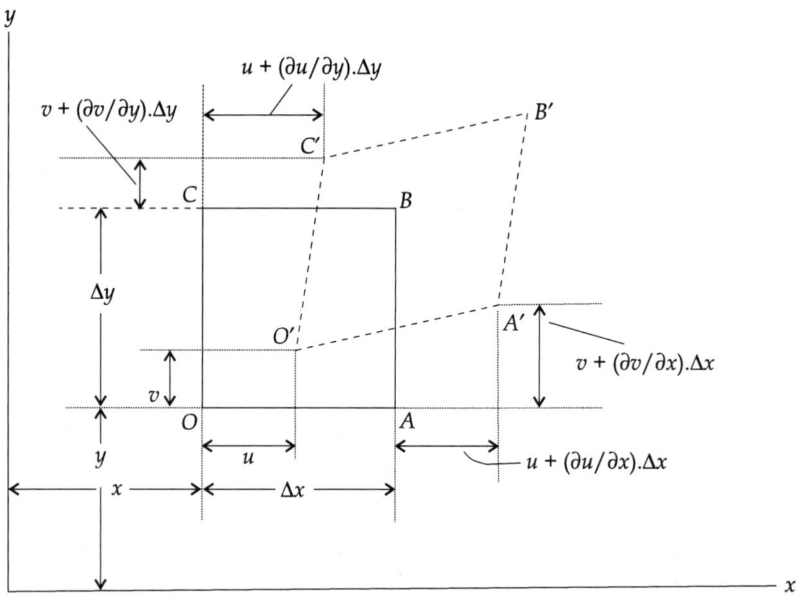

Fig. 4.3

Referring to Fig. 4.3 and using Eqn. 4.3, the strains, ε_x and ε_y, are written as

$$\varepsilon_x = \lim_{\Delta x \to 0} [\{\Delta x + u + (\partial u/\partial x)\cdot\Delta x - u\} - \Delta x]/\Delta x, \quad \text{or} \quad \varepsilon_x = \partial u/\partial x \tag{4.5}$$

Similarly, $$\varepsilon_y = \lim_{\Delta y \to 0} [\{\Delta y + v + (\partial v/\partial y)\cdot\Delta y - v\} - \Delta y]/\Delta y, \quad \text{or} \quad \varepsilon_y = \partial v/\partial y \tag{4.6}$$

To find an expression of shear strain, γ_{xy}, the deformed $\angle A'O'C'$ is pulled back as rigid body translation, such that O' coincides with O, as shown in Fig. 4.4.

For small values of angles, we have

$$\angle AOA' = \{(\partial v/\partial x)\cdot\Delta x\}/\Delta x = \partial v/\partial x$$

and $$\angle COC' = \{(\partial u/\partial y)\cdot\Delta y\}/\Delta y = \partial u/\partial y$$

Therefore, using Eqn. 4.4,

$$\gamma_{xy} = \lim_{\substack{\Delta x \to 0 \\ \Delta y \to 0}} (\angle AOC - \angle A'O'C')$$

$$= \lim_{\substack{\Delta x \to 0 \\ \Delta y \to 0}} [\pi/2 - \{\pi/2 - (\partial v/\partial x) - (\partial u/\partial y)\}]$$

$$\therefore \quad \gamma_{xy} = (\partial v/\partial x) + (\partial u/\partial y). \qquad (4.7)$$

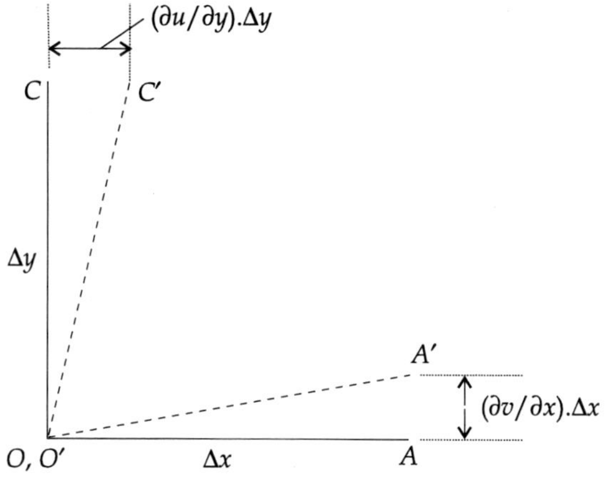

Fig. 4.4

Next, we can define average rotation of the element as a whole in the form of rotations of two mutually perpendicular line segments shown in Fig. 4.5. Such a rotation is denoted by ω_z.

Referring to Fig. 4.5, we have

$$(\omega_z)_{OA} = +(\partial v/\partial x), \quad (\omega_z)_{OC} = -(\partial u/\partial y).$$

$\therefore$ Average rotation at the point O,

$$\omega_z = (1/2)\cdot[(\omega_z)_{OA} + (\omega_z)_{OC}]$$

or

$$\omega_z = (1/2)\cdot[(\partial v/\partial x) - (\partial u/\partial y)] \qquad (4.8)$$

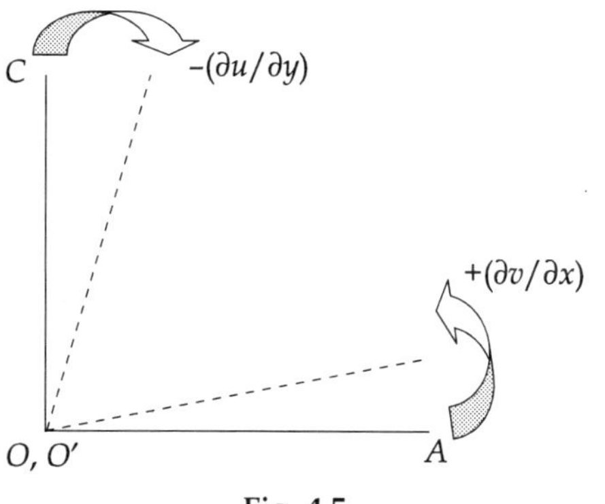

Fig. 4.5

4.4 TRANSFORMATION OF PLANE STRAIN

It was shown in the preceding section that the strain components at point O in x-y plane are obtained from two displacement components (u and v) by considering the deformation of infinitesimal element (at O) of sides Δx and Δy. As such, we obtain

$$\varepsilon_x = \partial u/\partial x,\ \varepsilon_y = \partial v/\partial y \text{ and } \gamma_{xy} = (\partial v/\partial x) + (\partial u/\partial y).$$

If another set of coordinates ($x' - y'$) on the same plane is chosen at an inclination (θ) with respect to x-y coordinates (see Fig. 4.6), the displacements (u' and v') at the same point O but in the directions x' and y' can be used to obtain the strain components in x'-y' coordinate system. The strain components are written as

$$\varepsilon_{x'} = \partial u'/\partial x',\ \varepsilon_{y'} = \partial v'/\partial y' \text{ and } \gamma_{x'y'} = (\partial v'/\partial x') + (\partial u'/\partial y').$$

Since x and y can be written as functions of x' and y', we have

$$\left.\begin{aligned} \varepsilon_{x'} &= (\partial u'/\partial x)\cdot(\partial x/\partial x') + (\partial u'/\partial y)\cdot(\partial y/\partial x'),\\ \varepsilon_{y'} &= (\partial v'/\partial x)\cdot(\partial x/\partial y') + (\partial v'/\partial y)\cdot(\partial y/\partial y'),\\ \text{and}\quad \gamma_{x'y'} &= \{(\partial v'/\partial x)\cdot(\partial x/\partial x') + (\partial v'/\partial y)\cdot(\partial y/\partial x')\} + \{(\partial u'/\partial x)\cdot(\partial x/\partial y') +\\ &\quad (\partial u'/\partial y)\cdot(\partial y/\partial y')\} \end{aligned}\right\} \tag{4.9}$$

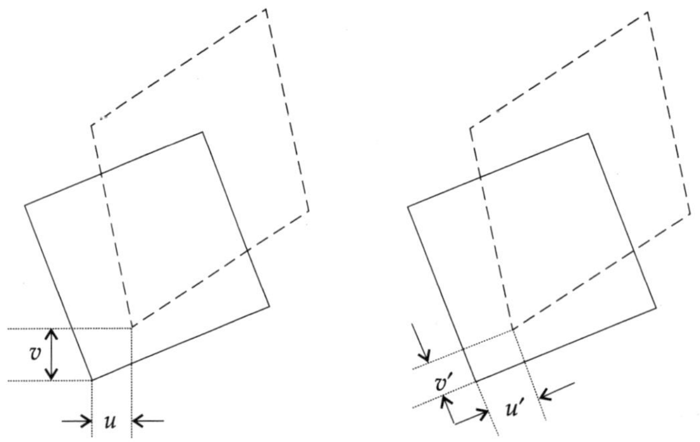

Fig. 4.6

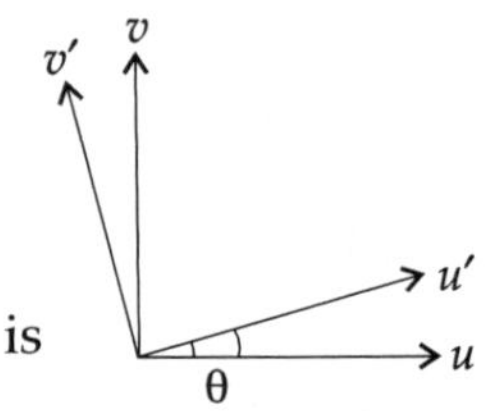

The displacements (u', v') and (u, v) are related according to

$$\begin{Bmatrix} u' \\ v' \end{Bmatrix} = \begin{pmatrix} \cos\theta & \sin\theta \\ -\sin\theta & \cos\theta \end{pmatrix} \begin{Bmatrix} u \\ v \end{Bmatrix},$$

in which the first matrix on right side of the above equation is known as **rotation matrix**.

The matrix equation given above can be written in the form of two equations, as

$$u' = u\cdot\cos\theta + v\cdot\sin\theta, \quad \text{and} \quad v' = -u\cdot\sin\theta + v\cdot\cos\theta \tag{4.10}$$

The above equations are used to write the following derivatives:

$$\left.\begin{aligned} \partial u'/\partial x &= (\partial u/\partial x)\cdot\cos\theta + (\partial v/\partial x)\cdot\sin\theta,\\ \partial u'/\partial y &= (\partial u/\partial y)\cdot\cos\theta + (\partial v/\partial y)\cdot\sin\theta,\\ \partial v'/\partial x &= (-\partial u/\partial x)\cdot\sin\theta + (\partial v/\partial x)\cdot\cos\theta,\\ \text{and}\quad \partial v'/\partial y &= (-\partial u/\partial y)\cdot\sin\theta + (\partial v/\partial y)\cdot\cos\theta. \end{aligned}\right\} \quad (4.11)$$

Also, the coordinates (x, y) and (x', y') are related as

$$\begin{Bmatrix} x \\ y \end{Bmatrix} = \begin{pmatrix} \cos\theta & -\sin\theta \\ \sin\theta & \cos\theta \end{pmatrix} \begin{Bmatrix} x' \\ y' \end{Bmatrix},$$

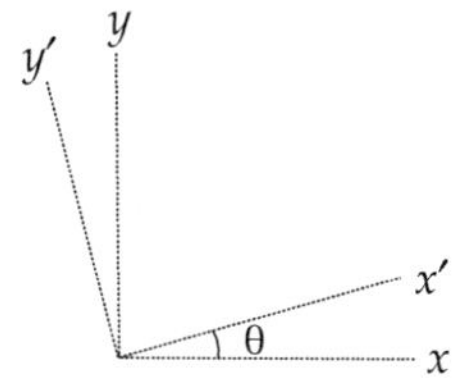

The above matrix can be split up into two equations,

$$x = x'\cdot\cos\theta - y'\cdot\sin\theta,$$

and $y = x'\cdot\sin\theta + y'\cdot\cos\theta.$

Therefore, using partial derivatives, we have

$$\partial x/\partial x' = \cos\theta,\ \partial x/\partial y' = -\sin\theta,\ \partial y/\partial x' = \sin\theta \text{ and } \partial y/\partial y' = \cos\theta \quad (4.12)$$

Next, Eqns. 4.11 and 4.12 are used to simplify Eqn. 4.9 to obtain

$$\varepsilon_{x'} = \{(\partial u/\partial x)\cdot\cos\theta + (\partial v/\partial x)\cdot\sin\theta\}\cdot(\cos\theta) + \{(\partial u/\partial y)\cdot\cos\theta + (\partial v/\partial y)\cdot\sin\theta\}\cdot(\sin\theta),$$

or $\varepsilon_{x'} = (\partial u/\partial x)\cdot\cos^2\theta + (\partial v/\partial y)\cdot\sin^2\theta + \{(\partial v/\partial x) + (\partial u/\partial y)\}\cdot\sin\theta\cdot\cos\theta,$

$$\therefore\quad \varepsilon_{x'} = \{(\varepsilon_x + \varepsilon_y)/2\} + \{(\varepsilon_x - \varepsilon_y)/2\}\cdot\cos 2\theta + (\gamma_{xy}/2)\cdot\sin 2\theta. \quad (4.13)$$

Similarly, it can be deduced that

$$\varepsilon_{y'} = \{(\varepsilon_x + \varepsilon_y)/2\} - \{(\varepsilon_x - \varepsilon_y)/2\}\cdot\cos 2\theta - (\gamma_{xy}/2)\cdot\sin 2\theta, \quad (4.14)$$

and $\gamma_{x'y'}/2 = -\{(\varepsilon_x - \varepsilon_y)/2\}\cdot\sin 2\theta + (\gamma_{xy}/2)\cdot\cos 2\theta.$ (4.15)

It is found from the above equations that with known strain components (ε_x, ε_y and γ_{xy}) at a point, the strains ($\varepsilon_{x'}$, $\varepsilon_{y'}$ and $\gamma_{x'y'}$) at an orientation of angle (θ) can be completely determined.

Example 4.1: A thin plate is rectangular in shape, as shown in Fig. 4.7. Uniformly distributed strains in the plate are $\varepsilon_x = 2000\ \mu$, $\varepsilon_y = 1000\ \mu$ and $\gamma_{xy} = -1000\ \mu$ rad., where, $\mu = 10^{-6}$. Determine the normal strain in the direction of unit vector (**n**) along the diagonal.

Solution: Given that,

$$\varepsilon_x = 2000\ \mu,\ \varepsilon_y = 1000\ \mu,\ \gamma_{xy} = -1000\ \mu \text{ rad.},\ \theta = 40°$$

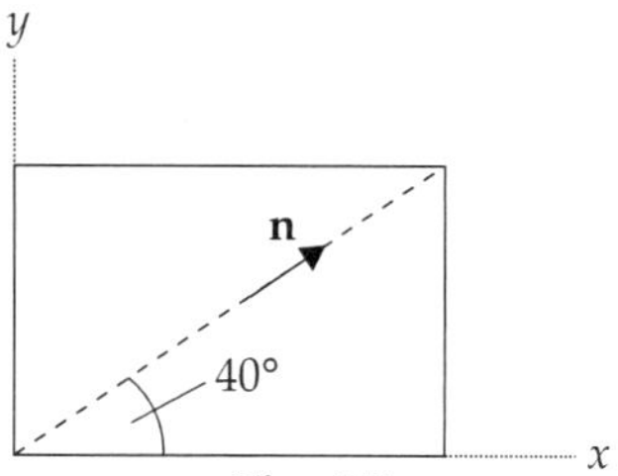

Fig. 4.7

The normal strain ($\varepsilon_{\mathbf{n}}$) along **n** is obtained according to Eqn. 4.13, as

$$\varepsilon_{\mathbf{n}} = \varepsilon_{x'} = \{(2000 + 1000)/2\} + \{(2000 - 1000)/2\}\cdot\cos(2\times 40°) + \{(-1000)/2\}\cdot\sin(2\times 40°)$$

or $\varepsilon_{\mathbf{n}} = 1094\ \mu$ **Ans.**

Example 4.2: Uniform deformation of a thin plate as shown in Fig. 4.8, results into strains $\varepsilon_x = -1200\ \mu$, $\varepsilon_y = +900\ \mu$ and $\gamma_{xy} = +800\ \mu$ rad. Find the normal strain ε_{BD} along the diagonal BD.

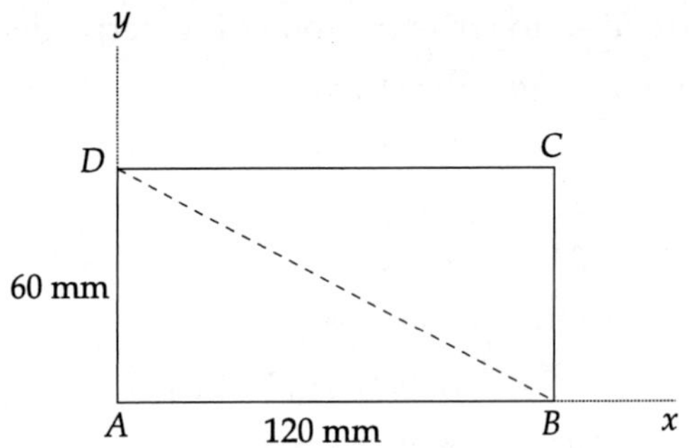

Fig. 4.8

Solution: Referring to Fig. 4.8,

$\Psi = \tan^{-1}(60/120) = 26.565°$

$\therefore \quad \theta = 180° - \Psi = 153.435°$

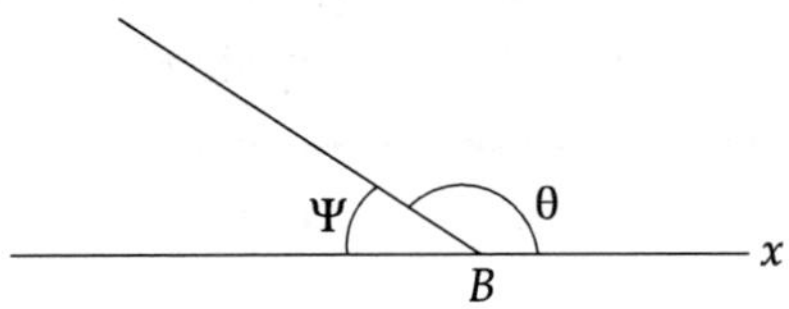

From Eqn. 4.13, $\varepsilon_{BD} = \varepsilon_{x'} = \{(-1200 + 900)/2\} + \{(-1200 - 900)/2\}\cdot\cos(2 \times 153.435°) + (800/2)\cdot\sin(2 \times 153.435°)$

or $\varepsilon_{BD} = -1100\ \mu$ **Ans.**

Example 4.3: A thin plate shown in Fig. 4.9 is uniformly deformed to contain the strains $\varepsilon_x = 1600\ \mu m/m$, $\varepsilon_y = -2000\ \mu m/m$, and $\gamma_{xy} = -800\ \mu$ rad. Determine the normal strain (ε_n) along n-axis and shear strain (γ_n) in the n-t plane.

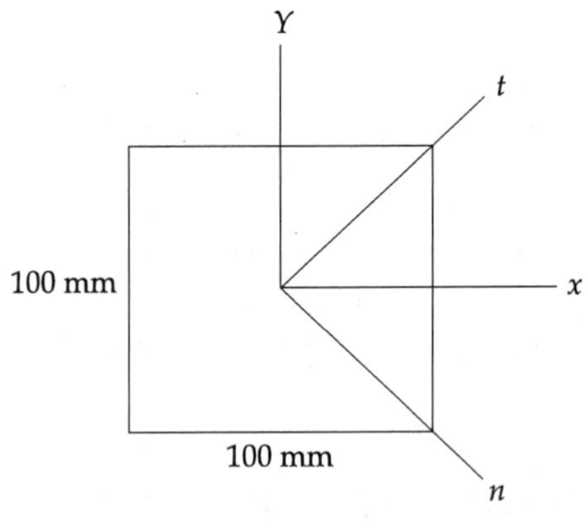

Fig. 4.9

Solution: Inclination of n-axis with respect to x-axis is $-45°$, or $\theta = -45°$.

$\therefore \quad 2\theta = -90°$, $\sin 2\theta = -1$ and $\cos 2\theta = 0$.

From Eqn. 4.13,

$\varepsilon_n = \varepsilon_{x'} = [\{1600 + (-2000)\}/2] + [\{1600 - (-2000)/2\}]\cdot 0 + \{(-800)/2\} \times (-1)$

$\therefore \quad \varepsilon_n = 200\ \mu m/m$ **Ans.**

Shear strain in the *n-t* plane,

$$\gamma_{nt}/2 = -[\{1600 - (-2000)\}/2]\cdot(-1) + \{(-800)/2\}\cdot 0$$

$$\gamma_{nt} = 3600\ \mu\ \text{rad.}$$ **Ans.**

Example 4.4: At a point in a body, the strain components under plane strain are given as $\varepsilon_x = 750\ \mu$, $\varepsilon_y = 350\ \mu$, and $\gamma_{xy} = -300\ \mu$ rad. Determine the strains (ε_n, ε_s and γ_{ns}) in the *n-s* coordinates shown in Fig. 4.10.

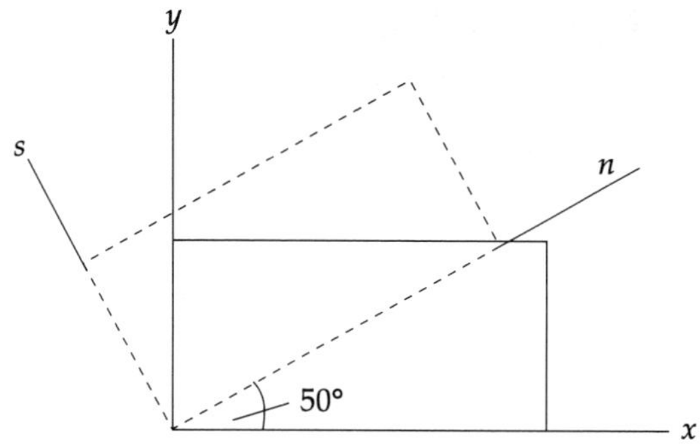

Fig. 4.10

Solution: Given that $\varepsilon_x = 750\ \mu$, $\varepsilon_y = 350\ \mu$, $\gamma_{xy} = -300\ \mu$, $\theta = 50°$ or $2\theta = 100°$

$$\varepsilon_n = \{(750 + 350)/2\} + \{(750 - 350)/2\}\cdot\cos(100°) + \{(-300)/2\}\cdot\sin(100°)$$

$$\varepsilon_n = 369\ \mu$$ **Ans.**

$$\varepsilon_s = \{(750 + 350)/2\} - \{(750 - 350)/2\}\cdot\cos(100°) - \{(-300)/2\}\cdot\sin(100°)$$

$$\varepsilon_s = 731\ \mu$$ **Ans.**

$$\gamma_{ns}/2 = -\{(750-350)/2\}\cdot\sin(100°) + \{(-300)/2\}\cdot\cos(100°)$$

$$\gamma_{ns} = -342\ \mu\ \text{rad.}$$ **Ans.**

Example 4.5: A thin plate shown in Fig. 4.11 contains uniformly distributed strains $\varepsilon_n = 1600\ \mu\text{m/m}$, $\varepsilon_s = 1300\ \mu\text{m/m}$, and $\varepsilon_x = 1200\ \mu\text{m/m}$. Find the shearing strain (γ_{ns}) and normal strain (ε_y) in the plate.

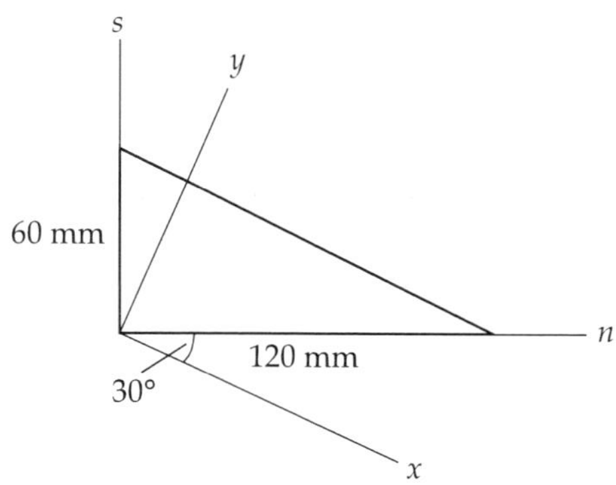

Fig. 4.11

Solution: Addition of Eqns. 4.13 and 4.14 yields to

$$\varepsilon_{x'} + \varepsilon_{y'} = \varepsilon_x + \varepsilon_y.$$

In the present case, $\varepsilon_n + \varepsilon_s = \varepsilon_x + \varepsilon_y$

$\therefore$ $$1600 + 1300 = 1200 + \varepsilon_y.$$

$$\varepsilon_y = 1700\ \mu\text{m/m}$$ **Ans.**

Again, subtracting Eqn. 4.14 from Eqn. 4.13,

$$\varepsilon_n - \varepsilon_s = (\varepsilon - \varepsilon_y)\cdot\cos 2\theta + \gamma_{xy}\cdot\sin 2\theta.$$

or $$1600 - 1300 = (1200 - 1700)\cdot\cos 60° + \gamma_{xy}\cdot\sin 60°$$

$$\gamma_{xy} = 635\ \mu\ \text{rad}.$$

$\therefore$ $$\gamma_{ns}/2 = -\{(1200 - 1700)/2\}\cdot\sin 60° + (635/2)\cdot\cos 60°$$

$$\gamma_{ns} = 750.5\ \mu\ \text{rad}.$$ **Ans.**

4.5 PRINCIPAL STRAINS AND MOHR'S CIRCLE FOR PLANE STRAIN

If the strain components (ε_x, ε_y and γ_{xy}) at a point are known in x-y coordinates, a set of mutually perpendicular axes can be obtained at the same point such that the normal strains are maximum and minimum along these axes. Such strains are known as **principal strains**. In the set of axes defined by the directions of normal strains, the shear strain is found to be zero.

To find principal strains, the expression of $\varepsilon_{x'}$ in Eqn. 4.13 is used to write

$$d\varepsilon_{x'}/d\theta = -\{(\varepsilon_x - \varepsilon_y)/2\}\cdot 2\cdot\sin 2\theta + (\gamma_{xy}/2)\cdot 2\cdot\cos 2\theta = 0, \text{ or } \tan 2\theta = \gamma_{xy}/(\varepsilon_x - \varepsilon_y) \quad (4.16)$$

or $\theta = \theta_1 = (1/2)\cdot\tan^{-1}\{\gamma_{xy}/(\varepsilon_x - \varepsilon_y)\}$, where, θ_1 gives the inclination of the first principal strain from x-axis.

From Eqn. 4.16, we have

$$\sin 2\theta = \gamma_{xy}/\sqrt{(\varepsilon_x - \varepsilon_y)^2 + \gamma_{xy}^2}\cdot \quad \cos 2\theta = (\varepsilon_x - \varepsilon_y)/\sqrt{(\varepsilon_x - \varepsilon_y)^2 + \gamma_{xy}^2}\cdot$$

Therefore, the first principal strain (ε_1) is written according to Eqn. 4.13, and after simplification becomes

$$\varepsilon_1 = \{(\varepsilon_x + \varepsilon_y)/2\} + \sqrt{\{(\varepsilon_x - \varepsilon_y)/2\}^2 + (\gamma_{xy}/2)^2} \quad (4.17)$$

The direction of other principal strain is found by replacing θ with $\theta + (\pi/2)$, so that $\sin\{2\cdot(\theta + \pi/2)\} = -\sin 2\theta$, and $\cos\{2\cdot(\theta + \pi/2)\} = -\cos 2\theta$.

and $$\sin 2\theta = -\gamma_{xy}/\sqrt{(\varepsilon_x - \varepsilon_y)^2 + \gamma_{xy}^2}, \text{ and } \cos 2\theta = -(\varepsilon_x - \varepsilon_y)/\sqrt{(\varepsilon_x - \varepsilon_y)^2 + \gamma_{xy}^2}$$

Hence, the principal strain (ε_2) is obtained by substituting for sin 2θ and cos 2θ in Eqn. 4.13 to write

$$\varepsilon_2 = \{(\varepsilon_x + \varepsilon_y)/2\} - \sqrt{\{(\varepsilon_x - \varepsilon_y)/2\}^2 + (\gamma_{xy}/2)^2} \quad (4.18)$$

Next, the maximum shearing strain is derived from Eqn. 4.15 as follows :

$$d(\gamma_{xy}/2)/d\theta = -(\varepsilon_x - \varepsilon_y)\cdot\cos 2\theta - \gamma_{xy}\cdot\sin 2\theta = 0$$

or $$\tan 2\theta = \tan 2\theta_s = -(\varepsilon_x - \varepsilon_y)/\gamma_{xy} \tag{4.19}$$

It is to be noted that $(\tan 2\theta_1)\cdot(\tan 2\theta_s) = -1$, which implies that the maximum shearing strain occurs at 45° angular displacement from the principal strains. An expression of maximum shearing strain is derived after writing equations of sin 2θ and cos 2θ from Eqn. 4.19 and using them in Eqn. 4.15. Therefore,

$$\gamma_{\max}/2 = \sqrt{\{(\varepsilon_x - \varepsilon_y)/2\}^2 + (\gamma_{xy}\ 2)^2}. \tag{4.20}$$

It may be observed from Eqn. 4.17 to Eqn. 4.20 that $\gamma_{xy}/2$ is included in the same way as τ_{xy} has its place for writing the principal stresses. Hence, we can conclude that the strain tensor under plane strain condition can be written in the form

$$\begin{Bmatrix} \varepsilon_x & \gamma_{xy}/2 \\ \gamma_{xy}/2 & \varepsilon_y \end{Bmatrix} \tag{4.21}$$

Since the above matrix (in Eqn. 4.21) represents a second order tensor, a Mohr's circle construction for plane strain can be accomplished in the same way as that for plane stress condition. To obtain principal strains (ε_1, ε_2) and the maximum shearing strain ($\gamma_{\max}$) a Mohr's circle is constructed using ε_x, ε_y and γ_{xy}, as shown in Fig. 4.12. To draw Mohr's circle the sign of shear strain (γ_{xy}) should be properly diploid. As γ_{xy} is written according to Eqn. 4.7, as

$$\gamma_{xy} = (\partial v/\partial x) + (\partial u/\partial y),$$

for the positive values of displacement-derivatives $\partial v/\partial x$ and $\partial u/\partial y$ the shear strain is positive. As such, there is an angular distortion of the element to attain an acute angle. This was described in the derivation of Eqn. 4.7. If an obtuse angle is obtained after angular distortion, γ_{xy} is taken to be negative.

In Fig. 4.12, the normal strain (ε) is measured along the ordinate and half of shear strain ($\gamma_{xy}/2$) is measured along the abscissa, with O as origin of the coordinate system. This selection of coordinates complies with the strain tensor given under Eqn. 4.21. The points X and Y are taken on ε-axis, such that $OX = \varepsilon_x$ and $OY = \varepsilon_y$. At the point X, a perpendicular $XX' = \gamma_{xy}/2$ is drawn. Similarly, another perpendicular, $YY' = -\gamma_{xy}/2$, is drawn at the point Y. Then $X'Y'$ is joined to intersect ε-axis at C, which gives the centre of Mohr's circle. With $CX' = CY'$ as radius, a circle is drawn to complete the construction. The circle intersects ε-axis at two points P_1 and P_2 so that $OP_1 = \varepsilon_1$ and $OP_2 = \varepsilon_2$ are the two principal strains.

It can be checked that $OP_1 = \varepsilon_1 = OC + CP_1 = OC + CX'$.

∴ $$\varepsilon_1 = \{(\varepsilon_x + \varepsilon_y)/2\} + \sqrt{\{(\varepsilon_x - \varepsilon_y)/2\}^2 + (\gamma_{xy}/2)^2}.$$

Also, it is found that $$OP_2 = \varepsilon_2 = \{(\varepsilon_x + \varepsilon_y)/2\} - \sqrt{\{(\varepsilon_x - \varepsilon_y)/2\}^2 + (\gamma_{xy}/2)^2},$$

and $$CP = \gamma_{max}/2 = \sqrt{\{(\varepsilon_x - \varepsilon_y)/2\}^2 + (\gamma_{xy}/2)^2}.$$

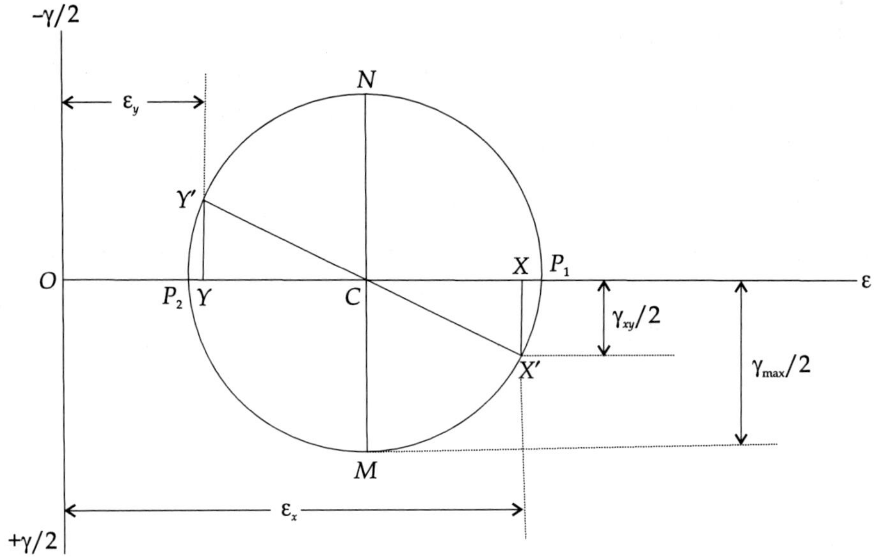

Fig. 4.12

Above-mentioned deductions are same as those obtained earlier in Eqns. 4.17, 4.18 and 4.20, respectively.

The following are observed in Mohr's circle of strain :

(a) The sum $\varepsilon_1 + \varepsilon_2 = \varepsilon_x + \varepsilon_y$.

(b) $CM = \gamma_{max}/2 = CP_1 = CP_2 = (\varepsilon_1 - \varepsilon_2)/2$

or $\gamma_{max} = \varepsilon_1 - \varepsilon_2$.

Example 4.6: An element of material in plane strain is subjected to strains $\varepsilon_x = 500 \times 10^{-6}$, $\varepsilon_y = 100 \times 10^{-6}$, and $\gamma_{xy} = 400 \times 10^{-6}$. Determine the principal strains and the maximum shear strain.

Solution: Given that $\varepsilon_x = 500\ \mu$, $\varepsilon_y = 100\ \mu$, $\gamma_{xy} = 400\ \mu$, where, $1\ \mu = 10^{-6}$.

Principal strains are obtained from Eqns. 4.17 and 4.18, as

$$\varepsilon_{1,2} = \{(500 + 100)/2\} \pm \sqrt{\{(500 - 100)/2\}^2 + (400/2)^2} = 300 \pm 282.8$$

$\therefore \quad \varepsilon_1 = 582.8\ \mu$ and $\varepsilon_2 = 17.2\ \mu$ or $\varepsilon_1 = 582.8 \times 10^{-6}$ and $\varepsilon_2 = 17.2 \times 10^{-6}$ **Ans.**

Maximum shear strain is written according to Eqn. 4.20, as

$\gamma_{max}/2 = \sqrt{\{(500 - 100)/2\}^2 + (400/2)^2}$, $\therefore \gamma_{max} = 566\ \mu$ or 566×10^{-6} **Ans.**

Alternatively, $\gamma_{max} = \varepsilon_1 - \varepsilon_2 = 582.8 - 17.2 \simeq 566\ \mu$.

The inclination of the first principal strain, $\theta_1 = (1/2)\cdot\tan^{-1}\{400/(500 - 100)\} = 22.5°$.

Hence, the maximum shearing strain occurs at $\theta_s = 22.5° - 45° = -22.5°$.

Principal strain deformation and maximum shearing strain distortion of elements are shown in the following sketch.

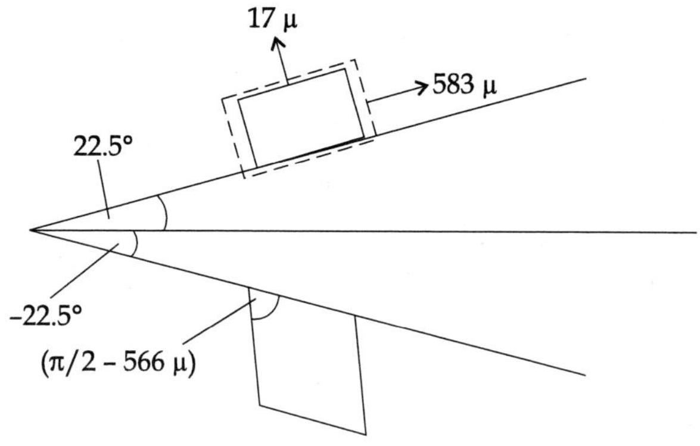

Fig. 4.13

Example 4.7: The strain components at a point under plane strain are ε_x = –325 μ, ε_y = –625 μ, and γ_{xy} = +380 μ. Use Mohr's circle construction to find the principal strains and the maximum shear strain at the point. Also, find the inclination of the first principal strain.

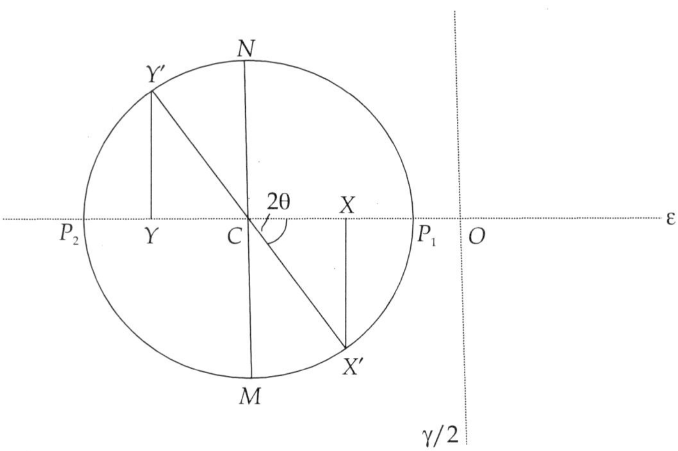

Fig. 4.14

Solution: Mohr's circle of plane strain is constructed, as given below:

From the origin O of ε – $\gamma/2$ system, OX = –325 and OY = –625 are measured on ε-axis to some scale. The centre of Mohr's circle is found from OC = {(–325) + (-625)}/2 = –475. XX' = 380/2 = 190 and YY' = –190 are the perpendiculars drawn at X and Y, respectively. The points X' and Y' are joined to intersect ε-axis at C. With C as centre and CX' or CY' as radius, the circle is drawn which meets ε-axis at two points, P_1 and P_2. Hence, $OP_1 = \varepsilon_1$ and $OP_2 = \varepsilon_2$ are the two principal strains.

Referring to the figure, $CX = 475 - 325 = 150$. Therefore, $CX' = \sqrt{(CX)^2 + (XX')^2}$ $= \sqrt{(150)^2 + (190)^2}$.

Or $CX' = 242 = CY'$, which gives the radius of Mohr's circle.

$\therefore \quad \varepsilon_1 = OP_1 = OC - CP_1 = OC - CX'$, or $\varepsilon_1 = -(475 - 242) = -233\ \mu$ **Ans.**

$\varepsilon_2 = OC + CP_2 = OC + CX'$, or $\varepsilon_2 = -(475 + 242) = -717\ \mu$ **Ans.**

$\gamma_{max}/2 = CM = CX' = 242$. $\therefore \gamma_{max} = 484$ μrad. **Ans.**

Alternatively, $\gamma_{max} = \varepsilon_1 - \varepsilon_2 = (-233) - (-717)$, or $\gamma_{max} = 484$ μrad.

Inclination of the first principal strain (θ_1) : $2\theta_1 = \tan^{-1}(XX'/CX) = \tan^{-1}(190/150)$.

$\therefore \theta_1 = 25.9°$ **Ans.**

Example 4.8: For the state of plane strain at a point, the strains, $\varepsilon_x = 500\ \mu$, $\varepsilon_y = -800\ \mu$, and the first principal strain, $\varepsilon_1 = 1000\ \mu$, are known. With a Mohr's circle construction, determine γ_{xy}, ε_2, γ_{max} and the inclination of ε_1 with respect to x-axis.

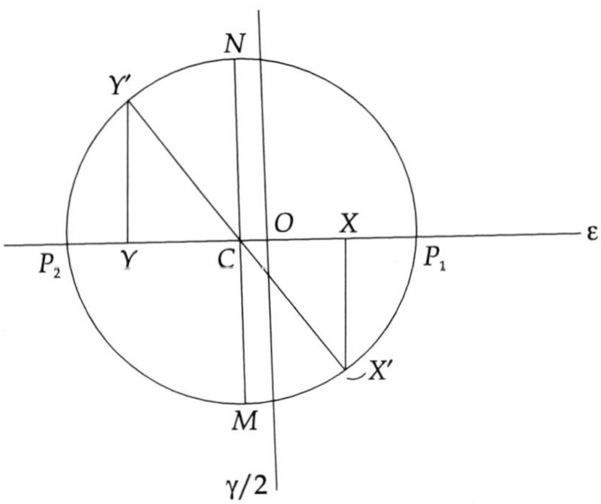

Fig. 4.15

Solution: Mohr's circle for the problem is drawn in which $OX = 500$, $OY = -800$, $OP_1 = 1000$.

$XX' = \gamma_{xy}$ is assumed to be positive.

$OC = \{500 + (-800)\}/2 = -150$, $CX' = CP_1 = 150 + 1000 = 1150$.

Also, $CX = 150 + 500 = 650$.

$\therefore \gamma_{xy}/2 = XX' = \sqrt{(CX')^2 + (CX)^2} = \sqrt{(1150)^2 + (650)^2}$, or $\gamma_{xy} = 1897\ \mu$

$\gamma_{xy} = \pm 1897\ \mu$ **Ans.**

Next, $OP_2 = OC + CP_2 = OC + CX' = 150 + 1150 = 1300$. $\therefore \varepsilon_2 = -1300\ \mu$ **Ans.**

Further, $CM = CX' = 1150$. $\therefore \gamma_{max}/2 = 1150$, or $\gamma_{max} = 2300\ \mu$ **Ans.**

Inclination of ε_1 : $2\theta_1 = \tan^{-1}(XX'/CX) = \tan^{-1}(948.68/650) = 55.6° \Rightarrow \theta_1 = 27.8°$ **Ans.**

4.6 STRAIN MEASUREMENTS: STRAIN ROSETTE

It was explained earlier that the stresses on the surface or interior of plates and simple structural components can be determined on the basis of stress analysis. However, the machine elements and structural components are generally complicated in shape and are subjected to various types of loads. As such, the experimental determination of stresses becomes imperative. The stress cannot be measured directly because it is a mathematical concept applied for the analytical convenience. It is a common practice to measure the normal (longitudinal) strain on the surface of long prismatic bars by using mechanical strain gauges, as in tension test of materials. For this, a large gauge length is available over which the change in length per unit of gauge length is measured. Because of limited space at the critical regions on the surface of components, the mechanical gauges are not found suitable. To overcome this difficulty, electrical resistance strain gauges are used which are small in size and sensitive enough to measure normal strains within acceptable range of accuracy. After finding the values of strains, the stresses can be obtained from the stress-strain relations.

An electrical resistance strain gauge is a widely used device. It consists of a thin wire filament or a filament of metallic foil mounted on small insulating paper or plastic sheet, as shown in Fig. 4.16. The filament of thin wire is coiled several times to accommodate a large length of wire within a short gauge length and it is embedded between paper or plastic sheets cemented together with resin. This arrangement is fixed on the surface of a structural component by means of some adhesive. The two ends of filament are connected to the terminals called leads which, in turn, are attached to an external measuring circuit.

The working principle of electrical resistance strain gauge is that there is a change in electrical resistance of filament wire due to a change in its length. The electrical resistance is measured by a suitable external circuit.

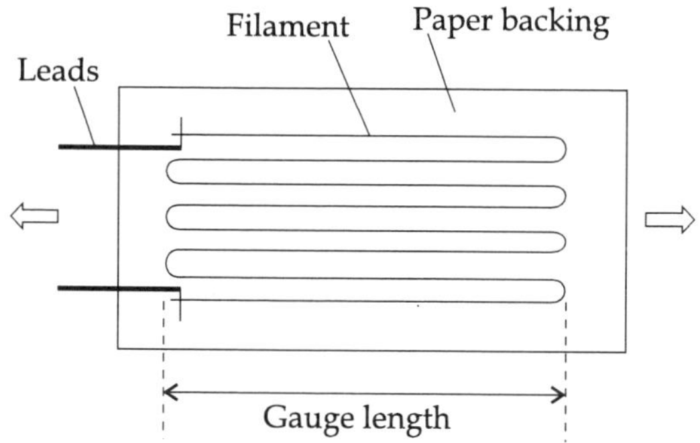

Fig. 4.16

It should be noted that a strain gauge described above can measure longitudinal strain only, and there is no device available to measure the shearing strain. In order to obtain the values of strains, ε_x, ε_y and γ_{xy}, at a point, three small strain gauges are placed very close to the point, as shown in Fig. 4.17. If the inclination of strain gauges are θ_a, θ_b and θ_c with respect to x-axis, and the readings of strains in them are ε_a, ε_b and ε_c respectively, the simplified form of Eqn. 4.13 can be used to write

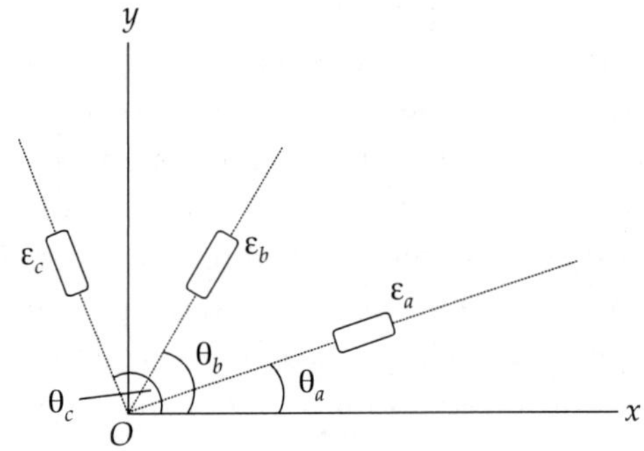

Fig. 4.17

$$\varepsilon_a = \varepsilon_x \cdot \cos^2\theta_\alpha + \varepsilon_y \cdot \sin^2\theta_\alpha + \gamma_{xy} \cdot \sin\theta_\alpha \cdot \cos\theta_\alpha,$$
$$\varepsilon_b = \varepsilon_x \cdot \cos^2\theta_b + \varepsilon_y \cdot \sin^2\theta_b + \gamma_{xy} \cdot \sin\theta_b \cdot \cos\theta_b, \quad (4.22)$$
and $$\varepsilon_c = \varepsilon_x \cdot \cos^2\theta_c + \varepsilon_y \cdot \sin^2\theta_c + \gamma_{xy} \cdot \sin\theta_c \cdot \cos\theta_c.$$

From the three equations given above, the three unknowns, ε_x, ε_y and γ_{xy}, are obtained. Such an arrangement of three strain gauges is called a **strain rosette**. Two common types of rosettes generally used are (1) 45° rosette, and (2) 60° rosette or Delta rosette. Schematic drawings of these rosettes are given in Fig. 4.18.

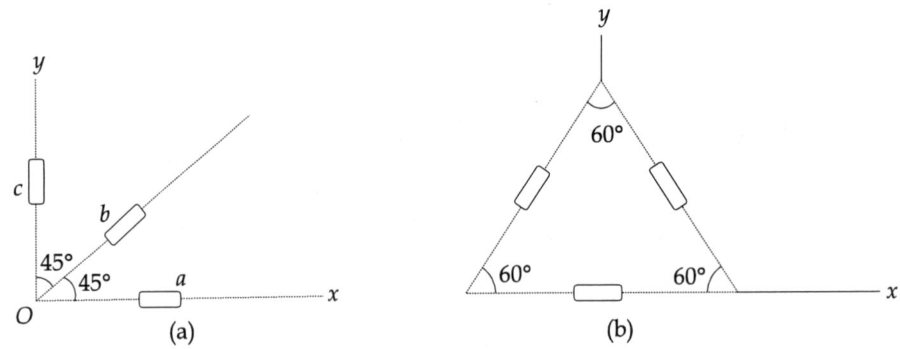

Fig. 4.18

Example 4.9: A 45° rosette mounted on the surface of an airplane wing is shown in Fig. 4.19. The strain gauge readings are $\varepsilon_A = 500 \times 10^{-6}$, $\varepsilon_B = 350 \times 10^{-6}$ and $\varepsilon_C = -100 \times 10^{-6}$. Determine the principal strains and the maximum shear strain.

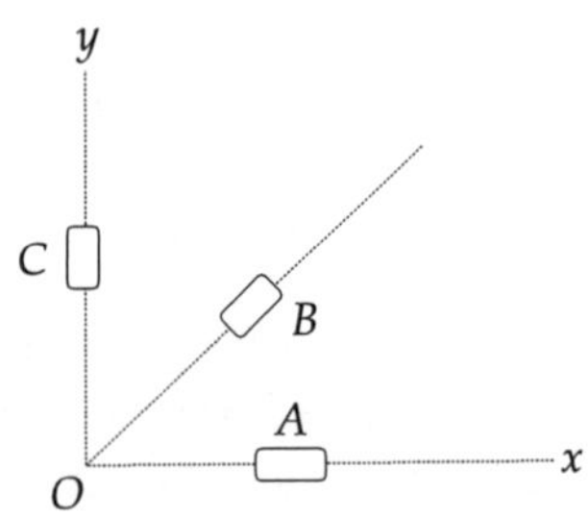

Fig. 4.19

Solution: Given that,

$$\varepsilon_A = 500\ \mu,\ \varepsilon_B = 350\ \mu,\ \varepsilon_C = -100\ \mu, \text{ where } 1\ \mu = 10^{-6}$$

For gauge A at $\theta = 0°$,

$$\varepsilon_A = 500 = \varepsilon_x \cdot \cos^2 0° + \varepsilon_y \cdot \sin^2 0° + \gamma_{xy} \cdot \sin 0° \cdot \cos 0°$$

or $\varepsilon_x = 500\ \mu$.

$$\varepsilon_B = 350 = \varepsilon_x \cdot \cos^2 45° + \varepsilon_y \cdot \sin^2 45° + \gamma_{xy} \cdot \sin 45° \cdot \cos 45°$$

or $350 = \{(500 + \varepsilon_y)/2\} + (\gamma_{xy}/2)$

$\therefore\ \gamma_{xy} + \varepsilon_y = 200$ ... (a)

For gauge C at $\theta = 90°$, $-100 = \varepsilon_x \cdot \cos^2 90° + \varepsilon_y \cdot \sin^2 90° + \gamma_{xy} \cdot \sin 90° \cdot \cos 90° \cdot\ \varepsilon_y = -100\ \mu$

Therefore, in (a), $\gamma_{xy} + (-100) = 200$. $\therefore\ \gamma_{xy} = 300\ \mu$.

Principal strains: $\varepsilon_{1,2} = \{(500 - 100)/2\} + \sqrt{\{(500 + 100)/2\}^2 + (300/2)^2}$

$\varepsilon_1 = 535.4\ \mu$, $\varepsilon_2 = -135.4\ \mu$ **Ans.**

Maximum shear strain: $\gamma_{max} = \varepsilon_1 - \varepsilon_2 = 535.4 - (-135.4) = 670.8\ \mu$ **Ans.**

Example 4.10: Three strain gauges, A, B and C, are mounted on the surface of a structural component as shown in Fig. 4.20. If the strains recorded in the three gauges are $\varepsilon_A = 1200 \times 10^{-6}$, $\varepsilon_B = 300 \times 10^{-6}$ and $\varepsilon_C = 300 \times 10^{-6}$, determine principal strains and the principal stresses. Take $E = 42$ GPa and $\nu = 0.35$ for the material of the component.

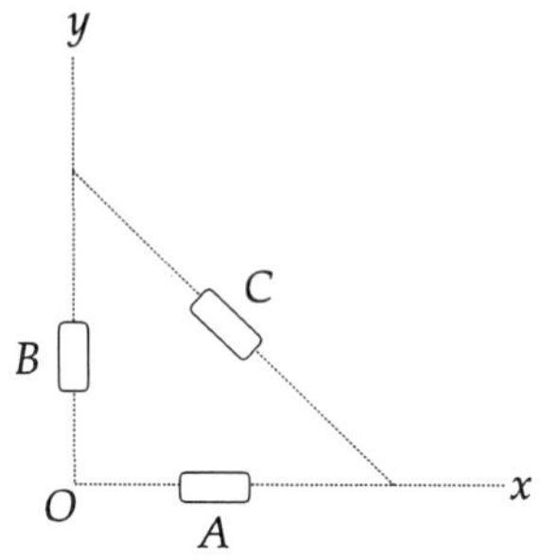

Fig. 4.20

Solution: $\varepsilon_A = 1200\ \mu$ at $\theta = 0°$, $\varepsilon_B = 300\ \mu$ at $\theta = 90°$, $\varepsilon_C = 300\ \mu$ at $\theta = 135°$

$\therefore$ $\varepsilon_A = 1200 = \varepsilon_x \cdot \cos^2 0° + \varepsilon_y \cdot \sin^2 0° + \gamma_{xy} \cdot \sin 0° \cdot \cos 0°$, or $\varepsilon_x = 1200\ \mu$.

$\varepsilon_B = 300 = \varepsilon_x \cdot \cos^2 90° + \varepsilon_y \cdot \sin^2 90° + \gamma_{xy} \cdot \sin 90° \cdot \cos 90°$, or $\varepsilon_y = 300\ \mu$.

$\varepsilon_C = 300 = 1200 \cdot \cos^2 135° + 300 \cdot \sin^2 135° + \gamma_{xy} \cdot \sin 135° \cdot \cos 135°$, $\therefore\ \gamma_{xy} = 900\ \mu$.

Principal strains:

$$\varepsilon_{1,2} = \{(1200 + 300)/2\} \pm \sqrt{\{(1200 + 300)/2\}^2 + (900/2)^2}$$

$$= 750 \pm 636.4$$

$\therefore$ $\varepsilon_1 = 1386.4\ \mu$, $\varepsilon_2 = 113.6\ \mu$ **Ans.**

Principal stresses:

From $\varepsilon_1 = (1/E) \cdot (\sigma_1 - \nu\sigma_2)$ and $\varepsilon_2 = (1/E) \cdot (\sigma_2 - \nu\sigma_1)$,

$$\sigma_1 = \{E/(1 - \nu^2)\} \cdot (\varepsilon_1 + \nu\varepsilon_2)$$

and $\sigma_2 = \{E/(1 - \nu^2)\} \cdot (\varepsilon_2 + \nu\varepsilon_1)$.

Hence, substituting for E, ν, ε_1 and ε_2 in the above equations,

$\sigma_1 = 68.3$ MPa and $\sigma_2 = 28.7$ MPa **Ans.**

Example 4.11: A Delta rosette as shown in Fig. 4.21 consists of three strain gauges, A, B and C. The readings in gauges are recorded as $\varepsilon_a = 800\ \mu$, $\varepsilon_b = 350\ \mu$ and $\varepsilon_c = -300\ \mu$, respectively. If the elastic constants for the material are known to be $E = 100$ GPa and $\nu = 0.3$, determine (a) principal strains and maximum shear strain in the plane of rosette, (b) the principal stresses on the same plane, and (c) principal strain in a direction perpendicular to the plane of rosette.

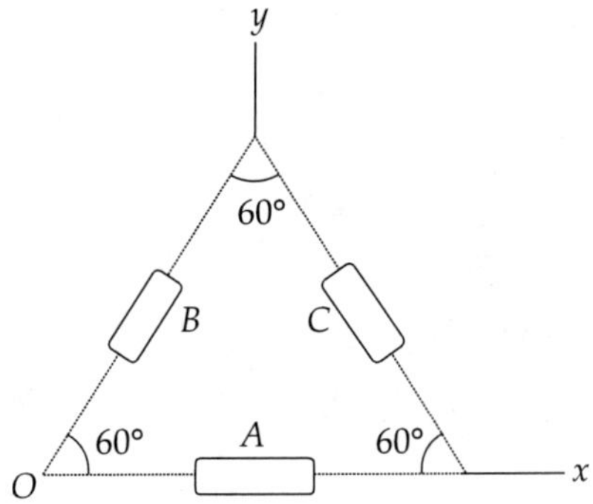

Fig. 4.21

Solution: Given that

$\varepsilon_a = 800\ \mu$ at $\theta = 0°$

$\varepsilon_b = 350\ \mu$ at $\theta = 60°$

$\varepsilon_c = -300\ \mu$ at $\theta = 120°$

$\therefore\quad \varepsilon_a = 800 = \varepsilon_x \cdot \cos^2 0° + \varepsilon_y \cdot \sin^2 0° + \gamma_{xy} \cdot \sin 0° \cdot \cos 0°,$

or $\quad \varepsilon_x = 800\ \mu.$

$\varepsilon_b = 350 = \varepsilon_x \cdot \cos^2 60° + \varepsilon_y \cdot \sin^2 60° + \gamma_{xy} \cdot \sin 60° \cdot \cos 60°$

or $\quad 350 = 0.25 \cdot \varepsilon_x + 0.75 \cdot \varepsilon_y + 0.433 \cdot \gamma_{xy}$ ... (i)

$\varepsilon_c = -300 = \varepsilon_x \cdot \cos^2 120° + \varepsilon_y \sin^2 120° + \gamma_{xy} \cdot \sin 120° \cdot \cos 120°$

or $\quad -300 = 0.25 \cdot \varepsilon_x + 0.75 \cdot \varepsilon_y - 0.433 \cdot \gamma_{xy}$ ... (ii)

Subtracting (ii) from (i) and simplifying, $\gamma_{xy} = 750.6\ \mu$.

Therefore from (i), $\varepsilon_y = -233.3\ \mu$.

(a) Principal strains: $\varepsilon_{1,2} = \{(800 - 233.3)/2\} \pm \sqrt{\{(800 + 233.3)/2\}^2 + (750.6/2)^2}$

$= 283.35 \pm 638.57$

$\therefore\ \varepsilon_1 = 922\ \mu,\ \varepsilon_2 = -355\ \mu$ **Ans.**

Maximum shearing strain $\gamma_{max} = 1277\ \mu$ **Ans.**

(b) Principal stresses: Substituting for the values of E, ν, ε_1 and ε_2 in $\sigma_1 = \{E/(1 - \nu^2)\} \cdot (\varepsilon_1 + \nu \cdot \varepsilon_2)$,

$\sigma_1 = 89.6$ MPa **Ans.**

Similarly, from $\sigma_2 = \{E/(1 - \nu^2)\} \cdot (\varepsilon_2 + \nu \cdot \varepsilon_1)$,

$\sigma_2 = -8.6$ MPa **Ans.**

(c) As $\sigma_3 = 0$, $\varepsilon_3 = (1/E) \cdot [-\nu.(\sigma_1 + \sigma_2)] = (1/100 \times 10^9) \cdot [-0.3 \cdot (89.6 - 8.6) \cdot 10^6]$

or $\varepsilon_3 = -243\ \mu$ **Ans.**

PROBLEMS

1. A thin plate shown in Fig. 4.22 is subjected to uniform strain given by $\varepsilon_x = 2000$ μm/m, $\varepsilon_y = -1500$ μm/m and $\gamma_{xy} = 1200$ μrad. Find the normal strain (ε_n) along the unit vector, **n**, in the plate.

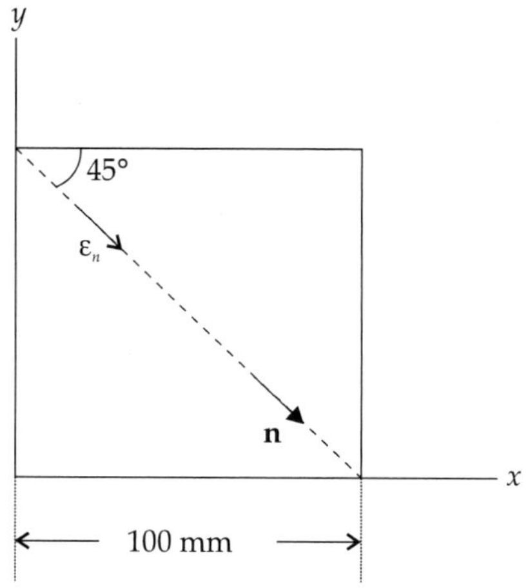

Fig. 4.22

Ans. $\varepsilon_n = -350$ μm/m

2. A thin plate (Fig. 4.23) is uniformly deformed so that $\varepsilon_x = -850$ μ, $\varepsilon_y = +660$ μ, and $\gamma_{xy} = 1000$ μ. Determine the normal strain along the diagonal (*PR*) of the plate.

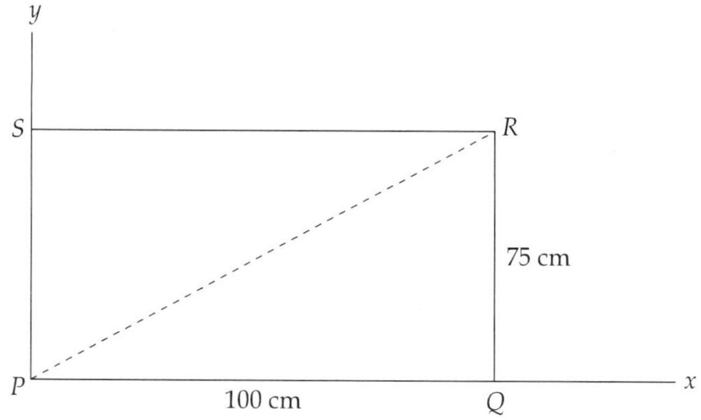

Fig. 4.23

Ans. 174 μ

3. A square plate as shown in Fig. 4.24 is subjected to uniform strains $\varepsilon_x = 3000$ μm/m, $\varepsilon_y = 1400$ μm/m and $\gamma_{xy} = 1000$ μrad. Find the values of normal strain ε_n and shear strain γ_{ns} in the plate.

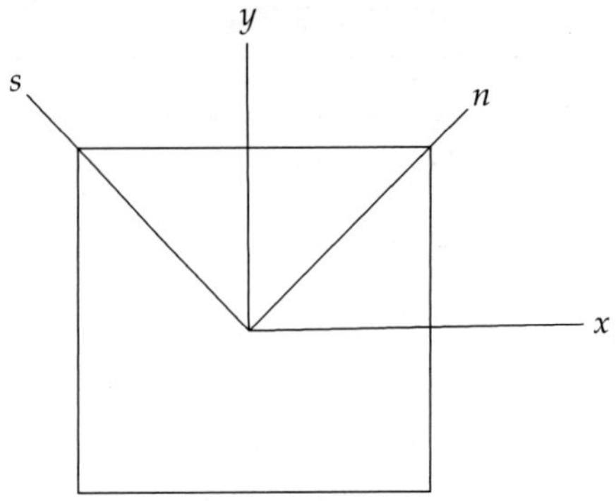

Fig. 4.24

Ans. $\varepsilon_n = 2700\ \mu m/m$, $\gamma_{ns} = -1600\ \mu rad$.

4. At a point *P* in a stressed body, the strain components are given as $\varepsilon_x = -850\ \mu$, $\varepsilon_y = +600\ \mu$ and $\gamma_{xy} = -900\ \mu$. If another coordinate system (n – t) is at an inclination 40° with respect to *x*-*y* coordinates, determine the strain components, ε_n, ε_t and γ_{nt} (see Fig. 4.25).

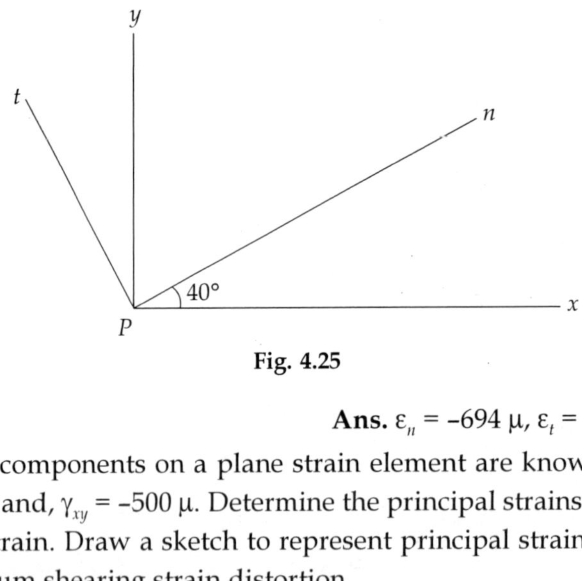

Fig. 4.25

Ans. $\varepsilon_n = -694\ \mu$, $\varepsilon_t = +444\ \mu$, $\gamma_{nt} = 1272\ \mu$

5. The strain components on a plane strain element are known to be $\varepsilon_x = 950\ \mu$, $\varepsilon_y = -300\ \mu$ and, $\gamma_{xy} = -500\ \mu$. Determine the principal strains and the maximum shearing strain. Draw a sketch to represent principal strain deformations and the maximum shearing strain distortion.

Ans. $\varepsilon_1 = 998\ \mu$, $\varepsilon_2 = -348\ \mu$, $\gamma_{max} = 1346\ \mu$.

6. At a point under plane strain, $\varepsilon_x = 400\ \mu m/m$, $\varepsilon_y = -1000\ \mu m/m$ and the first principal strain $\varepsilon_1 = 1600\ \mu m/m$,determine the shearing strain (γ_{xy}), the other principal strain (ε_2), the maximum shearing strain (γ_{max})and the inclination of ε_1 relative to *x*-axis.

Ans. $\gamma_{xy} = 3533\ \mu$, $\varepsilon_2 = -2200\ \mu$, $\gamma_{max} = \pm 3800\ \mu$, $\theta_1 = 34.2°$

7. The strain components at a point in x-y plane are given as $\varepsilon_x = +1000$ μm/m, $\varepsilon_y = -230$ μm/m, and $\gamma_{xy} = +280$ μrad. Using Mohr's circle construction, find the principal strains and the maximum shearing strain at the point. What is the inclination of the first principal strain with respect to x-axis ?

Ans. $\varepsilon_1 = 1016$ μm/m, $\varepsilon_2 = -246$ μm/m, $\gamma_{max} = 1261$ μrad., $\theta_1 = 6.4°$

8. If $\varepsilon_x = -700$ μ, $\varepsilon_y = 300$ μ and the first principal strain, $\varepsilon_1 = 400$ μ, are known at a point in plane strain condition, use Mohr's construction to obtain γ_{xy}, ε_2, γ_{max} and θ_1 at the point.

Ans. $\gamma_{xy} = 663$ μ, $\varepsilon_2 = -800$ μ, $\gamma_{max} = 1200$ μ, $\theta_1 = -16.8°$

9. Three strain gauges are arranged very close to a point on the surface of a machine element, as shown in Fig. 4.26. The readings in the gauges are $\varepsilon_a = 680$ μ, $\varepsilon_b = 400$ μ and $\varepsilon_c = 850$ μ. Determine:

(a) ε_x, ε_y and γ_{xy} at the point

(b) The principal strains and maximum shear strain at the point on the surface

(c) The principal strain perpendicular to the surface of machine element, assuming $\nu = 0.30$ for the material.

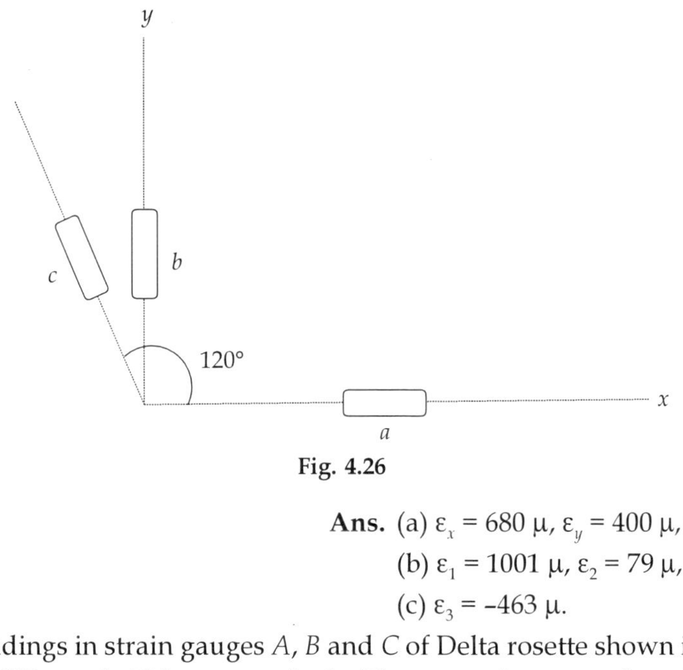

Fig. 4.26

Ans. (a) $\varepsilon_x = 680$ μ, $\varepsilon_y = 400$ μ, $\gamma_{xy} = -877.6$ μ.
(b) $\varepsilon_1 = 1001$ μ, $\varepsilon_2 = 79$ μ, $\gamma_{max} = 921$ μ
(c) $\varepsilon_3 = -463$ μ.

10. The readings in strain gauges A, B and C of Delta rosette shown in Fig. 4.27 are 1100 μ, 800 μ and –700 μ, respectively. The rosette is mounted on the free surface of a material for which $E = 150$ GPa and $\nu = 0.35$. Determine (a) the principal strains on the surface and also the third principal strain perpendicular to the surface, (b) the maximum shearing strain on the surface, and (c) the principal stresses at a point close to the rosette.

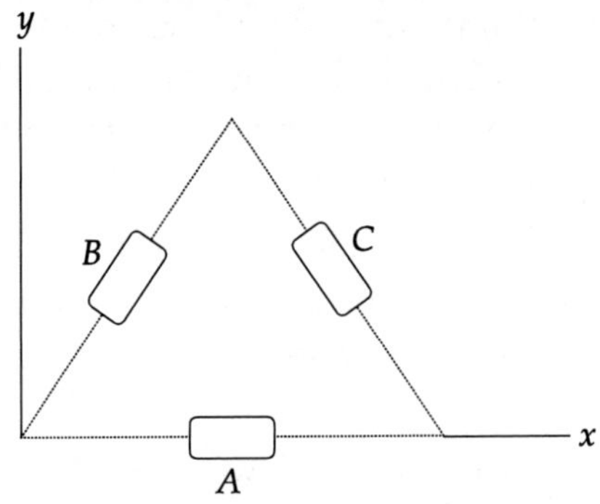

Fig. 4.27

Ans. (a) $\varepsilon_1 = 1513.5\ \mu$, $\varepsilon_2 = -713.5\ \mu$, $\varepsilon_3 = -430.8\ \mu$.

(b) $\gamma_{max} = 2227\ \mu$.

(c) $\sigma_1 = 216$ MPa, $\sigma_2 = -31.4$ MPa.

5
Relation between Stress and Strain

5.1 INTRODUCTION

It is understood from discussions in chapter 2 that the proportionality between deformation of a member and the force acting on it according to Hooke's law is extended in the form of stress-strain proportionality, provided that the stress remains within the elastic limit of the material. A linear relation between stress and strain was obtained as $\sigma = E \cdot \varepsilon$, for the case of uniaxial loading of bars. The term "modulus of elasticity" was introduced as a material property which is denoted by E.

Poisson's ratio was introduced to obtain transverse strain that appears due to the development of longitudinal or axial strain. In the following sections of this chapter the two-dimensional and three-dimensional stress-strain relations will be derived. The constitutive equations, i.e., equations relating stress and strain will be written by including elasticity modulus € and Poisson's ratio (ν). The engineering materials like steel, aluminium, etc., are assumed to be homogeneous and isotropic with only two elastic constants (E and ν) to define the elastic properties for each material. The stress-strain relations obtained with such assumption are generally found to agree with experimental observation and practical applications in industries.

5.2 STRESS-STRAIN RELATIONS IN THREE DIMENSIONS

It is assumed that a body made from a homogeneous and isotropic material is loaded externally such that there exists a point inside the body which is in a state of tri-axial stresses. An infinitely small cubical element can be assumed to enclose the point and the stresses acting on the element in x-y-z coordinates are shown in Fig. 5.1. A positive normal strain along x-direction due to the application of σ_x is given by σ_x/E. Since the normal strain (σ_y/E) is simultaneously acting in y-direction, a ransverse strain (negative) along x-direction becomes $-\nu \cdot \sigma_y/E$, where, ν = Poisson's ratio. Similarly, another normal strain (σ_z/E) in z-direction results into a negative strain ($\nu \cdot \sigma_z/E$) along x-direction. After adding all these strains algebraically along x-axis, we have the net normal strain in x-direction as

$$\varepsilon_x = \sigma_x/E - \nu\cdot\sigma_y/E - \nu\cdot\sigma_z/E$$

which is rearranged in the form

$$\varepsilon_x = (1/E)\cdot[\sigma_x - \nu\cdot(\sigma_y + \sigma_z)]. \quad (5.1)$$

Similarly, the normal strains in y- and z-directions are

$$\left.\begin{aligned} \varepsilon_y &= (1/E)\cdot[\sigma_y - \nu\cdot(\sigma_x + \sigma_z)], \\ \text{and} \quad \varepsilon_z &= (1/E)\cdot[\sigma_z - \nu\cdot(\sigma_x + \sigma_y)]. \end{aligned}\right\} \quad (5.2)$$

For isotropic materials, the shear strain acting on a plane remains unaffected by other shear strains on the orthogonal planes. Hence, the relations among shear strains and shear stresses on x-y, x-z and y-z planes respectively, are

$$\gamma_{xy} = \tau_{xy}/G, \gamma_{xz} = \tau_{xz}/G, \gamma_{yz} = \tau_{yz}/G \quad (5.3)$$

where, G = shear modulus of the material.

In Eqns. 5.1 and 5.2, the normal strains are written in terms of normal stresses. These expressions can be used to obtain the normal stresses in terms of normal strains also. For this, the above equations are rearranged as

$$\left.\begin{aligned} \sigma_x - \nu\cdot\sigma_y - \nu\cdot\sigma_z &= E\cdot\varepsilon_x, \\ \sigma_y - \nu\cdot\sigma_x - \nu\cdot\sigma_z &= E\cdot\varepsilon_y, \\ \sigma_z - \nu\cdot\sigma_x - \nu\cdot\sigma_y &= E\cdot\varepsilon_z. \end{aligned}\right\} \quad (5.4)$$

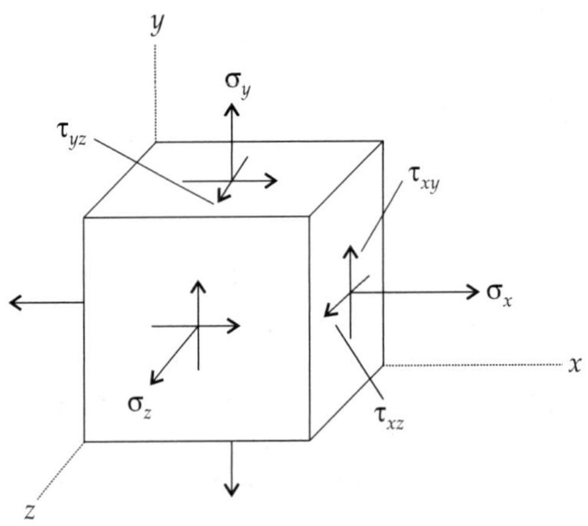

Fig. 5.1

Next, eliminating σ_z from the first two equations in Eqn. 5.4, we have

$$\sigma_x - \sigma_y = \{E/(1 + \nu)\}\cdot(\varepsilon_x - \varepsilon_y). \quad (5.5)$$

Also, from the first and third equations in Eqn. 5.4,

$$\{(1 - \nu)/\nu\}\cdot\sigma_x - \sigma_y = [E/\{\nu\cdot(1 - \nu)\}]\cdot(\varepsilon_x + \nu\cdot\varepsilon_z) \quad (5.6)$$

After subtracting Eqn. 5.5 from Eqn. 5.6 and rearranging of the terms,

$$\sigma_x = [E\cdot(1 - \nu)/\{(1 + \nu)\cdot(1 - 2\nu)\}]\cdot\varepsilon_x + [\nu\cdot E/\{(1 + \nu)\cdot(1 - 2\nu)\}].\varepsilon_y + [\nu.E/\{(1 + \nu)\cdot(1 - 2\nu)\}]\cdot\varepsilon_z,$$

so that a further simplification yields to

$$\sigma_x = [\nu\cdot E/\{(1 + \nu)\cdot(1 - 2\nu)\}]\cdot(\varepsilon_x + \varepsilon_y + \varepsilon_z) + [E/(1 + \nu)]\cdot\varepsilon_x. \quad (5.7)$$

The above can be written in the form

$$\sigma_x = \lambda\cdot(\varepsilon_x + \varepsilon_y + \varepsilon_z) + 2\cdot G\cdot\varepsilon_x,$$

where, $\lambda = \nu\cdot E/\{(1+\nu)\cdot(1-2\nu)\}$, which is known as Lame's constant, and $E/(1+\nu) = 2G$, where, G = shear modulus of the material which is also known as modulus of rigidity (a relation between E and G will be derived later on).

Therefore the three normal stresses are written in terms of three normal strains as

$$\left.\begin{aligned} \sigma_x &= (\lambda + 2G)\cdot\varepsilon_x + \lambda\cdot\varepsilon_y + \lambda\cdot\varepsilon_z, \\ \sigma_y &= \lambda\cdot\varepsilon_x + (\lambda + 2G)\cdot\varepsilon_y + \lambda\cdot\varepsilon_z, \\ \text{and} \quad \sigma_z &= \lambda\cdot\varepsilon_x + \lambda\cdot\varepsilon_y + (\lambda + 2G)\cdot\varepsilon_z. \end{aligned}\right\} \tag{5.8}$$

Also, the shear stresses are written in terms of shear strains as

$$\tau_{xy} = G\cdot\gamma_{xy},\ \tau_{xz} = G\cdot\gamma_{xz}\ \tau_{yz} = G\cdot\gamma_{yz}, \tag{5.9}$$

which completes the solution.

5.3 STATE OF PLANE STRESS

In a state of plane stress, the stresses act on a plane only and there is no stress associated with the direction perpendicular to that plane. If x-y plane is assumed as the said plane, then the plane stress condition implies that σ_x, σ_y and τ_{xy} are the stresses acting in the plane x-y and the stresses $\sigma_z = 0$, and $\tau_{xz} = \tau_{yz} = 0$. A thin plate has thickness very small compared to its length and breadth. When a thin plate is subjected to loads acting in its plane, its contraction or extension along the thickness remains unrestricted and hence, no stress develops in the direction perpendicular to the plane of the plate. Thus, a condition of plane stress is attained. A thin plate under uniform stress distribution is shown in Fig. 5.2. As $\sigma = 0$, the normal strains, ε_x and ε_y along x- and y-axes, are written from Eqn. 5.1 and Eqn. 5.2, as

$$\left.\begin{aligned} \varepsilon_x &= (1/E)\cdot(\sigma_x - \nu\cdot\sigma_y), \\ \text{and} \quad \varepsilon_y &= (1/E)\cdot(\sigma_y - \nu\cdot\sigma_x). \end{aligned}\right\} \tag{5.10}$$

Similarly, the shear strain is written according to Eqn. 5.9, as

$$\gamma_{xy} = \tau_{xy}/G. \tag{5.11}$$

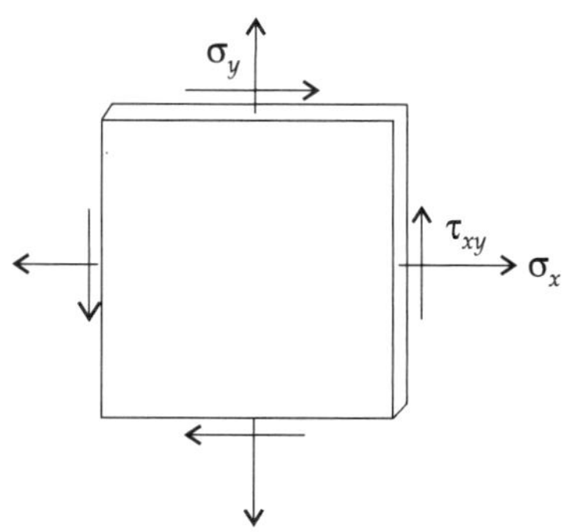

Fig. 5.2

After rearranging the terms in Eqn. 5.10, the normal stresses (σ_x and σ_y) are written as

$$\left.\begin{aligned} \sigma_x &= \{E/(1-\nu^2)\}\cdot(\varepsilon_x + \nu\cdot\varepsilon_y), \\ \text{and} \qquad \sigma_y &= \{E/(1-\nu^2)\}\cdot(\varepsilon_y + \nu\cdot\varepsilon_x), \end{aligned}\right\} \tag{5.12}$$

and the shear stress as $\tau_{xy} = G\cdot\gamma_{xy}$. (5.13)

It is important to note that in plane stress the stress in the third direction (*z*-axis) is zero, but the normal strain (ε_z) along *z*-axis is not necessarily zero. The strain (ε_z) in z-direction is obtained under the condition $\sigma_z = 0$ in Eqn. 5.2 and becomes

$$\varepsilon_z = -(\nu/E)\cdot(\sigma_x + \sigma_y). \tag{5.14}$$

Using expressions of σ_x and σ_y from Eqn. 5.12, the above equation is modified as

$$\varepsilon = -\{\nu/(1-\nu)\}\cdot(\varepsilon_x + \varepsilon_y) \tag{5.15}$$

which implies that when ε_x and ε_y are positive (causing extension), ε_z is negative indicating contraction.

It is also important to note that on the surface of a thick plate or a structural element the stress in a direction perpendicular to the surface is absent and a plane stress condition is obtained.

5.4 STATE OF PLANE STRAIN

A plane strain condition is attained when the normal and shear strains are acting on a plane and there is no strain attached to the third direction which is perpendicular to the plane. When a thick plate is subjected to external loads, a state of plane strain is attained at a point sufficiently inside the plate material. Due to internal resistance within the material, the displacement of points along the thickness direction are constrained and the strain along the thickness is zero. Only the strains in the plane containing length and breadth of the plate may be present. However, because of constraint caused by internal resistance there is development of normal stress in the thickness direction.

Therefore, we have $\varepsilon_z = 0$ and $\gamma_{xz} = \gamma_{yz} = 0$, whereas the strains, ε_x, ε_y and $\gamma_{xy,}$ are present on *x-y* plane. With $\varepsilon_z = 0$ in Eqn. 5.2

$$\sigma_z = \nu\cdot(\sigma_x + \sigma_y) \tag{5.16}$$

which gives the normal stress perpendicular to *x-y* plane as discussed above.

To find ε_x and ε_y in *x-y* plane, the expression of σ_z from from Eqn. 5.16 is used in Eqn. 5.1 and Eqn. 5.2 to obtain

$$\varepsilon_x = \{(1-\nu^2)/E\}\cdot[\sigma_x - \{\nu/(1-\nu)\}\cdot\sigma_y]. \tag{5.17}$$

Similarly, we have $\varepsilon_y = \{(1-\nu^2)/E\}\cdot[\sigma_y - \{\nu/(1-\nu)\}\cdot\sigma_x]$. (5.18)

The shear strain is written similar to plane stress state, as $\gamma_{xy} = \tau_{xy}/G$.

The stress-strain relations in Eqns. 5.17 and 5.18 can be written in the form of Eqn. 5.10 of plane stress with the assumption $E' = E/(1-\nu^2)$ and $\nu' = \nu/(1-\nu)$, so that

$$\left.\begin{aligned}\varepsilon_x &= (1/E')\cdot(\sigma_x - \nu'\cdot\sigma_y),\\ \text{and}\qquad \varepsilon_y &= (1/E')\cdot(\sigma_y - \nu'\cdot\sigma_x).\end{aligned}\right\} \tag{5.19}$$

Further, the stresses can be written in terms of strains on putting $\varepsilon_z = 0$ in Eqn. 5.8, so that

$$\sigma_x = [E/\{(1 + \nu)\cdot(1 - 2\nu)\}]\cdot[(1 - \nu)\cdot\varepsilon_x + \nu\cdot\varepsilon_y],$$

$$\text{and}\qquad \sigma_y = [E/\{(1 + \nu)\cdot(1 - 2\nu)\}]\cdot[(1 - \nu)\cdot\varepsilon_y + \nu\cdot\varepsilon_x]. \tag{5.20}$$

Also, σ_z with the help of Eqn. 5.16 becomes

$$\sigma_z = [\nu\cdot E/\{(1 + \nu)\cdot(1 - 2\nu)\}]\cdot(\varepsilon_x + \varepsilon_y). \tag{5.21}$$

Example 5.1: A machine element is made of steel having E = 210 GPa and ν = 0.3. At a point inside the machine element, the state of stress is given by σ_x = 100 MPa, σ_y = 35 MPa, σ_z = 50 MPa, τ_{xy} = 40 MPa, τ_{xz} = 22 MPa and τ_{yz} = 35 MPa. What is the state of strain at this point?

Solution: Referring to Eqns. 5.1 and 5.2, the normal stresses are as follows:

$\varepsilon_x = (1/210 \times 10^9)\cdot[100 - 0.3\cdot(35 + 50)]\cdot10^6$, or $\varepsilon_x = 355\ \mu\text{m/m}$ **Ans.**

$\varepsilon_y = (1/210 \times 10^9)\cdot[35 - 0.3\cdot(100 + 50)]\cdot10^6$. or $\varepsilon_y = -48\ \mu\text{m/m}$ **Ans.**

$\varepsilon_z = (1/210 \times 10^9)\cdot[50 - 0.3\cdot(100 + 35)]\cdot10^6$ or $\varepsilon_z = 45\ \mu\text{m/m}$ **Ans.**

Also, the shear strain $\gamma_{xy} = \tau_{xy}/G = \{2.(1 + \nu)/G\}.\tau_{xy},$

$\therefore\qquad \gamma_{xy} = \{2\cdot(1 + 0.3)/210 \times 10^9\}\cdot(40 \times 10^6) = 495\ \mu\text{rad.}$ **Ans.**

Similarly, $\gamma_{xz} = \{2\cdot(1 + 0.3)/210 \times 10^9\}\cdot(22 \times 10^6) = 272\ \mu\text{rad.}$ **Ans.**

$\Gamma_{yz} = \{2\cdot(1 + 0.3)/210 \times 10^9\}\cdot(35 \times 10^6) = 433\ \mu\text{rad.}$ **Ans.**

Example 5.2: At a point inside a machine part made of brass (E = 100 GPa, ν = 0.3), the strain components are known to be: ε_x = 1825 μm/m, ε_y = –1910 μ/m, ε_z = 458 μm/m, γ_{xy} = 1275 μrad., γ_{xz}= 2769 μrad, and γ_{yz} = 1749 μrad. What is the state of stress at this point?

Solution: The normal stresses are obtained from Eqn. 5.7 or Eqn. 5.8.

Lame's constant, $\lambda = 0.3 \times 100/[(1 + 0.3)\cdot\{1 - (2 \times 0.3)\}] = 57.69$ GPa

Shear modulus, $G = E/\{2\cdot(1 + \nu)\}$, $\therefore\ 2G = 100/(1 + 0.3) = 76.92$ GPa

Therefore, $\sigma_x = \lambda\cdot(\varepsilon_x + \varepsilon_y + \varepsilon_z) + 2G\cdot\varepsilon_x,$

$= 57.69 \times 10^9\cdot[1825 + (-1910) + 458]\cdot10^{-6} + 76.92 \times 10^9\cdot1825 \times 10^{-6}\ \text{N/m}^2$

or $\qquad \sigma_x = 162$ MPa **Ans.**

Similarly, $\sigma_y = 57.69 \times 10^9\cdot[1825 + (-1910) + 458]\cdot10^{-6} + 76.92 \times 10^9\cdot(-1910 \times 10^{-6})\ \text{N/m}^2$

or $\qquad \sigma_y = -125$ MPa **Ans.**

$\sigma_z = 57.69 \times 10^9\cdot[1825 + (-1910) + 458]\cdot10^{-6} + 76.92 \times 10^9\cdot458 \times 10^{-6}\ \text{N/m}^2$

or $\qquad \sigma_z = 57$ MPa **Ans.**

Shear stresses: G = 38.46 GPa

$\tau_{xy} = (38.46 \times 10^9)\cdot(1275 \times 10^{-6}) = 49$ MPa

$\tau_{xz} = (38.46 \times 10^9)\cdot(2769 \times 10^{-6}) = 106$ MPa

$\tau_{yz} = (38.46 \times 10^9)\cdot(1749 \times 10^{-6}) = 67$ MPa **Ans.**

Example 5.3: At a point on the surface of a structural component, the strains are $\varepsilon_x = 950\ \mu m/m$, $\varepsilon_y = -400\ \mu m/m$ and $\gamma_{xy} = -500\ \mu rad$. If $E = 200$ GPa and $\nu = 0.30$ for the material, determine σ_x, σ_y and τ_{xy} at the point.

Solution: There is a state of plane stress on the surface.

$$\therefore \quad \sigma_x = \{E/(1-\nu^2)\}\cdot(\varepsilon_x + \nu.\varepsilon_y)$$

$$= [200 \times 10^9/\{1-(0.3)^2\}]\cdot[950 + 0.3\cdot(-400)],\ 10^{-6}\ N/m^2$$

or $\sigma_x = 182$ MPa **Ans.**

Similarly, $\sigma_y = [200 \times 10^9/\{1-(0.3)^2\}]\cdot[-400 + 0.3\cdot(950)]\cdot 10^{-6}\ N/m^2$

or $\sigma_y = -25$ MPa **Ans.**

Also, $\tau_{xy} = G\cdot\gamma_{xy} = [E/\{2\cdot(1+\nu)\}]\cdot\gamma_{xy} = [200 \times 10^9/\{2\cdot(1+0.3)\}]\cdot(-500 \times 10^{-6})\ N/m^2$

or $\tau_{xy} = -38.5$ MPa **Ans.**

Example 5.4: A 200 mm × 150 mm × 10 mm aluminum plate (Fig. 5.3) is subjected to uniform biaxial stresses σ_x and σ_y. Two strain gauges *A* and *B* are attached to the surface of the plate as shown. If the readings in strain gauges are $\varepsilon_A = 200 \times 10^{-6}$ and $\varepsilon_B = 285 \times 10^{-6}$, what are the values of σ_x and σ_y ? What is the reduction in thickness of the plate as a result of loading ?

Take $E = 75$ GPa and $\nu = 0.33$ for aluminium.

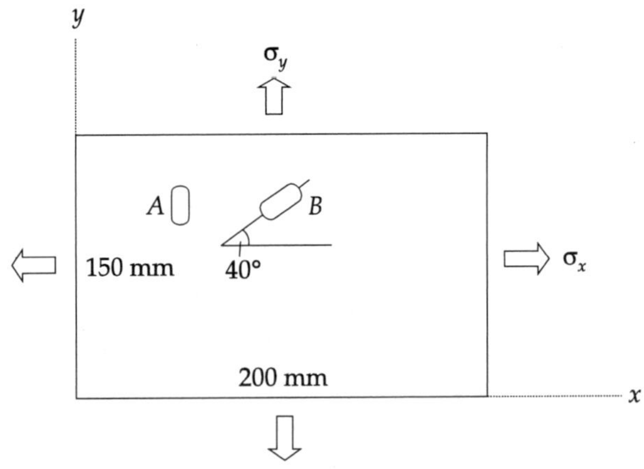

Fig. 5.3

Solution: $\varepsilon_A = 200 \times 10^{-6} = \varepsilon_x\cdot\cos^2 90° + \varepsilon_y\cdot\sin^2 90°$. $\therefore\ \varepsilon_y = 200 \times 10^{-6}$.

$\varepsilon_B = 285 \times 10^{-6} = \varepsilon_x\cdot\cos^2 40° + (200 \times 10^{-6})\cdot\sin^2 40°$. $\therefore\ \varepsilon_x = 344.85 \times 10^{-6}$.

Next, using Eqn. 5.10 for plane stress,

$$\varepsilon_x = 344.85 \times 10^{-6} = (1/75 \times 10^9)\cdot(\sigma_x - 0.33.\sigma_y),$$

or $\sigma_x - 0.33\cdot\sigma_y = 25.863$ MPa (a)

Also, $\varepsilon_y = 200 \times 10^{-6} = (1/75 \times 10^9)\cdot(\sigma_y - 0.33.\sigma_x)$,

or $\sigma_y - 0.33\cdot\sigma_x = 15.0$ MPa (b)

From (a) and (b), $\sigma_x = 34.6$ MPa, $\sigma_y = 26.4$ MPa **Ans.**

Change in thickness: $\Delta t = t \cdot \varepsilon_z = t[(-\nu/E)\cdot(\sigma_x + \sigma_y)$

$= -\{10\cdot(0.33)/(75 \times 10^9)\}\cdot(34.6 + 26.4)\cdot 10^6$ mm

$= -2684 \times 10^{-6}$ mm

Reduction in thickness = 2684×10^{-6} mm **Ans.**

Example 5.5: A 100 mm × 100 mm × 10 mm aluminium plate is subjected to uniform plane stress in *x-y* plane, as shown in Fig. 5.4. The strain gauges *A* and *B* mounted on the surface of the plate read 450×10^{-6} and -300×10^{-6}, respectively. Determine (a) the normal stresses σ_x and σ_y, and (b) the change in volume of the plate. Take E = 75 GPa and ν = 0.33.

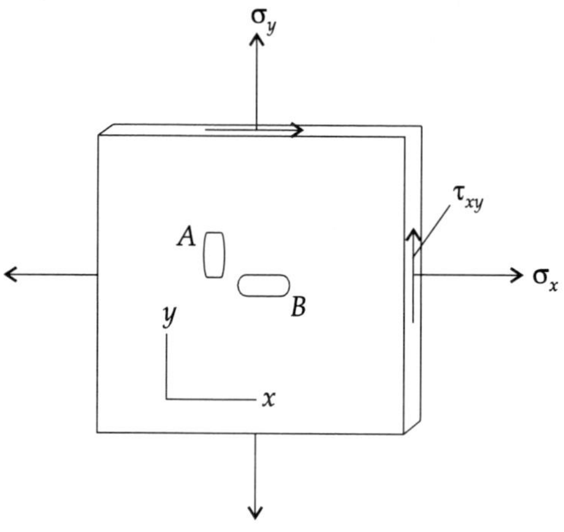

Fig. 5.4

Solution: (a) $\varepsilon_A = 450 \times 10^{-6} = \varepsilon_x \cdot \cos^2 90° + \varepsilon_y \cdot \sin^2 90° + \gamma_{xy} \cdot \sin 90° \cdot \cos 90°$,

$\varepsilon_y = 450 \times 10^{-6}$

$\varepsilon_B = -300 \times 10^{-6} = \varepsilon_x \cdot \cos^2 0° + \varepsilon_y \cdot \sin^2 0° + \gamma_{xy} \cdot \sin 0° \cdot \cos 0°$,

$\varepsilon_x = -300 \times 10^{-6}$

Hence, using Eqn. 5.12,

$\sigma_x = [75 \times 10^9/\{1-(0.33)^2\}]\cdot\{(-300) + 0.33\cdot(450)\}\cdot 10^{-6}$ N/m²

or σ_x = 12.8 MPa (c) **Ans.**

$\sigma_y = [75 \times 10^9/\{1-(0.33)^2\}]\cdot\{(450) + 0.33\cdot(-300)\}\cdot 10^{-6}$ M/m²

or σ_y = 29.5 MPa (T) **Ans.**

(b) Volume strain : If *a*, *b* and *c* are the length of sides of a parallelepiped,

Volume, $V = a \cdot b \cdot c$

Because of strains ε_x, ε_y and ε_z, the changed volume is V' and

$V' = [a\cdot(1 + \varepsilon_x)]\cdot[b\cdot(1 + \varepsilon_y)]\cdot[c\cdot(1 + \varepsilon_z)] = a\cdot b\cdot c.(1 + \varepsilon_x + \varepsilon_y + \varepsilon_z) = V\cdot(1 + \varepsilon_x + \varepsilon_y + \varepsilon_z)$.

(since the strain components are small, their products are ignored)

$\therefore$ Volume strain, $\varepsilon_V = (V' - V)/V = \varepsilon_x + \varepsilon_y + \varepsilon_z$.

But, $\varepsilon_z = -\{\nu/(1-\nu)\}\cdot(\varepsilon_x + \varepsilon_y)$, $\therefore$ $\varepsilon_V = \{(1-2\nu)/(1-\nu)\}\cdot(\varepsilon_x + \varepsilon_y)$.

Change in volume of the plate,

$\Delta V = V\cdot\varepsilon_V = (100)\cdot(100)\cdot(10)\cdot[\{(1 - 2.0.33)/(1-0.33)\}\cdot(-300 + 450)\cdot 10^{-6}]$ mm^3

or $\Delta V = 7.6$ mm^3 (increase) **Ans.**

Example 5.6: A steel plate is subjected to biaxial stresses (σ_x and σ_y only), as shown in Fig. 5.5. There is a strain gauge fixed to the surface at an angle $\phi = 45°$ with respect to x-axis. If $\sigma_x = 130$ MPa and reading in strain gauge is $\varepsilon = 500 \times 10^{-6}$, what is *t* he maximum shear stress (τ_{max}) and maximum shear strain (γ_{max}) in the plane x-y ? What are maximum shear strains in x-z plane and y-z plane ?

Take $E = 200$ GPa and $\nu = 0.30$ for steel.

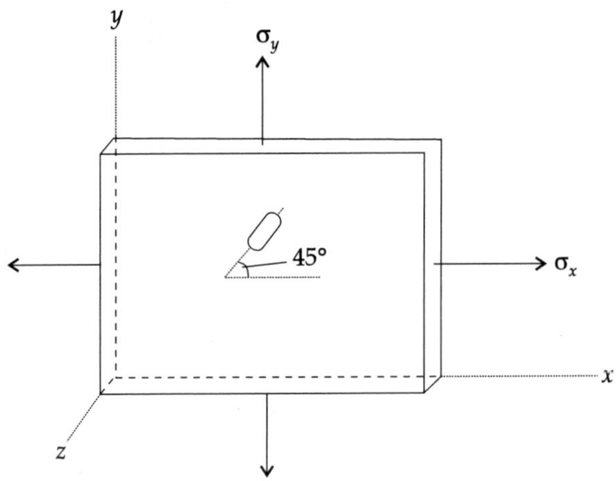

Fig. 5.5

Solution: For the plane stress condition, $\varepsilon_x = (1/E)\cdot(\sigma_x - \nu\cdot\sigma_y)$, and $\varepsilon_y = (1/E)\cdot(\sigma_y - \nu\cdot\sigma_x)$

The strain measured in the gauge, $\varepsilon = \varepsilon_x\cdot\cos^2\phi + \varepsilon_y\cdot\sin^2\phi$. (since $\gamma_{xy} = 0$)

or $\varepsilon = (1/E)\cdot[(\sigma_x - \nu\cdot\sigma_y)\cdot\cos^2\phi + (\sigma_y - \nu\cdot\sigma_x)\cdot\sin^2\phi]$

$= (1/E)\cdot[\sigma_x\cdot(\cos^2 45° - 0.3\cdot\sin^2 45°) + \sigma_y\cdot(\sin^2 45° - 0.3\cdot\cos^2 45°)]$

or $\varepsilon = (1/E)\cdot(0.35\cdot\sigma_x + 0.35\cdot\sigma_y)$,

$\therefore$ $500 \times 10^{-6} = \{0.35/(200 \times 10^9)\}\cdot(130 + \sigma_y)\cdot 10^6$. $\sigma_y = 155.7$ MPa.

As σ_x and σ_y are principal stresses, the maximum shear stress on x-y plane,

$\tau_{max} = (\sigma_x - \sigma_y)/2 = (130 - 155.7)/2 = -12.85$. or $\tau_{max} = 12.85$ MPa **Ans.**

The principal strains are obtained as follows :

$\varepsilon_x = \{1/(200 \times 10^9)\}\cdot[130 - (0.3 \times 155.7)]\cdot 10^6 = 416.45 \times 10^{-6}$,

$\varepsilon_y = \{1/(200 \times 10^9)\}\cdot[155.7 - (0.3 \times 130)]\cdot 10^6 = 583.50 \times 10^{-6}$.

(Check: $\varepsilon = 416.45 \times 10^{-6}\cdot\cos^2 45° + 583.50 \times 10^{-6}\cdot\sin^2 45° = 499.975 \times 10^{-6} \simeq 500 \times 10^{-6}$)

Maximum shear strain in x-y plane, $(\gamma_{max})_{x-y} = (416.45 - 583.50)\cdot 10^{-6}$

$= -167 \times 10^{-6}$ **Ans.**

Normal strain in z-direction, $\varepsilon_z = \{-0.3/(1 - 0.3)\}\cdot(416.45 + 583.50)\cdot 10^{-6}$

$= -428.55 \times 10^{-6}$.

$\therefore$ Maximum shear strain in x-z plane, $(\gamma_{max})_{x-z} = \varepsilon_x - \varepsilon_z = [416.45 - (-428.55)]\cdot 10^{-6}$

$= 845 \times 10^{-6}$ **Ans.**

Maximum shear strain in y-z plane, $(\gamma_{max})_{y-z} = \varepsilon_y - \varepsilon_z = [583.50 - (-428.55)]\cdot 10^{-6}$

$= 1012 \times 10^{-6}$ **Ans.**

5.5 RELATION BETWEEN ELASTIC CONSTANTS

Many equations pertaining to stress-strain relations include parameters such as elasticity modulus (E), shear modulus (G) and bulk modulus (K), which are known as elastic constants. However, a homogeneous and isotropic material is identified by two independent parameters. These are elasticity modulus (E) and Poisson's ratio (ν), known as fundamental elastic constants. It can be shown that the shear modulus (G) and bulk modulus (K) can be expressed in terms of E and ν only.

5.5.1 Relation between G and E

If a plane element in x-y plane is subjected to pure shear, a Mohr's circle is drawn using shear stress, $\tau_{xy} = \tau_s$, as in Fig. 5.6. From the Mohr's circle drawn against pure shear, it is found that

1. The centre of Mohr's circle coincides with the origin of $\sigma - \tau$ coordinate system and the radius of Mohr's circle is equal to the shear stress τ_s,
2. The two principal stresses, σ_1 and σ_2, and the maximum shear stress are all of equal magnitude (τ_s), and
3. Algebraic sum of normal stresses is zero for any inclination of plane element.

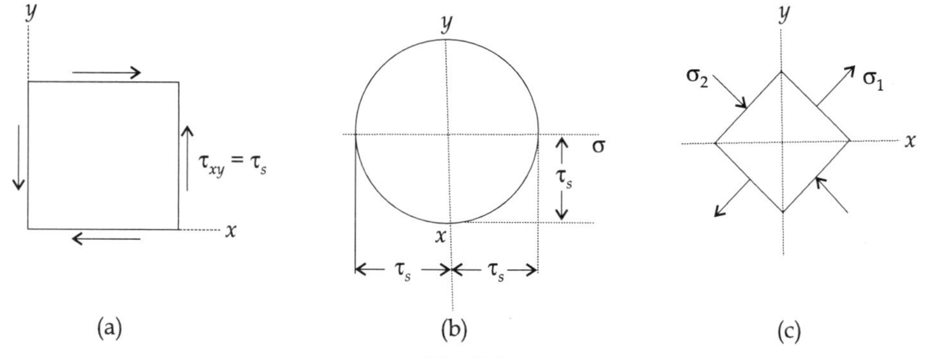

Fig. 5.6

Referring to Mohr's circle in Fig. 5.6(b), $\sigma_1 = \tau_s$, $\sigma_2 = -\tau_s$, so that the principal strains are

$$\varepsilon_1 = (1/E)\cdot(\sigma_1 - \nu\cdot\sigma_2) = (1/E)\cdot[\tau_s - \nu\cdot(-\tau s_s)], \quad \text{or } \varepsilon_1 = \{(1 + \nu)/E\}\cdot\tau_s.$$

Similarly, $\varepsilon_2 = (1/E)\cdot[(-\tau_s) - \nu\cdot\tau_s]$, or $\varepsilon_2 = -\{(1+\nu)/E\}\cdot\tau_s$.

$\therefore$ Maximum shear strain, $\gamma_{max} = \varepsilon_1 - \varepsilon_2 = [2\cdot(1+\nu)/E]\cdot\tau_s$ (5.22)

If G = modulus of rigidity or shear modulus, we have

$$\gamma_{max} = \tau_{max}/G = \tau_s/G. \quad (5.23)$$

Therefore, from Eqns. 5.22 and 5.23,

$\tau_s/G = [2\cdot(1+\nu)/E]\cdot\tau_s$, or $G = E/\{2\cdot(1+\nu)\}$. (5.24)

It means that shear modulus (G) is not independent but it can be obtained from known values of E and ν.

5.5.2 Relation between K and E

The bulk modulus (K) of a material is obtained when the average normal stress acting on a body made from this material is divided by the volume strain of the body. A small element in the shape of a parallelepiped of sides, measuring Δx, Δy and Δz, is shown in Fig. 5.7. It is subjected to normal compressive stress ($-p$) on all of its faces. This type of stress is called hydrostatic stress by drawing analogy with uniform pressure on the entire surface area of a body which is immersed deep in water.

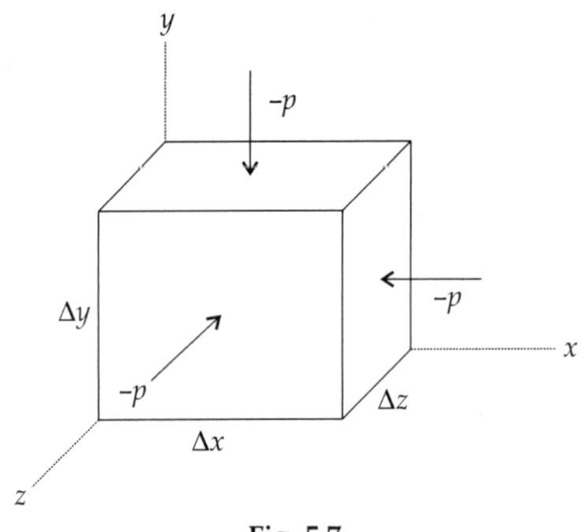

Fig. 5.7

In this case, the normal stresses are $\sigma_x = \sigma_y = \sigma_z = -p$.

The volume strain of the body is defined as the change in volume divided by its original volume, and it is written as $\delta V/V$, where, δV is the change in volume and V is the original volume. Hence, the volume strain is obtained as follows :

As $V = \Delta x\cdot\Delta y\cdot\Delta z$, $\ln V = \ln\Delta x + \ln\Delta y + \ln\Delta z$,

where, ln denotes Napierian logarithm.

From differentials on both sides of the above, $\delta V/V = \delta(\Delta x)/\Delta x + \delta(\Delta y)/\Delta y + \delta(\Delta z)/\Delta z$.

$\therefore \quad \varepsilon_V = \varepsilon_x + \varepsilon_y + \varepsilon_z$.

But, $\varepsilon_x = (1/E)\cdot[-p - \nu(-p-p)]$ (from Eqn. 5.1). or $\varepsilon_x = -\{p\cdot(1-2\nu)\}/E$.

Similarly, $\varepsilon_y = \varepsilon_z = -\{p\cdot(1-2\nu)\}/E$.

Therefore, $\varepsilon_V = -\{3p\cdot(1-2\nu)\}/E$.

According to the definition of bulk modulus, $K = -p/\varepsilon_V = [-p/\{-3p\cdot(1-2\nu)\}]\cdot E$.

or $$K = E/\{3\cdot(1-2\nu)\}, \quad (5.25)$$

which indicates that the bulk modulus depends on E and ν.

It can be checked in Eqn. 5.25 that the value of K becomes equal to infinity as $\nu \rightarrow 0.5$, which is valid only for a perfectly rigid body. Hence, for all the isotropic engineering materials, the value of $\nu < 0.5$.

Five elastic constants λ, E, ν, G and K have been discussed so far. Lame's constant (λ) has no physical significance and it is applied only for mathematical simplicity in stress-strain relations. The constant (G) has both mathematical and physical significance. And it is used extensively in torsion problems. The constant (K) is used to find the volume change in a body that is subjected to hydrostatic stresses. Elastic constants, E and ν, are widely used in almost all areas of stress analyses.

Example 5.7: An element of a material is shown in Fig. 5.8. The stresses acting on it are σ_x = 35 MPa (tension), σ_y = 32 MPa (compression) and σ_z = 20 MPa (compression). Also, the normal strains are found as $\varepsilon_x = 710 \times 10^{-6}$ (elongation) and $\varepsilon_y = -500 \times 10^{-6}$ (contraction). What is the bulk modulus (K) for the material?

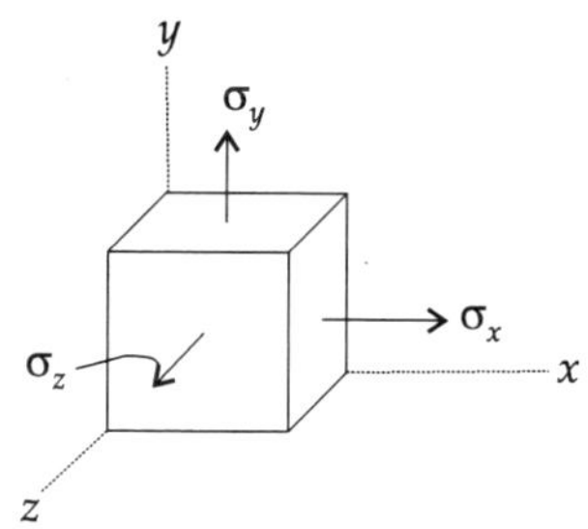

Fig. 5.8

Solution: From Eqns. 5.1 and 5.2,

$$710 \times 10^{-6} = (1/E)\cdot[35 - \nu\cdot(-32-20)],\ 10^6 \text{ or } 35 + 52.\nu = 710 \times 10^{-12}\cdot E. \quad \text{(a)}$$

Also, $-500 \times 10^{-6} = (1/E)\cdot[-32 - \nu\cdot(35-20)].\ 10^6$ or $32 + 15.\nu = 500 \times 10^{-12}\cdot E$. (b)

Eliminating E from (a) and (b), $\nu = 0.34$ is obtained.

Therefore, in (a), $35 + (52 \times 0.34) = 710 \times 10^{-12}\cdot E$.

or $E = 74.197 \times 10^9$ N/m². $E = 74.197$ GPa .

$\therefore$ Bulk modulus $K = E/\{3\cdot(1-2\nu)\} = 74.197/\{3\cdot(1-2\times 0.34)\}$

$= 77.288$ GPa or $K = 77.3$ GPa **Ans.**

An alternative solution is:

$$\sigma_{av} = (\sigma_x + \sigma_y + \sigma_z)/3 = [\{35 + (-32) + (-20)\}/3]\cdot 10^6 = -(17/3)\cdot 10^6 \text{ N/m}^2$$

$$\varepsilon_z = (1/74.197 \times 10^9)\cdot[-20 - 0.34\cdot(35-32)]\cdot 10^6.\ \therefore\ \varepsilon_z = -283.299 \times 10^{-6}.$$

Volume strain, $\varepsilon_V = \varepsilon_x + \varepsilon_y + \varepsilon_z$. or $\varepsilon_V = [710 + (-500) + (-283.299)]\cdot 10^{-6}$.

$$\varepsilon_V = -73.299 \times 10^{-6}.$$

Bulk modulus, $K = \sigma_{av}/\varepsilon_V$, or $K = (17/3) \times 10^6/(73.299 \times 10^{-6})$

$$= 77.3 \times 10^9 \text{ N/m}^2$$

$\therefore$ $K = 77.3$ GPa **Ans.**

Example 5.8: A rubber cylinder of length L and cross-sectional area A is put inside a steel cylinder and compressed by a force P such that a uniformly distributed pressure acts at the top of rubber cylinder. This is shown by a sectional view in Fig. 5.9.

(a) Find the expression of lateral pressure p between rubber and steel. Ignore friction between rubber and steel and assume the steel cylinder to be rigid compared to the rubber cylinder.

(b) Find the shortening of rubber cylinder (δ).

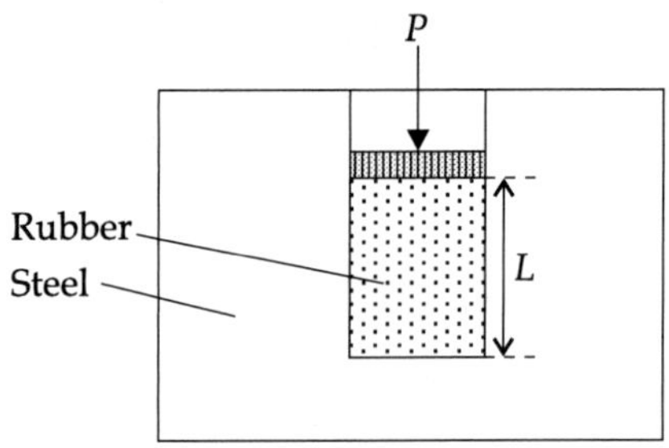

Fig. 5.9

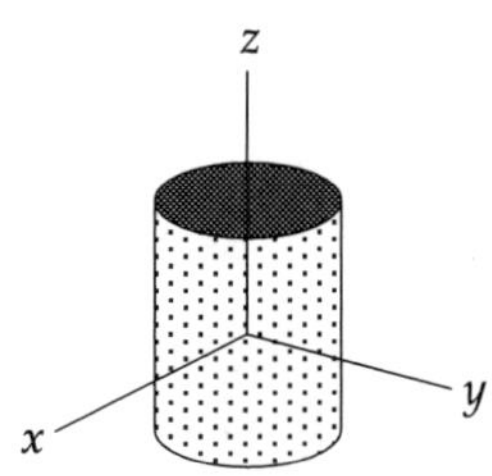

Solution: Assuming the axis of rubber cylinder coinciding with z-axis, $\sigma_z = -P/A$, which is a compressive stress. It should be noted that the rubber cylinder can deform along z-axis, but its deformations along x- and y-directions are restricted.

Hence, $\varepsilon_z \neq 0$, but $\varepsilon_x = \varepsilon_y = 0$.

Also rubber cylinder can deform symmetrically about z-axis, so that the stresses developed in x- and y-directions are equal in magnitude, i.e., $\sigma_x = \sigma_y = -p$ (assumed).

(a) For $\varepsilon_x = 0$, $(1/E)\cdot[\sigma_x - \nu\cdot(\sigma_y + \sigma_z)] = 0$, or $\sigma_x - \nu\cdot\sigma_y - \nu\cdot\sigma_z = 0$.

$\therefore -p - \nu\cdot(-p) - \nu\cdot(-P/A) = 0$, Therefore, $p = [\nu/(1-\nu)]\cdot(P/A)$ **Ans.**

(b) $\varepsilon_z = (1/E)\cdot[\sigma_z - \nu\cdot(\sigma_x + \sigma_y)] = (1/E)\cdot[(-P/A) - \nu\cdot\{(-2\nu P/A)/(1-\nu)\}] = \{-P/(A\cdot E)\}\cdot[1 - \{2\nu^2/(1-\nu)\}]$, which can be simplified to obtain

$\varepsilon_z = -\{P/(A\cdot E)\}\cdot[\{(1+\nu)\cdot(1-2\nu)\}/(1-\nu)]$

$\therefore$ Change in length of rubber cylinder, $\Delta L = L\cdot\varepsilon_z = -\{(P\cdot L)/(A\cdot E)\}\cdot[\{(1+\nu)\cdot(1-2\nu)\}/(1-\nu)]$, which is shortening of rubber cylinder. **Ans.**

Example 5.9: A rubber block is placed between the parallel walls of a steel block (Fig. 5.10) and a force (F) is applied at the top of the rubber block to produce a uniformly distributed pressure (p_0) on it. Determine the pressure between the rubber block and the walls of steel block. If V is the initial volume of the rubber block, what is the change in volume, δV? Neglect any friction between the rubber and steel. The steel can be assumed to be rigid relative to rubber.

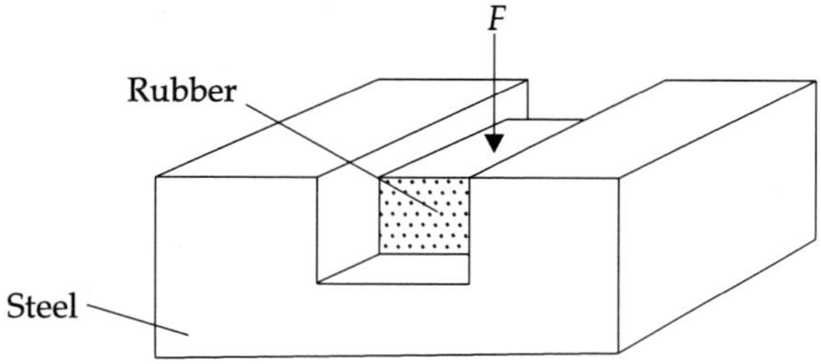

Fig. 5.10

Solution: A coordinate system, x-y-z, is attached with the rubber block. Note that the rubber block is confined within steel walls and cannot expand horizontally in x-direction. Due to this constraint, a stress (σ_x) develops along x-axis. The rubber block can expand freely along z-axis and there is no normal stress along z-direction because of the absence of any constraint. There is a uniform pressure from y-direction and the block can deform along y-axis.

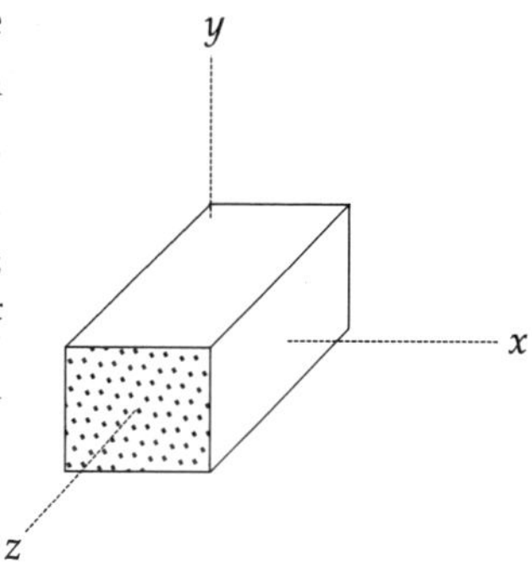

Therefore, we have

$$\sigma_x \neq 0, \quad \varepsilon_x = 0,$$
$$\sigma_y = -p_0, \quad \varepsilon_y \neq 0,$$
$$\sigma_z = 0, \quad \varepsilon_z \neq 0.$$

$\therefore \quad \varepsilon_x = (1/E)\cdot[\sigma_x - \nu\cdot(\sigma_y + \sigma_z)] = (1/E)\cdot[\sigma_x - \nu\cdot(-p_0 + 0)] = (1/E)\cdot[\sigma_x + \nu\cdot p_0)]$.

As $\quad \varepsilon_x = 0, \sigma_x = -\nu\cdot p_0$. Pressure between rubber and steel = $\nu\cdot p_0$ **Ans.**

Next, $\varepsilon_y = (1/E)\cdot[\sigma_y - \nu(\sigma_x + \sigma_z)] = (1/E)\cdot[(-p_0) - \nu\cdot\{(-\nu\cdot p_0) + 0\}] = -(p_0/E)\cdot(1 - \nu^2)$.

Also, $\varepsilon_z = (1/E)\cdot[\sigma_z - \nu\cdot(\sigma_x + \sigma_y)] = (1/E)\cdot[0 - \nu\cdot\{(-\nu\cdot p_0) + (-p_0)\}] = (\nu\cdot p_0/E)\cdot(1 + \nu)$.

Hence, the volume strain $\varepsilon_V = 0 + \{-(p_0/E)\cdot(1 - \nu^2)\} + (\nu p_0/E)\cdot(1 + \nu)$,

Which after simplification becomes $\varepsilon_V = -(p_0/E)\cdot(1 + \nu)\cdot(1 - 2\nu)$.

$\therefore$ Change in volume, $\delta V = -(p_0\cdot V/E)\cdot(1 + \nu)\cdot(1 - 2\nu)$ **Ans.**

Example 5.10: A solid aluminium sphere, 25 cm in diameter, is lowered at a depth of 3 km into ocean. Determine the decrease (δD) in diameter and the decrease (δV) in volume of the sphere.

Take E = 75 GPa and ν = 0.33 for aluminium.

Solution: Bulk modulus, $K = E/\{3\cdot(1 - 2\nu)\}$, or $K = 75/\{3\cdot(1 - 2 \times 0.33)\} = 73.5294$ GPa.

Hydrostatic pressure: $p = \rho\cdot g\cdot h = (1000\ \text{kg/m}^3)\cdot(9.81\ \text{m/s}^2)\cdot(3000\ \text{m})$,

or $p = 29.43 \times 10^6\ \text{N/m}^2$.

$\therefore$ Volume strain, $\varepsilon_V = -p/K$, or $\varepsilon_V = -(29.43 \times 10^6)/(73.5294 \times 10^9) = -0.0004$.

Volume of sphere, $V = (4/3)\cdot\pi\cdot(25/2)^3 = 8181.23$ cm^3 .

Change in volume, $\delta V = V\cdot\varepsilon_V$. $\delta V = (8181.23)\cdot(-0.0004)$ cm^3 = -3.27 cm^3 (decrease)

Ans.

Change in diameter, $\delta D = (1/3)\cdot D\cdot\varepsilon_V\cdot$ or $\delta D = (1/3)\cdot(25)\cdot(-0.0004)$ cm

$= -0.0033$ cm (decrease) **Ans.**

Example 5.11: A solid steel cube of 25 cm side is subjected to uniformly distributed pressure of 210 MPa on its four faces with their outward normal pointing to *x*- and *y*-directions. Because of rigid constraints, the deformation of cube in *z*-direction is restricted to 0.005 cm. Determine the normal stress (σ_z) that develops as the pressure is applied. Take $E = 210$ GPa and $\nu = 0.30$ for steel.

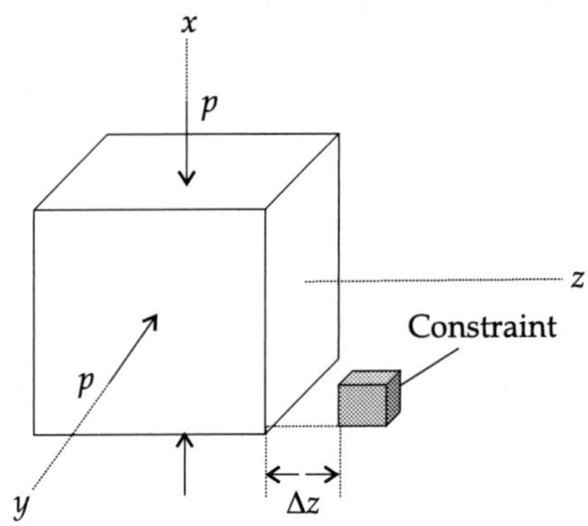

Fig. 5.11

Solution: Length of each side, $x = y = z = 25$ cm, full pressure = 210 MPa, $\Delta z = 0.005$ cm, $E = 210$ GPa, $\nu = 0.3$.

Let us assume that the pressure is gradually increased from zero to a certain value (p') till there is free expansion of cube in *z*-direction up to an amount $\Delta z = 0.005$ cm. The stress, $\sigma_z = 0$ till the expansion occurs up to the constraint.

Strain, $\varepsilon_z = \Delta z/z = 0.005\ /25 = 2 \times 10^{-4}$.

$\therefore$ at the stage when the constraint (obstacle) is just reached,

$2 \times 10^{-4} = \{1/(210 \times 10^9)\}\cdot[0 - 0.3(-p' - p')]$, (as $\sigma_z = 0$, $\sigma_x = \sigma_y = -p'$). or $0.6\text{x}p' = 42 \times 10^6$ N/m^2

$\therefore p' = 70$ MPa.

After this, $\varepsilon = 0$ and applied pressure further increases by $p - p' = 210 - 70$

$= 140$ MPa.

Because of constraint, σ_z develops as the pressure attains its full value.

Hence, $\varepsilon_z = 0 = \{1/(210 \times 10^9)\}\cdot[\sigma_z - 0.3\cdot(-140 - 140)]$, or $\sigma_z = -84$ MPa.

$\therefore \sigma_z = 84$ MPa (C) **Ans.**

Example 5.12: A 10 mm × 10 mm block of synthetic rubber is 25.4 mm long, and it is pushed inside a 10 mm × 10 mm × 25 mm hole in steel, as shown in Fig. 5.12. The hole is well lubricated to avoid any friction between the rubber and steel. A force (P) is applied at the top of rubber block through a rigid washer, as shown. Determine the value of P required to push the block into the hole so that the top surface of rubber is in same horizontal level with the top surface of steel. Take E = 1.50 GPa and ν = 0.42 for synthetic rubber.

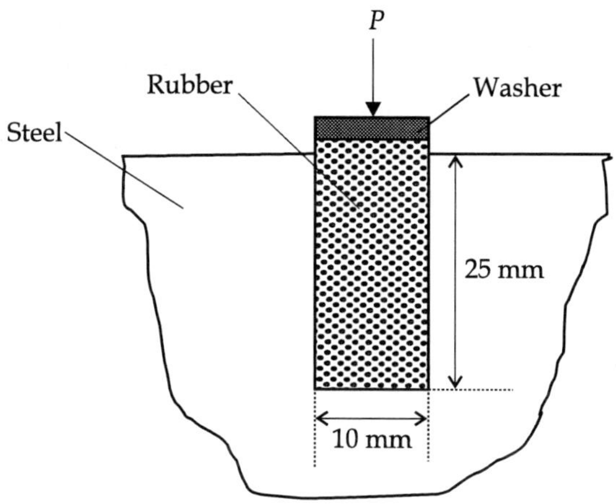

Fig. 5.12

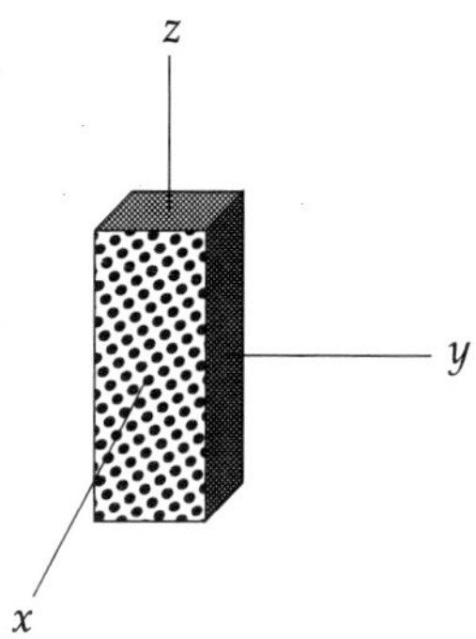

Solution: The coordinates (x-y-z) are attached with the rubber block.

$$\sigma_z = -P/A = -p_0 \,.$$

Since there is symmetry about z-axis, $\sigma_x = \sigma_y$. Also, due to constraints in x- and y-directions, $\varepsilon_x = 0$, $\varepsilon_y = 0$.

As $\varepsilon_x = 0$, $(1/E)\cdot[\sigma_x - \nu\cdot\sigma_y - \nu\cdot\sigma_z] = 0$

or $(1/E)\cdot[\sigma_x - \nu\cdot\sigma_x - \nu\cdot(-p_0)] = 0$

$\therefore$ $\sigma_x = \sigma_y = -\{\nu/(1-\nu)\}\cdot p_0.$

Strain in z-direction, $\varepsilon_z = -(0.4/25.4) = -0.015748.$ ...(a)

But $\varepsilon_z = (1/E)\cdot[\sigma_z - \nu\cdot(\sigma_x + \sigma_y)] = (1/E)\cdot[(-p_0) - \nu\cdot\{2\cdot(-\nu p_0/1-\nu)\}]$

or $\varepsilon_z = -(p_0/E)\cdot[(1+\nu)\cdot(1-2\nu)/(1-\nu)].$ ...(b)

Therefore, from (a) and (b), $-0.015748 = -(p_0/1.5\times10^9)\cdot[(1+0.42)\cdot(1-2\times0.42)/(1-0.42)]$

After simplification, $p_0 = 0.0603\times10^9$ N/m^2.

$\therefore$ Force, $P = A\cdot p_0 = \{(10\times10)\times10^{-6}\}\cdot(0.0603\times10^9)$ N = 6030 N or P = 6.03 kN **Ans.**

5.6 GENERALIZED HOOKE'S LAW

It was stated that the stress-strain relations for engineering materials are obtained by assuming isotropic behaviour of these materials. For an isotropic material, the elasticity modulus is same in every direction within the material. There is single value of Poisson's ratio and the normal stresses are in no way dependent on shear strains. However, such limitations are not valid for the materials which are non-isotropic or anisotropic in nature. For the last five decades, a large number of materials such as fibrous and particulate composites have been developed in laboratories which found wide applications in space technology, aircraft industries and even in the area of sports and games. Such materials are anisotropic in the sense that their elasticity modulus values are different in different directions, there are more than one Poisson's ratio and in certain cases the normal stresses are functions of shear strains also.

For a generally anisotropic material, the six stress components and six strain components are related according to Hooke's law written in the form of the following matrix:

$$\begin{Bmatrix} \sigma_x \\ \sigma_y \\ \sigma_z \\ \tau_{xy} \\ \tau_{xz} \\ \tau_{yz} \end{Bmatrix} = \begin{pmatrix} K_{11} & K_{12} & K_{13} & K_{14} & K_{15} & K_{16} \\ K_{21} & K_{22} & K_{23} & K_{24} & K_{25} & K_{26} \\ K_{31} & K_{32} & K_{33} & K_{34} & K_{35} & K_{36} \\ K_{41} & K_{42} & K_{43} & K_{44} & K_{45} & K_{46} \\ K_{51} & K_{52} & K_{53} & K_{54} & K_{55} & K_{56} \\ K_{61} & K_{62} & K_{63} & K_{64} & K_{65} & K_{66} \end{pmatrix} \cdot \begin{Bmatrix} \varepsilon_x \\ \varepsilon_y \\ \varepsilon_z \\ \gamma_{xy} \\ \gamma_{xz} \\ \gamma_{yz} \end{Bmatrix} \tag{5.26}$$

The set of equations given above are known as **constitutive equations**, which relate stresses and strains. In the above matrix form $\{\sigma\} = [K]\cdot\{\varepsilon\}$, the matrix $[K]$ is known as **stiffness matrix**. An alternative to above is written in the form $\{\varepsilon\} = [C]\cdot\{\sigma\}$, where the matrix $[C]$ having same number of elements as $[K]$ is known as **compliance matrix**.

From Eqn. 5.26, the normal stress σ_x along x-direction will be written as

$$\sigma_x = K_{11}\cdot\varepsilon_x + K_{12}\cdot\varepsilon_y + K_{13}\cdot\varepsilon_z + K_{14}\cdot\gamma_{xy} + K_{15}\cdot\gamma_{xz} + K_{16}\cdot\gamma_{yz}\,. \tag{5.27}$$

It is found from the above equation that the normal stress is not only related to normal strains along and perpendicular to x-axis, but also to shear strains on the three mutually perpendicular planes. This is in contradiction with experimentally observed behaviour of common engineering materials. In fact, Eqn. 5.26 represents a set of constitutive equations for anisotropic material and $[K]$ matrix consists of 36 elements or elastic constants. A material having same composition throughout is called a homogeneous material. Therefore, a linear elastic, homogeneous and anisotropic body has 36 elastic constants to establish stress-strain relations below the proportional limit. However, unlike general anisotropic material, most of materials used in engineering applications have mechanical properties (stress-strain relations) that remain same in more than one direction. Orthotropic and isotropic materials fall under such categories.

It can be shown that for isotropic materials the 36 elastic constants in Eqn. 5.26 degenerate into only two elastic constants. For this, x-y-z axes can be suitably rotated to new positions x'-y'-z' and three-dimensional stress and strain transformation rules are applied in definite steps as explained below.

The stresses in x'-y'-z' coordinates are obtained from the stresses in x-y-z coordinates by transformation rules written in abbreviated form as

$$\sigma_{mn} = \sigma_{ij} \cdot \cos(m, i) \cdot \cos(n, j);\ m, n = x', y', z' \text{ and } i, j = x, y, z \tag{5.28}$$

where, $\cos(m, i)$ stands for cosine of angle between m-axis and i-axis.

Also, $\sigma_{mn} = \sigma_m$ or σ_n, if $m = n$, $\sigma_{ij} = \sigma_i$ or σ_j, if $i = j$,

and $\sigma_{mn} = \tau_{mn}$ if $m \neq n$, $\sigma_{ij} = \tau_{ij}$ if $i \neq j$.

Equation 5.28 can be used to write

$$\begin{aligned}\sigma_{x'} = {} & \sigma_x \cdot \cos(x', x) \cdot \cos(x', x) + \sigma_y \cdot \cos(x', y) \cdot \cos(x', y) + \sigma_z \cdot \cos(x', z) \cdot \cos(x', z) + \tau_{xy} \cdot \\ & \cos(x', x) \cdot \cos(x', y) + \tau_{yx} \cdot \cos(x', y) \cdot \cos(x', x) + \tau_{yz} \cdot \cos(x', x) \cdot \cos(x', z) + \tau_{zx} \cdot \\ & \cos(x', z) \cdot \cos(x', x) + \tau_{yz} \cdot \cos(x', y) \cdot \cos(x', z) + \tau_{zy} \cdot \cos(x', z) \cdot \cos(x', y).\end{aligned}$$

Since $\tau_{ij} = \tau_{ji}$, $i, j = x, y, z$, the above expression is simplified as

$$\begin{aligned}\sigma_{x'} = {} & \sigma_x \cdot \cos^2(x', x) + \sigma_y \cdot \cos^2(x', y) + \sigma_z \cdot \cos^2(x', z) + 2 \cdot \tau_{xy} \cdot \cos(x', x) \cdot \cos(x', y) + 2 \cdot \tau_{xz} \cdot \\ & \cos(x', x) \cdot \cos(x' \cdot z) + 2 \cdot \tau_{yz} \cdot \cos(x', y) \cdot \cos(x', z).\end{aligned} \tag{5.29}$$

Similarly, we have

$$\begin{aligned}\tau_{x'y'} = {} & \sigma_x \cdot \cos(x', x) \cdot \cos(y', x) + \sigma_y \cdot \cos(x', y) \cdot \cos(y', y) + \sigma_z \cdot \cos(x', z) \cdot \cos(y', z) \\ & + \tau_{xy} \cdot [\cos(x', x) \cdot \cos(y', y) + \cos(x', y) \cdot \cos(y', x)] \\ & + \tau_{xz} \cdot [\cos(x', x) \cdot \cos(y', z) + \cos(x', z) \cdot \cos(y', x)] \\ & + \tau_{yz} \cdot [\cos(x', y) \cdot \cos(y', z) + \cos(x', z) \cdot \cos(y', y)].\end{aligned} \tag{5.30}$$

A similar transformation rule applies for strains also.

In order to modify Eqn. 5.26 and to obtain Hooke's law for orthotropic material, the following steps are considered.

Step 1: The coordinate system (x-y-z) is rotated about z-axis through 180° and a new set of axes (x'-y'-z') is found , as shown in Fig. 5.13. Cosine of angle between respective axes are given in Table 5.1.

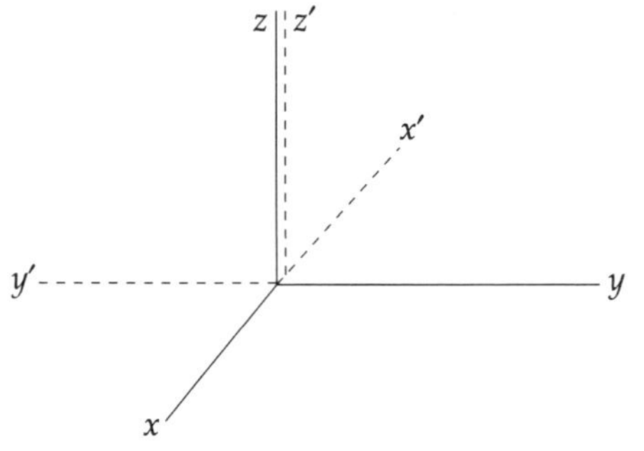

Fig. 5.13

Table 5.1

	x'	y'	z'
x	-1	0	0
y	0	-1	0
z	0	0	1

Some examples of cosine values in Eqn. 5.28, found from Table 5.1 are:

$\cos(x', x) = \cos(x, x') = -1$, $\cos(x, z') = \cos(z', x) = 0$, $\cos(y, z') = \cos(z', y) = 0$, $\cos(z, z') = \cos(z', z) = 1$, etc.

Therefore, referring to Fig. 5.13 and Table 5.1, the stress transformation rules from Eqn. 5.28 gives $\sigma_{x'} = \sigma_x$, $\sigma_{y'} = \sigma_y$, $\sigma_{z'} = \sigma_z$, $\tau_{x'y'} = \tau_{xy}$, $\tau_{x'z'} = -\tau_{xz}$, and $\tau_{y'z'} = -\tau_{yz}$. (5.31)

Similarly, the strain transformation gives

$$\varepsilon_{x'} = \varepsilon_x,\ \varepsilon_{y'} = \varepsilon_y,\ \varepsilon_{z'} = \varepsilon_z,\ \gamma_{x'y'} = \gamma_{xy},\ \gamma_{x'z'} = -\gamma_{xz},\ \text{and } \gamma_{y'z'} = \gamma_{yz}. \quad (5.32)$$

In the new coordinates (x'-y'-z') the stress, $\sigma_{x'}$, is written as

$$\sigma_{x'} = K_{11}\cdot\varepsilon_{x'} + K_{12}\cdot\varepsilon_{y'} + K_{13}\cdot\varepsilon_{z'} + K_{14}\cdot\gamma_{x'y'} + K_{15'}\gamma_{x'z'} + K_{16}\cdot\gamma_{y'z'},$$

which can be written with the help of Eqns. 5.31 and 5.32 as

$$\sigma_x = K_{11}\cdot\varepsilon_x + K_{12}\cdot\varepsilon_y + K_{13}\varepsilon_z + K_{14}\cdot\gamma_{xy} - K_{15}\cdot\gamma_{xz} - K_{16}\cdot\gamma_{yz}. \quad (5.33)$$

From Eqns. 5.27 and 5.33, we have

$$2\cdot K_{15}\cdot\gamma_{xz} + 2\cdot K_{16}\cdot\gamma_{yz} = 0$$

As γ_{xz} and γ_{yz}, in general, are not zero, $K_{15} = K_{16} = 0$.

Similarly, with $\sigma_{y'} = \sigma_y$, $\sigma_{z'} = \sigma_z$, and $\tau_{x'y'} = \tau_{xy}$, it is found that $K_{25} = K_{26} = 0$, $K_{35} = K_{36} = 0$ and $K_{45} = K_{46} = 0$.

Next, from $\tau_{x'z'} = -\tau_{xz}$, and $\tau_{y'z'} = -\tau_{yz}$, it follows that $K_{51} = K_{52} = K_{53} = K_{54} = 0$ and $K_{61} = K_{62} = K_{63} = K_{64} = 0$.

Thus, after completing Step 1 the stiffness matrix of Eqn. 5.26 reduces to

$$[K] = \begin{pmatrix} K_{11} & K_{12} & K_{13} & K_{14} & 0 & 0 \\ K_{21} & K_{22} & K_{23} & K_{24} & 0 & 0 \\ K_{31} & K_{32} & K_{33} & K_{34} & 0 & 0 \\ K_{41} & K_{42} & K_{43} & K_{44} & 0 & 0 \\ 0 & 0 & 0 & 0 & K_{55} & K_{56} \\ 0 & 0 & 0 & 0 & K_{65} & K_{66} \end{pmatrix} \quad (5.34)$$

Step 2: Assuming that the material behaviour remains same as the conditions imposed in Step 1, the coordinate system (x'-y'-z') is now rotated about y'-axis through 180°, as shown in Fig. 5.14.

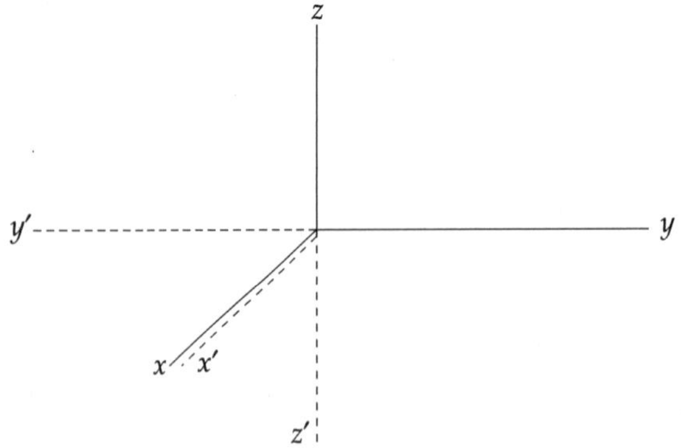

Fig. 5.14

Table 5.2

	x'	y'	z'
x	1	0	0
y	0	-1	0
z	0	0	-1

Referring to Fig. 5.14, cosines of angles between the axes are given in Table 5.2. The results of transformation of stresses and strains are:

$\sigma_{x'} = \sigma_x,\ \ \sigma_{y'} = \sigma_y,\ \ \sigma_{z'} = \sigma_z,\ \ \tau_{x'y'} = -\tau_{xy},\ \ \tau_{x'z'} = -\tau_{xz},\ \ \tau_{y'z'} = \tau_{yz}$, and

$\varepsilon_{x'} = \varepsilon_x,\ \ \varepsilon_{y'} = \varepsilon_y,\ \ \varepsilon_{z'} = \varepsilon_z,\ \ \gamma_{x'y'} = -\gamma_{xy},\ \ \gamma_{x'z'} = -\gamma_{xz},\ \ \gamma_{y'z'} = \gamma_{yz}$.

Hence, we have $\sigma_{x'} = K_{11}\cdot\varepsilon_{x'} + K_{12}\cdot\varepsilon_{y'} + K_{13}\cdot\varepsilon_{z'} + K_{14}\cdot\gamma_{x'y'}$

or
$$\sigma_x = K_{11}\cdot\varepsilon_x + K_{12}\cdot\varepsilon_y + K_{13}\cdot\varepsilon_z - K_{14}\cdot\gamma_{xy}. \tag{5.35}$$

But, by using $[K]$ matrix of Eqn. 5.34, σ_x is written as

$$\sigma_x = K_{11}\cdot\varepsilon_x + K_{12}\cdot\varepsilon_y + K_{13}\cdot\varepsilon_z + K_{14}\cdot\gamma_{xy}. \tag{5.36}$$

On comparing Eqn. 5.35 with Eqn. 5.36, $K_{14} = 0$.

In the same way, from the expressions of $\sigma_{y'}$, $\sigma_{z'}$, $\tau_{x'y'}$, $\tau_{x'z'}$ and $\tau_{y'z'}$ it is found that $K_{24} = 0$, $K_{34} = 0$, and $K_{41} = K_{42} = K_{43} = K_{56} = K_{65} = 0$.

Therefore, the $[K]$ matrix of Eqn. 5.34 is further reduced, so that the stress-strain relations can be expressed in the form

$$\begin{Bmatrix} \sigma_x \\ \sigma_y \\ \sigma_z \\ \tau_{xy} \\ \tau_{xz} \\ \tau_{yz} \end{Bmatrix} = \begin{pmatrix} K_{11} & K_{12} & K_{13} & 0 & 0 & 0 \\ K_{21} & K_{22} & K_{23} & 0 & 0 & 0 \\ K_{31} & K_{32} & K_{33} & 0 & 0 & 0 \\ 0 & 0 & 0 & K_{44} & 0 & 0 \\ 0 & 0 & 0 & 0 & K_{55} & 0 \\ 0 & 0 & 0 & 0 & 0 & K_{66} \end{pmatrix} \cdot \begin{Bmatrix} \varepsilon_x \\ \varepsilon_y \\ \varepsilon_z \\ \gamma_{xy} \\ \gamma_{xz} \\ \gamma_{yz} \end{Bmatrix} \tag{5.37}$$

The matrix equation given above is Hooke's law for orthotropic material. Using Steps 1 and 2 the same stress-strain relation is established in three mutually perpendicular directions, which is characteristic to an orthotropic material.

Following the similar procedure explained under the two steps but by rotating the coordinate system through 90° about suitably chosen axes, it can be summarized for isotropic material that

$K_{11} = K_{22} = K_{33}$, $K_{12} = K_{21}$, $K_{13} = K_{31}$, $K_{23} = K_{32}$, and $K_{44} = K_{55} = K_{66}$.

Finally, with the use of elastic constants the Hooke's law for isotropic material is written as

$$\begin{Bmatrix} \sigma_x \\ \sigma_y \\ \sigma_z \\ \tau_{xy} \\ \tau_{xz} \\ \tau_{yz} \end{Bmatrix} = \begin{pmatrix} (\lambda+2G) & \lambda & \lambda & 0 & 0 & 0 \\ \lambda & (\lambda+2G) & \lambda & 0 & 0 & 0 \\ \lambda & \lambda & (\lambda+2G) & 0 & 0 & 0 \\ 0 & 0 & 0 & G & 0 & 0 \\ 0 & 0 & 0 & 0 & G & 0 \\ 0 & 0 & 0 & 0 & 0 & G \end{pmatrix} \cdot \begin{Bmatrix} \varepsilon_x \\ \varepsilon_y \\ \varepsilon_z \\ \gamma_{xy} \\ \gamma_{xz} \\ \gamma_{yz} \end{Bmatrix}, \tag{5.38}$$

where, $\lambda = \nu E/\{(1+\nu)\cdot(1-2\nu)\}$ is Lame's constant and $G = E/\{2\cdot(1+\nu)\}$ is shear modulus.

From Eqn.5. 38, the set of constitutive equations are separately written which were given in Eqns. 5.7 and 5.8.

5.7 THEORIES OF ELASTIC FAILURE

Strength theories predict failure conditions of materials under combined stresses, assuming that the behaviour of materials in a simple tension or compression test is known. They have been developed to determine the suitable allowable stresses for complicated stress conditions which occur in practical designs.

Failure of the material means either yielding or actual rupture, whichever occurs first.

Let us consider a body as shown in the following figure where, σ_1, σ_2 and σ_3 are principal stresses. Tension is regarded as positive and compression as negative, further the axes are chosen such that the relations between the algebraic values of the principal stresses are:

$\sigma_1 > \sigma_2 > \sigma_3$.

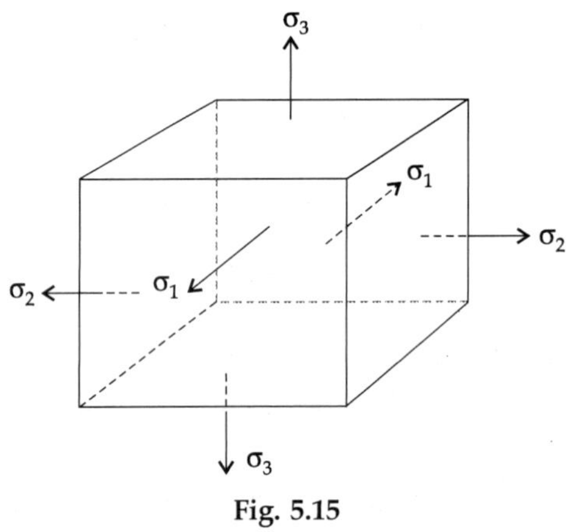

Fig. 5.15

5.7.1 Maximum Stress Theory

The maximum stress theory considers the maximum or minimum principal stress as the criterion for strength. For ductile materials this means that yielding begins in an element of a stressed body when either the maximum stress reaches the yield point in simple tension or the minimum stress reaches the yield point in simple compression.

$$\sigma_1 = \sigma_y \text{ (tensile) or } \sigma_3 = \sigma_y \text{ (compressive)}.$$

For a simple body under tension sliding occurs along planes inclined at 45° to the axis of specimen and failure here is caused by shear stresses. So this theory is not sufficient to determine the conditions for yielding of the material or its fracture.

5.7.2 Maximum Strain Theory

According to this theory a ductile material begins to yield either when the maximum strain (elongation) equals the yield point strain in simple tension or when the minimum strain (shortening) equals the yield point in simple compression.

We know that

$\varepsilon_1 = [\sigma_1 - \mu(\sigma_2 + \sigma_3)]/E$

$\varepsilon_2 = [\sigma_2 - \mu(\sigma_1 + \sigma_3)]/E$

$\varepsilon_3 = [\sigma_3 - \mu(\sigma_1 + \sigma_2)]/E$

Using these we get

$\sigma_1 - \mu(\sigma_3 + \sigma_2)] = \sigma$ (tensile)

$\sigma_3 - \mu(\sigma_1 + \sigma_2)] = \sigma$ (compressive)

5.7.3 Maximum Shear Theory

Maximum shear theory states that yielding begins when the maximum shear stress in the material becomes equal to the maximum shear stress at the yield point in a simple tension test.

$$\tau_{max} = (\sigma_1 - \sigma_3)/2 = \sigma_y/2$$

Thus, $(\sigma_1 - \sigma_3) = \sigma_y$.

5.7.4 Deformation Energy Theory

This theory uses strain energy of deformation per unit volume of the material as a criterion for predicting failure in the given specimen.

Strain energy per unit volume is

$$u = (\varepsilon_1\sigma_1 + \varepsilon_2\sigma_2 + \varepsilon_3\sigma_3)/2$$

Substituting the values of ε_1, ε_2 and ε_3, we get

$$u = \left[\sigma_1^2 + \sigma_2^2 + \sigma_3^2 - 2\mu(\sigma_1\sigma_2 + \sigma_2\sigma_3 + \sigma_3\sigma_1)\right]/2E$$

Materials can withstand very large hydrostatic pressure without failure. Hence, it is reasonable to resolve the total strain energy as

(1) The strain energy of uniform tension or compression

(2) The strain energy of distortion

We have

$$\varepsilon_1 + \varepsilon_2 + \varepsilon_3 = \frac{1 - 2\mu}{E}(\sigma_1 + \sigma_2 + \sigma_3)$$

Let us consider

$$\frac{\sigma_1 + \sigma_2 + \sigma_3}{3} = p, \text{ then}$$

$\sigma_1 = p + \sigma_1'$, $\sigma_2 = p + \sigma_2'$ and $\sigma_3 = p + \sigma_3'$.

$\sigma_1' + \sigma_2' + \sigma_3' = 0$. (These stresses produce only distortion.)

Let us consider a particular case where $\sigma_2 = 0, \sigma_3 = 0$

$$\frac{\sigma_1}{3} = p, \frac{2\sigma_1}{3} = \sigma_1', \quad \sigma_2' = \sigma_3' = -\frac{\sigma_1}{3}.$$

Substituting, $\sigma_1 = \sigma_2 = \sigma_3 = \dfrac{\sigma_1 + \sigma_2 + \sigma_3}{3}$ in strain energy equation, we get

$$u = \frac{1-2\mu}{6E}(\sigma_1 + \sigma_2 + \sigma_3)^2$$

The strain energy of distortion,

$$u = \frac{1}{2E}[\sigma_1^2 + \sigma_2^2 + \sigma_3^2 - 2\mu(\sigma_1\sigma_2 + \sigma_2\sigma_3 + \sigma_3\sigma_1)] - \frac{1-2\mu}{E}(\sigma_1 + \sigma_2 + \sigma_3)^2$$

$$= \frac{1+\mu}{6E}[(\sigma_1 + \sigma_2)^2 + (\sigma_2 - \sigma_3)^2 + (\sigma_3 - \sigma_1)^2]$$

As per the theory yielding occurs when $\sigma_1 = \sigma_y, \sigma_2 = \sigma_3 = 0$,

Thus, $u = \dfrac{1+\mu}{6E}\sigma_y^2$

Then the condition for yielding based on distortion energy theory is

$(\sigma_1 - \sigma_2)^2 + (\sigma_2 - \sigma_3)^2 + (\sigma_3 - \sigma_1)^2 = 2\sigma_y^2$.

This theory is currently accepted as valid for ductile materials, and it is assumed that the material begins to yield when the strain energy of distortion reaches a definite value.

5.7.5 Maximum Strain Energy Theory

This theory assumes that yielding or fracture occurs when total strain energy exceeds the strain energy.

$$\frac{[\sigma_1^2 + \sigma_2^2 + \sigma_3^2 - 2\mu(\sigma_1\sigma_2 + \sigma_2\sigma_3 + \sigma_3\sigma_1)]}{2E} = \frac{\sigma^2}{2E}$$

Example 5.13: The principal stresses at a point in an elastic material are 2σ (tensile), σ (tensile) and 0.5σ (compressive). If the elastic limit in tension is 200 MPa and $\mu = 0.25$. Calculate the value of failure (σ) computed with different theories of failure.

Solution:

1. According to maximum principal stress theory

   $\sigma_1 \le \sigma_e \Rightarrow +2\sigma \le 200 \Rightarrow \sigma \le 100$ MPa

2. According to maximum shear stress theory

   $$\text{Max}\left[\left|\frac{\sigma_1 - \sigma_2}{2}\right|, \left|\frac{\sigma_2 - \sigma_3}{2}\right|, \left|\frac{\sigma_3 - \sigma_1}{2}\right|\right] \le \frac{\sigma_e}{2}$$

$$\text{Max}\left[\left|\frac{2\sigma-\sigma}{2}\right|,\left|\frac{\sigma+0.5\sigma}{2}\right|,\left|\frac{-0.5\sigma-2\sigma}{2}\right|\right]\leq\frac{\sigma_e}{2}$$

$$\text{Max}\,(0.5\sigma,\,0.75\sigma,\,1.25\sigma)\leq\frac{\sigma_e}{2}$$

$$\Rightarrow 1.25\sigma\leq\frac{200}{2}\Rightarrow\sigma\leq 80\text{ MPa}$$

3. According to maximum principal strain theory

$$\frac{\sigma_1}{E}-\mu\frac{\sigma_2}{E}-\mu\frac{\sigma_3}{E}\leq\frac{\sigma_e}{E}\Rightarrow\sigma_1-\mu\sigma_2-\mu\sigma_3\leq\sigma_e$$

$$2\sigma-0.25(\sigma)-0.25(-0.5\sigma)\leq 200$$

$$\Rightarrow 1.875\,\sigma\leq 200\Rightarrow\sigma\leq 106.67\text{ MPa}$$

4. According to maximum strain energy theory

$$\sigma_1^2+\sigma_2^2+\sigma_3^2-2\mu(\sigma_1\sigma_2+\sigma_2\sigma_3+\sigma_3\sigma_1)\leq\sigma_e^2$$

$$\Rightarrow (2\sigma)^2+(\sigma)^2+(-0.5\sigma)^2$$

$$-2\times 0.25[(2\sigma\times\sigma)+(\sigma\times -0.5\sigma)+(-0.5\sigma\times 2\sigma)]\leq 200^2$$

$$\Rightarrow 5\sigma^2\leq 200^2\Rightarrow\sigma\leq 89.4\text{ MPa}$$

5. According to maximum shear strain energy theory

$$(\sigma_1-\sigma_2)^2+(\sigma_2-\sigma_3)^2+(\sigma_3-\sigma_1)^2\leq 2\sigma_e^2\qquad (2\sigma-\sigma)^2+(\sigma+0.5\sigma)^2+(-0.5\sigma-2\sigma)^2\leq 2\sigma_e^2$$

$$\Rightarrow 9.6\,\sigma^2\leq 2\times 200^2\Rightarrow\sigma\leq 91.76\text{ MPa}$$

PROBLEMS

1. A steel plate is subjected to uniformly distributed biaxial stresses, with σ_x = 125 MPa and σ_y = –65 MPa. The normal strains are found as $\varepsilon_x = 688\times 10^{-6}$ and $\varepsilon_y = -488\times 10^{-6}$. Determine the modulus of elasticity € and Poisson's ratio (ν) for steel.

Ans. E = 210 GPa, ν = 0.3

2. A 30 cm × 25 cm × 25 cm steel plate is subjected to normal stresses, σ_x = 66 MPa (tension) and σ_y = 22 MPa (compression), as shown in Fig. 5.16. Determine (a) the maximum shearing strain (γ_{max})in the plane of plate, (b) change in thickness of the plate, and (c) the change in volume of the plate. Take E = 210 GPa and ν = 0.3.

Ans. (a) $\gamma_{max} = 545\times 10^{-6}$,
(b) $\Delta t = -157\times 10^{-6}$cm (decrease),
(c) ΔV = 0.156263 cm^3 (increase)

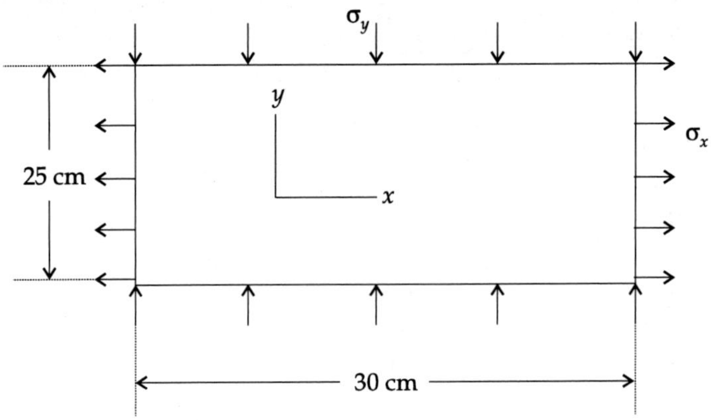

Fig. 5.16

3. A plate of dimensions $L \times b \times h$, is shown in Fig. 5.17. A force P is applied which produces a uniform normal stress σ_x along x-direction. Find the strains ε_x, ε_y and ε_z if E and ν are elastic constants for the material of the plate. What is change in volume of the plate?

Ans. $P/(bhE)$, $-\nu P/(bhE)$, $-\nu P/(bhE)$
$\Delta V = (1 - 2\nu).PL/E$

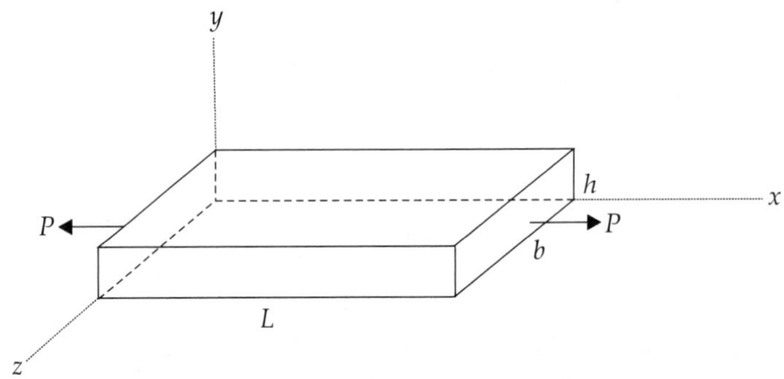

Fig. 5.17

4. A granite cube of 60 mm side is subjected to tri-axial stresses. The measurements of normal strains give $\varepsilon_x = -600 \times 10^{-6}$ and $\varepsilon_y = \varepsilon_z = -250 \times 10^{-6}$. If $E = 60$ GN/m^2 and $\nu = 0.25$ for granite, determine (a) normal stresses σ_x, σ_y and σ_z acting on the faces of cube with outward normal pointing to x-, y- and z-directions, respectively, (b) the maximum shear stress (τ_{max}) in the material, and (c) the change in volume (ΔV) of the cube.

Ans. (a) $\sigma_x = 55.2$ MPa (C), $\sigma_y = \sigma_z = 38.4$ MPa (C)
(b) $\tau_{max} = 8.4$ MPa
(c) $\Delta V = 237.6$ mm^3

5. The stresses, $\sigma_x = 5$ MPa (C), $\sigma_y = 4$ MPa (C) and $\sigma_z = 2$ MPa (C), are acting on an element of nylon (Fig. 5.18). Also, the normal strains in x- and y-directions are found as $\varepsilon_x = -720 \times 10^{-6}$ and $\varepsilon_y = -300 \times 10^{-6}$. What is the bulk modulus (K) for nylon?

Ans. $K = 7.64$ GPa

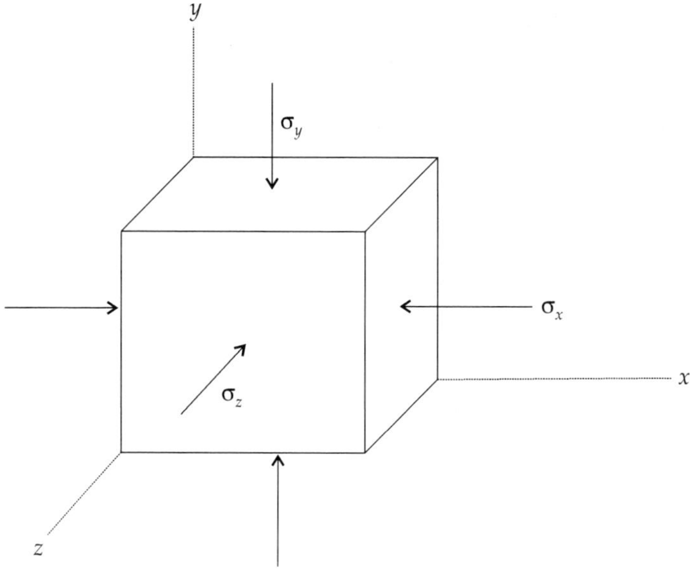

Fig. 5.18

6. A solid sphere of brass ($E = 100$ GPa, $\nu = 0.34$) is tested to hydraulic pressure such that its decrease in volume per unit volume is 0.4%. What is the bulk modulus of brass and the applied pressure?

Ans. $K = 104$ GPa, $p = 521$ MPa

7. A sphere made of steel is taken deep into ocean up to a point where its volume reduces by 19 mm^3. If the diameter of the sphere is 60 cm, what is the depth to which it is immersed?

Ans. 3 km

8. A bar of 5 cm × 5 cm cross section and 50 cm long is made of aluminium. It is snug-fit in the well lubricated square hole of a rigid cover, as shown in Fig. 5.19. If a compressive force, $P = 4$ kN, is applied at the ends of the bar to produce uniform pressure over its cross section, what is the change in length of the bar? What is the pressure developed between the bar and rigid cover?

Take $E = 73$ GPa and $\nu = 0.33$ for aluminium.

Ans. 0.0074 mm (decrease), 0.8 MPa

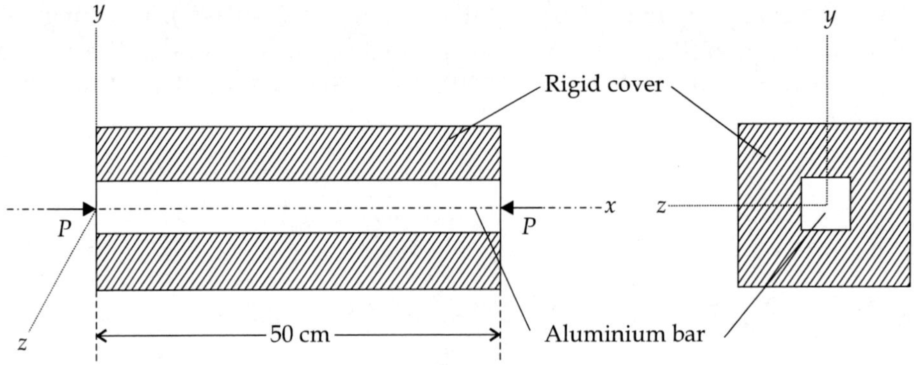

Fig. 5.19

9. The strain matrix at a point in a body is given by

$$\begin{Bmatrix} 350 & -180 & 0 \\ -180 & -100 & 50 \\ 0 & 50 & 200 \end{Bmatrix} \times 10^{-6}$$

If $E = 73$ GPa and $\nu = 0.33$ for the material, determine the stress matrix.

Ans. $\begin{Bmatrix} 52.88 & -6.92 & 0 \\ -6.92 & 18.27 & 1.92 \\ 0 & 1.92 & 41.34 \end{Bmatrix}$ MPa

10. If the shear modulus of a material is $G = 56$ GPa and its bulk modulus is, $K = 167$ GPa, what is the elasticity modulus € of the material?

Ans. $E = 151$ GPa

6
Thin-walled Pressure Vessels

6.1 INTRODUCTION

Pressure vessels are containers of regular shape which are used to store liquids and gasses under pressure. Common types of thin-walled vessels are thin cylindrical or spherical shells. These vessels are usually constructed with metal plates. Non-metallic materials are also used in certain applications. Due to the internal pressure of fluids, the vessels tend to burst and hence the tensile stresses develop in its wall. It is found with thin cylinders that there is almost negligible variation in stresses along the thickness of their wall if thickness of the wall is less than $1/10^{th}$ of its radius. The same is true for other thin-walled vessels. Boiler shells, gas storage tanks and the pipes in which some fluid flows under pressure, fall in the category of thin-walled vessels. On the other end, the gun barrels, some containers in chemical plants and pipes for hydraulic press, etc., are treated as thick cylinders.

It can be established that in thin-walled cylinders there is little variation of circumferential strain (and hence, the stress) along the thickness of the wall. For this, let us consider a cylinder of inner radius, r_i , outer radius, r_0, and thickness, $t = r_0 - r_i$. The cross section of the cylinder is shown in Fig. 6.1(a). When there is a pressure (p) of fluid inside the cylinder, it deforms symmetrically about its axis such that a circular section remains circular in shape and there is no warping anywhere in the cylinder. As a result of deformation, there are changes in the inner and outer radii but the thickness remains unaffected by overall deformation of the cylinder. The deformed cross section of cylinder is shown in Fig. 6.1(b) with R_i = inner radius, R_0 = outer radius and thickness $t = R_0 - R_i$.

Let ε_i and ε_0 be the circumferential strains at the inner and outer fibre, respectively. Then we have at inner fibre,

$$\varepsilon_i = (2\pi R_i - 2\pi r_i)/(2\pi r_i) = (R_i - r_i)/r_i. \tag{6.1}$$

Similarly, the strain at the outer fibre,

$$\varepsilon_0 = (R_0 - r_0)/r_0 = \{(R_i + t) - (r_i + t)\}/(r_i + t)$$

or

$$\varepsilon_0 = (R_i - r_i)/(r_i + t). \tag{6.2}$$

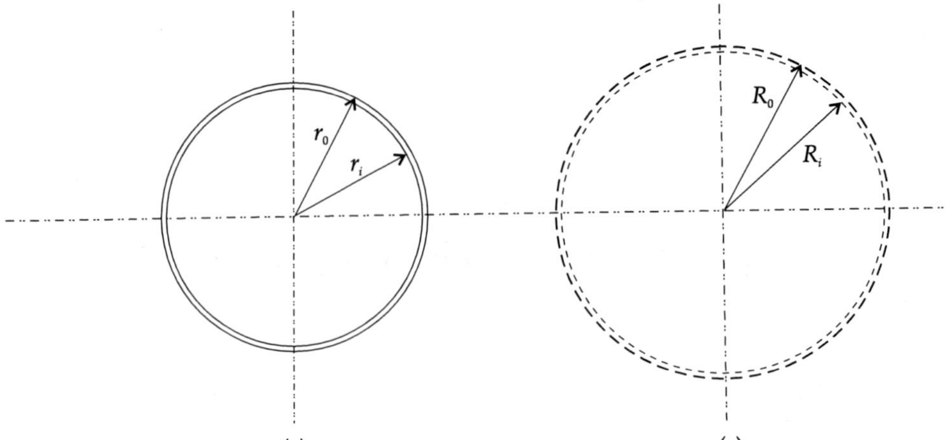

Fig. 6.1

For the thickness lying in the range $(1/20)\cdot r_i < t < (1/10)\cdot r_i$, $r_i + t \simeq r_i$.

Hence, Eqn. 6.2 becomes $\varepsilon_0 = (R_i - r_i)/r_i$, or $\varepsilon_0 = \varepsilon_i$ (from Eqn. 6.1).

As the strain remains unchanged through the thickness of the cylinder, the stress also remains constant and the thin cylinder is defined accordingly.

6.2 STRESSES IN THIN CYLINDER

When a thin cylinder is subjected to internal pressure, there are two major types of stresses produced at the cylinder's wall, which are (1) the circumferential stress, and (2) the longitudinal stress.

1. **Circumferential stress:** It is a normal stress which can be denoted by σ_C. It acts normal to the longitudinal plane (Fig. 6.2 (a)) of the cylinder. A longitudinal plane passes through the axis of the cylinder and divides it longitudinally into two halves. A cut-section showing one-half is given in Fig. 6.2 (b), which is in equilibrium under the forces produced due to the stress (σ_C) and the x-component of the bursting force caused by internal pressure, p.

Let r = radius of the cylinder (Fig. 6.2 (b))

D = diameter of the cylinder (=$2r$)

t = thickness of the cylinder

p = internal pressure in the cylinder

σ_C = circumferential stress

L = length of the cylinder measured parallel to its axis

Then, the force due to σ_C along x-axis, $F_C = -2\cdot\sigma_C\cdot t\cdot L$ (referring to Fig. 6.2 (b)).

To find the x-component of force due to pressure, p, an element of length, $r\cdot d\theta$, is chosen to get an elementary surface area (inner) = $L\cdot r\cdot d\theta$. The force of pressure over this elementary area is $dF_p = p \cdot L \cdot r \cdot d\theta$, which acts in a direction making an angle θ with the x-axis.

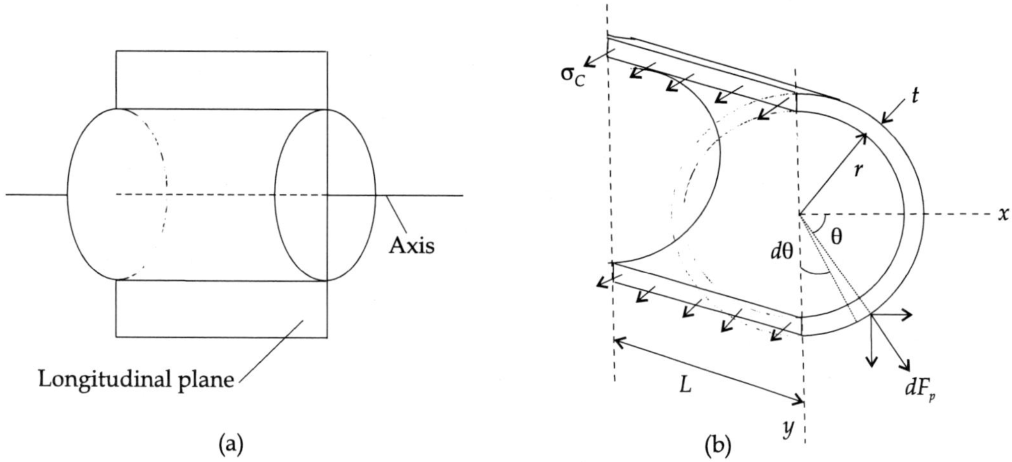

Fig. 6.2

$\therefore$ the x-component of this force, $dF = dF_p \cdot \cos\theta = (p \cdot L \cdot r \cdot d\theta) \cdot \cos\theta$.

Therefore, the total force due to pressure over the entire inner surface of cut-section,

$$F = 2 \cdot \int_0^{\pi/2} p \cdot L \cdot r \cdot \cos\theta \cdot d\theta, \text{ or } F = 2 \cdot p \cdot L \cdot r.$$

For equilibrium along x-axis, $F_C + F = 0$, or $-2 \cdot \sigma_C \cdot t \cdot L + 2 \cdot p \cdot L \cdot r = 0$.

Hence, the circumferential stress,

$$\left.\begin{aligned} \sigma_C &= p \cdot r / t, \\ \text{or} \quad \sigma_C &= p \cdot D / (2 \cdot t). \end{aligned}\right\} \tag{6.3}$$

The circumferential stress is also known as "**hoop stress**".

2. **Longitudinal stress:** When a cylinder is intersected by a transverse plane which is perpendicular to its axis, the cross section of the cylinder is exposed (Fig. 6.3 (a)). The normal stress (σ_L) acting on the cross section is called the longitudinal stress. Stress (σ_L) acts parallel to the axis of the cylinder which is shown in Fig. 6.3 (b). The force due to longitudinal stress on the cross section is balanced by the bursting force of pressure (p) acting over the area at the end of the cylinder.

Referring to Fig. 6.3(b), $\sum F_x = -\pi \cdot r^2 \cdot p + \sigma_L \cdot 2\pi \cdot r \cdot t = 0$.

$$\left.\begin{aligned} \therefore \quad \sigma_L &= p \cdot r / (2 \cdot t), \\ \text{or} \quad \sigma_L &= p \cdot D / (4 \cdot t). \end{aligned}\right\} \tag{6.4}$$

It is found from Eqns. 6.3 and 6.4 that the circumferential, as well as longitudinal stress are both directly proportional to pressure (p) and diameter (D), and inversely proportional to thickness (t) of the cylinder.

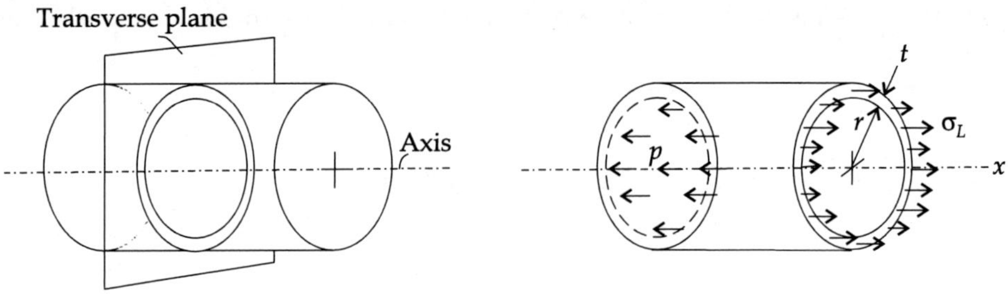

Fig. 6.3

When a pressure (p) is applied, both types of stresses (σ_C and σ_L) are produced simultaneously at the wall of cylinder. While the circumferential stress (σ_C) acts on the longitudinal section, the longitudinal stress (σ_L) acts on the transverse section. These two normal stresses are perpendicular to each other, as shown in Fig. 6.4. At a certain value of pressure (p), the stress (σ_C) is two times in magnitude than the stress, σ_L. Hence, on increasing pressure (p) the stress (σ_C) will first attain the level of permissible stress of the plate material of the cylinder. As such, the failure will occur by tearing of plate along the longitudinal section. This is the reason when a thin cylinder with closed ends is constructed by joining the sheets of a certain thickness (t), the longitudinal joint (see Fig. 6.5) is made two times stronger than the girth joint. If longitudinal joint is not exactly twice stronger than the girth joint, then the permissible pressure in the cylinder is decided according to the strength of the longitudinal joint.

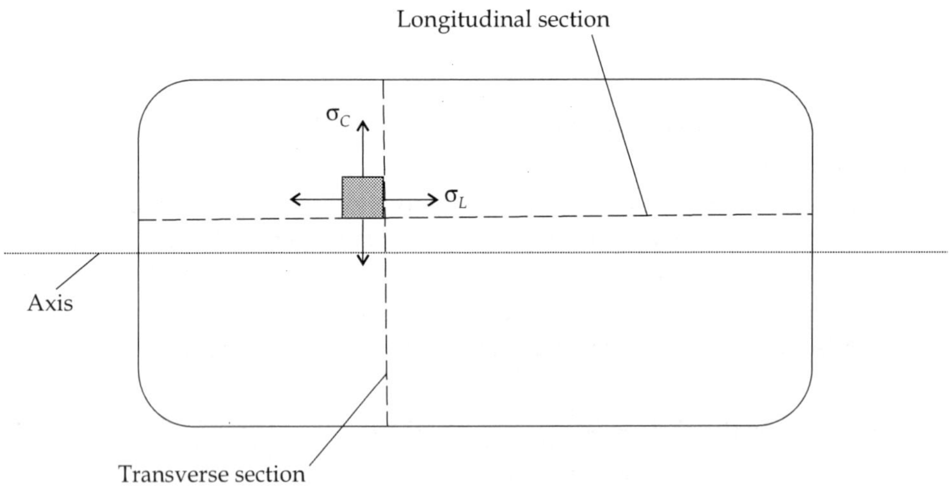

Fig. 6.4

It should be noted that there is also a third stress due to pressure (p) which is normal to the inner cylindrical surface. However, it is not included in the analysis because $|p| << |p \cdot D/(4 \cdot t)| < |p \cdot D/(2 \cdot t)|$.

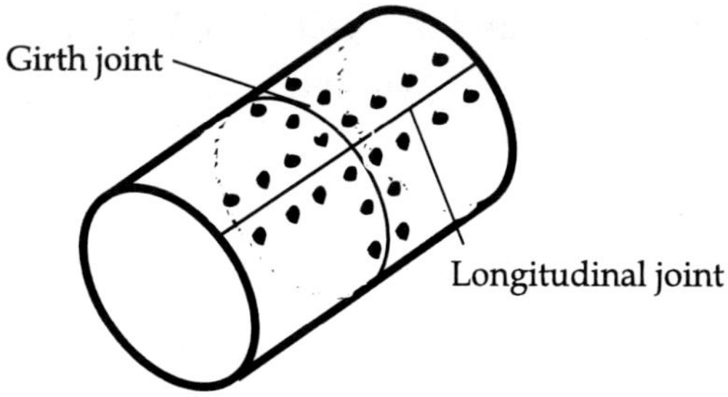

Fig. 6.5

6.2.1 Analyses of Stresses and Strains in Thin Cylinders

For a thin cylinder subjected to internal pressure, a small element of thickness (t) is chosen on its cylindrical surface and the two normal stresses (σ_C and σ_L) acting on the element are shown in Fig. 6.6. The circumferential stress and longitudinal stress are lying in a plane. Both are acting normal to planes free from any shear stress. Hence, these are principal stresses and shown by Mohr's circle construction in Fig. 6.6(c).

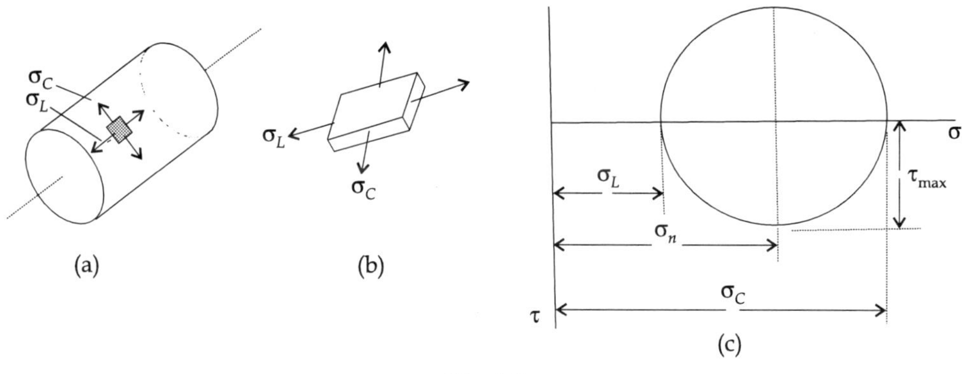

Fig. 6.6

$$\tau_{max} = (\sigma_C - \sigma_L)/2 = (1/2)\cdot[(p\cdot D/2\cdot t) - (p\cdot D/4\cdot t)].\ \therefore\ \tau_{max} = p\cdot D/(8\cdot t), \qquad (6.5)$$

which acts on a plane at 45° to the principal directions.

The normal stress on the plane containing maximum shear stress is written as

$$\sigma_n = (\sigma_C + \sigma_L)/2 = (1/2)\cdot[(p\cdot D/2\cdot t) + (p\cdot D/4\cdot t)], \text{ or } \sigma_n = (3/4)(p\cdot D/2\cdot t). \qquad (6.6)$$

All the stresses at a point in the wall of the thin cylinder lie in a plane. Therefore, under a plane stress condition the circumferential strain (ε_c) and longitudinal strain (ε_L) are written as

$$\left.\begin{aligned} \varepsilon_c &= (1/E)\cdot[\sigma_C - \nu\sigma_L], \\ \text{and} \quad \varepsilon_L &= (1/E)\cdot[\sigma_L - \nu\sigma_C], \end{aligned}\right\} \qquad (6.7)$$

where, E = elasticity modulus, and ν = Poisson's ratio of the plate material of the cylinder.

Using the relations $\sigma_C = p{\cdot}D/(2{\cdot}t)$ and $\sigma_L = p{\cdot}D/(4.t)$, Eqn. 6.7 becomes

$$\varepsilon_c = \{p{\cdot}D/(4{\cdot}t{\cdot}E)\}{\cdot}(2 - \nu),$$

and $$\varepsilon_L = \{p{\cdot}D/(4{\cdot}t{\cdot}E)\}{\cdot}(1 - 2\nu). \quad (6.8)$$

Change in the volume of the cylinder

If the diameter (D) and the length (L) of the cylinder change to D' and L' respectively, after loading with pressure (p), then

$$\varepsilon_c = (\text{change in circumference/original circumference}) = (\pi{\cdot}D' - \pi{\cdot}D)/(\pi{\cdot}D)$$
$$= (D' - D)/D = \Delta D/D.$$

Hence, using Eqn. 6.8, we have the change in diameter, $\Delta D = D{\cdot}\varepsilon_c$, so that

$$\Delta D = \{p{\cdot}D^2/(4{\cdot}t{\cdot}E)\}{\cdot}(2 - \nu). \quad (6.9)$$

Also, the change in length, $\Delta L = L' - L = L{\cdot}\varepsilon_L$,

or $$\Delta L = \{p{\cdot}D{\cdot}L/(4{\cdot}t{\cdot}E)\}{\cdot}(1 - 2\nu) \quad (6.10)$$

Volume of the cylinder, $V = (\pi/4){\cdot}D^2{\cdot}L$, so that $\ln V = \ln(\pi/4) + 2{\cdot}\ln D + \ln L$.

After taking differential on both sides of the above equation, the volume strain is obtained in the form

$$dV/V = 2{\cdot}(dD/D) + dL/L, \text{ or } \Delta V/V = 2{\cdot}(\Delta D/D) + \Delta L/L.$$

∴ From Eqn.6.8, the volume strain, $\Delta V/V = 2{\cdot}\{p{\cdot}D/(4{\cdot}t{\cdot}E)\}{\cdot}(2 - \nu) + \{p{\cdot}D/(4{\cdot}t{\cdot}E)\}{\cdot}(1 - 2\nu)$.

or $$\Delta V/V = \{p{\cdot}D/(4{\cdot}t{\cdot}E)\}{\cdot}(5 - 4\nu) \quad (6.11)$$

Therefore, the change in volume is written as

$$\Delta V = \{(\pi/4){\cdot}D^2{\cdot}L\}{\cdot}\{p{\cdot}D/(4{\cdot}t{\cdot}E)\}{\cdot}(5 - 4\nu) \quad (6.12)$$

Example 6.1: A thin cylinder of 50 cm radius is made from 3 mm steel plates. If a gas is filled in the cylinder at a pressure of 750 kN/m^2 (gauge), determine:

(a) the circumferential stress in the wall of the cylinder,
(b) the longitudinal stress,
(c) the maximum shear stress in the plane of the above two stresses,
(d) change in the diameter of the cylinder due to the pressure of gas,
(e) change in the length for 2.5 m length of the cylinder,
(f) change in the volume of the cylinder.

Take E = 210 GPa and ν = 0.3 for steel.

Solution: Given that r = 50 cm, t = 0.3 cm and $p = 750 \times 10^3/\text{m}^2$.

(a) Circumferential stress, $\sigma_C = (750 \times 10^3 \times 50)/0.3$ N/m^2 (from Eqn. 6.3)

$$= 125 \times 10^6 \text{ N/m}^2$$

or σ_C = 125 MPa **Ans.**

(b) Longitudinal stress, $\sigma_L = (750 \times 10^3 \times 50)/(2 \times 0.3)$ N/m^2 (from Eqn. 6.4)

or σ_L = 62.5 MPa **Ans.**

(c) Maximum shear stress, $\tau_{max} = \{750 \times 10^3{\cdot}(2 \times 50)\}/(8 \times 0.3)$ N/m^2 (from Eqn. 6.5)

$= 31.25$ MPa **Ans.**

(d) Change in diameter, $\Delta D = [\{750 \times 10^3 \times (2 \times 0.5)^2\}/(4 \times 3 \times 10^{-3} \times 210 \times 10^9)]$.

$$(2 - 0.3) \text{ m (from Eqn. 6.9)}$$

$$= 0.506 \times 10^{-3} \text{ m}$$

or $\Delta D = 0.5$ mm **Ans.**

(e) Change in length, $\Delta L = [\{750 \times 10^3 \times (1.0) \times (2.5)\}/(4 \times 3 \times 10^{-3} \times 210 \times 10^9)]$.

$$\{1 - (2 \times 0.3)\} \text{ m (from Eqn. 6.10)}$$

or $\Delta L = 0.3$ mm **Ans.**

(f) Change in volume, $\Delta V = \{(\pi/4) \times 1^2 \times 2.5\} \cdot \{(750 \times 10^3 \times 1)/(4 \times 3 \times 10^{-3} \times 21 \times 10^9)\} \cdot$

$$\{5 - (4 \times 0.3)\} \text{ m}^3 \text{ (from Eqn. 6.12)}$$

or $\Delta V = 0.0022 \text{ m}^3$ **Ans.**

Example 6.2: A 25 cm diameter steel pipe is used to carry a flow of steam at a pressure of 5.5 MN/m^2.

Because of a longitudinal weld in the pipe, the hoop stress is limited to 70 MPa. What should be the minimum thickness of the pipe?

Solution: Given that D = 25 cm, p = 5.5 MN/m^2, σ_C = 70 MPa (allowable).

$\therefore$ $p \cdot D/(2 \cdot t) \le \sigma_C$, or $t \ge p \cdot D/(2 \cdot \sigma_C)$. $t \ge 5.5 \times 25/(2 \times 70)$ or $t \ge 0.98$ cm

$\therefore$ $t = 1.0$ cm **Ans.**

Example 6.3: A vertical steel tank, 4 m in diameter and 15 m tall, is used to store water. If the hoop stress in the tank is not to exceed 35 MN/m^2, find the minimum plate thickness of the pipe. The density of water, ρ = 1000 kg/m^3.

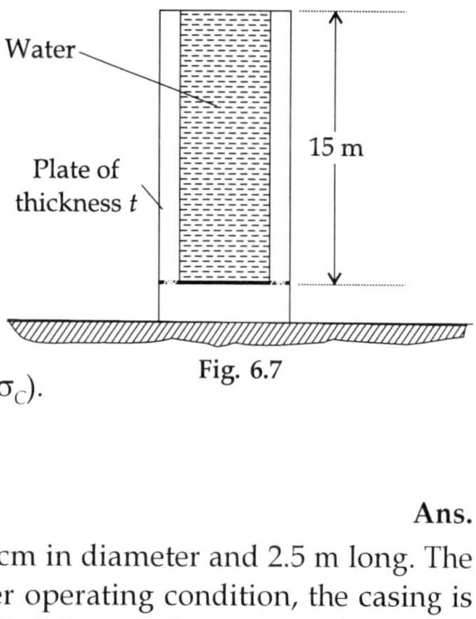

Fig. 6.7

Solution: The maximum pressure of water is present at the bottom and given by $p = \rho \cdot g \cdot h$, where h = 15 m.

$\therefore$ $p = 1000 \times 9.81 \times 15$ N/m^2.

For $p \cdot D/(2.t) \le \sigma_C$ (allowable stress), $t \ge p \cdot D/(2 \cdot \sigma_C)$.

or $t \ge (1000 \times 9.81 \times 15 \times 4)/(2 \times 35 \times 10^6)$ m

$t \ge 8.4 \times 10^{-3}$ m

$\therefore$ $t = 8.5$ mm (safe value) **Ans.**

Example 6.4: The casing of a rocket motor is 178 cm in diameter and 2.5 m long. The thickness of this alloy steel casing is 2 mm. Under operating condition, the casing is subjected to a longitudinal tensile force of 560 kN. If the tensile strength of the plate material is limited to 1450 MPa, what is the maximum value of operating pressure?

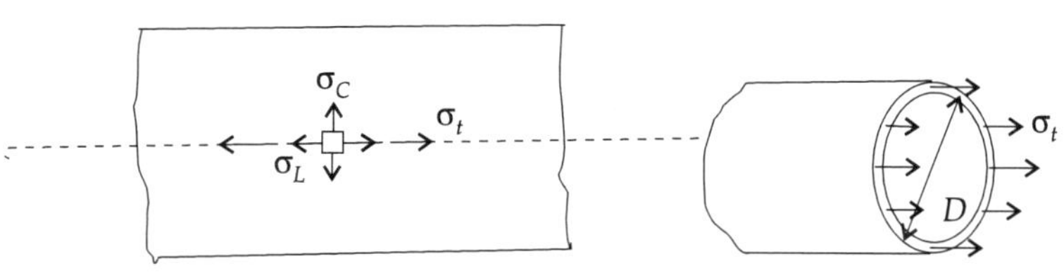

Fig. 6.8

Solution: Apart from the hoop stress (σ_C) and longitudinal stress (σ_L), there will be an additional tensile stress (σ_t) parallel to the axis of the casing.

Let p = operating pressure (in N/m²).

Also, $\sigma_w = 1450 \times 10^6$ N/m².

Tensile stress (σ_t) on the cross section of the cylinder:

$\sigma_t.(\pi.D.t) = F$

$$\sigma_t = 560 \times 10^3/(\pi \times 178 \times 2 \times 10^{-5})$$

$$\therefore \quad \sigma_t = 50 \times 10^6 \text{ N/m}^2.$$

According to the given condition, either $\sigma_C \le \sigma_w$, or $\sigma_L + \sigma_t \le \sigma_w$.

For $\sigma_C \le \sigma_w$, $p \cdot 178/(2 \times 0.2) \le 1450 \times 10^6$ N/m² or $p \le 3.26$ MPa.

Again, for $\sigma_L + \sigma_t \le \sigma_w$, $\{p \cdot 178/(4 \times 0.2)\} + 50 \times 10^6 \le 1450 \times 10^6$ N/m² or $p \le 6.29$ MPa.

Lower of the two values of pressure given above will be the permissible pressure. If the pressure exceeds (p = 3.26 MPa), the motor casing will fail by tearing of plate along the longitudinal section.

Hence, p = 3.26 MPa **Ans.**

Example 6.5: Plates are welded along a spiral seam to construct a thin cylindrical tank as shown in Fig. 6.9. If the diameter of the tank is 85 cm and it contains a fluid at a pressure of 1.4 MN/m²,what normal and shearing forces will be acting per cm length of the weld?

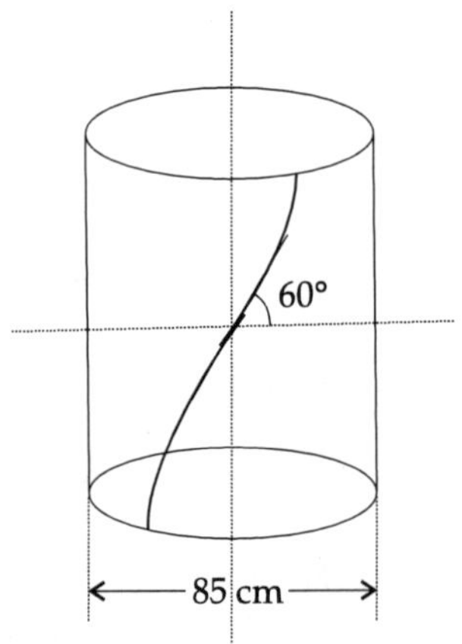

Fig. 6.9

Solution: Let t = thickness of plate (in cm).

Circumferential stress, $\sigma_C = p \cdot D/(2 \cdot t) = (1.4 \times 10 \times 85)/(2 \cdot t)$ N/m²,

or $\sigma_C = 59.5 \times 10^6/t$ N/m².

Longitudinal stress, $\sigma_L = (1/2) \cdot \sigma_C = 29.75 \times 10^6/t$ N/m².

Normal stress (σ_n) and shear stress (τ) at a point on the seam is obtained from stress transformation given in chapter 3.

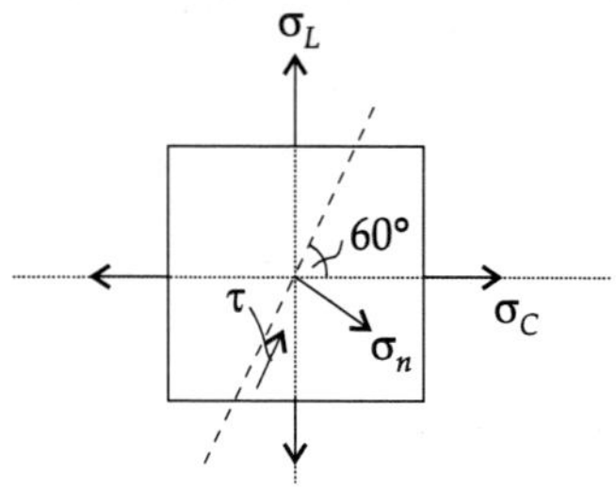

$$\sigma_n = (\sigma_C + \sigma_L)/2 + \{(\sigma_C - \sigma_L)/2\}\cdot\cos(-60°)$$

Using the expressions of σ_C and σ_L given above,

$$\sigma_n = 52.0625 \times 10^6/t \text{ N/m}^2.$$

Similarly, $\tau = \{(\sigma_L - \sigma_C)/2\}\cdot\sin(-60°)$

$\therefore$ $\tau = 12.882 \times 10^6/t \text{ N/m}^2$.

Area of 1 cm length of weld, $A = 1 \times \text{t cm}^2 = \tau \times 10^{-4} \text{ m}^2$.

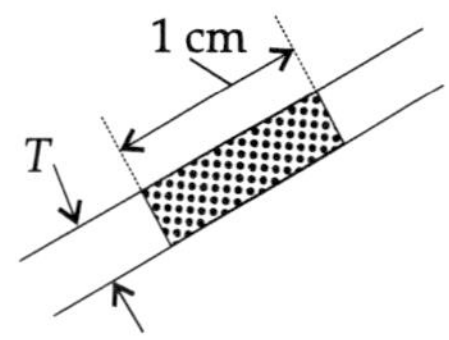

$\therefore$ Normal force, $F_n = \sigma_n \cdot A = (52.0625 \times 10^6/t)\cdot(t \times 10^{-4})$ N

$F_n = 5206$ N or $F_n = 5.2$ kN **Ans.**

Shearing force, $F_s = \tau \cdot A = (12.882 \times 10^6/t)\cdot(t \times 10^{-4})$ N

$\therefore$ $F_s = 1288$ N or $F_s \simeq 1.3$ kN **Ans.**

Example 6.6: Two strain gauges are mounted at right angle to each other on the surface of a cylindrical pressure vessel, as shown in Fig. 6.10. The gauges measure the normal strains $\varepsilon_1 = 633$ μm/m and $\varepsilon_2 = 330$ μm/m along 1- and 2-directions, respectively. The diameter of the pressure vessel, $D = 50$ cm and its plate thickness, $t = 3$ mm. Determine (a) the normal stresses (σ_1 and σ_2) along 1- and 2-directions, respectively, (b) internal pressure (p) in the vessel, and (c) factor of safety based on failure due to hoop stress if UTS for the plate material is 580 MPa.

Take $E = 210$ GPa and $\nu = 0.3$ for the plate material.

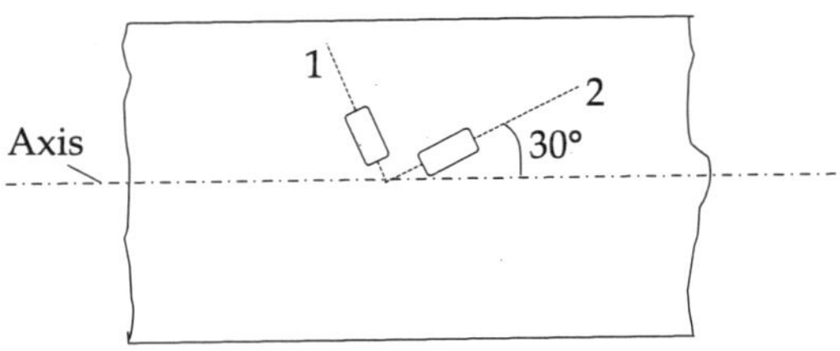

Fig. 6.10

Solution:

(a) From Eqn. 5.2.

$\sigma_1 = [210 \times 10^9/\{1 - (0.3)^2\}]\cdot[633 + (0.3 \times 330)]\cdot 10^{-6}$ N/m².

$= 168.9 \times 10^6$ N/m²

$\sigma_1 = 169$ MPa **Ans.**

Similarly,

$\sigma_2 = [210 \times 10^9/(1 - (0.3)^2\}]\cdot[330 + (0.3 \times 633)]\cdot 10^{-6}$ N/m².

$= 120$ MPa **Ans.**

(b) Assume that the longitudinal stress, $\sigma_L = \sigma$, so that the circumferential stress, $\sigma_C = 2\sigma$.

Therefore, for $\sigma_2 = 120$ MPa at 30° from the longitudinal axis,

$\sigma_2 = (\sigma + 2\sigma)/2 + \{(\sigma - 2\sigma)/2\}\cdot\cos(2 \times 30°)$, $\therefore$ $\sigma_2 = 1.25\cdot\sigma$.

$\sigma = \sigma_2/1.25 = 120/1.25 = 96$ MPa.

From Eqn. 6.4, $96 \times 10^6 = p\cdot 50/(4 \times 0.3)$

$p = 2300$ kPa **Ans.**

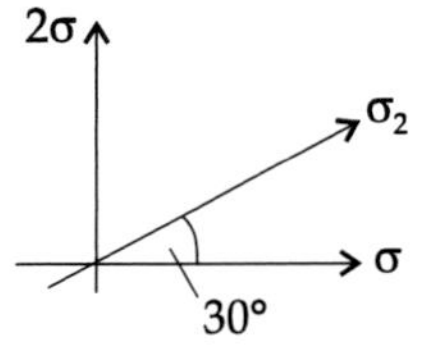

(c) Hoop stress, $\sigma_C = 2\sigma = 2 \times 96 = 192$ MPa.

Factor of safety = 580/192 = 3 **Ans.**

6.3 STRESSES IN SPHERICAL VESSELS

Like thin cylindrical tanks, the spherical vessels are also used as storage tanks for liquids and gases under pressure. The wall thickness of spherical shells is very small compared to its diameter. As a general practice, a spherical shell is considered to be thin-walled if its wall thickness is less than up to 1/10th of its radius. When a gas or liquid is filled under pressure, there is tendency of vessel to bulge out which is prevented by tensile stresses developed at the wall of the vessel. A soap bubble is a good example of thin-walled spherical shell. The pressure of air trapped within a bubble is in excess over atmospheric pressure and the force of surface tension in its membrane prevents the bubble from bursting. Since a spherical shell is constructed with metal plates, it is essential to determine the stress developed in its wall due to the internal pressure of a fluid which is in excess over the atmospheric pressure outside.

To obtain normal stress in the wall, a spherical shell (with internal pressure) can be assumed to be intersected by a diametral plane passing through its centre, as shown in Fig. 6.11(a). The diametral plane is a plane of symmetry and divides the spherical shell into two halves. One half of the shell is considered to be in equilibrium (Fig. 6.11 (b)). The stress (σ_t) is shown acting normal to the exposed cross section.

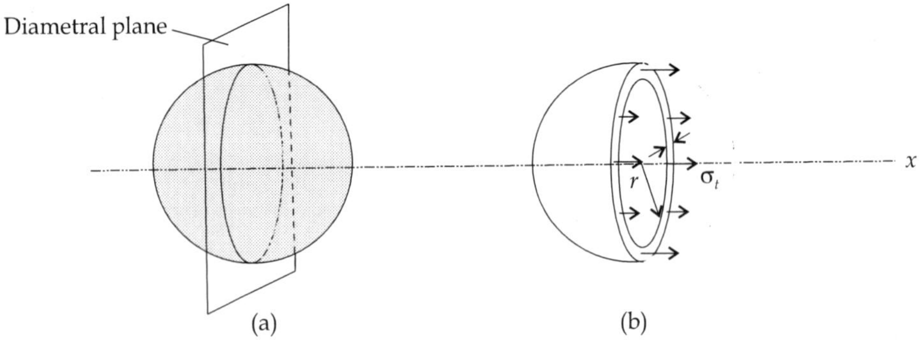

Fig. 6.11

Let r = inner radius of spherical shell and t = plate thickness, such that $t < (r/10)$. Then, the area over which the normal stress (σ_t) acts is given by $A = 2\pi\cdot r\cdot t$. Hence, the force due to σ_t along x-axis,

$$F_t = 2\pi \cdot r \cdot t \cdot \sigma_t \tag{6.13}$$

The pressure (p) of gas or liquid acts normal to the inner surface area, at every point. It can be shown that the net force due to pressure (p) has a component towards negative of x-axis. A sectional view of hemispherical shell is given in Fig. 6.12. At an angle θ, an elementary area ($r \cdot d\theta \cdot ds$) is taken, where ds belongs to a ring-shaped elementary area of radius, $r \cdot \sin\theta$ (see Fig. 6.12 (b)). Therefore, the elementary force, $dF = p \cdot r \cdot d\theta \cdot ds$, and the component of dF along x-axis $dF_x = p \cdot r \cdot d\theta \cdot ds \cdot \cos\theta$.

$\therefore$ The component due to force distribution over the ring-shaped area

$dF_p = p \cdot r \cdot \cos\theta \cdot d\theta \cdot (2\pi \cdot r \cdot \sin\theta)$, or $dF_p = p \cdot \pi \cdot r^2 \cdot \sin 2\theta \cdot d\theta$.

$$\therefore \text{ Net force along } x\text{-axis, } F_p = \int_0^{\pi/2} p \cdot \pi \cdot r^2 \cdot \sin 2\theta \cdot d\theta = p \cdot \pi \cdot r^2 \tag{6.14}$$

For equilibrium of hemispherical shell, Eqns. 6.13 and 6.14 are included to establish

$$\sum F_x = 0, \text{ or } -p \cdot \pi \cdot r^2 + 2\pi \cdot r \cdot t \cdot \sigma_t = 0.$$

$$\therefore \quad \sigma_t = p \cdot r / (2 \cdot t). \tag{6.15}$$

If $D = 2r$ is the diameter, the above expression can be written as

$$\sigma_t = p \cdot D / (4 \cdot t). \tag{6.16}$$

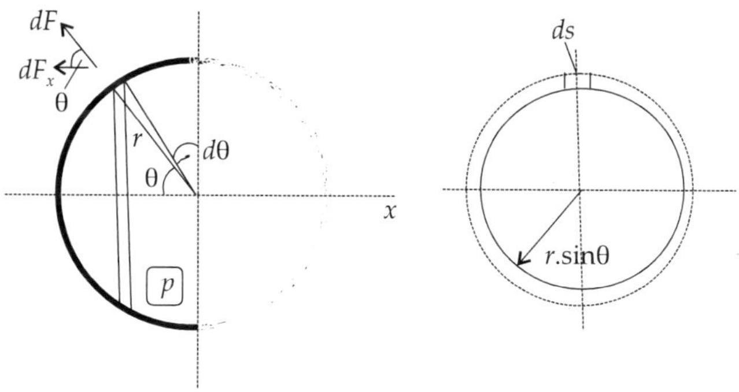

Fig. 6.12

Since the diametral sectional plane at any inclination is the plane of symmetry for the spherical shell, it implies that the normal stress given by Eqn. 6.15 or Eqn. 6.16 is same in every direction. This is shown in Fig. 6.13 with arbitrarily chosen plane elements on the surface of the spherical shell.

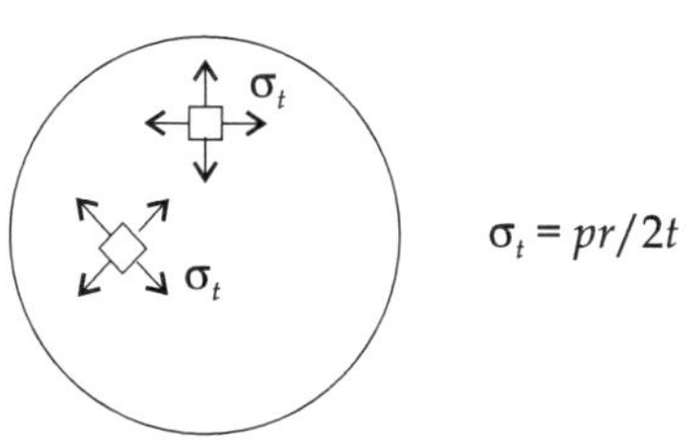

Fig. 6.13

It should be noted that the normal stresses acting at the surface on mutually perpendicular planes are tensile in nature and are equal in magnitude. Since these stresses act on planes that are free from shear, these are principal stresses. Also, the stress, $\sigma_t = pr/2t$, is a circumferential stress or hoop stress acting on the spherical shell.

Referring to Fig. 6.13, the principal stresses at any point are found as $\sigma_1 = \sigma_2 = pr/2t$. Hence, the maximum shearing stress at the point, $\tau_{max} = (\sigma_1 - \sigma_2)/2 = 0$. However, the outer surface is free from any normal stress, so that $\sigma_3 = 0$ (see Fig. 6.14). Therefore the out of plane maximum shearing stress is obtained as

$$\tau_{max} = (\sigma_1 - 0)/2 = (\sigma_2 - 0)/2 = p{\cdot}r/(4{\cdot}t). \tag{6.17}$$

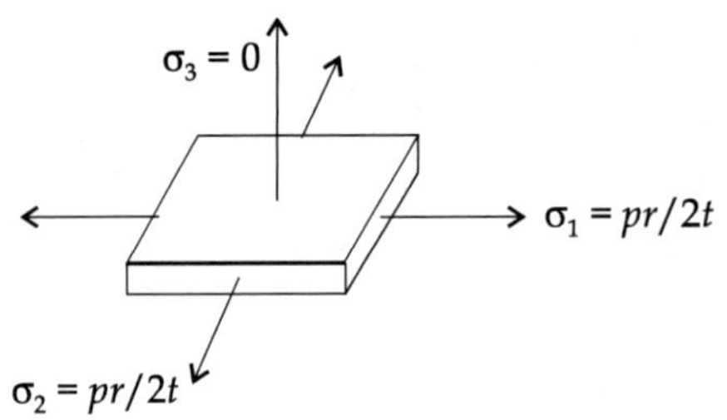

Fig. 6.14

At the inner surface of shell, the maximum shearing stress, $\tau_{max} = \{\sigma_1 - (-p)\}/2$,

or $\tau_{max} = \{p{\cdot}r/(4{\cdot}t)\}{\cdot}\{1 + (2t/r)\}$. (6.18)

If $t/r = 1/10$, it is found from Eqn. 6.18 given above that τ_{max} at inner surface is 20% higher than that at the outer surface. But, usually $t/r < 1/10$ is adapted and the permissible stress in shear is decided by selecting higher factor of safety. Hence, from the standpoint of design, $\tau_{max} = pr/4t$. is accepted to be same through the plate thickness and it is safe for practical purposes. Also, it should be noted that the spherical shells used in chemical industries or other applications must have inlet and outlet points for handling liquids or gases. The pipes with flanges, etc., are fitted at such points which present the regions of stress concentrators. Analyses of higher stresses at these regions are beyond the scope of the present study and hence, these are omitted. Moreover, it should be kept into notice that in the derivation of Eqn. 6.15 or Eqn. 6.16 the self-weight of any fluid inside a spherical vessel is ignored. These are the only limitations in stress analysis of vessels.

Change in volume of spherical vessel: At a point on the surface of spherical vessel subjected to pressure, the mutually perpendicular normal stresses are

$\sigma_1 = \sigma_2 = \sigma = pr/2t$.

$\therefore$ The normal strain, $\varepsilon_x = \varepsilon_y = (1/E){\cdot}(\sigma - \nu{\cdot}\sigma) = (\sigma/E){\cdot}(1 - \nu)$.

The change in radius of vessel, $\Delta r = r{\cdot}\varepsilon_x = (\sigma/E){\cdot}(1 - \nu){\cdot}r$.

Volume of spherical shell, $V = (4/3){\cdot}\pi{\cdot}r^3$, $\therefore \Delta V/V = 3{\cdot}(\Delta r/r) = 3{\cdot}(\sigma/E){\cdot}(1 - \nu){\cdot}r{\cdot}(1/r)$.

$\therefore$ Change in volume, $\Delta V = (3\sigma/E){\cdot}(1 - \nu){\cdot}V = \{3{\cdot}p{\cdot}r/(2{\cdot}t{\cdot}E)\}{\cdot}(1 - \nu){\cdot}\{(4/3){\cdot}(\pi{\cdot}D^3/8)\}$.

or $\Delta V = \{\pi{\cdot}p{\cdot}D^4/(8{\cdot}t{\cdot}E)\}{\cdot}(1 - \nu)$. (6.19)

Example 6.7: A spherical shell is made from structural plate to withstand an internal pressure of 500 kPa. The yield strength of steel is 250 MPa. If the diameter of the shell is 8 m and the factor of safety 2.5 is taken with respect to failure by yielding, determine the minimum thickness of the plate.

Solution: Given that $p = 500 \times 10^3$ N/m², $\sigma_{yld} = 250 \times 10^6$ N/m², $D = 8$m, $n = 2.5$.

Safe tensile stress in the wall, $\sigma_w = \sigma_{yld}/\text{F.S.} = 250 \times 10^6/2.5$ N/m².

Circumferential stress, $\sigma = p{\cdot}D/(4{\cdot}t)$, or $\sigma = 500 \times 10^3 \times 8/(4{\cdot}t)$ N/m², where t is in metre.

For $\sigma \le \sigma_w$, $\{500 \times 10^3 \times 8/(4{\cdot}t)\} \le \{250 \times 10^6/2.5\}$, $\therefore\ t \ge 10 \times 10^{-3}$ m or $t = 10$ mm **Ans.**

Example 6.8: A hemispherical window (Fig. 6.15) is 80 cm in diameter and 25 mm thick. A pressure of 620 kPa is applied in its concave side as shown. The window is attached to the wall of a chamber by 20 bolts. Find the tensile stress developed in the material of hemispherical window. What is the tensile force on each bolt?

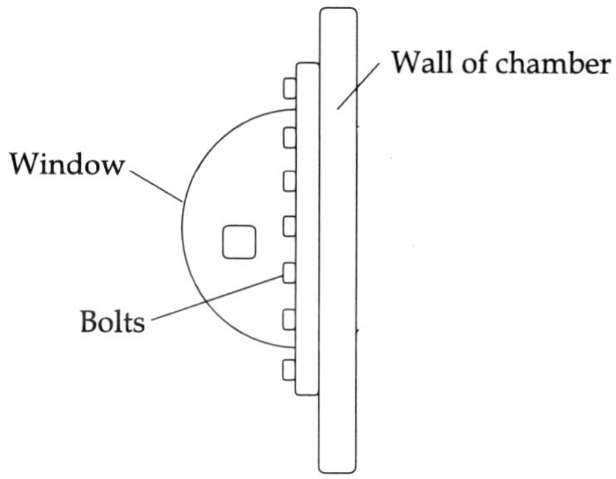

Fig. 6.15

Solution: Tensile stress (hoop stress), $\sigma = p{\cdot}D/(4{\cdot}t) = 620 \times 10^3 \times 80/(4 \times 2.5)$ N/m².

$\therefore\ \sigma = 4960 \times 10^3$ N/m², or $\sigma = 4.96$ MPa **Ans.**

It may be observed in Fig. 6.13 that the bursting force of pressure towards the left is balanced by the tensile forces in bolts. If F = tensile force in each bolt and $N = 20$ is the number of bolts,

$$F{\cdot}N = p{\cdot}(\pi{\cdot}D^2/4),\ \therefore\ F = p{\cdot}\pi{\cdot}D^2/(4{\cdot}N) = 620 \times 10^3 \times \pi \times (80)^2 \times 10^{-4}/(4 \times 20)\ \text{N}$$

$$\therefore\ F = 15.58\ \text{kN or } F = 15.6\ \text{kN}$$ **Ans.**

Example 6.9: An inflated rubber ball contains air at a pressure of 75 kPa. Its diameter is 215 mm and wall thickness is 1.5 mm. Determine the hoop stress and the maximum normal strain developed in the ball, if $E = 3.5$ MPa and $\nu = 0.45$ for rubber.

Solution: Given that $p = 75 \times 10^3$ N/m², $D = 215$ mm, $t = 1.5$ mm, $E = 3.5 \times 10^6$ N/m², $\nu = 0.45$.

Hoop stress, $\sigma = p{\cdot}D/(4{\cdot}t) = 75 \times 10^3 \times 215/(4 \times 1.5)$. $\sigma = 2.687$ MPa **Ans.**

Normal strain, $\varepsilon = \{(1 - \nu)/E\}{\cdot}\sigma = (1 - 0.45) \times 2.687/3.5$. $\varepsilon = 0.422$ **Ans.**

Example 6.10: A spherical vessel of 500 mm diameter and 8 mm thickness is made of steel. It is coated with a brittle coating material in which crack will develop (see Fig. 6.16) if the strain exceeds 225×10^{-6}. Determine internal pressure in the vessel which will cause the coating to develop cracks.

Take $E = 210$ GPa and $\nu = 0.3$ for steel.

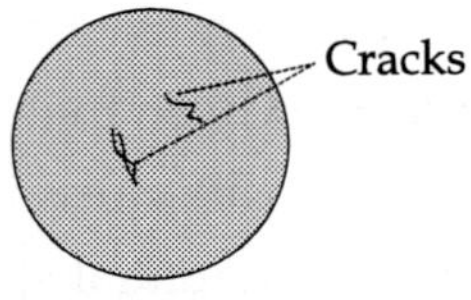

Fig. 6.16

Solution: Normal strain, $\varepsilon = \{(1 - \nu)/E\} \cdot \{p \cdot D/(4 \cdot t)\}$

or pressure, $p = 4 \cdot t \cdot E \cdot \varepsilon / \{(1 - \nu) \cdot D\}$

$$p = 4 \times 8 \times 210 \times 10^9 \times 225 \times 10^{-6} / \{(1 - 0.3) \times 500\} \text{ N/m}^2$$

$\therefore \quad p = 4.32$ MPa **Ans.**

Example 6.11: A spherical steel tank of 40 cm diameter is used to store a gas at a pressure of 20 MPa.

For the steel used to construct the tank, the following properties are known: yield stress in tension = 970 MPa, yield stress in shear = 450 MPa, elasticity modulus = 210 GPa and Poisson's ratio = 0.3.

Factor of safety = 3.0 is to apply with respect to yielding in tension, as well as in shear. Also, the maximum normal strain should not exceed 980×10^{-6}. Determine the minimum plate thickness for the tank.

Solution: The minimum plate thickness for tank can be obtained after considering the three different modes of failure, which are (1) failure in tension, (2) failure in shear, and (3) failure due to excessive deformation or strain.

(1) Filure in tension: It is known that hoop stress, $\sigma = pD/4t$, and to prevent failure, $\sigma \le \sigma_{yld}/n$, where, σ_{yld} is the yield stress in tension and n = factor of safety.

$\therefore\ p \cdot D/(4 \cdot t) \le \sigma_{yld}/n$, or $t \ge p \cdot D \cdot n / (4 \cdot \sigma_{yld})$. $t \ge 20 \times 40 \times 3/(4 \times 970)$ or $t \ge 0.6186$ cm.

(2) Failure in shear: Maximum shear stress, $\tau_{max} = \sigma/2$ (from Eqn. 6.17) $= p \cdot D/(8 \cdot t)$.

But, $\tau_{max} \le \tau_y/n$, where, τ_y is yield stress in shear.

$\therefore\ p \cdot D/(8 \cdot t) \le \tau_y/n$, or $t \ge p \cdot D \cdot n/(8 \cdot \tau_y)$

$t \ge 20 \times 40 \times 3/(8 \times 450)$ cm, or $t \ge 0.6667$ cm.

(3) Failure under strain: Normal strain, $\varepsilon = \{(1 - \nu)/E\} \cdot \sigma$.

For $\varepsilon \le \varepsilon_{max}$, $\{(1 - \nu)/E\} \cdot \{p \cdot D/(4 \cdot t)\} \le \varepsilon_{max}$.

$\therefore\ t \ge (1 - \nu) \cdot p \cdot D/(4 \cdot E \cdot \varepsilon_{max})$, or $t \ge (1 - 0.3) \times 20 \times 10^6 \times 40/(4 \times 210 \times 10^9 \times 980 \times 10^{-6})$.

or $t \ge 0.6803$ cm.

After comparing the three values of thickness (t) obtained above, $t = 0.681$ cm can be adapted.

$\therefore\ t = 6.81$ mm **Ans.**

(With $t = 6.81$ mm, failure will not occur in any of the three modes discussed above.)

6.4 GENERAL EXPRESSION OF STRESSES IN PRESSURE VESSELS

In this section, let us consider stress analysis in thin-walled pressure vessels which are in the form of surface of revolution having single axis of symmetry. A pressure vessel in this category is shown in Fig. 6.17. There are parallel circles of different radii on the surface of the vessel which are concentric to the axis of revolution. Also, on the surface there are longitudinal curves called meridians each of which lies on the same plane passing through the axis of revolution.

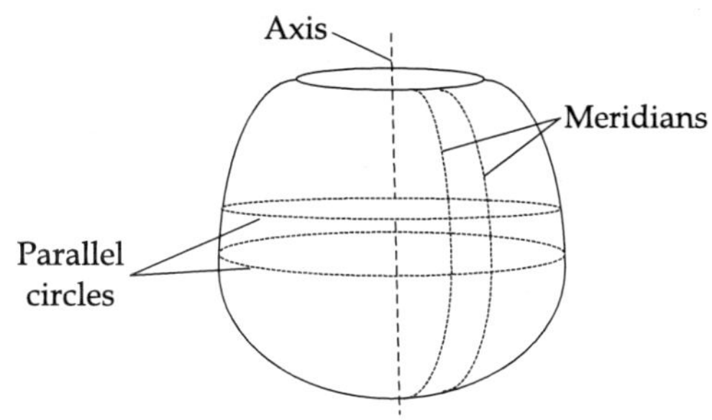

Fig. 6.17

When there is an internal pressure in the vessel, two normal stresses develop at any point in the wall of the vessel. One is hoop stress, which acts tangential to periphery or along the parallel circle and the other one is meridional stress acting along the meridian. These stresses are derived with the following assumptions:

1. There should be no discontinuity or sharp bend in meridional curves. As such, there should be no stress concentration at any point.
2. The normal stresses developed in the wall do not vary through the thickness of the wall of the vessel.

A thin-walled vessel in the form of surface of revolution is shown in Fig. 6.18. A very small element between two meridians and two parallel circles is taken with side (ds_1) along the circumference of parallel circle and side, ds_2 along meridian (see Fig. 6.18 (b)). The radius of curvature of the surface along the parallel circle is denoted by r_1 and that of surface along meridian by r_2. The angles subtended by sides ds_1 and ds_2 about the respective centres of curvature are denoted as $d\alpha_1$ and $d\alpha_2$, such that

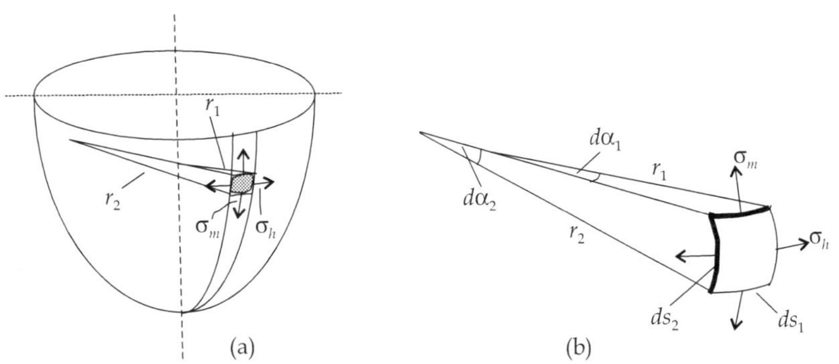

Fig. 6.18

$d\alpha_1 = ds_1/r_1$ and $d\alpha_2 = ds_2/r_2$. Because of internal pressure (p), the normal stresses occurring on the sides of the element are:

σ_h = hoop stress which is tangential to parallel circle and acts on the side, ds_2,

σ_m = meridional stress which is along the meridian and acts on the side, ds_1.

Also, t = thickness of the element in Fig. 6.18 (b), which is obviously the thickness of the vessel.

The force acting on the element due to pressure (p) acts perpendicular to the plane of the element along its outward normal, which is balanced by components of forces due to the normal stresses (σ_h and σ_m) acting on the sides of the element.

Hence, the force due to pressure (p) on the element = $p \cdot ds_1 \cdot ds_2$.

Force due to hoop stress, σ_h (see Fig. 6.19 (a)),

$$F_h = 2\cdot(\sigma_h \cdot t \cdot ds_2)\cdot(d\alpha_1/2),$$

since $\sin(d\alpha_1/2) \simeq d\alpha_1 2$, (for small value of $d\alpha_1$)

or $\quad F_h = \sigma_h \cdot t \cdot ds_1 \cdot ds_2/r_1$.

Similarly, the force due to meridional stress, σ_m,

$$F_m = \sigma_m \cdot t \cdot ds_1 \cdot ds_2/r_2.$$

It can be checked from the drawings that both F_h and F_m are acting in the direction opposite to F_p.

$\therefore F_h + F_m = F_p$ for equilibrium.

or $(\sigma_h \cdot t \cdot ds_1 \cdot ds_2/r_1) + (\sigma_m \cdot t \cdot ds_1 \cdot ds_2/r_2) = p \cdot ds_1 \cdot ds_2$

$$\therefore (\sigma_h/r_1) + (\sigma_m/r_2) = (p/t) \qquad (6.20)$$

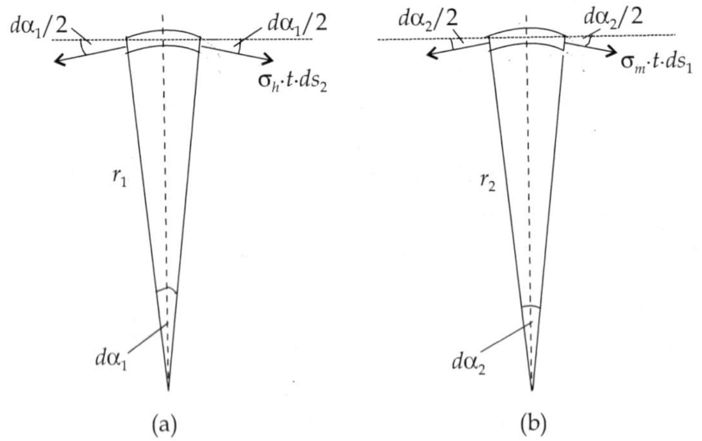

Fig. 6.19

There are instances when the radii (r_1 and r_2) are not on the same side of element. It implies that the force component (F_m) will be in same direction as F_p, and equilibrium will be satisfied according to $F_h - F_m = F_p$. For solving problems with vessels of revolution, the meridional stress (σ_m) is obtained from the given configuration and type of loading, which is followed by application of Eqn. 6.20 to find the tangential stress, σ_h.

Equation 6.20 derived above can be verified in the following cases:

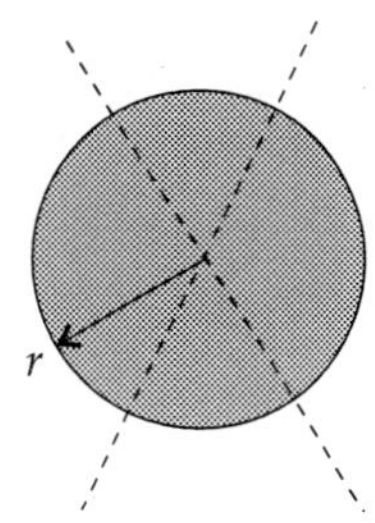

1. Stresses produced in spherical shell: Because of symmetry, $\sigma_h = \sigma_m = \sigma$, and $r_1 = r_2 = r$, so that
   $(\sigma/r) + (\sigma/r) = (p/t)$
   or $\sigma = p{\cdot}r/(2{\cdot}t) = p{\cdot}D/(4{\cdot}t)$, where, $D = 2r$.
2. Stresses produced in thin-walled cylinder: In this case,
   $\sigma = \sigma$ = hoop stress,
   $r_1 = r$ = radius of cylinder,
   and $r_2 = \infty$.
   $\therefore$ $(\sigma/r) = (p/t)$, or $\sigma = p{\cdot}r/t = p{\cdot}D/(2{\cdot}t)$.
   As $r_2 = \infty$ offers a particular condition, the longitudinal stress is obtained from the bursting pressure acting at the closed ends of the cylinder.

Example 6.12: A conical storage tank with apex angle $2\alpha = 60°$ is shown in Fig. 6.20. The tank is made of 10 mm steel plates and contains water up to a height, $h = 4$ m. Determine the meridional stress and tangential stress at a height, $y = 1.5$ m. Also, find the maximum values of meridional and tangential stresses developed in the wall of the tank. Density of water, $\rho = 1000$ kg/m^3.

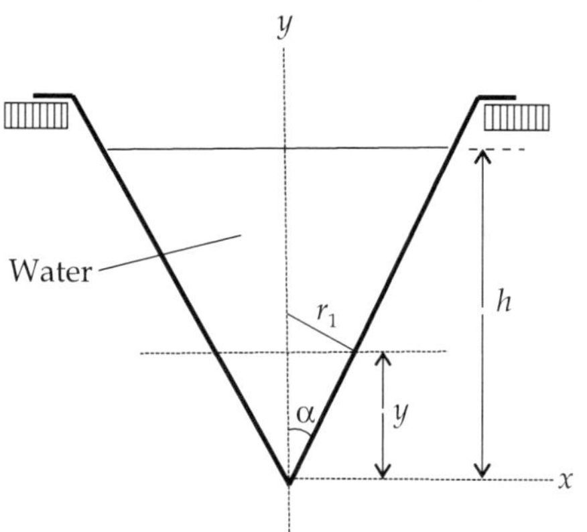

Fig. 6.20

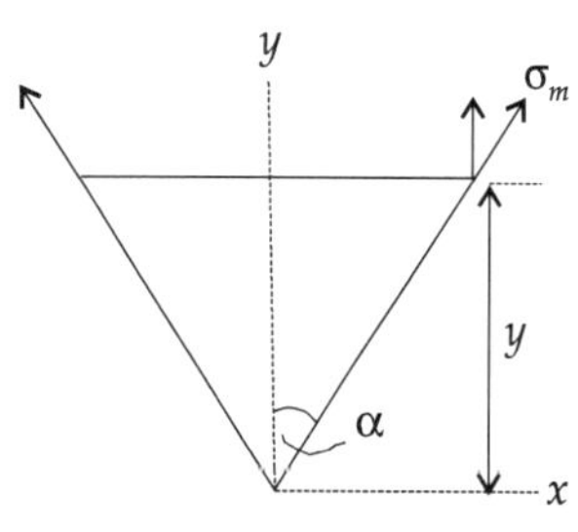

Solution: Pressure of water at a height y, $p = \rho{\cdot}g{\cdot}(h - y)$.

$\therefore$ $(\sigma_h/r_1) + (\sigma_m/r_2) = \rho{\cdot}g{\cdot}(h - y)/t$,

Where, t = wall thickness.

But, the radius of curvature of a meridian, $r_2 = \infty$.

$\therefore$ $\sigma_h/r_1 = \rho{\cdot}g{\cdot}(h - y)/t$, or $\sigma_h = \rho{\cdot}g{\cdot}r_1{\cdot}(h - y)/t$.

From Fig. 6.20, $r_1 = y \cdot \tan\alpha/\cos\alpha$.

$\therefore \quad \sigma_h = \rho \cdot g \cdot \tan\alpha \cdot y \cdot (h - y)/(t \cdot \cos\alpha)$ (a)

or $\quad \sigma_h = 1000 \times 9.81 \times \tan 30° \times 1.5 \times (4 - 1.5)/(10 \times 10^{-3} \times \cos 30°)$ N/m^2

$\sigma_h = 2.453$ MPa **Ans.**

The y-component of force due to σ_m is opposed by force of water pressure at the depth $(h - y)$ plus the weight of water up to level y.

For $\sum F_x = 0$, $\sigma_m \cdot (2\pi \cdot y \cdot \tan\alpha \cdot t) \cdot \cos\alpha = \rho \cdot g \cdot (h - y) \cdot \pi \cdot (y \cdot \tan\alpha)^2 + \rho g \cdot (1/3) \cdot \pi \cdot (y \cdot \tan\alpha)^2 \cdot y \cdot$

$\therefore \quad \sigma_m = \rho \cdot g \cdot y \cdot \{h - (2y/3)\} \cdot \tan\alpha/(2t \cdot \cos\alpha)$ (b)

or $\quad \sigma_m = 1000 \times 9.81 \times 1.5 \times \{4 - \{(2/3) \cdot (1.5)\} \cdot \tan 30°/(2 \times 10 \times 10^{-3} \times \cos 30°)$ N/m^2

$= 1471500$ N/m^2

$\sigma_m = 1.472$ MPa **Ans.**

Maximum tangential stress

For the tangential stress, σ_h to be maximum, $d\sigma_h/dy = 0$.

$\therefore$ from (a), $\{\rho \cdot g \cdot \tan\alpha/(t \cdot \cos\alpha)\} \cdot (h - 2y) = 0$, or $y = h/2$.

With $y = h/2$, $(\sigma_h)_{max} = \rho \cdot g \cdot h^2 \cdot \tan\alpha/(4 \cdot t \cdot \cos\alpha)$

Therefore, $\quad (\sigma_h)_{max} = 1000 \times 9.81 \times (4)^2 \times \tan 30°/(4 \times 10 \times 10^{-3} \times \cos 30°)$ N/m^2

$= 2.616$ MPa **Ans.**

Maximum meridional stress

For the meridional stress to be maximum, $d\sigma_m/dy = 0$.

$\therefore$ from (b), $\{\rho \cdot g \cdot \tan\alpha/(2t \cdot \cos\alpha)\} \cdot \{h - (4y/3)\} = 0$ or $y = 3h/4$.

With $y = 3h/4$, $(\sigma_m)_{max} = 3 \cdot \rho \cdot g \cdot h^2 \cdot \tan\alpha/(16 \cdot t \cdot \cos\alpha)$

$(\sigma_m)_{max} = 3 \times 1000 \times 9.81 \times (4)^2 \times \tan 30°/(16 \times 10 \times 10^{-3} \times \cos 30°)$ N/m^2

$= 1.962$ MPa **Ans.**

Example 6.13: If the conical tank in Example 6.12 is tightly fitted with a cover plate at the top and a gas is filled in it at a pressure of 20 kPa, determine the stresses, σ_m and σ_h, at a height $y = 2$ m.

Solution: Using Eqn. 6.20 with $r_2 = \infty$, $\sigma_h/(y \cdot \tan\alpha/\cos\alpha) = p/t$,

$\therefore \sigma_h = p \cdot y \cdot \tan\alpha/(t \cdot \cos\alpha)$

or $\sigma_h = 20 \times 10^3 \times 2 \times \tan 30°/(10 \times 10^{-3} \times \cos 30°)$ N/m^2

$= 2.666$ MPa **Ans.**

It is observed that at cross section of tank at $y = 2$ m, the y-component of force due to σ_m is balanced by the force generated by gas pressure in opposite direction.

Therefore, from $\sum F_y = 0$, $\sigma_m \cdot (2\pi \cdot y \cdot \tan\alpha) \cdot t \cdot \cos\alpha - p \cdot \pi \cdot (y \cdot \tan\alpha)^2 = 0$,

Hence, the meridional stress, $\sigma_m = p \cdot y \cdot \tan\alpha/(2 \cdot t \cdot \cos\alpha)$

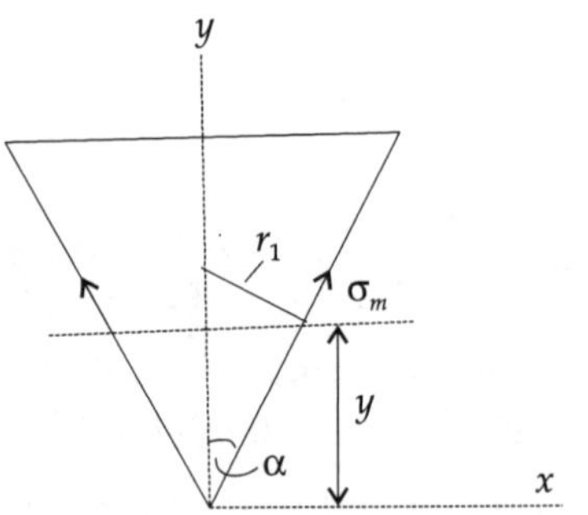

Fig. 6.21

$\therefore\ \sigma_m = 20 \times 10^3 \times 2 \times \tan 30°/(2 \times 10 \times 10^{-3} \times \cos 30°\ \text{N/m}^2$

$= 1.333$ MPa **Ans.**

Example 6.14: A pressure vessel in the shape of a paraboloid and covered at the top with a tightly fitted cover-plate is shown in Fig. 6.22. The vessel is made of 6 mm steel plate. The generating parabola is represented by $y = x^2/3$, where, x and y are in cm. If there is gas pressure of 1800 kPa in the vessel, determine the meridional stress (σ_m) and tangential stress (σ_h) in the shell at a point

$Y = 27$ cm above the bottom of the vessel.

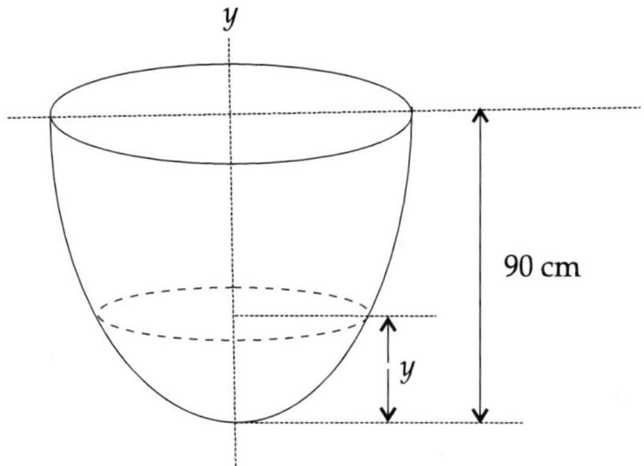

Fig. 6.22

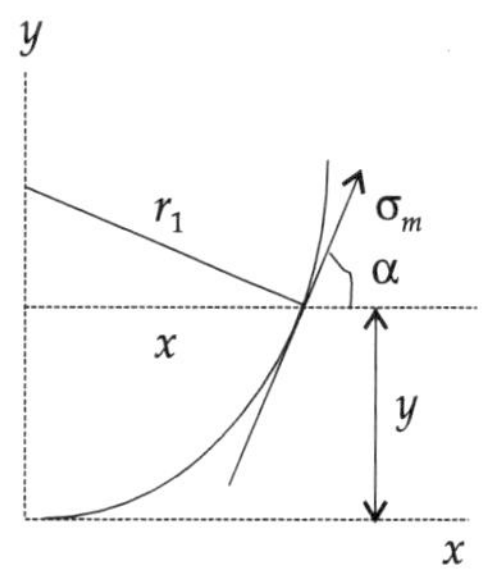

Solution: For $y = x^2/3$, $dy/dx = 2x/3$

At $y = 27$ cm, $x = 9$ cm and $dy/dx = 6 = \tan\alpha$

$\therefore$ $\sin\alpha = 6/\sqrt{37}$. Also, $d^2y/dx^2 = 2/3$.

Meridional stress, σ_m: $-p \cdot \pi \cdot x^2 + \sigma_m \cdot 2\pi \cdot x..\sin\alpha = 0$, $\therefore\ \sigma_m = p \cdot x / (2t \cdot \sin\alpha)$.

or $\sigma_m = 1800 \times 10^3 \times 9/\{2 \times 0.6 \times (6/\sqrt{37})\}\ \text{N/m}^2$

$= 13.7$ MPa **Ans.**

Tangential stress, σ_h : Radius of curvature of meridian at y, $r_2 = [1 + (dy/dx)^2]^{3/2}/d^2y/dx^2$.

or $r_2 = [1 + 6^2]^{1.5}/(2/3) = 337.59$ cm

Radius, $r_1 = x/\sin\alpha = 9 \times \sqrt{37}/6 = 9.124$ cm

Applying Eqn. 6.20,

$[\sigma_h/(9.124 \times 10^{-2})] + [13.7 \times 10^6/(337.59 \times 10^{-2})] = 1800 \times 10^3/(6 \times 10^{-3})$

$\therefore\ \sigma_h = 27 \times 10^6\ \text{N/m}^2$ or 27 MPa **Ans.**

Example 6.15: A closed hourglass-shaped small container is made of 0.5 mm aluminium sheet, as shown in Fig. 6.23. The equation of generating curve is given as $x^2 = (1/4) + y^2$, where, x and y are in cm. Internal pressure in the container is 10 kPa. Find the merdional stress and tangential stress at $Y = 3$ cm.

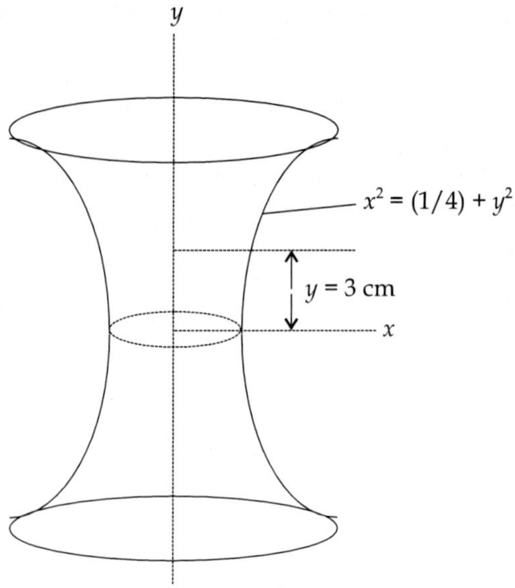

Fig. 6.23

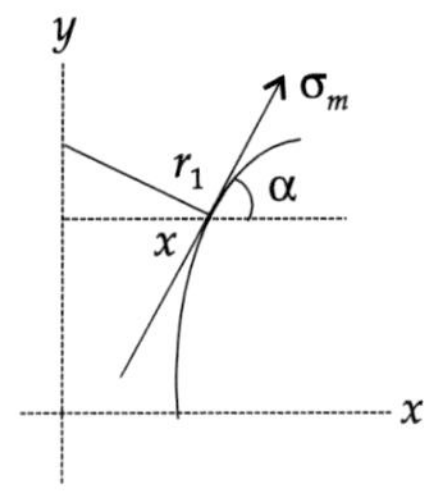

Solution: With $x^2 = (1/4) + y^2$: at $y = 3$ cm, $x = 3.04$ cm

Also, $2x = 2y\cdot(dy/dx)$, $\therefore$ $dy/dx = x/y = 3.04/3 = 1.0133$.

Hence, $\tan\alpha = 1.0133$, so that $\sin\alpha = 0.7118$

Meridional stress:

For $\sum F_y = 0$, $\sigma_m\cdot 2\pi\cdot x\cdot t\cdot\sin\alpha - p\cdot\pi\cdot x^2 = 0$

$\therefore$ $\sigma_m = p\cdot x/(2t\cdot\sin\alpha)$

or $\sigma_m = 10\times 10^3\times 3.04/(2\times 0.05\times 0.7118)\ \text{N/m}^2$

$= 427$ kPa **Ans.**

Tangential stress:

Radius, $r_1 = x/\sin\alpha = 3.04/0.7118 = 4.27$ cm

From $x = y\cdot(dy/dx)$, $1 = y\cdot(d^2y/dx^2) + (dy/dx)^2$, or $1 = 3\cdot(d^2y/dx^2) + (1.0133)^2$.

$\therefore$ $d^2y/dx^2 = -0.008926$.

Radius, $r_2 = [1 + (1.0133)^2]^{3/2}/(-0.008926)$, or $r_2 = -323.26$ cm

It is to note that r_1 and r_2 lie on opposite sides of the wall of the container and hence the force components of σ_h and σ_m are also opposite in sense. Therefore, applying Eqn. 6.20,

$(\sigma_h/4.27) + \{427\times 10^3/(-323.26)\} = 10\times 10^3/0.05$

$\therefore$ $\sigma_h = 859.6\times 10^3$ kPa, or $\sigma_h = 860$ kPa **Ans.**

Example 6.16: A rubber tube in the form of torus is made from 3 mm thick rubber sheet. A sectional view of torus is given in Fig. 6.24. The tube is inflated to an internal pressure of 180 kPa. Determine the meridional stress and tangential stress at the point *A* of the torus shown.

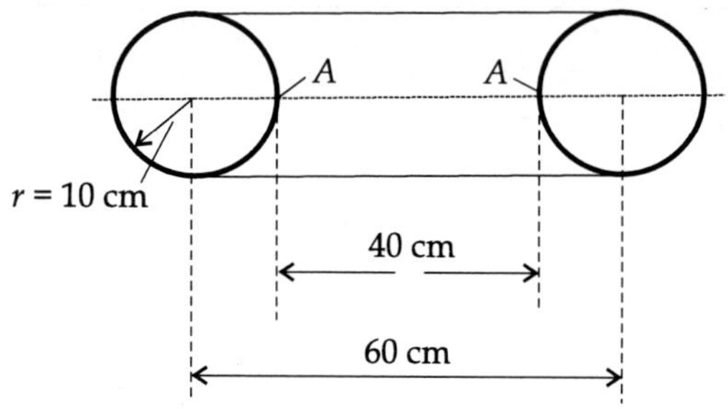

Fig. 6.24

Solution: Given that $p = 180 \times 10^3$ N/m^2, $t = 0.3$ cm, $r = 10$ cm

Meridional stress: $\sigma_m \cdot 2\pi \cdot r \cdot t = p \cdot \pi \cdot r^2$, or $\sigma_m = p \cdot r/(2.t)$

$$\sigma_m = 180 \times 10^3 \times 10/(2 \times 0.3) \text{ N/m}^2$$

$$= 3 \text{ MPa}$$ **Ans.**

Tangential stress: $r_1 = 10$ cm, $r_2 = 40/2 = 20$ cm.

It may be observed at the point A that the radii r_1 and r_2 are not on the same side of the wall or on the same side of an element at A. While r_1 is positive, r_2 is negative.

$\therefore (\sigma_h/10) + \{3 \times 10^6/(-20)\} = 180 \times 10^3/0.3$

$\sigma_h = 7.5$ MPa **Ans.**

PROBLEMS

1. A 3.5 m diameter cylindrical tank is made from 25 mm steel plate and contains a gas under pressure. If the circumferential stress is limited to 100 MPa and longitudinal strain is not to exceed 80×10^{-6}, what should be the maximum value of pressure in the cylinder?

   Take $E = 210$ GPa and $\nu = 0.30$ for steel.

   Ans. 1.27 MPa

2. A cylindrical tank of air compressor is 1 m in diameter, 1 m in length and contains air at a pressure of 2 MN/m^2. The tank is made from 25 mm steel plate ($E = 210$ GPa and $\nu = 0.30$). Determine (a) the maximum shearing stress in the wall of the tank, and (b) the change in volume after air is filled in the tank.

   Ans. (a) 10 MPa, (b) 74.8 cm^3

3. A cylindrical pressure vessel is constructed with a strip of plastic of thickness (t) and width (B) by rolling it into a helix, as shown in Fig. 6.25. A continuous fused joint is maintained along the helix. When there is pressure of a fluid in the cylinder, the normal tensile stress on the fused helical joint is not to exceed 80 per cent of the maximum normal stress in the parent plastic in cylinder's wall. Find the maximum permissible width (B) of the strip.

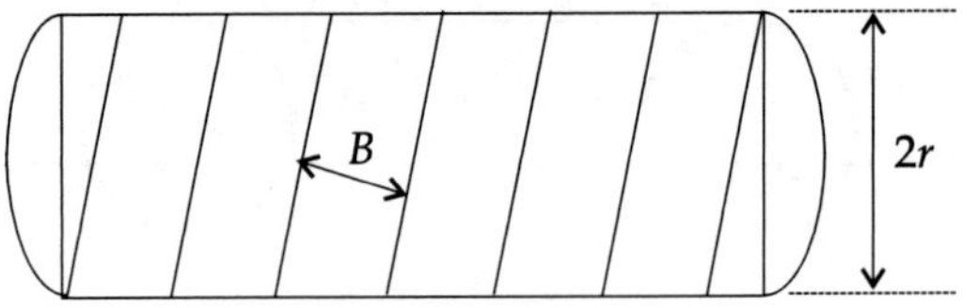

Fig. 6.25

Ans. $B = 4.9$ r

4. A cylindrical tank 1.5 m in diameter and 2 m long, is made from 10 mm thick steel plate. The tank is filled with a liquid at the atmospheric pressure. If an additional 25 cm^3 of liquid is now pumped in, what will be the pressure inside the tank and hoop stress developed in its wall? Ignore the compressibility of the liquid. Take $E = 210$ GPa and $\nu = 0.30$ for steel.

Ans. 10.4 kPa, 391 kPa

5. Two hemispherical shells, each of 0.5 m radius and 8 mm thick, are secured tightly by the arrangement of flanges and 16 bolts to form a spherical shell, as shown in Fig. 6.26. Determine the force developed in each bolt if a fluid is introduced in the shell at a pressure of 55 kPa.

Ans. 2.7 kN

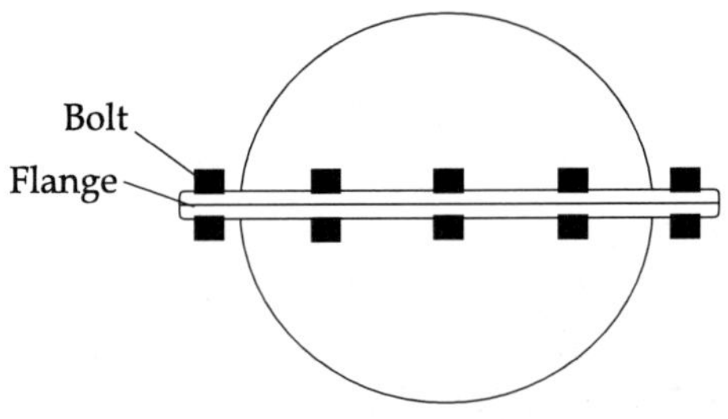

Fig. 6.26

6. A spherical vessel 1.8 m in diameter is made from 16 mm thick steel plate. It is filled with water at atmospheric pressure. What additional volume of water must be pumped into the vessel so that the internal pressure becomes 2 MPa ? The bulk modulus of water, $K = 2.2$ GPa and take $E = 210$ GPa, $\nu = 0.30$ for steel.

Ans. 4495×10^{-6} m^3

7. A 240 mm diameter ball has 1.5 mm wall thickness after it is inflated to a pressure of 80 kPa. What is the maximum normal stress and the maximum shear stress in the ball ?

Ans. 3.2 MPa, 1.6 MPa

8. The container shown in Fig. 6.27 is fabricated from 3 mm steel plate. If the internal pressure in the container is p = 850 kPa, determine the longitudinal stress (σ_L) and hoop stress (σ_c) on the curved side of container.

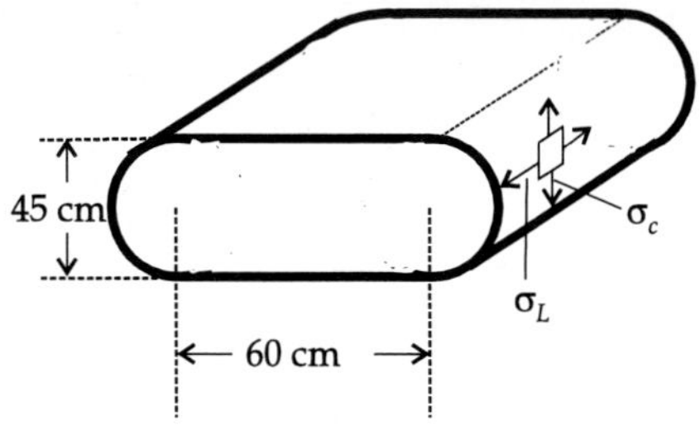

Fig. 6.27

Ans. $\sigma_L = 46.5$ MPa, $\sigma_c = 148.75$ MPa

9. A conical tank of 3 mm wall thickness is supported on a horizontal base, as shown in Fig. 6.28. It is subjected to internal gas pressure, $p = 50$ kPa. Neglecting self-weight of conical tank, find the stresses, σ_1 and σ_2, at the level $h = 1.5$ m below the apex. The semi apex angle is 20° as shown.

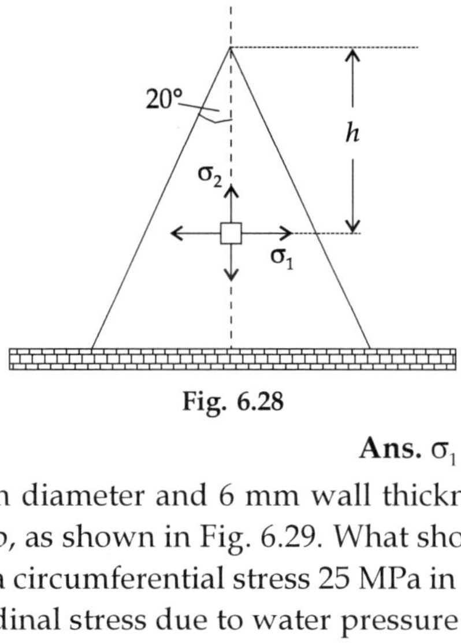

Fig. 6.28

Ans. $\sigma_1 = 9.68$ MPa, $\sigma_2 = 4.84$ MPa

10. A tall pipe of 1.8 m diameter and 6 mm wall thickness is fixed to the ground and open at the top, as shown in Fig. 6.29. What should be the height of water level (h) to induce a circumferential stress 25 MPa in the wall of the pipe? What will be the longitudinal stress due to water pressure at this point?

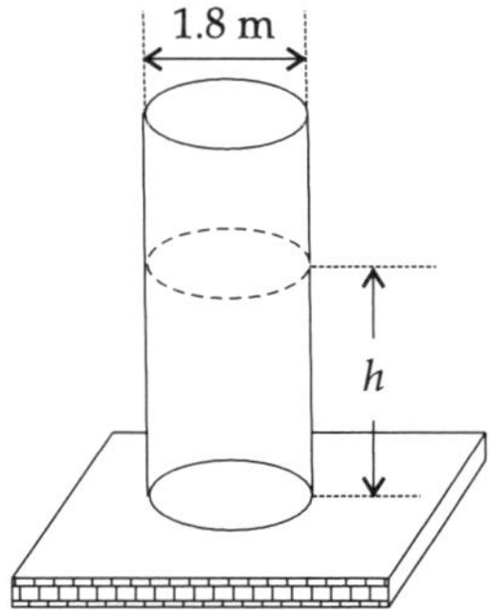

Fig. 6.29

Ans. $H = 17$ m, zero

7

Axisymmetric Problems: Thick Cylinders and Rotating Disks

7.1 INTRODUCTION

There are certain material bodies such as thick cylinders or circular disks, which are in the form of solids of revolution that are symmetric about their axes. The deformation of this type of bodies under the effect of external forces are also found to be symmetric about their axes of revolution. Hence, the stresses and strains in axisymmetric bodies can be more conveniently analyzed in cylindrical polar coordinates, rather than in Cartesian coordinates. Regarding this, the two types of normal stresses are dominant, which are (1) the radial stress (σ_r) acting in radial direction or in a direction perpendicular to the axis of revolution, and (2) the circumferential stress (σ_θ) which acts along tangential direction and remains fixed at a fixed radial distance from the axis. These stresses are included in the differential equation of equilibrium discussed in the following section.

7.2 STRESSES AND STRAINS IN CYLINDRICAL POLAR COORDINATES

Let us consider a general axisymmetric body shown in Fig. 7.1, in which the position of a point is defined by the coordinates r, θ and z. While r is the radial distance from z-axis, θ is the angular position with respect to some base direction x and the distance z is measured along z-axis from the base plane containing x-axis.

When external forces act, the deformation of body is symmetric about z-axis and there is no warping or distortion in r-θ, r-z or θ-z plane. As such, only the normal stresses (σ_r, σ_θ or σ_z) are induced and the shear stresses will be absent.

Let us consider an element lying on a plane normal to z-axis, as shown in Fig. 7.2. It is assumed that the plane element is very small in size, and is lying at a distance (r) z-axis and subtends an angle ($\Delta\theta$) about the point O. The side of this element nearer to the point O is of length $r\cdot\Delta\theta$ and the length of side farther from O is $(r + \Delta r)\cdot\Delta\theta$. The radial stress ($\sigma_r$) acts on the side of length, $r\cdot\Delta\theta$. Assuming a first degree variation of σ_r along radial direction, a stress $[\sigma_r + (d\sigma_r/dr)\cdot\Delta r]$ acts on the side of length $(r + \Delta r)\cdot\Delta\theta$.

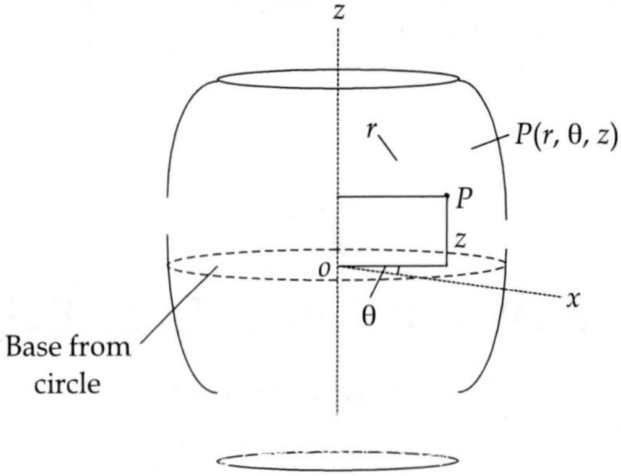

Fig. 7.1

Other two faces of the element are of length Δr on which the circumferential stress (σ_θ) is acting as shown. It should be noted that there is no variation in σ_θ along tangential direction, which is obvious for an axisymmetric case. Assuming the thickness of the element as unity (1), equilibrium condition for the plane stress case is satisfied according to $\sum F_r = 0$. Since $\sum F_\theta = 0$ is satisfied uniquely for any orientation of the element, it is not considered in the analysis.

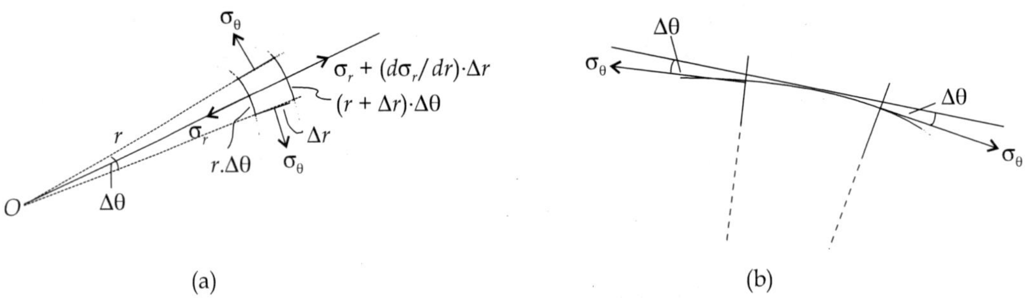

Fig. 7.2

Therefore, from $\sum F_r = 0$, $\{\sigma_r + (d\sigma_r/dr).\Delta r\}\cdot(r + \Delta r)\cdot\Delta\theta - \sigma_r\cdot r\cdot\Delta\theta - 2\cdot\sigma_\theta\cdot\Delta r\cdot(\Delta\theta/2) + V_r = 0$, (7.1)

where, V_r is the body force component along positive r (radial) direction. Also, Fig. 7.2 can be referred for the component of force due to σ_θ towards negative of r-direction. Equation 7.1 can be simplified to write

$$(d\sigma_r/dr) + \{(\sigma_r - \sigma_\theta)/r\} + \gamma_r = 0 \tag{7.2}$$

where, γ_r = body force component per unit volume along the radial direction.

It should be noted that in the simplification of Eqn. 7.1, the terms containing $(\Delta r)^2$, etc., are ignored because Δr and $\Delta\theta$ are very small quantities. The above expression is known as differential equation of equilibrium in cylindrical polar coordinates under the plane stress condition. When Eqn. 7.2 is applied for obtaining stresses in a thick cylinder, the body force component (γ_r) is ignored because the self-weight of a thick cylinder is generally small compared to the applied forces of pressure. On the other hand, γ_r is retained as the outcome of centrifugal action in the stress analysis of disks rotating with angular speeds.

Radial strain and circumferential strain in polar coordinates

The circumferential and radial strains at a point on a cross-sectional plane are obtained from displacement at that point in circumferential and radial directions, respectively. For this, let us consider a plane in circular shape as shown in Fig. 7.3. The centre is at the point O through which the axis of symmetry is perpendicular to the plane of paper. Two points A and B separated by a distance (Δr) are chosen on the radial axis, r. If the displacement at A in radial direction is denoted by u, displacement at the point B in radial direction is $u + (\partial u/\partial r)\cdot\Delta r$. The points A and B assume new positions A' and B' as a result of deformation, such that $AA' = u$ and $BB' = u + (\partial u/\partial r)\cdot\Delta r$. Therefore, the radial strain is

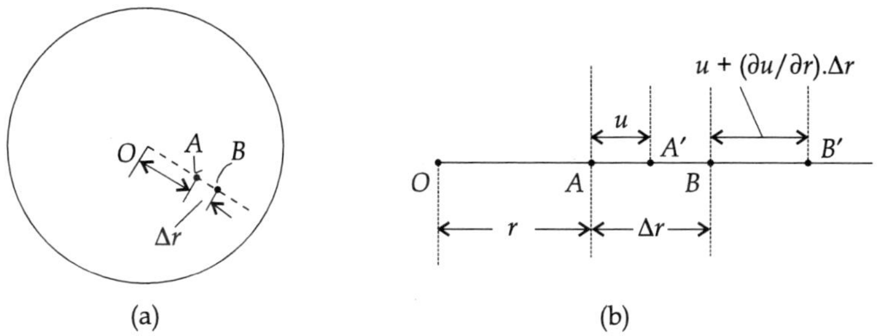

Fig. 7.3

written as $\varepsilon_r = (A'B' - AB)/AB = [(\Delta r - u) + \{u + (\partial u/\partial r)\cdot\Delta r\} - \Delta r]/\Delta r$

or $$\varepsilon_r = \partial u/\partial r \tag{7.3}$$

Next, with the arcs AA and $A'A'$ at distances r and $r + u$, respectively, from centre at O (see Fig. 7.4), the circumferential stress is written as

$$\varepsilon_\theta = (A'A' - AA)/AA = \{(r + u)\cdot\Delta\theta\} - r\cdot\Delta\theta\}/r\cdot\Delta\theta,$$

or $$\varepsilon_\theta = u/r \tag{7.4}$$

In the analyses of axisymmetric cases, the strains (ε_r and ε_θ) are related to stresses σ_r and σ_θ with the help of appropriate constitutive equations.

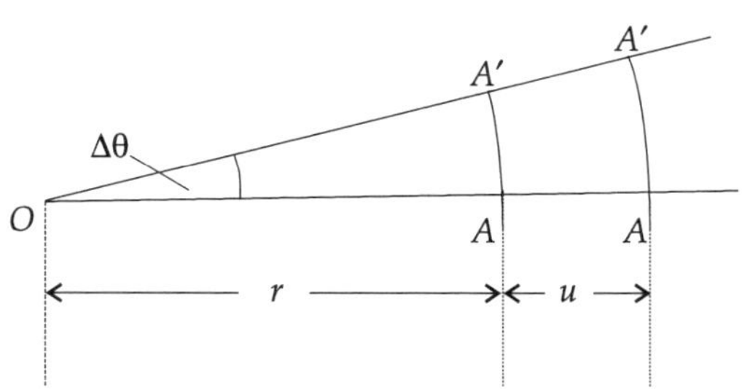

Fig. 7.4

7.3 THICK CYLINDERS

Thick cylinders are widely used in industries to carry steam, water, air or any fluid under pressure. Also, the barrels of light and heavy duty guns are designed on the basis of stress/strain analyses of thick cylinders. Unlike thin cylinders, the radial and circumferential stresses in thick cylinders are found to vary along wall-thickness of cylinder.

A cylinder of inner radius (r_i) and outer radius (r_o) is shown in Fig. 7.5. There is a uniform pressure (p_i) acting on the inner surface and a uniform pressure (p_o) on the outer surface. The stresses developed at any point on the crosssection of cylinder can be obtained either under the plane stress condition ($\sigma_z = 0$) or under plane strain condition ($\varepsilon_z = 0$).

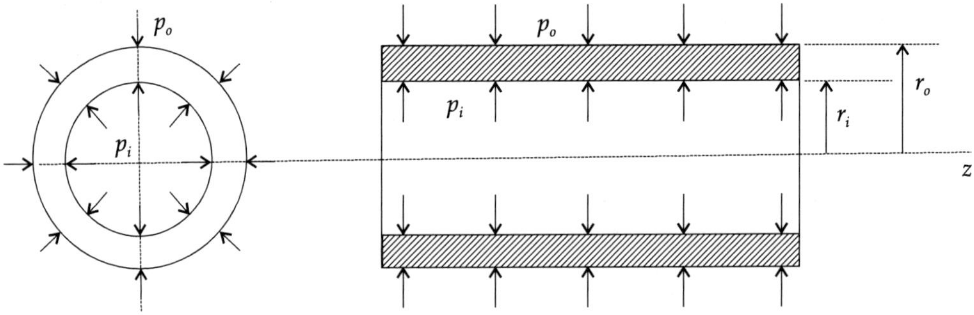

Fig. 7.5

7.3.1 Plane Stress

If the two ends of cylinder are not constrained, i.e., the cylinder is free to expand or contract along its axis (z-direction), there will be no stress in axial direction after the pressure p_i and p_o are applied. Only the radial stress and circumferential stress will be produced at the cross section and it is a plane stress problem. Since the loads due to pressures are symmetric about the axis, there will be no distortion or warping anywhere in the cylinder. Hence it is found that $\tau_{rz} = \tau_{\theta z} = 0$.

With $\sigma_z = 0$ and neglecting body-force per unit volume ($\gamma_r = \gamma_z = 0$), equilibrium equation is written according to Eqn. 7.2 as

$$(d\sigma_r/dr) + \{(\sigma_r - \sigma_\theta)/r\} = 0$$

Which can be rearranged in the form

$$\{d(r\cdot\sigma_r)/dr\} - \sigma_\theta = 0 \tag{7.5}$$

Next, using the stress-strain relation for the plane stress condition,

$$\varepsilon_r = (1/E)\cdot(\sigma_r - \nu\sigma_\theta),$$

and $$\varepsilon_\theta = (1/E)\cdot(\sigma_\theta - \nu\sigma_r).$$

The above can be rearranged as $\sigma_r = \{E/(1 - \nu^2)\}\cdot(\varepsilon_r + \nu\varepsilon_\theta)$,

and $$\sigma_\theta = \{E/(1 - \nu^2)\}\cdot(\varepsilon_\theta + \nu\varepsilon_r). \tag{7.6}$$

With $\varepsilon_r = du/dr$ and $\varepsilon_\theta = u/r$, we have

$$\sigma_r = \{E/(1 - \nu^2)\}\cdot\{(du/dr) + \nu\cdot(u/r)\},$$

and
$$\sigma_\theta = \{E/(1 - \nu^2)\}\cdot\{(u/r) + \nu\cdot(du/dr)\}. \quad (7.7)$$

From Eqn. 7.5 and Eqn. 7.7, we have

$$(d/dr)\cdot\{r\cdot(du/dr) + \nu\cdot u\} - \{(u/r) + \nu\cdot(du/dr)\} = 0,$$

or
$$r\cdot(d^2u/dr^2) + (du/dr) - (u/r) = 0. \quad (7.8)$$

A solution to the above differential equation is written as $u = C_1 + (C_2/r)$, where C_1 and C_2 are constants.

With $u = C_1 + (C_2/r)$, $u/r = C_1 + (C_2/r^2)$ and $du/dr = C_1 - (C_2/r^2)$, so that Eqn. 7.7 becomes

$$\sigma_r = \{E/(1 - \nu^2)\}\cdot[C_1\cdot(1 + \nu) - (C_2/r^2)\cdot(1 - \nu)],$$

$$\sigma_\theta = \{E/(1 - \nu^2)\}\cdot[C_1\cdot(1 + \nu) + (C_2/r^2)\cdot(1 - \nu)]. \quad (7.9)$$

The constants C_1 and C_2 are obtained by applying boundary conditions, as given below:

At $r = r_i$, $\sigma_r = -p_i$ and at $r = r_o$, $\sigma_r = -p_o$ (see Fig. 7.6).

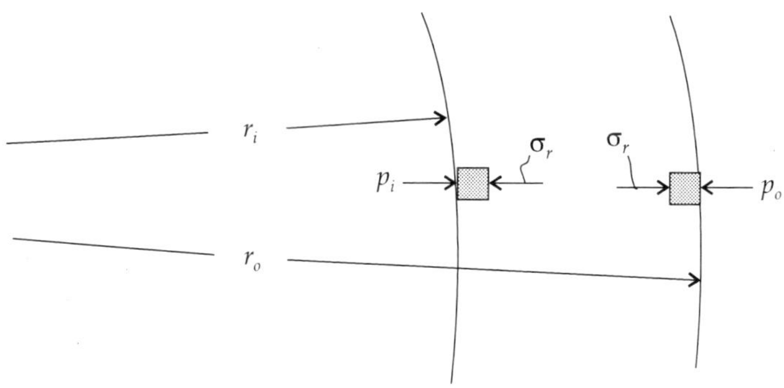

Fig. 7.6

Using the above conditions in Eqn. 7.9 and simplifying,

$$\left.\begin{aligned} C_1 &= \{(1 - \nu)/E\}\cdot\{(p_i\cdot r_i^2 - p_o\cdot r_o^2)/(r_o^2 - r_i^2)\}, \\ \text{and} \quad C_2 &= \{(1 + \nu)/E\}\cdot\{r_o^2\cdot r_i^2/(r_o^2 - r_i^2)\}\cdot(p_i - p_o). \end{aligned}\right\} \quad (7.10)$$

Therefore, substituting for constants C_1 and C_2 from Eqn. 7.10, the radial stress σ_r and circumferential stress σ_θ are obtained from Eqn. 7.9 and written as

$$\left.\begin{aligned} \sigma_r &= [(p_i\cdot r_i^2 - p_o\cdot r_o^2)/(r_o^2 - r_i^2)] - \{r_o^2\cdot r_i^2/(r_o^2 - r_i^2)\}\cdot\{(p_i - p_o)/r^2\}, \\ \text{and} \quad \sigma_\theta &= [(p_i\cdot r_i^2 - p_o\cdot r_o^2)/(r_o^2 - r_i^2)] + \{r_o^2\cdot r_i^2/(r_o^2 - r_i^2)\}\cdot\{(p_i - p_o)/r^2\}. \end{aligned}\right\} \quad (7.11)$$

The stresses σ_r and σ_θ are derived as functions of r, where r is the radial distance from the axis of the cylinder. In Fig. 7.7, these stresses are shown acting on a small element at the cross section of cylinder.

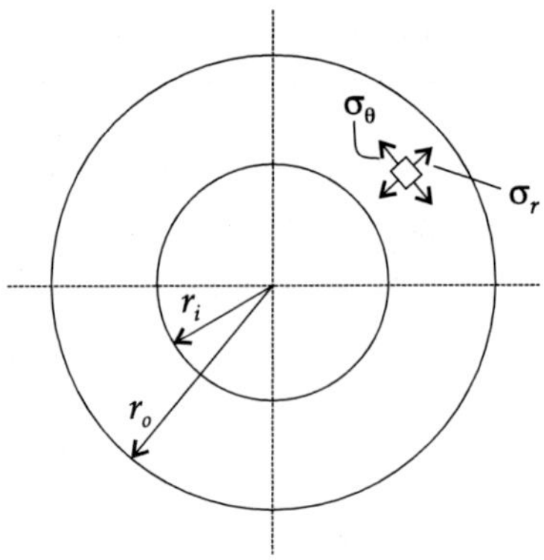

Fig. 7.7

It is found from Eqn. 7.11 that the sum $(\sigma_r + \sigma_\theta)$ is independent of 'r'. Since the strain in z- direction is given by $\varepsilon_z = -(\nu/E)\cdot(\sigma_r + \sigma_\theta)$, it implies that the extension or contraction in z-direction is same at every point of a cross section and a plane cross section remains plane after the forces due to pressures p_i and p_o are applied.

Further, using the expressions of C_1 and C_2 from Eqn. 7.10, the radial displacement at a point on the cross section is also written as a function of 'r' as

$$u = \{(1 - \nu)/E\}\cdot\{(p_i\cdot r_i^2 - p_o\cdot r_o^2)/(r_o^2 - r_i^2)\}\cdot r + \{(1 + \nu)/E\}\cdot\{r_o^2\cdot r_i^2/(r_o^2 - r_i^2)\}\cdot\{(p_i - p_o)/r\} \tag{7.12}$$

where E = elasticity modulus and ν = Poisson's ratio for the material of the cylinder.

The expressions of stresses were obtained in Eqn. 7.11 when both internal pressure p_i and external pressure p_o are acting on the cylinder. Let us now inspect the cases when $p_i \neq 0$, $p_o = 0$, and $p_i = 0$, $p_o \neq 0$, because these are the cases of practical importance.

Case I A cylinder subjected to internal pressure only

Assuming $p_o = 0$ and $p_i = p$, the radial and circumferential stresses are obtained from Eqn. 7.11 and written as

$$\sigma_r = \{p\cdot r_i^2/(r_o^2 - r_i^2)\}\cdot\{1 - (r_o^2/r^2)\},$$

and

$$\sigma_\theta = \{p\cdot r_i^2/(r_o^2 - r_i^2)\}\cdot\{1 + (r_o^2/r^2)\}. \tag{7.13}$$

Since $r \leq r_o$, the radial stress $\sigma_r \leq 0$ or always a compressive stress, and the circumferential stress $\sigma_\theta > 0$ or always a tensile stress. According to Eqn. 7.13 it is found that both σ_r and σ_θ vary hyperbolically across the thickness of cylinder. Also, it is found for the inner and outer surfaces of cylinder that

$\sigma_r = -p$ at $r = r_i$, and $\sigma_r = 0$ at $r = r_o$:

$\sigma_\theta = p\cdot\{(r_o^2 + r^2)/(r_o^2 - r_i^2)\}$ at $r = r_i$, and $\sigma_\theta = 2\cdot p\cdot r^2/(r_o^2 - r_i^2)$ at $r = r_o$.

Variation of radial and circumferential stresses across the thickness of cylinder is shown in Fig. 7.8 against the internal pressure p acting at the inner surface. It should be noted that in such a case the stress σ_θ is maximum at the inner surface of cylinder and radial stress σ_r is equal to pressure p in magnitude (see Fig. 7.9). At inner surface,

σ_θ is always greater than the magnitude of pressure p and it increases with the value of outer radius r_o. It cannot be reduced below p, whatever be the larger value of r_o.

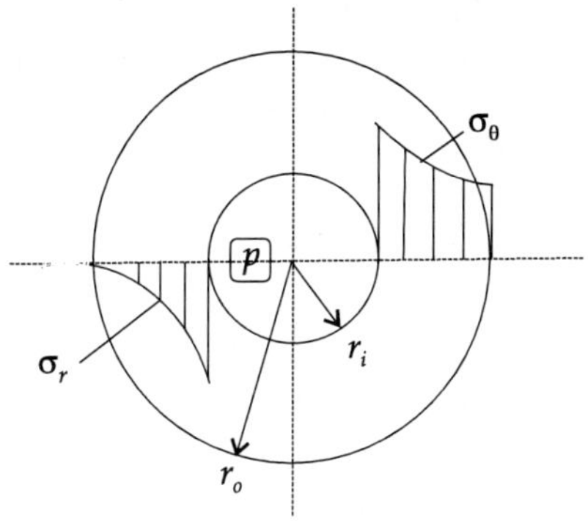

Fig. 7.8

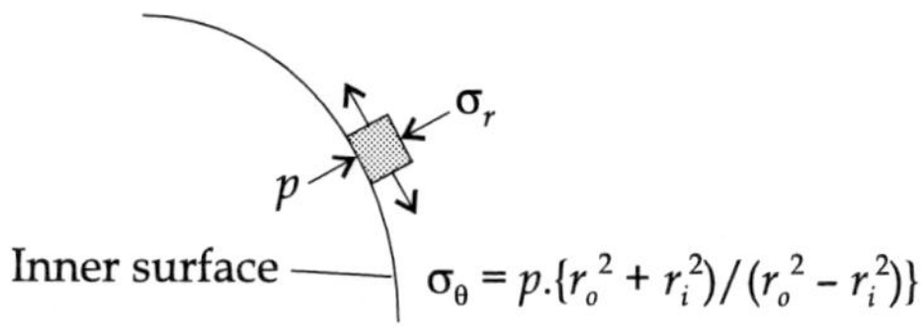

Fig. 7.9

Case II A cylinder subjected to external pressure only

If $p_i = 0$ and $p_o = p$, Eqn. 7.11 yields to

$$\sigma_r = -\{p\cdot r_o^2/(r_o^2 - r_i^2)\}\cdot\{1 - (r_i^2/r^2)\},$$

and $$\sigma_\theta = -\{p\cdot r_o^2/(r_o^2 - r_i^2)\}\cdot\{1 + (r_i^2/r^2)\}. \tag{7.14}$$

It is observed from the above equations that

$\sigma_r = 0$ at $r = r_i$, and $\sigma_r = -p$ at $r = r_o$;

$\sigma_\theta = -2\cdot p\cdot r_o^2/(r_o^2 - r_i^2)$ at $r = r_i$ and $\sigma_\theta = -p\cdot\{(r_o^2 + r_i^2)/(r_o^2 - r_i^2)\}$ at $r = r_o$.

Variation of σ_r and σ_θ are shown in Fig. 7.10. In this case, σ_r and σ_θ, are both compressive in nature.

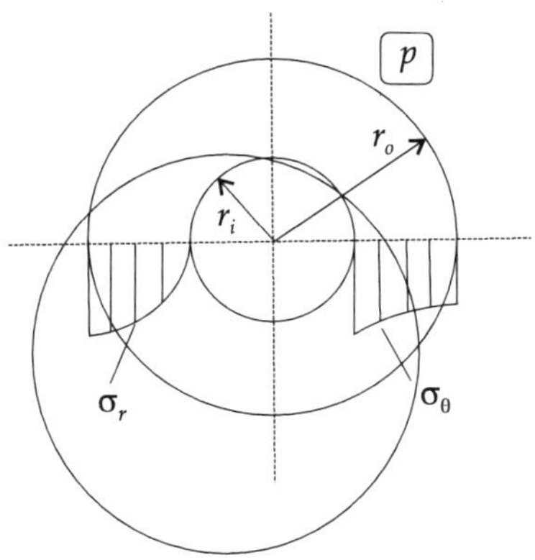

Fig. 7.10

If there is a solid cylinder ($r_i = 0$, $r_o \neq 0$), and an external pressure ($p_o = p$) is applied to it, it is found from Eqn. 7.14 that $\sigma_r = -p$ and $\sigma_\theta = -p$, and these stresses are uniformly distributed in the cylinder.

7.3.2 Plane Strain

If a cylinder is constrained at the two ends and the forces due to pressures (p_i and p_o) are applied to it, there is no displacement in z-direction at the points far from the ends. At such points a cross section is free from strain in axial (z-) direction and a plane strain condition is attained. Therefore, for any point on the cross section, the stress-strain relation is given by

$$\left.\begin{aligned} \sigma_r &= [E/\{(1-2\nu)\cdot(1+\nu)\}]\cdot[(1-\nu)\cdot\varepsilon_r + \nu\cdot\varepsilon_\theta], \\ \text{and} \quad \sigma_\theta &= [E/\{(1-2\nu)\cdot(1+\nu)\}]\cdot[(1-\nu)\cdot\varepsilon_\theta + \nu\cdot\varepsilon_\rho]. \end{aligned}\right\} \tag{7.15}$$

From Eqns. 7.5-.7.15 and using the relations ($\varepsilon_r = du/dr$, $\varepsilon_\theta = u/r$), a differential equation same as Eqn. 7.8 is obtained. Next, the solution of differential equation, as in plane stress case, leads to the expressions of radial stress (σ_r) and circumferential stress (σ_θ) which are same as in Eqn. 7.11. Also, a stress (σ_z) in axial direction is obtained under plane strain condition as

$$\sigma_z = \nu.(\sigma_r + \sigma_\theta) = 2\nu\cdot[(p_i\cdot r_i^2 - p_o\cdot r_o^2)/(r_o^2 - r_i^2)]. \tag{7.16}$$

Failure of thick cylinder in shear

Failure of a cylinder due to high internal pressure can be predicted according to maximum shear stress theory, written as $\tau_{max} = (\sigma_1 - \sigma_3)/2$, where, σ_1 and σ_3 are the maximum and minimum principal stresses at a point. When only an internal pressure is applied, the circumferential stress (σ_θ) is tensile (positive) in nature and radial stress (σ_r) is compressive (negative) in nature. Also, the magnitudes of both, σ_θ and σ_r, are maximum at the inner surface of the cylinder. Under plane stress condition, $\sigma_z = 0$, so that $\sigma_1 = \sigma_\theta$ and $\sigma_3 = \sigma_r$. As such, the maximum shear stress at any point on the inner surface becomes $\tau_{max} = (\sigma_\theta - \sigma_r)/2$. On the other hand, if there is a plane strain condition, the stress in axial direction, $\sigma_z = \nu.(\sigma_\theta + \sigma_r)$. The value of σ_z is found in between σ_θ and σ_r and hence $\tau_{max} = (\sigma_\theta - \sigma_r)/2$ still holds for the plane strain condition.

With $\quad \sigma_\theta = p\cdot[(r_o^2 + r_i^2)/(r_o^2 - r_i^2)]$ and $\sigma_r = -p$ at $r = r_i$, we have

$$(\sigma_\theta - \sigma_r)/2 = (1/2)\cdot[p\cdot\{(r_o^2 + r_i^2)/(r_0^2 - r_i^2)\} - (-p)]$$
$$= p\cdot r_o^2/(r_o^2 - r_i^2).$$

As $(\sigma_\theta - \sigma_r)/2 \le \tau_{max}$. $\quad p\cdot r_o^2/(r_o^2 - r_i^2) \le \tau_{max}$, $\quad$ (7.17)

which gives a necessary condition that the cylinder does not fail at its inner surface.

Example 7.1: A thick cylinder with inner diameter of 100 mm and outer diameter of 200 mm is made from a material that can withstand a maximum shearing stress of 165 MN/m^2. Determine the maximum internal pressure that can be applied to the cylinder. What is the maximum tensile stress developed in cylinder if 80 per cent of permissible internal pressure is applied?

Solution: Given that $r_i = 5$ cm, $r_o = 10$ cm, $\tau_{max} = 165 \times 10^6$ N/m^2.

From Eqn. 7.17, $p\cdot 10^2/(10^2 - 5^2) = 165 \times 10^6$, or $p = 123.75$ MPa **Ans.**

The circumferential stress under the given condition is tensile in nature and maximum tensile stress occurs at the inner surface, it is given by

$\sigma_\theta = p \cdot [(r_o^2 + r_i^2)/(r_o^2 - r_i^2)$,

$= (0.80 \times 123.75) \cdot \{(10^2 + 5^2)/(10^2 - 5^2)\}$ MPa

$= 165$ MPa **Ans.**

Example 7.2: A thick-walled cylindrical shell of inside diameter 180 mm and outside diameter of 350 mm is subjected to an internal pressure of 150 MN/m². Determine:

(a) the maximum shearing stress in the cylinder, and

(b) the radial and tangential stresses at a point equidistant from inside and outside surfaces.

Solution: Given that $r_i = 9.0$ cm, $r_o = 17.5$ cm, and $p_i = p = 150 \times 10^6$ N/m².

(a) The maximum shearing stress (from Eqn. 7.17)

$\tau_{max} = p \cdot r_o^2/(r_o^2 - r_i^2) = 150 \times 10^6 \times (17.5)^2/(17.5^2 - 9^2)$

$= 203.94 \times 10^6$ N/m² or 204 MPa **Ans.**

(b) The radial distance of a point midway between inner and outer surfaces

$r = (17.5 + 9.0)/2 = 13.25$ cm

$\therefore$ the radial stress, $\sigma_r = \{150 \times 10^6 \times 9^2/(17.5^2 - 9^2)\} \cdot \{1 - (17.5/13.25)^2\}$ N/m² (from Eqn. 7.13)

$\therefore \sigma_r = -40.15 \times 10^6$ N/m² $= 40.15$ MPa (compressive) **Ans.**

The tangential stress, $\sigma_\theta = \{150 \times 10^6 \times 9^2/(17.5^2 - 9^2)\} \cdot \{1 + (17.5/13.25)^2\}$ N/m² (from Eqn. 7.13)

$= 148$ MPa (tensile) **Ans.**

Example 7.3: A gun barrel is in the form of thick cylinder with 400 mm bore. If during an operation it develops internal pressure of 140 MN/m². Determine (a) outside diameter if the maximum tensile stress in gun barrel is limited to 250 MPa, and (b) the change in inner diameter when the gun is fired if the outside diameter is taken according to that found in part (a). Take $E = 207$ GPa and $G = 80$ GPa for the material of the gun.

Solution: $r_i = d_i/2 = 200$ mm or 20 cm, $p_i = p = 140 \times 10^6$ N/m².

(a) For $(\sigma_\theta)_{max}$ 250×10^6 N/m², $250 \times 10^6 = 140 \times 10^6 \cdot \{(r_o^2 + 20^2)/(r_o^2 - 20^2)\}$, (at $r = r_i$)

or $r_o = 37.77$ cm ≈ 378 mm **Ans.**

(b) For $E = 2G.(1 + \nu)$, $\nu = (E/2G) - 1$. $\therefore \nu = \{207/(2 \times 80)\} - 1$ or $\nu = 0.294$.

Hence, using Eqn. 7.12 the radial displacement at the inner surface,

$u = [p.r_i/\{E.(r_o^2 - r_i^2)\}] \cdot [(1 - \nu) \cdot r_i^2 + (1 + \nu) \cdot r_o^2]$

$= [140 \times 10^6 \times 20 / \{207 \times 10^9 \cdot (37.8^2 - 20^2)\}] \cdot [(1 - 0.294) \cdot (20^2) + (1 + 0.294) \cdot (37.8^2)]$

$= 28 \times 10^{-3}$ cm or 0.28 mm (increase)

$\therefore$ Change in the inner diameter, $\Delta d_i = 2 \times 0.28 = 0.56$ mm **Ans.**

7.4 COMPOSITE CYLINDERS

It was discussed earlier that the internal pressure in a single thick cylinder is restricted due to maximum shearing stress value of the material (see Eqn. 7.17). In other words, a high internal pressure, such as in gun barrels, may cause damage by shear failure. To eradicate this problem a composite cylinder is constructed which can withstand much higher internal pressure. A composite cylinder is formed by fitting a cylinder over another cylinder by shrink fit.

To construct a composite cylinder a cylinder of inner radius (r_i) and outer radius (r_c) is taken, as shown in Fig. 7.11(a). Another cylinder of inner radius [$r_c - \Delta$ ($\Delta << r_c$)] and outer radius (r_o) is arranged as shown in Fig. 7.11(b). This larger cylinder is heated so that its inner diameter increases due to thermal expansion and overcomes the outer diameter of smaller cylinder. It is then slipped on the smaller cylinder and allowed to cool down to room temperature. Finally, the larger cylinder fits over the smaller one by shrink fit (Fig. 7.11 (c)).

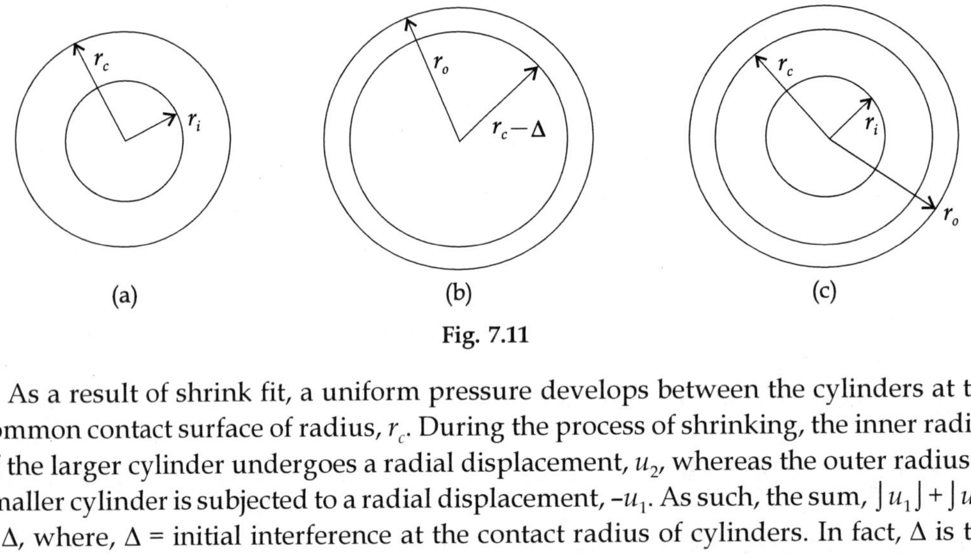

Fig. 7.11

As a result of shrink fit, a uniform pressure develops between the cylinders at the common contact surface of radius, r_c. During the process of shrinking, the inner radius of the larger cylinder undergoes a radial displacement, u_2, whereas the outer radius of smaller cylinder is subjected to a radial displacement, $-u_1$. As such, the sum, $|u_1| + |u_2| = \Delta$, where, Δ = initial interference at the contact radius of cylinders. In fact, Δ is the radial interference because according to system of tolerances (interference, clearance, etc.) as per IS: 2709-1982 and IS: 919-1993 interference is measured on the basis of diameter, so that initial interference = $2 \cdot \Delta$. However, for simplicity in calculations, only Δ based on radius is considered at this stage. A relation between pressure (p_c) at the contact surface and the interference (Δ) is derived as follows:

Let the smaller cylinder be made from a material with elastic constants ν_1 and E_1. Its boundary conditions are

at $r = r_i$, $p_i = 0$ and $r = r_c$, $p_o = p_c$ (the contact pressure).

Hence, Eqn. 7.12 is used to write the radial displacement at the outer surface of the smaller cylinder as

$$u_1 = \{(1 - \nu_1)/E_1\}\cdot\{(-p_c\cdot r_c^2)/(r_c^2 - r_i^2)\}\cdot r_c + \{(1 + \nu_1)/E_1\}\cdot\{r_c^2\cdot r_i^2/(r_c^2 - r_i^2)\}\cdot(-p_c/r_c),$$

or $$u_1 = -[p_c\cdot r_c/\{E_1\cdot(r_c^2 - r_i^2)\}]\cdot[(r_c^2 + r_i^2) - \nu_1\cdot(r_c^2 - r_i^2)] \quad (7.18)$$

For the larger cylinder made from a material with elastic constants ν_2 and E_2, the boundary conditions are

At $r = r_c$, $p_i = p_c$ and $r = r_o$, $p_o = 0$.

Therefore, from Eqn. 7.12, the radial displacement at the inner surface of larger cylinder becomes

$$u_2 = \{(1 - \nu_2)/E_2\}\cdot\{(p_c\cdot r_c^2)/(r_o^2 - r_c^2)\}\cdot r_c + \{(1 + \nu_2)/E_2\}\cdot\{r_o^2\cdot r_c^2/(r_o^2 - r_c^2)\}\cdot(p_c/r_c),$$

or $$u_2 = [p_c\cdot r_c/\{E_2\cdot(r_o^2 - r_c^2)\}]\cdot[(r_o^2 + r_c^2) + \nu_2\cdot(r_o^2 - r_c^2)] \quad (7.19)$$

As $|u_1| + |u_2| = \Delta$, we have from Eqns. 7.18 and 7.19,

$$[p_c\cdot r_c/\{E_1\cdot(r_c^2 - r_i^2)\}]\cdot[(r_c^2 + r_i^2) - \nu_1\cdot(r_c^2 - r_i^2)] + [p_c\cdot r_c/\{E_2\cdot(r_o^2 - r_c^2)\}]\cdot[(r_o^2 + r_c^2) + \nu_2\cdot(r_o^2 - r_c^2)] = \Delta. \quad (7.20)$$

If both smaller and larger cylinders are made from same material, putting $\nu_1 = \nu_2 = \nu$, and $E_1 = E_2 = E$ in the above equation the following is obtained.

$$p_c = \{E\cdot\Delta/(2\cdot r_c^3)\}\cdot\{(r_c^2 - r_i^2)\cdot(r_o^2 - r_c^2)/(r_o^2 - r_i^2)\}. \quad (7.21)$$

The above equation indicates that the contact pressure (p_c) is proportional to the radial interference, Δ. As a result of uniform pressure developed at the outer surface of the smaller cylinder and inner surface of larger cylinder, the radial and circumferential stresses are induced in both the cylinders. These stresses can be obtained from Eqns. 7.13 and 7.14 as functions of radial distance (r) from the common axis of cylinders. The distribution of these stresses at cross sections of cylinders are presented in Fig. 7.12.

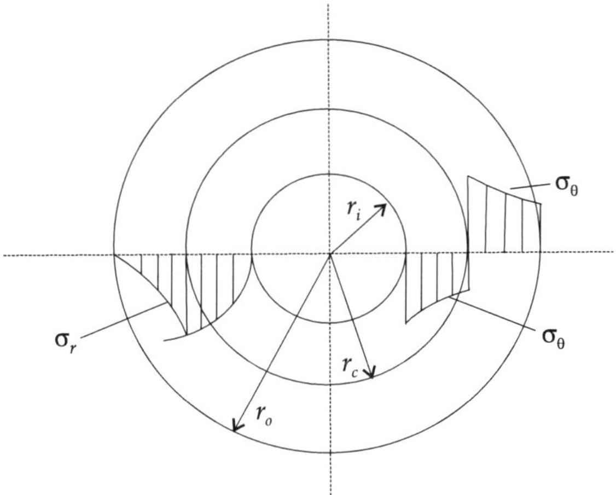

Fig. 7.12

It is found that the radial stress (σ_r) is compressive in both the cylinders. Circumferential stress (σ_θ) is compressive in the smaller cylinder, while it is tensile in the larger cylinder (Fig. 7.12). Both, σ_r and σ_θ, are residual stresses present in the cylinders due to shrink fit. When a composite cylinder is subjected to internal pressure (p_i) it behaves an angle cylinder and some additional stresses (radial and tangential) are produced which superimpose with the residual stresses already present at the cross section. The distribution of stresses due to internal pressure (p_i) and due to the contact pressure (p_c) are shown together in Fig. 7.13.

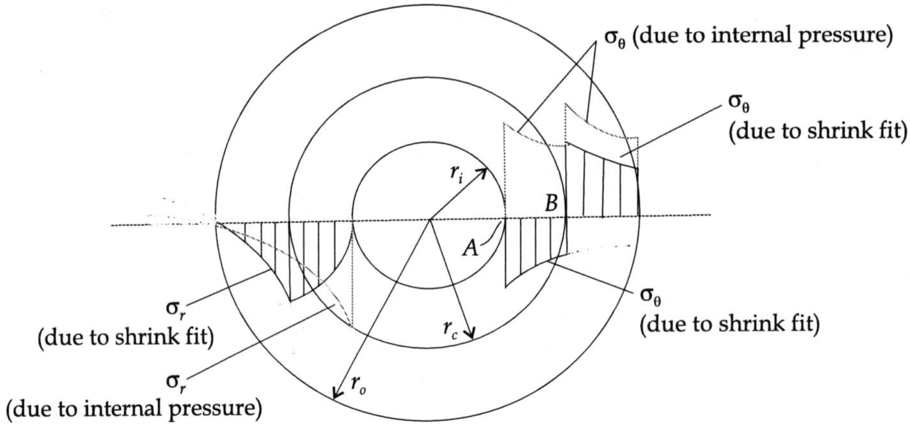

Fig. 7.13

It may be observed that at the point *A* (of inner surface) in Fig. 7.13, that σ_θ due to internal pressure is tensile in nature, which gets reduced due to already present residual compressive stress. Hence, a higher internal pressure can be applied so that the shearing stress can be kept below the maximum permissible value as per Eqn. 7.17. Further, it is found at a point like *B* (at the interface between two cylinders) that σ_θ due to internal pressure adds up with residual stress, since both are tensile in nature. Thus, the peak values of σ_θ occur at the point *A* ($r = r_i$) and at the point *B* ($r = r_c$).

From the standpoint of maximum shear stress theory, it is the concern of designers to maintain equal values of ($\sigma_\theta - \sigma_r$) at the points like *A* and *B*, such that ($\sigma_\theta - \sigma_r$) is minimum. This is accomplished according to the analysis given below.

At the point *A* ($r = r_i$),

$$(\sigma_\theta)_p = \{(r_o^2 + r_i^2)/(r_o^2 - r_i^2)\}\cdot p, \text{ and } (\sigma_r)_p = -p, \text{ due to the pressure } p.$$

$$\therefore\ (\sigma_\theta - \sigma_r)_p = \{(r_o^2 + r_i^2)/(r_o^2 - r_i^2)\}\cdot p - (-p) = \{2\cdot r_o^2/((r_o^2 - r_i^2)\}\cdot p.$$

Also $(\sigma_\theta)_c = \{-2\cdot r_c^2/(r_c^2 - r_i^2)\}\cdot p_c$, and $(\sigma_r)_c = 0$, due to contact pressure p_c.

$$\therefore\ (\sigma_\theta - \sigma_r)_c = \{-2\cdot r_c^2/(r_c^2 - r_i^2)\}\cdot p_c.$$

Therefore, at *A*, $(\sigma_\theta - \sigma_r)_A = (\sigma_\theta - \sigma_r)_p + (\sigma_\theta - \sigma_r)_c$.

$$\text{or } (\sigma_\theta - \sigma_r)_A = \{2\cdot r_o^2/((r_o^2 - r_i^2)\}\cdot p - \{2\cdot r_c^2/(r_c^2 - r_i^2)\}\cdot p_c \tag{7.22}$$

Similarly, at the point *B* ($r = r_c$),

$$(\sigma_\theta)_p = \{p\cdot r_i^2/(r_o^2 - r_i^2)\}\cdot\{1 + (r_o^2/r_c^2)\}, \text{ and } (\sigma_r)_p = \{p\cdot r_i^2/(r_o^2 - r_i^2)\}\cdot\{1 - (r_o^2/r_c^2)\}.$$

$$\therefore\ (\sigma_\theta - \sigma_r)_p = [2\cdot r_o^2\cdot r_i^2/\{r_c^2\cdot(r_o^2 - r_i^2)\}]\cdot p.$$

Also, $(\sigma_\theta)_c = \{(r_o^2 + r_c^2)/(r_o^2 - r_c^2)\}\cdot p_c$, and $(\sigma_r)_c = -p_c$.

$\therefore \quad (\sigma_\theta - \sigma_r)_c = \{2\cdot r_o^2/((r_o^2 - r_c^2)\}\cdot p_c$.

or $\quad (\sigma_\theta - \sigma_r)_B = [2\cdot r_o^2\cdot r_i^2/\{r_c^2\cdot(r_o^2 - r_i^2)\}]\cdot p + \{2\cdot r_o^2/(r_o^2 - r_c^2)\}\cdot p_c$. $\quad$ (7.23)

For $(\sigma_\theta - \sigma_r)_A = (\sigma_\theta - \sigma_r)_B$, Eqns. 7.22 and 7.23 can be equated and simplified further to obtain

$$(r_o^2/r_c^2)\cdot\{(r_c^2 - r_i^2)/(r_c^2 - r_i^2)\}\cdot p = [\{r_o^2\cdot(r_c^2 - r_i^2) + r_c^2\cdot(r_o^2 - r_c^2)\}/\{(r_o^2 - r_c^2)\cdot(r_c^2 - r_i^2)\}]\cdot p_c,$$

which can be rearranged to write

$$p_c = p\cdot(r_o^2/r_c^2)\cdot[\{(r_o^2 - r_c^2)\cdot(r_c^2 - r_i^2)\}/\{r_o^2\cdot(r_c^2 - r_i^2) + r_c^2\cdot(r_o^2 - r_c^2)\}] \quad (7.24)$$

Next, substituting for p_c in terms of p in Eqn. 7.22 and rearranging further, we have

$$(\sigma_\theta - \sigma_r)_A = p\cdot\{2\cdot r_o^2/((r_o^2 - r_i^2)\}\cdot\{1 - (1/\psi)\} \quad (7.25)$$

where, $\quad \psi = \{r_o^2/(r_o^2 - r_c^2)\} + \{r_c^2/(r_c^2 - r_i^2)\}$

Implying the condition that $(\sigma_\theta - \sigma_r)_A$ is to be minimum, the term (ψ) in the above expression should be minimum. Therefore, with $d\psi/dr_c = 0$ the following is written

$$\{2\cdot r_o^2\cdot r_c^2/(r_o^2 - r_c^2)^2\} - \{2\cdot r_c^2\cdot r_i^2/(r_c^2 - r_i^2)\} = 0,$$

so that the simplification of the above equation yields to

$$r_c = \sqrt{r_i r_o}. \quad (7.26)$$

With $r_c = \sqrt{r_i r_o}$, the composite cylinder will be subjected to minimum stresses, complying with the maximum shear stress theory. Using this expression of r_c in Eqn. 7.25, we have

$$(\sigma_\theta - \sigma_r)_{minimum} = p\cdot\{r_o/(r_o - r_i)\} \quad (7.27)$$

In order to find an expression of radial interference Δ, the relation between p_c and Δ from Eqn. 7.21 can be used with Eqn. 7.24 to write

$$\Delta = (2\cdot p/E)\cdot[r_c\cdot r_o^2\cdot(r_c^2 - r_i^2)/\{r_o^2\cdot(r_c^2 - r_i^2) + r_c^2\cdot(r_o^2 - r_i^2)\}] \quad (7.28)$$

Further, with $r_c = \sqrt{r_i r_o}$, the above equation becomes

$$\Delta = (p/E)\cdot\sqrt{r_i r_o}. \quad (7.29)$$

To maintain the minimum shearing stress at the cross section, the radial interference is chosen according to internal pressure applied to a composite cylinder.

Example 7.4: A composite cylinder is fabricated by shrinking a steel cylinder with inner diameter of 199.796 mm and outer diameter of 300 mm on to another steel cylinder having inner diameter of 100 mm and outer diameter of 200 mm. Determine the maximum internal pressure that can be applied if the circumferential stress at the inner surface of the composite cylinder is limited to 210 MPa. Take E = 210 GPa and ν = 0.30 for steel.

Solution: Given that r_o = 15 cm, r_c = 10 cm, r_i = 5 cm, $(\sigma_\theta)_{max}$ = 210 MPa at $r = r_i$, E = 210 GPa, ν = 0.30.

Radial interference, Δ = (200 – 199.796)/2 = 0.102 mm

From Eqn. 7.21, the contact pressure is written as

$$p_c = \{210 \times 10^9 \times 0.102 \times 10^{-3} \times (10^2 - 5^2) \times 10^{-4} \times (15^2 - 10^2) \times 10^{-4}\}/\{2 \times 10^3 \times 10^{-6} \times (15^2 - 5^2) \times 10^{-4}\} \text{ N/m}^2$$

or $p_c = 50.2 \times 10^6$ N/m² $\quad \therefore p_c = 50.2$ MPa.

The tangential stress at the inner surface, $\sigma_\theta = \{(r_o^2 + r_i^2)/(r_o^2 - r_i^2)\}\cdot p - \{2\cdot r_c^2/(r_c^2 - r_i^2)\}\cdot p_c$.

$\therefore \{(15^2 + 5^2)/(15^2 - 5^2)\}\cdot p - \{2 \times 10^2/(10^2 - 5^2)\} \times 50.2 \le 210$, or $p \le 275$ MPa **Ans.**

Example 7.5: A composite cylinder is formed by shrinking a steel cylinder on to another steel cylinder such that its dimensions are: inner diameter = 160 mm, outer diameter = 320 mm and the diameter at junction = 228 mm. The contact pressure at the interface is 15 MN/m². Determine (a) the shrinkage allowance applied to radius, (b) the minimum temperature through which the outer cylinder should be heated so as to slip down over the inner cylinder, and (c) the resultant tangential stresses in the composite cylinder if an internal fluid pressure of 20 MN/m² is introduced. Take the elasticity modulus, E = 200 GPa and coefficient of linear expansion, $\alpha = 0.6 \times 10^{-5}$/°C.

Solution: Given that $d_c = 228$ mm, $\therefore r_c = 11.4$ cm,

$d_o = 320$ mm, $\therefore r_o = 16$ cm,

$d_i = 160$ mm, $\therefore r_i = 8$ cm,

$p_c = 15$ MPa, $\quad p = 20$ MPa.

(a) Shrinkage allowance (Δ)

For $p_c = \{E\cdot\Delta/(2\cdot r_c^3)\}\cdot\{(r_c^2 - r_i^2)\cdot(r_o^2 - r_c^2)/(r_o^2 - r_i^2)\}$,

$15 \times 10^6 = \{200 \times 10^9\cdot\Delta/(2 \times (11.4)^3\}\cdot\{(11.4^2 - 8^2)\cdot(16^2 - 11.4^2)/(16^2 - 8^2)\}$

$\therefore \quad \Delta = 5.13 \times 10^{-3}$ cm

$= 0.0513$ mm **Ans.**

(b) For $\alpha\cdot\theta\cdot r_c = \Delta$, $0.6 \times 10^{-5} \times \theta\cdot 114 = 5.13 \times 10^{-2}$. $\quad \therefore \theta = 75°$C **Ans.**

(c) **Shrinkage stresses:**

Inner cylinder: At $r = r_i$, $\sigma_\theta = -\{2.r_c^2/(r_c^2 - r_i^2)\}.p_c$

$= -\{2 \times (11.4)^2/(11.4^2 - 8^2)\} \times 15$

$= -59.108$ MPa

At $r = r_c$, $\sigma_\theta = -\{(r_c^2 + r_i^2)/(r_c^2 - r_i^2)\}\cdot p_c$

$= -\{(11.4^2 + 8^2)/(11.4^2 - 8^2)\} \times 15$

$= -44.109$ MPa

Outer cylinder: At $r = r_c$, $\sigma_\theta = \{(r_o^2 + r_c^2)/(r_o^2 - r_c^2)\}\cdot p_c$

$= \{(16^2 + 11.4^2)/(16^2 - 11.4^2)\} \times 15$

$= +45.933$ MPa

At $r = r_o$, $\sigma_\theta = \{2.r_c^2/(r_o^2 - r_c^2)\}\cdot p_c$

$= \{2 \times (11.4)^2/(16^2 - 11.4^2)\} \times 15$

$= +30.933$ MPa

Stresses due to internal pressure:

Inner cylinder: At $r = r_i$, $\sigma_\theta = \{(r_o^2 + r_i^2)/(r_o^2 - r_i^2)\}\cdot p$

$= \{(16^2 + 8^2)/(16^2 - 8^2)\} \times 20$

$= +33.333$ MPa

At $r = r_c$, $\sigma_\theta = \{p.r_i^2/(r_o^2 - r_i^2)\}\cdot\{1 + (r_o^2/r_c^2)\}$

$= \{20 \times 8^2/(16^2 - 8^2)\}\cdot\{1 + (16^2/11.4^2)\}$

$= +19.799$ MPa

Outer cylinder: At $r = r_c$, $\sigma_\theta = +19.799$ MPa

At $r = r_o$, $\sigma_\theta = \{2.r_i^2/(r_o^2 - r_i^2)\}.p$

$= \{2 \times 8^2/(16^2 - 8^2)\} \times 20$

$= +13.333$ MPa

Resultant tangential stresses

Radius	Cylinder	Shrinkage stresses (MPa)	Stresses due to fluid pressure (MPa)	Resultant stress (MPa)
8.0	Inner	-59.108	+33.333	25.78 (C)
11.4	Inner	-44.109	+19.799	24.31 (C)
11.4	Outer	+45.933	+19.799	65.73 (T)
16.0	Outer	+30.933	+13.333	44.27 (T)

Example 7.6: A steel cylinder of 320 mm outside diameter and 210 mm inside diameter is shrunk over another steel cylinder of 120 mm inside diameter to form a composite cylinder. When an internal pressure of 150 MN/m^2 is applied, the maximum tensile stress developed in the smaller cylinder is 145 MPa. Find the initial shrinkage pressure and maximum tensile stress in the larger cylinder.

Solution: Given that $r_o = 16$ cm, $r_c = 10.5$ cm, $r_i = 6$ cm, $p = 150$ MPa.

Maximum tensile stress (tangential stress) in smaller cylinder, $(\sigma_\theta)_{max} = (\sigma_\theta)_p + (\sigma_\theta)_c$.

$\therefore\ 145 = \{(16^2 + 6^2)/(16^2 - 6^2)\} \times 150 + \{-2 \times 10.5^2/(10.5^2 - 6^2)\} \times p_c$

$\therefore\ p_c = 18.2$ MPa.

Initial shrinkage pressure = 18.2 MPa **Ans.**

Maximum tensile stress in the larger cylinder, $(\sigma_\theta)_{max} = (\sigma_\theta)_p + (\sigma_\theta)_c$.

$(\sigma_\theta)_{max} = \{150 \times 6^2/(16^2 - 6^2)\}\cdot\{1 + (16^2/10.5^2)\} + \{(16^2 + 10.5^2)/(16^2 - 10.5^2)\} \times 18.2$ MPa

$= 127$ MPa **Ans.**

7.5 ROTATING DISK

When a thin disk of uniform thickness rotates about its axis, the deformation of disk takes place due to centrifugal action. Equal radial displacements are observed for points lying at the same radial distance from the axis of rotation and the displacement varies with radial distance of the point. The net result is deformation of the disk which is symmetric about its axis. Hence, the radial and circumferential stresses are induced along with deformation of the disk. In this section, the stresses and displacements in rotating disks are derived as functions of radial distance.

Let us consider a disk rotating at an angular speed ω, as shown in Fig. 7.14. It is assumed that radial and circumferential stresses do not vary along the thickness of thin disk. Also, a plane stress condition implies that the stress (σ_z) in axial direction (z-axis) is zero. If we consider the dynamic equilibrium of an element of mass (m) at a radial distance (r) from the axis of rotation, the centrifugal force ($m\cdot r\cdot\omega^2$) is balanced by the forces due to radial and tangential stresses. Hence, a body force per unit volume in the radial direction is $m\cdot r\cdot\omega^2/V = \rho$, where, V is volume of elementary mass and ρ is the density of the material of the disk. Therefore, the equilibrium equation is written by including this body force, as

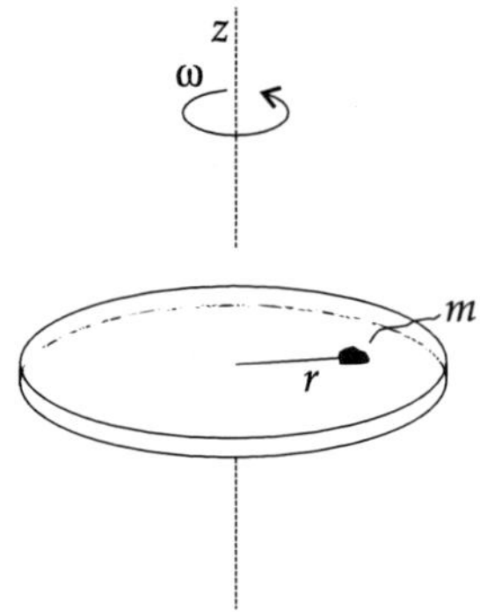

Fig. 7.14

$$d\sigma_r/dr + \{(\sigma_r - \sigma_\theta)/r\} + \rho r\omega^2 = 0, \qquad (7.30)$$

which can be rearranged as

$$d(r\cdot\sigma_r)/dr - \sigma_\theta + \rho r^2\omega^2 = 0. \qquad (7.31)$$

For plane stress condition at every point of the disk, the radial and tangential strains are written in terms of stresses, as

$$\left.\begin{aligned} \varepsilon_r &= (1/E)\cdot(\sigma_r - \nu\cdot\sigma_\theta), \\ \text{and} \quad \varepsilon_\theta &= (1/E)\cdot(\sigma_\theta - \nu\cdot\sigma_r), \end{aligned}\right\} \qquad (7.32)$$

where, σ_r is radial stress and σ_θ is tangential stress.

For tangential strain, $\varepsilon_\theta = u/r$, $u = r\cdot\varepsilon_\theta$. Hence, the radial strain, $\varepsilon_r = du/dr = d(r\cdot\varepsilon_\theta)/dr$.

From Eqn. 7.32, $\varepsilon_r = (d/dr)\cdot\{(1/E)\cdot(r\cdot\sigma_\theta - \nu\cdot r\cdot\sigma_r)\}$,

or $$\varepsilon_r = (1/E)\cdot\{(d/dr)\cdot(r\cdot\sigma_\theta - \nu\cdot r\cdot\sigma_r)\}. \qquad (7.33)$$

Next, using Eqn. 7,32, $\varepsilon_r = (1/E)\cdot(\sigma_r - \nu\cdot\sigma_\theta) = (1/E)\cdot\{(d/dr)\cdot(r\cdot\sigma_\theta - \nu\cdot r\cdot\sigma_r)$

or $$\sigma_r - \nu\cdot\sigma_\theta = (d/dr)\cdot(r\cdot\sigma_\theta - \nu\cdot r\cdot\sigma_r) \qquad (7.34)$$

Assuming $r\cdot\sigma_r = y$, we have $\sigma_r = y/r$ and $\sigma_\theta = (dy/dr) + \rho r^2\omega^2$ (from Eqn. 7.31)

Therefore, in Eqn. 7.34, $(y/r) - \nu\cdot(dy/dr) - \nu\,\rho r^2\omega^2 = (d/dr)\cdot\{r\cdot(dy/dr) + \rho r^3\omega^2\} - \nu\cdot(dy/dr)$,

Which can be simplified to write

$$r\cdot(d^2y/dr^2) + r\cdot(dy/dr) - y + (3 + \nu)\cdot\rho r^3\omega^2 = 0 \qquad (7.35)$$

The solution to the above differential equation is found in the form

$$y = C_1\cdot r + C_2\cdot(1/r) - \{(3 + \nu)/8\}\cdot\rho r^3\omega^2, \qquad (7.36)$$

where, C_1 and C_2 are the arbitrary constants.

Also, for $\sigma_r = y/r$, $\sigma_r = C_1 + C_2\cdot(1/r^2) - \{(3 + \nu)/8\}\cdot\rho r^2\omega^2$. (7.37)

Further, we have $\sigma_\theta = (dy/dr) + \rho r^2\omega^2 = C_1 - C_2\cdot(1/r^2) - \{3\cdot(3 + \nu)/8\}\cdot\rho r^2\omega^2 + \rho r^2\omega^2$

$$\therefore \quad \sigma_\theta = C_1 - C_2\cdot(1/r^2) - \{(1 + 3\nu)/8\}\cdot\rho r^2\omega^2 \qquad (7.38)$$

The constants, C_1 and C_2, in Eqns. 7.37 and 7.38 are determined according to boundary conditions in a rotating disk problem.

Case I: Solid disk

A thin solid disk of radius R does not contain hole at its centre (see Fig. 7.14). It should be noted in this case that the constant C_2 in Eqns. 7.37 and 7.38 must be absent because as $r \to 0$, the stresses (σ_r and σ_θ) cannot be defined. Since the radial stress, $\sigma_r = 0$ at the periphery ($r = R$), we have (from Eqn. 7.37),

$$C_1 - C = 0 \text{ or } C_1 = \{(3 + \nu)/8\} \cdot \rho r^2 \omega^2 .$$

The constant (C_1) obtained above is used to write the expressions of radial and tangential stresses in a solid rotating disk (from Eqns. 7.37 and 7.38) as

$$\sigma_r = \{(3 + \nu)/8\} \cdot \rho.\omega^2 \cdot (R^2 - r^2),$$

$$\sigma_\theta = \{(3 + \nu)/8\} \cdot \rho.\omega^2 \cdot R^2 - \{(1 + 3\nu)/8\} \cdot \rho \cdot \omega^2 \cdot r^2. \tag{7.39}$$

It can be checked that at $r = 0$, $\sigma_r = \sigma_\theta = \{(3 + \nu)/8\} \cdot \rho r^2 \omega^2$,

and $r = R$, $\sigma_r = 0$ and $\sigma_\theta = \{(1 - \nu)/4\} \cdot \rho R^2 \omega^2$.

It comes out that the stresses attain their maximum values at the centre of the disk.

Case II: Disk with a hole at its centre

Let us consider a thin rotating disk of outer radius r_o, containing a hole of radius r_i at its centre (see Fig. 7.15). The boundary conditions in this case are: at $r = r_i$, $\sigma_r = 0$ and at $r = r_o$, $\sigma_r = 0$. The constants (C_1 and C_2) are obtained from Eqns. 7.37 and 7.38, as

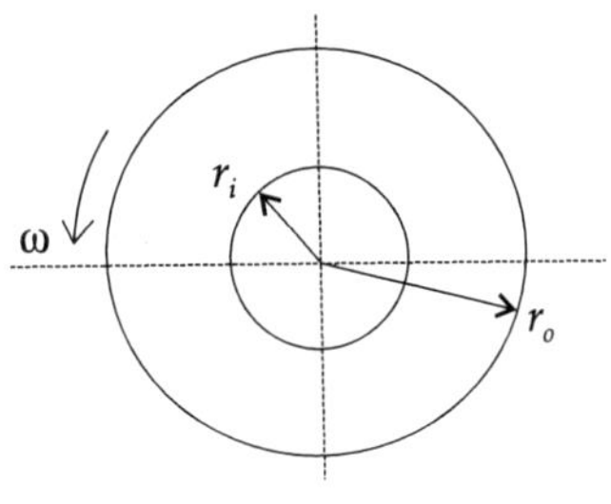

Fig. 7.15

$$C_1 = \{(3 + \nu)/8\} \cdot \rho.\omega^2 \cdot (r_o^2 + r_i^2),$$

and $$C_2 = -\{(3 + \nu)/8\} \cdot \rho.\omega^2 \cdot (r_o^2 \cdot r_i^2).$$

Therefore, equations of radial and tangential stresses become

$$\sigma_r = \{(3 + \nu)/8\} \cdot \rho \cdot \omega^2 \cdot [r_o^2 + r_i^2 - (r_o^2 \cdot r_i^2 / r^2) - r^2],$$

$$\sigma_\theta = \{(3 + \nu)/8\} \cdot \rho \cdot \omega^2 \cdot [r_o^2 + r_i^2 + (r_o^2 \cdot r_i^2 / r^2) - \{(1 + 3\nu)/(3 + \nu)\} r^2], \tag{7.40}$$

Maximum radial stress

It is found from Eqn. 7.40 that both, σ_r and σ_θ, are functions of radial distance (r) from the axis of rotation, and these stresses change with r. For maximum radial stress, $d\sigma_r / dr = 0$

or $(2r_o^2 r_i^2 / r^3) - 2r = 0$ (from Eqn. 7.40). Hence, $r = \sqrt{r_o r_i}$.

Using $r = \sqrt{r_o r_i}$, the maximum radial stress is obtained as

$$(\sigma_r)_{\max} = \{(3 + \nu)/8\} \cdot \rho \cdot \omega^2 \cdot (r_o - r_i)^2. \tag{7.41}$$

Maximum tangential stress

It can be observed in Eqn. 7.40 that σ_θ is maximum at $r = r_i$, so that

$$(\sigma_\theta)_{\max} = \{(3 + \nu)/4\} \cdot \rho \cdot \omega^2 \cdot [r_o^2 + \{(1 - \nu)/(3 + \nu)\} r_i^2]. \tag{7.42}$$

After comparing Eqn. 7.41 with Eqn. 7.42, it is found that $(\sigma_\theta)_{\max}$ is larger in magnitude than $(\sigma_r)_{\max}$.

Also referring to Eqn. 7.42, it is found that as $r_i \to 0$, $(\sigma_\theta)_{\max} = \{(3 + \nu)/4\} \cdot \rho \cdot \omega^2 \cdot r_o^2$, which is two times larger than $(\sigma_\theta)_{\max}$ obtained at the centre of a solid rotating disk (see

Eqn. 7.39). It implies that by introducing even a very small hole at the centre of a solid rotating disk the circumferential stress becomes two times larger.

Example 7.7: A thin circular disk of outer radius 25 cm and inner radius 4 cm is rotating at a speed of 2000 r.p.m. Determine:

(a) the radial and hoop stresses as functions of radial distance r in the disk,

(b) maximum hoop stress in the disk,

(c) maximum radial stress, and

(d) show that the variation of σ_r and σ_θ with r in a tabular form.

Take density, $\rho = 7800$ kg/m^3 and $\nu = 0.30$ for the material of the disk.

Solution: Given that $r_o = 25$ cm, $r_i = 4$ cm, $N = 2000$ r.p.m.

Angular speed, $\omega = 2\pi \times 2000/60$ rad/s = 209.4 rad/s.

(a) The radial stress (from Eqn. 7.40),

$$\sigma_r = \{(3 + 0.3)/8\} \times 7800 \times (209.4)^2 \times [25^2 + 4^2 - \{(25 \times 4)^2/r^2\} - r^2]\ \text{N/m}^2$$

or $\sigma_r = 1.41 \times 10^4 \times \{641 - (10^4/r^2) - r^2)\ \text{N/m}^2$ **Ans.**

The hoop stress (from Eqn. 7.40),

$$\sigma_\theta = \{(3 + 0.3)/8\} \times 7800 \times (209.4)^2 \times [25^2 + 4^2 + \{(25 \times 4)^2/r^2\} - \{(1 + 3 \times 0.3)/(3 + 0.3)\} \times r^2]\ \text{N/m}^2$$

or $\sigma_\theta = 1.41 \times 10^4 \times \{641 + (10^4/r^2) - 0.576\, r^2)\ \text{N/m}^2$ **Ans.**

(b) At $r = r_i = 4$ cm, maximum value of hoop stress $(\sigma_\theta)_{max} = 17.72 \times 10^6$ N/m^2 = 17.72 MPa **Ans.**

(c) At $r = \sqrt{r_o r_i} = \sqrt{25 \times 4} = 10$ cm, maximum radial stress is $(\sigma_r)_{max} = 6.2 \times 10^6$ N/m^2 = 6.2 MPa **Ans.**

(d) The stresses σ_r and σ_θ are calculated from the results in part (a) and given in the following Table.

r(cm)	4	10	15	20	25
σ_r(MPa)	0	6.2	5.24	3.05	0
σ_θ(MPa)	17.72	9.64	7.84	6.14	4.19

Example 7.8: A steel disk of 25 cm outer diameter and 5 cm inner diameter is shrunk on a steel shaft so that the pressure between the shaft and disk at standstill is 45 MN/m^2. Assuming the shaft to be rigid, find at what speed the disk will loosen from the shaft. Density, $\rho = 7800$ kg/m^3 and $\nu = 0.30$ for steel.

Solution: Given that $r_o = 12.5$ cm, $r_i = 2.5$ cm, $p_c = 45 \times 10^6$ N/m^2.

To achieve shrink-fit, the disk is heated to an extent that its inner diameter becomes slightly larger than the shaft's diameter. The disk is then inserted on the shaft and allowed to cool to room temperature so that it fits over the rigid shaft and its inner diameter becomes same as of the shaft diameter (see Fig. 7.16). As such, there is a residual circumferential strain at the inner surface of the disk. At stationary stage this strain is written as $\varepsilon_\theta = (1/E)\cdot(\sigma_\theta - \nu.\sigma_r)$, where $\sigma_r = -p_c = -45$ MPa, and

$$\sigma_\theta = \{(r_o^2 + r_i^2)/(r_o^2 - r_i^2)\}\cdot p_c$$
$$= \{(12.5^2 + 2.5^2)/(12.5^2 - 2.5^2)\}\cdot 45$$
$$= 48.75 \text{ MPa}$$

$$\therefore \quad \varepsilon_\theta = (1/E)\cdot\{48.75 - 0.3\cdot(-45)\} \times 10^6$$
$$= (1/E)\cdot 62.25 \times 10^6 \qquad \text{(a)}$$

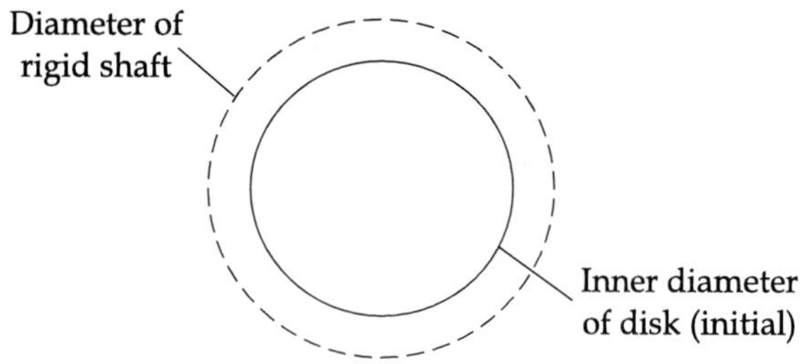

Fig. 7.16

When the disk rotates at the required angular speed (ω), inner diameter of the disk increases in size due to centrifugal action and it becomes equal to the diameter of the shaft. Then the contact pressure between shaft and the disk becomes zero and the disk is loose over the shaft. Under such condition the circumferential strain at the inner surface of the rotating disk is equal to the residual circumferential strain at the stationary stage.

At $r = r_i = 2.5$ cm, $\sigma_r = 0$ and $\sigma_\theta = \{(3 + \nu)/4\}.\ \rho\cdot\omega^2\cdot[r_o^2 + \{(1 - \nu)/(3 + \nu)\}r_i^2]$.

or $\quad \sigma_\theta = \{(3 + 0.3)/4\}\cdot(7800)\cdot\omega^2\cdot[12.5^2 + \{(1 - 0.3)/(3 + 0.3)\}\cdot 2.5^2] \times 10^{-4} \text{ N/m}^2$
$$= 101.4\cdot\omega^2 \text{ N/m}^2$$

$$\therefore \quad \varepsilon_\theta = (1/E)\cdot[101.4\cdot\omega^2 - 0] = 101.4\cdot\omega^2 \qquad \text{(b)}$$

Hence, equating (a) and (b), $101.4\cdot\omega^2/E = 62.25 \times 10^6/E$.

$\therefore \quad \omega = 783.5$ rad/s. Speed $N = 60\cdot\omega/2\pi = 7482$ r.p.m. **Ans.**

Alternative method: After shrink-fit is attained, there is an initial (residual) radial displacement at the inner radius of the disk. When the disk is in motion, there is a radial displacement at the inner radius given by $u = r\cdot\varepsilon_\theta$, where, ε_θ = circumferential strain at inner radius in rotating condition. As soon as both types of displacements are equal, the disk is loose over the shaft.

Initial radial displacement at inner radius of disk due to shrink-fit,

$$u_2 = \Delta = \{p_c\cdot r_c/E\cdot(r_o^2 - r_c^2)\}\cdot[(r_o^2 + r_c^2) + \nu\cdot(r_o^2 - r_c^2)],$$

or $\quad u_2 = \{45 \times 10^6 \times 2.5 \times 10^{-2}/E\cdot(12.5^2 - 2.5^2) \times 10^{-4}\}\cdot[(12.5^2 + 2.5^2) + 0.3\cdot(12.5^2 - 2.5^2)] \times 10^{-4}$ m
$$= 15.56 \times 10^5/\text{E m}.$$

For $u_2 = u$, where, $u = r\cdot\varepsilon_\theta$, $15.56 \times 10^5/E = 2.5 \times 10^{-2} \times (101.4\cdot\omega^2/E)$

$\omega = 783.5$ rad/s and N = 7482 r.p.m.

Example 7.9: A 750 mm diameter steel disk is shrunk and fitted on a solid steel shaft of 150 mm diameter, such that the contact pressure between them at stand still is 90 MN/m^2. Calculate:

(a) interference allowance and (b) the speed at which the disk will become loose over the shaft. Take

E = 210 GPa, ν = 0.30 and ρ = 7800 kg/m^3 for steel.

Solution:

(a) With $r_i = 0$ (solid sphere) in Eqn. 7.21, the contact pressure is written as

$p_c = \{E.\Delta/(2.r_c)\}\cdot\{1 - (r_c^2/r_o^2)\}$, where, $r_c = r$, the radius of shaft,

$\therefore\ 90 \times 10^6 = \{210 \times 10^9\cdot\Delta/(2 \times 7.5)\}\cdot\{1 - (7.5^2/37.5^2)\}$. or $\Delta = 0.067$ mm **Ans.**

(b) It should be noted that:

(i) The shaft is deformable and its radial dimension is affected by the centrifugal action. It has not been assumed to be rigid in the given problem.

(ii) When the disk becomes loose over the shaft, the contact between them ceases and $\sigma_r = 0$ for both of them at the interface.

Shaft

$\sigma_\theta = \{(1 - \nu)/4\}\cdot \rho\cdot\omega^2\cdot r_c^2$ (from Eqn. 7.39).

$\therefore\ \sigma_\theta = \{(1 - 0.3)/4\}\cdot (7800)\cdot\omega^2\cdot 7.5^2 \times 10^{-4}$ N/m^2 = $7.678\cdot\omega^2$ N/m^2,

and $\sigma_r = 0$.

$\therefore\ \varepsilon_\theta = (1/E)\cdot(7.678\cdot\omega^2 - 0.3 \times 0) = 7.678\cdot\omega^2/(210 \times 10^9)$

The radial displacement of the shaft at its outer surface (at a speed ω),

$U_s = r.\ \varepsilon_\theta = (7.5 \times 7.678\cdot\omega^2)/(210 \times 10^9)$ cm $= 57.585\omega^2/(210 \times 10^9)$ cm

Disk

$\sigma_\theta = \{(3 + 0.3)/4\}\cdot 7800\cdot\omega^2\cdot[37.5^2 + \{(1 - 0.3)/(3 + 0.3)\}7.5^2]$ N/m^2 (from Eqn. 7.42).

or $\sigma_\theta = 912.6\cdot\omega^2$ N/m^2,

and $\sigma_r = 0$.

$\therefore\ \varepsilon_\theta = (1/E)\cdot(\sigma_\theta - \nu\sigma_r) = 912.6\cdot\omega^2/(210 \times 10^9)$

Radial displacement at the inner surface of the disk, $u_d = r\cdot\varepsilon_\theta = 7.5 \times 912.6\cdot\omega^2/(210 \times 10^9)$ cm

or $u_d = 6844.5\cdot\omega^2/(210 \times 10^9)$ cm

If the shaft is deformable, its radial displacement (u_s) occurs due to centrifugal action, also. Hence, the contact between the shaft and the disk ceases at a certain speed (ω) such that

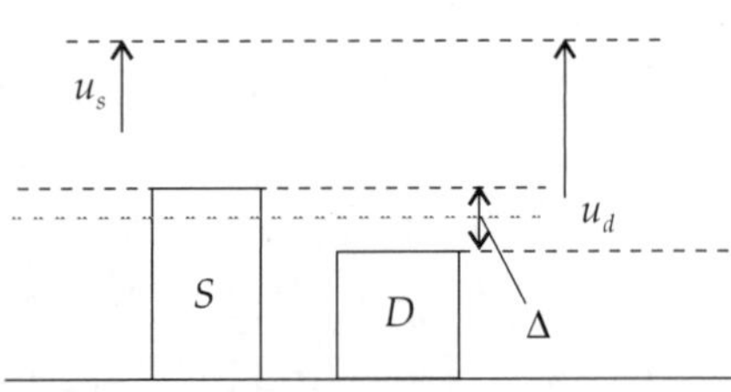

$u_d - u_s = \Delta$.

Therefore, with $u_d - u_s = \Delta$, $\{(6844.5 - 57.585)\cdot\omega^2\}/(210 \times 10^9) = 0.0067$

or $\omega = 455$ rad/s

$\therefore$ $N = 60 \cdot \omega / 2\pi = 4345$ r.p.m. **Ans.**

PROBLEMS

1. A thick cylinder of 200 mm inner diameter is subjected to an internal pressure of 50 N/mm^2. If the tensile stress in the cylinder is not to exceed 120 MPa, what should be the thickness of the cylinder? **Ans.** 55 mm
2. A thick cylinder of 50 mm inner diameter and 110 mm outer diameter is subjected to an internal pressure of 60 MPa and external pressure of 30 MPa. Find the maximum radial and tangential stresses in the cylinder.
   Ans. –60 MPa, 15.6 MPa
3. A hydraulic press consists of a cylinder having 200 mm inside diameter. What should be the thickness of the cylinder against an internal pressure of 50 MN/m^2 if the shearing stress in the cylinder is limited to 90 MPa? **Ans.** 50 mm
4. A cylinder of 160 mm outside diameter and 115 mm inside diameter is shrunk on to another cylinder of 80 mm inside diameter, so that a contact pressure of 30 N/mm^2 develops at the junction. Determine the tangential stresses induced at the inner and outer surfaces of the compound cylinder. Also, find the tangential stresses produced at the junction.
   Ans. At inner surface, $\sigma_\theta = 116$ MPa (C)
   At outer surface, $\sigma_\theta = 64$ MPa (T)
   At the junction:
   $\sigma_\theta = 86$ MPa (C) in the inner cylinder
   $\sigma_\theta = 94$ MPa (T) in the outer cylinder
5. A gear-hub of 120 mm outside diameter and 50 mm in length is shrunk on a shaft, 60 mm in diameter. If $\mu = 0.377$ at the contact surface between the hub and shaft and the tangential stress in the gear-hub at the junction is not to exceed 25 MPa, find the maximum torque that can be transmitted before the gear slips over the shaft. **Ans.** 1.6 kNm
6. A steel cylinder of inside diameter 212 mm and outside diameter 300 mm is shrunk on to another steel cylinder of inside diameter 150 mm to form a composite cylinder. The contact pressure due to shrink-fit is 30 MN/m^2. Determine (a) initial interference required to produce the contact pressure, and (b) maximum tensile stress developed in the composite cylinder when an internal pressure of 180 MPa is applied. Take $E = 200$ GPa and $\nu = 0.30$ for steel.
   Ans. 270 MPa
7. A thin disk of 50 cm outside diameter contains a circular hole of 20 cm diameter at its centre. The disk is running at 1500 r.p.m. Determine the values of maximum

radial and hoop stresses in the disk. At what radius the radial stress is maximum? Take density, $\rho = 7800$ kg/m^3 and $\nu = 0.30$ for disk material.

Ans. 1.78 MPa, 10.25 MPa, 15.8 cm

8. A tube of 200 mm inner diameter and 30 mm thickness is shrunk on to another tube of 200 mm outer diameter and 30 mm thickness. Both tubes are made from same material. The pressure at junction due to shrink fit is 15 N/mm^2. If the compound tube is subjected to an internal pressure of 90 N/mm^2, determine the resultant hoop stresses at the inner surface, the junction and the outer surface of compound tube.

Ans.

Radius (cm)	Cylinder	Shrinkage stress (MPa)	Stress due to fluid pressure (MPa)	Resultant stress (MPa)
7.0	Inner	-58.82	+163.50	104.68 (T)
10.0	Inner	-43.82	+98.86	55.04 (T)
10.0	Outer	+58.48	+98.86	157.34 (T)
13.0	Outer	+43.48	+73.50	116.98 (T)

9. A thin solid disk of 80 cm diameter is rotating at 3000 r.p.m. If $\rho = 7800$ kg/m^3 and $\nu = 0.30$ for disk material, what are the radial and tangential stresses at the centre and at the periphery of the disk? If a hole of 0.2 mm radius is drilled at the centre, what will be the tangential stress near the centre ?

Ans. At the centre, $\sigma_r = \sigma_\theta = 50.8$ MPa

At periphery, $\sigma_r = 0$ and $\sigma_\theta = 21.6$ MPa

101.6 MPa (near the centre)

10. A steel disk of 80 cm outside diameter with a 20 cm diameter circular hole at its centre is shrunk over a solid steel shaft. The radial allowance is 1 in 1000 parts.
    (a) At what r.p.m. the disk will become loose over the shaft?
    (b) What are the maximum radial and tangential stresses in the disk rotating at a speed calculated in part (a) ?
    (c) What is maximum tangential stress in the disk at stand still ?

Take $E = 200$ GPa, $\nu = 0.30$ and $\rho = 7800$ kg/m^3.

Ans. (a) 4219 r.p.m. (b) 56.5 MPa, 203 MPa (c) 106 MPa.

8

Torsion of Shafts

8.1 INTRODUCTION

A prismatic bar subjected primarily to twisting moment is known as a shaft. Shafts of circular cross section are widely used for the different practical applications. A major use of shafts is in the transfer of mechanical power from one point to another. Also a twisted shaft can be used as a spring of prescribed stiffness with respect to rotation. Since most of the shafts are cylindrical , it is necessary to discuss torsion of shafts of circular cross section. In the transmission of power by a shaft, it is required to determine the stresses produced due to the torque acting on the shaft. When a shaft is used as a torsional spring, the relation between the applied twisting moment and the resulting angular twist of shaft becomes important. The overall force-deformation relation is accomplished from the distribution of stress and strain in the entire shaft. In order to understand the effect of twisting on a shaft, it is convenient to begin the analysis by examining geometric behaviour of the twisted shaft with its angular deformation. Next, the strains, stresses and their relations are incorporated, and finally, the equilibrium condition is used to include the twisting moment applied to the shaft.

8.2 TORSIONAL SHEAR STRAIN IN CIRCULAR SHAFTS

To find the strain due to torsion, it is required to investigate the nature of deformation of a shaft under torsion. For this, a circular shaft of radius R and length L is considered, as shown in Fig. 8.1. A line AB on the surface is parallel to the axis of shaft. Two cross-sectional planes, circular in shape are taken at distances z and $z + dz$ from the left end. Two lines ab and cd are parallel to line AB and the ends of these lines lie on the periphery of circular sections. Hence, the lines, ab and cd, are parallel to each other. This shaft is now subjected to equal and opposite twisting moments (T) at its ends, which results into deformation of the shaft as shown in Fig. 8.2. The changes found in the deformed shaft are supported by experimental observation.

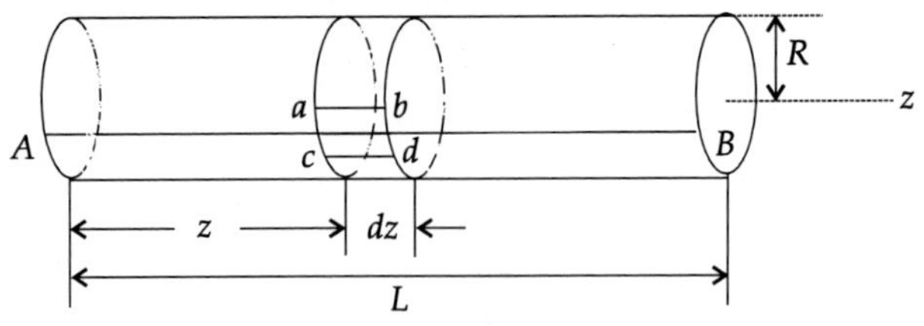

Fig. 8.1

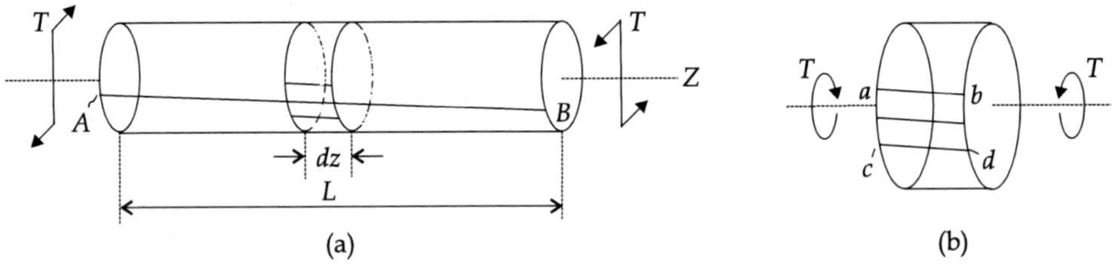

Fig. 8.2

After deformation, the line *AB* which is initially parallel to shaft-axis takes the form of a helix on the surface of the shaft. The two circular sections separated by *dz* remain transverse to the axis of shaft but rotate relative to each other such that the lines *ab* and *cd* also become parts of two helices (Fig. 8.2 (b)). Also, these lines remain parallel to each other. There is no distortion either of lines *ab* and *cd*, or periphery of circular sections. There is no change in separation, *dz*. As it is shown in Fig. 8.2(b), the portion of shaft between two circular sections is in equilibrium under the two equal and opposite twisting moments. In fact, the shaft can be assumed to be split up into a number of such elements each of which is subjected to same type of equilibrium condition and undergoes similar deformation as shown in Fig. 8.3.

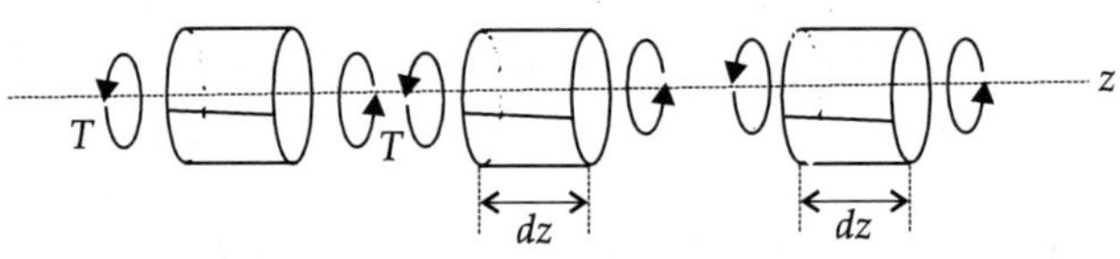

Fig. 8.3

After twisting moment is applied, the different cross sections successively rotate relative to one another and an initially straight line parallel to shaft-axis takes the shape of a helix. This is given in Fig. 8.4.

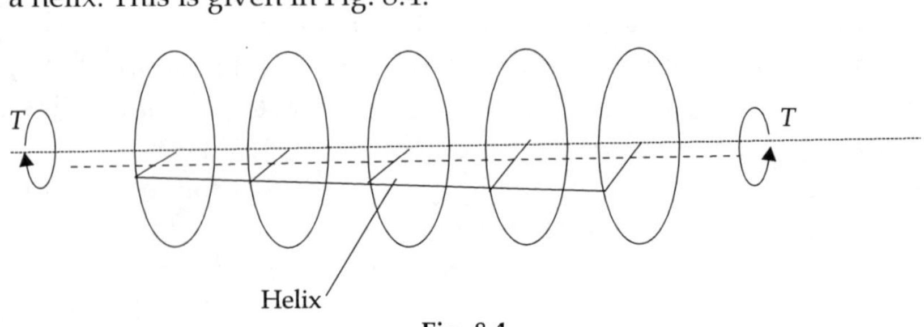

Fig. 8.4

Let us take a single element of length dz which is separated by two circular cross sections, as in Fig. 8.5. If the left section undergoes an angular displacement (φ) relative to the left end of shaft where, $\varphi = 0$, the section on the right side undergoes an angular displacement ($\varphi + d\varphi$). As such, the relative angular displacement between the two sections is $d\varphi$. According to the definition of helix, $d\varphi/dz$ = constant, which is consistently valid for every element shown in Fig. 8.3.

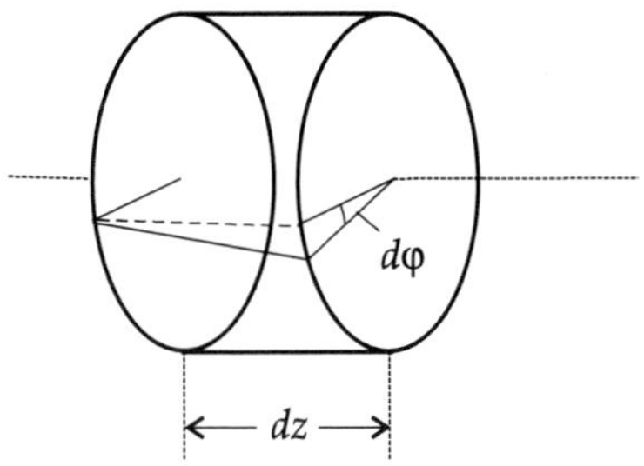

Fig. 8.5

From the discussions given above, some important conclusions regarding torsional configuration are as given below.

1. A plane cross section transverse to the axis of circular bar remains plane and perpendicular to the axis even after the bar is twisted.
2. The neighbouring cross sections rotate relative to one another throughout the length of the shaft such that the rate of change of angular displacement with respect to the change in length is constant.
3. A rectangular plane element between two circular sections and parallel to shaft-axis is subjected to pure shear. There is no significant axial deformation in such element.
4. The radius of a circular cross section remains straight even after twisting of the bar.

Experimental observation, provided that the amount of twist (angular deformation) is small. It is so for most structural metals. In case of materials like rubber for which there is possibly a large amount of twist, the assumptions given above are to be re-examined.

Since there is no radial or longitudinal dilatation and no warping due to twisting of the bar, all the extensional strains are zero, i.e., $\varepsilon_r = \varepsilon_\theta = \varepsilon_z = 0$. The only mode of deformation is one in which the adjacent cross sections remain un-deformed but rotate relative to each other. An element of radius r ($r < R$) between two cross sections is shown in deformed state in Fig. 8.6, in which $oa = oa' = o'b = o'b' = r$ ($r < R$), $ab = a'b' = dz$, $\angle aoa' = \angle bo'b' = \varphi$, and $\angle b'ob'' = d\varphi$. Also, oa' is along radial direction, $a'\varphi$ is tangential to the circle of radius r, and $a'b'$ is along z-direction.

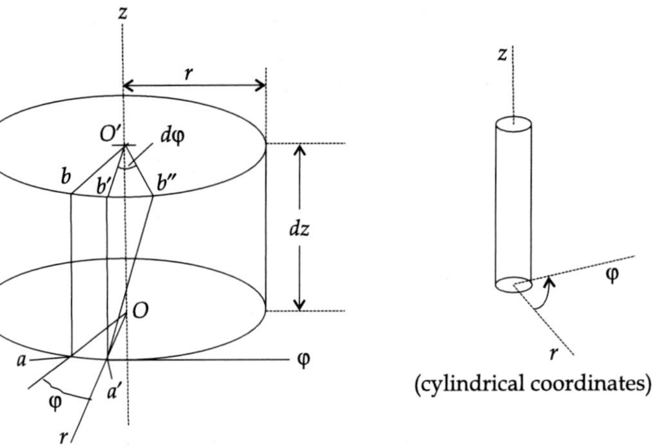

Fig. 8.6

Since $\angle oa'b' = \pi/2$ and there is no change of this angle, the shear strain $\gamma_{rz} = 0$. Also, for $\angle oa'\varphi = \pi/2$ and it remains unchanged, the shear strain $\gamma_{r\varphi} = 0$. But, $\angle b'a'\varphi = \pi/2$ and the change of this angle is $\angle b'a'b''$, the shear strain, $\gamma_{\varphi z} = \angle b'a'b''$ or $\gamma_{\varphi z} = b'b''/a'b' = r.d\varphi/dz$.

$$\therefore \qquad \gamma_{\varphi z} = r\cdot(d\varphi/dz). \tag{8.1}$$

It should be noted that $(d\varphi/dz)$ is constant according to the definition of helix. Hence, the shear strain, $\gamma_{\varphi z} \propto r$. At $r = 0$, $\gamma_{\varphi z} = 0$ and $\gamma_{\varphi z}$ is maximum at the surface of the shaft, where, $r = R$. The maximum shear strain manifests in the form of deformation of rectangular element such as *abcd* (Fig. 8.1). As explained with the help of Fig. 8.6, each slice of shaft of length *dz* deforms in same manner such that the rate of angular twist $d\varphi/dz$ is same for each of them. It implies that the same shear strain $(\gamma_{\varphi z})$ distribution is found throughout the length of the shaft.

8.3 TORSIONAL SHEAR STRESS AND TORSION FORMULA

Assuming that the homogeneous and isotropic material of cylindrical bar is linearly elastic and a small strain due to applied torque (T) results into stress within elastic limit, the stress-strain relations according to Hooke's law are applied in cylindrical coordinates to obtain

$$\left.\begin{aligned} \sigma_r = \sigma_\varphi = \sigma_z = \tau_{r\varphi} = \tau_{rz} = 0, \\ \tau_{\varphi z} = G\cdot\gamma_{\varphi z} = G\cdot r\cdot(d\varphi/dz). \end{aligned}\right\} \tag{8.2}$$

It is found that on applying twisting moment the only stress developed in the bar is shear stress, $\tau_{\varphi z}$. Like shear strain, the shear stress also varies proportionally with the radial distance (r) from the axis of the bar. The shear stress distribution for a slice of length (dz) is shown in Fig. 8.7. It is shown that the shear stress neither varies with φ nor with z, but only varies with r. At shaft surface, $\tau_{\varphi z} = \tau$ is maximum. Therefore, we have

$$(\tau_{\varphi z}/r) = (\tau/R)\,. \tag{8.3}$$

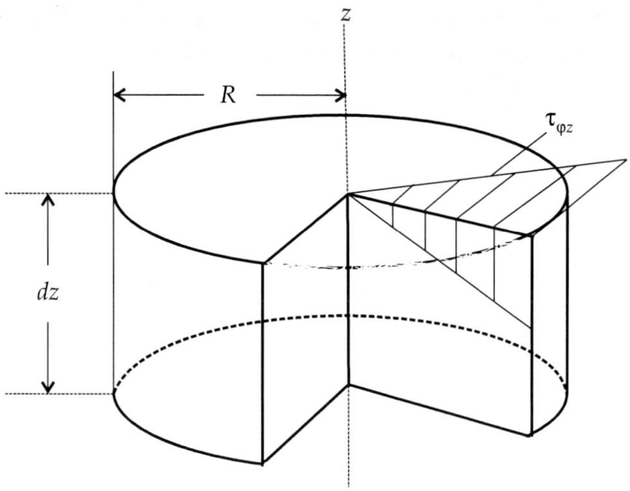

Fig. 8.7

Equilibrium requirement: A bar with twisting moment applied on it (as in Fig. 8.2) can be divided by a cross section into two parts and the part on the right end can be considered for equilibrium, as given in Fig. 8.8. The reaction at the exposed cross section can be found in the form of shearing stress distribution. Considering a small area (dA) on the cross section at a distance (r) from centre, a small force acting on this area will be $dF = \tau_{\varphi z} \cdot dA$ and an elementary torque will be $(\tau_{\varphi z} \cdot dA) \cdot r$.

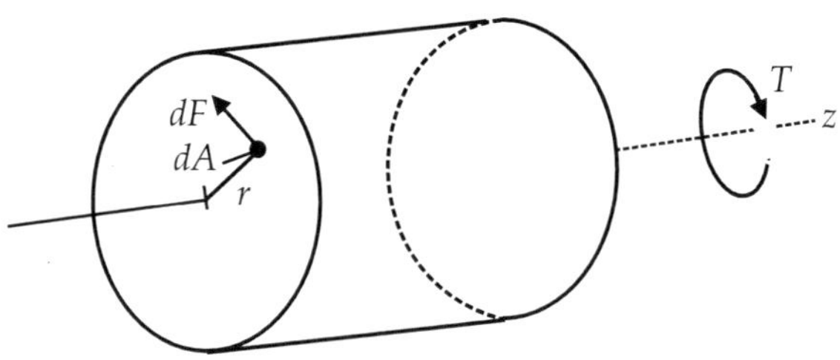

Fig. 8.8

Hence, for equilibrium of the part, we have $T - \int_A \tau_{\varphi z} \cdot dA \cdot r = 0$.

$$\therefore\ T = \int_A G \cdot r^2 \cdot (d\varphi/dz) \cdot dA = G \cdot (d\varphi/dz) \cdot \int_A r^2 \cdot dA,$$

$$\text{or } T = G \cdot (d\varphi/dz) \cdot I_p, \qquad (8.4)$$

where, $I_p = \int r^2 \cdot dA = \pi \cdot R^4/2 = (\pi/32) \cdot d^4$ is the polar moment of inertia of circular crosssection and $R = d/2$ with d = diameter of shaft.

It follows according to Eqn. 8.4 that a value of twisting moment (T) is obtained against a given radius) of the shaft, as the quantity $G \cdot (d\varphi/dz)$ remains fixed.

Torsion formula: For a bar of radius R and length L (as in Fig. 8.2) in which the total angular displacement of the cross section at right end is φ relative to $\varphi = 0$ at the left end, Eqn. (8.4) can be rearranged to write

$$\int_0^{\varphi} d\varphi = (T/(G{\cdot}I_p))\cdot\int_0^{L} dZ$$

or $\varphi = (T/(G{\cdot}I_p)){\cdot}L$, which is rearranged as

$$(T/I_p) = (G{\cdot}\varphi/L). \tag{8.5}$$

Also, from Eqn. (8.2), we have $(\tau_{\varphi z}/r)\cdot\int_0^{L} dZ = G\cdot\int_0^{\varphi} d\varphi$

or
$$(\tau_{\varphi z}/r) = (G\varphi/L) \tag{8.6}$$

Therefore, from Eqn. (8.3), Eqn. (8.5) and Eqn. (8.6), the following is obtained.

$$T/I_p = G{\cdot}\varphi/L = \tau_{\varphi z}/r = \tau/R, \tag{8.7}$$

which is known as "torsion formula" to establish relations between applied torque (T), radius of shaft (R), length (L) and shear modulus (G) of the material used. Torsion formula is used to obtain shear stress on the surface of a shaft. For example, a circular shaft of radius (R) and length (L) is shown in Fig. 8.9. A small square element is chosen on the surface of the shaft. The shear stresses (τ) acting on the element give rise to a state of pure shear. It should be noted that the shear stresses act on planes perpendicular to the surface of the shaft and very close to the surface, but these stresses are not acting directly on the surface. Because of pure shear, there are principal stresses, $\sigma_1 = \tau$ and $\sigma_2 = -\tau$, which are tensile and compressive stresses, respectively and act at 45° to shaft-axis.

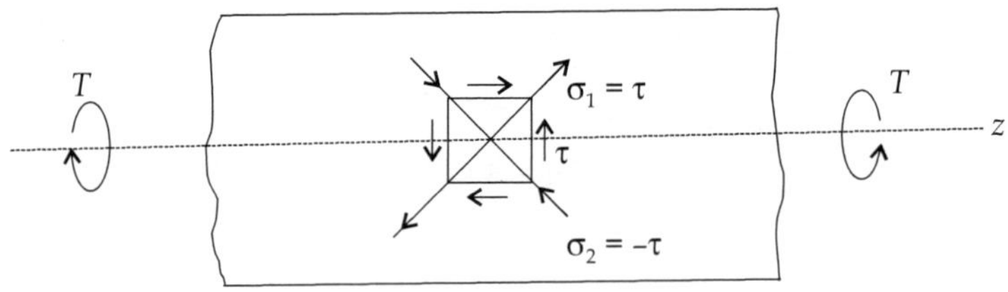

Fig. 8.9

Suppose a shaft is made from mild steel which is weaker in shear than in tension. If the magnitude of applied torque (T) is increased gradually, the shear stress also increases according to the relation $\tau = (T/I_p){\cdot}R$ till the failure occurs due to shear. Then the fracture surface will be almost perpendicular to the axis of the shaft. On the other end, if a shaft is of a brittle material like cast-iron which is weaker in tension than in shear, the failure will be due to tension and the fracture surface is found close to 45° to the shaft-axis. This is shown in Fig. 8.10.

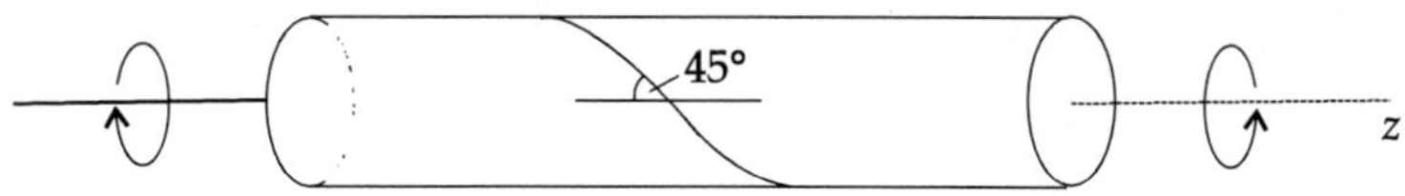

(failure of shaft of brittle material)

Fig. 8.10

Torsion formula can also be applied to hollow cylindrical bars. For a hollow circular bar of cross section with inner radius (r_i) and outer (r_o), the polar moment of inertia is written as $(\pi/2)\cdot(r_o^4 - r_i^4)$ and the shear stresses at inner and outer radii are obtained from Eqn. (8.7). The distribution of shear stress on the cross section of a hollow bar is given below in Fig. 8.11.

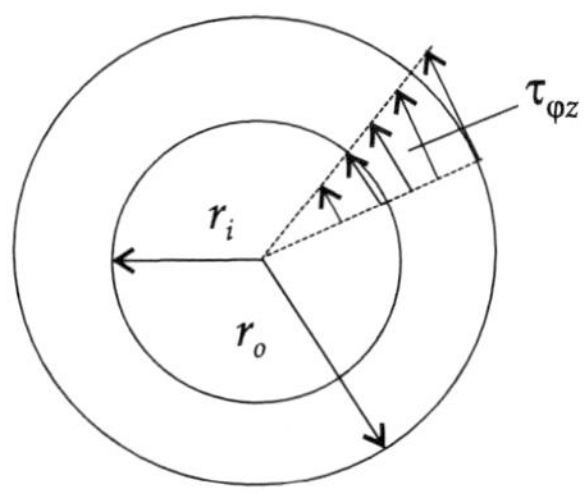

Fig. 8.11

8.4 TORSIONAL STIFFNESS

Force per unit axial deformation of a helical spring is known as its stiffness and is written as $k = F/\delta$, where, F is axial force and δ is axial deformation of the spring. Similarly, the torsional stiffness or torsional rigidity of a shaft is defined as torque per unit angular deformation. If a shaft of length L is subjected to twisting moment T so that the relative angular displacement between the two ends of shaft is φ, then the torsional formula in Eqn. (8.7) is used to write the torsional stiffness as

$$T/\varphi = G\cdot I_p/L, \tag{8.8}$$

which has unit in twisting moment per radian

When the length and the material of a shaft is fixed, torsional stiffness is proportional to its polar moment of inertia.

Example 8.1: A solid circular steel shaft transmits 350 hp at 3000 rpm. If the allowable stress in shear for the shaft is 40 N/mm^2, determine the diameter of the shaft.

Solution: Torque (T)

From $hp = T\cdot 2\pi\cdot N/(60 \times 746)$, $350 = T \times 2\pi \times 3000/(60 \times 746)$

$\therefore$ T = 831 Nm.

From torsion formula, $T = (\pi/16)\cdot d^3\cdot\tau_w$, where, d = dia. of shaft and τ_w = allowable stress

In shear ($\tau_w = 40 \times 10^6$ N/m²),

$$d = [16 \cdot T/(\pi \cdot \tau_w)]^{1/3} = [16 \times 831/(\pi \times 40 \times 10^6\,]^{1/3},$$

or $d = 4.7 \times 10^{-2}$ m (d = 4.7 cm) **Ans.**

Example 8.2: A 20 mm diameter steel shaft is running at 1000 rpm. If the shear stress in the shaft is not to exceed 35 N/mm², what is the maximum hp that can be transmitted at the given speed?

Solution: Given that d = 20 mm, N = 1000 rpm, $\tau_w = 35 \times 10^6$ N/m².

$$\therefore \text{ Torque } T = (\pi/16) \cdot d^3 \cdot \tau_w = (\pi/16) \times 20^3 \times 10^{-9} \times 35 \times 10^6 \text{ Nm}$$
$$= 54.978 \text{ Nm}.$$

$$\text{Power transmitted, } hp = T \cdot 2\pi \cdot N/(60 \times 746)$$
$$= 54.978 \times 2\pi \times 1000/(60 \times 746)$$
$$= 7.7$$

Example 8.3: A solid circular steel shaft is 50 mm in diameter and 2 m in length and it is subjected to a torque of 2 kNm. Find (a) maximum shear stress developed in the shaft, and (b) relative angle of twist between the two ends of the shaft. Take G = 80 GPa for steel.

Solution: Given that d = 50 mm, L = 2 m, $T = 2 \times 10^3$ Nm, $G = 80 \times 10^9$ N/m².

(a) Polar moment of inertia of the shaft

$I_p = (\pi/32) \times 5^4 \times 10^{-8}\ \text{m}^4 = 61{,}357 \times 10^{-8}\ \text{m}^4$

$\therefore$ Applying torsion formula (Eqn. 8.7),

$2 \times 10^3/(61.359 \times 10^{-8}) = \tau_{max}/((5/2) \times 10^{-2})$, or $\tau_{max} \approx 81.5$ MPa **Ans.**

(b) Also, from torsion formula, $T/I_p = G.\varphi/L$.

$\therefore\ 2 \times 10^3/(61.359 \times 10^{-8}) = 80 \times 10^9 \times \varphi/(2)$ or $\varphi = 0.08149$ rad.

$= 4.7°$ **Ans.**

Example 8.4: Five gears are rigidly attached to a shaft and torques transmitted through the gears are shown in Fig. 8.12.

(a) Find the torques carried by cross sections in parts AB, BC, CD and DE.

(b) Draw a torque diagram for the shaft.

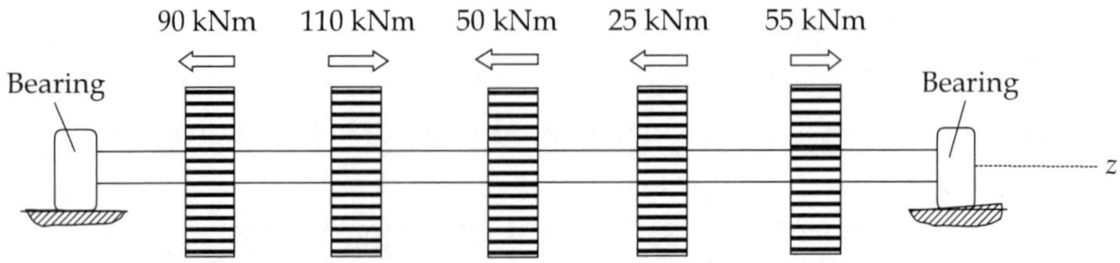

Fig. 8.12

Solution: It should be noted that the torques are shown by double-line arrows which is according to the right hand rule of vectors, as explained in the adjacent figure.

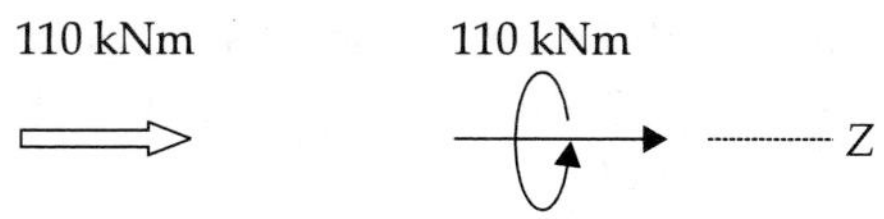

(a) Since the entire shaft is in equilibrium, from FBD up to a cross section in part AB and assuming torque () on the exposed cross section of *AB* in positive sense,

As $\sum T_z = 0, T - 90 = 0. \therefore T = 90$ kNm

or $T_{AB} = 90$ kNm **Ans.**

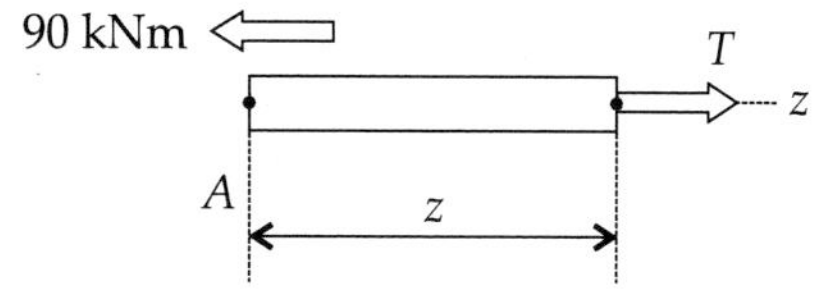

From FBD of part up to a cross section in *BC*,

$T + 110 - 90 = 0$, or $T = -20$ kNm

or $T_{BC} = -20$ kNm **Ans.**

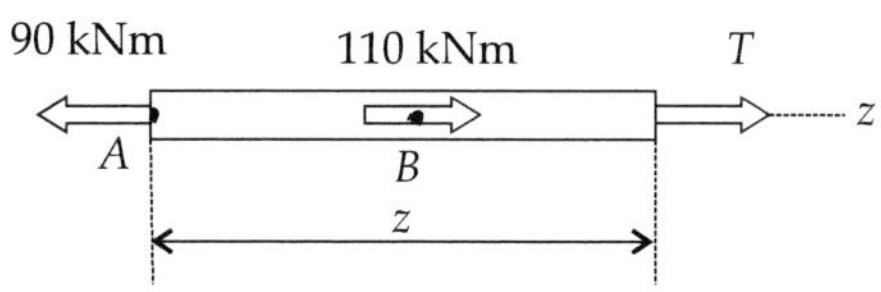

Similarly, in part *CD*, $T - 90 + 110 - 50 = 0$, or $T = 30$ kNm.

or $T_{CD} = 30$ kNm **Ans.**

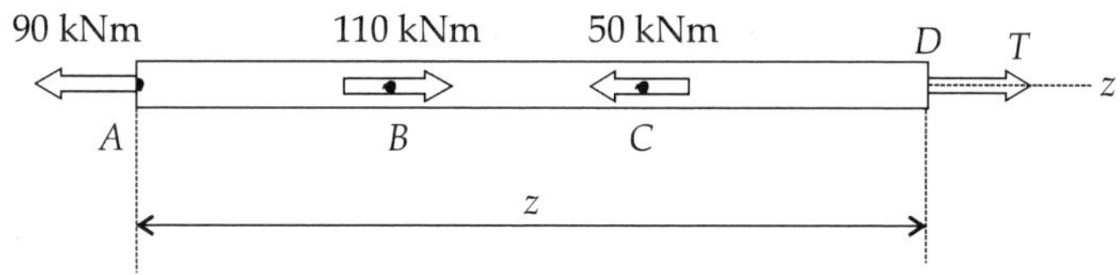

In part *DE*, $T - 90 + 110 - 50 - 25 = 0, \therefore T = 55$ kNm.

or $T_{DE} = 55$ kNm **Ans.**

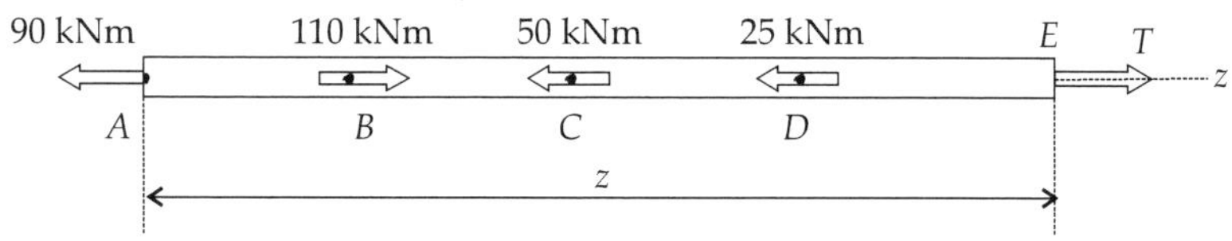

(b) Torque diagram

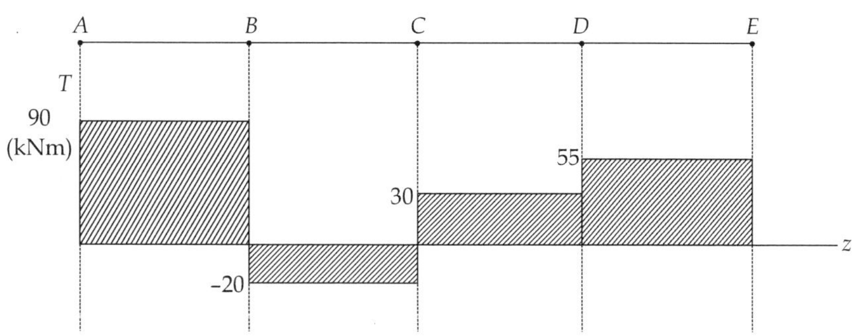

Example 8.5: A 10 mm diameter solid steel bar is built in at its two ends and a disk is rigidly attached to the bar at C, as shown in Fig. 8.13. What is the maximum angle through which the disk can be rotated about the axis of the bar, if the allowable shear stress in bar is 70 N/mm² ? Take shear modulus, G = 80 GPa for steel.

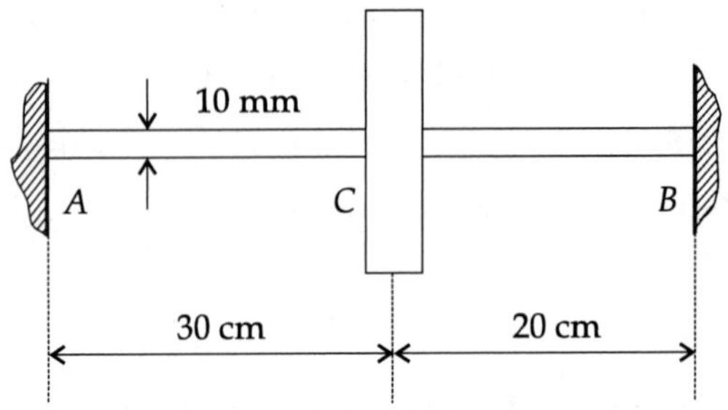

Fig. 8.13

Solution: As the disk at C is rotated by an angle φ, a twisting moment (T_0) develops at C which is balanced by opposing twisting moments at the ends A and B. Hence, considering FBD of the bar in equilibrium,

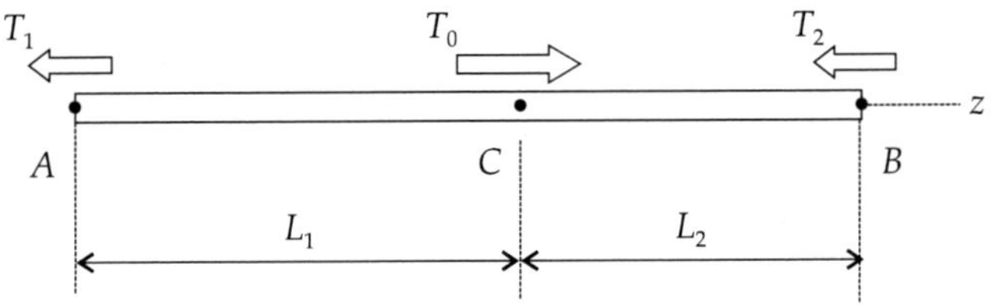

(L_1 = 30 cm, L_2 = 20 cm)

From $\sum T_z = 0$, $T_0 - T_1 - T_2 = 0$

$\therefore \quad T_1 + T_2 = T_0$ (a)

But, $T_1/I_p = G \cdot \varphi / L_1$,

and $T_2/I_p = G \cdot \varphi / L_2$, (b)

where, $I_p = \pi \cdot d^4/32$, the polar moment of inertia.

From (b), $G \cdot \varphi \cdot I_p = T_1 \cdot L_1 = T_2 \cdot L_2$ (c)

$\therefore$ From (a) and (c), $T_1 = T_0 \cdot L_2/(L_1 + L_2)$ and $T_2 = T_0 \cdot L_1/(L_1 + L_2)$,

Where, T_1 and T_2 are torques developed in parts AC and CB, respectively.

After comparing T_1 and T_2 above, it is found that $T_2 > T_1$ since $L_1 > L_2$.

Again from Eqn. (8.7), if τ_1 and τ_2 are shear stresses induced near the surfaces in parts Ab and CB, we have $T_1/I_p = \tau_1 /(d/2)$, or $T_1 = (\pi/16) \cdot d^3 \cdot \tau_1$, and similarly, $T_2 = (\pi/16) \cdot d^3 \cdot \tau_2$.

As $T_2 > T_1$, $\tau_2 > \tau_1$, which means that for same rotation of disk at C, a larger shear stress develops in part CB which is 20 cm in length. By increasing the angle of rotation (φ) of disk at C a stage is attained when $\tau_2 = \tau_w$ = 70 MPa.

$\therefore$ from Eqn. (b) above, $G \cdot \varphi / L_2 = T_2 / I_p = \tau_w / (d/2)$, or $\varphi = 2 \cdot \tau_w \cdot L_2 / (G \cdot d)$

or $$\varphi = 2 \times 70 \times 10^6 \times 20 \times 10^{-2} / (80 \times 10^9 \times 10 \times 10^{-3})$$
$$= 0.035 \text{ rad}$$
$$= 2^\circ$$ **Ans.**

Example 8.6: A hollow steel shaft of 200 mm outer diameter, 140 mm inner diameter and 2.5 m in length is subjected to a torque of 40 kNm. The shear modulus of steel, G = 80 GPa. Determine:

(a) Shear stress on the cross section at the outside surface of the shaft, (b) shear stress on crosssection at the inside surface of shaft, (c) the magnitude of angle of twist between the ends of the shaft, (d) maximum compressive stress at the surface of shaft, and (e) the magnitude of angle of twist between the ends of a 2.5 m long solid shaft which is of same weight as the hollow shaft.

Solution: Given that, $D = 200$ mm, $d = 140$ mm, $L = 2.5$ m, $T = 40$ kNm, $G = 80 \times 10^9$ N/m^2.

(a) $I_p = (\pi/32) \cdot (20^4 - 14^4) \times 10^{-8}$ m^4, or $I_p = 11936.48 \times 10^{-8}$ m^4. (refer to Fig. 8.11)

$\therefore$ Applying torsion formula, $40 \times 10^3 / (11936.48 \times 10^{-8}) = \tau / ((20/2) \times 10^{-2})$

or $\tau = 33.5 \times 10^6$ N/m^2 = 33.5 MPa **Ans.**

(b) Similarly, for shear stress at the inside surface,

$40 \times 10^3 / (11936.48 \times 10^{-8}) = \tau / ((14/2) \times 10^{-2})$

or $\tau = 23.46 \times 10^6$ N/m$^2 \approx 23.5$ MPa **Ans.**

(c) $40 \times 10^3 / (11936.48 \times 10^{-8}) = 80 \times 10^9 \cdot \varphi / (2.5)$.

$\therefore \varphi = 0.0147$ rad. $= 0.6^0$ **Ans.**

(d) Referring to Fig. 8.9, the maximum compressive stress at the surface of the shaft will be equal in magnitude to the shear stress at the outside surface.

$\therefore$ maximum compressive stress = –33.5 MPa **Ans.**

(e) Let d_0 = diameter of solid shaft of same weight,

ρ = density of material.

For same weight, $\rho \cdot g \cdot (\pi/4) \cdot d_0^2 . L = \rho \cdot g \cdot (\pi/4) \cdot (D^2 - d^2) \cdot L$.

or $$d_0 = \sqrt{D^2 - d^2} = \sqrt{20^2 - 14^2}$$
$$d_0 = 17.32 \text{ cm}$$

$\therefore I_p = (\pi/32) \cdot (17.32)^4 \cdot 10^{-8} = 8834.69 \times 10^{-8}$ m^4

$40 \times 10^3 / (8834.69 \times 10^{-8}) = 80 \times 10^9 \cdot \varphi / 2.5$. or $\varphi = 0.014148$ rad

$$\varphi = 0.8^\circ$$

Example 8.7: A circular shaft carries four gears attached to it and the torques transmitted through gears are shown in Fig. 8.14. The permissible shear stress in the shaft is 75 MPa.

Find (a) the required diameter (d) of the shaft if it has solid cross section, and (b) the outside diameter (d) of the shaft if it is a hollow shaft with 45 mm inner diameter.

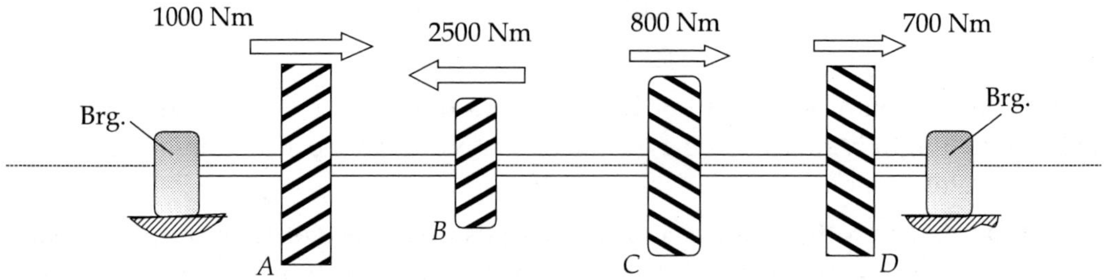

Fig. 8.14

Solution:

(a) From equilibrium consideration and the torques produced at cross sections in steps AB, BC and CD, it can be checked that the maximum torque develops in part BC which is obtained as follows:

$$\sum T_z = 0, \mathrm{T} + 1000 - 2500 = 0$$

$$T = 1500 \text{ Nm}$$

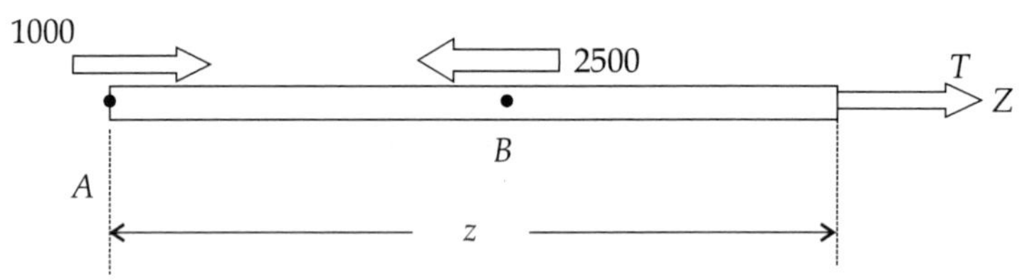

Therefore, the uniform shaft diameter is obtained against this maximum torque $T = 1500$ Nm

From $T = (\pi/16)\cdot d^3\cdot\tau_w$, we have $1500 = (\pi/16)\cdot d^3 \times 75 \times 10^6$

$\therefore \quad d = 4.67 \times 10^{-2}$ m, or $d = 4.67$ cm **Ans.**

(b) Let d = outside diameter of hollow shaft (in m), so that

$1500/[(\pi/32)\cdot(d^4 - 0.045^4)] = 75 \times 10^6/(d/2)$.

After solving by trial & error, $d = 0.056$ m, or $d = 5.6$ cm

Example 8.8: A solid brass rod of 65 mm diameter with a steel jacket of 7.5 mm thickness rigidly fixed over it is shown in Fig. 8.15. If the permissible shear stresses in brass rod and steel jacket are $\tau_b = 53$ MPa and $\tau_s = 83$ MPa respectively, find the safe torque that can be applied to this compound shaft. What is the ratio of safe torque for compound shaft to the safe torque that the brass rod alone can carry? Take the shear modulus, $G_b = 35$ GPa for brass and $G_s = 80$ GPa for steel.

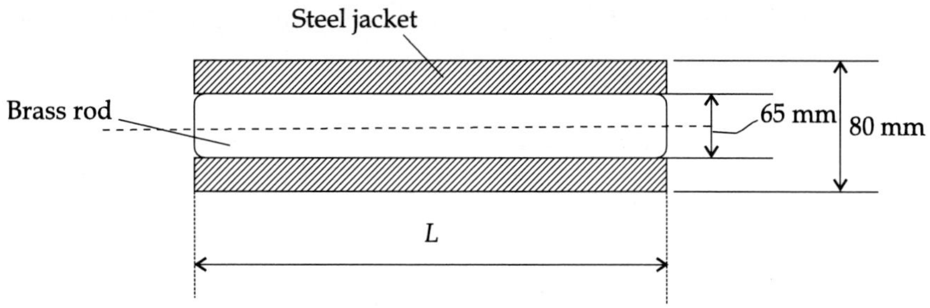

Fig. 8.15

Solution: Given that,

$\tau_b = 53$ MPa, $G_b = 35$ GPa,

$\tau_s = 83$ MPa, $G_s = 80$ GPa.

Also, polar moment of inertia, $I_b = (\pi/32) \times (6.5)^4 \times 10^{-8}\ \text{m}^4 = 175.248 \times 10^{-8}\ \text{m}^4$

and $I_s = (\pi/32) \times (8^4 - 6.5^4) \times 10^{-8}\ \text{m}^4 = 226.876 \times 10^{-8}\ \text{m}^4$.

When a torque (T) is applied to a compound shaft, it is the sum of torques carried by brass rod (T_b) and steel jacket (T_s), such that $T = T_b + T_s$. It should be noted that the angular displacement (φ) between the two ends of compound shaft is common to both brass rod and steel jacket.

Therefore, the following relation is satisfied:

$$\varphi/L = T_b/(I_b \cdot G_b) = T_s/(I_s \cdot G_s)$$

$\therefore$ the ratio $T_s/T_b = (G_s \cdot I_s)/(G_b \cdot I_b) = 80 \times 226.876 \times 10^{-8}/(35 \times 175.248 \times 10^{-8})$

or $T_s/T_b = 2.959$, which is the ratio of torque shared by steel and brass.

Torque carrying capacity of brass rod,

$$T_b = 175.248 \times 10^{-8} \times 53 \times 10^6/\{(65/2) \times 10^{-3}\}\ \text{Nm} = 2858\ \text{Nm},$$

and the torque carrying capacity of steel jacket,

$$T_s = 226.876 \times 10^{-8} \times 83 \times 10^6/\{(80/2) \times 10^{-3}\}\ \text{Nm} = 4708\ \text{Nm}.$$

Because the ratio of torques (2.959) is fixed according to the common angular displacement, either the brass rod or steel jacket may carry maximum torque while the ratio remains intact.

If it is assumed that $T_b = 2858$ Nm, $T_s = 2858 \times 2.959 = 8457$ Nm, which exceeds the safe value of torque as 4708 in steel jacket.

Therefore, with $T_s = 4708$ Nm the torque in brass rod $T_b = 4708/2.959$, or

$T_b = 1591$ Nm which is within the safe limit.

Therefore, the safe torque for the compound shaft,

$T = T_s + T_b = 4708 + 1591 = 6299$ Nm

Also, the ratio $T/T_b = 6299/2858 = 2.2$

Example 8.9: A circular steel shaft 6 m in length is subjected to torsions as shown in Fig. 8.16.

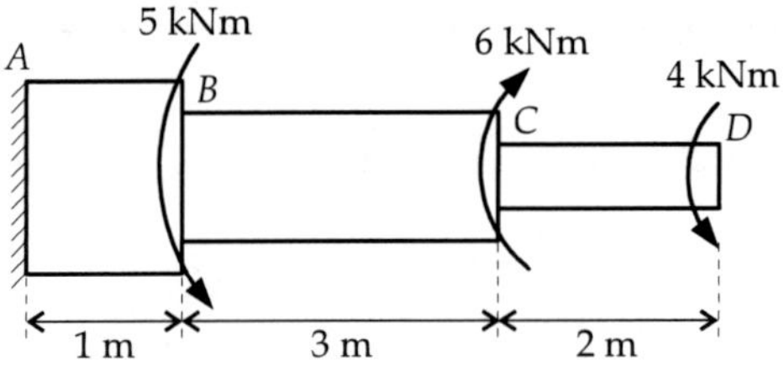

Fig. 8.16

The torsional rigidity of portions *AB*, *BC* and *CD* are 750, 500 and 250 kNm²/rad, respectively. Find the angular twist at the free end and the strain energy stored in the shaft.

Solution:

Torque in portion *CD* $T_{CD} = 4$ kNm

Torque in portion *BC* $T_{BC} = 4 - 6 = -2$ kNm

Torque in portion AB $T_{AB} = 4 - 6 + 5 = 3$ kNm

Angle of twist

$$\theta_{CD} = \frac{T_{CD}\,L_{CD}}{GI_{p(CD)}} = \frac{4 \times 2}{250} = 0.032 \text{ rad}$$

$$\theta_{BC} = \frac{T_{BC}\,L_{BC}}{GI_{p(BC)}} = \frac{-2 \times 3}{500} = -0.012 \text{ rad}$$

$$\theta_{AB} = \frac{T_{AB}\,L_{AB}}{GI_{p(AB)}} = \frac{3 \times 1}{750} = 0.004 \text{ rad}$$

Hence, total angle of twist at free end is

$\theta_T = \theta_{AB} + \theta_{BC} + \theta_{CD} = 0.004 - 0.012 + 0.032 = 0.024$ rad

For strain energy stored in the shaft since $T_{AB} = 3$ kNm, $T_{BC} = -2$ kNm and $T_{CD} = 4$ kNm

$$\text{Strain energy } U = \sum \frac{T^2 L}{2GI_p} = \frac{3^2 \times 1}{2 \times 750} + \frac{(-2)^2 \times 3}{2 \times 500} + \frac{4^2 \times 2}{2 \times 250} = 0.082 \text{ kNm} = 82 \text{ Nm}$$

Example 8.10: A circular shaft 5 m in length is subjected to torsions as shown in Fig. 8.17.

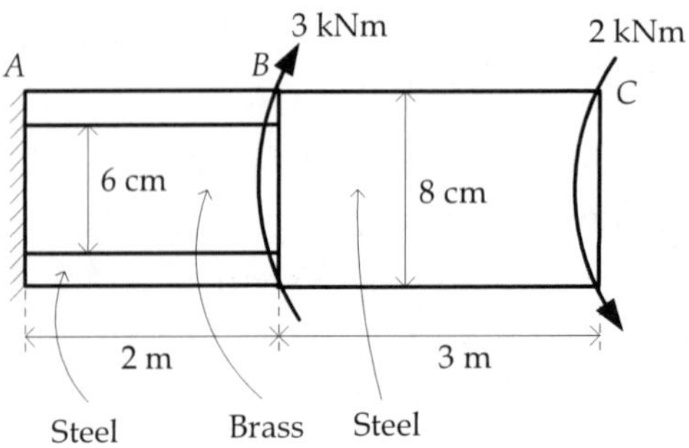

Fig. 8.17

AB is composite shaft with outer material as steel and inner as brass, while *BC* is solid shaft of steel. Find the angular twist at the free end. (Given G_{brass} = 50 GPa and G_{steel} = 100 GPa)

Solution:

Torque in portion *BC* T_{BC} = 2 kNm

Torque in portion *AB* T_{AB} = 2 –3 = –1 kNm

Polar moment of inertia

$$I_{P(BC)} = \frac{\pi D^4{}_{BC}}{32} = \frac{\pi \times (0.08)^4}{32} = 4 \times 10^{-6}\ \text{m}^4$$

$$I_{P(brass)} = \frac{\pi D^4{}_{brass}}{32} = \frac{\pi \times 0.06^4}{32} = 1.27 \times 10^{-6}\ \text{m}^4$$

$$I_{P(steel)} = \frac{\pi(D^4{}_{steel,\,0} - D^4{}_{steel,\,i})}{32} = \frac{\pi \times (0.08^4 - 0.06^4)}{32} = 2.75 \times 10^{-6}\ \text{m}^4$$

$$\theta_{BC} = \frac{T_{BC}L_{BC}}{GI_{p(BC)}} = \frac{2 \times 3}{100 \times 10^6 \times 4 \times 10^{-6}} = 0.015\ \text{rad}$$

Since portion *AB* is composite of two materials therefore torque act in these two materials will be different while rotation of both the shaft is same.

Let T_b and T_s are the torques act on brass and steel, respectively, then

$\theta_{AB} = \theta_b = \theta_s$

$$T_{AB} = T_b + T_s = \frac{G_{brass}I_{p(brass)}\theta_{AB}}{L_{AB}} + \frac{G_{steel}I_{p(steel)}\theta_{AB}}{L_{AB}} = -1$$

$$\Rightarrow \theta_{AB} = \frac{-1 \times 2}{G_{brass}I_{p(brass)} + G_{steel}I_{p(steel)}}$$

$$= \frac{-2}{(50 \times 10^6 \times 1.27 \times 10^{-6} + 100 \times 10^6 \times 2.75 \times 10^{-6})}$$

$\Rightarrow \theta_{AB}$ = –0.006 rad

Hence, total angle of twist at free end is

$$\theta_T = \theta_{AB} + \theta_{BC} = 0.015 - 0.006 = 0.009\ \text{rad}$$

Example 8.11: A circular steel shaft is subjected to linearly varying torsion from free end to fixed end as shown in Fig. 8.18. (Torsional rigidity of shaft is *GI*)

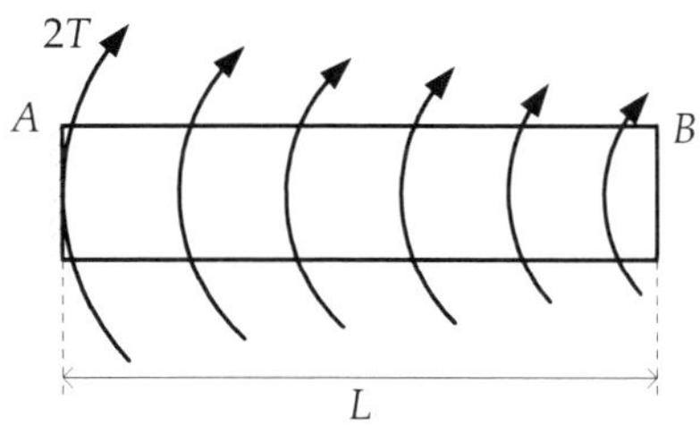

Fig. 8.18

Find the angular twist at the free end and the strain energy stored in the shaft.

Solution:

Torque at a distance x from B is $T_x = -T\left(1+\frac{x}{L}\right)$

$$\theta_{AB} = \int_0^L \frac{T_{AB}}{GI}dx = \int_0^L \frac{-T\left(1+\frac{x}{L}\right)}{GI}dx = \frac{-3TL}{2GI}$$

For the strain energy stored in the shaft, since

$$T_x = -T\left(1+\frac{x}{L}\right)$$

$$U_{AB} = \int_0^L \frac{T_{AB}^2}{2GI}dx = \int_0^L \frac{T^2\left(1+\frac{x}{L}\right)^2}{2GI}dx = \frac{7T^2L}{6GI}$$

Example 8.12: A steel shaft is to be manufactured either as a *solid circular shaft* or as a *hollow circular shaft.* A shaft *(either solid circular* or a *hollow circular)* is required to transmit a torque of 1000 Nm without exceeding an allowable shear stress of 50 MPa and an allowable rate of twist of 1.0°/m. The shear modulus of elasticity of the steel is 78 GPa.

Find the required diameter D of the solid shaft, the required outer diameter D_o of the hollow shaft. (assume thickness of hollow shaft $t = 0.1\ D_o$) and the ratio of weights of the hollow and solid shafts.

Solution: Given that: $T = 1000$ Nm, $G = 78$ GPa

$\lambda_{\text{allowable}} = 50$ MPa, $\phi = \theta_{\text{allowable}}/L = 1.0°/\text{m}$

Since the allowable shear stress $\lambda_{\text{allowable}}$ and allowable rate of twist $\theta_{\text{allowable}}$ are given. Hence, two separate relationships have to be used for both the cases, i.e.

Case-1 Stress and torque relationship: $\tau = \frac{Tr}{I_P}$

Case-2 Twist and torque relationship: $\phi = \frac{\theta}{L} = \frac{T}{GI_P}$

Solid shaft:

$I_P = \frac{\pi D^4}{32}$ for solid shaft then

Diameter D for maximum shear stress is obtained as

$\tau_{\max} \le 50\ \text{MPa} \Rightarrow \frac{TR}{I_P} \le 50$ MPa putting the value of T, R and I_P, we get $D \ge 46.7$ mm

Diameter D for maximum twist is obtained as:

$\phi_{\max} = \frac{\theta_{\max}}{L} = \frac{T}{GI_P} \le 1° = \frac{\pi}{180}$ rad putting the value of T, G and I_P, we get $D \ge 52.3$ mm

52.3 mm diameter required for solid shaft.

Hollow shaft:

$D_i = D_o - 2t = D_o - 2(0.1\ D_o) = 0.8\ D_o \quad I_P = \dfrac{\pi\left(D_o^4 - (0.8D_o)^4\right)}{32} = 0.05796\ D_o^4$

Diameter D for maximum shear stress is

$\tau_{max} \le 50\ \text{MPa} \Rightarrow \dfrac{TR_o}{I_P} \le 50\ \text{MPa}$ putting the value of T, R_o and I_P, we get $D_o \ge 55.67$ mm

Diameter D for maximum twist is obtain as:

$\phi_{max} = \dfrac{\theta_{max}}{L} = \dfrac{T}{GI_P} \le 1° = \dfrac{\pi}{180}$ rad putting the value of T, G and I_P, we will get $D_o \ge$ 55.67 mm

59.6 mm outer diameter required for hollow shaft.

Now since $\dfrac{D_o}{D} \ge 1.14$ mm, i.e., hollow design is 14% bulkier, but the ratio of weights between hollow & solid design $\dfrac{W_{hollow}}{W_{solid}} = \dfrac{A_{hollow}L}{A_{solid}L} = \dfrac{\pi\left(D_o^2 - (0.8D_o)^2\right)/4}{\pi D^2/4} = 46.8\%$, i.e., hollow design is more efficient (saves over 50% of material cost) than solid shaft.

PROBLEMS

1. A circular steel shaft 5 m in length is subjected to torsions as shown in Fig. 8.19.

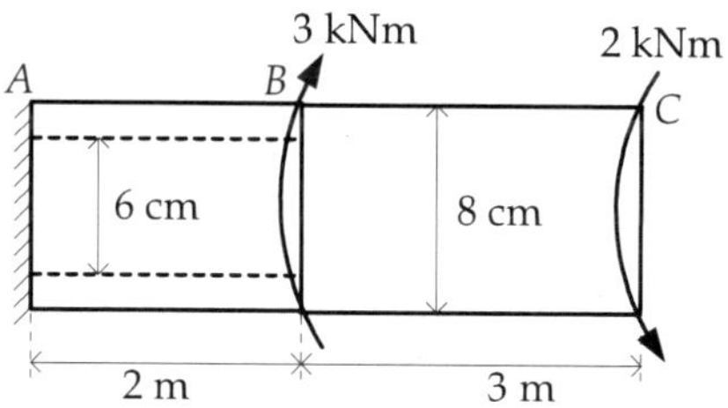

Fig. 8.19

AB is hollow shaft while BC is solid shaft. Find the angular twist at the free end. (Given G = 100 GPa) **Ans.** 0.008 rad

2. A circular steel shaft 5 m in length is subjected to torsions as shown in Fig. 8.20. AB is hollow shaft while BC is solid shaft. Find the resisting torque exerted by the fixed supports. (Given G = 100 GPa)

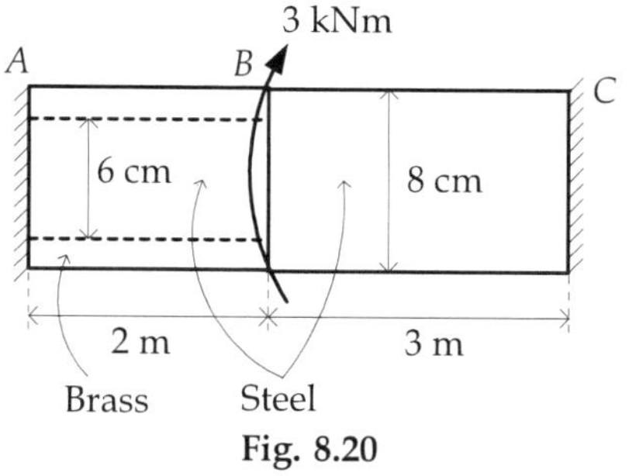

Fig. 8.20

Ans. T_A = 1.52 kNm and T_C = 1.48 kNm

3. A circular steel shaft with linearly varying diameter is subjected to torsion as shown in the Fig. 8.21. Find the angular twist at the free end.

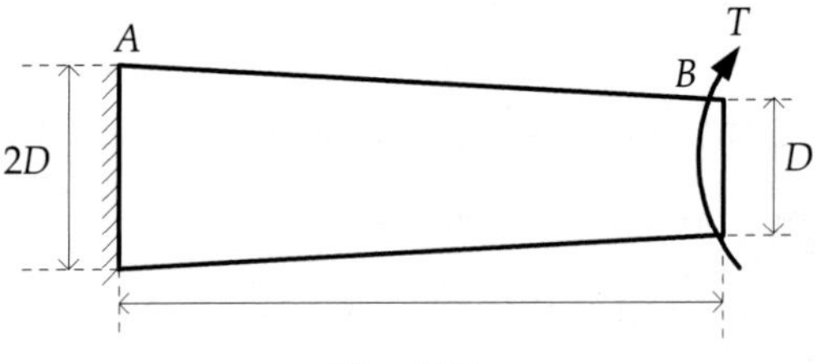

Fig. 8.21

Ans. $\theta_{AB} = \dfrac{28TL}{3\pi GD^4}$

4. Find the ratio of diameters of two solid shafts to transmit 100 horsepower each without exceeding a shear stress of 500 kg/cm^2, when one of these shafts operates at 10 rpm and the other at 10,000 rpm. **Ans.** 10
5. The internal diameter of steel shaft is 75% of the external diameter. The shaft is to transmit 3000 kW at 200 rpm. If the maximum allowable stress in the shaft material is 50 MPa, then find the external diameter of the shaft and if the length of the shaft is 5 m and G = 80 MPa, then find the maximum twist in the shaft.

Ans. 0.0223 radian or 1.3°

6. A thin tube of circular cross section with outer diameter 104 mm, thickness 4 mm and length 1000 mm is fixed at one end. It is subjected to a twisting moment of 1 kNm at the free end. Find the shear stress in the wall of the tube and the angle of twist at the free end. (G = 80 MPa)

Ans. Shear stress, τ = 16 N/mm^2 and angle of twist, θ = 0.004 radian

7. Find the ratio of power transmission of equal length solid shaft of diameter D, and a hollow shaft of external diameter D and internal diameter 0.75 D of same material. **Ans.** 1.47
8. A horizontal shaft 6 m in length is fixed at ends as shown in Fig. 8.22.

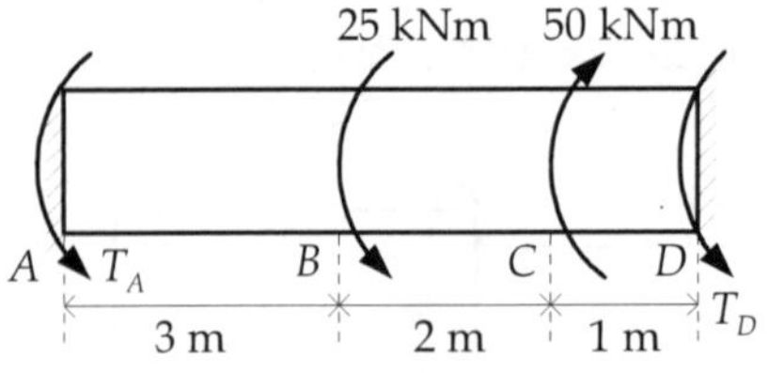

Fig. 8.22

When viewed from its left end axial couples of 25 kN-m clockwise and 50 kN-m counterclockwise act at 3 m and 5 m from the left end respectively. Find the position from left end where the shaft suffers no angular twist. **Ans.** 3.6 m.

9. A fixed circular steel shaft 6 m in length is subjected to torsions as shown in Fig. 8.23.

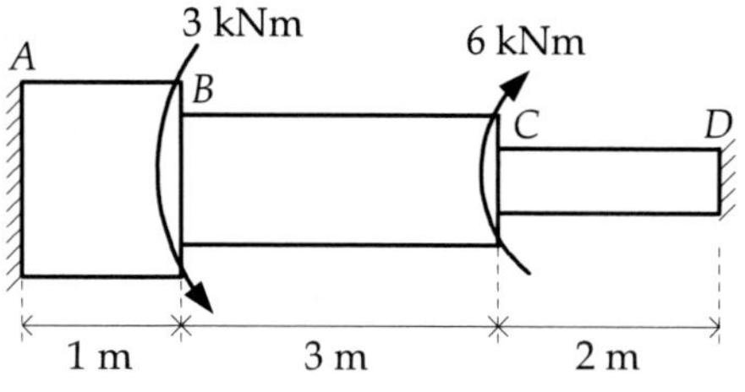

Fig. 8.23

The torsional rigidity of portions *AB*, *BC* and *CD* are 750, 500 and 250 kNm^2/rad, respectively. Find the resisting torque exerted by the fixed supports.

Ans. T_A = 0.4 kNm and T_D = 2.6 kNm

10. A circular shaft 5 m in length is subjected to torsions as shown in Fig. 8.24.

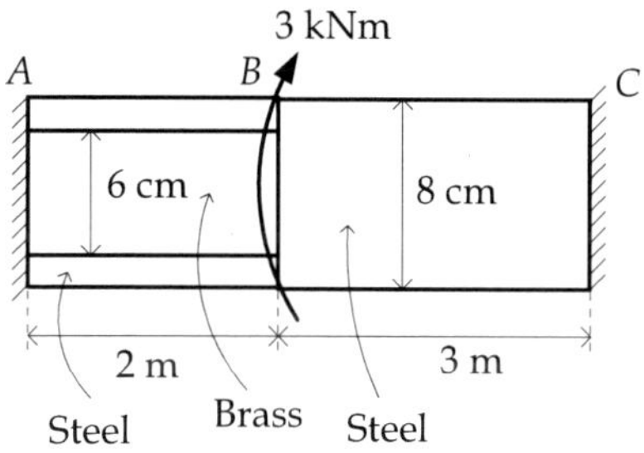

Fig. 8.24

AB is composite shaft with outer material as steel and inner as brass, while *BC* is solid shaft of steel. Find the resisting torque exerted by the fixed supports. (Given G_{brass} = 50 GPa and G_{steel} = 100 GPa)

Ans. T_A = 1.67 kNm and T_C = 1.33 kNm

9

Bending Moments and Shear Forces

9.1 INTRODUCTION

When a structural bar is subjected to transverse forces, it deforms in the plane of the applied load. To resist the deformation of the bar due to external load, two important types of loading actions are set up at any section of the bar; that is, bending moment and shear force.

A long slender bar is called a beam if the external loads are applied transverse to the axis of the bar and the length of the bar is large as compared to its cross-sectional dimension. When a beam is subjected to transverse loading, it bends and moment is developed at any section of the beam to resist bending of the beam. This internal resistance of the beam is known as **bending moment** which is equal to external moment acting at that point.

Analysis of a beam for different support condition and external applied loading conditions are discussed in this chapter. The beam is to be analyzed using equations of equilibrium that incorporate the applied forces and reaction forces at its support. The beam is completely analyzed when the bending moment and shear force diagram are found knowing the external applied forces and reaction forces in support. Internal forces and moments give rise to stresses, strains or deformations which help in deciding the load carrying capacity of a beam.

9.2 CLASSIFICATION OF SUPPORTS

A beam must be supported and the reactions provided by the supports must balance the applied forces to maintain equilibrium. The support conditions are also known as boundary conditions. Common supports and external reactions for two-dimensional loading are shown in Fig. 9.1.

9.2.1 Roller Support

A roller support will only produce a reaction force (only one unknown R_{Ay}) perpendicular to the plane. (Fig. 9.1 (a)). In a roller supported beam the axial displacements and rotations (which cause slope changes) at the supports are unrestrained (i.e., the beam is free to move along the x-axis and rotate about the z-axis).

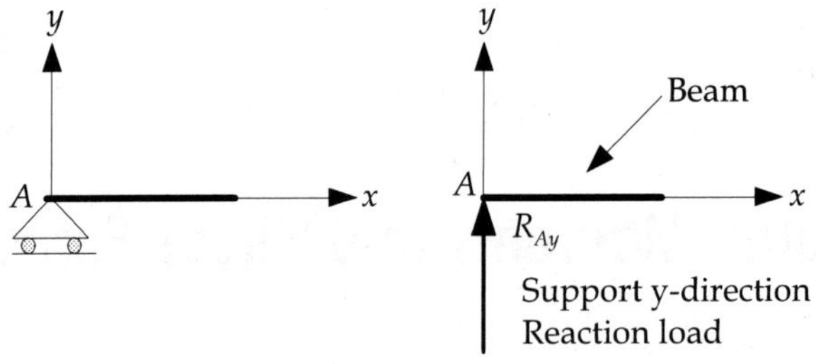

Fig. 9.1 (a)

9.2.2 Hinged or Pinned End support

Hinged or pinned supports provide similar support to roller supports with the addition of support in the plane (two unknowns, R_{Ax} and R_{Ay}), i.e., displacements in the axial direction are prevented (Fig. 9.1 (b)) the beam is only free to rotate about the z-axis).

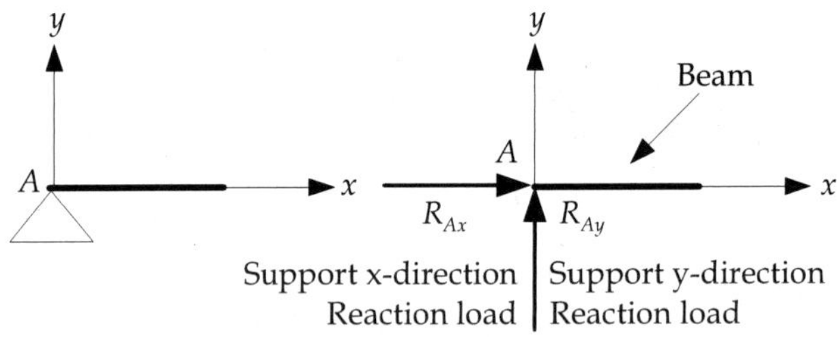

Fig. 9.1 (b)

9.2.3 Fixed or Built-in End Support

Fixed end supports support moments (three unknowns, R_{Ax}, R_{Ay} and M_A) in addition to lateral and axial forces (Fig. 9.1 (c)). No axial, lateral or rotational movements are possible at the built-in end (i.e., the end is not able to move in either the x- or y-direction nor can it rotate about the z-axis

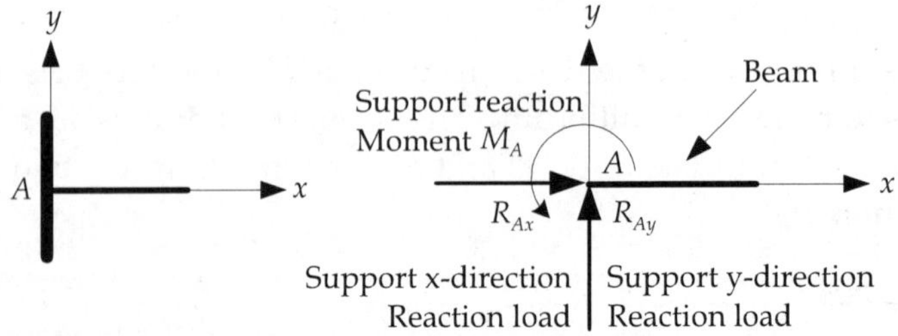

Fig. 9.1 (c)

9.3 CLASSIFICATION OF BEAMS

Beams are often classified on the types of supports they have. Beams are classified into two broad categories, which are (1) statically determinate beams, and (2) statically indeterminate beams. With equilibrium consideration of a beam, the reaction forces at the supports are calculated from the applied external load. To achieve this, the equations of statics are used, depending on the force-system to which the beam is subjected. The statically determinate beams are those in which the number of support reactions are less than or equal to the number of equations available for solving them. After calculating the reaction forces a further analysis to obtain the internal forces or moments becomes possible.

There are commonly three types of determinate beams which come under this category. These are (a) simply supported beams, (b) cantilever beams, and (c) overhanging beams, as shown distinctly in Fig. 9.2.

In a simply supported beam, one support is hinged support with supposedly frictionless pin and the other support is on the roller. As such, any movement at the hinge in vertical or horizontal direction is restricted but a free rotation about the pin fitted to the hinge is permitted. At the roller support, only the movement (downward) in vertical direction is constrained.

In a cantilever beam, one end of the beam is fixed to the wall or may be welded to a vertical structural member and the other end is free. At fixed support, neither the movement in vertical plane nor any rotation is permitted.

The overhanging beam is similar to a simply supported beam with the difference that its length extends beyond one or both the supports. The distance between supports in the first and the third types is known as **span** while there is free length of the cantilever beam measured from the fixed end.

Simply supported beam

Beam supported freely at both the ends (Fig. 9.2 (a)).

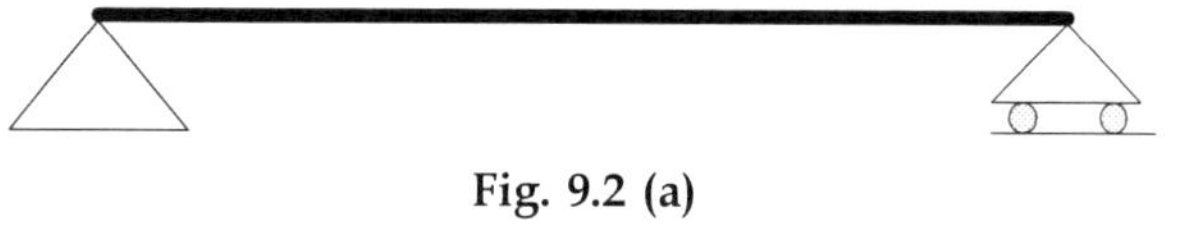

Fig. 9.2 (a)

Cantilever beam

Beam with one end fixed and the other free (Fig. 9.2 (b)).

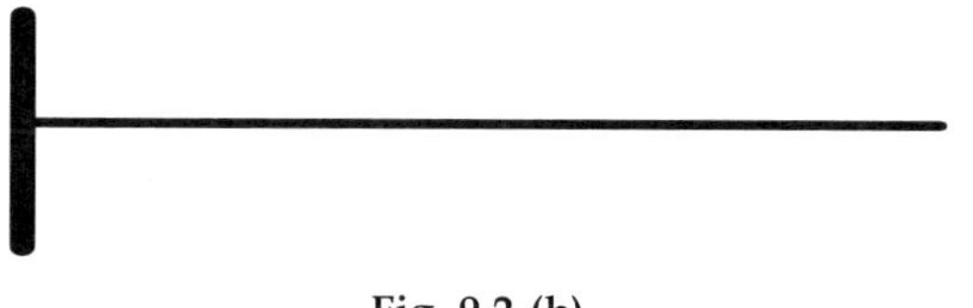

Fig. 9.2 (b)

Overhang beam

Beam which has part that extends beyond a support either on one side or on both the sides (Fig. 9.2 (c)).

Fig. 9.2 (c)

In statically indeterminate beams the number of unknown reaction forces are greater than the number of equilibrium equations available and these cannot be solved directly by applying equations of statics. The common types of statically indeterminate beams are shown in Fig. 9.2. In fact, a complete solution can be obtained by considering the deformation of such a beam and using techniques related to it. In the process some additional equations are obtained.

Propped cantilever beam

A cantilever beam provided with an additional simple support on the other end (Fig. 9.2 (d)).

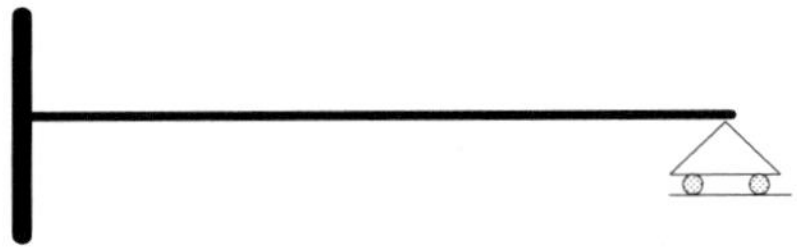

Fig. 9.2 (d)

Fixed beam

Beam with fixed support at both the ends (Fig. 9.2 (e)).

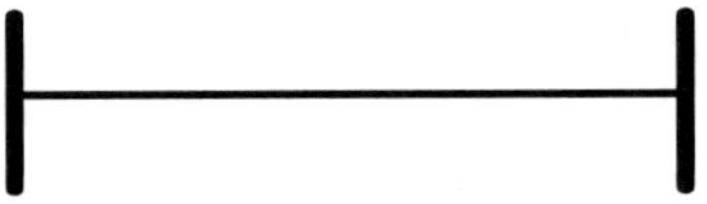

Fig. 9.2 (e)

Continuous beam

Beam with more than two supports (Fig. 9.2 (f))

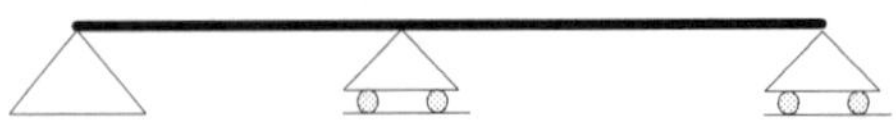

Fig. 9.2 (f)

9.4 CLASSIFICATION OF LOADS

Generally, the external forces acting on a beam lie in a vertical plane and hence the reaction forces at the supports also lie in the same plane. In other words a beam is a two-dimensional structure and all the forces acting on it are coplanar. There are two types of external loads (forces) applied on a beam which are (1) concentrated or point load, and (2) the distributed load (Fig. 9.3). Normally, a load acting on a beam covers a limited amount of surface area on the top surface of the beam. It means the load is distributed within a small region. Further calculations regarding the reaction forces at supports, etc., will then invite unnecessary complications. Hence, for simplicity of analysis the load is idealized as a point load or concentrated load. On the other hand, a distributed load covers an appreciable length along the axis of the beam and it is measured in the units of force per unit length. The distributed load on a beam commonly falls under two categories, which are (1) uniformly distributed load, and (2) uniformly varying load.

Concentrated or point loads

Load acts at a point and its SI unit is kN.

Uniformly distributed loads

Load distributed over some length of the beam with constant magnitude and its SI unit is kN/m.

Non-uniform distributed loads

Load distributed over some length of the beam with varying magnitude and its SI unit is kN/m.

Moments (or couple)

Beam loaded by an applied couple and its SI unit is kNm.

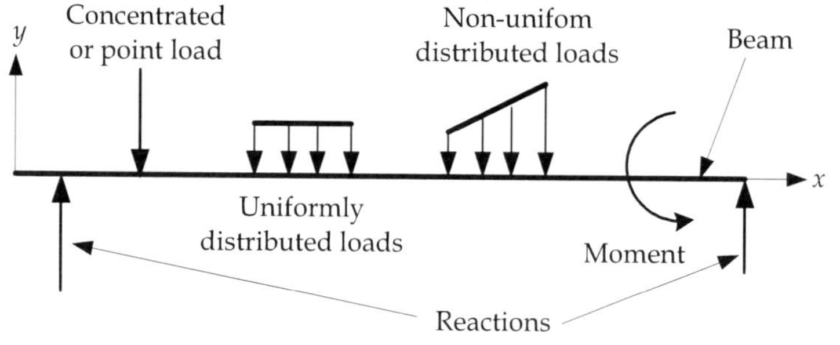

Fig. 9.3

9.5 DISTRIBUTED LOADS TO EQUIVALENT POINT LOAD

To simplify the analysis of a distributed load it is usually easier to replace the distributed load with a point load acting at an appropriate location.

Let w (force per unit length) be the distributed loads; R is its equivalent point load, l (from a to b) is the length over which the load is distributed and $\bar{x}$ is its point of application (Fig. 9.4).

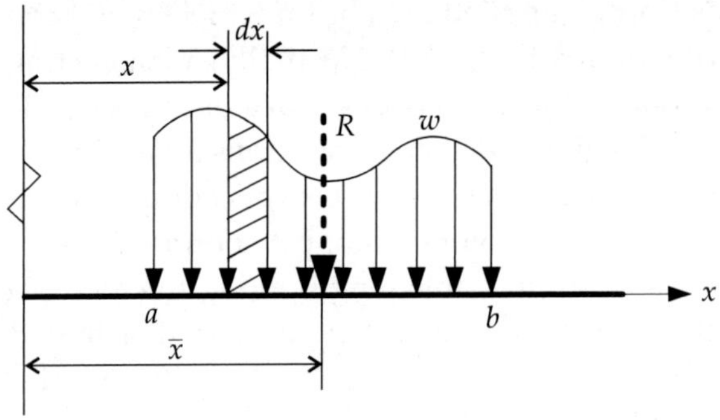

Fig. 9.4

Equivalent force (R) is equal to the area formed by the intensity (w) and the length (L) over which the force is distributed. The equivalent force passes through the centroid of this area.

Differential increment of force, $dR = w\,dx$. The total equivalent force is then the sum of the differential forces, i.e.

$$\boxed{R = \int_a^b w\,dx}$$

The equivalent (R) is located at the centroid of the area under consideration. The x-coordinate of this centroid (point of application) is found by the principle of moments, i.e.,

$$\boxed{\bar{x} = \frac{\int_a^b w\,x\,dx}{R}}$$

Some Special Cases

Rectangular case

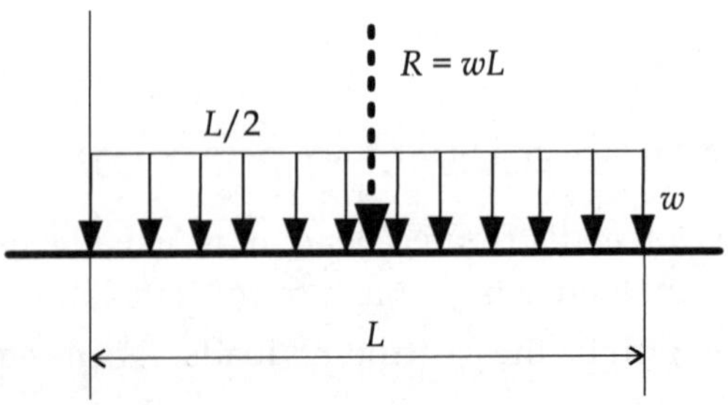

Fig. 9.4 (a)

Triangular case

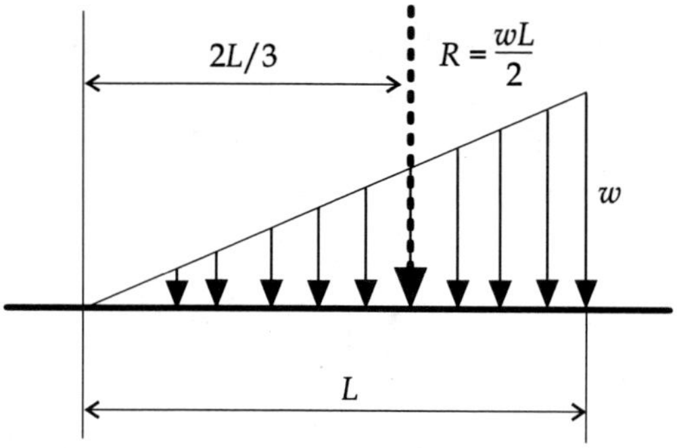

Fig. 9.4 (b)

Trapezoidal case

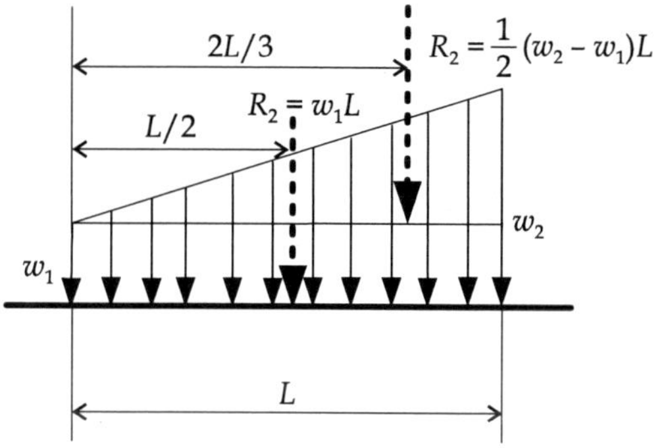

Fig. 9.4 (c)

9.6 INTERNAL LOADINGS OF BEAMS

Internal loading at a specific point or section (let it be along section *aa*, Fig. 9.5) in a beam consists of normal (or axial) force (N), shear force (V) and bending moment (M).

M = Sum of moments due to all forces to one side of the section *aa* (including support forces) is called the *bending moment* acting on the vertical face of the beam at position *aa*.

N = Sum of all x-direction (axial) forces to one side of the section (*aa*) is called the *axial force* acting on the vertical face of the beam at section, *aa*.

V = Sum of all the y-direction (transverse or shear) forces to one side of the section *aa* is called the *shear force* acting on the vertical face of the beam at section, *aa*.

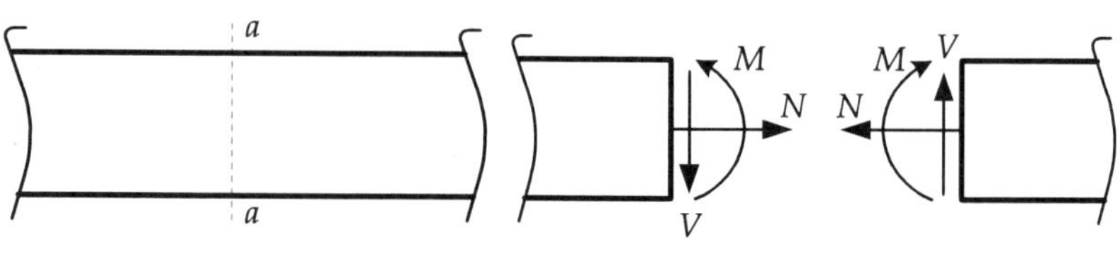

Fig. 9.5

9.6.1 Sign Convention

Sign convention widely accepted is shown in Fig. 9.6, i.e., on the left part of the beam normal force (N) acts *towards right*, the internal shear force (V) acts *downwards* and the moment (M) acts *clockwise*. Similarly, by Newton's third law, right part of the beam gives equal and opposite normal force, shear force and moment (since the whole beam is in equilibrium).

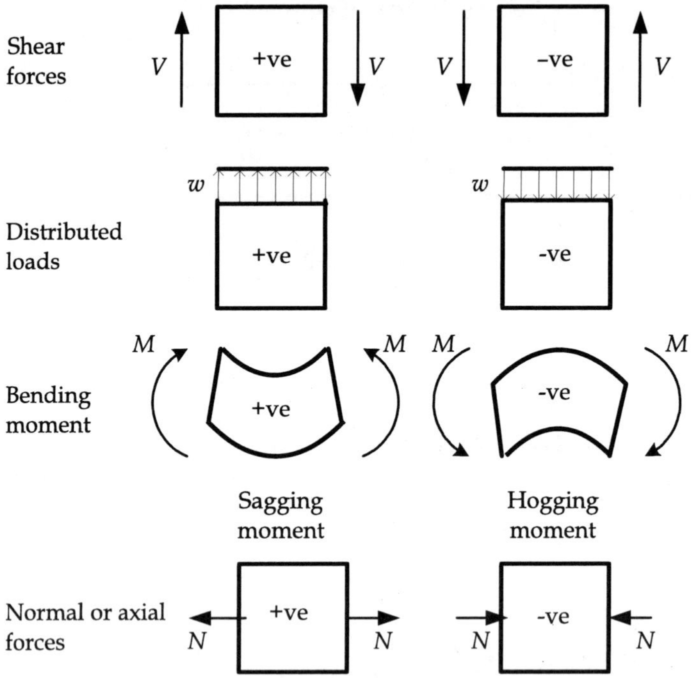

Fig. 9.6

9.7 ANALYSIS OF THE SUPPORT REACTIONS OF THE BEAM

- **Sign conventions:** Horizontal forces along *rightward direction* (→), vertical forces along *upward direction* (↑) and moment in *clockwise direction* (↻) should be taken as positive.
- Draw the free-body diagram (FBD) of the beam by assuming the direction of all the forces and moments of the reactions supports to be positive. If any of the forces calculated are negative then they act in the opposite sense to that assumed in the FBD.
- Apply the conditions of equilibrium of forces and moments on the whole beam, i.e.,

$$\sum F_x = 0, \quad \sum F_y = 0 \text{ and } \sum M = 0$$

Analysis of the internal loadings

- First determine the support reactions by using equations of equilibrium.
- Draw the FBD of the beam showing all the external loadings and support reactions.
- For analyzing the internal loadings, cut the beam at the section where the internal loadings is to be determined.
- For equilibrium we require equal and opposite forces (*N* and *V*) as well as equal and opposite moments (*M*), acting at the severed sections of the two parts of the beam.
- Draw the FBD of each section of the beam showing all the external loadings, support reactions and require equal and opposite forces (*N* and *V*) as well as equal and opposite moment (*M*), acting at the severed sections.
- Apply the equations of equilibrium in the section that has least number of loads on it. Calculate the required normal (or axial) force (*N*), shear force (*V*) and bending moment (*M*) by indicating the unknowns (*N*, *V* and *M*) acting in their positive directions.
- If the answer is positive, then the assumed direction is correct, otherwise it is in the opposite direction to that of shown in the FBD.

9.8 SHEAR FORCE DIAGRAMS (SFDs) AND BENDING MOMENT DIAGRAMS (BMDs)

The graphical representation of change of shear force along the beam axis is known as shear force diagram (SFD) and graphical representation of bending moment variation along the length of beam is known as bending moment diagram (BMD).

Consider a beam with load intensity, w, (Fig. 9.7). Draw the FBD of an element of length, dx.

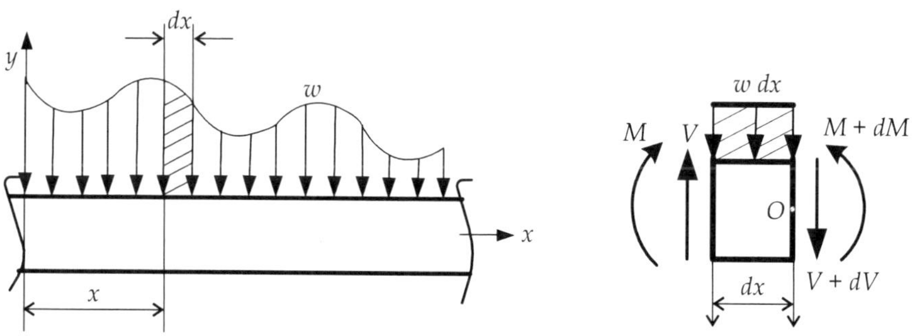

Fig. 9.7

Since the element is in equilibrium, the algebraic sum of vertical forces and moments on the element is zero, i.e.,

$$\sum F_y = 0 \Rightarrow V - w\,dx - (V + dV) = 0 \Rightarrow dV = -\,w\,dx$$

$$\sum M_o = 0 \Rightarrow M + V\,dx - w\,dx\frac{dx}{2} - (M + dM) = 0$$

$$\Rightarrow dM = V\,dx \text{ (assume } dx^2 << dx\text{)}$$

$$\boxed{\begin{aligned}\frac{dV}{dx} &= -w\\ \frac{dM}{dx} &= V\end{aligned}}$$

On integrating the above equations, we get:

$$\boxed{\begin{aligned}dV &= -\int w\,dx\\ dM &= -\int V\,dx\end{aligned}}$$

Note:

- $\frac{dV}{dx} = w$; indicates that slope of the shear diagram at a point is equal to the negative intensity of the distributed load at that point.
- $\frac{dM}{dx} = V$; indicates that slope of the moment diagram at a point is equal to the intensity of the shear at that point.
- $dV = -\int w\,dx$; indicates that the change in the shear between any two points on a beam equals the negative area under the distributed loading diagram between the points.
- $dM = -\int V\,dx$; indicates that the change in the moment between any two points on a beam equals the area under the shear diagram between the points.

Procedure for analysis

- First calculate the support reactions by considering force and moment equilibrium conditions.
- Divide the beam in subsections starting from left end (one can start from right end also) where concentrated loads/moments are applied or where distributed loads start or finish.
- Determine the shear force and bending moment relationships with distance (x) along the length of the beam for each section of the beam.
- Plot the shear force (V) and bending moment (M) diagrams along the length of the beam.

- Check consistency of shear force and bending moment expressions between sections and at locations where concentrated loads/moments are applied or where distributed loads start or finish.

Some Important Points

With our sign convention either one can start drawing shear force and bending moment diagram from left end or form right end of the section. Only the sign of shear force diagram and moment diagram will change while the magnitude will remain the same.

- The magnitudes of the shear force and bending moments at the end points of the beam are equal to the magnitudes of the end reaction load and moment.
- Whenever there is a concentrated load or bending moment applied to a beam, the corresponding shear force and bending moment diagrams should show a step of the same magnitude.
- The shear force and bending moment at a free ends are zero.
- Bending moment is maximum (or minimum) at the section where shear force is zero or it changes its sign.
- If w is a curve of degree n, then V is of degree $n+1$ and M is of degree $n+2$.

Nature of SFD and BMD Curve for different loading

Loading	SFD	BMD
Concentrated moment	Zero	Constant
Concentrated load	Constant	Linear
Uniformly distributed moment	Constant	Linear
Uniformly distributed load	Linear	Parabolic
Linearly varying distributed load	Parabolic	Cubic
n^{th} degree varying distributed load	$(n + 1)$ degree	$(n + 2)$ degree

EXAMPLES

Example 9.1: Consider the simply supported beam AB of length L subjected to a concentrated load P as shown in Fig. 9.8. Draw shear force and bending moment diagrams.

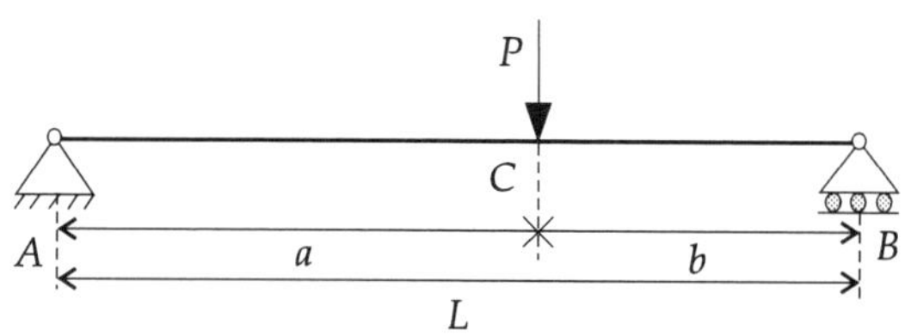

Fig. 9.8

Solution: The first step is to find the reactions, R_A and R_B, by considering two equations of equilibrium, i.e., moments about ends A and B equal to zero.

Free-body diagram

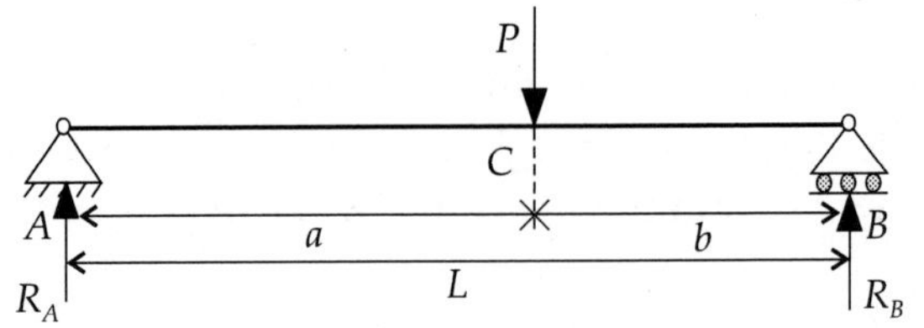

Equations of equilibrium

Moment of forces about B, $M_B = 0 \Rightarrow R_A \times L - P \times b = 0 \Rightarrow R_A = \dfrac{Pb}{L}$

Moment of forces about A, $M_A = 0 \Rightarrow -R_B \times L + P \times a = 0 \Rightarrow R_B = \dfrac{Pa}{L}$

Now, the beam is to be cut at two cross sections, one between A and C and other between C and B.

Free-body diagram (When Section is cut at a point between A and C)

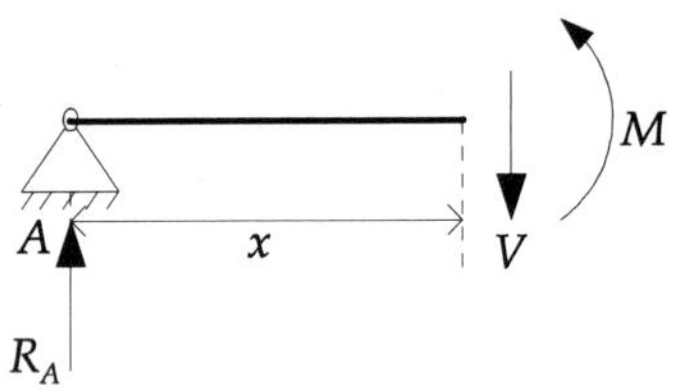

The free-body diagram (FBD) of either half of the beam is drawn, let left half of the beam whose FBD is shown in the figure above. The unknown shear force and bending moment will be obtained by considering two equilibrium equations, i.e., summation of vertical forces and summation of moment of forces.

$$\sum F_y = 0 \Rightarrow R_A - V = 0 \Rightarrow V = R_A = \frac{Pb}{L}$$

$$\sum M = 0 \Rightarrow R_A \times x - M = 0 \Rightarrow M = \frac{Pb}{L}x$$

Thus, shear force will be constant between A and C, while the moment varies linearly.

Free-body diagram (when section is cut at a point between C and B)

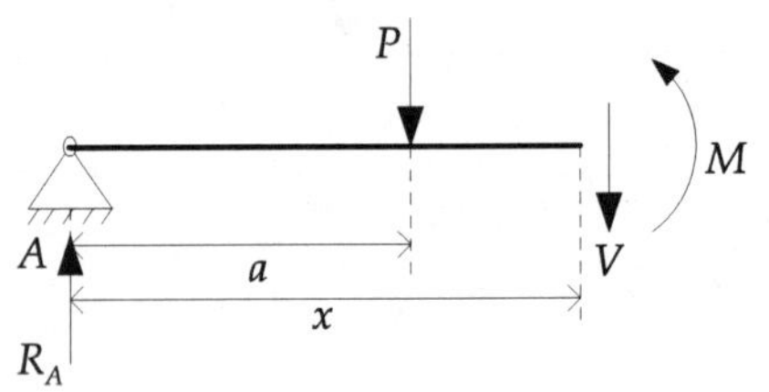

Considering two equilibrium equations, i.e., summation of vertical forces and summation of moment of forces, we get

$$\sum F_y = 0 \Rightarrow R_A - P - V = 0 \Rightarrow V = R_A - P = -\frac{Pa}{L}$$

$$\sum M = 0 \Rightarrow R_A \times x - P \times (x - a) - M = 0 \Rightarrow M = \frac{Pa(L-x)}{L}$$

Thus, shear force will be constant between C and B, while the moment varies linearly.

Beam section	Range of x	Shear force	Bending moment
A to C	$0 < x < a$	$\frac{Pb}{L}$	$\frac{Pb}{L}x$
C to B	$a < x < L$	$-\frac{Pa}{L}$	$\frac{Pa(L-x)}{L}$

At point C

At just left of C ($x = a$) $\quad V = \frac{Pb}{L}$ and $M = \frac{Pb}{L}x = \frac{Pab}{L}$

At just right of C ($x = a$) $\quad V = -\frac{Pb}{L}$ and $M = \frac{Pa(L-x)}{L} = \frac{Pa(L-a)}{L} = \frac{Pab}{L}$

Therefore, due to concentrated force (P at C, shear is discontinuous at C and moment is continuous at C. Moment at C will change its slope at C $\left(\frac{dM}{dx} = V\right)$.

Note: The slope of bending moment diagram is equal to the shear force $\left(\frac{dM}{dx} = V\right)$

The maximum value of bending moment occurs at the cross section where the shear force is zero.

SFD and BMD

SFD and BMD diagrams are shown in Fig. 9.8 (a). Here we started from left side of the section, but one can also start from the right side of the section the difference will come only in signs while the values at every point will remain same as shown in Fig. 9.8 (b).

Special case: ($a = b = L/2$)

Beam section	Range of x	Shear force	Bending moment
A to C	$0 < x < L/2$	$\frac{P}{2}$	$\frac{P}{2}x$
C to B	$L/2 < x < L$	$-\frac{P}{2}$	$\frac{P}{2}(L-x)$

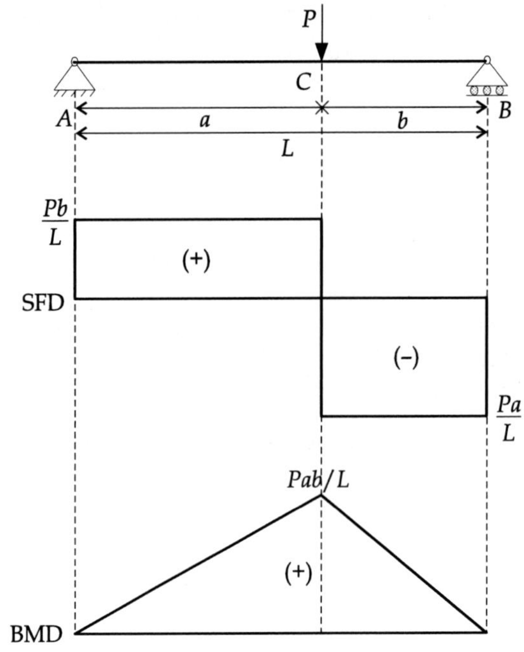

Fig. 9.8 (a)

SFD and BMD

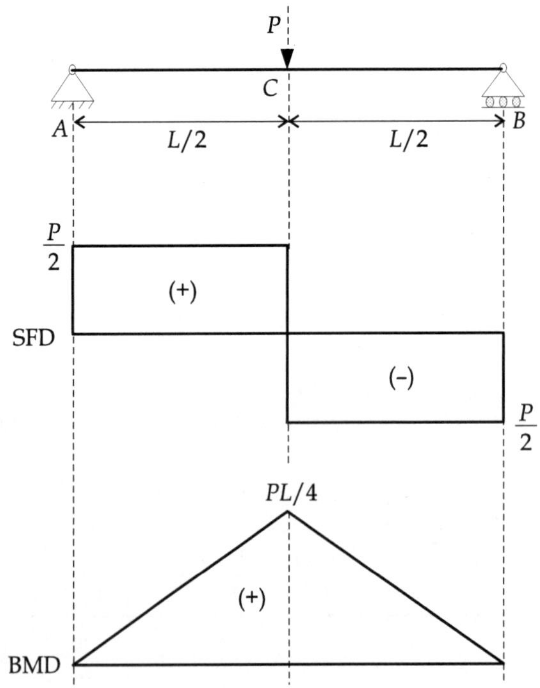

Fig. 9.8 (b)

Example 9.2: Consider the simply supported beam *AB* of length *L* subjected to a uniformly distributed load *w* as shown in Fig. 9.9. Draw shear force and bending moment diagrams.

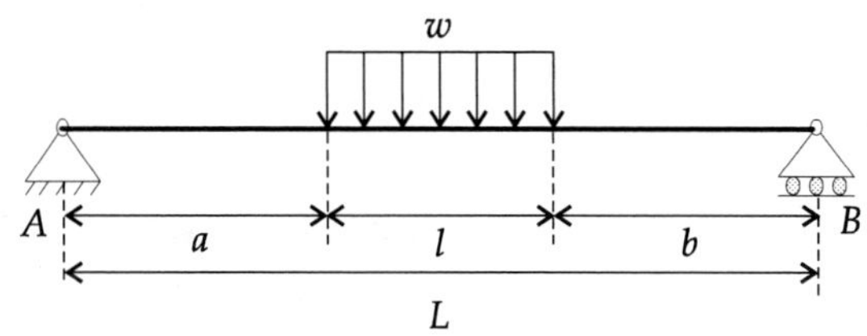

Fig. 9.9

Solution: The first step is to find the reactions R_A and R_B by considering two equations of equilibrium, i.e., moments about end *A* and *B* equal to zero.

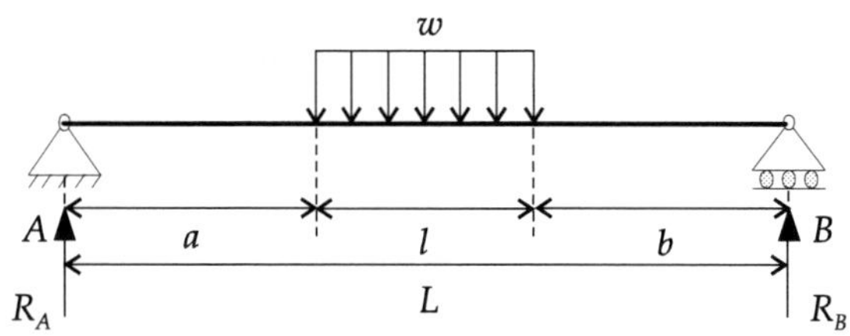

Equations of equilibrium

Moment of forces about *B*,

$$M_B = 0 \Rightarrow R_A \times L - wl \times \left\{b + \frac{l}{2}\right\} = 0 \Rightarrow R_A = \left\{\frac{wl(2b+l)}{2L}\right\}$$

Moment of forces about *A*,

$$M_A = 0 \Rightarrow -R_B \times L + wl \times \left\{a + \frac{l}{2}\right\} = 0 \Rightarrow R_B = \left\{\frac{wl(2b+l)}{2L}\right\}$$

Free-body diagram

Now, the beam is to be cut at three cross sections, i.e., between *A* and *C*; between *C* and *D* and between *D* and *B*.

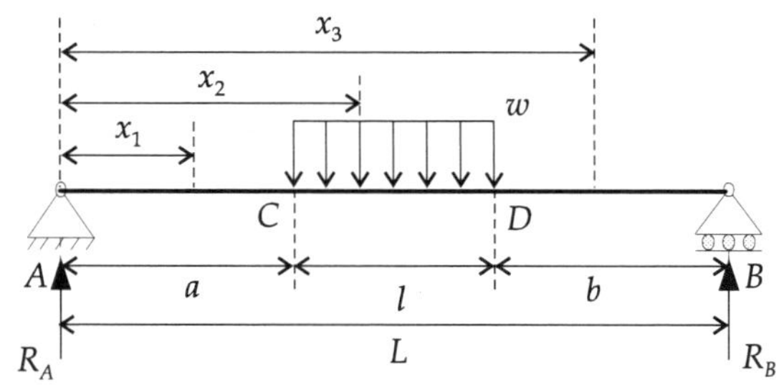

Free-body diagram (when section is cut at a point between *A* and *C*)

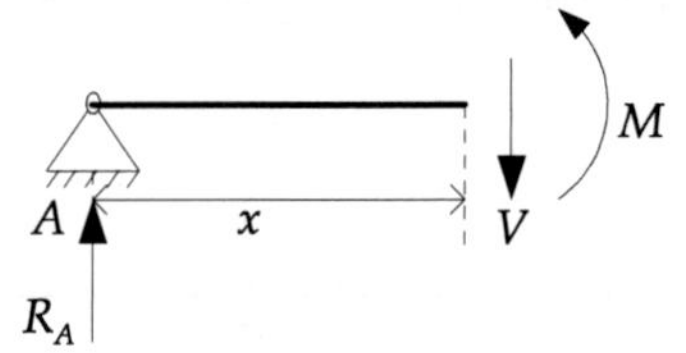

The free-body diagram of either half of the beam is drawn, let left half of the beam whose free-body diagram is shown in the figure above. The unknown shear force and bending moment will be obtained by considering two equilibrium equations, i.e., summation of vertical forces and summation of moment of forces.

$$\sum F_y = 0 \Rightarrow R_A - V = 0 \Rightarrow V = R_A$$

$$\sum M = 0 \Rightarrow R_A \times x - M = 0 \Rightarrow M = R_A x$$

Thus, shear force will be constant between A and C, while moment varies linearly.

Free-body diagram (when section is cut at a point between *C* and *D*)

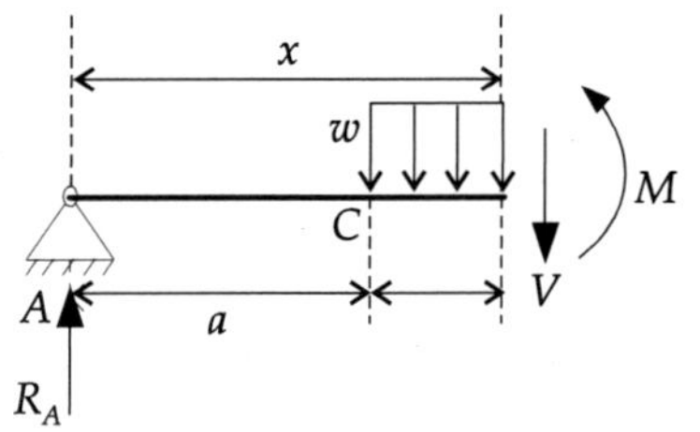

Considering two equilibrium equations, i.e., summation of vertical forces and summation of moment of forces, we get

$$\sum F_y = 0 \Rightarrow R_A - w(x - a) - V = 0 \Rightarrow V = R_A - w(x - a)$$

$$\sum M = 0 \Rightarrow R_A x - w\frac{(x-a)^2}{2} - M = 0 \Rightarrow M = R_A x - w\frac{(x-a)^2}{2}$$

Thus, shear force will vary linearly between *C* and *D*, while moment varies parabolically.

Free-body diagram (when section is cut at a point between *D* and *B*)

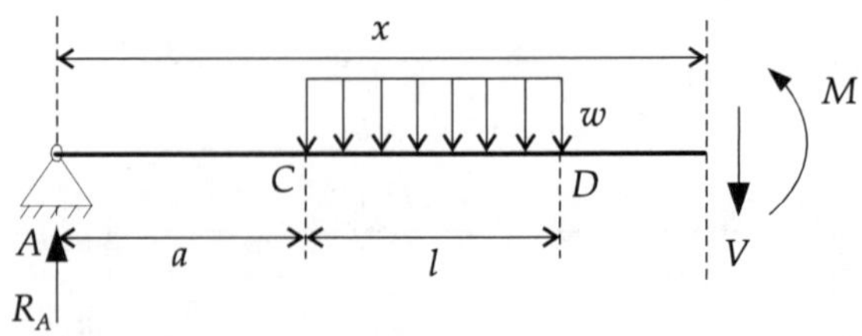

Considering two equilibrium equations, i.e., summation of vertical forces and summation of moment of forces, we get

$$\sum F_y = 0 \Rightarrow R_A - wl - V = 0 \Rightarrow V = R_A - wl$$

$$\sum M = 0 \Rightarrow R_A x - wl(x - a - l/2) - M = 0$$

$$\Rightarrow M = R_A x - wl(x - a - l/2)$$

Thus, shear force will be constant between D and B, while the moment varies linearly.

Beam section	Range of x	Shear force	Bending moment
A to C	$0 < x < a$	R_A	$R_A x$
C to D	$a < x < L - b$	$R_a - w(x - a)$	$R_A x - w\dfrac{(x-a)^2}{2}$
D to B	$L - b < x < L$	$R_A - wl = -R_B$	$R_A x - wl(x - a - l/2)$

Since uniformly distributed load is applied between points C and D, therefore, SFD and BMD changes its sign or slope at point C and D, thus it is necessary to find the value of SFD and BMD at theses points.

At point *C*

At just left of C, put ($x = a$) in beam section, AC

$V = R_A \qquad M = R_A x = R_A x$

At just right of C, put ($x = a$) in beam section, CD

$$V = R_A - w(x - a) = R_A$$

$$M = R_A x - w\frac{(x-a)^2}{2} = R_A a$$

Since, value of shear and moment at just left of C and just right of C are same, therefore, shear and moment both are continuous at C.

Similarly, at point *D*

At just left of D, put ($x = a + l$) in beam section CD

$$V = R_A - w(x - a) = R_A - wl$$

$$M = R_A x - w\frac{(x-a)^2}{2} = R_A(a+l) - w\frac{l^2}{2}$$

At just right of D, put ($x = a + l$) in beam section DB

$$V = R_A - wl$$

$$M = R_A x - wl(x - a - l/2) = R_A(a+l) - w\frac{l^2}{2}$$

Since, value of shear and moment at just left of D and just right of D are same, therefore, shear and moment both are continuous at D.

Note: Student must check that moment at D $M = R_A(a+l) - w\dfrac{l^2}{2} = -R_B\, b$

Note: The slope of bending moment diagram is equal to the shear force $\left(\frac{dM}{dx} = V\right)$

The maximum value of bending moment occurs at the cross section where the shear force is zero.

Maximum moment will occur in beam section BC where the shear force is zero.

$$V = R_A - w(x - a) = R_A - w(x - a) = 0 \Rightarrow x = \frac{R_A}{w} + a$$

Therefore, maximum bending moment

$$M_{max} = R_A\left(\frac{R_A}{w} + a\right) - \frac{(R_A)^2}{2w} = \frac{(R_A)^2}{2w} + R_A a$$

SFD and BMD

SFD and BMD diagrams are shown in Fig. 9.9 (a).

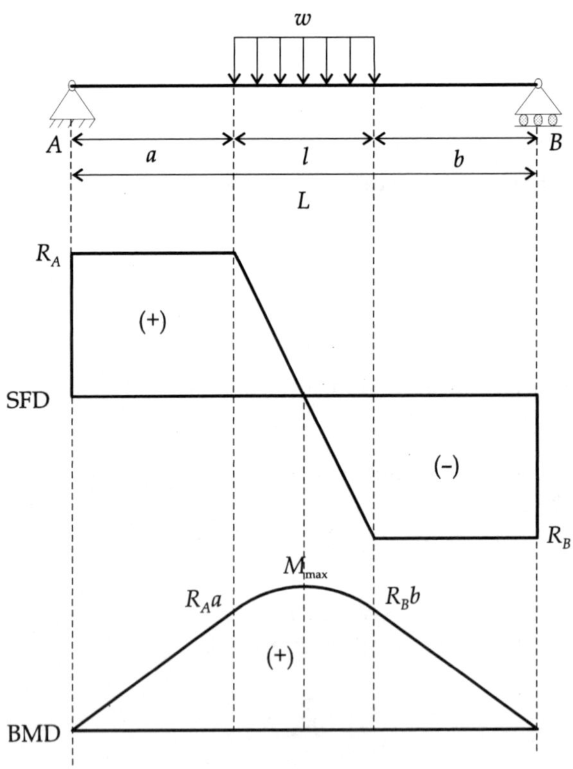

Fig. 9.9 (a)

Special case-I: $(a = b = l = L/3)$

$$R_A = \left\{\frac{wl(2b + l)}{2L}\right\} = \frac{wL}{6} \qquad R_B = \left\{\frac{wl(2a + l)}{2L}\right\} = \frac{wL}{6}$$

Beam section	Range of x	Shear force	Bending moment
A to C	$0 < x < \frac{L}{3}$	$R_A = \frac{wL}{6}$	$R_A x = \frac{wL}{6}x$
C to D	$\frac{L}{3} < x < \frac{2L}{3}$	$R_A - w(x - a)$ $= \frac{wL}{6} - w\left(x - \frac{L}{3}\right)$	$R_A x - w\frac{(x-a)^2}{2}$ $= \frac{wL}{6}x - w\frac{(x-L/3)^2}{2}$
D to B	$\frac{2L}{3} < x < L$	$R_A - wl = R_B = -\frac{wL}{6}$	$R_A x - wl(x - a - l/2)$ $= \frac{wL}{6}x - w\frac{L}{3}(x - L/2)$

At point C ($x = a = L/3$)

$$V = R_A = \frac{wL}{6}$$

$$M = R_A a = \frac{wL^2}{18}$$

At point D ($x = a + l = 2L/3$)

$$V = R_A - w(x - a) = R_A - wl = \frac{wL}{6} - \frac{wL}{3} = -\frac{wL}{6}$$

$$M = R_A(a+l) - w\frac{l^2}{2} = \frac{wL}{6}(2L/3) - w\frac{L^2}{18} = \frac{wL}{18}$$

Maximum moment will occur in beam section BC where the shear force is zero.

$$V = R_A - w(x - a) = R_A - w(x - a) = 0 \Rightarrow x = \frac{R_A}{w} + a = \frac{wL}{6w} + \frac{L}{3} = \frac{L}{2}$$

Therefore, maximum bending moment at $x = \frac{L}{2}$

$$M_{\max} = R_A x - w\frac{(x-a)^2}{2}$$

$$= \frac{wL}{6}x - w\frac{(x-L/3)^2}{2} = \frac{wL}{6}\frac{L}{2} - w\frac{(L/2 - L/3)^2}{2} = \frac{5wL^2}{72}$$

SFD and BMD

SFD and BMD diagrams are shown in Fig. 9.9 (b).

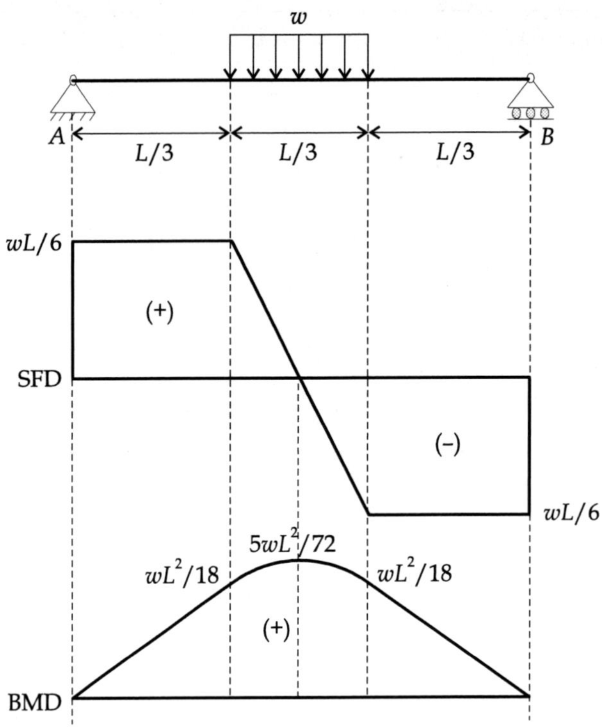

Fig. 9.9 (b)

Special case-II: $(b = 0 \Rightarrow l = L - a)$

$$R_A = \left\{\frac{wl(2b+l)}{2L}\right\} = \frac{w(L-a)^2}{2L}$$

$$R_B = \left\{\frac{wl(2a+l)}{2L}\right\} = \frac{w(L-a)(L+a)}{2L}$$

Beam section	Range of x	Shear force	Bending moment
A to C	$0 < x < a$	$R_A = \dfrac{w(L-a)^2}{2L}$	$R_A x = \dfrac{w(L-a)^2}{2L}x$
C to B	$a < x < L$	$R_A - w(x-a)$ $= \dfrac{w(L-a)^2}{2L} - w(x-a)$	$R_A x - w\dfrac{(x-a)^2}{2}$ $= \dfrac{w(L-a)^2}{2L}x - w\dfrac{(x-a)^2}{2}$

At point C $(x = a)$

$$V = R_A = \frac{w(L-a)^2}{2L} \qquad M = R_A a = \frac{wa(L-a)^2}{2L}$$

Maximum moment will occur in beam section CB where the shear force is zero.

$$V = \frac{w(L-a)^2}{2L} - w(x-a) = 0 \Rightarrow x = \frac{L^2 + a^2}{2L}$$

Therefore, maximum bending moment at $x = \dfrac{L^2 + a^2}{2L}$

$$M_{\max} = \frac{w(L-a)^2}{2L}\left(\frac{L^2+a^2}{2L}\right) - w\frac{\left\{\left(\frac{L^2+a^2}{2L}\right) - a\right\}}{2}$$

$$= \frac{w(L-a)^2}{2L}\left(\frac{L^2+a^2}{2L}\right) - w\frac{w(L-a)^4}{8L^2} = \frac{w(L^2-a^2)^2}{8L^2}$$

SFD and BMD

SFD and BMD diagrams are shown in Fig. 9.9 (c).

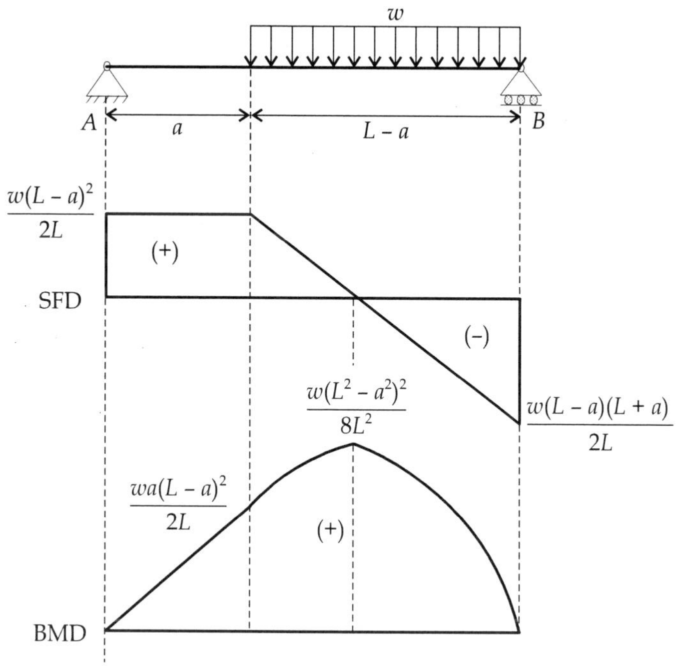

Fig. 9.9 (c)

Special case-III: $(a = b = 0 \Rightarrow l = L)$

$$R_A = \left\{\frac{wl(2b+l)}{2L}\right\} = \frac{wL}{2}$$

$$R_B = \left\{\frac{wl(2a+l)}{2L}\right\} = \frac{wL}{2}$$

Beam section	Range of x	Shear force	Bending moment
A to B	$0 < x < L$	$R_A - w(x - a)$ $= \frac{wL}{2} - wx$	$R_A x - w\frac{(x-a)^2}{2}$ $= \frac{wL}{2}x - w\frac{x^2}{2}$

Maximum moment will occur in beam where shear force (at centre) is zero.

$$V = R_A - w(x - a) = \frac{wL}{2} - wx = 0 \Rightarrow x = \frac{L}{2}$$

Therefore, maximum bending moment at $x = \frac{L}{2}$

$$M_{\max} = R_A x - w\frac{(x-a)^2}{2} = \frac{wL}{2}x - w\frac{x^2}{2} = \frac{wL}{2}\frac{L}{2} - w\frac{(L/2)^2}{2} = \frac{wL^2}{8}$$

SFD and BMD

SFD and BMD diagrams are shown in Fig. 9.9 (d).

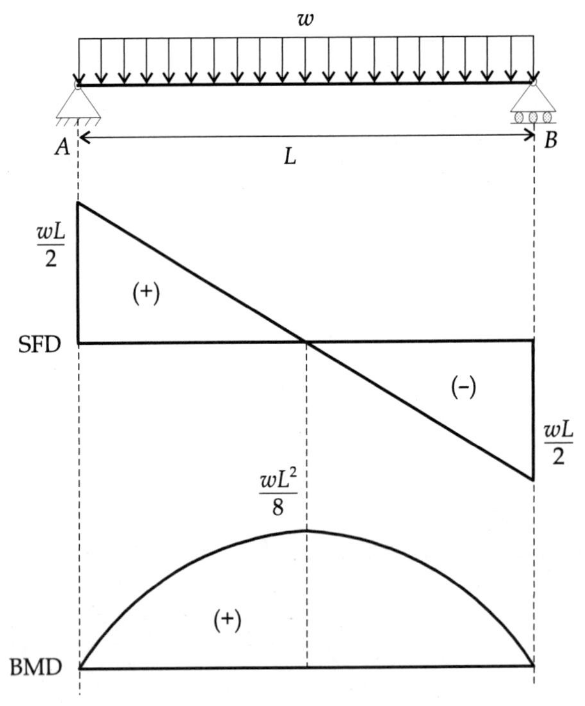

Fig. 9.9 (d)

Example 9.3: Consider the simply supported beam AB of length L subjected to a several concentrated loads as shown in Fig. 9.10. Draw shear force and bending moment diagrams.

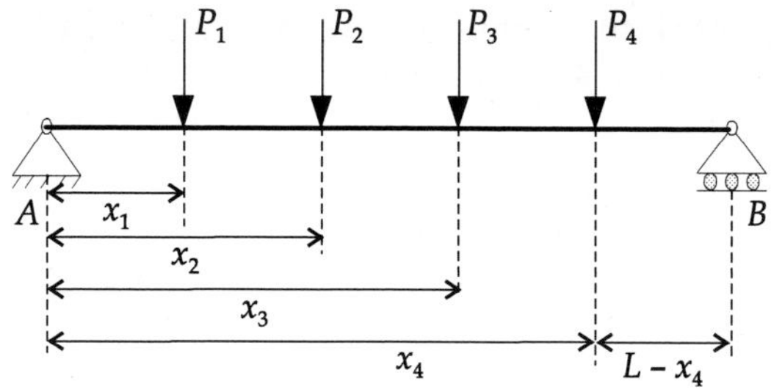

Fig. 9.10

Solution: The first step is to find the reactions R_A and R_B by considering two equations of equilibrium, i.e., moments about end A and B equal to zero.

Free-body diagram

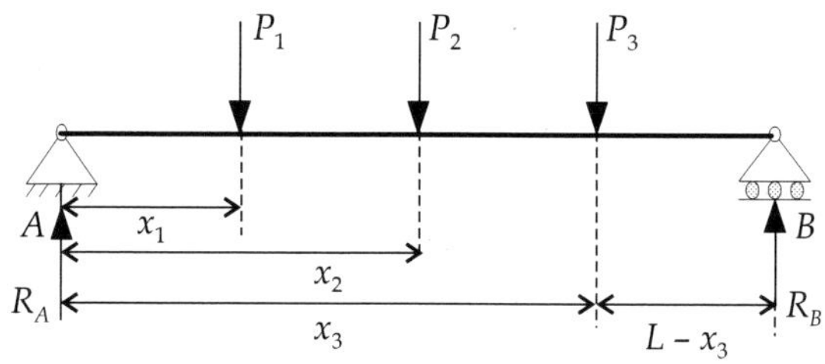

Equations of equilibrium

Moment of forces about A,

$$M_A = 0 \Rightarrow -R_B \times L + P_1 \times x_1 + P_2 \times x_2 + P_3 \times x_3 = 0$$

$$\Rightarrow R_B = \frac{P_1 \times x_1 + P_2 \times x_2 + P_3 \times x_3}{L}$$

Moment of forces about B,

$$M_B = 0 \Rightarrow R_A \times L - P_1 \times (L - x_1) + P_2 \times (L - x_2) + P_3 \times (L - x_3) = 0$$

$$\Rightarrow R_A = \frac{P_1 \times (L - x_1) + P_2 \times (L - x_2) + P_3 \times (L - x_3)}{L}$$

Now, the beam is to be cut at four cross sections, before and after each concentrated load.

Free-body diagram ($0 < x < x_1$)

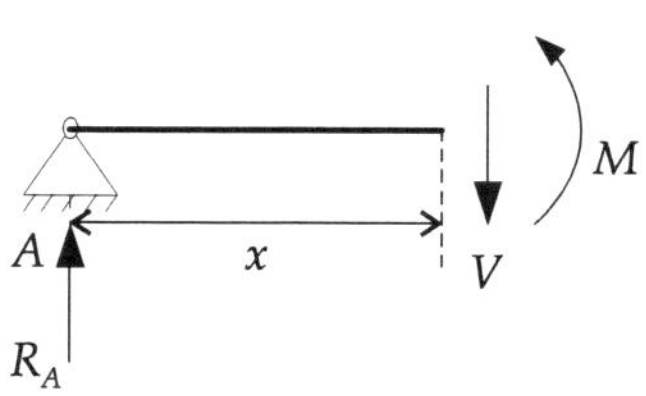

Considering two equilibrium equations, i.e., summation of vertical forces and summation of moment of forces, we get

$$\sum F_y = 0 \Rightarrow R_A - V = 0 \Rightarrow V = R_A$$

$$\sum M = 0 \Rightarrow R_A \times x - M = 0 \Rightarrow M = R_A x$$

Thus, shear force will be constant, while moment varies linearly.

Free-body diagram ($x_1 < x < x_2$)

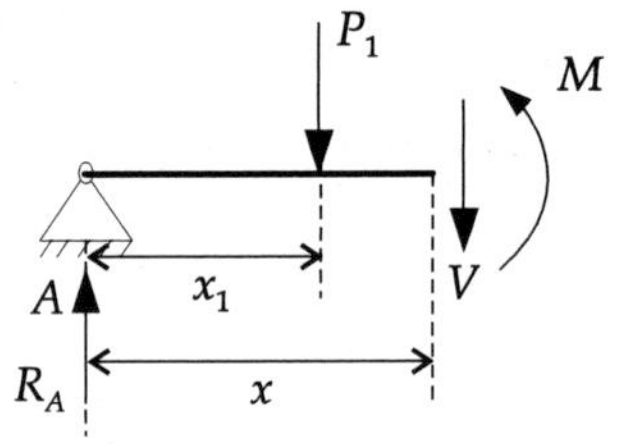

Considering two equilibrium equations, i.e., summation of vertical forces and summation of moment of forces, we get

$$\sum F_y = 0 \Rightarrow R_A - P_1 - V = 0 \Rightarrow V = R_A - P_1$$

$$\sum M = 0 \Rightarrow R_A \times x - P_1 \times (x - x_1) - M = 0 \Rightarrow M = R_A \times x - P_1 \times (x - x_1)$$

Thus, shear force will be constant, while moment varies linearly.

Free-body diagram ($x_2 < x < x_3$)

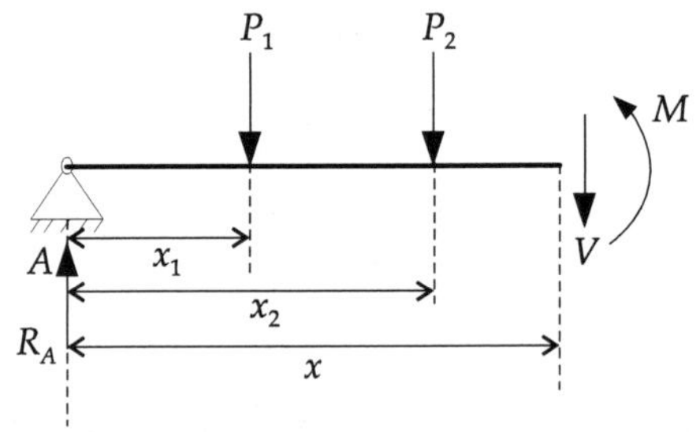

Considering two equilibrium equations, i.e., summation of vertical forces and summation of moment of forces, we get

$$\sum F_y = 0 \Rightarrow R_A - P_1 - P_2 - V = 0 \Rightarrow V = R_A - P_1 - P_2$$

$$\sum M = 0 \Rightarrow R_A \times x - P_1 \times (x - x_1) - P_2 \times (x - x_2) - M = 0$$

$$\Rightarrow M = R_A \times x - P_1 \times (x - x_1) - P_2 \times (x - x_2)$$

Thus, shear force will be constant, while moment varies linearly.

Free-body diagram ($x_3 < x < L$)

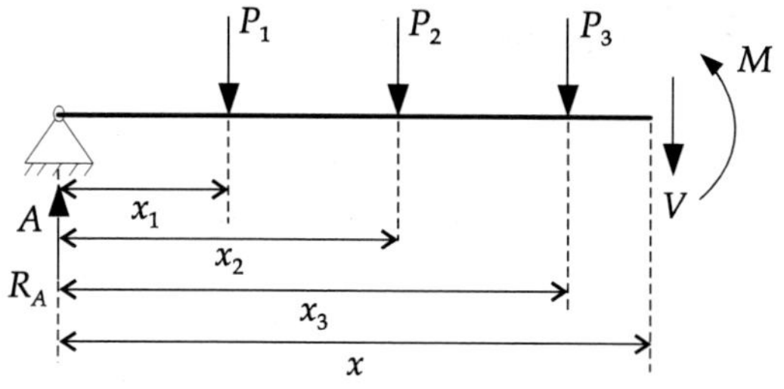

Considering two equilibrium equations, i.e., summation of vertical forces and summation of moment of forces, we get

$\sum F_y = 0 \Rightarrow R_A - P_1 - P_2 - P_3 - V = 0 \Rightarrow V = R_A - P_1 - P_2 - P_3$

$\sum M = 0 \Rightarrow R_A \times x - P_1 \times (x - x_1) - P_2 \times (x - x_2) - P_3 \times (x - x_3) - M = 0$

$\Rightarrow M = R_A \times x - P_1 \times (x - x_1) - P_2 \times (x - x_2) - P_3 \times (x - x_3)$

Thus, shear force will be constant, while moment varies linearly.

Under load P_1 ($x = x_1$)

At just left of P_1

$V_1 = R_A$ and $M_1 = R_A x_1$

At just right of P_1

$V_1 = R_A - P_1$ and $M_1 = R_A x_1$

Therefore, due to concentrated force (P_1, shear is discontinuous while moment is continuous.

Under load P_2 ($x = x_2$)

At just left of P_2

$V_2 = R_A - P_1$ and $M_2 = R_A x_2 - P_1(x_2 - x_1)$

At just right of P_2

$V_2 = R_A - P_1 - P_2$ and $M_2 = R_A x_2 - P_1(x_2 - x_1)$

Therefore, due to concentrated force P_2, Shear is discontinuous while moment is continuous.

Under load P_3 ($x = x_3$)

At just left of P_3

$V_3 = R_A - P_1 - P_2$ and $M_3 = R_A x_3 - P_1(x_3 - x_1) - P_2(x_3 - x_2)$

At just right of P_3

$V_3 = R_A - P_1 - P_2 - P_3$ and $M_3 = R_A x_3 - P_1(x_3 - x_1) - P_2(x_3 - x_2)$

Therefore, due to concentrated force P_3, shear is discontinuous while moment is continuous.

SFD and BMD

SFD and BMD diagrams are shown in Fig. 9.10 (a). Under each concentrated load we get the jump in shear force equal to load at that point.

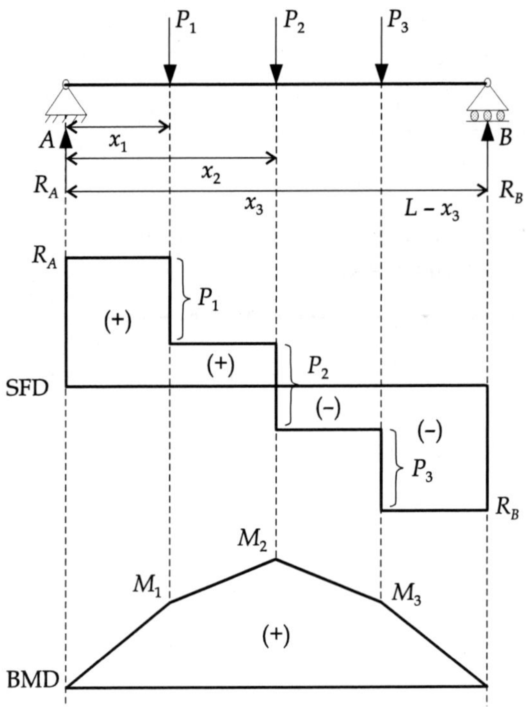

Fig. 9.10 (a)

Example 9.4: Consider the simply supported beam AB of length L subjected to a varying distributed load as shown in Fig. 9.11. Draw shear force and bending moment diagrams.

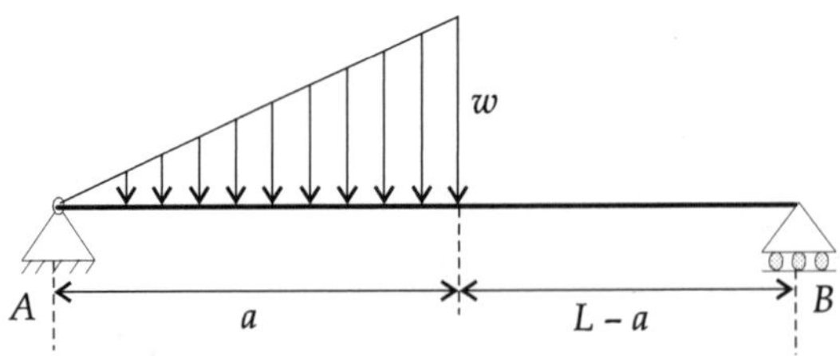

Fig. 9.11

Solution: The first step is to find the reactions R_A and R_B by considering two equations of equilibrium, i.e., moments about end A and B equal to zero.

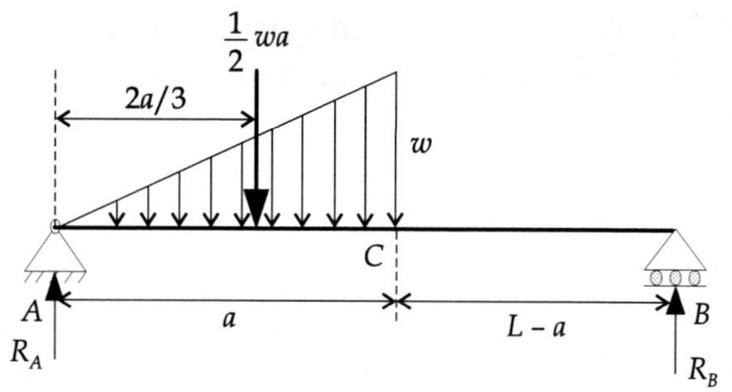

Equations of equilibrium

Moment of forces about B,

$$M_B = 0 \Rightarrow R_A \times L - \frac{1}{2} w a \times (L - 2a/3) = 0 \Rightarrow R_A = \frac{wa(3L - 2a)}{6L}$$

Moment of forces about A,

$$M_A = 0 \Rightarrow -R_B \times L + \frac{1}{2} w a \times (2a/3) = 0 \Rightarrow R_B = \frac{wa^2}{3L}$$

Free-body diagram

Now, the beam is to be cut at two cross sections, i.e., between A and C; and between C and B

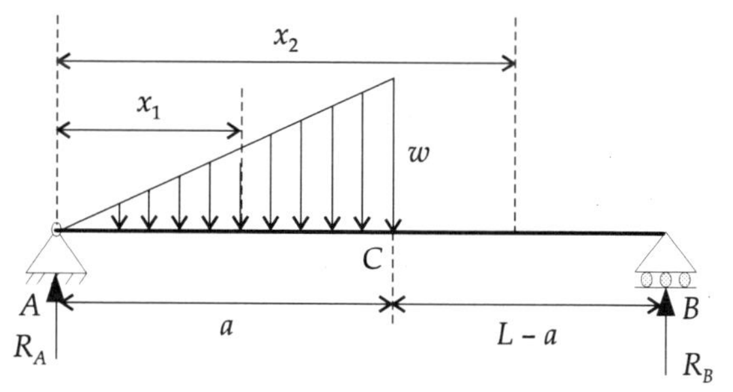

Free-body diagram (when section is cut at a point between A and C)

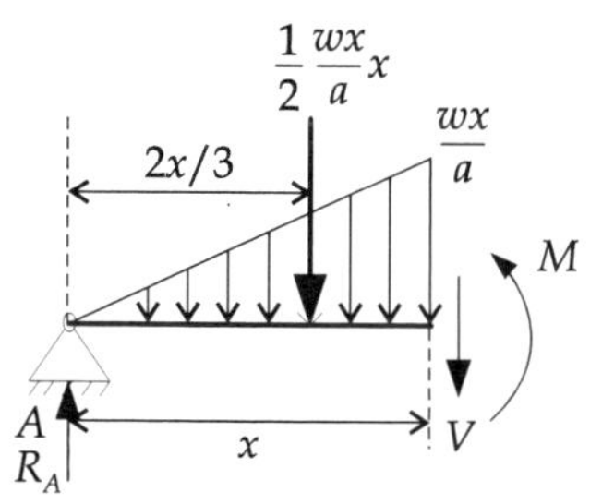

The free-body diagram of either half of the beam is drawn, let left half of the beam whose free-body diagram is shown in the figure above. The unknown shear force and bending moment will be obtained by considering two equilibrium equations, i.e., summation of vertical forces and summation of moment of forces.

$$\sum F_y = 0 \Rightarrow R_A - \frac{wx^2}{2a} - V = 0 \Rightarrow V = R_A - \frac{wx^2}{2a}$$

$$\sum M = 0 \Rightarrow R_A \times x - \frac{wx^2}{2a}(x/3) - M = 0 \Rightarrow M = R_A \times x - \frac{wx^3}{6a}$$

Thus, shear force vary parabolically between A to C, while moment varies cubically.

Free-body diagram (when section is cut at a point between C and B)

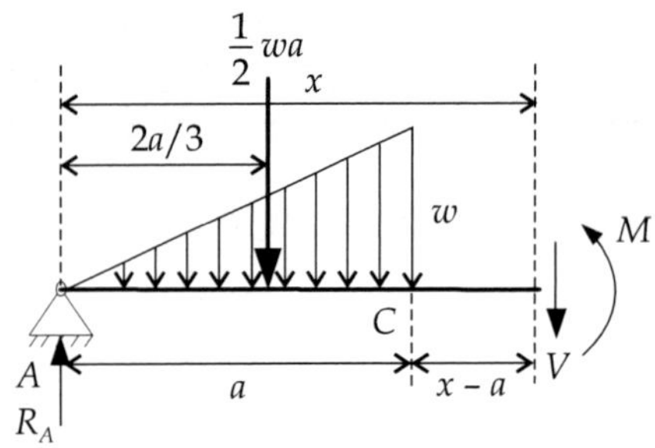

Considering two equilibrium equations, i.e., summation of vertical forces and summation of moment of forces, we get

$$\sum F_y = 0 \Rightarrow R_A - \frac{wa}{2} - V = 0 \Rightarrow V = R_A - \frac{wa}{2}$$

$$\sum M = 0 \Rightarrow R_A x - \frac{wa}{2}(x - 2a/3) - M = 0 \Rightarrow M = R_A x - \frac{wa(3x - 2a)}{6}$$

Thus, shear force will be constant between C and B, while moment varies linearly.

Beam section	Range of x	Shear force	Bending moment
A to C	$0 < x < a$	$R_A - \frac{wx^2}{2a}$	$R_A \times x - \frac{wx^3}{6a}$
C to B	$a < x < L$	$R_A \times \frac{wa}{2}$	$R_A x \times \frac{wa(3x - 2a)}{6}$

Since varying distributed load is applied between points A and C, therefore SFD and BMD changes its sign or slope at point C, thus it is necessary to find the value of SFD and BMD at theses points.

At point C

At just left of C, put $(x = a)$ in beam section AC

$$V = R_A - \frac{wa}{2} \qquad M = R_A a - \frac{wa^2}{6}$$

At just right of C, put ($x = a$) in beam section CB

$$V = R_A - \frac{wa}{2} \qquad M = R_A a - \frac{wa^2}{6}$$

Since, value of shear and moment at just left of C and just right of C are same, therefore, shear and moment both are continuous at C.

Thus at point C

$$V = R_A - \frac{wa}{2} = \frac{wa(3L-2a)}{6L} - \frac{wa}{2} = -\frac{wa^2}{3L} = -R_B$$

$$M = R_A a - \frac{wa^2}{6} = \frac{wa(3L-2a)a}{6L} - \frac{wa^2}{6} = \frac{wa^2(L-a)}{3L}$$

Note: The maximum value of bending moment occurs at the cross section where the shear force is zero.

Maximum moment will occur in beam section AC where shear force is zero, i.e.,

$$V = R_A - \frac{wx^2}{2a} = \frac{wa(3L-2a)}{6L} - \frac{wx^2}{2a} = 0 \Rightarrow x = \sqrt{\frac{(3L-2a)a^2}{3L}}$$

SFD and BMD

SFD and BMD diagrams are shown in Fig. 9.11 (a).

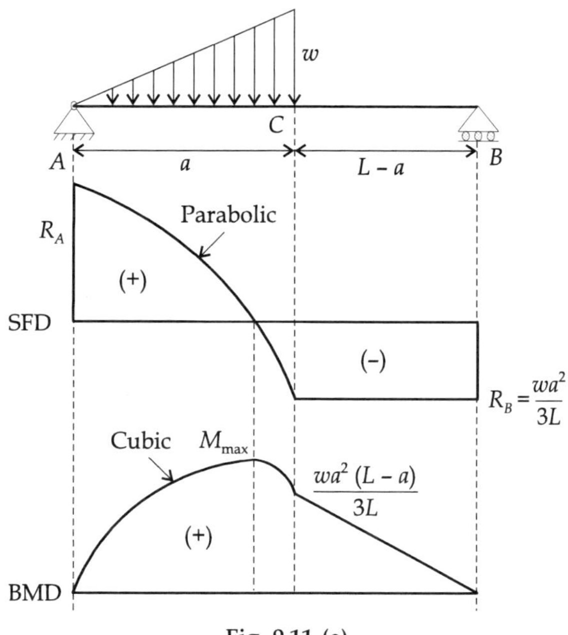

Fig. 9.11 (a)

Special case: ($a = L$)

$$R_A = \frac{wL}{6}$$

$$R_B = \frac{wL}{3}$$

Shear force and bending moment, $0 < x < L$

$$V = R_A - \frac{wx^2}{2a} = \frac{wL}{6} - \frac{wx^2}{2L}$$

$$M = R_A \times x - \frac{wx^3}{6a} = \frac{wL}{6}x - \frac{wx^3}{6L}$$

Thus, shear force varies parabolically, while moment varies cubically.

Maximum moment will occur in beam at a section where shear force is zero.

$$V = R_A - \frac{wx^2}{2a} = \frac{wL}{6} - \frac{wx^2}{2L} = 0 \Rightarrow x = \sqrt{1/3}\ L$$

Therefore, maximum bending moment at $x = \sqrt{1/3}\ L$

$$M_{\max} = \frac{wL}{6}x - \frac{wx^3}{6L} = \frac{wL}{9\sqrt{3}}$$

SFD and BMD

SFD and BMD diagrams are shown in Fig. 9.11 (b).

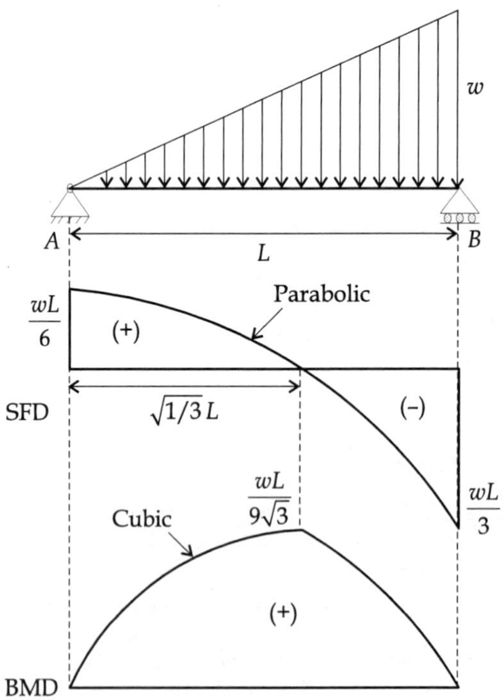

Fig. 9.11 (b)

Example 9.5: Consider the cantilever beam AB of length L subjected to several concentrated loads as shown in Fig. 9.12. Draw shear force and bending moment diagrams.

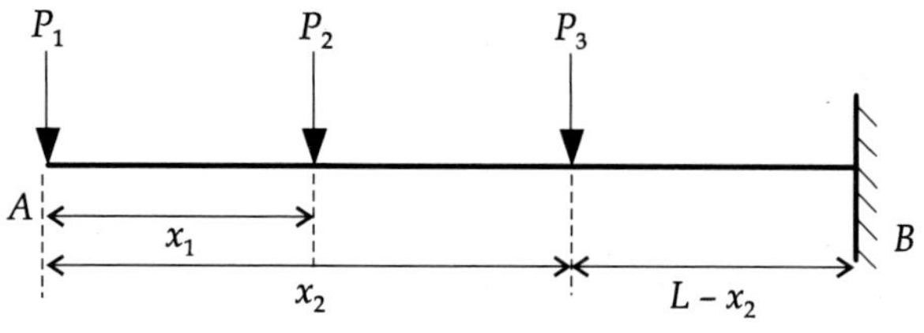

Fig. 9.12

Solution: Consider the three cross sections from left, after each concentrated load.

Free-body diagram $(0 < x < x_1)$

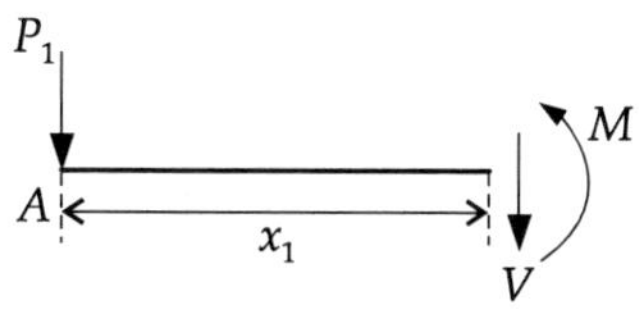

Considering two equilibrium equations, i.e., summation of vertical forces and summation of moment of forces, we get

$$\sum F_y = 0 \Rightarrow -P_1 - V = 0 \Rightarrow V = -P_1$$

$$\sum M = 0 \Rightarrow -P_1 \times x_1 - M = 0 \Rightarrow M = -P_1 \times x_1$$

Thus shear force will be constant (negative), while moment varies linearly (negative).

Free-body diagram $(x_1 < x < x_2)$

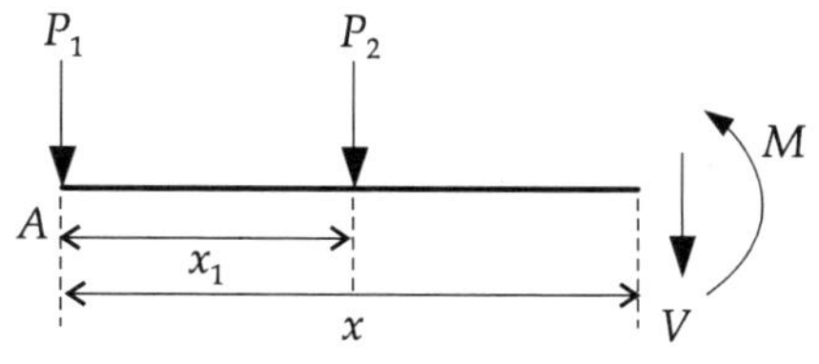

Considering two equilibrium equations, i.e., summation of vertical forces and summation of moment of forces, we get

$$\sum F_y = 0 \Rightarrow -P_1 - P_2 - V = 0 \Rightarrow V = -P_1 - P_2$$

$$\sum M = 0 \Rightarrow -P_1 \times x - P_2 \times (x - x_1) - M = 0 \Rightarrow M = -P_1 \times x - P_2 \times (x - x_1)$$

Thus shear force will be constant (negative), while moment varies linearly (negative).

Near fixed end B $(x = L)$ $\quad V = -P_1 - P_2$ and $M = -P_1 \times L - P_2 \times (L - x_1)$

SFD and BMD SFD and BMD diagrams are shown in Fig. 9.12 (a).

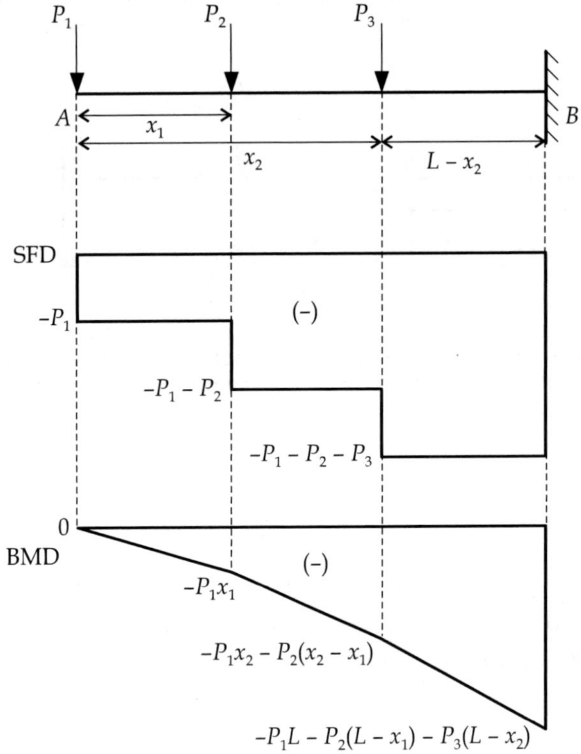

Fig. 9.12 (a)

Example 9.6: Consider the cantilever beam AB of length L subjected to a uniformly distributed load (w) as shown in Fig. 9.13. Draw shear force and bending moment diagrams.

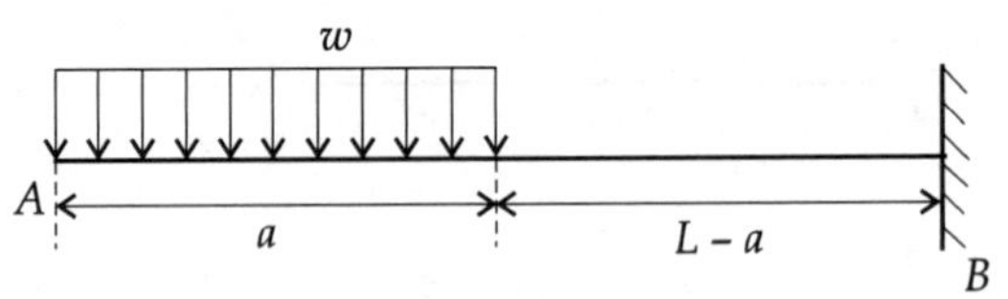

Fig. 9.13

Solution: Consider the two cross sections of the beam, one in between uniformly distributed load and another after uniformly distributed load.

Free-body diagram ($0 < x < a$)

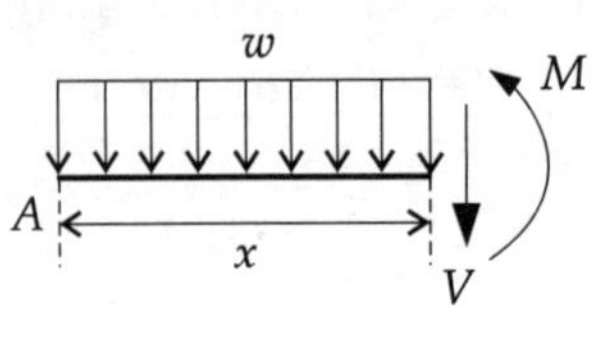

Considering two equilibrium equations, i.e., summation of vertical forces and summation of moment of forces, we get

$$\sum F_y = 0 \Rightarrow -wx - V = 0 \Rightarrow V = -wx$$

$$\sum M = 0 \Rightarrow -(wx)\frac{x}{2} - M = 0 \Rightarrow M = \frac{wx^2}{2}$$

Thus, shear force will vary linearly, while moment varies parabolically between 0 and a.

Free-body diagram ($a < x < L$)

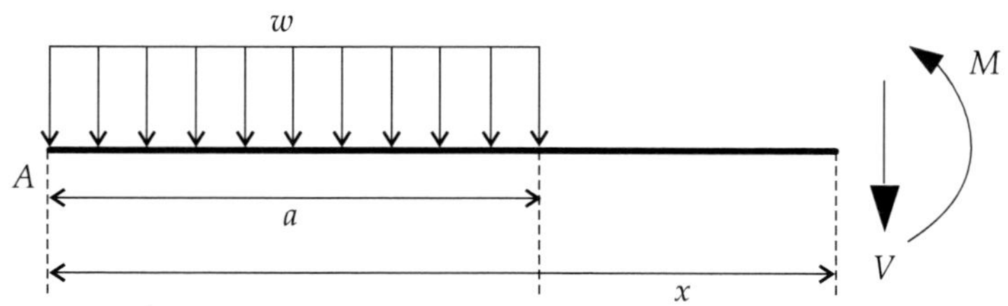

Considering two equilibrium equations, i.e., summation of vertical forces and summation of moment of forces, we get

$$\sum F_y = 0 \Rightarrow -wa - V = 0 \Rightarrow V = -wa$$

$$\sum M = 0 \Rightarrow -(wa)\left(x - \frac{a}{2}\right) - M = 0 \Rightarrow M = -\frac{wa(2x-a)}{2}$$

Thus shear force will be constant, while moment varies linearly between a to L.

At point ($x = a$)

$$V = -wa \quad \text{and} \quad M = -\frac{wa^2}{2}$$

Near fixed end B ($x = L$)

$$V = -wa \quad \text{and} \quad M = -\frac{wa(2x-a)}{2} = -\frac{wa(2L-a)}{2}$$

SFD and BMD

SFD and BMD diagrams are shown in Fig. 9.13 (a).

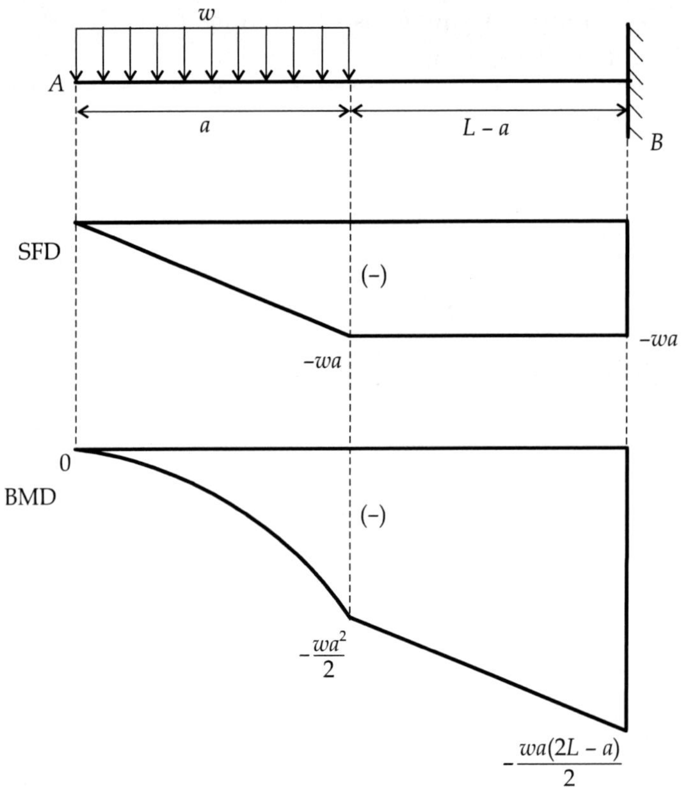

Fig. 9.13 (a)

Special case: ($a = L$)

Free-body diagram ($0 < x < L$)

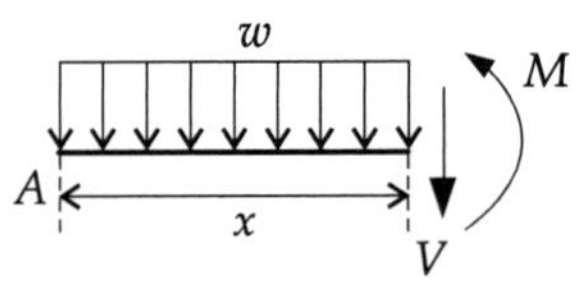

Considering two equilibrium equations, i.e., summation of vertical forces and summation of moment of forces, we get

$$\sum F_y = 0 \Rightarrow -wx - V = 0 \Rightarrow V = -wx$$

$$\sum M = 0 \Rightarrow -(wx)\frac{x}{2} - M = 0 \Rightarrow M = \frac{wx^2}{2}$$

Thus, shear force will vary linearly, while moment varies parabolically.

Near fixed end B ($x = L$) $\quad V = -wL$ and $M = -\frac{wx^2}{2} = -\frac{wL^2}{2}$

SFD and BMD

SFD and BMD diagrams are shown in Fig. 9.13 (b).

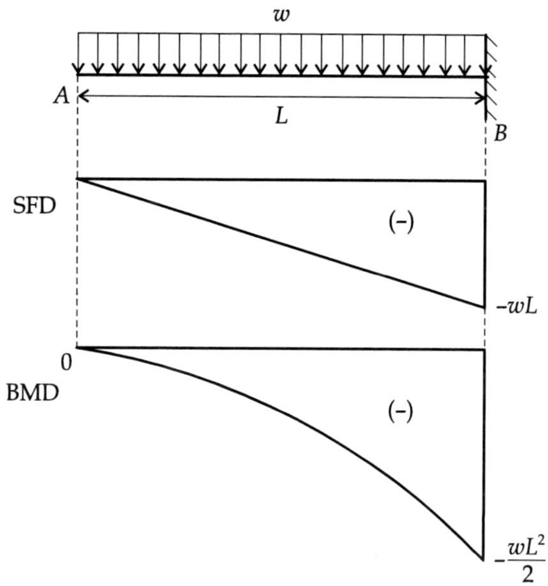

Fig. 9.13 (b)

Example 9.7: Consider the cantilever beam *AB* of length *L* subjected to a varying distributed load as shown in Fig. 9.14. Draw shear force and bending moment diagrams.

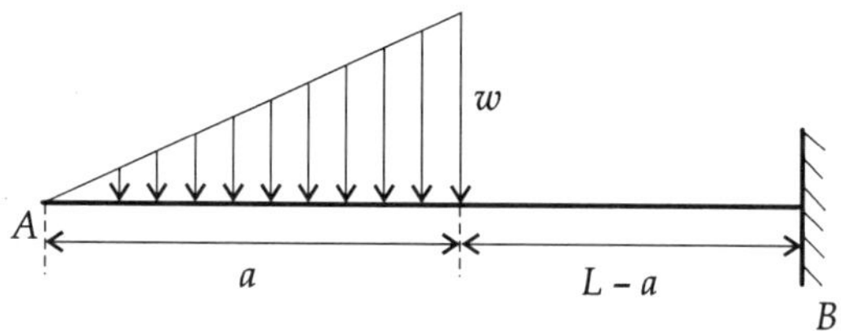

Fig. 9.14

Solution:

Free-body diagram

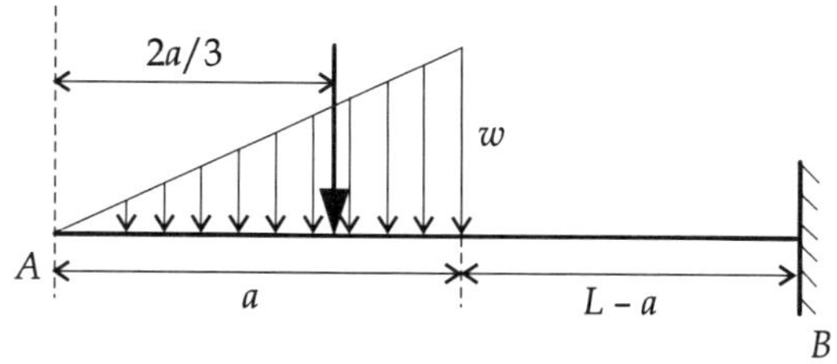

Now, the beam is to be cut at two cross sections, i.e., between *A* and *C*; and between *C* and *B*.

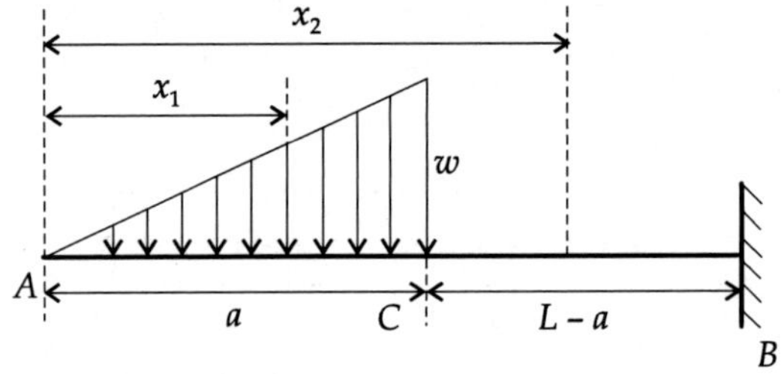

Free-body diagram (when section is cut at a point between *A* and *C*)

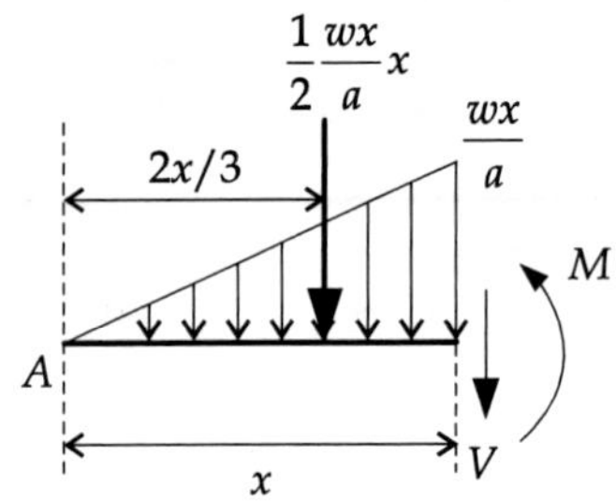

The free-body diagram of either half of the beam is drawn, let left half of the beam whose free-body diagram is shown in the figure above. The unknown shear force and bending moment will be obtained by considering two equilibrium equations, i.e., summation of vertical forces and summation of moment of forces.

$$\sum F_y = 0 \Rightarrow -\frac{wx^2}{2a} - V = 0 \Rightarrow V = -\frac{wx^2}{2a}$$

$$\sum M = 0 \Rightarrow -\frac{wx^2}{2a}(x/3) - M = 0 \Rightarrow M = -\frac{wx^3}{6a}$$

Thus, shear force vary parabolically between *A* and *C*, while moment varies cubically.

Free-body diagram (when section is cut at a point between *C* and *B*)

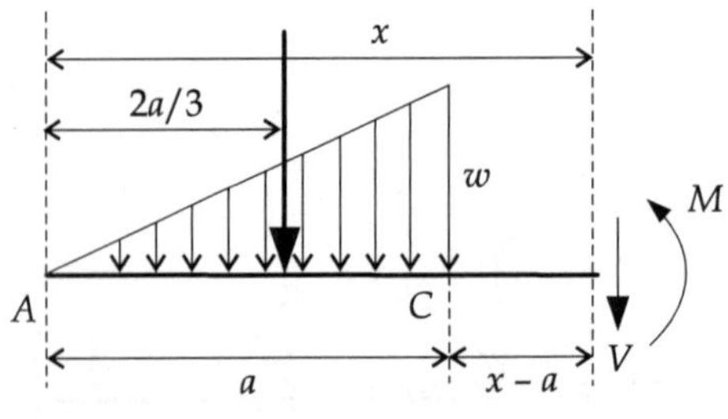

Considering two equilibrium equations, i.e., summation of vertical forces and summation of moment of forces, we get

$$\sum F_y = 0 \Rightarrow -\frac{wa}{2} - V = 0 \Rightarrow V = -\frac{wa}{2}$$

$$\sum M = 0 \Rightarrow -\frac{wa}{2}(x - 2a/3) - M = 0 \Rightarrow M = -\frac{wa(3x - 2a)}{6}$$

Thus, shear force will be constant between C and B, while moment varies linearly.

Beam section	Range of x	Shear force	Bending moment
A to C	$0 < x < a$	$-\frac{wx^2}{2a}$	$-\frac{wx^3}{6a}$
C to B	$a < x < L$	$-\frac{wa}{2}$	$-\frac{wa(3x-2a)}{6}$

Since varying distributed load is applied between points A and C, therefore SFD and BMD changes its sign or slope at point C, thus it is necessary to find the value of SFD and BMD at theses points.

At point C

At just left of C, put $(x = a)$ in beam section AC

$$V = -\frac{wa}{2}$$

$$M = -\frac{wa^2}{6}$$

At just right of C, put $(x = a)$ in beam section CB

$$V = -\frac{wa}{2}$$

$$M = -\frac{wa^2}{6}$$

Since, value of shear and moment at just left of C and just right of C are same, therefore, shear and moment both are continuous at C.

Thus at point C

$$V = -\frac{wa}{2}$$

$$M = -\frac{wa^2}{6}$$

Near fixed support B $(x = L)$

$$V = -\frac{wa}{2}$$

$$M = \frac{wa(3x - 2a)}{6} = -\frac{wa(3L - 2a)}{6}$$

SFD and BMD

SFD and BMD diagrams are shown in Fig. 9.14 (a).

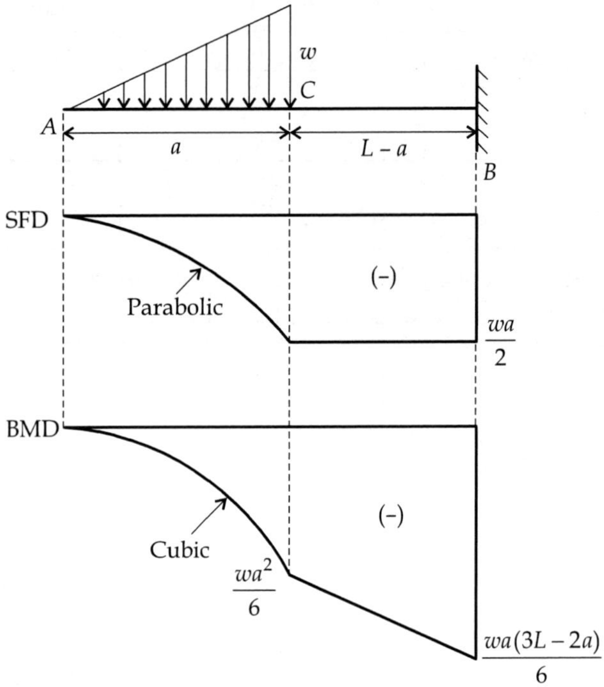

Fig. 9.14 (a)

Special case: ($a = L$)

Shear force and bending moment $0 < x < L$

$$V = -\frac{wx^2}{2L}$$

$$M = -\frac{wx^3}{6L}$$

Thus shear force varies parabolically, while moment varies cubically. Maximum shear will occur near fixed end at $x = L$

$$V = -\frac{wL}{2}$$

Maximum moment will occur near fixed end at $x = L$

$$M_{max} = -\frac{wL^2}{6}$$

SFD and BMD

SFD and BMD diagrams are shown in Fig. 9.14 (b).

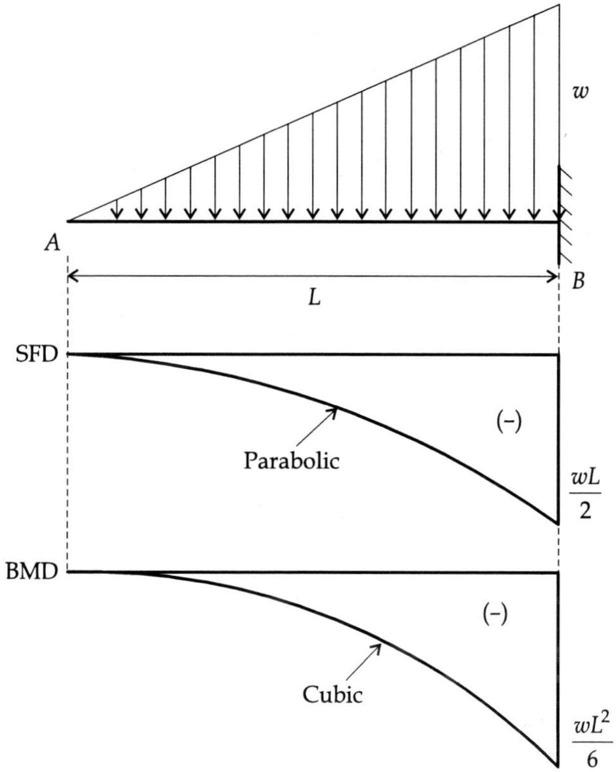

Fig. 9.14 (b)

Example 9.8: The frame *AB* is circular quadrant arch with radius *R* as shown in the figure Fig. 9.15. Draw axial force diagram, shear force diagram and bending moment diagram of the frame.

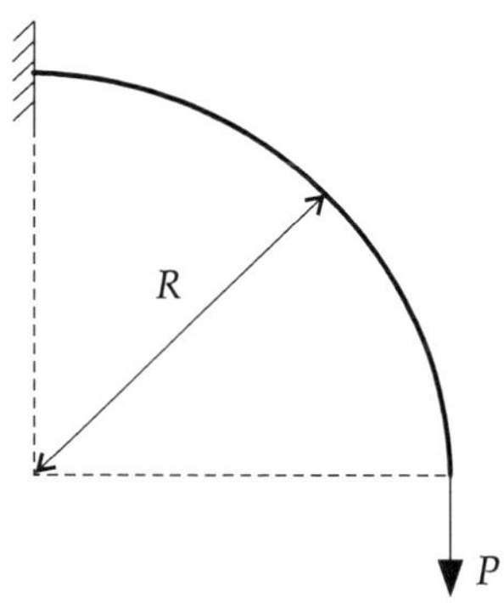

Fig. 9.15

Solution:

Free-body diagram

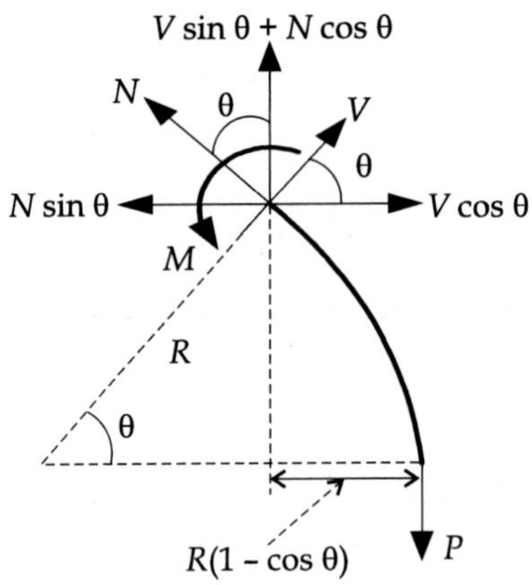

The unknown axial force, shear force and bending moment will be obtained by considering three equilibrium equations, i.e., summation of horizontal forces, vertical forces and summation of moment of forces.

$$\sum F_x = 0 \Rightarrow V\cos\theta - N\sin\theta = 0 \Rightarrow V = \frac{N\sin\theta}{\cos\theta}$$

$$\sum F_y = 0 \Rightarrow N\cos\theta + V\sin\theta - P = 0$$

Solving the above two equations, we get

$N = P\cos\theta$, $\quad V = P\sin\theta$

$$\sum M = 0 \Rightarrow M = PR(1 - \cos\theta)$$

Thus, axial force, shear force and bending moment vary in trigonometric function.

Frame section	At A	At B
θ	0	$\pi/2$
Axial force (N)	P	0
Shear force (V)	0	P
Bending moment (M)	0	PR

Note: The maximum value of axial force will occur at free end, and maximum value of shear force and bending moment occurs at the fixed end, i.e.,

$N_{max} = N(0) = P$, $\quad V_{max} = V(\pi/2) = P$, $\quad M_{max} = M(\pi/2) = PR$

SFD and BMD

SFD and BMD diagrams are shown in Fig. 9.15 (a).

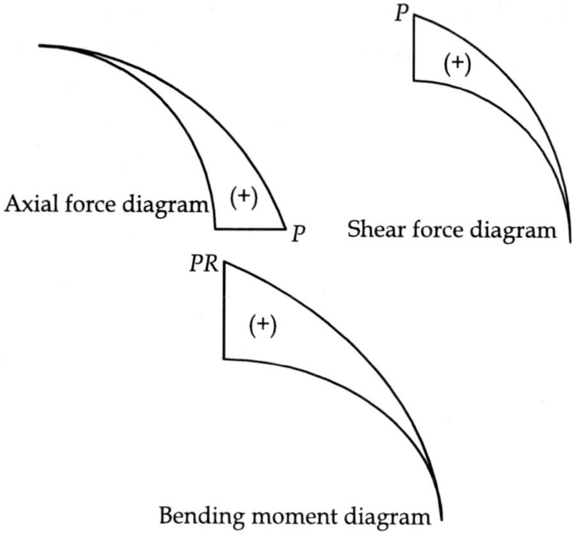

Fig. 9.15 (a)

Example 9.9: Consider the beam *AB* as shown in Fig. 9.16. Draw shear force and bending moment diagrams of beam *AB*.

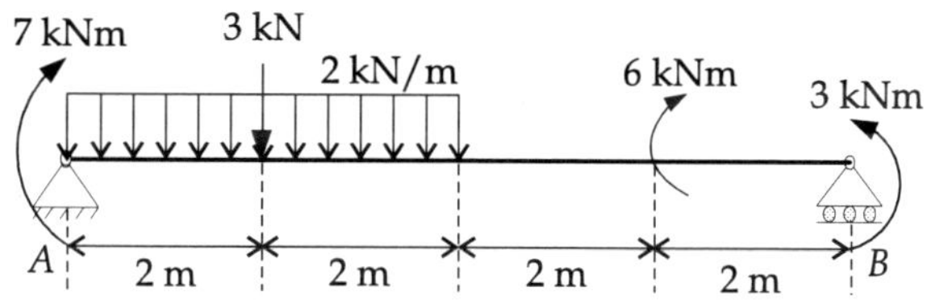

Fig. 9.16

Solution: The first step is to find the external reactions at *A* and *B*, using equations of equilibrium.

Free-body diagram

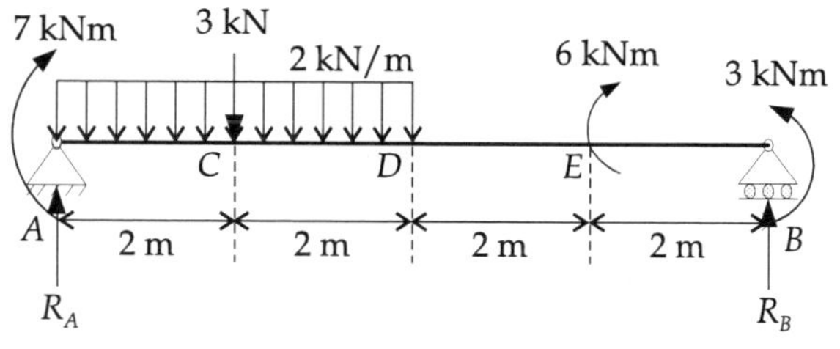

Equations of equilibrium

$$\sum M_B = 0 \Rightarrow R_A \times 8 + 7 - (2 \times 4) \times (2 + 4) - 3 \times 6 + 6 - 3 = 0 \Rightarrow R_A = 7 \text{ kN}$$

$$\sum F_y = 0 \Rightarrow R_A - 2 \times 4 - 3 + R_B = 0 \Rightarrow R_B = 4 \text{ kN}$$

Due to uniformly distributed load between A and D, shear force diagram will be linear while bending moment diagram will be parabolic in beam section AD. In beam section DB shear force diagram will be constant while bending moment diagram varies linearly.

Due to concentrated load at C there will be jump in shear force diagram at C.

Due to couple at A, E and B, there will be jump in bending moment diagram at these points. Normally, when there is no external couple at simply supports then the bending moment at supports are zero, but in this case external couple gives jump in bending moments at supports.

Now calculate the shear and bending moments at each critical point and join them accordingly.

At point *A*

At just right of A ($x = 0^+$)

$V = R_A = 7 \text{ kN}$ and $M = 7 \text{ kNm}$

At point *C*

At just left of C ($x = 2^-$)

$$V = R_A - 2x = 3 \text{ kN} \quad \text{and} \quad M = 7 + R_A x - 2x\frac{x}{2} = 17 \text{ kNm}$$

At just right of C ($x = 2^+$)

$$V = R_A - 2x - 3 = 0 \quad \text{and} \quad M = 7 + R_A x - 3(x-2) - 2x\frac{x}{2} = 17 \text{ kNm}$$

At point *D*

At just left of D ($x = 4^-$)

$$V = R_A - 2x - 3 = -4 \text{ kN} \quad \text{and} \quad M = 7 + R_A x - 3(x-2) - 2x\frac{x}{2} = 13 \text{ kNm}$$

At just right of D ($x = 4^+$)

$$V = R_A - 2\times 4 - 3 = -4 \text{ kN} \quad \text{and} \quad M = 7 + R_A x - 3(x-2) - (2\times 4)\times\frac{4}{2} = 13 \text{ kNm}$$

At point *E*

At just left of E ($x = 6^-$)

$V = R_A - 2\times 4 - 3 = -4 \text{ kN}$ and $M = 7 + R_A x - 3(x - 2) - (2\times 4)\times(x - 2) = 5 \text{ kNm}$

At just right of E ($x = 6^+$)

$V = R_A - 2\times 4 - 3 = -4 \text{ kN}$ and $M = 7 + R_A x - 3(x - 2) - (2\times 4)\times(x - 2) + 6 = 11 \text{ kNm}$

At point *B*

At just left of B ($x = 8^-$)

$V = R_A - 2\times 4 - 3 = -4 \text{ kN}$ and $M = 7 + R_A x - 3(x - 2) - (2\times 4)\times(x - 2) + 6 = 3 \text{ kNm}$

Note: Student can check that the shear just left of B must be equal to negative of reaction at B, and moment at just left of B must be equal to negative of moment at B, such that the summation of forces and moments must equal to zero Fig. 9.16 (a).

SFD and BMD

Point	Shear force		Bending moment	
	Just left of	Just right of	Just Left of	Just Right of
A	-	7 kN	-	7 kNm
C	3 kN	0	17 kNm	17 kNm
D	-4 kN	-4 kN	13 kNm	13 kNm
E	-4 kN	-4 kN	5 kNm	11 kNm
B	-4 kN	-	3 kNm	-

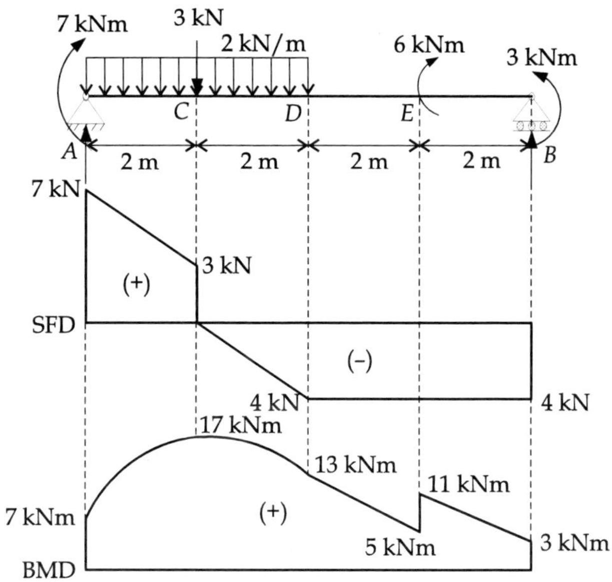

Fig. 9.16 (a)

Example 9.10: Consider the cantilever beam *AB* as shown in the Fig. 9.17. Draw shear force and bending moment diagrams of beam *AB*.

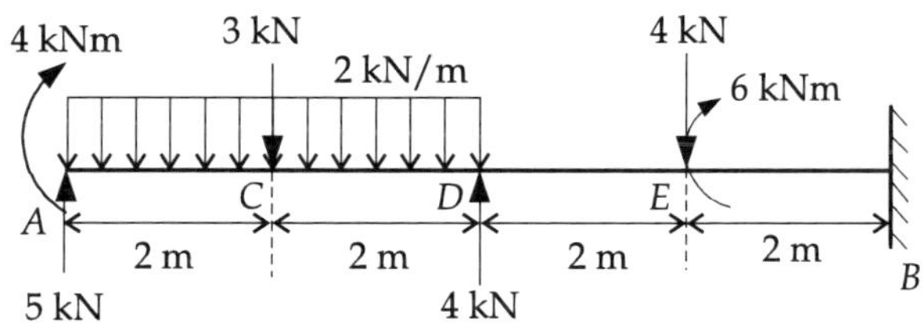

Fig. 9.17

Solution: Due to uniformly distributed load between *A* and *D*, shear force diagram will be linear while bending moment diagram will be parabolic in beam section *AD*. In beam section *DB* shear force diagram will be constant while bending moment diagram varies linearly.

Due to concentrated load at A, C, D and E there will be jump in shear force diagram at these points. Normally, when there is no concentrated load at free end then the shear at free end is zero, but in this case external concentrated load at free end gives jump in shear at free end.

Due to couple at A and E, there will be jump in bending moment diagram at these points. Normally, when there is no external couple at free end then the bending moment at free end is zero, but in this case external couple gives jump in bending moments at the free end.

Now calculate the shear and bending moments at each critical point and join them accordingly.

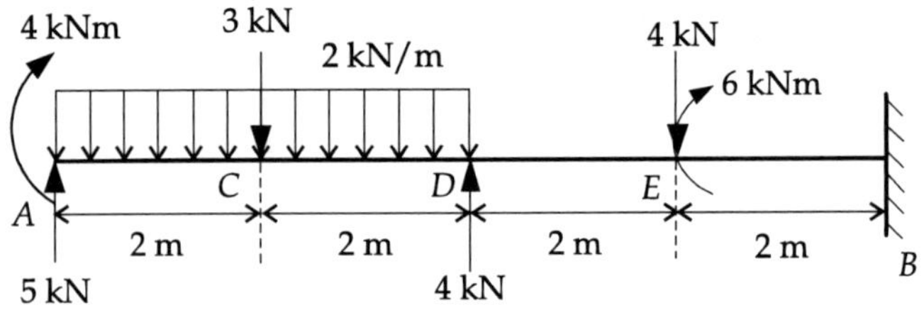

At point A

At just right of A ($x = 0^+$)

$V = 5$ kN and $M = 4$ kNm

At point C

At just left of C ($x = 2^-$)

$$V = 5 - 2x = 1 \text{ kN} \quad \text{and} \quad M = 4 + 5x - 2x\frac{x}{2} = 10 \text{ kNm}$$

At just right of C ($x = 2^+$)

$$V = 5 - 2x - 3 = -2 \text{ kN} \quad \text{and} \quad M = 4 + 5x - 3(x-2) - 2x\frac{x}{2} = 10 \text{ kNm}$$

At point D

At just left of D ($x = 4^-$)

$$V = 5 - 2x - 3 = -6 \text{ kN} \quad \text{and} \quad M = 4 + 5x - 3(x-2) - 2x\frac{x}{2} = 2 \text{ kNm}$$

At just right of D ($x = 4^+$)

$$V = 5 - 2x - 3 + 4 = -2 \text{ kN} \quad \text{and} \quad M = 4 + 5x - 3(x-2) + 4(x-4) - 2x\frac{x}{2} = 2 \text{ kNm}$$

At point E

At just left of E ($x = 6^-$)

$V = 5 - 2 \times 4 - 3 + 4 = -2$ kN and

$M = 4 + 5x - 3(x - 2) + 4(x - 4) - (2 \times 4) \times (x - 2) = -2$ kNm

At just right of E ($x = 6^+$)

$V = 5 - 2 \times 4 - 3 + 4 - 4 = -6$ kN and

$M = 4 + 5x - 3(x - 2) + 4(x - 4) - (2 \times 4) \times (x - 2) - 4(x - 6) + 6 = 4$ kNm

At point *B*

At just left of B ($x = 8^-$)

$V = 5 - 2 \times 4 - 3 + 4 - 4 = -6$ kN and

$M = 4 + 5x - 3(x - 2) + 4(x - 4) - (2 \times 4) \times (x - 2) - 4(x - 6) + 6 = 4$ kNm

Note: Student can check that the shear just left of B must be equal to negative of reaction at B, and moment at just left of B must be equal to negative of moment at B, such that the summation of forces and moments must equal to zero Fig. 9.17(a).

SFD and BMD

Point	Shear force		Bending moment	
	Just left of	Just right of	Just left of	Just right of
A	-	5 kN	-	4 kNm
C	1 kN	-2 kN	10 kNm	10 kNm
D	-6 kN	-2 kN	2 kNm	2 kNm
E	-2 kN	-6 kN	-2 kNm	4 kNm
B	-6 kN	-	-6 kNm	-

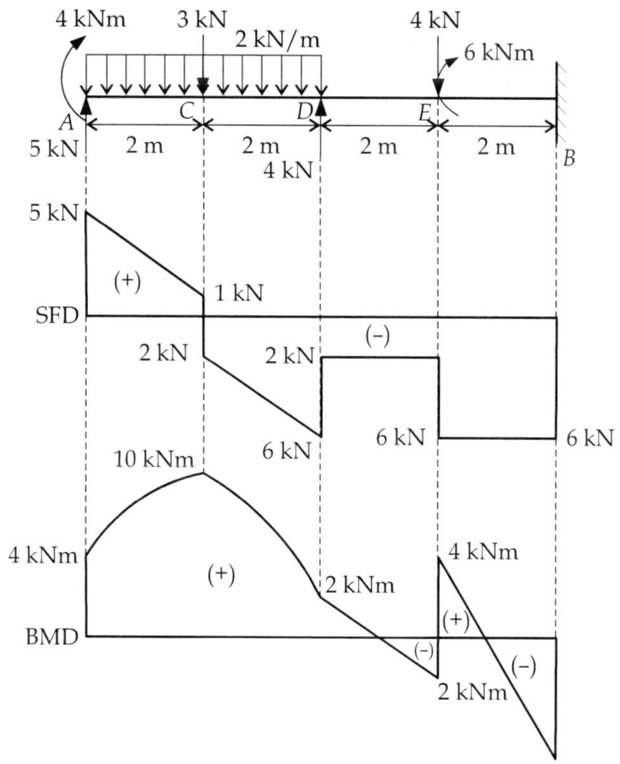

Fig. 9.17 (a)

Example 9.11: Consider the simply supported beam AB of length $L = \pi/2$ subjected to a varying distributed load $q(x) = \sin x + \cos x$ as shown in Fig. 9.18. Draw shear force and bending moment diagrams.

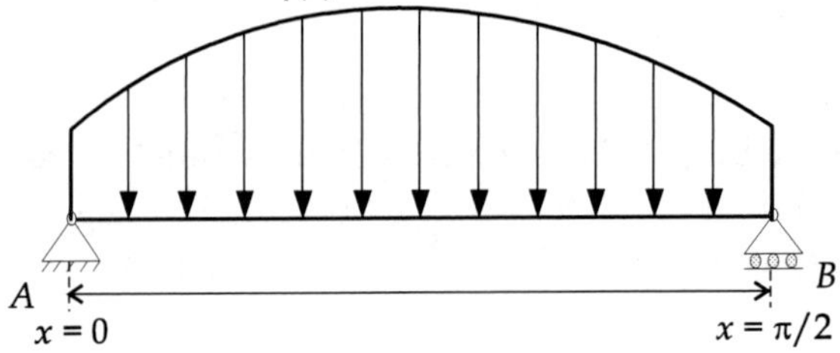

Fig. 9.18

Solution: Consider left support A at $x = 0$ and right support B at $x = \pi/2$. Since the load is symmetrical, therefore reactions $R_A = R_B$.

$$R_A = \frac{1}{2}\int_0^{\pi/2} q(x)dx = \frac{1}{2}\int_0^{\pi/2} (\sin x + \cos x)\,dx = \frac{1}{2}(-\cos x + \sin x)\Big|_0^{\pi/2} \quad R_A = R_B = 1 \text{ kN}$$

Free-body diagram (when section is cut at a point between A and B)

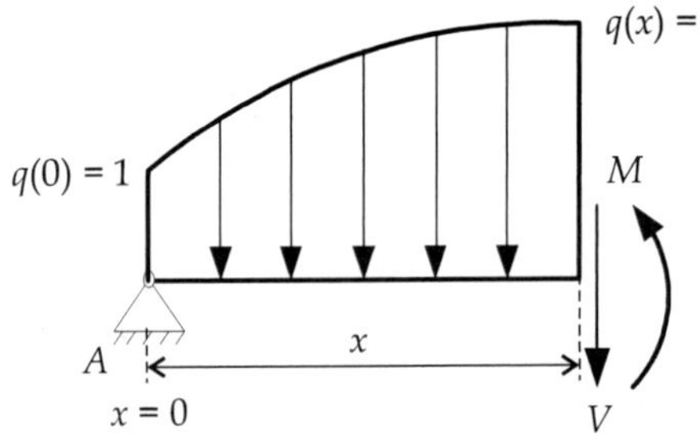

The free-body diagram of either half of the beam is drawn, let left half of the beam whose free-body diagram is shown in the figure above. The unknown shear force and bending moment will be obtained by considering two equilibrium equations, i.e., summation of vertical forces and summation of moment of forces.

$$\sum F_y = 0 \Rightarrow V = R_A - \int_0^x q(x)dx = 1 - \int_0^x (\sin x + \cos x)dx$$

$$\Rightarrow V = 1 - (-\cos x + \sin x)\Big|_0^x = \cos x - \sin x$$

$$\sum M = 0 \Rightarrow = R_A \times x - \int_0^x q(x).xdx - M = 0 \Rightarrow M = 1.x - \int_0^x q(x).xdx$$

$$= 1 - \int_0^x x(\sin x + \cos x)dx = x - [(x+1)\sin x - (x-1)\cos x]_1^x$$

$$\Rightarrow M = x - (x+1)\sin x + (x-1)\cos x + 1$$

Thus, shear force and bending moment both vary in trigonometric function.

Beam section	x	Shear force	Bending moment
At A	0	1	0
At B	$\pi/2$	-1	0
At centre (C)	$\pi/4$	0	$1-\sqrt{2}+\pi/4$

Note: The maximum value of bending moment occurs at the cross section where the shear force is zero, i.e., at centre

$$M_{max} = M(\pi/4) = 1-\sqrt{2}+\pi/4$$

SFD and BMD

SFD and BMD diagrams are shown in Fig. 9.18 (a).

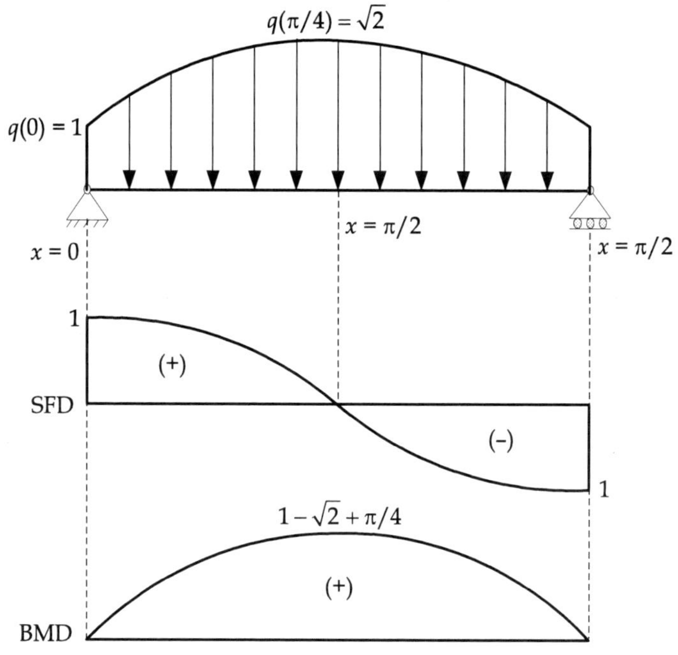

Fig. 9.18 (a)

Example 9.12: Consider the cantilever beam AB of length $L = 1$ unit subjected to a uniformly distributed load $q(x) = e^x$ as shown in Fig. 9.19. Draw shear force and bending moment diagrams.

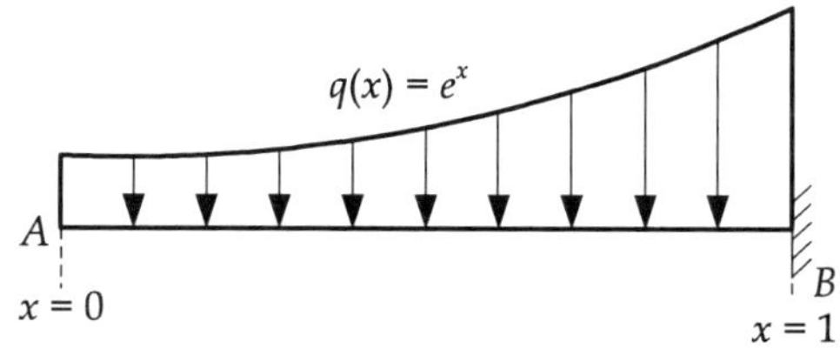

Fig. 9.19

Solution: Consider the crosssections of the beam at a distance x from free end.

Free-body diagram

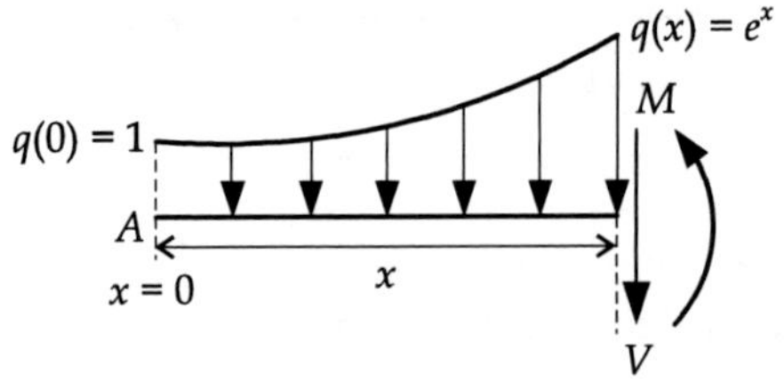

Considering two equilibrium equations, i.e., summation of vertical forces and summation of moment of forces, we get

$$\sum F_y = 0 \Rightarrow V = \int_0^x q(x)dx = \int_0^x (e^x)dx = (e^x)\Big|_0^x = e^x - 1$$

$$\sum M = 0 \Rightarrow M = \int_0^x q(x).xdx = \int_0^x (e^x\ x)dx = \left[e^x(x-1)\right]\Big|_0^x = e^x(x-1)+1$$

Thus, shear force and bending moment both vary in exponential function.

Beam section	x	Shear force	Bending moment
At A	0	0	0
At B	1	$e - 1$	1

Note: The maximum value of shear force and bending moment occurs at the fixed end, i.e.,

$$V_{max} = V(1) = e - 1, \qquad M_{max} = M(1) = 1$$

SFD and BMD

SFD and BMD diagrams are shown in Fig. 9.19 (a).

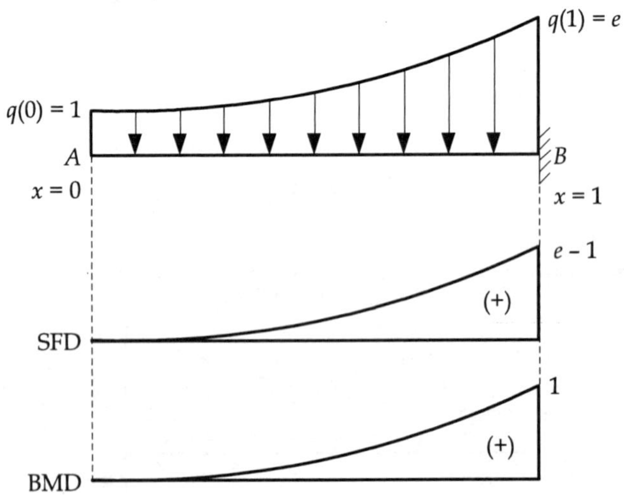

Fig. 9.19 (a)

PROBLEMS

1. Consider the arrangement of simply supported beams as given below, calculate the shear and moment in beam AB at point C (2 m from A).

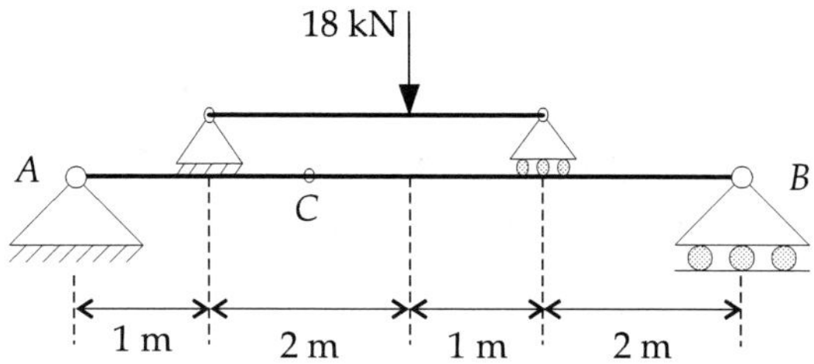

Ans. $V_C = 3$ kN and $M_C = 12$ kNm

2. Draw the shear force diagram and bending moment diagram of the beam given below.

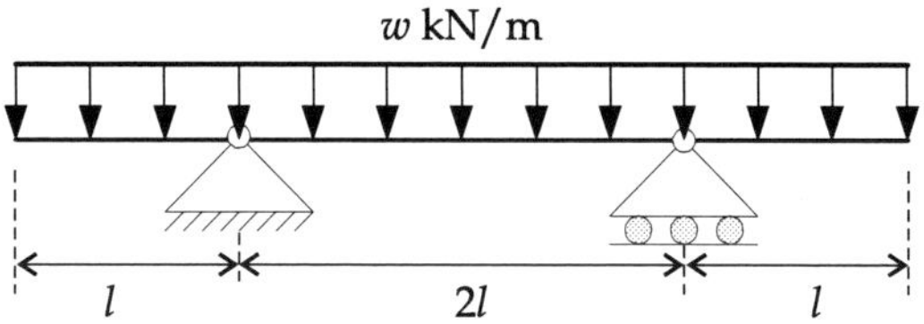

Ans.

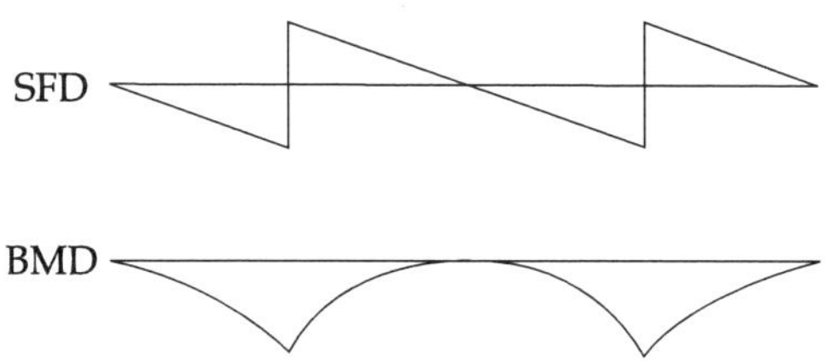

3. Consider the two beams and their shear force diagram and bending moment, match the correct shear force diagram and bending moment diagram.

(I)

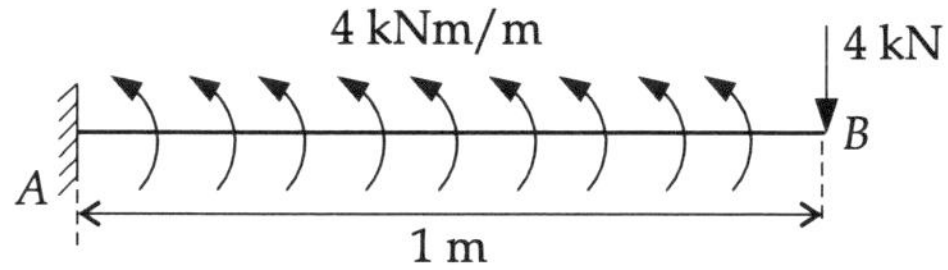

(II)

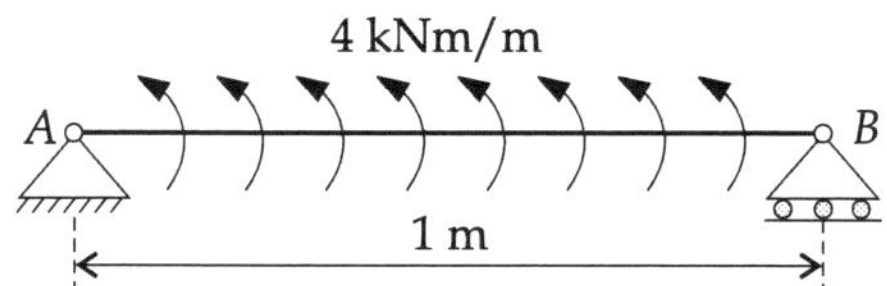

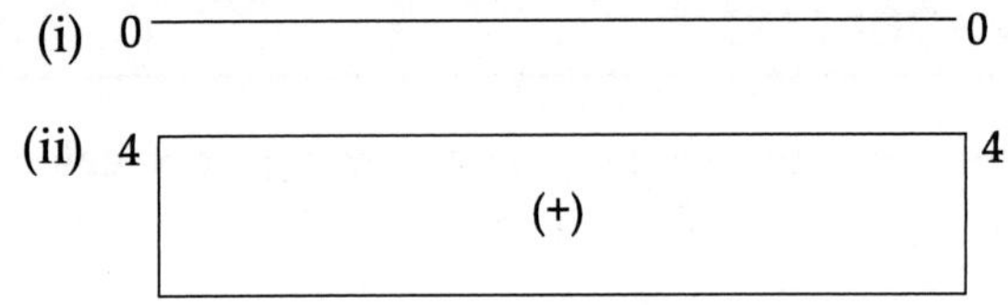

Ans. (ii) represents SFD of both beams I and II and
(i) represents BMD of both beams I and II.

4. Draw the axial force diagram, shear force diagram and bending moment diagram of the beam-pulley arrangement given below.

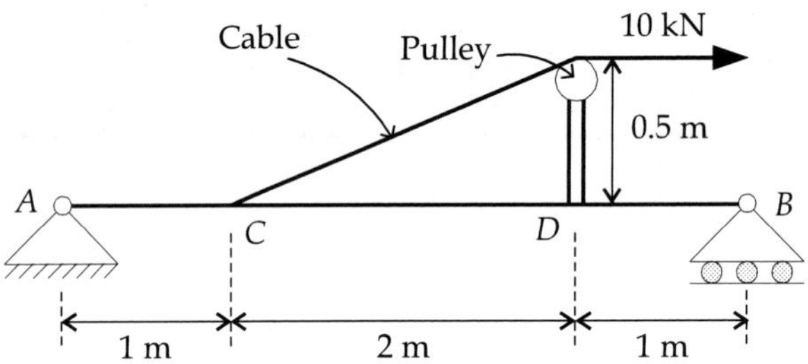

Ans. (i)

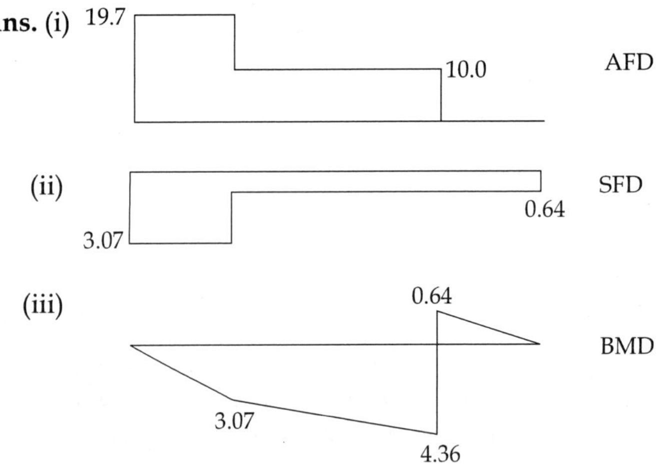

5. Consider the simply supported beam AB of length L subjected to a couple M_0 as shown in figure below, draw shear force and bending moment diagrams.

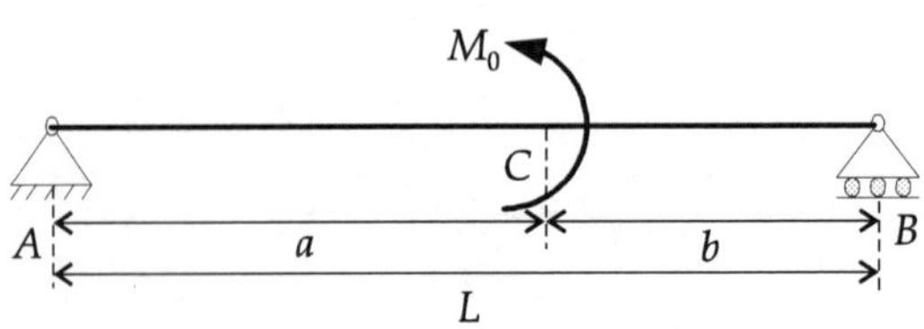

Ans.

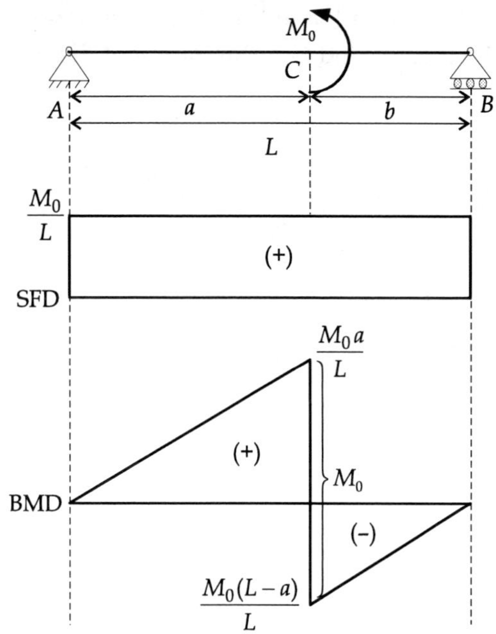

6. Draw shear force and bending moment diagrams of beam AB.

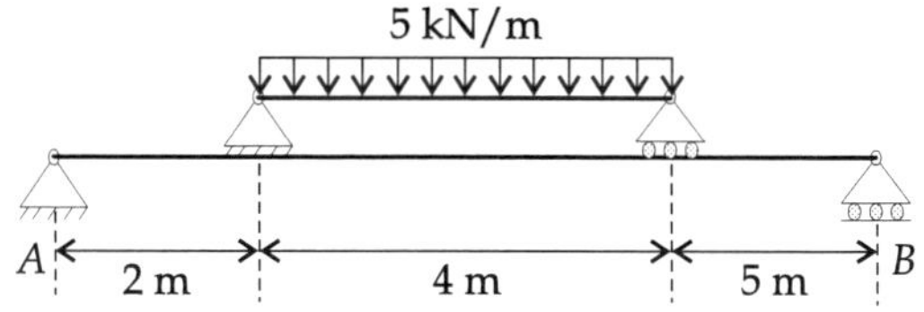

Ans.

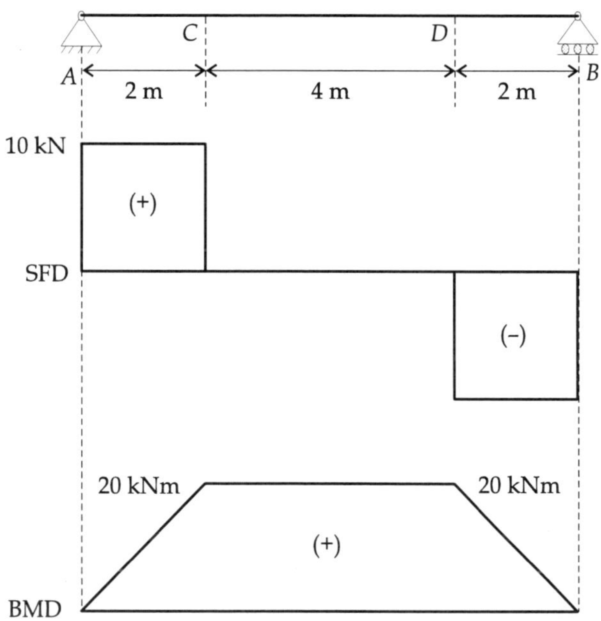

7. Consider the beam *AB* with a loaded hook at *C* as shown in the figure below, draw axial force, shear force and bending moment diagrams of beam *AB*.

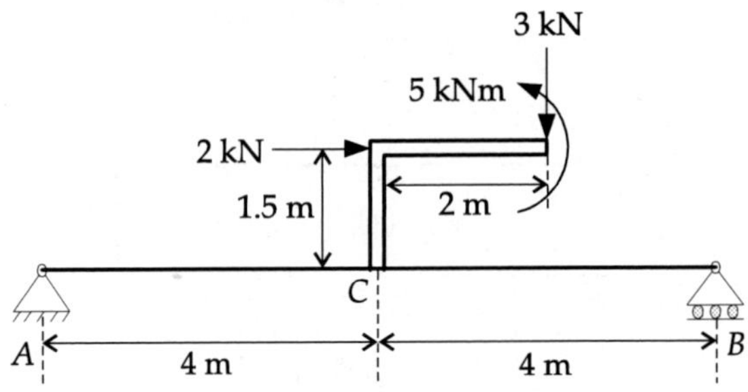

Ans.

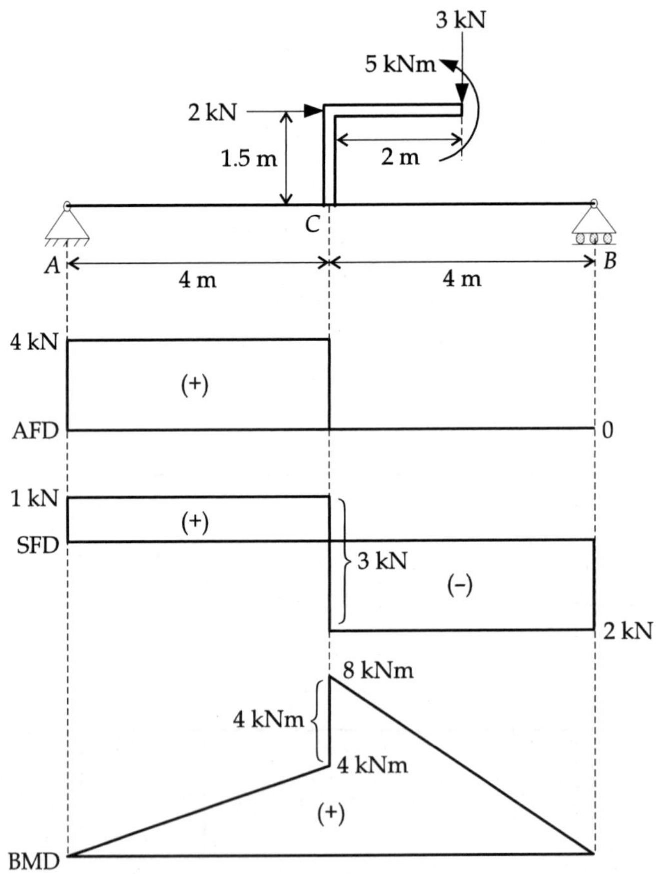

8. Draw shear force and bending moment diagrams of the beam given below.

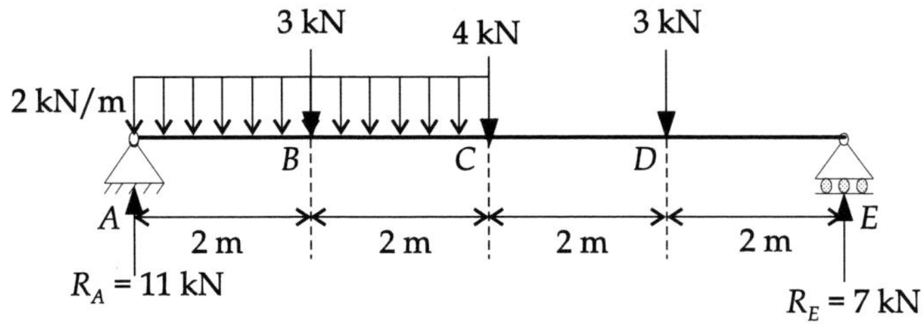

Ans.

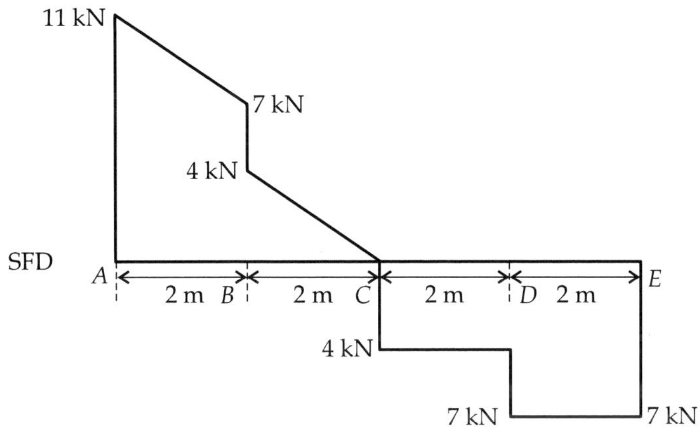

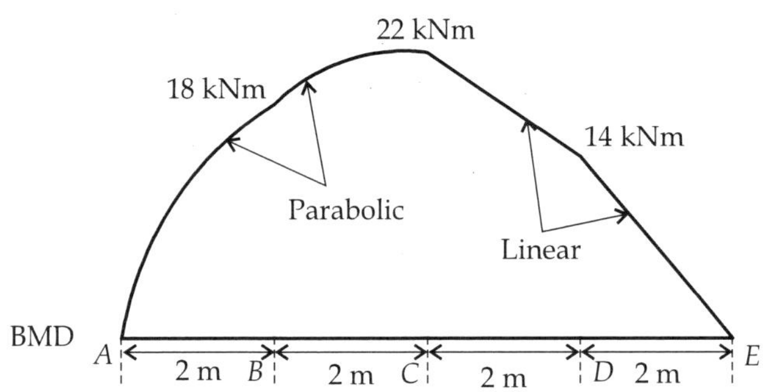

10

Slope and Deflection of Beams

10.1 INTRODUCTION

The originally straight beam deflects when transverse loads is applied on it. If axial force and shear deformation of the beam is neglected, the deformation of the beam depends on the bending moment. The deflected shape of the beam after the applied load is known as **elastic curve** or the **deflection curve**. The deformation response at any point of the beam is measured on elastic curve in terms of vertical deflection and slope. The deflection diagram or elastic curve of a simply supported beam for a concentrated load applied at the centre of the beam is shown in Fig. 10.1.

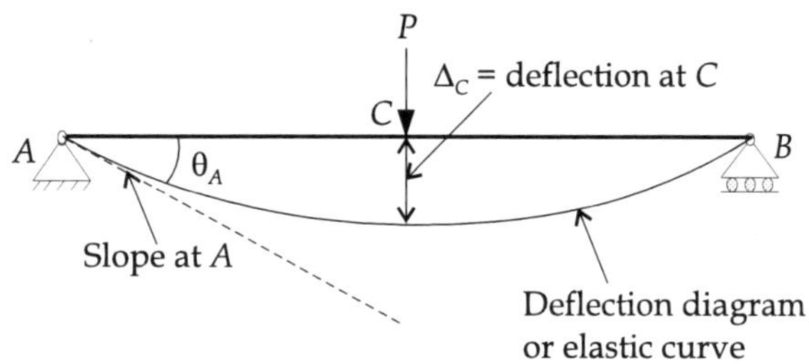

Fig. 10.1

The slope at any point of the beam is defined as the angle between the original beam axis and the tangent at the point on the elastic curve. It is measured in radian. The deflection at any point of the beam is defined as the distance between the originally straight beam and the point on the elastic curve measured in the perpendicular direction of the originally straight beam.

Table 10.1: Slope and deflection at the different type of supports are given in the table below.

Support condition	Slope/deflection
Roller support 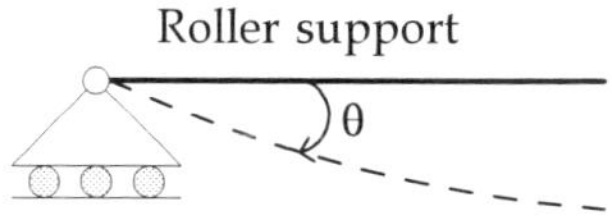	Slope (θ) $\neq 0$ Deflection (Δ) $= 0$

Hinge support 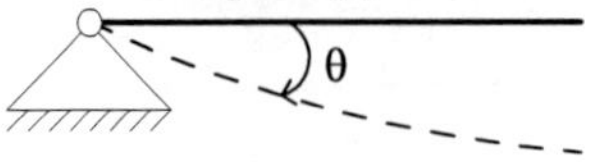	Slope (θ) ≠ 0 Deflection (Δ) = 0
Fixed support 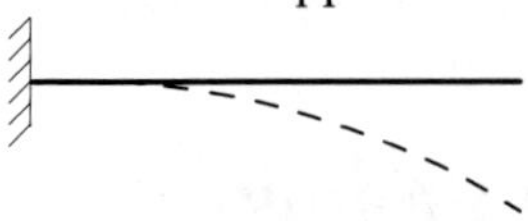	Slope (θ) = 0 Deflection (Δ) = 0

10.2 POINT OF CONTRAFLEXURE (OR INFLECTION POINT)

Inflection point is the point where elastic curve changes its slope or bending moment changes its sign. In inflection point the magnitude of bending moment is zero. Figure 10.2 shows bending moment diagram, deflection curve and inflection points of some beams.

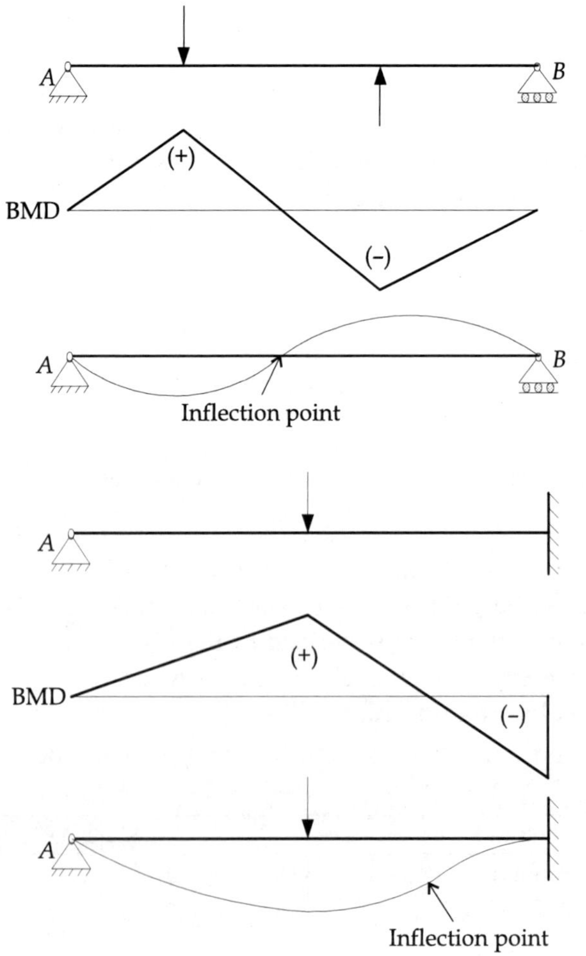

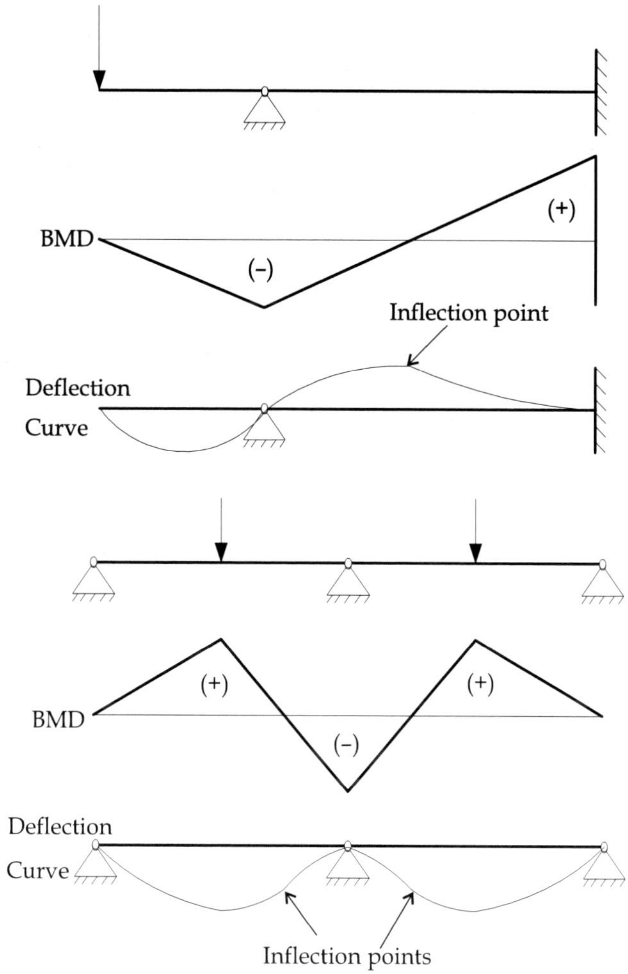

Fig. 10.2

Four well known methods to find slope and deflection for a statically determinate beam are discussed here.

1. Double integration method
2. Moment area method
3. Conjugate beam method
4. Castigliano's theorem for beam deflection (partial derivative method)

10.3 DOUBLE INTEGRATION METHOD

When a straight beam of uniform cross section subjected to end couples moment (M) applied about the principal axis, it bends like a circular arc. The relation between the radius of curvature (ρ) and bending moment (M) is

$$\frac{1}{\rho} = \frac{M}{EI}$$

where, *EI* is the product of Young's modulus (*E*) and the second moment of area *I* about the neutral axis.

The differential equation of curvature in *X-Y* coordinate system

$$\frac{1}{\rho} = -\frac{\dfrac{d^2y}{dx^2}}{\left[1 + \left(\dfrac{dy}{dx}\right)^2\right]^{\frac{3}{2}}}$$

The magnitude of $\left(\dfrac{dy}{dx}\right)^2$ can be negligible if the value of y is very small.

$$\frac{1}{\rho} = -\frac{d^2y}{dx^2}$$

Let a straight beam be subjected to lateral load and get deflected shape as shown in Fig 10.3. Any two points separated by *dx* length on the original straight beam is subtended a small angle ($d\theta$) in the center of the circle after bends by the small arc length, *ds*. For small deflection, it can be assumed that the length of the original beam is equal to the length of elastic curve at neutral plane.

From Fig. 10.3, we can write $ds = \rho d\theta \Rightarrow \dfrac{d\theta}{ds} = \dfrac{1}{\rho}$

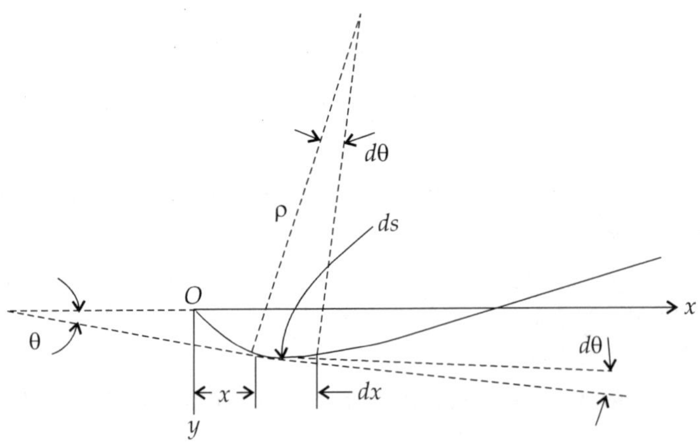

Fig. 10.3

For small deflection

$$ds \approx dx \text{ and } \theta \approx \frac{dy}{dx}$$

Thus,

$$\frac{d^2y}{dx^2} = -\frac{M}{EI} = \frac{1}{\rho}$$

We get the moment curvature relation

$$\frac{d^2y}{dx^2} = -\frac{M}{EI}$$

If the load has a non-uniform distribution the following equations are used to determine the equation of elastic curve.

$$EI\frac{d^3y}{dx^3} = -\frac{dM}{dx} = -V$$

$$EI\frac{d^4y}{dx^4} = -\frac{dV}{dx} = w$$

where, V is the shear force at any cross section, and w is the intensity of distributed load.

10.3.1 Procedure for Analysis

In this method $EI\frac{d^2y}{dx^2} = -M(x)$ equation is used to get slope after 1st integration and deflection after 2nd integration with boundary conditions.

$$EI\int\frac{d^2y}{dx^2} = -\int M(x)$$

$$\text{Slope } (\theta) = \frac{dy}{dx} = -\frac{1}{EI}\int M(x)dx + C_1$$

$$\text{Deflection } (\Delta) = y = -\frac{1}{EI}\int\int M(x)dx + \int C_1 dx + C_2$$

Constants C_1 and C_2 can be determined by applying boundary conditions.

Example 10.1: Find the equation of elastic curve for a simple supported beam of length L due to a concentrated load P applied at centre of the beam as shown in Fig. 10.4. Also find the maximum deflection and slope. EI is uniform throughout the beam.

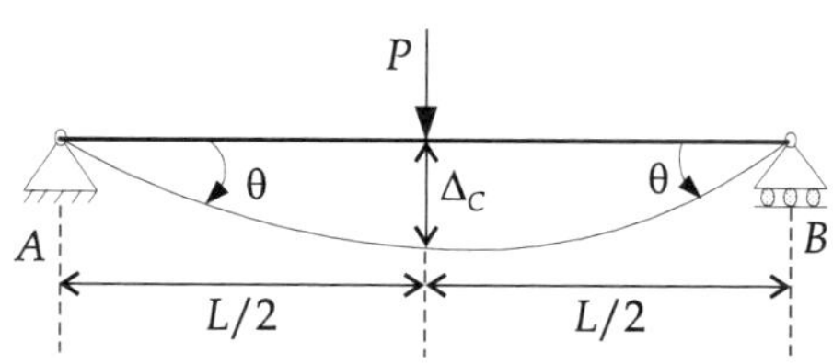

Fig. 10.4

Solution: The reaction force at the support, $R_A = R_B = P/2$

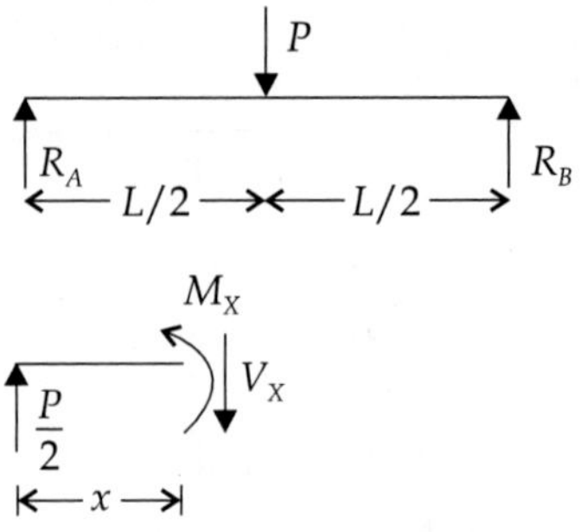

Fig. 10.4 (a)

The bending moment at any section, $M_x = \frac{P}{2}x$ when $0 < x < L/2$

We know

$$EI\frac{d^2y}{dx^2} = -M_x$$

Therefore,

$$EI\frac{d^2y}{dx^2} = -\frac{P}{2}x$$

Integrating ones

$$EI\frac{dy}{dx} = -\frac{P}{2}.\frac{x^2}{2} + C_1$$

Again integrating

$$EIy = -\frac{P}{4}.\frac{x^3}{3} + C_1x + C_2$$

Boundary condition

The deflection at support is zero and slope at the middle point of the beam is zero as it is parallel to the original straight beam

$y = 0$ at $x = 0$

$\frac{dy}{dx} = 0$ at $x = \frac{L}{2}$

The constants of integration (C_1 and C_2) can be found from the boundary condition

Using the second boundary condition, $0 = -\frac{P}{2}.\frac{L^2}{8} + C_1$

$$C_1 = \frac{PL^2}{16}$$

At $x = 0$,

$$EI\frac{dy}{dx} = C_1$$

The equation becomes

$$EIy = -\frac{P}{12}x^3 + \frac{PL^2}{16}x + C_2$$

Putting the first boundary condition $C_2 = 0$

The equation of elastic curve,

$$EIy = -\frac{P}{12}x^3 + \frac{PL^2}{16}x$$

Hence, the equation of elastic curve,

$$y = \frac{PL^3}{48EI}\left(3\left(\frac{x}{L}\right) - 4\left(\frac{x}{L}\right)^3\right)$$

The maximum deflection occurs at $x = L/2$

$$y_{\max} = \frac{PL^3}{48EI}$$

Maximum slope at x = 0

$$\theta_{\max} = \frac{PL^2}{16EI}$$

Example 10.2: A simple supported beam is subjected to udl (w) throughout its length, L as shown in Fig. 10.5. The flexural rigidity (EI) of the beam is constant. Determine the equation of elastic curve, maximum deflection and the slope.

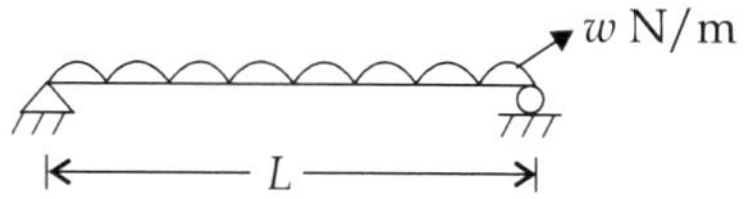

Fig 10.5

Solution:

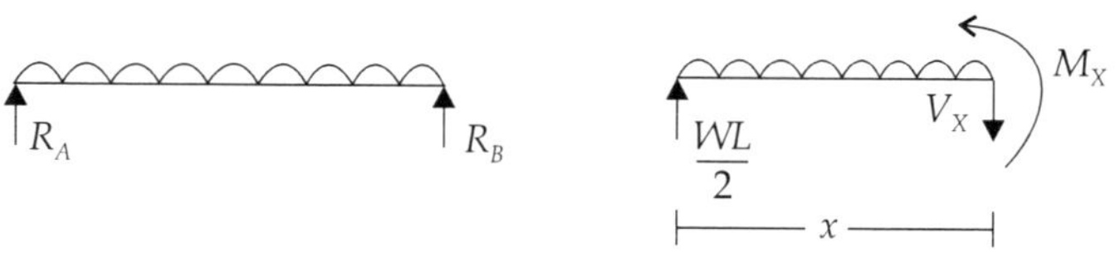

Fig. 10.5 (a)

The reaction force at support,

$$R_A = R_B = \frac{wL}{2}$$

The moment at a distance x from the left support,

$$M_x = \frac{wL}{2}x - \frac{wx^2}{2}$$

Now, we know

$$EI\frac{d^2y}{dx^2} = -M_x$$

$$EI\frac{d^2y}{dx^2} = -\frac{wLx}{2} + \frac{wx^2}{2}$$

Integrating

$$EI\frac{dy}{dx} = -\frac{wLx^2}{2} + \frac{w}{2}.\frac{x^3}{6} + C_1$$

Again integrating

$$EIy = -\frac{wLx^3}{12} + \frac{wx^4}{24} + C_1x + C_2$$

Integration constants (C_1 and C_2) can be found from the boundary condition.

Boundary condition

$y = 0$ at $x = 0$

$\frac{dy}{dx} = 0$ at $x = \frac{L}{2}$

Putting the boundary condition

$\frac{dy}{dx} = 0$ at $x = \frac{L}{2}$

$C_1 = \frac{wL^3}{24}$

$y = 0$ at $x = 0$

$C_2 = 0$

Thus, putting the value of constants

$$EIy = -\frac{wLx^3}{12} + \frac{w}{24}x^4 + \frac{wL^3}{24}x$$

The equation of elastic curve

$$y = \frac{wL^4}{24EI}\left[\left(\frac{x}{L}\right)^4 - 2\left(\frac{x}{L}\right)^3 + \left(\frac{x}{L}\right)\right]$$

At $x = L/2$ deflection is maximum

$$y_{max} = \frac{5wL^4}{384EI}$$

Maximum slope occurs at $x = 0$ and $x = L/2$

$$\theta_{max} = \frac{wL^3}{24EI}$$

Example 10.3: A cantilever beam of uniform flexural rigidity (EI) is carrying a concentrated load (P) at its free end as shown in Fig. 10.6. Find the equation of elastic curve and maximum deflection.

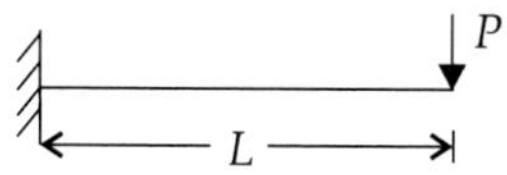

Fig. 10.6

Solution:

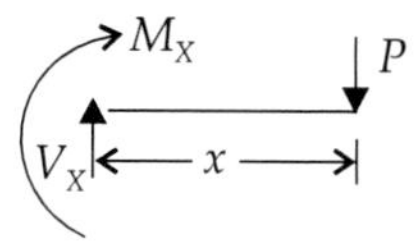

Fig. 10.6 (a)

The moment taken from the free end at a distance x

$$M_x = -Px$$

Then

$$EI\frac{d^2y}{dx^2} = Px$$

Integrating

$$EI\frac{dy}{dx} = P\frac{x^2}{2} + C_1$$

Again integrating

$$EIy = P\frac{x^3}{2} + C_1x + C_2$$

Boundary condition

Here, x is measured from the free end. The deflection and slope are zero at clamped edge.

$y = 0$ at $x = L$

$\frac{dy}{dx} = 0$ at $x = L$

From the second boundary condition

$$C_1 = -\frac{PL^2}{2}$$

From the first boundary condition

$$C_1 = -\frac{PL^3}{3}$$

The equation of elastic curve

$$EIy = \frac{P}{6}x^3 - \frac{PL^2}{2}x + \frac{PL^3}{3}$$

Hence,

$$y = \frac{PL^3}{6EI}\left[\left(\frac{x}{L}\right)^3 - 3\left(\frac{x}{L}\right) + 2\right]$$

Maximum deflection occurs at the free end.

Put $x = 0$ in the above equation

$$y_{max} = \frac{PL^3}{3EI}$$

and slope at the free end

$$\theta_{max} = -\frac{PL^2}{2EI}$$

Example 10.4: A cantilever beam carrying an udl (w) in its whole span as shown in Fig. 10.7. Flexural rigidity (EI) is uniform. Determine maximum deflection of the beam.

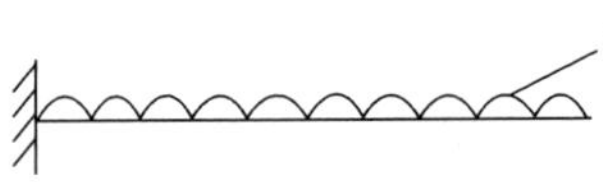

Fig. 10.7

Solution:

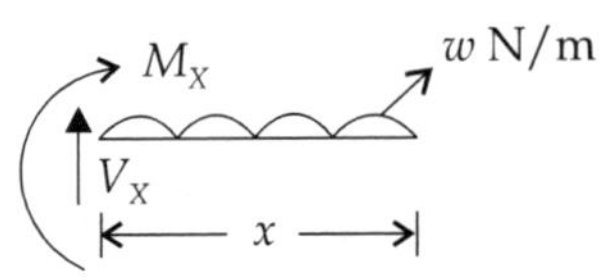

Fig. 10.7 (a)

The moment is taken at a distance x from the free end.

The moment at any section, $0 < x < L$

$$M_x = -\frac{wx^2}{2}$$

The differential equation for bending

$$EI\frac{d^2y}{dx^2} = \frac{wx^2}{2}$$

Integrating

$$EI\frac{dy}{dx} = w\frac{x^3}{6} + C_1$$

Again integrating

$$EIy = w\frac{x^4}{24} + C_1x + C_2$$

Putting the boundary condition

$\frac{dy}{dx} = 0$ at $x = L$

We get

$$C_1 = \frac{-wL^3}{6}$$

Deflection zero at the fixed end

$y = 0$ at $x = L$

We get

$$C_2 = \frac{-wL^4}{8}$$

The equation of elastic curve

$$EIy = \frac{w}{24}x^4 - \frac{wL^3}{6}x + \frac{wL^4}{8}$$

$$y = \frac{wL^4}{24EI}\left[\left(\frac{x}{L}\right)^4 - 4\left(\frac{x}{L}\right)^3 + 3\right]$$

The maximum deflection occurs at the free end.

Putting $x = 0$ in the elastic curve equation

$$y_{\max} = \frac{wL^4}{8EI}$$

And maximum slope

$$\theta_{\max} = -\frac{wL^3}{6EI}$$

Example 10.5: Consider the simply supported beam AB of length L subjected to a concentrated load P as shown in Fig. 10.8. Using the double integration method, determine displacement at point C and slope at points A and B.

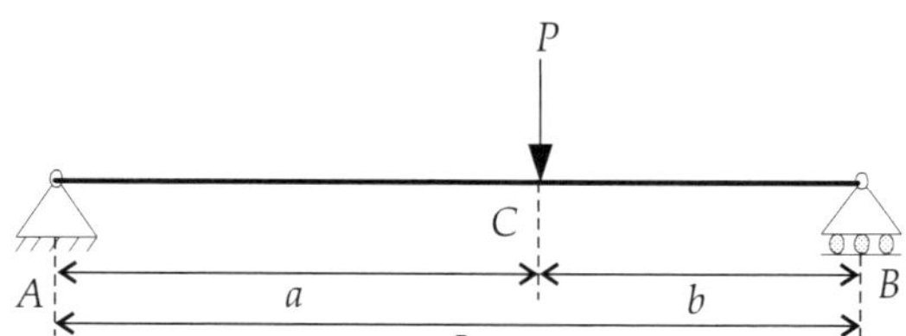

Fig. 10.8

Solution: The deflected elastic curve of the beam is shown in Fig. 10.9.

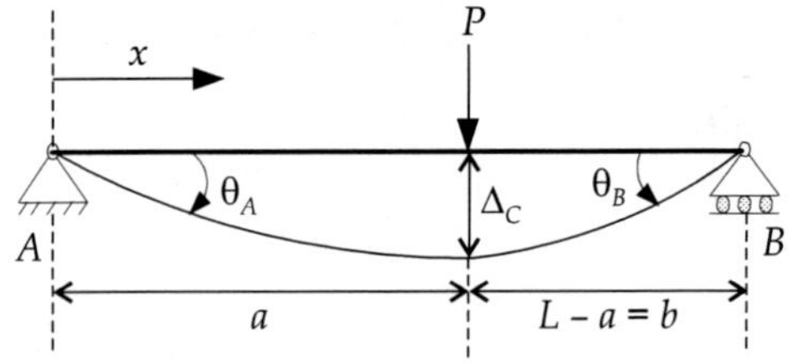

Fig. 10.9

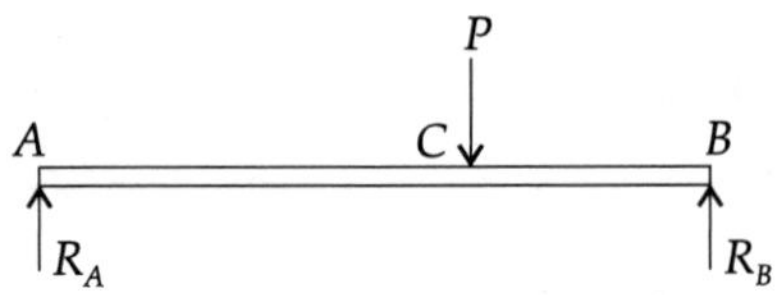

$$R_A + R_B = P$$

Taking moment about point B

$$R_A \times L = P \times b$$

$$R_A = \frac{Pb}{L}$$

The support reaction at A and B

$$R_A = \frac{Pb}{L} \quad \text{and} \quad R_B = \frac{Pa}{L}$$

We get two moment functions for $0 \le x \le a$

$$M_x = \frac{Pb}{L}x$$

and for $a \le x \le L$

$$M_x = \frac{Pb}{L}x - P(x-a) = Pa\left(1 - \frac{x}{L}\right)$$

Putting the value of moment

$$EI\frac{d^2y}{dx^2} = -M_x$$

For $0 \le x \le a$ integration elastic curve equation, we get

$$EI\frac{d^2y}{dx^2} = -\frac{Pb}{L}x$$

$$EI\frac{dy}{dx} = -\frac{Pb}{L}\frac{x^2}{2} + C_1 \tag{1}$$

$$EIy = -\frac{Pb}{L}\frac{x^3}{6} + C_1x + C_2 \tag{2}$$

And for $a \le x \le L$ integration elastic curve equation, we get

$$EI\frac{d^2y}{dx^2} = -Pa\left(1-\frac{x}{L}\right)$$

$$EI\frac{dy}{dx} = -Pa\left(x-\frac{x^2}{2L}\right)+C_3 \quad (3)$$

$$EI\,y = -Pa\left(\frac{x^2}{2}-\frac{x^3}{6L}\right)+C_3x+C_4 \quad (4)$$

Boundary conditions

Two unknown constants are determined using two boundary conditions, i.e., $y = 0$ at $x = 0$ and $y = 0$ at $x = L$

Continuity equations

Two unknown constants are determined using two continuity equations, i.e., slope and deflection at just left and right of point C ($x = a$) must be equal, $\left(\frac{dy}{dx}\right)_{C^-} = \left(\frac{dy}{dx}\right)_{C^+}$ and $y_{C^-} = y_{C^+}$

Applying the first boundary condition $y = 0$ at $x = 0$ in equation (2), we get

$$0 = -0 + 0 + C_2 \Rightarrow C_2 = 0 \quad (5)$$

Applying the second boundary condition $y = 0$ at $x = L$ in equation (4), we get

$$0 = -Pa\frac{L^2}{3}+C_3L+C_4 \quad (6)$$

Applying the first continuity equation $\left(\frac{dy}{dx}\right)_{C^-} = \left(\frac{dy}{dx}\right)_{C^+}$ at $x = a$ in equations (1) and (3), we get

$$-\frac{Pb}{L}\frac{a^2}{2}+C_1 = -Pa\left(a-\frac{a^2}{2L}\right)+C_3$$

or
$$C_1 - C_3 = -\frac{Pa^2}{2} \quad (7)$$

Applying the second continuity equation $y_{C^-} = y_{C^+}$ at $x = a$ in equations (2) and (4), we get

$$-\frac{Pb}{L}\frac{a^3}{6}+C_1a = -Pa\left(\frac{a^2}{2}-\frac{a^3}{6L}\right)+C_3a+C_4$$

or
$$(C_1 - C_3)a - C_4 = -\frac{Pa^3}{3} \quad (8)$$

Equating equations (7) and (8), we get

$$C_4 = -\frac{Pa^3}{6}$$

Putting value of C_4 in equation (6), we get

$$0 = -Pa\frac{L^2}{3} + C_3L - \frac{Pa^3}{6}$$

or

$$C_3 = \frac{Pa(2L^2 + a^2)}{6L}$$

Putting value of C_3 in equation (7), we get

$$C_1 = \frac{Pa(2L^2 + a^2)}{6L} - \frac{Pa^2}{2} = \frac{Pab(L+b)}{6L}$$

Putting the values of C_1, C_2, C_3 and C_4 in equations (1) to (4), we get the elastic curve equations of slope and deflection.

For $0 \le x \le a$

Slope, $$\frac{dy}{dx} = \frac{1}{EI}\left(-\frac{Pb}{L}\frac{x^2}{2} + \frac{Pab(L+b)}{6L}\right) \quad (9)$$

Deflection, $$y = \frac{1}{EI}\left\{-\frac{Pb}{L}\frac{x^3}{6} + \left(\frac{Pab(L+b)}{6L}\right)x\right\} \quad (10)$$

For $a \le x \le L$

Slope, $$\frac{dy}{dx} = \frac{1}{EI}\left\{-Pa\left(x - \frac{x^2}{2L}\right) + \frac{Pa(2L^2 + a^2)}{6L}\right\} \quad (11)$$

Deflection, $$y = \frac{1}{EI}\left\{-Pa\left(\frac{x^2}{2} - \frac{x^3}{6L}\right) + \left(\frac{Pa(2L^2 + a^2)}{6L}\right)x - \frac{Pa^3}{6}\right\} \quad (12)$$

Slope near support *A* put $x = 0$ in equation (9)

$$\theta_A = \left.\frac{dy}{dx}\right|_{x=0} = \frac{1}{EI}\left(\frac{Pab(L+b)}{6L}\right) \text{ (Clockwise: positive)}$$

Slope near support *B* put $x = L$ in equation (11)

$$\theta_B = \left.\frac{dy}{dx}\right|_{x=L} = \frac{-1}{EI}\left(\frac{Pab(L+b)}{6L}\right) \text{ (anticlockwise: negative)}$$

For deflection at *C* put $x = a$ in equation (10) or (12)

$$\Delta_C = y|_{x=a} = \frac{1}{EI}\left(\frac{Pa^2b^2}{3L}\right) \text{ (downward: positive)}$$

If $a > b$

The maximum deflection occurs at from the left support

$$x = \sqrt{\frac{L^2 - b^2}{3}}$$

And the magnitude of maximum deflection

$$y_{max} = \frac{Pb}{9\sqrt{3}EIL}\left(\sqrt{(L^2 - b^2)^3}\right)$$

Special case (Fig. 10.10): When load is at the centre ($a = b = L/2$)

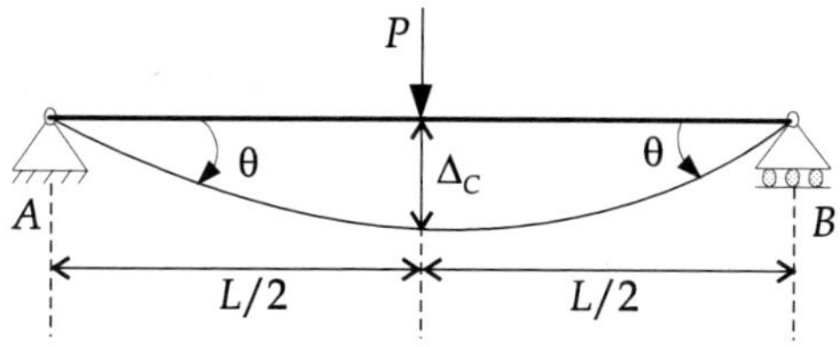

Fig. 10.10

For slope near support A put $x = 0$ in equation (9)

$$\theta_A = \left.\frac{dy}{dx}\right|_{x=0,\, a=L/2} = \frac{1}{EI}\left(\frac{Pab(L+b)}{6L}\right) = \frac{PL^2}{16EI} \text{ (Clockwise)}$$

For slope near support B put $x = L$ in equation (11)

$$\theta_B = \left.\frac{dy}{dx}\right|_{x=L,\, a=L/2} = \frac{-1}{EI}\frac{Pab(L+a)}{6L} = -\frac{PL^2}{16EI} \text{ (Anticlockwise)}$$

Since this is a symmetrical case therefore slope at A and B will be equal but opposite in sign, $\theta_A = \theta_B = \theta = \frac{PL^2}{16EI}$, i.e., at A positive (clockwise direction) and at B negative (anticlockwise direction).

For deflection at C put $x = a = b = L/2$ in equation (10) or (12)

$$\Delta_C = y|_{x=a=L/2} = \frac{1}{EI}\left(\frac{Pa^2b^2}{3L}\right) = \frac{PL^3}{48EI} \text{ (Downward)}$$

Example 10.6: Consider the cantilever beam AB of length L subjected to a concentrated load P as shown in Fig. 10.11. Using the double integration method, determine the slope and displacement at free end A and loading point C.

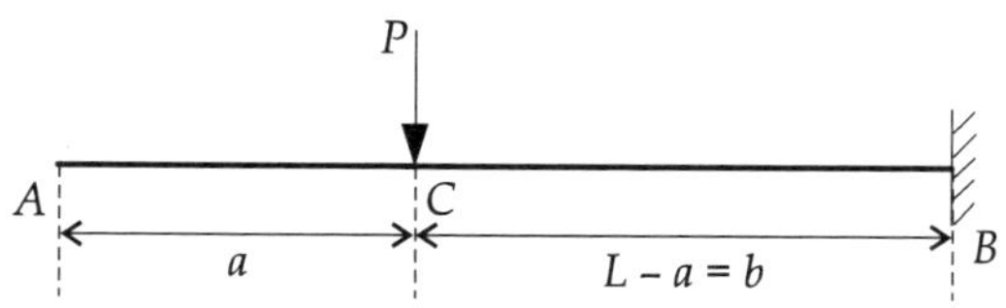

Fig. 10.11

Solution: Free-body diagram

The deflected elastic curve of the beam is shown in Fig. 10.12.

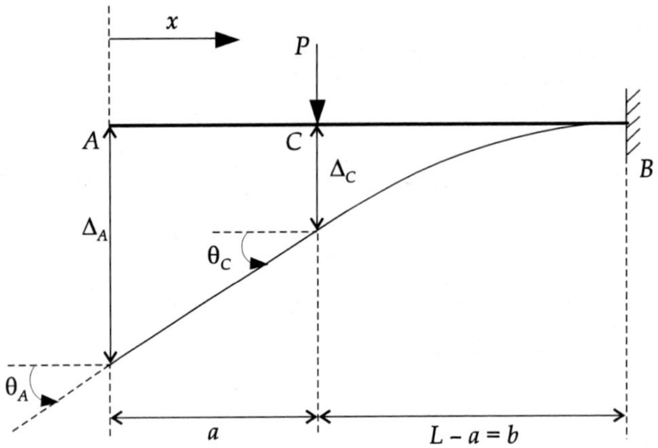

Fig. 10.12

Taking x from free end A, we get two moment functions

For $0 \leq x \leq a$

$$M_x = 0$$

For $a \leq x \leq L$

$$M_x = -P(x - a)$$

Elastic curve equation

$$EI\frac{d^2y}{dx^2} = -M_x$$

For $0 \leq x \leq a$ integration elastic curve equation we, get

$$EI\frac{d^2y}{dx^2} = 0$$

$$EI\frac{dy}{dx} = C_1 \tag{1}$$

$$EI\,y = C_1x + C_2 \tag{2}$$

For $a \leq x \leq L$ integration elastic curve equation, we get

$$EI\frac{d^2y}{dx^2} = P(x - a)$$

$$EI\frac{dy}{dx} = \frac{Px^2}{2} - Pax + C_3 \tag{3}$$

$$EI\,y = \frac{Px^3}{6} - \frac{Pax^2}{2} + C_3x + C_4 \tag{4}$$

Boundary conditions

Two unknown constants are determined using the two-boundary conditions, i.e., $\frac{dy}{dx} = 0$ at $x = L$ and $y = 0$ at $x = L$.

Continuity equations

Two unknown constants are determined using two continuity equations, i.e., the slope and deflection at just left and right of point C ($x = a$) must be equal, $\left(\frac{dy}{dx}\right)_{C^-} = \left(\frac{dy}{dx}\right)_{C^+}$ and $y_{C^-} = y_{C^+}$

Applying the first boundary condition $\frac{dy}{dx} = 0$ at $x = L$ in equation (3), we get

$$C_3 = PaL - \frac{PL^2}{2} = \frac{PL(a-b)}{2} \tag{5}$$

Applying the second boundary condition $y = 0$ at $x = L$ in equation (4), we get

$$0 = \frac{PL^3}{6} - \frac{PaL^2}{2} + C_3L + C_4 \tag{6}$$

Applying the first continuity equation $\left(\frac{dy}{dx}\right)_{C^-} = \left(\frac{dy}{dx}\right)_{C^+}$ at $x = a$ in equations (1) and (3), we get

$$C_1 = -\frac{Pa^2}{2} + C_3 \tag{7}$$

Applying the second continuity equation $y_{C^-} = y_{C^+}$ and $x = a$ in equations (2) and (4), we get

$$C_1a + C_2 = -\frac{Pa^2}{3} + C_3a + C_4 \tag{8}$$

Putting value of C_3 in equation (6), we get

$$C_4 = \frac{PL^2(2b-a)}{6}$$

Putting value of C_3 in equation (7), we get

$$C_1 = -\frac{Pa^2}{2} + \frac{PL(2b-a)}{6} = -\frac{Pb^2}{2}$$

Putting value of C_1, C_3 and C_4 in equation (8), we get

$$C_2 = \frac{Pa^3}{6} - \frac{PL^2a}{2} + \frac{PL^3}{3} = \frac{Pb^2(2L+a)}{6}$$

Putting the values of C_1, C_2, C_3 and C_4 in equations (1) to (4), we get the elastic curve equations of the slope and deflection.

For $0 \le x \le a$

$$\text{Slope, } \frac{dy}{dx} = \frac{-1}{EI}\frac{Pb^2}{2} \tag{9}$$

$$\text{Deflection, } y = \frac{1}{EI}\left\{\left(\frac{-Pb^2}{2}\right)x + \frac{Pb^2(2L+a)}{6}\right\} \tag{10}$$

For $a \le x \le L$

$$\text{Slope, } \frac{dy}{dx} = \frac{1}{EI}\left(\frac{Px^2}{2} - Pax + \frac{PL(a-b)}{2}\right) \tag{11}$$

$$\text{Deflection, } y = \frac{1}{EI}\left\{\frac{Px^3}{6} - \frac{Pax^2}{2} + \left(\frac{PL(a-b)}{2}\right)x + \frac{PL^2(2b-a)}{6}\right\} \tag{12}$$

For slope at free end A put $x = 0$ in equation (9)

$$\theta_A = \left.\frac{dy}{dx}\right|_{x=0} = -\frac{1}{EI}\frac{Pb^2}{2} \text{(Anticlockwise: Negative)}$$

For slope at C put $x = a$ in equation (9) or (11)

$$\theta_B = \left.\frac{dy}{dx}\right|_{x=a} = -\frac{1}{EI}\frac{Pb^2}{2} \text{(Anticlockwise: Negative)}$$

Note: The deflected curve between A to C is linear since there is no load between A to C, therefore slope at A and C will be same, and since slope is anticlockwise direction therefore it should be negative.

For deflection at free end A put $x = 0$ in equation (10)

$$\Delta_A = y|_{x=0} = \frac{1}{EI}\frac{Pb^2(2L+a)}{6} \text{ (Downward: Positive)}$$

For deflection at C put $x = a$ in equation (10) or (12)

$$\Delta_C = y|_{x=a} = \frac{1}{EI}\frac{Pb^2(L+a)}{3} \text{ (Downward: Positive)}$$

To check: $\Delta_A = \Delta_C - \theta_C a$ (Negative sign is taken to make slope positive)

$$\text{RHS: } \Delta_C - \theta_C a = \frac{1}{EI}\left(\frac{Pb^2(L-a)}{3} + \frac{Pb^2}{2}a\right) = \frac{1}{EI}\left(\frac{Pb^2(2L+a)}{6}\right) = \Delta_A = \text{LHS}$$

Special case: When load is at free end A ($a = 0$)

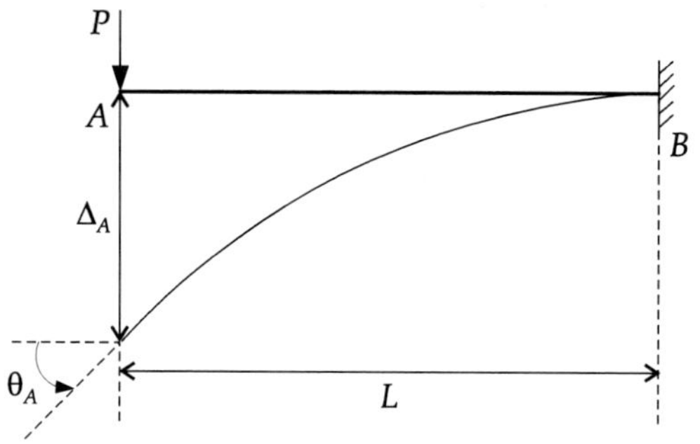

Fig. 10.13

In this special case (Fig. 10.13) since $a(= 0) \leq x \leq L(= b)$ therefore equations (3), (4), (11) and (12) will be under consideration.

Since $a(= 0) \leq x \leq L(= b)$ therefore,

Slope,

$$\frac{dy}{dx} = \frac{1}{EI}\left(\frac{Px^2}{2} - Pax + \frac{PL(a-b)}{2}\right) = \frac{1}{EI}\left(\frac{Px^2}{2} - \frac{PL^2}{2}\right)$$

Deflection, $y = \frac{1}{EI}\left\{\frac{Px^3}{6} - \frac{Pax^2}{2} + \left(\frac{PL(a-b)}{2}\right)x + \frac{PL^2(2b-a)}{6}\right\}$

$$y = \frac{1}{EI}\left\{\frac{Px^3}{6} - \frac{Pax^2}{2} + \left(\frac{PL(a-b)}{2}\right)x + \frac{PL^2(2b-a)}{6}\right\}$$

$$= \frac{1}{EI}\left\{\frac{Px^3}{6} - \frac{PL^2}{2}x + \frac{PL^3}{3}\right\}$$

For slope at free end A put $x = 0$ in slope equation

$$\theta_A = \left.\frac{dy}{dx}\right|_{x=0} = \frac{1}{EI}\left(\frac{Px^2}{2} - \frac{PL^2}{2}\right) = \frac{PL^2}{2EI} \text{ (Anticlockwise)}$$

For deflection at free end A put $x = 0$ in deflection equation

$$\Delta_A = y|_{x=0} = \frac{1}{EI}\left(\frac{Px^3}{6} - \frac{PL^2}{2}x + \frac{PL^3}{3}\right) = \frac{PL^3}{3EI} \text{ (Downward)}$$

Example 10.7: Consider a simply supported beam AB of length L subjected to uniformly distributed load as shown in Fig. 10.14. Using the double integration method, determine displacement at point C and slope at points A and B.

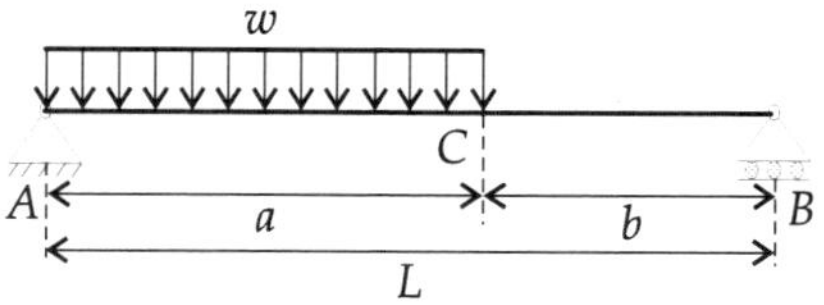

Fig. 10.14

Solution: Free-body diagram

The deflected elastic curve of the beam is shown in Fig. 10.15.

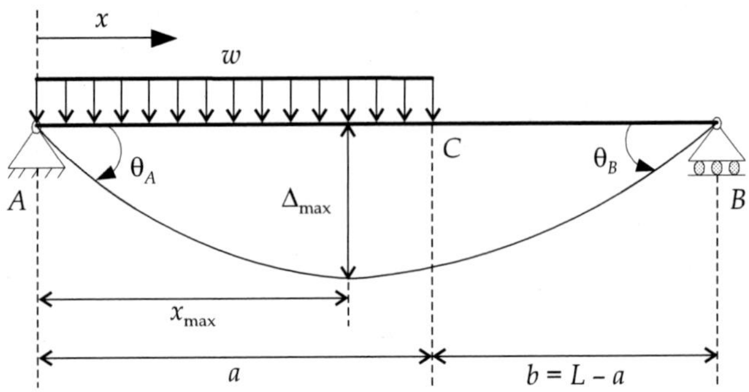

Fig. 10.15

Using equations of equilibrium, we get

$R_A = wa\left(1-\frac{a}{2L}\right)$ and $R_B = \frac{wa^2}{2L}$

Now by taking x from end A, we get two moment functions

For $0 \le x \le a$

$$M_x = R_A x - \frac{wx^2}{2}$$

For $a \le x \le L$

$$M_x = R_A x - wa(x - a/2) = \frac{wx^2}{2}\left(1-\frac{x}{L}\right)$$

Elastic curve equation

$$EI\frac{d^2y}{dx^2} = -M_x$$

For $0 \le x \le a$ integration elastic curve equation, we get

$$EI\frac{d^2y}{dx^2} = \frac{wx^2}{2} - R_A x$$

$$EI\frac{dy}{dx} = \frac{wx^2}{6} - R_A\frac{x^2}{2} + C_1 \quad (1)$$

$$EI\,y = \frac{wx^4}{24} - R_A\frac{x_3}{6} + C_1 x + C_2 \quad (2)$$

For $a \le x \le L$ integration elastic curve equation, we get

$$EI\frac{d^2y}{dx^2} = -\frac{wa^2}{2}\left(1-\frac{x}{L}\right)$$

$$EI\frac{dy}{dx} = -\frac{wa^2}{2}\left(x - \frac{x^2}{2L}\right) + C_3 \tag{3}$$

$$EI\,y = -\frac{wa^2}{2}\left(\frac{x^2}{2} - \frac{x^3}{6L}\right) + C_3x + C_4 \tag{4}$$

Boundary conditions

Two unknown constants are determined using two boundary conditions, i.e., $y = 0$ at $x = 0$ and $y = 0$ at $x = L$

Continuity equations

Two unknown constants are determined using two continuity equations, i.e., slope and deflection at just left and right of point C ($x = a$) must be equal, $\left(\frac{dy}{dx}\right)_{C^-} = \left(\frac{dy}{dx}\right)_{C^+}$ and $y_{C^-} = y_{C^+}$

Applying the first boundary condition $y = 0$ at $x = 0$ in equation (2), we get

$$0 = 0 - 0 + 0 + C_2 \Rightarrow C_2 = 0 \tag{5}$$

Applying second boundary condition $y = 0$ at $x = L$ in equation (4), we get

$$0 = -\frac{wa^2L^2}{6} + C_3L + C_4 \tag{6}$$

Applying the first continuity equation $\left(\frac{dy}{dx}\right)_{C^-} = \left(\frac{dy}{dx}\right)_{C^+}$ at $x = a$ in equations (1) and (3), we get

$$\frac{wx^3}{6} - R_A\frac{x^2}{2} + C_1 = -\frac{wa^2}{2}\left(x - \frac{x^2}{2L}\right) + C_3$$

or
$$C_1 - C_3 = -\frac{wa^3}{6} \tag{7}$$

Applying the second continuity equation $y_{C^-} = y_{C^+}$ at $x = a$ in equations (2) and (4), we get

$$\frac{wx^4}{24} - \left(wa\left(1 - \frac{a}{2L}\right)\right)\frac{x^3}{6} + C_1x = -\frac{wa^2}{2}\left(\frac{x^2}{2} - \frac{x^3}{6L}\right) + C_3x + C_4$$

or
$$(C_1 - C_3)a - C_4 = -\frac{wa^4}{8} \tag{8}$$

Equating equations (7) and (8), we get

$$C_4 = -\frac{wa^4}{24}$$

Putting value of C_4 in equation (6), we get

$$C_3 = \frac{wa^2}{24L}(4L^2 + a^2)$$

Putting value of C_3 in equation (7), we get

$$C_1 = \frac{wa^2}{24L}(4L^2 + a^2 - 4aL)$$

Putting the values of C_1, C_2, C_3 and C_4 in equations (1) to (4) we get the elastic curve equations of slope and deflection.

For $0 \le x \le a$

$$\text{Slope, } \frac{dy}{dx} = \frac{1}{EI}\left\{\frac{wx^3}{6} - R_A\frac{x^2}{2} + \frac{wa^2}{24L}(4L^2 + a^2 - 4aL)\right\} \quad (9)$$

$$\text{Deflection, } y = \frac{1}{EI}\left\{\frac{wx^4}{24} - R_A\frac{x^3}{2} + \frac{wa^2 x}{24L}(4L^2 + a^2 - 4aL)\right\} \quad (10)$$

For $a \le x \le L$

$$\text{Slope, } \frac{dy}{dx} = \frac{1}{EI}\left\{-\frac{wa^2}{2}\left(x - \frac{x^2}{2L}\right) + \frac{wa^2}{24L}(4L^2 + a^2)\right\} \quad (11)$$

$$\text{Deflection, } y = \frac{1}{EI}\left\{-\frac{wa^2}{2}\left(\frac{x^2}{2} - \frac{x^3}{6L}\right) + \frac{wa^2}{24L}(4L^2 + a^2) - \frac{wa^4}{24}\right\} \quad (12)$$

For slope near support A put $x = 0$ in equation (9)

$$\theta_A = \left.\frac{dy}{dx}\right|_{x=0} = \frac{1}{EI}\left\{\frac{wa^2}{24L}(4L^2 + a^2 - 4aL)\right\} \text{(Clockwise: Positive)}$$

For slope near support B put $x = L$ in equation (11)

$$\theta_B = \left.\frac{dy}{dx}\right|_{x=L} = \frac{1}{EI}\left\{-\frac{wa^2L}{12} + \frac{wa^4}{24L}\right\} \text{(Anticlockwise: Negative)}$$

Maximum deflection

Maximum deflection will occur at a point where slope will be zero, let assume it between $0 \le x \le a$, i.e. for slope equation (9)

$$\frac{1}{EI}\left\{\frac{wx^3}{6} - R_A\frac{x^2}{2} + \frac{wa^2}{24L}(4L^2 + a^2 - 4aL)\right\} = 0$$

Putting value of R_A and solving we get x_{max} and then putting value of x_{max} in equation (10) we get maximum deflection Δ_{max}

Special case-I (Fig. 10.16): When uniformly distributed load is over half beam ($a = b = L/2$)

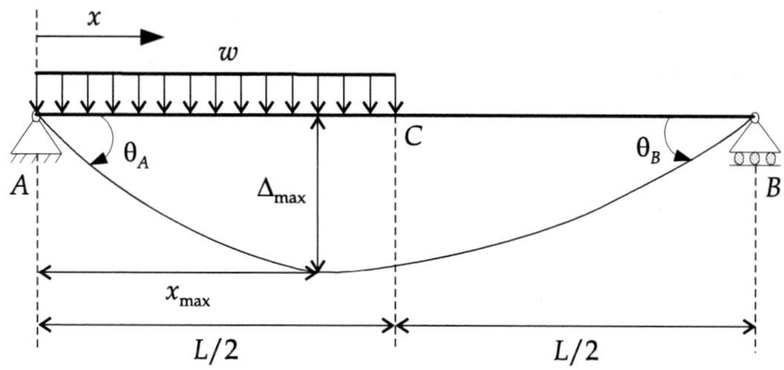

Fig. 10.16

For slope near support A put $x = 0$ in equation (9)

$$\theta_A = \left.\frac{dy}{dx}\right|_{x=0,\, a=L/2} = \frac{3wL^3}{128EI} \text{ (Clockwise: Positive)}$$

For slope near support B put $x = L$ in equation (11)

$$\theta_B = \left.\frac{dy}{dx}\right|_{x=L,\, a=L/2} = -\frac{7wL^3}{384EI} \text{ (Anticlockwise: Negative)}$$

Special Case-II (Fig. 10.17): When uniformly distributed load is over full beam ($a = L$)

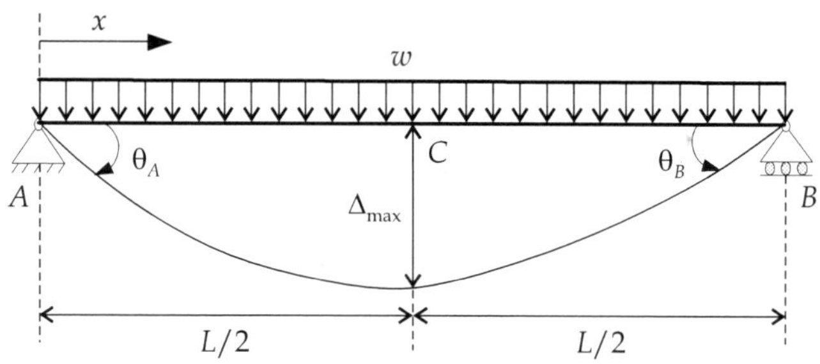

Fig. 10.17

For slope near support A put $x = 0$ in equation (9)

$$\theta_A = \left.\frac{dy}{dx}\right|_{x=0,\, a=L/2} = \frac{wL^3}{24EI} \text{ (Clockwise: Positive)}$$

For slope near support B put $x = L$ in equation (11)

$$\theta_B = \left.\frac{dy}{dx}\right|_{x=L,\, a=L/2} = -\frac{wL^3}{24EI} \text{ (Anticlockwise: Negative)}$$

Since this is a symmetrical case therefore slope at A and B will be equal but opposite in sign, $\theta_A = \theta_B = \theta = \frac{wL^3}{24EI}$, i.e., at A positive (clockwise direction) and at B negative (anticlockwise direction).

Maximum deflection

Maximum deflection will occur at centre, i.e., at $x = L/2$

Putting $a = L$ and $x = L/2$ in equation (10) or (12), we get

$$\Delta_{max} = y|_{x=L/2} = \frac{wx}{24EI}(x^3 - 2Lx^2 + L^3) = \frac{5wL^4}{384EI} \text{ (Downward)}$$

Example 10.8: Consider the cantilever beam AB of length L subjected to uniformly distributed load as shown in Fig. 10.18. Using the double integration method, determine the slope and displacement at free end A.

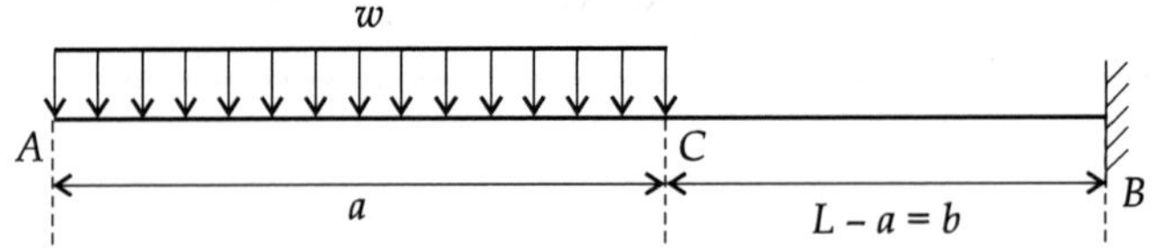

Fig. 10.18

Solution: Free-body diagram

The deflected elastic curve of the beam is shown in Fig. 10.19.

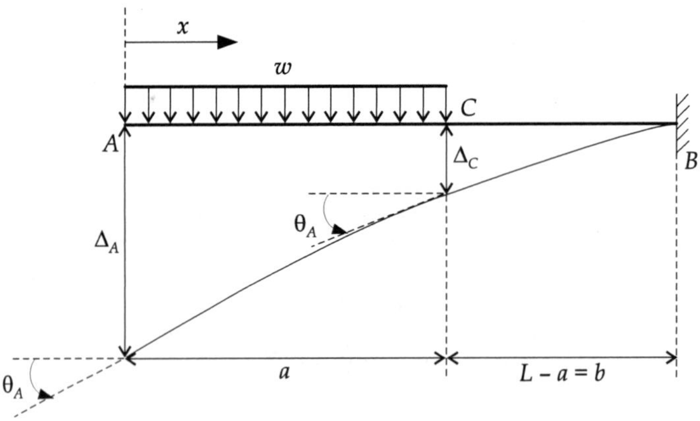

Fig. 10.19

Taking x from free end A, we get two moment functions

For $0 \leq x \leq a$

$$M_x = -\frac{wx^2}{2}$$

For $a \leq x \leq L$

$$M_x = -wa(x - a/2)$$

Elastic curve equation

For $0 \le x \le a$ integration elastic curve equation, we get

$$EI\frac{d^2y}{dx^2} = \frac{wx^2}{2}$$

$$EI\frac{dy}{dx} = -\frac{wx^3}{6} + C_1 \tag{1}$$

$$EI\,y = -\frac{wx^4}{24} + C_1x + C_2 \tag{2}$$

For $a \le x \le L$ integration elastic curve equation, we get

$$EI\frac{d^2y}{dx^2} = wa(x - a/2)$$

$$EI\frac{dy}{dx} = \frac{wax^2}{2} - \frac{wa^2x}{2} + C_3 \tag{3}$$

$$EI\,y = \frac{wax^2}{6} - \frac{wa^2x^2}{4} + C_3x + C_4 \tag{4}$$

Boundary conditions

Two unknown constants are determined using two boundary conditions, i.e., $\frac{dy}{dx} = 0$ at $x = L$ and $y = 0$ at $x = L$.

Continuity equations

Two unknown constants are determined using two continuity equations, i.e., slope and deflection at just left and right of point C ($x = a$) must be equal, $\left(\frac{dy}{dx}\right)_{C^-} = \left(\frac{dy}{dx}\right)_{C^+}$ and $y_{C^-} = y_{C^+}$

Applying the first boundary condition $\frac{dy}{dx} = 0$ at $x = L$ in equation (3), we get

$$C_3 = \frac{waL(a - L)}{2} \tag{5}$$

Applying the second boundary condition $y = 0$ at $x = L$ in equation (4), we get

$$0 = \frac{waL^3}{6} - \frac{wa^2L^2}{4} + C_3L + C_4 \tag{6}$$

Applying the first continuity equation $\left(\frac{dy}{dx}\right)_{C^-} = \left(\frac{dy}{dx}\right)_{C^+}$ at $x = a$ in equations (1) and (3), we get

$$C_1 = -\frac{wa^3}{6} + C_3 \tag{7}$$

Applying the second continuity equation $y_{C^-} = y_{C^+}$ at $x = a$ in equations (2) and (4), we get

$$C_1 a + C_2 = -\frac{wa^4}{8} + C_3 a + C_4 \quad (8)$$

Putting value of C_3 in equation (6), we get

$$C_4 = \frac{waL^3}{3} - \frac{wa^2L^2}{4}$$

Putting value of C_3 in equation (7), we get

$$C_1 = -\frac{wa^3}{6} + \frac{waL(a-L)}{2}$$

Putting value of C_1, C_3 and C_4 in equation (8), we get

$$C_2 = \frac{wa^4}{24} + \frac{waL^3}{3} - \frac{wa^2L^2}{4}$$

Putting the values of C_1, C_2, C_3 and C_4 in equations (1) to (4), we get the elastic curve equations of slope and deflection.

For $0 \le x \le a$

$$\text{Slope, } \frac{dy}{dx} = \frac{1}{EI}\left(\frac{wx^3}{6} - \frac{wa^3}{6} + \frac{waL(a-L)}{2}\right) \quad (9)$$

$$\text{Deflection, } y = \frac{1}{EI}\left\{\frac{wx^4}{24} + \left(\frac{waL(a-L)}{2} - \frac{wa^3}{6}\right)x + \frac{wa^4}{24} + \frac{waL^3}{3} - \frac{wa^2L^2}{4}\right\} \quad (10)$$

For $a \le x \le L$

$$\text{Slope, } \frac{dy}{dx} = \frac{1}{EI}\left(\frac{wax^2}{2} - \frac{wa^2x}{2} + \frac{waL(a-L)}{2}\right) \quad (11)$$

$$\text{Deflection, } y = \frac{1}{EI}\left(\frac{wax^3}{6} - \frac{wa^2x^2}{4} + \frac{waL(a-L)}{2}x + \frac{waL^3}{3} - \frac{wa^2L^2}{4}\right) \quad (12)$$

For slope at C put $x = a$ in equation (9)

$$\frac{dy}{dx} = \frac{1}{EI}\left(\frac{wx^3}{6} - \frac{wa^3}{6} + \frac{waL(a-L)}{2}\right)$$

$$\theta_C = \left.\frac{dy}{dx}\right|_{x=a} = \frac{-1}{EI}\left(\frac{waL(L-a)}{2}\right) \text{(Anticlockwise: Negative)}$$

For deflection at C put $x = a$ in equation (10)

$$y = \frac{1}{EI}\left\{\frac{wx^4}{24} + \left(\frac{waL(a-L)}{2} - \frac{wa^3}{6}\right)x + \frac{wa^4}{24} + \frac{waL^3}{3} - \frac{wa^2L^2}{4}\right\}$$

$$\Delta_C = y\big|_{x=a} = \frac{w}{EI}\left\{\frac{-a^4}{12} + \frac{-a^3L}{2} + \frac{aL^3}{3} - \frac{3a^2L^2}{4}\right\} \text{ (Downward: Positive)}$$

For slope at free end A put $x = 0$ in slope equation (9)

$$\frac{dy}{dx} = \frac{1}{EI}\left\{\frac{wx^3}{6} - \frac{wa^3}{6} + \frac{waL(a-L)}{2}\right\}$$

$$\theta_A = \frac{dy}{dx}\bigg|_{x=0} = \frac{1}{EI}\left\{-\frac{wa^3}{6} + \frac{waL(a-L)}{2}\right\} \text{ (Anticlockwise)}$$

For deflection at free end A put $x = 0$ in deflection equation (10)

$$y = \frac{1}{EI}\left\{\frac{wa^4}{24} + \left(\frac{waL(a-L)}{2} - \frac{wa^3}{6}\right)x + \frac{wa^4}{24} + \frac{waL^3}{3} - \frac{wa^2L^2}{4}\right\}$$

$$\Delta_A = y\big|_{x=0} = \frac{1}{EI}\left(\frac{wa^4}{24} + \frac{waL^3}{3} - \frac{wa^2L^2}{4}\right) \text{ (Downward)}$$

Special Case (Fig. 10.20): When uniformly distributed load is over full beam ($a = L$)

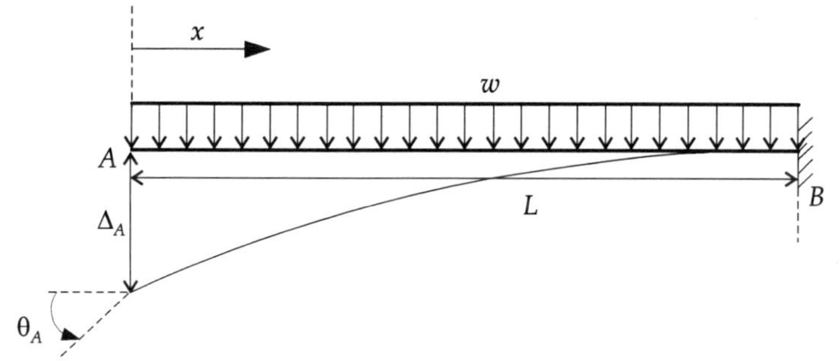

Fig. 10.20

In this special case since $a = L$ therefore, equations (1), (2), (9) and (10) will be under consideration.

Slope, $\dfrac{dy}{dx} = \dfrac{1}{EI}\left(\dfrac{wx^3}{6} - \dfrac{wL^2}{6}\right)$

Deflection, $y = \dfrac{1}{EI}\left(\dfrac{wx^4}{24} - \dfrac{wL^3}{6}x + \dfrac{wL^4}{8}\right)$

For slope at free end A put $x = 0$ in slope equation

$$\theta_A = \frac{dy}{dx}\bigg|_{x=0} = \frac{1}{EI}\left(\frac{wx^3}{6} - \frac{wL^3}{6}\right) = -\frac{wL^3}{6EI} \text{ (Anticlockwise)}$$

For deflection at free end A put $x = 0$ in deflection equation

$$\Delta_A = y\big|_{x=0} = \frac{1}{EI}\left(\frac{wx^4}{24} - \frac{wL^3}{6}x + \frac{wL^4}{8}\right) = \frac{wL^4}{8EI} \text{ (Downward)}$$

10.4 MOMENT AREA METHOD

Theorem 1: The change of angle θ in radians between tangents at any two points *A* and *B* on the elastic curve of an originally straight beam is equal to the total area of the bending moment diagram between *A* and *B*, divided by *EI*.

Theorem 2: The deflection of a point *B* on the elastic curve from the tangent at another point *A* on the elastic curve is equal to the statistical moment with respect to *B* of the bending moment area between *A* and *B*, divided by *EI*.

$$\delta = \int_A^B x\,d\theta = \int_A^B \frac{Mxdx}{EI}$$

In this method $\frac{M}{EI}$ diagram is used to get slope and deflection as

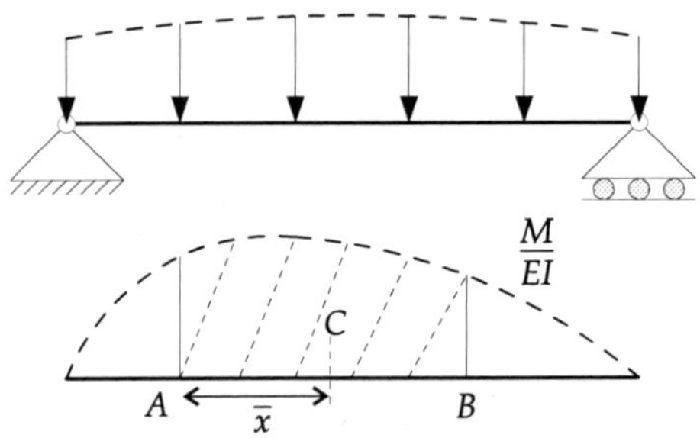

C is centroid of area under *AB*

Fig. 10.21

Slope at *B* with respect to *A* and deflection at *A* with respect to *B* are given by

$$\theta_{BA} = \int_A^B \frac{M}{EI}dx \text{ and } \Delta_{AB} = \bar{x}\int_A^B \frac{M}{EI}dx$$

There may be an inflection point in the elastic line between *A* and *B* as shown in Fig. 10.22.

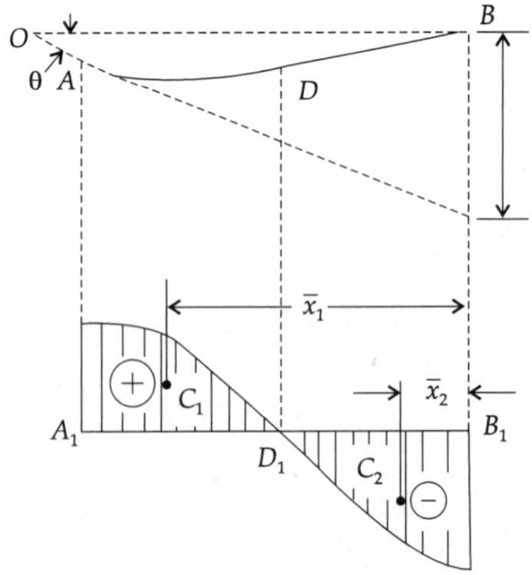

Fig. 10.22

In such cases the angle θ between the tangents at *A* and *B*

$$\theta = \int_A^D \frac{M\,dx}{EI} - \int_D^B \frac{M\,dx}{EI}$$

Similarly,

$$\delta = \int_A^D \frac{Mx\,dx}{EI} - \int_D^B \frac{Mx\,dx}{EI}$$

$\delta = \left[\dfrac{\text{Area } M - x \text{ graph}}{EI}\right]_A^D . \bar{x}_1 - \left[\dfrac{\text{Area } M - x \text{ graph}}{EI}\right]_D^B . \bar{x}_2$, where $\bar{x}_1$ and $\bar{x}_2$ are the centroidal distances from *B*.

10.4.1 Procedure for Analysis

- In general, the moment-area method does not give a direct solution for the slope or displacement at a point on the beam, instead it can only be used to determine the angles or deviations between two tangents on the beam's elastic curve.
- Determine the support reactions and draw the beam's *M*/*EI* diagram.
- Draw an inflated view of the beam's elastic curve. Considering that points of zero slope and zero displacement occur at a fixed support, and zero displacement occurs at the pin and roller supports.
- Draw the general shape of the elastic curve using the moment (or *M*/*EI*) diagram. When the beam is subjected to a positive moment the beam bends concave up, whereas negative moment bends the beam concave down. An inflection point or change in curvature occurs where the moment in the beam (or *M*/*EI*) is zero.
- Indicate the displacement and slope to be determined on the curve. Since the moment-area theorems apply only between two tangents, therefore tangents should be constructed on those points so that the angles or deviations between them lead to the solution of the problem. One must take the tangents at the points of unknown slope and displacement and at the supports, since the beam usually has zero displacement and/or zero slope at the supports.
- Apply Theorem 1 to determine the angle between two tangents, and Theorem 2 to determine vertical deviations between these tangents.
- Theorem 2, in general, will not yield the displacement of a point on the elastic curve. It will only give the vertical distance or deviation of a tangent at point *A* on the elastic curve from the tangent at *B*.
- After applying either Theorem 1 or Theorem 2, the slope and deflection can be calculated using simple geometry.

Example 10.9: Consider the simply supported beam *AB* of length *L* subjected to a concentrated load *P* as shown in Fig. 10.23. Using the moment area method, determine the displacement at point C and the slope at points A and B.

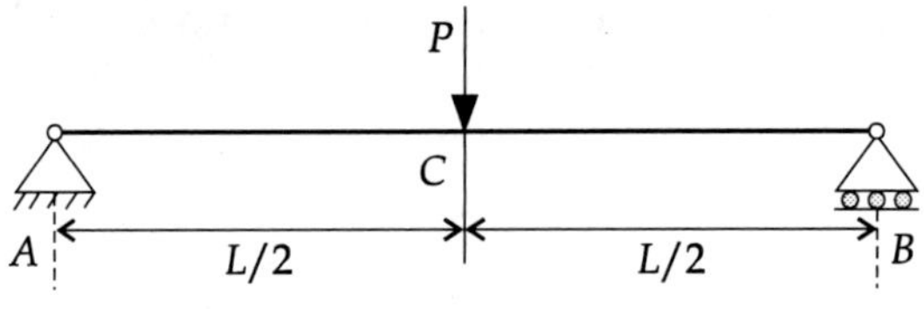

Fig. 10.23

Solution:

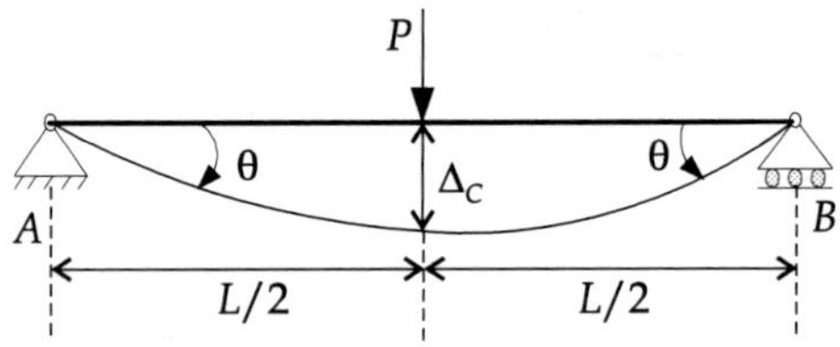

The load is applied at the centre of the beam. From the deflected curve we can see it is a symmetrical case. Therefore, slope at A and B will be equal but opposite in sign, $\theta_A = \theta_B = \theta$, and at centre point of the elastic curve, the slope will be zero.

Bending moment diagram and *M/EI* diagram (Fig. 10.24)

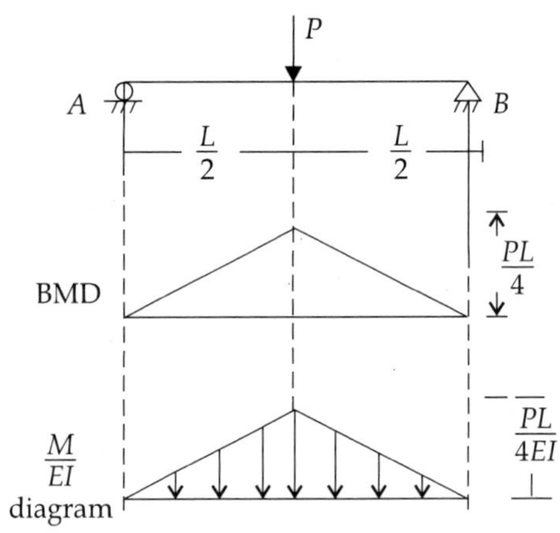

Fig. 10.24

Using Theorem 1 of the moment area method:

$$\theta_{AC} = \int_A^C \frac{Mdx}{EI} = M/EI \text{ area between } A \text{ and } C.$$

Also, $\theta_{AC} = \theta_A - \theta_C = \theta_A$ $(\because \theta_C = 0)$

Therefore, slope at A,

$$\theta_A = M/EI \text{ area between } A \text{ and } C = \frac{1}{EI}\left(\frac{1}{2}\frac{L}{2}\frac{PL}{4}\right) = \frac{PL^2}{16EI} \quad \text{(Clockwise: Positive)}$$

$$\theta_B = -\theta_A = -\frac{PL^2}{16EI} \text{ (Anticlockwise: Negative)}$$

Using Theorem 2 of the moment area method:

$$\Delta_{AC} = \int_A^C \frac{M\bar{x}\,dx}{EI} = \text{moment of } M/EI \text{ area between } A \text{ and } C \text{ about } A.$$

Also, $\Delta_{AC} = \Delta_A - \Delta_C = -\Delta_C$ $\qquad (\because \Delta_A = 0)$

We will find the deflection of point A with respect to point C as the slope of the beam is zero at the centre which is parallel to the original un-deflected straight beam.

Take moment of the area about the point A of the M/EI diagram.

$-\Delta_C$ = Moment of M/EI area between A and C about A

$$= \left(\frac{1}{2} \times \frac{L}{2} \times \frac{PL}{4EI}\right) \times \left(\frac{2}{3}\frac{L}{2}\right) = \frac{PL^3}{48EI}$$

$$\Delta_C = \frac{PL^3}{48EI} \text{ (Downward)}$$

Example 10.10: Consider the simply supported beam AB of length L subjected to uniformly distributed load as shown in Fig. 10.25. Using the moment area method, determine the maximum displacement and slope at points A and B.

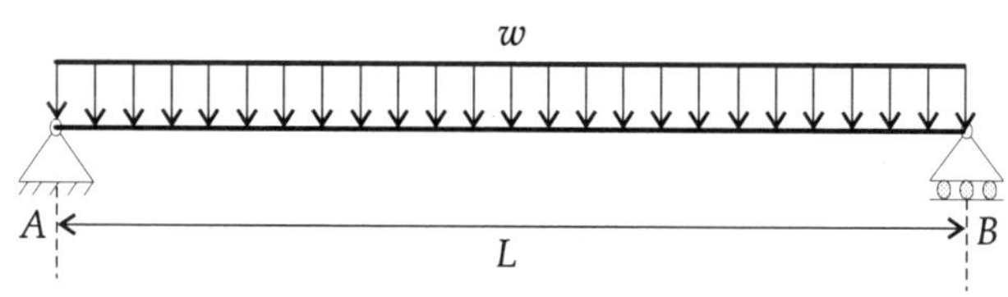

Fig. 10.25

Solution: Slope deflection diagram (Fig. 10.26)

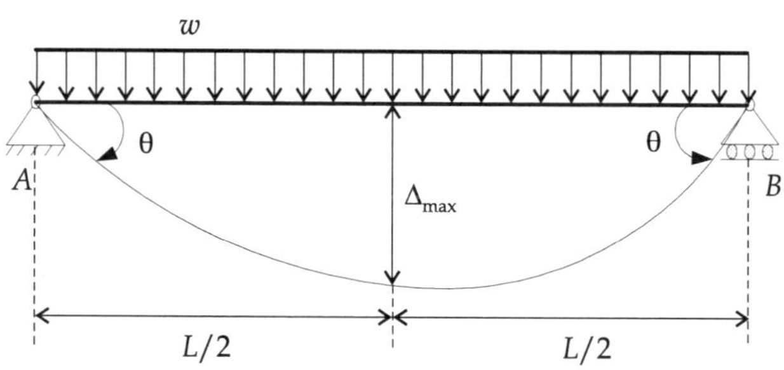

Fig. 10.26

Since this is a symmetrical case therefore slope at A and B will be equal but opposite in sign, $\theta_A = \theta_B = \theta$, and at the centre the slope will be zero and deflection will be maximum.

Bending moment diagram (Fig. 10.27)

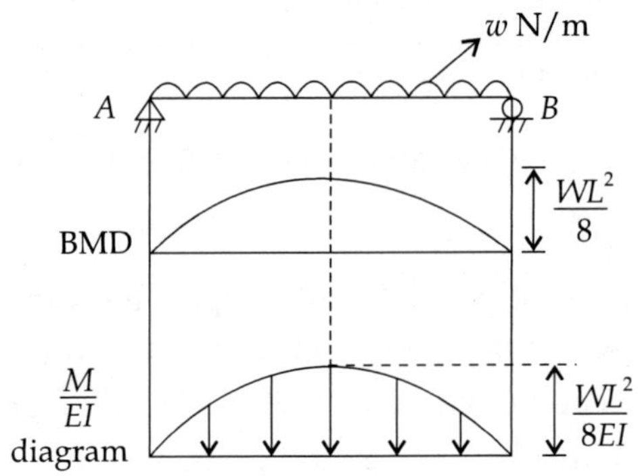

Fig. 10.27

Using Theorem 1 of the moment area method:

$$\theta_{AC} = \int_A^C \frac{M\,dx}{EI} = M/EI \text{ area between } A \text{ and } C$$

Also, $\theta_{AC} = \theta_A - \theta_C = \theta_A$ $\quad (\because \theta_C = 0)$

Therefore, slope at A,

$$\theta_A = M/EI \text{ area between } A \text{ and } C = \frac{1}{EI}\left(\frac{2}{3}\frac{L}{2}\frac{wL^2}{8}\right) = \frac{wL^3}{24EI} \text{ (Clockwise: Positive)}$$

$$\theta_B = -\theta_A = -\frac{wL^3}{24EI} \text{ (Anticlockwise: Negative)}$$

Using Theorem 2 of the moment area method:

$$\Delta_{AC} = \int_A^C \frac{M\bar{x}\,dx}{EI} = \text{Moment of } M/EI \text{ area between } A \text{ and } C \text{ about } A$$

Also, $\Delta_{AC} = \Delta_A - \Delta_C = -\Delta_C$ $\quad (\because \Delta_A = 0)$

Therefore, deflection at C,

$$-\Delta_C = \text{Moment of } M/EI \text{ area between } A \text{ and } C \text{ about } A = \left(\frac{wL^3}{24EI}\right) \times \left(\frac{5}{3}\frac{L}{2}\right) = \frac{5wL^4}{384EI}$$

$$\Rightarrow \Delta_{max} = \Delta_C = \frac{5wL^4}{384EI} \text{ (Downward)}$$

Example 10.11: Find the maximum deflection and slope of a cantilever beam under concentrated load at the free end as shown in Fig 10.28.

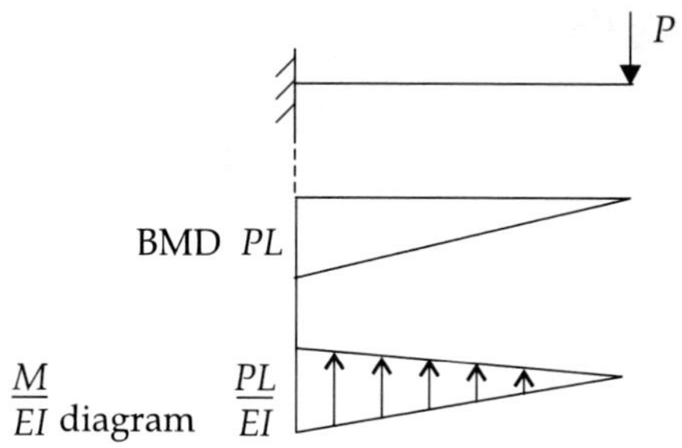

Fig. 10.28: Cantilever beam

The deflection of point B with respect to A

Take the moment of the M/EI area about the point B, give deflection

$$\Delta = \left(\frac{1}{2} \times L \times \frac{PL}{EI}\right) \times \left(\frac{2}{3}L\right)$$

Slope = Area of the M/EI diagram $\frac{1}{2} \times L \times \frac{PL}{EI}$

Example 10.12: Maximum deflection and slope of a cantilever beam under a moment M at free end

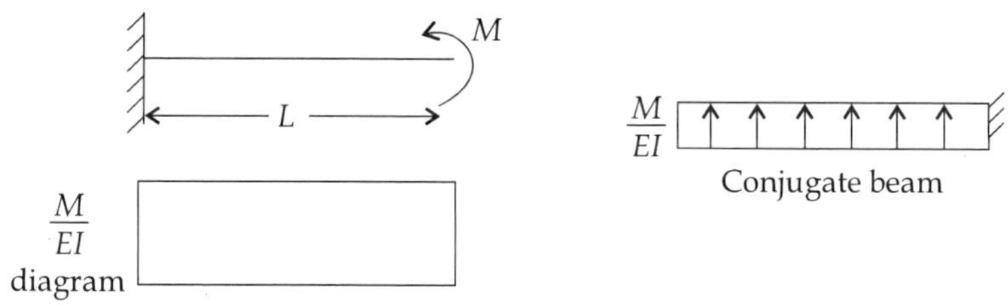

Fig. 10.29

The deflection of point B from tangent at A

Deflection, $\Delta = \left(\frac{M}{EI} \times L\right) \times \left(\frac{1}{2}L\right)$

The tip deflection due to moment M is $\frac{1}{2}\frac{ML^2}{EI}$

Maximum slope, $\frac{ML}{EI}$

Example 10.13: A cantilever beam subjected to moment and concentrated load

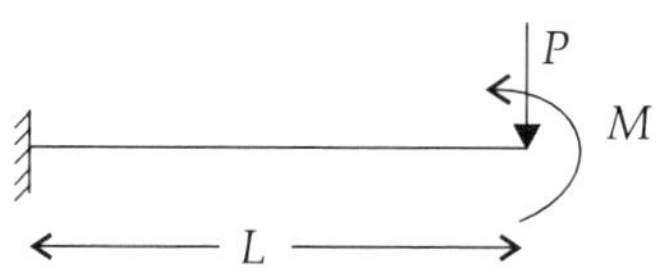

Fig. 10.30

By the method of superposition

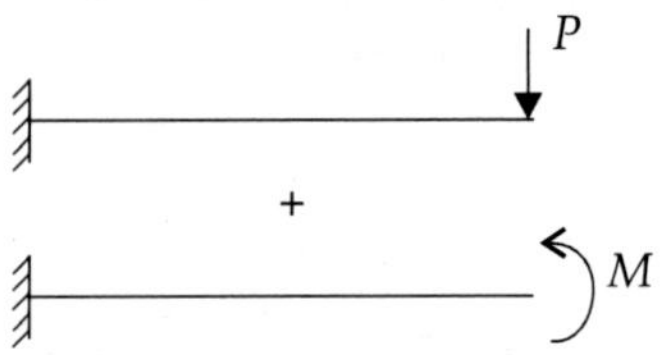

Fig. 10.31

For linear elastic structure the tip deflection,

$$\Delta = \frac{PL^3}{3EI} + \frac{ML^2}{2EI}$$

Example 10.14: The deflection of a cantilever beam due to udl, moment (M) and concentrated load (P)

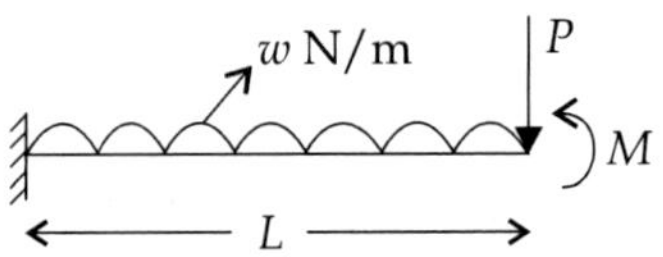

Fig. 10.32

The moment at support, $\frac{wL^2}{2EI}$

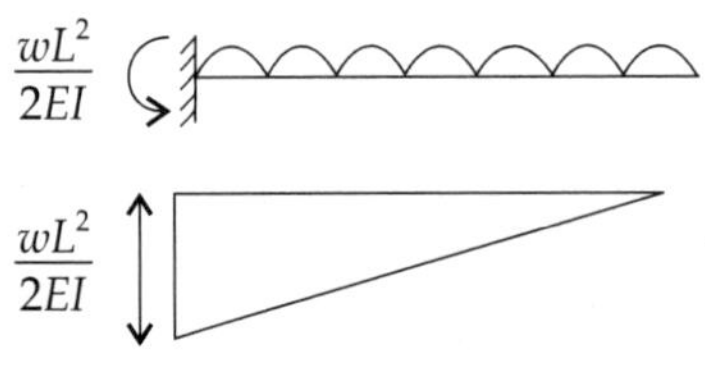

Fig. 10.32 (a)

The area of the moment diagram $= \frac{1}{3} \times L \times \frac{wL^2}{2}$

The moment of the area $= \left(\frac{1}{3} \times L \times \frac{wL^2}{2EI}\right) \times \left(\frac{3}{4} \times L\right)$

The maximum deflection due to udl acting throughout the beam, $\frac{wL^4}{8EI}$

Total deflection of the beam due to udl, moment and concentrated load acting at the free end can be obtained by the method of superposition.

Maximum deflection at free end, $\Delta = \frac{wL^4}{8EI} + \frac{ML^2}{2EI} + \frac{PL^3}{3EI}$

Example 10.15: A simply supported beam AB is loaded with a concentrated load P at the centre as shown in Fig. 10.33 below, the segments AB have flexural rigidity (EI) and the segment CB have flexural rigidity, $2EI$. Using the moment area method, determine the maximum slope and deflection.

Solution: Draw M/EI diagram of simply supported beam

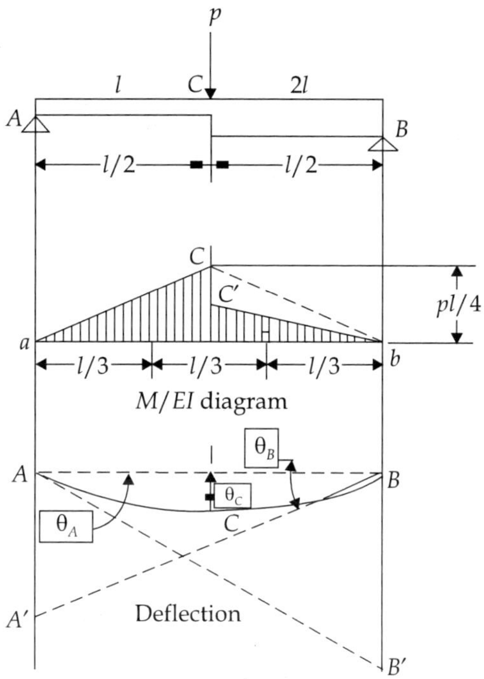

Fig. 10.33

Using the second moment area theorem

The deflection of the point A away from the tangent at B is

$$\delta_{A/B} = AA' = \frac{pl}{4EI}\cdot\frac{l}{4}\cdot\frac{l}{3} + \frac{pl}{8EI}\cdot\frac{l}{4}\cdot\frac{2l}{3} = \frac{pl^3}{24EI}$$

Then using geometry of figure,

$$\theta_B = \frac{AA'}{l} = \frac{pl^2}{24EI}$$

Similarly, deflection of the point B away from the tangent at A is

$$\delta_{B/A} = BB' = \frac{pl}{8EI}\cdot\frac{l}{4}\cdot\frac{l}{3} + \frac{pl}{4EI}\cdot\frac{l}{4}\cdot\frac{2l}{3} = \frac{5pl^3}{96EI}$$

$$\theta_A = \frac{BB'}{l} = \frac{5pl^2}{96EI}$$

Since, ratio $\theta_A/\theta_B = \dfrac{5}{4}$

Now, deflection of point C away from the tangent at B is

$$\delta_{C/B} = \frac{pl}{8EI}\cdot\frac{l}{4}\cdot\frac{l}{6} = \frac{pl^3}{192EI}$$

Therefore, from the geometry of figure,

$$\text{Maximum deflection, } \delta_C = \frac{AA'}{2} - \delta_{C/B} = \frac{pl^3}{48EI} - \frac{pl^3}{192EI} = \frac{pl^3}{64EI}$$

Since, $\theta_A/\theta_B = \dfrac{5}{4} > 1$, therefore maximum slope should be at A, $\theta_A = \dfrac{5pl^2}{96EI}$

Example 10.16: A simply supported beam $ABCD$ is loaded with a concentrated load P at centre as shown in Fig. 10.34 below, the segments AB and CD have flexural rigidity EI and the segment BC is rigid.

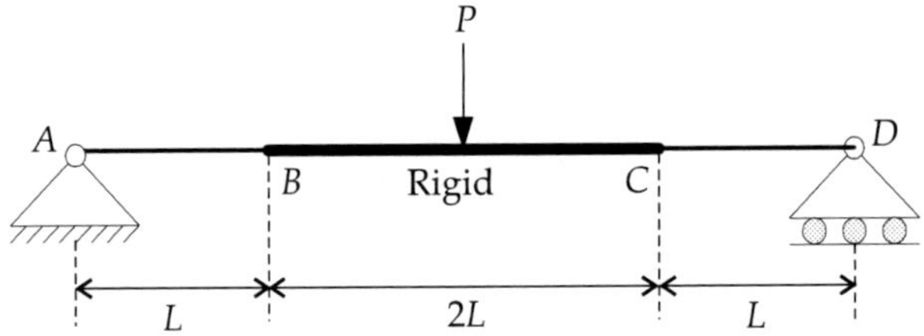

Fig. 10.34

Consider the following statements:

(i) The beam will be straight between B and C.

(ii) The magnitude of maximum deflection at mid span and deflection at B be equal to $\dfrac{PL^3}{6EI}$

(iii) The magnitude of slope at point support A is $\dfrac{PL^2}{4EI}$

Solution: Deflection diagram (Fig. 10.35)

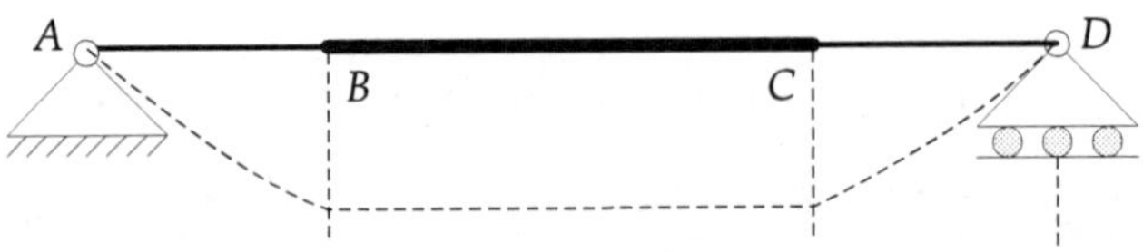

Fig. 10.35

From B to C the beam will be straight due to rigidity.

The M/EI diagram (Fig. 10.36) of simply supported beam (assuming flexural rigidity EI throughout the length) is given below

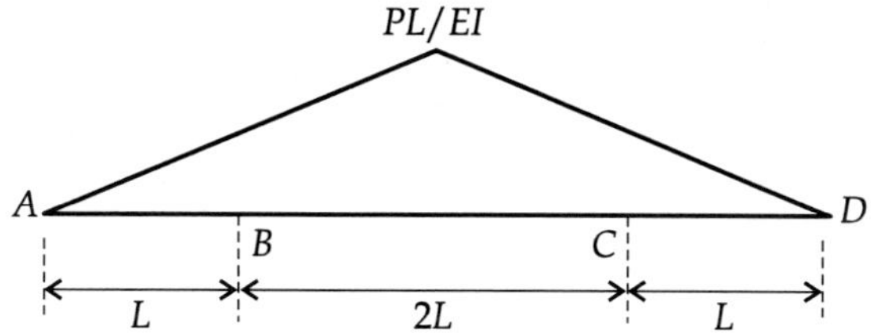

Fig. 10.36

Since the segment BC is rigid, i.e., $EI = \infty$, therefore, the net M/EI diagram (Fig. 10.37) of beam is

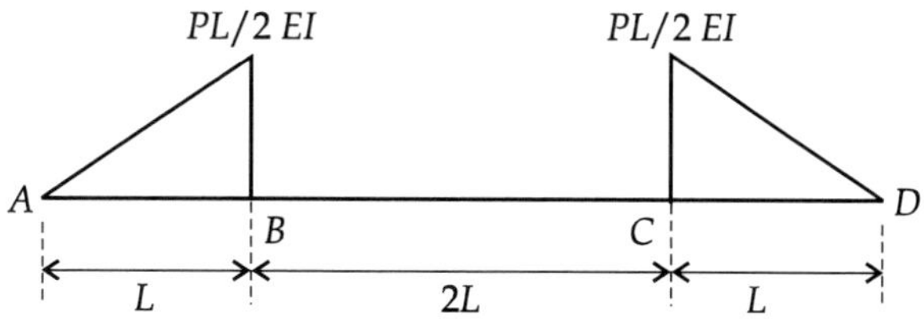

Fig. 10.37

Slope at A,

$$\theta_A = \text{Area } (AB) = \frac{1}{2} \times L \times \frac{PL}{2EI} = \frac{PL^2}{4EI}$$

Deflection at B = Maximum deflection at the centre

$$\Delta_B = \Delta_{max} = \text{Area}(AB) \times \bar{x} = \frac{PL^2}{4EI} \times \frac{2}{3}L = \frac{PL^3}{6EI}$$

Problem 10.17: A circular cantilever beam AB having diameter 1 m at free end and 2 m at the fixed end is loaded with a transverse load of 3 kN at the free end, if the length of the beam is 4 m as shown in Fig. 10.38, calculate the deflection at free end B.

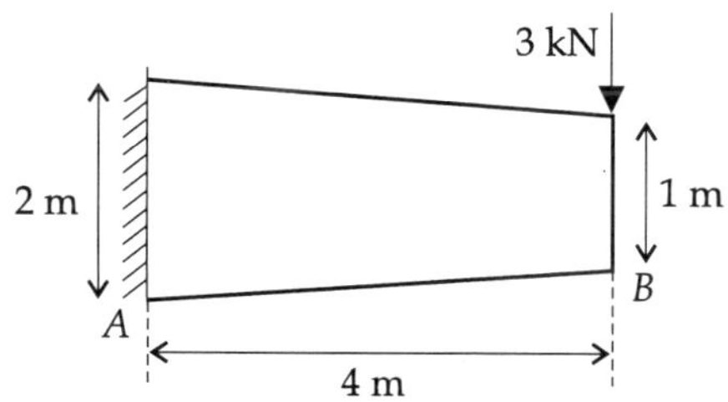

Fig. 10.38

Solution: Extend the beam up to zero diameter and take x from extreme free end, then using the double integration method (Fig. 10.39)

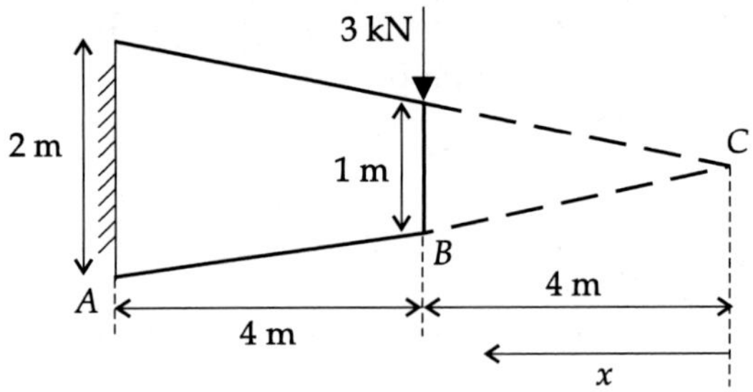

Fig. 10.39

$$d_x = \frac{2}{8} \times x = \frac{x}{4} \text{ and } I_x = \frac{\pi}{64} \times (d_x)^4 = \frac{\pi x^4}{64 \times 4^4}$$

$$EI\frac{d^2y}{dx^2} = -M(x) \Rightarrow E\frac{d^2y}{dx^2} = \frac{M(x)}{I} = \frac{3 \times 64 \times 4^4 (4 - x)}{\pi x^4}$$

$$= \frac{3 \times 64 \times 4^4}{\pi}(4x^{-4} - x^{-3})$$

After first integration,

$$\text{Slope} = \frac{dy}{dx} = \frac{3 \times 64 \times 4^4}{E\pi}\left(\frac{4x^{-3}}{-3} - \frac{x^{-2}}{-2} + C_1\right)$$

After second integration

$$\text{Deflection} = y = \frac{3 \times 64 \times 4^4}{E\pi}\left(\frac{4x^{-2}}{6} - \frac{x^{-1}}{2} + C_1 x + C_2\right)$$

Since

$$\text{At } x = 8 \Rightarrow \frac{dy}{dx} = y = 0 = C_1 = \frac{-2.67}{8^3} \text{ and } C_2 = \frac{48}{8^3}$$

Therefore, deflection at free end, $x = 4$ is

$$y(4) = \frac{3 \times 64 \times 4^4}{E\pi}\left(\frac{4 \times 4^{-2}}{6} - \frac{4^{-1}}{2} - \frac{2.67 \times 4}{8^3} + \frac{48}{8^3}\right) = \frac{-512}{E\pi}$$

10.5 CONJUGATE BEAM METHOD

In the conjugate beam method, the deflection and slope at any point of a beam is found by introducing another fictitious beam having the same length as the real beam. The imaginary beam is known as conjugate beam and it is analyzed for elastic load which is the same as the $\frac{M}{EI}$ diagram of the real beam. The conjugate beam concept makes use of the analogy between the slope and shear between deflection and bending moment of conjugate beam and real beam. The support conditions of conjugate beam are changed suitably to make it an analogy of real beam.

Theorem 1: The clockwise slope at any point in a real beam is numerically equal to the positive shear force at the corresponding point in a conjugate beam.

Theorem 2: The downward deflection at any point of a real beam is numerically equal to the positive moment at the corresponding point in the conjugate beam.

The support conditions are to be changed during analyze the conjugate beam as follows:

Table 10.1: Real conjugate beam supports conditions

Real beam	Conjugate beam
Roller support $\theta \neq 0;\quad \Delta = 0$	Roller support $V \neq 0;\quad M = 0$
Pin support $\theta \neq 0;\quad \Delta = 0$	Pin support $V \neq 0;\quad M = 0$
Fixed support $\theta = 0;\quad \Delta = 0$	Free end $V = 0;\quad M = 0$
Free end $\theta \neq 0;\quad \Delta \neq 0$	Fixed support $V \neq 0;\quad M \neq 0$
Internal roller support $\theta \neq 0;\quad \Delta = 0$	Internal hinge $V \neq 0;\quad M = 0$
Internal pin support $\theta \neq 0;\quad \Delta = 0$	Internal hinge $V \neq 0;\quad M = 0$
Internal hinge $\theta \neq 0;\quad \Delta \neq 0$	Internal roller support $V \neq 0;\quad M \neq 0$

10.5.1 Procedure for Analysis

- First, draw the bending moment (M) diagram of the real beam subjected to actual load acts on the beam.
- Then draw the conjugate beam for the real beam. The conjugate beam has the same length as the real beam.

- Provide the support condition of the conjugate beam as per Table 10.1. The conjugate beam is to be loaded with the real beam's M/EI diagram. This loading is assumed to be distributed over the conjugate beam and is directed upwards when M/EI is positive and downward when M/EI is negative. In other words, the loading always acts away from the beam.
- Using the equations of equilibrium, determine the support reactions of conjugate beam due to the above M/EI loads applied.
- Determine the unknown shear force (V') and moment (M') using the equations of equilibrium at the desired point of the conjugate beam. The shear force (V') of the conjugate beam will be the slope and moment (M') will be deflection for the real beam at the corresponding point.

Example 10.18: A simply supported beam is subjected to concentrated load at the centre (Fig 10.40). Find the maximum deflection and slope of the beam using the conjugate beam method. EI is constant.

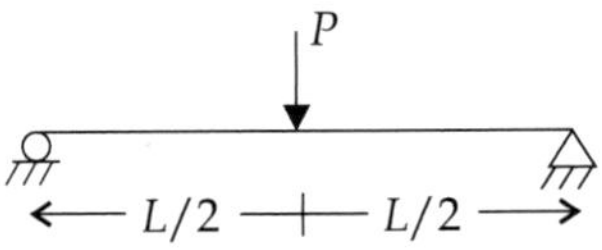

Fig. 10.40

The beam is subjected to concentrated load at the mid span of the simply supported beam of length L.

The M/EI diagram of the beam is shown in Figure 10.41.

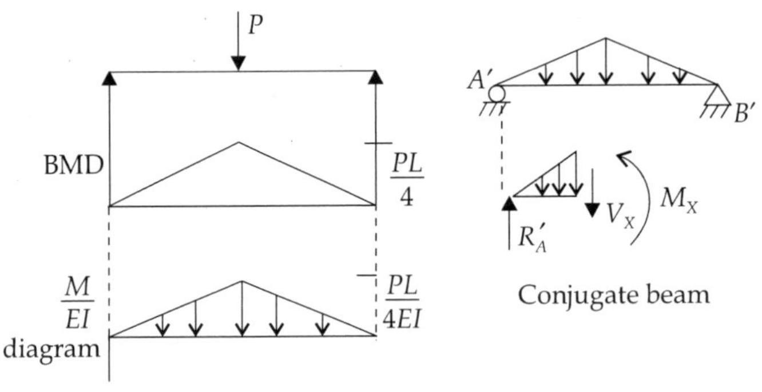

Fig. 10.41

The reaction force due to M/EI load on the conjugate beam,

$$R_A = \frac{1}{2} \times \frac{L}{2} \times \frac{PL}{4EI}$$

$$= \frac{PL^2}{16EI}$$

Taking moment about the centre point of the conjugate beam,

$$M_C = R_A \times \frac{L}{2} - \frac{1}{2} \times \frac{L}{2} \times \frac{PL}{4EI} \times \left(\frac{1}{3} \cdot \frac{L}{2}\right)$$

$$= \frac{PL^3}{32EI} - \frac{1}{96}\frac{PL^3}{EI}$$

By the conjugate method, the deflection at any point is the bending moment of the conjugate beam at that point.

$$\text{Central deflection, } \delta = \frac{PL^3}{48EI}$$

Example 10.19: A cantilever beam is subjected to concentrated load at the free end. Find the maximum deflection of the beam using conjugate beam method.

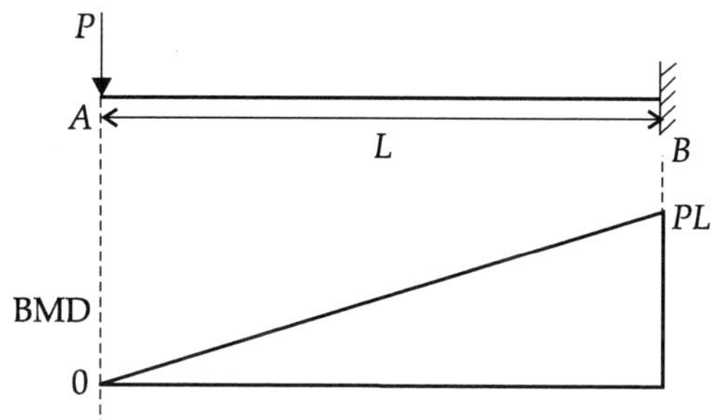

Fig. 10.42

Conjugate beam with M/EI loading (Fig. 10.43)

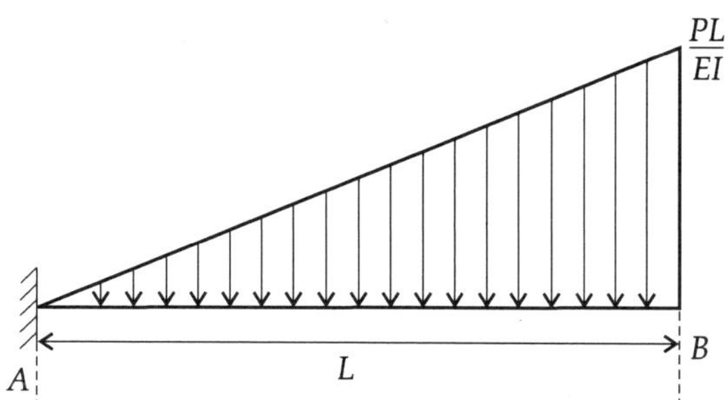

Fig. 10.43

Slope at A in real beam is equal to shear at A in conjugate beam

$$\theta_A = V_A = \frac{1}{2}L\frac{PL}{EI} = \frac{PL^2}{2EI}$$

Deflection at A in real beam is equal to moment at A in conjugate beam

$$\Delta_A = M_A = \frac{1}{2}L\frac{PL}{EI} \times \left(\frac{2L}{3}\right) = \frac{PL^3}{3EI}$$

Example 10.20: Find the deflection and slope when uniformly distributed load is over the full length of the cantilever beam.

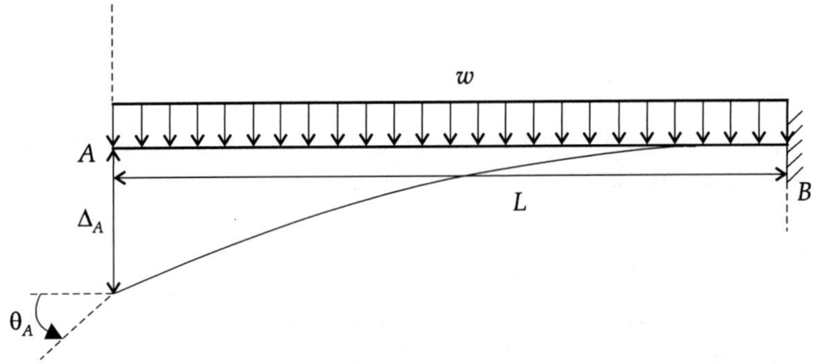

Fig. 10.44

Bending moment diagram (Fig. 10.45)

Draw the bending moment diagram

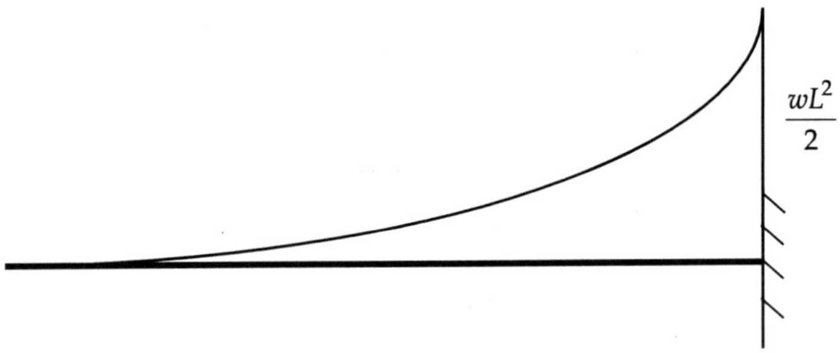

Fig. 10.45

Conjugate beam with M/EI loading (Fig. 10.46)

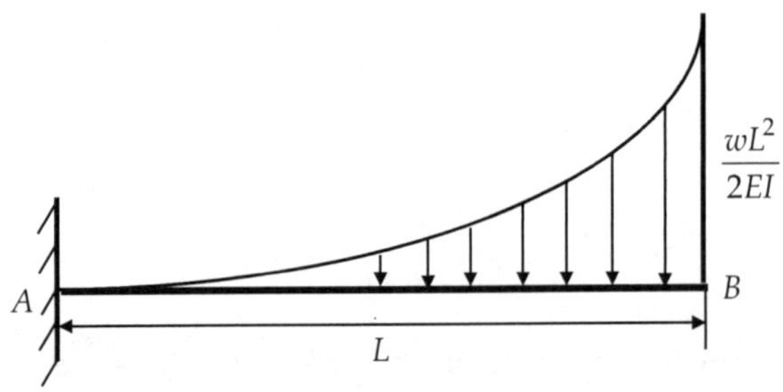

Fig. 10.46

Slope at A in real beam is equal to shear at A in conjugate beam

$$\theta_A = V_A = \frac{1}{3}\frac{wL^2}{2EI} \times L = \frac{wL^3}{6EI}$$

Deflection at A in real beam is equal to moment at A in conjugate beam

$$\Delta_A = M_A = \frac{3}{4}\frac{wL^3}{6EI} \times L = \frac{wL^4}{8EI}$$

Example 10.21: Consider the cantilever beam AB of length L subjected to a concentrated load P as shown in Fig. 10.47. Using the conjugate beam method, determine the slope and displacement at free end A.

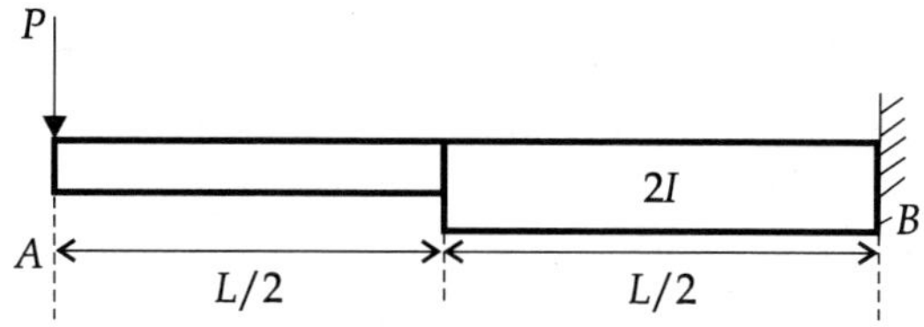

Fig. 10.47

Solution: Bending moment diagram

Draw M/EI diagram considering moment of inertia I for total beam.

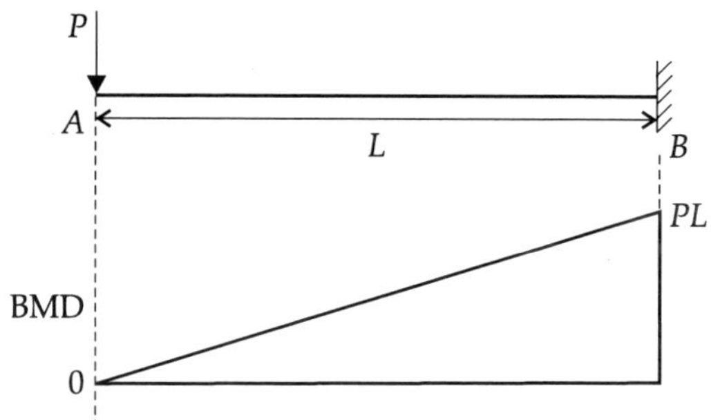

Fig. 10.48

Conjugate beam with M/EI loading (Fig. 10.49)

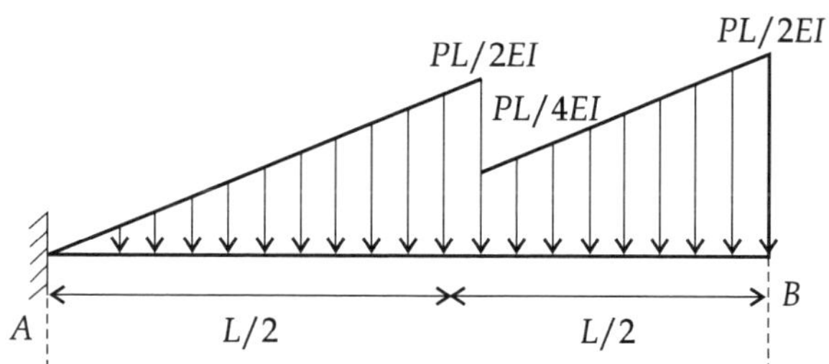

Fig. 10.49

Slope at A in real beam is equal to shear at A in conjugate beam

$$\theta_A = V_A = \frac{1}{2}\frac{L}{2}\frac{PL}{2EI} + \frac{1}{2}\frac{L}{2}\frac{PL}{4EI} + \frac{L}{2}\frac{PL}{4EI} = \frac{5PL^2}{16EI}$$

Deflection at A in real beam is equal to moment at A in conjugate beam

$\Delta_A = M_A = \theta_A = V_A$

$$= \frac{1}{2}\frac{L}{2}\frac{PL}{2EI} \times \left(\frac{2}{3}\frac{L}{2}\right) + \frac{1}{2}\frac{L}{2}\frac{PL}{4EI} \times \left(\frac{L}{2} + \frac{2}{3}\frac{L}{2}\right) + \frac{L}{2}\frac{PL}{4EI} \times \left(\frac{L}{2} + \frac{1}{2}\frac{L}{2}\right)$$

$$= \frac{PL^3}{24EI} + \frac{43PL^3}{48EI} + \frac{3PL^3}{32EI} = \frac{99PL^3}{96EI} = \frac{33PL^3}{32EI}$$

Example 10.22: A simply supported beam *ABCD* is loaded with a concentrated load *P* at the centre as shown in Fig. 10.50 below, the segments *AB* and *CD* have flexural rigidity *EI* and the segment *BC* have flexural rigidity 2*EI*. Using the conjugate beam method, determine the slope at support end A and displacement at centre *O*.

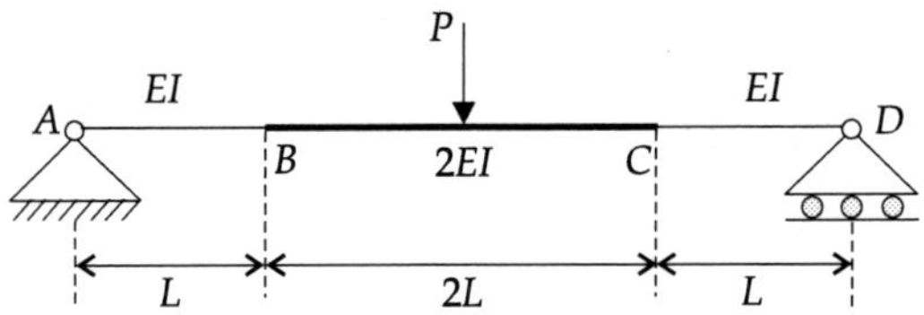

Fig. 10.50

Solution: Draw bending moment diagram of the simply supported beam.

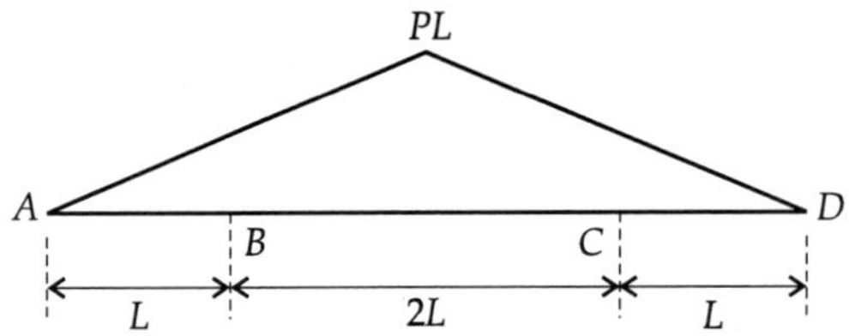

Fig. 10.51

Then net M/EI diagram of beam is

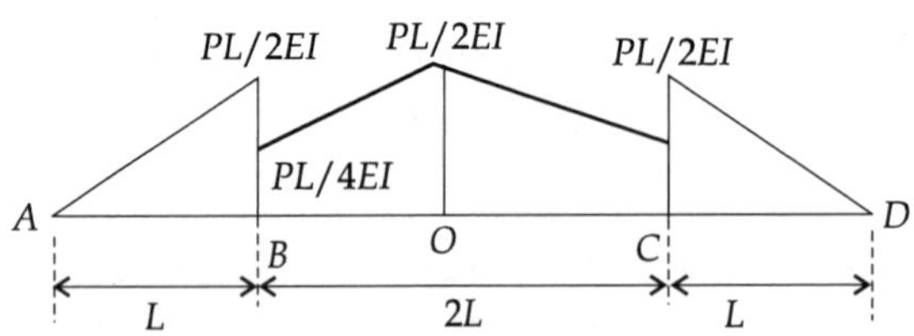

Fig. 10.52

Due to symmetry of diagram

Slope at A

θ_A = Area of beam section (AO)

$$= \frac{1}{2} \times L \times \frac{PL}{2EI} + \frac{1}{2} \times L \times \frac{PL}{4EI} + L \times \frac{PL}{4EI} = \frac{5PL^2}{8EI}$$

Deflection at O = Maximum deflection at centre

$$\Delta_O = \Delta_{max} = \text{Area } (AO) \times \bar{x}$$

$$= \frac{1}{2} \times L \times \frac{PL}{2EI} \times \left(\frac{2}{3}L\right) + \frac{1}{2} \times L \times \frac{PL}{4EI} \times \left(L \times \frac{2}{3}L\right) + L \times \frac{PL}{4EI} \times \left(L \times \frac{1}{2}L\right)$$

$$= \frac{PL^3}{6EI} + \frac{5PL^3}{24EI} + \frac{3PL^3}{8EI} = \frac{6PL^3}{8EI}$$

Example 10.23: Consider the cantilever beam AB of length L subjected to a concentrated load P as shown in Fig. 10.53. Using conjugate beam method, determine the slope and displacement at free end A and loading point C.

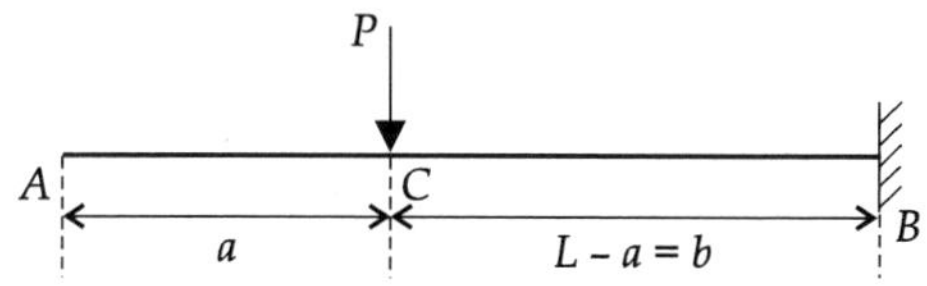

Fig. 10.53

Solution: Free-body diagram (Fig. 10.54)

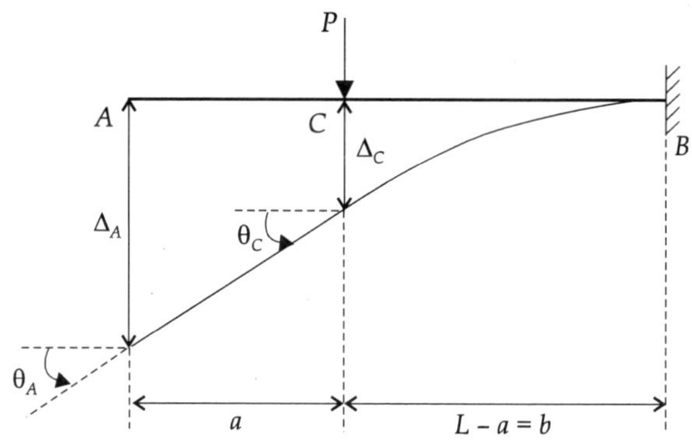

Fig. 10.54

Bending moment diagram (Fig. 10.55)

Draw the bending moment diagram

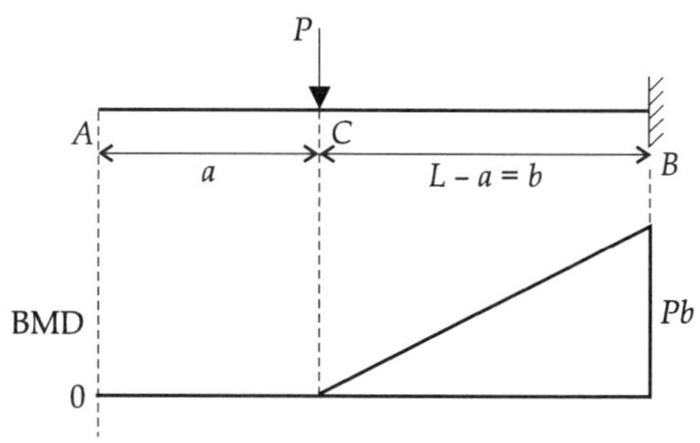

Fig. 10.55

Conjugate beam with M/EI loading (Fig. 10.56)

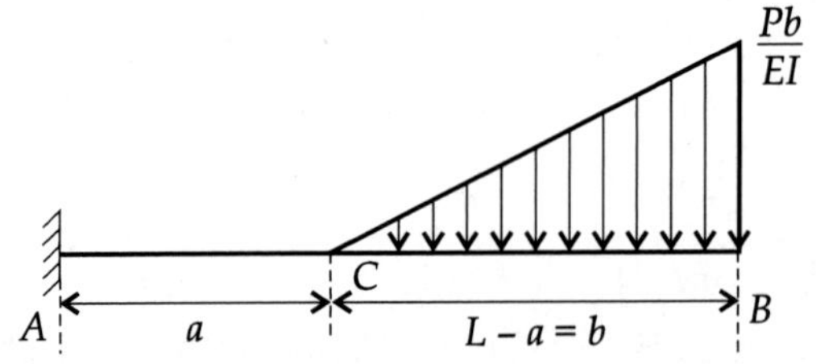

Fig. 10.56

Slope at C in real beam is equal to shear at C in conjugate beam

$$\theta_C = V_C = \frac{1}{2} b \frac{Pb}{EI} = \frac{Pb^2}{2EI}$$

Slope at A in real beam is equal to shear at A in conjugate beam

$$\theta_A = V_A = \frac{1}{2} b \frac{Pb}{EI} = \frac{Pb^2}{2EI}$$

Note: The deflected curve between A to C is linear since there is no load between A to C, therefore slope at A and C will be same.

Deflection at C in real beam is equal to moment at C in conjugate beam

$$\Delta_C = M_C = \frac{1}{2} b \frac{Pb}{EI} \times \frac{2b}{3} = \frac{Pb^3}{3EI}$$

Deflection at A in real beam is equal to moment at A in conjugate beam

$$\Delta_A = M_A = \frac{1}{2} b \frac{Pb}{EI} \times \left(a + \frac{2b}{3} \right) = \frac{Pb^2(2L + a)}{6EI}$$

Special case (Fig. 10.57): When load is at free end A ($a = 0$)

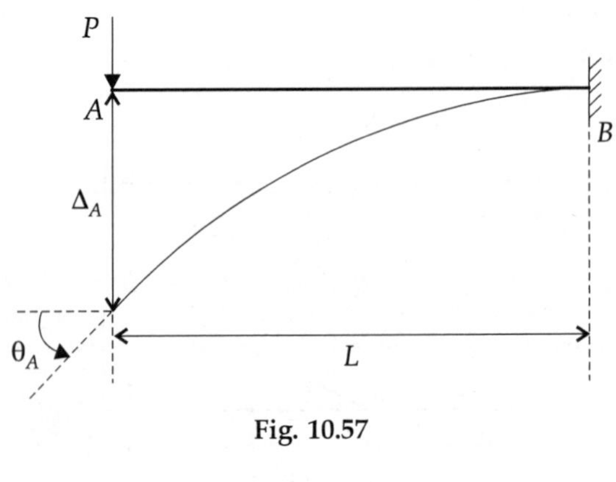

Fig. 10.57

Example 10.24: Consider the cantilever beam AB of length L subjected to uniformly distributed load as shown in Fig. 10.58. Using the conjugate beam method, determine the slope and displacement at free end A.

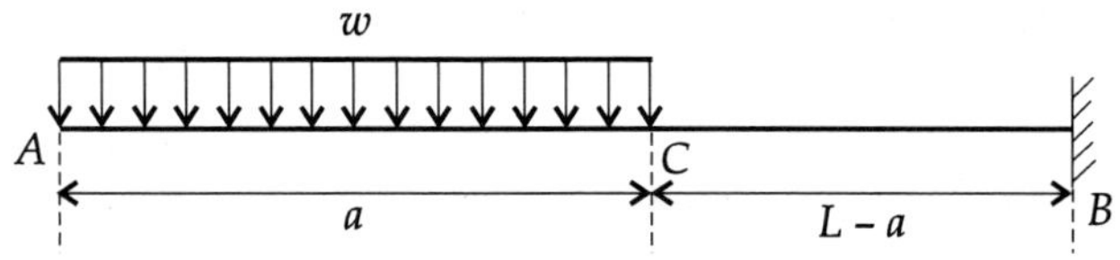

Fig. 10.58

Solution: Free-body diagram (Fig. 10.59)

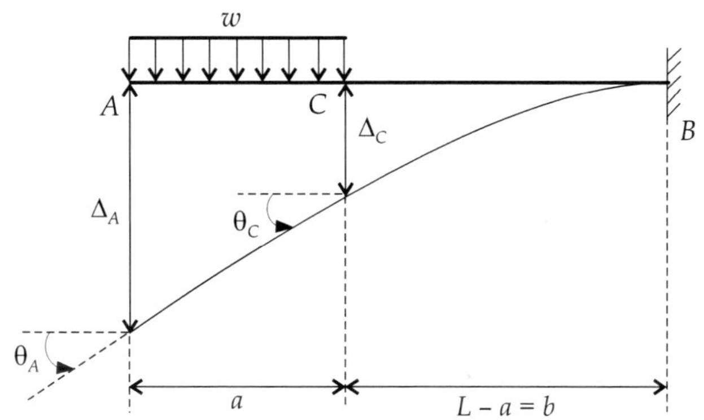

Fig. 10.59

Bending moment diagram (Fig. 10.60)

Draw the bending moment diagram

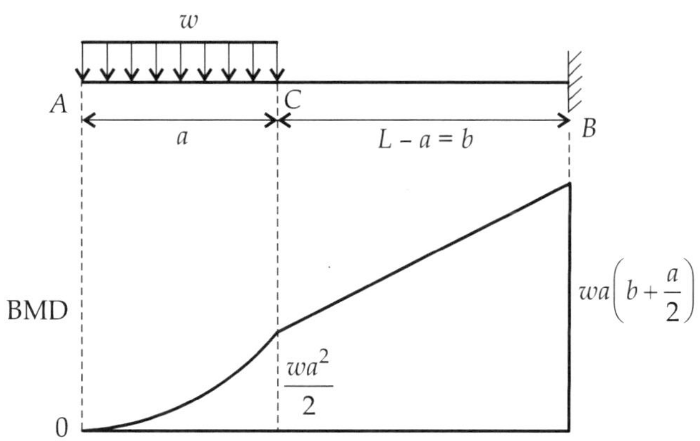

Fig. 10.60

Conjugate beam with M/EI loading (Fig. 10.61)

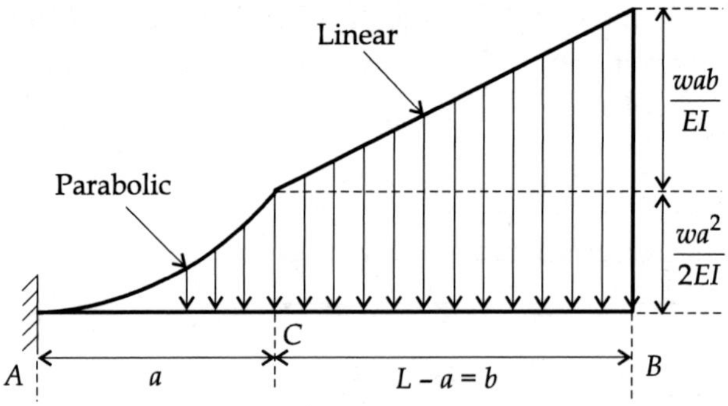

Fig. 10.61

Slope at C in real beam is equal to shear at C in conjugate beam

$$\theta_C = V_C = \frac{wa^2}{2EI} \times b + \frac{1}{2} \times \frac{wab}{EI} \times b = \frac{wab(a+b)}{2EI}$$

Slope at A in real beam is equal to shear at A in conjugate beam

$$\theta_A = V_A = \frac{wab(a+b)}{2EI} + \frac{1}{3}\frac{wa^2}{2EI} \times a = \frac{w\left[3ab(a+b)+a^3\right]}{6EI}$$

Note: The deflected curve between A to C is linear since there is no load between A to C, therefore slope at A and C will be same.

Deflection at C in real beam is equal to moment at C in conjugate beam,

$$\Delta_C = M_C = \frac{wa^2b}{2EI} \times \frac{b}{2} + \frac{wab^2}{2EI} \times \frac{2}{3}b = \frac{wab^2(3a+4b)}{12EI}$$

Deflection at A in real beam is equal to moment at A in conjugate beam,

$$\Delta_C = M_A = \frac{wa^2b}{2EI} \times \left(a + \frac{b}{2}\right) + \frac{wab^2}{2EI} \times \left(a + \frac{2}{3}b\right) + \frac{wa^3}{6EI} \times \frac{3}{4}a$$

$$= \frac{wab}{2EI}\left(a^2 + \frac{3ab}{2} + \frac{2b^2}{3}\right) + \frac{wa^4}{4EI}$$

10.6 CASTIGLIANO'S THEOREM FOR BEAM DEFLECTION (PARTIAL DERIVATIVE METHOD)

Castigliano's second theorem which forms the basis of the partial derivative method for finding the slope and deflection at any point of the elastic curve of the beam.

Castigliano's second theorem

The deflection at a point of a beam is the first partial derivative of the total internal energy of the beam with respect to external load applied at that point.

The Castigliano's theorem can be applied when the supports of the structure are unyielding and the temperature of the structure is constant.

For an elastic structure if the strain energy stored in the structure U due to external load P, then the deflection of the structural system in the direction of the external load will be $\delta = \dfrac{\partial U}{\partial P}$

Example 10.25: Find the free end displacement of the cantilever beam subjected to a concentrated load P at the tip. The flexural rigidity of the beam is EI uniform throughout the beam of length L.

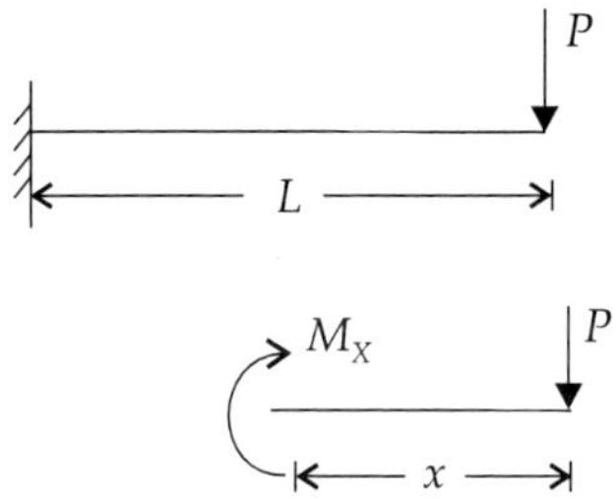

Fig. 10.62

The strain energy stored in the beam, $U = \int_0^L \dfrac{M^2 ds}{2EI}$

Moment at a distance x from the free end, $M = -Px$

Strain energy stored in the beam due to this loading, $U = \int_0^L \dfrac{P^2x^2\,dx}{2EI} = \dfrac{P^2L^3}{6EI}$

AS per the Castiliagon's theorem, the first partial derivatives of the strain energy with respect to external force will give the deflection of that point in the direction of the load acting.

Here, $$U = \frac{P^2L^3}{6EI}$$

$$\frac{\partial U}{\partial P} = \frac{PL^3}{3EI}$$

The deflection of beam at the free end, $\delta = \dfrac{PL^3}{3EI}$

Example 10.26: Determine tip deflection of the cantilever beam due to a uniform distributed load (w) as shown in Figure 10.63. The flexural rigidity of the beam is EI uniform.

The bending moment at any section at a distance x form the free end, $M = -Qx - wx$

$$M = -\frac{wx^2}{2} - Qx$$

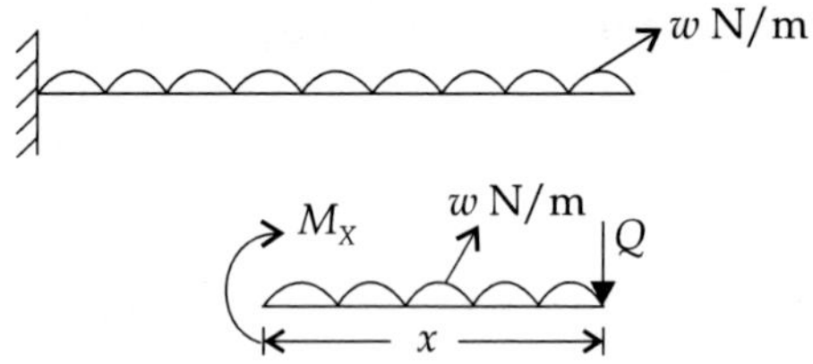

Fig. 10.63

The strain energy stored,

$$U = \int_0^L \frac{M^2 ds}{2EI}$$

The strain energy stored,

$$U = \int_0^L \frac{\left(-\frac{wx^2}{2} - Qx\right)^2}{2EI} dx$$

Take the terms containing only first order of Q and put $Q = 0$ in higher order terms of Q

Hence, $$U = \frac{QwL^4}{8EI}$$

$$\frac{\partial U}{\partial Q} = \frac{wL^4}{8EI}$$

The tip deflection of the cantilever beam due to udl w, $\delta = \dfrac{wL^4}{8EI}$

Example 10.27: A cantilever beam of uniform flexural rigidity EI is in the shape of a quadrant of a circle of radius R is loaded with P as shown in the Fig. 10.64. Find the vertical and horizontal deflections of the sector due to concentrated load, P.

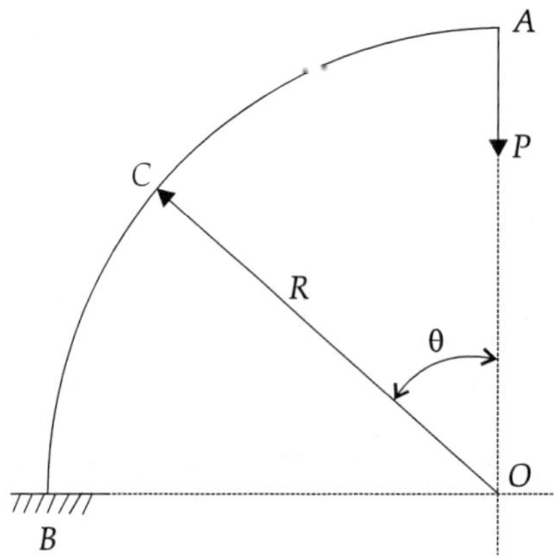

Fig. 10.64

The coordinate at point C taking O as origin $(-R \sin \theta, -R \cos \theta)$

The coordinate of point C, taking O as origin, $x = -R \sin \theta$ and $y = -R \cos \theta$

The bending moment at any point C, $M = PR \sin \theta$

Strain energy stored in the curved beam, $U = \int \frac{M^2 ds}{2EI}$

Here, $ds = R d\theta$

$$U = \int_0^{\frac{\pi}{2}} \frac{(P^2 R^2 \sin^2 \theta)}{2EI} R d\theta = \frac{P^2 R^3}{2EI} \int_0^{\frac{\pi}{2}} \sin^2 \theta \, d\theta$$

Strain energy,

$$\int_0^{\frac{\pi}{2}} \sin^2 \theta \, d\theta = \int_0^{\frac{\pi}{2}} \frac{1}{2}[1 - \cos 2\theta] d\theta = \frac{\pi}{4}$$

The strain energy stored, $U = \frac{\pi P^2 R^3}{8EI}$

The vertical deflection of the point A in the direction of the load P, $\frac{\partial U}{\partial P} = \frac{\pi P R^3}{4EI}$

For horizontal deflection at point A, introduce an imaginary load Q at A.

Now the bending moment at C is

$M = -PR \sin \theta - QR(1 - \cos \theta)$

The strain energy stored,

$$U = \int_0^{\frac{\pi}{2}} \frac{M^2 dx}{2EI} = \int_0^{\frac{\pi}{2}} \frac{(-PR \sin \theta - QR(1 - \cos \theta))^2 R d\theta}{2EI}$$

Now $\int_0^{\frac{\pi}{2}} \frac{(P^2 R^2 \sin^2 \theta)}{2EI} R d\theta = \frac{\pi P^2 R^3}{8EI}$

And $\int_0^{\frac{\pi}{2}} \frac{2PQR^3 \sin \theta (1 - \cos \theta)}{2EI} d\theta = \frac{PQR^3}{2EI}$

$$\int_0^{\frac{\pi}{2}} \frac{Q^2 R^3 (1 - \cos \theta)^2 d\theta}{2EI} = \frac{Q^2 R^3}{2EI}\left(\frac{3\pi}{4} - 2\right)$$

$$U = \frac{\pi P^2 R^3}{8EI} + \frac{PQR^3}{2EI} + \frac{Q^2 R^3}{2EI}\left(\frac{3\pi}{4} - 2\right)$$

$$\frac{\partial U}{\partial Q} = \frac{PR^3}{2EI} + \frac{QR^3}{EI}\left(\frac{3\pi}{4} - 2\right)$$

As there is no external load in the horizontal direction, hence put $Q = 0$

Horizontal deflection, $\delta_h = \dfrac{PR^3}{2EI}$

10.7 SLOPE-DEFLECTION OF COMMON BEAMS

Beam loading	Slope/deflection
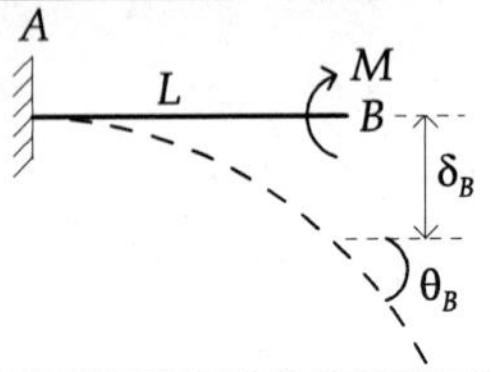	$\theta_B = \dfrac{ML}{EI}$ $\delta_B = \dfrac{ML^2}{2EI}$
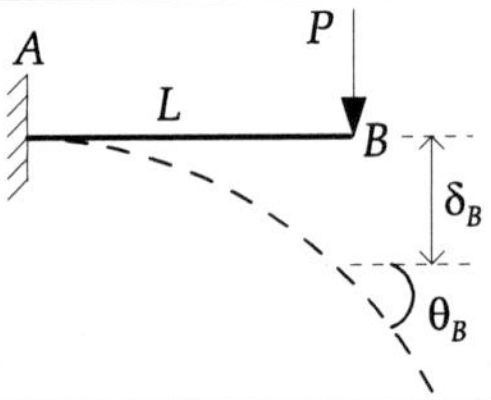	$\theta_B = \dfrac{PL^2}{2EI}$ $\delta_B = \dfrac{PL^3}{3EI}$
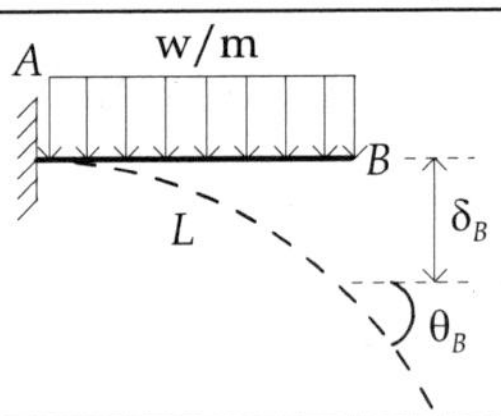	$\theta_B = \dfrac{wL^3}{6EI}$ $\delta_B = \dfrac{wL^4}{8EI}$
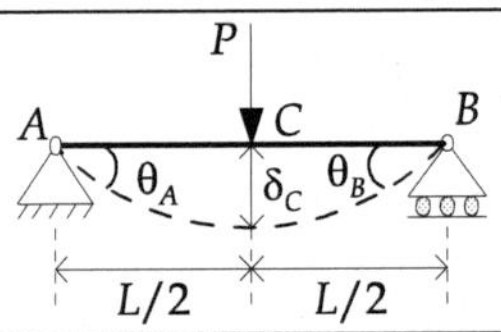	$\theta_A = \theta_B = \dfrac{PL^2}{16EI}$ $\delta_C = \dfrac{PL^3}{48EI}$
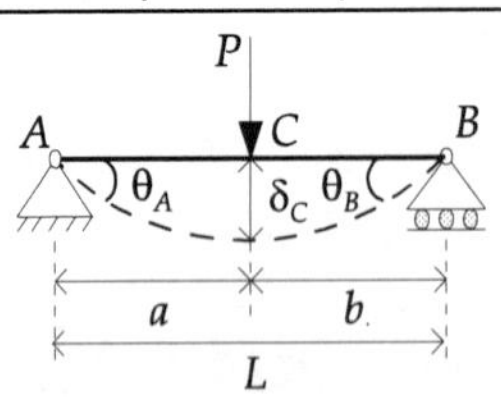	$\theta_A = \dfrac{Pab(2b+a)}{6LEI}$ $\theta_B = \dfrac{Pab(2a+b)}{6LEI}$ $\delta_C = \dfrac{Pa^2b^2}{3LEI}$
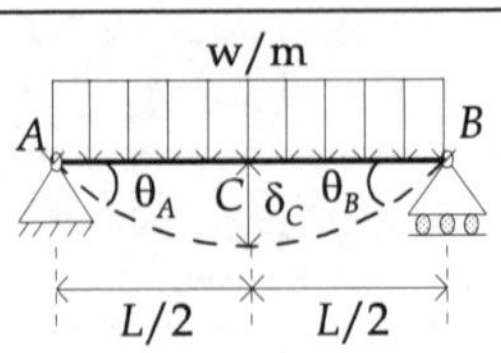	$\theta_A = \theta_B = \dfrac{wL^3}{24EI}$ $\delta_C = \dfrac{5wL^4}{384EI}$

PROBLEMS

1. Calculate the distance of the point of contraflexure from the free end *C*.

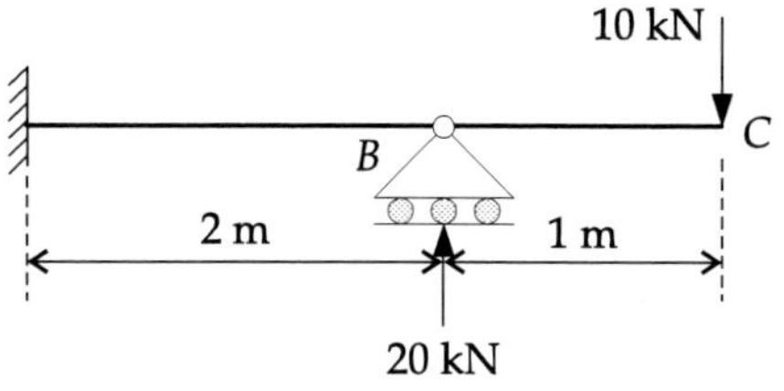

Fig. 10.65

Ans. Since inflection point is the point where bending moment is zero, let it be at a distance x from point C, then $10 \times x = 20 \times (x - 1) \Rightarrow x = 2$ m

2. How many inflection points are there in the beam given below.

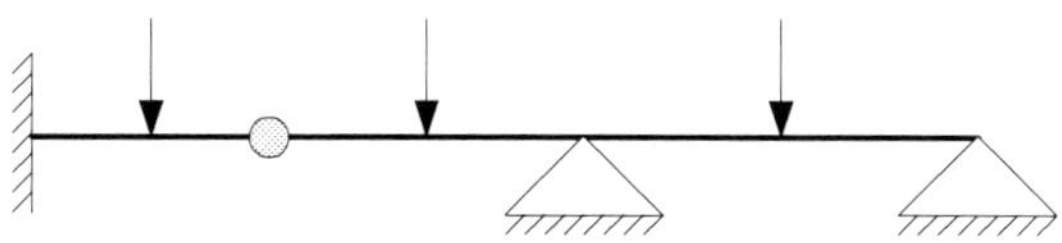

Fig. 10.66

Ans. (3) Inflection point is the point where bending moment is zero and deflection curve changes its sign.

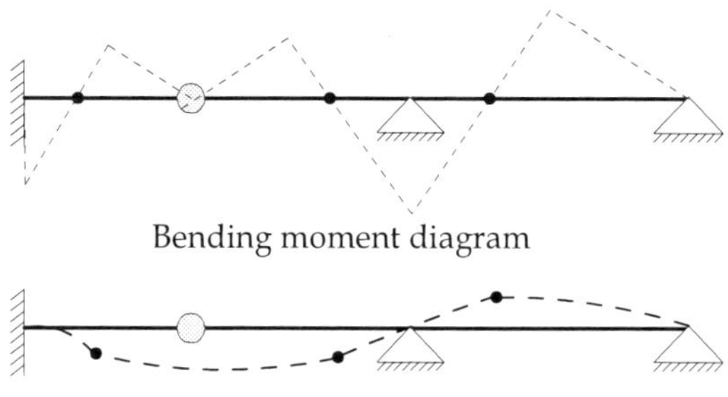

Fig. 10.67

3. Calculate the slope and deflection at point *D* in the beam given below.

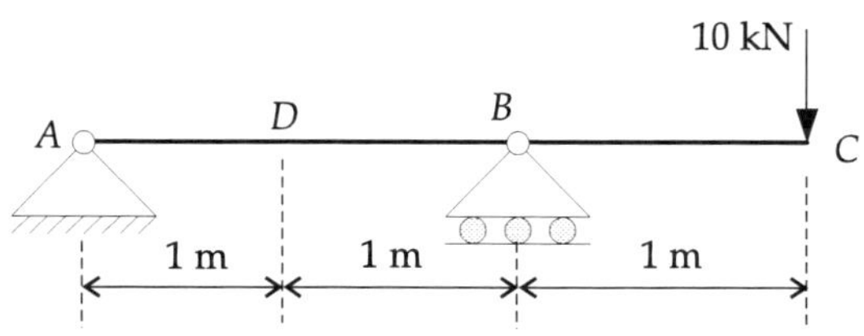

Fig. 10.68

Ans. $\theta_D = \dfrac{-5}{6EI}$ $\Delta_D = \dfrac{-2.5}{EI}$

4. Consider the frame in which AB is a straight beam of length L and BC is circular quadrant arch with radius ($R = \frac{L}{2}$) as shown in the figure below. Calculate the vertical deflection of point C.

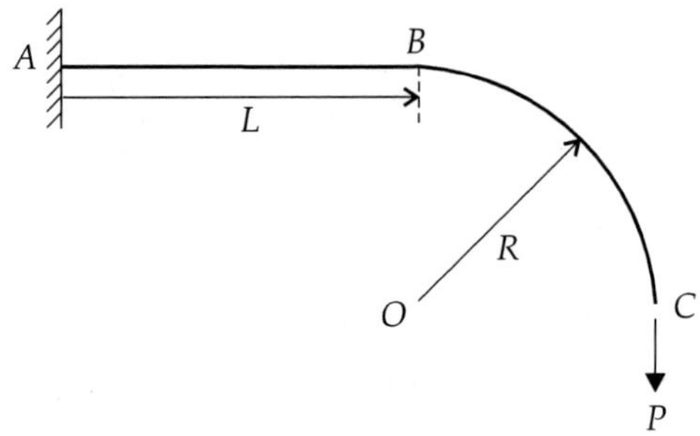

Fig. 10.69

Ans. $\Delta_{C(\text{vertical})} = \frac{PR^3}{EI}\left(\frac{3\pi}{4} + \frac{20}{3}\right)$

5. Consider the arrangement of simply supported beams as given below, calculate the vertical deflection at point R in beam AB.

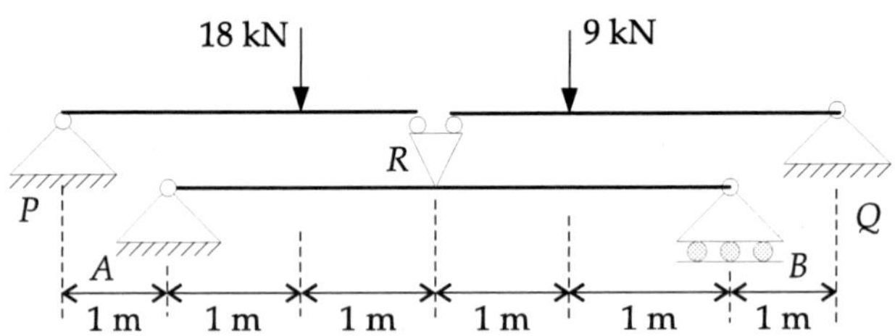

Fig. 10.70

Ans. $\delta_R = \frac{24}{EI}$

6. Consider the beam as shown in the figure below, calculate the vertical deflection of support B.

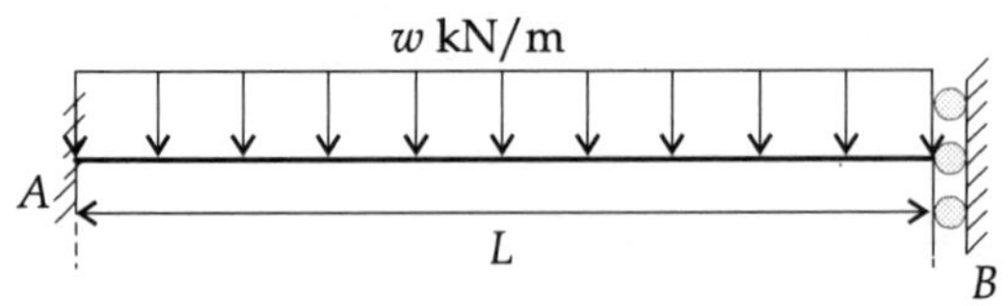

Fig. 10.71

Ans. $\Delta_B = \frac{-wL^4}{24EI}$

7. Consider the propped beam as shown below, if the propped support is rigid then calculate the magnitude of vertical reaction at propped support B.

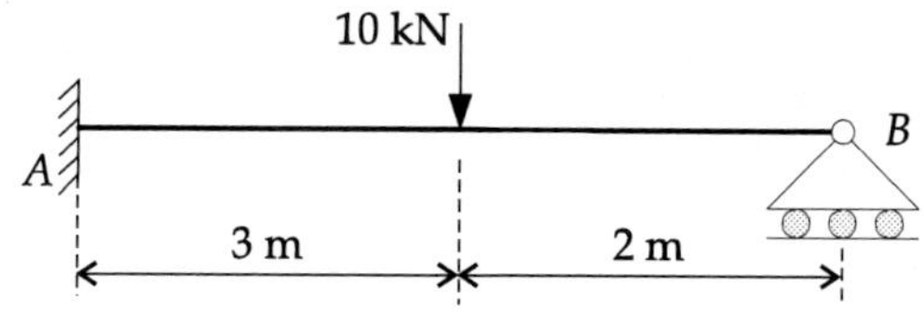

Fig. 10.72

Ans. $R_B = \dfrac{108}{25}$

8. A cantilever beam *ABCD* is loaded with a concentrated load *P* at free end as shown in the figure below, the segments *AB* and *CD* have flexural rigidity *EI* and the segment *BC* is rigid. Calculate the slope and deflection at *C*.

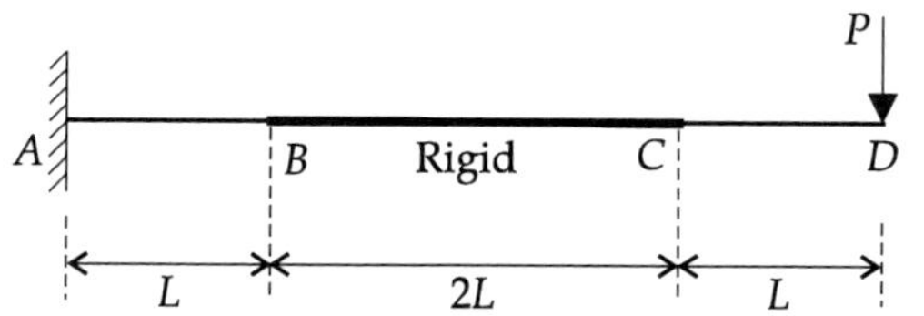

Fig. 10.73

Ans. Slope and deflection at *C* is $\dfrac{7PL^2}{2EI}$ and $\dfrac{53PL^3}{6EI}$

9. Consider the simply supported beam loaded with two end couples, as shown in the figure below, calculate the slope of beam at supports *A* and *B*.

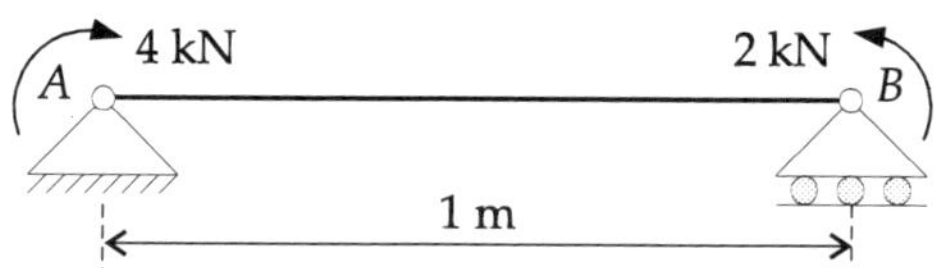

Fig. 10.74

Ans. $\theta_A = \dfrac{-5}{3EI}$ and $\theta_B = \dfrac{4}{3EI}$

11
Pure Bending and Bending Stresses

11.1 INTRODUCTION

Beam is a structural member which is primarily subjected to bending due to externally applied loads. A long slender prismatic bar is called a beam if the external loads are applied transverse to the axis of the bar. As a result of loading the forces and moments develop at the cross sections of a beam which may vary along the length (axis) of the beam. These internal forces and moments give rise to stresses, strains or deformations which produce bending in the beams.

The loads acting on a beam produce bending moment which tends to bend the beam along the section, and produces internal stresses called **bending stresses**. To accomplish this, a beam is analyzed for equilibrium using equations of statics that incorporate the applied forces and reaction forces at its support, and help in deciding the load carrying capacity of a beam.

The cross section of the beam is symmetrical about the plane of bending (i.e., plane in which the forces and moments act (Fig. 11.1).

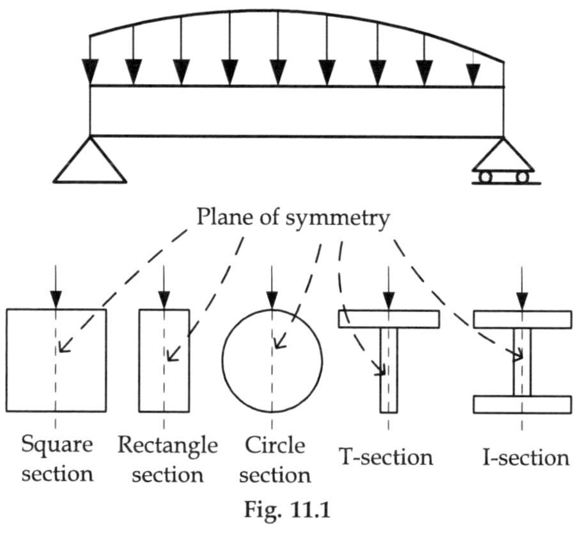

Fig. 11.1

Assumptions

To determine the stresses in beams certain fundamentals assumptions are made as follows:

- The beam is made of linear-elastic, isotropic homogeneous and the material follows Hooke's law.
- The beam is initially straight and unstressed.
- A transverse section of the beam which is plane before bending remains plane after bending.
- The loads are applied in the plane of bending.
- Young's modulus is same in tension and compression.
- Each layer is free to expand or contract without hindrance from adjacent layers.
- The lateral surface stresses are negligible. The lateral stress within the beam and the shear stresses between adjacent "layers" throughout the depth of the beam are ignored.

11.2 PURE BENDING OF BEAMS

A special case when the loads (P) are applied on the beam as shown in Fig. 11.2 at equal distance from the support. The shear force diagram and bending moment diagram (BMD) are drawn for this loading. From the BMD it is obvious that the bending moment is constant and shear force is zero in the middle portion of the beam (*bc* portion). As in the middle portion, there is no shear force present so the beam bends only for bending moment in that portion. This is an example of pure bending.

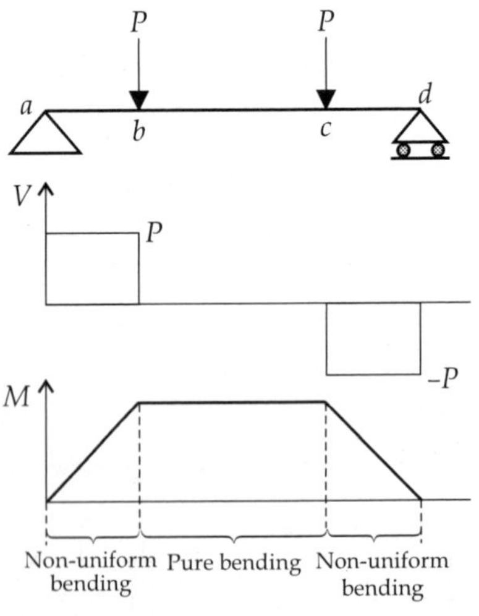

Fig. 11.2

Pure bending refers (Fig. 11.2) to flexure of a beam under a constant bending moment, that is, shear force will be zero (beam section *c*-*d*). Non-uniform bending refers to flexure in the presence of shear forces or variable bending moment (beam sections *a-b* and *c-d*).

In pure bending only normal stresses act on the cross section while in non-uniform bending both normal and shear stresses are developed in the beam. In case of pure bending, the deflected shape of the beam is taken as circular arc.

Neutral axis

The nature of stresses in fibres depends upon the nature of bending (Fig. 11.3). So under the action of sagging bending moment the beam bends concave upwards, then the layers at the top of the beam will be shortened (in compression) and the bottom layers will extend (in tension).

On the other hand, if the beam is under the action of hogging bending moment it bends concave downwards, then the layers at the bottom of the beam will be compressed (in compression) and the top layers will extend (in tension).

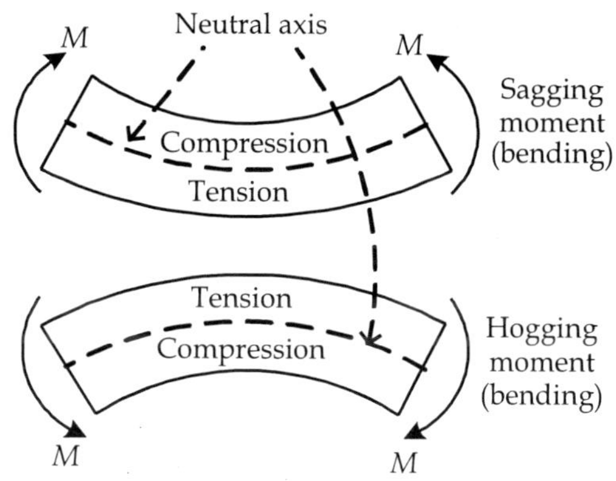

Fig. 11.3

Neutral surface: Somewhere between the top and bottom surfaces of the beam there is a surface, where there is no deformation and consequently no stress is called **neutral surface**. For small deflection of a beam, the length of the deformed beam measured at neutral surface along the length of the beam is same as the original straight beam, as no deformation takes place there.

Neutral axis: The perpendicular line on the longitudinal axis of the beam in the neutral surface is known as neutral axis.

Moment of resistance: The sum of the moments of the internal forces at any section must be equal to the external moment so that the beam is in equilibrium. This total moment developed internally is called moment of resistance. Moment of resistance normally refer of maximum stresses for a given condition of loading.

General Bending Equation

Bending moment generally varies along the length of the beam. However, it is reasonable to assume that bending moment is constant over a very small (infinitesimally small) length of the beam. So the case of such a small length dx of a beam subjected to a constant bending moment along its length may be regarded as pure bending.

Consider a small length (dx) of a beam subjected to a constant bending moment (M) over the length, dx. Consider a strip (MN) at a distance (y) from the neutral surface as shown in Fig. 11.4.

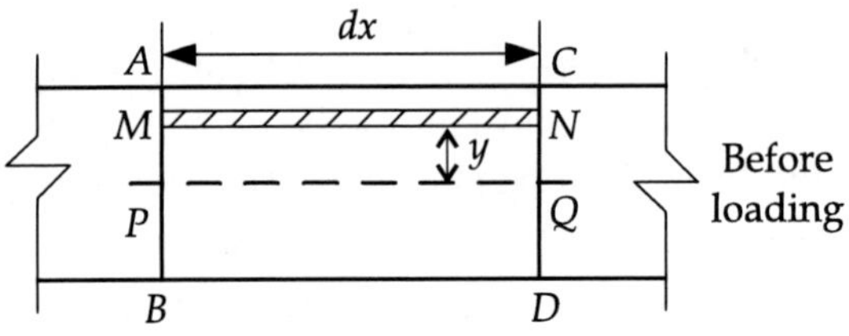

Fig. 11.4

As a result of the moment M, this small length of beam PQ bends into an arc $P'Q'$ of a circle with centre at O as shown in Fig. 11.5. Let plane $A'B'$ and $C'D'$ subtend an angle $d\theta$ at the centre of curvature O and R be the radius of curvature at the neutral surface $P'Q'$. The layer (MN) at a distance (y) from the neutral surface will take up the position ($M'N'$) after bending.

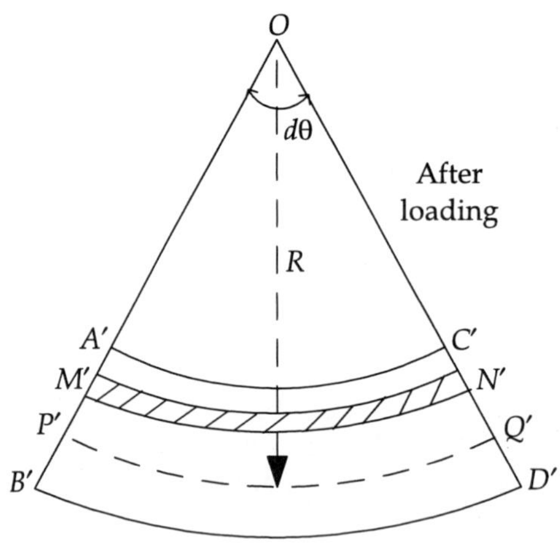

Fig. 11.5

Now, the arc length, $P'Q' = Rd\theta$.

Since no deformation occurs along the neutral surface, the length of neutral surface before bending and after bending is equal.

Length of neutral surface, $PQ = P'Q'$ and $MN = PQ$

After deformation length of $M'N' = (R - y)d\theta$

Therefore, change of length $= MN - M'N' = Rd\theta - (R - y)d\theta$

The strain of the fibre, $MN = yd\theta$

$$\varepsilon = \frac{\text{Change in length of } MN}{\text{Original length}} = \frac{yd\theta}{Rd\theta} = \frac{y}{R}$$

We know stress, σ = (strain) X (Young's modulus)

Hence, we get relation between radius of curvature and stress

$$\sigma = \frac{y}{R}E$$

Bending stresses at any point (y) across a section can be found using the above relation. The variation of bending stress of the section is shown in Fig. 11.6. At neutral axis the stress is zero, above the neutral axis stress is compressive and below it, tensile. The maximum stresses occur at the extreme fibre.

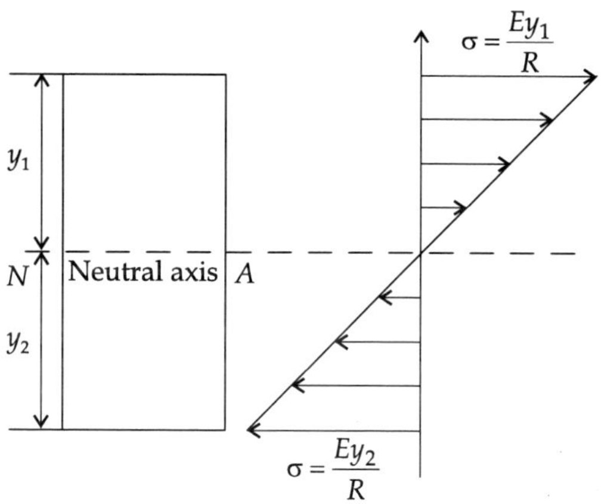

Fig. 11.6: A beam and its cross section has to be drawn

Let a beam be subjected to sagging moment M and the cross section of the beam at the point is as shown in Fig. 11.7. Now consider a strip of area (dA) at a distance (y) from the neutral axis parallel to neutral axis, NA.

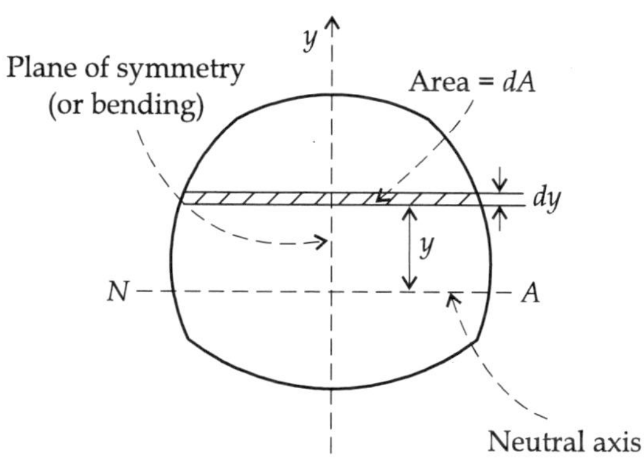

Fig. 11.7

The total compressive force acting in the element = Stress x Area of the strip

The bending stress at a distance y, $\sigma = \frac{y}{R}E$

Force acting in the element $dF = \sigma \times dA = \frac{E}{R} ydA$

Taking moment of the elemental force about the neutral axis $dM = y \times dF$

Considering a similar strip at a distance (y) at the bottom side from the neutral axis, we get the same expression.

The total couple moment acting in the section must be equal to the external moment, M.

Total couple acting in the section at a distance (y) from the neutral axis,

$$\int dM = \int_0^y y \frac{E}{R} ydA + \int_{-y}^{0} y \frac{E}{R} ydA$$

Total moment of the whole section, $M = \int_{y_1}^{y_2} y \frac{E}{R} ydA = \frac{E}{R}\int_{y_1}^{y_2} y^2 dA$

Now the integral, $I = \int y^2 dA$, is the second moment of inertia about the neutral axis.

So we can write $\frac{M}{I} = \frac{E}{R}$

Since $\frac{E}{R} = \frac{\sigma}{y}$

Therefore, the equation of general bending can be written as

$$\frac{M}{I} = \frac{\sigma}{y} = \frac{E}{R}$$

The bending stress at a distance (y) from the neutral axis,

$$\sigma = \frac{M}{I} y$$

Neutral axis of the section passes through centroid of the cross section

The total compressive force acting in the surface between the neutral axis and the top fibre,

$$F_c = \int_0^{y_2} \frac{Ey}{R} dA$$

Similarly, if we consider a strip below the neutral axis at a distance, y, then the total tensile force on the face between the neutral axis and the bottom fibre,

$$F_t = \int_0^{y_2} \frac{Ey}{R} dA$$

Now for equilibrium the total force on any cross section must be zero.

The total force of the surface of the cross section, $F = F_c + F_t$

Compressive force is negative and tensile force is positive.

Therefore,

$$-\int_0^{y_2} \frac{Ey}{R} dA + \int_0^{y_2} \frac{Ey}{R} dA = 0$$

$$\frac{E}{R} \int_{y_2}^{y_1} y dA = 0$$

But $\int_{y_1}^{y_2} y dA = 0$ is the first moment of area about the neutral axis and hence replaced by $A\bar{y}$, where, A is the cross-sectional area, and $\bar{y}$ is the distance between the neutral axis and the centroid. Hence,

$$\frac{E}{R} A\bar{y} = 0$$

Since $\frac{E}{R}$ and A cannot be zero, $\bar{y}$ must be zero.

Therefore, the neutral axis always passes through the centroid of the beam's cross section.

11.3 MAXIMUM BENDING STRESSES IN BEAMS

The bending equation may be used to calculate the maximum tensile and compressive stresses in a beam (Fig. 11.8).

For positive (or sagging) bending moment maximum tensile stress occurs at the bottom and maximum compressive stress occurs at the top edge of the beam. For a positive bending moment this corresponds to:

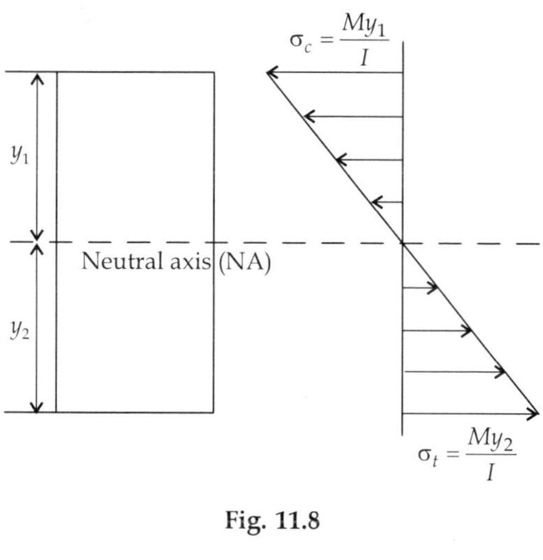

Fig. 11.8

$$\sigma_{max}\ (\text{compressive}) = \frac{M}{I_{xx}} y_1$$

$$\sigma_{max}\ (\text{tensile}) = \frac{M}{I_{xx}} y_2$$

Note:

- Normal stresses calculated above are applicable for both pure bending and non-uniform bending cases, because in non-uniform bending the impact of shear stresses is insignificant on flexural stresses.
- The flexural formula is applicable only when the shape of the beam is regular and loading is continuous. Discontinuity in loading and irregularity in beam's shape produces localized stresses called **stress concentrations** which are mush greater than flexural stresses.
- Neutral axis of a simple beam must pass through the beam;s cross section centroid.
- The magnitude of stress depends only on the moment (loads and their locations on the beam) and geometry (cross section) of the beam. It does not depend on the material from which the beam is made.
- To determine the stresses in a beam subjected to known bending moments, the only parameters needed are the cross section (shape) of the beam, the position of its centroid (to know the position of the neutral axis and to know y_1 and y_2) and the second moment of area, I_{xx}.
- *EI* (product of the modulus of elasticity of the material and the second moment of area of the cross section of the beam) is called the **flexural stiffness** of the beam. Compare this to the **torsional rigidity (GJ)** for a shaft under torsion. Both indicate the resistance of the beam/shaft to bending/torsion deflections.
- The strain and hence the stress distribution through the thickness of the beam are linear and have zero values at the neutral axis.

Section modulus

The relation for finding the stresses on the extreme fibres of the section, $M = \sigma \frac{I}{y}$, since maximum stresses occur at extreme ends, therefore, $M = \sigma_{max} \frac{I}{y_{max}} = \sigma_{max} Z$, this,

$Z = \frac{I}{y_{max}}$, is called **section modulus**.

Note:

- Strength or maximum stress of a beam for a given bending moment depends on the section modulus, i.e., the larger the value of the section modulus, stronger is the beam.

- For the sections with same section modulus and strength the section with smaller area is more economical.

Section modulus of some common cross sections (Fig. 11.9 to Fig. 11.15)

1. **Square**

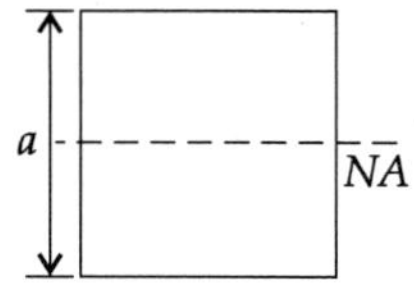

Fig. 11.9

Moment of inertia of the section about the neutral axis, $I = \dfrac{bd^3}{12} = \dfrac{a^4}{12}$

Distance to the extreme fibre, $y_{max} = \dfrac{d}{2} = \dfrac{a}{2}$

Therefore section modulus, $Z = \dfrac{I}{y_{max}} = \dfrac{a^3}{6}$

2. **Solid rectangle**

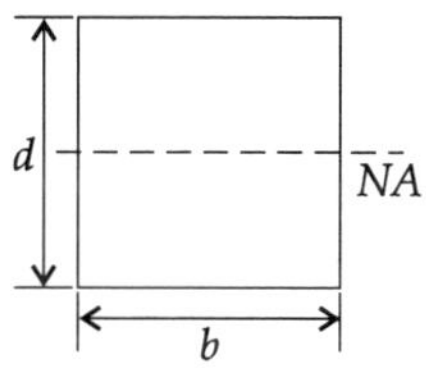

Fig. 11.10

Moment of inertia of the section about the neutral axis, $I = \dfrac{bd^3}{12}$

Distance to the extreme fibre, $y_{max} = \dfrac{d}{2}$

Therefore, section modulus, $Z = \dfrac{I}{y_{max}} = \dfrac{bd^2}{6}$

3. **Hollow rectangle**

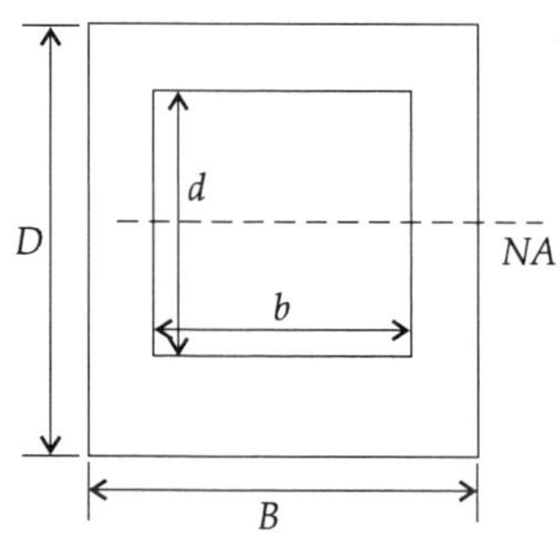

Fig. 11.11

Moment of inertia of the section about the neutral axis, $I = \frac{BD^3}{12} - \frac{bd^3}{12}$

Distance to the extreme fibre, $y_{max} = \frac{D}{2}$

Therefore, section modulus, $Z = \frac{I}{y_{max}} = \frac{BD^3 - bd^3}{6D}$

4. Solid circle

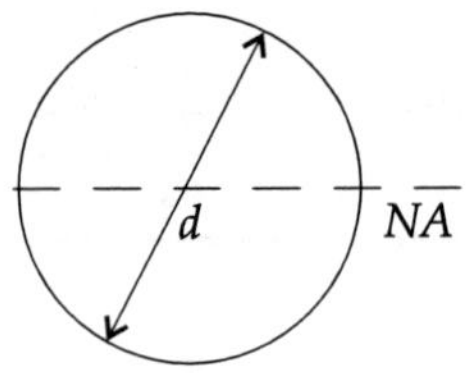

Fig. 11.12

Moment of inertia of the section about the neutral axis, $I = \frac{\pi d^4}{64}$

Distance to the extreme fibre, $y_{max} = \frac{d}{2}$

Therefore, section modulus, $Z = \frac{I}{y_{max}} = \frac{\pi d^3}{32}$

5. Hollow circle

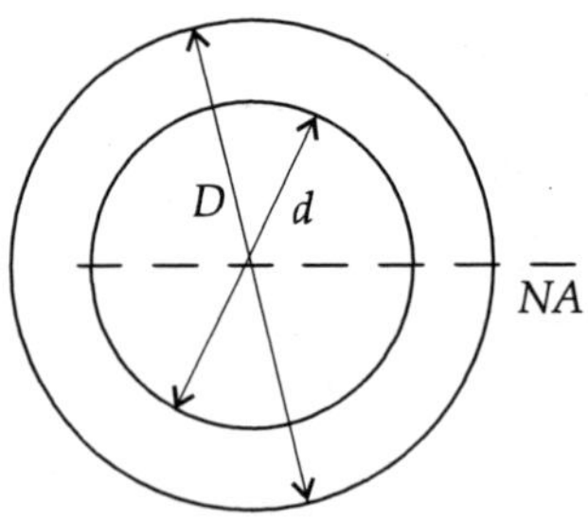

Fig. 11.13

Moment of inertia of the section about the neutral axis, $I = \frac{\pi D^4}{64} - \frac{\pi d^4}{64}$

Distance to the extreme fibre, $y_{max} = \frac{D}{2}$

Therefore, section modulus, $Z = \frac{I}{y_{max}} = \frac{\pi(D^4 - d^4)}{32D}$

6. Triangle

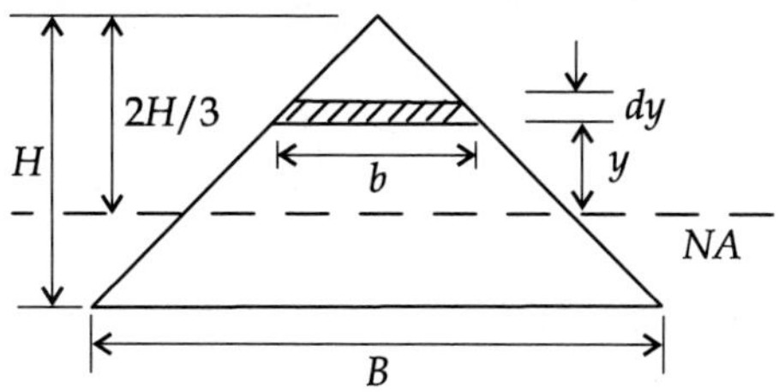

Fig. 11.14

Consider an elementary strip of thickness (dy) at an distance (y) from neutral axis. Let the width of the strip be b, then

$$\frac{b}{B}=\frac{2H/3-y}{H}\Rightarrow b=\frac{B}{H}(2H/3-y)$$

Moment of inertia of the section about the neutral axis,

$$I=\int_{-H/3}^{2H/3}(bdy)(y^2)=\int_{-H/3}^{2H/3}\frac{B}{H}\left(\frac{2H}{3}-y\right)y^2dy\Rightarrow I=\frac{BH^3}{36}$$

Distance to the extreme fibre, $y_{max}=\dfrac{2H}{3}$

Therefore, section modulus, $Z=\dfrac{I}{y_{max}}=\dfrac{BH^2}{24}$

7. Square placed diagonally

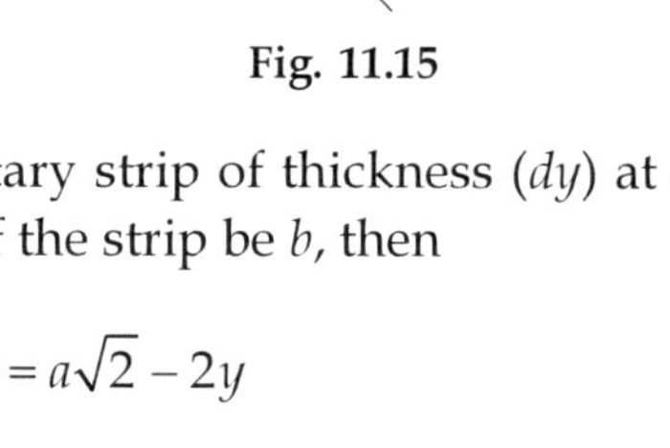

Fig. 11.15

Consider an elementary strip of thickness (dy) at an distance (y) from neutral axis. Let the width of the strip be b, then

$$\frac{b}{a\sqrt{2}}=\frac{a/\sqrt{2}-y}{a/\sqrt{2}}\Rightarrow b=a\sqrt{2}-2y$$

Moment of inertia of the section about the neutral axis,

$$I=2\int_{0}^{y}(bdy)(y^2)=2\int_{0}^{a/\sqrt{2}}(a\sqrt{2}-2y)y^2dy\Rightarrow I=\frac{a^4}{12}$$

Distance to the extreme fibre, $y_{max} = \frac{d}{2} = \frac{a}{\sqrt{2}}$

Therefore,section modulus, $Z = \frac{I}{y_{max}} = \frac{a^3}{6\sqrt{2}}$

Composite beams

A beam consists of two different materials is called composite or flitched beam (Fig. 11.16).

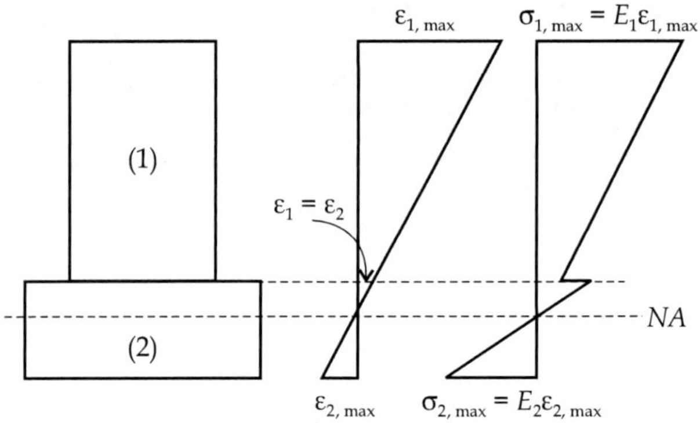

Fig. 11.16

Following points should be considered for the analysis of composite beams:

- Let E_1 and E_2 be the moduli of elasticity of two materials. Then, modular ratio, $m = \frac{E_1}{E_2}$
- Beam has common neutral axis.
- Strain at the junction of two materials must be same, i.e., $\varepsilon_1 = \varepsilon_2 \Rightarrow \frac{\sigma_1}{\sigma_2} = \frac{E_1}{E_2}$
- Strain is directly proportional to distance from neutral axis, i.e., $\frac{\varepsilon_{1,max}}{\varepsilon_{2,max}} = \frac{y_1}{y_2}$
- Total moment resist by the beam is equal to the sum of moments resist by two composite materials, i.e., $M_T = M_1 + M_2 = \frac{\sigma_{1,max}}{y_1} I_1 + \frac{\sigma_{2,max}}{y_2} I_2$

11.4 SHEAR STRESS IN BEAM

Whenever a beam is subjected to bending moment, bending normal stress (σ) is developed at any layer of the beam from the neutral plane along the longitudinal direction of the beam. The unbalanced force developed between two layers parallel to neutral plane due to bending stress must be counterbalanced by developing a shear stress.

Let us consider a beam subjected to bending moments (M and $M + dM$) at two planes (AA' and BB'), that is, (dx) distance apart along the longitudinal direction of the beam, Fig. 11.17. Consider a small strip (dA) at yz plane at a distance (y) from the neutral axis. Where, $dA = bdy$ and b is the width of the beam at location, y.

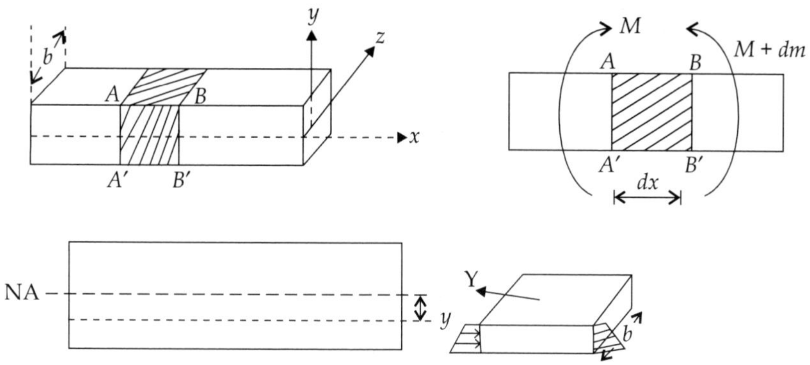

Fig. 11.17

To balance this force shear stress (τ) is developed at a distance (y) from neutral axis along the xz plane that is parallel to the neutral plane.

We have, $\sigma = \dfrac{M}{I}y$ and $\sigma + d\sigma = \dfrac{M + dM}{I}y$

Resultant shear force along the layers at a distance y,

$$(\sigma + d\sigma)dA - \sigma dA = \tau bdx$$

The end fibre distance is C_1 from neutral axis

$$\int_y^{c_1} \frac{(M + dM)}{I} ybdy - \int_y^{c_1} \frac{(M)}{I} ybdy = \tau bdx$$

$$\frac{dM}{I} b \int_y^{c_1} ydy = \tau bdx$$

$$\tau = \frac{1}{Ib}\frac{dM}{dx}\int_y^{c_1} ydA$$

This shear stress must be accompanied by a complementary shear stress along the yz plane

Again we know for shear force, $V = dM/dx$

So we can write

$$\tau = \frac{V}{Ib}\int_y^{c_1} ydA$$

$$\tau = \frac{VQ}{Ib},$$

where, q is statically moment area.

$$Q = \int_y^{c_1} ydA$$

11.4.1 Shear Stress Distribution of Some Common Beam Sections (Fig. 11.18 to Fig. 11.21)

Solid rectangle

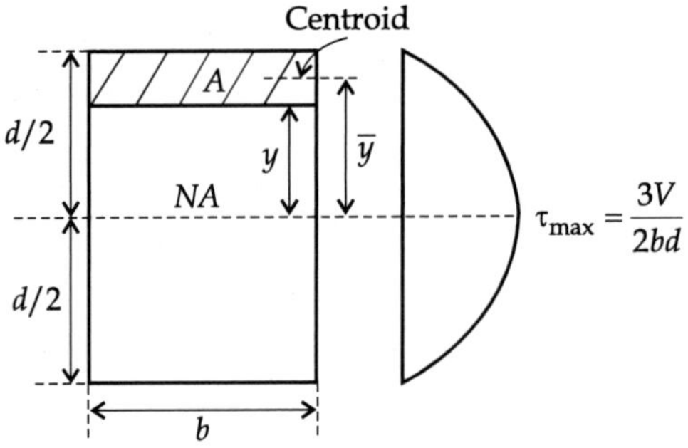

Fig. 11.18

$$\tau = \frac{V}{Ib}\int_{y}^{d/2} y\,dA = \frac{VA\bar{y}}{Ib}$$

where, τ is shear stress in the beam.

I and b are the moments of inertia about NA and width of the beam, respectively.

A is the area above y, and $\bar{y}$ is the centroid of area A from neutral axis.

$A = b(d/2 - y)$

$$\bar{y} = y + \frac{(d/2 - y)}{2} = \frac{y + d/2}{2}$$

$$I = \frac{bd^2}{12}$$

$$\tau = \frac{VA\bar{y}}{Ib} = \frac{6V(d^2/4 - y^2)}{bd^3}; \text{ parabolic}$$

At $y = d/2 \Rightarrow \tau_{min} = 0$

At $y = 0 \Rightarrow \tau_{max} = \dfrac{3V}{2bd}$

$$\tau_{average} = \frac{V}{bd} \Rightarrow \frac{\tau_{max}}{\tau_{average}} = 1.5$$

Solid circle

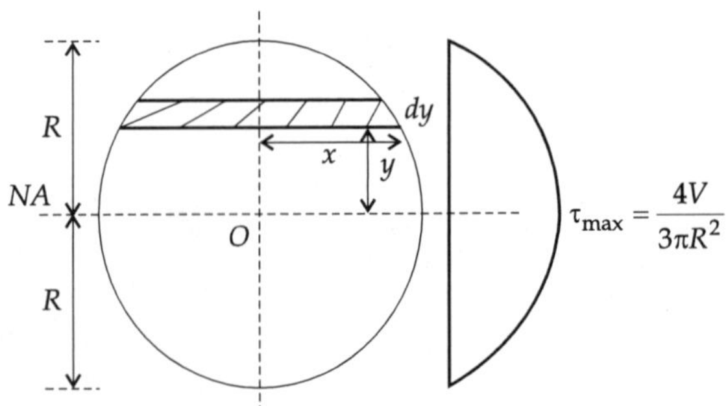

Fig. 11.19

Consider an elementary strip of thickness (dy) at an distance (y) from the neutral axis. Let the width of the strip be b, then $b = 2(R^2 - y^2)^{1/2}$

$$I = \frac{\pi R^4}{4}$$

$$dA = 2x dy$$

$$\bar{y} = y$$

$$\tau = \frac{V}{Ib}\int_y^R y dA = \frac{4V}{3\pi R^4}(R^2 - y^2)$$

At $y = R \Rightarrow \tau_{min} = 0$

At $y = 0 \Rightarrow \tau_{max} = \dfrac{4V}{3\pi R^4}$

$$\tau_{average} = \frac{V}{\pi R^2} \Rightarrow \frac{\tau_{max}}{\tau_{average}} = 1.33$$

Solid triangle

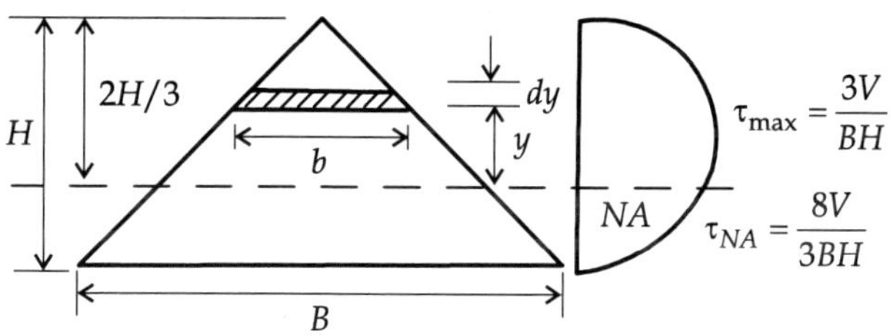

Fig. 11.20

Consider an elementary strip of thickness (dy) at an distance (y) from neutral axis. Let the width of the strip be b, then

$$\frac{b}{B} = \frac{2H/3 - y}{H} \Rightarrow b = \frac{B}{H}(2H/3 - y)$$

$$I = \frac{BH^3}{36}$$

$$dA = b dy$$

$$\bar{y} = y$$

$$\tau = \frac{V}{Ib}\int_y^{2H/3} y dA = \frac{12V}{BH^3}\left(\frac{2H}{3} - y\right)\left(\frac{H}{3} + y\right)$$

At $y = 2H/3$ and $y = -H/3 \Rightarrow \tau_{min} = 0$

At $y = 0 \Rightarrow \tau_{NA} = \dfrac{8V}{3BH}$

$$\text{At } y = H/6 \Rightarrow \tau_{max} = \frac{3V}{BH}$$

$$\tau_{average} = \frac{V}{1/2\,BH} \Rightarrow \frac{\tau_{max}}{\tau_{average}} = 1.5$$

Square placed diagonally

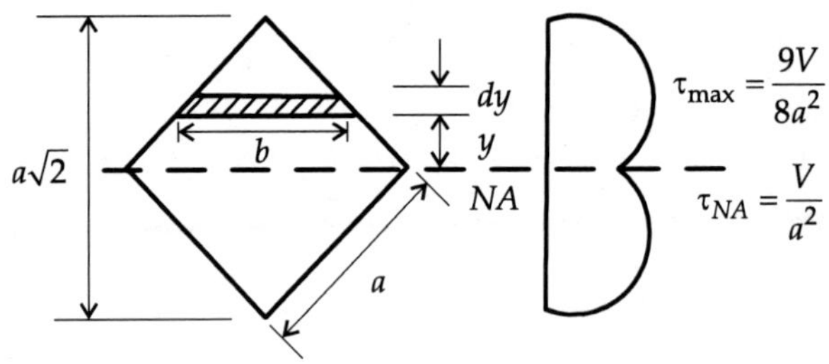

Fig. 11.21

Consider an elementary strip of thickness (dy) at a distance (y) from neutral axis. Let the width of the strip be b, then

$$\frac{b}{a\sqrt{2}} = \frac{a/\sqrt{2} - y}{a/\sqrt{2}} \Rightarrow b = a\sqrt{2} - 2y$$

Moment of inertia of the section about the neutral axis,

$$I = 2\int_0^y (bdy)(y^2) = 2\int_0^{a/\sqrt{2}} (a\sqrt{2} - 2y)y^2 dy \Rightarrow I = \frac{a^4}{12}$$

Distance to the extreme fibre, $y_{max} = \frac{d}{2} = \frac{a}{\sqrt{2}}$

$dA = bdy$ and $\bar{y} = y$

$$\tau = \frac{V}{Ib}\int_y^{a/\sqrt{2}} ydA = \frac{2V}{a^4}\left(\frac{a}{\sqrt{2}} - y\right)\left(\frac{a}{\sqrt{2}} + 2y\right)$$

$$\text{At } y = \frac{a}{\sqrt{2}} \Rightarrow \tau_{min} = 0$$

$$\text{At } y = 0 \Rightarrow \tau_{NA} = \frac{V}{a^2} = \tau_{average}$$

$$\text{At } y = a/4\sqrt{2} \Rightarrow \tau_{max} = \frac{9V}{8a^2} \Rightarrow \frac{\tau_{max}}{\tau_{average}} = 1.125$$

Schematic diagram of shear stress distribution of common beam sections

1. Symmetrical I-Section

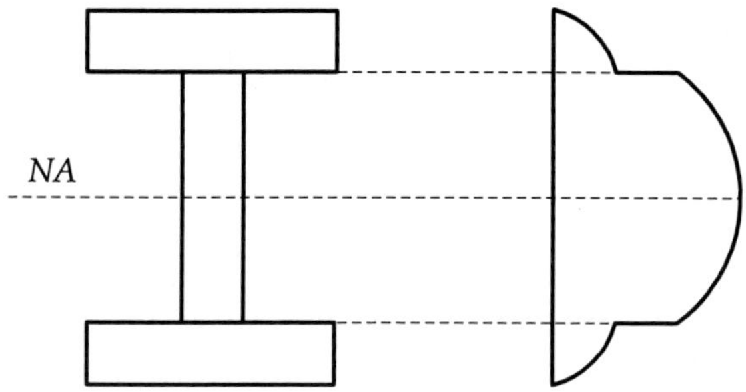

2. Unsymmetrical I-Section

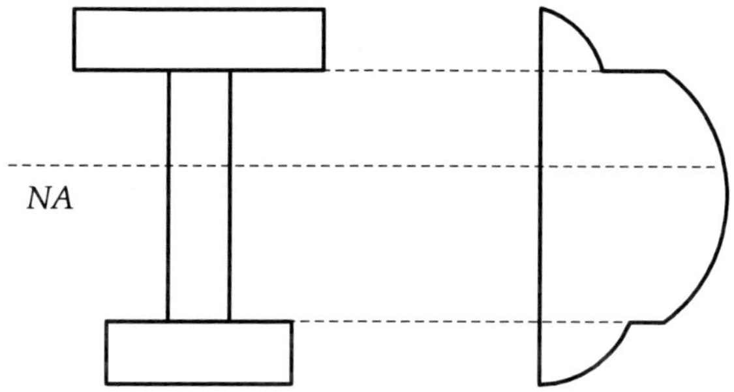

3. T-Section

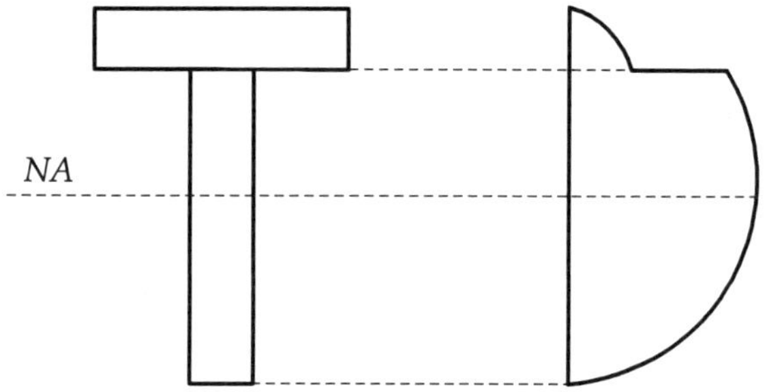

4. Inverted H-Section

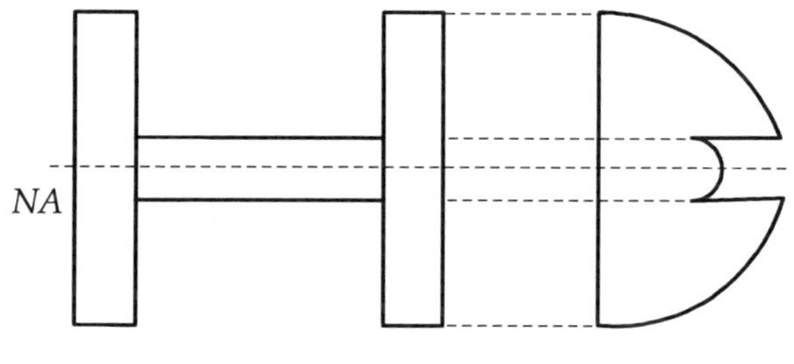

5. U-Section

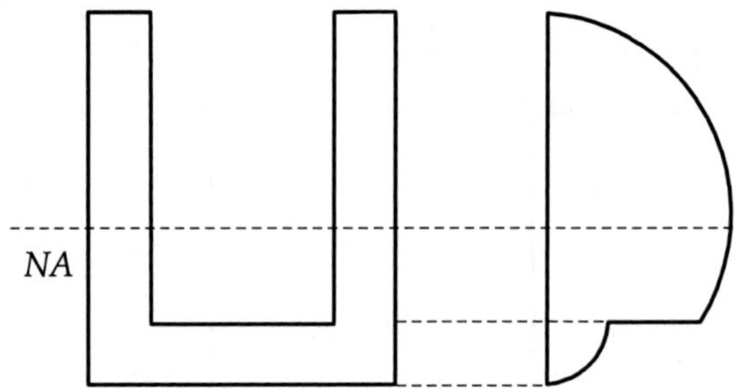

6. Plus-section

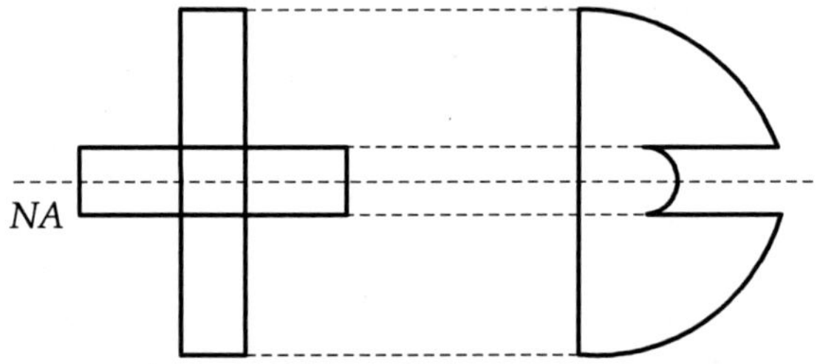

Centre line shear stress distribution

Since the thickness of sections is very less as compared to other dimensions, therefore shear distribution is assumed to be flow along a centreline of section's thickness, as shown in Fig. 11.22.

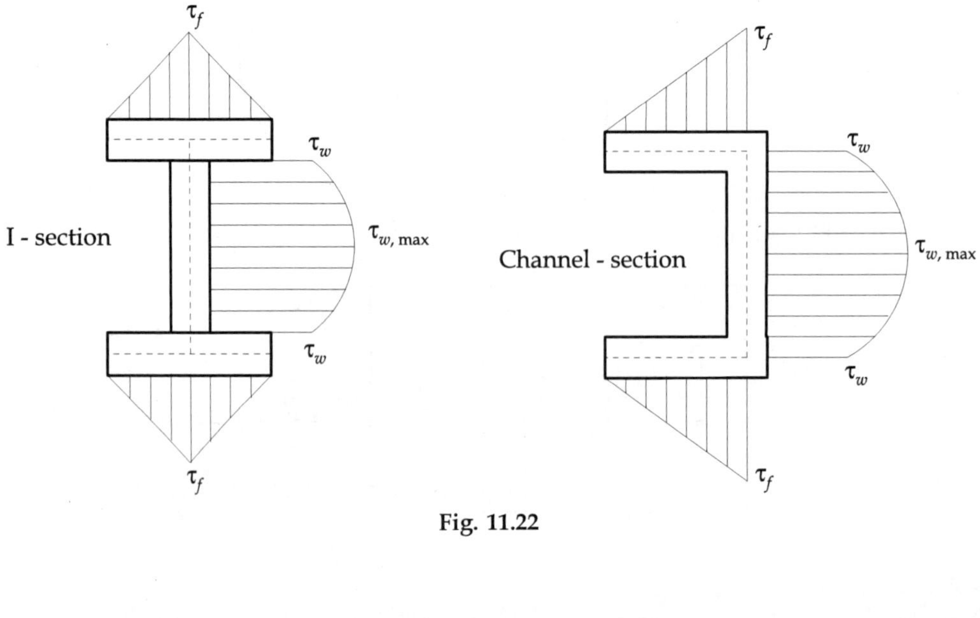

Fig. 11.22

11.5 UNSYMMETRICAL BENDING OF BEAM

Unsymmetrical bending occurs when the loads do not act in the plane of symmetry or when cross sections are not symmetrical or both (Fig. 11.23).

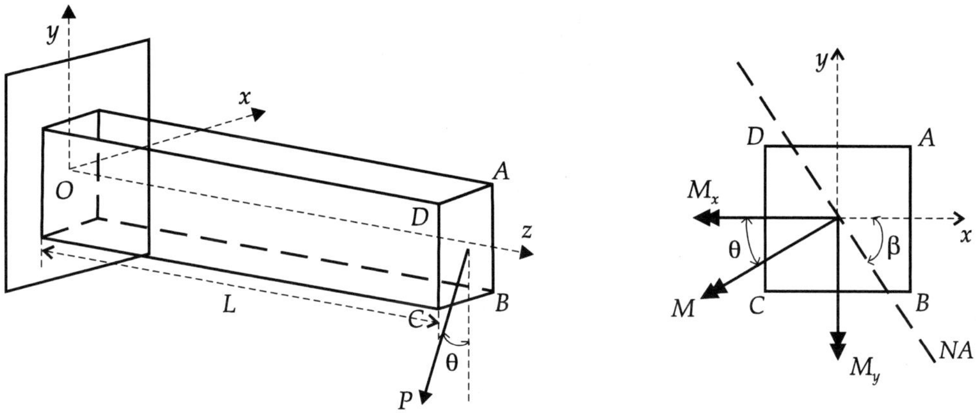

Fig. 11.23

$M_x = P\cos\theta \times L$ and $M_y = P\sin\theta \times L$

Neutral axis (NA) makes an angle β with *x*-*x* axis $\tan\beta = \frac{I_x}{I_y}\tan\theta$

Bending stress

$$\sigma = \pm\frac{M_x}{I_x}\times y \pm \frac{M_y}{I_y}\times x$$

Due to vertical component of load stresses at corners, *A* and *D*, will be tensile while stresses at corners *B* and *C* will be compressive.

Due to horizontal component of load stresses at corners, *A* and *B*, will be tensile while stresses at corners, *C* and *D*, will be compressive.

$$\sigma_A = +\frac{M_x}{I_x}\times y + \frac{M_y}{I_y}\times x$$

$$\sigma_B = -\frac{M_x}{I_x}\times y + \frac{M_y}{I_y}\times x$$

$$\sigma_C = -\frac{M_x}{I_x}\times y - \frac{M_y}{I_y}\times x$$

$$\sigma_D = +\frac{M_x}{I_x}\times y - \frac{M_y}{I_y}\times x$$

Resultant deflection

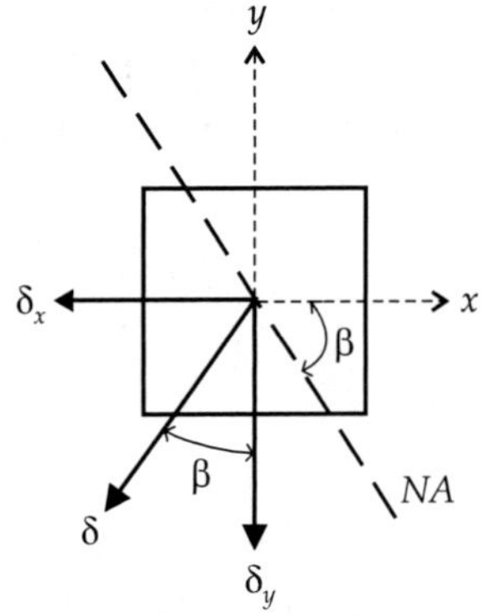

Fig. 11.24

Resultant deflection, is always perpendicular to neutral axis and at an angle of tan $\beta = \frac{I_x}{I_y}\tan\theta$.

Shear centre

- Load acting on shear centre produces only bending without any twisting.
- Normally, shear centre does not coincide with the centroid, except doubly symmetrical sections.
- In single symmetric section both centroid and shear centre lie on the axis of symmetry.
- Any force not acting on shear centre can be replaced by parallel force acting on shear centre and twisting couple equal to load into distance between points of load and shear centre.
- For shear centre moment of inertia is calculated by using centreline dimensions.

Shear centre (S) and centroid (C) of common beam sections

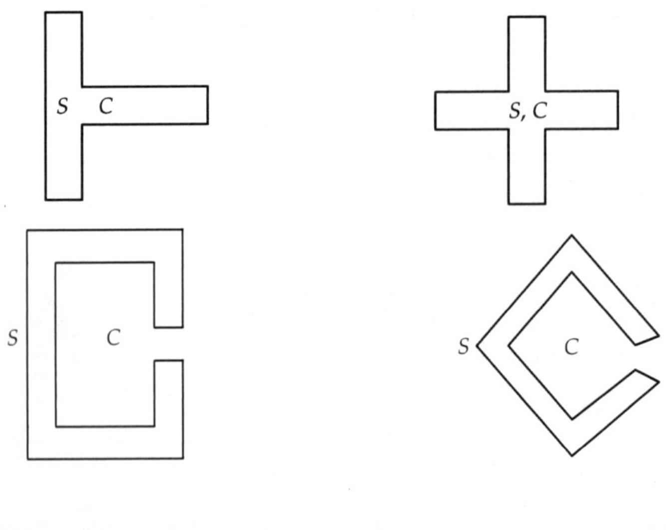

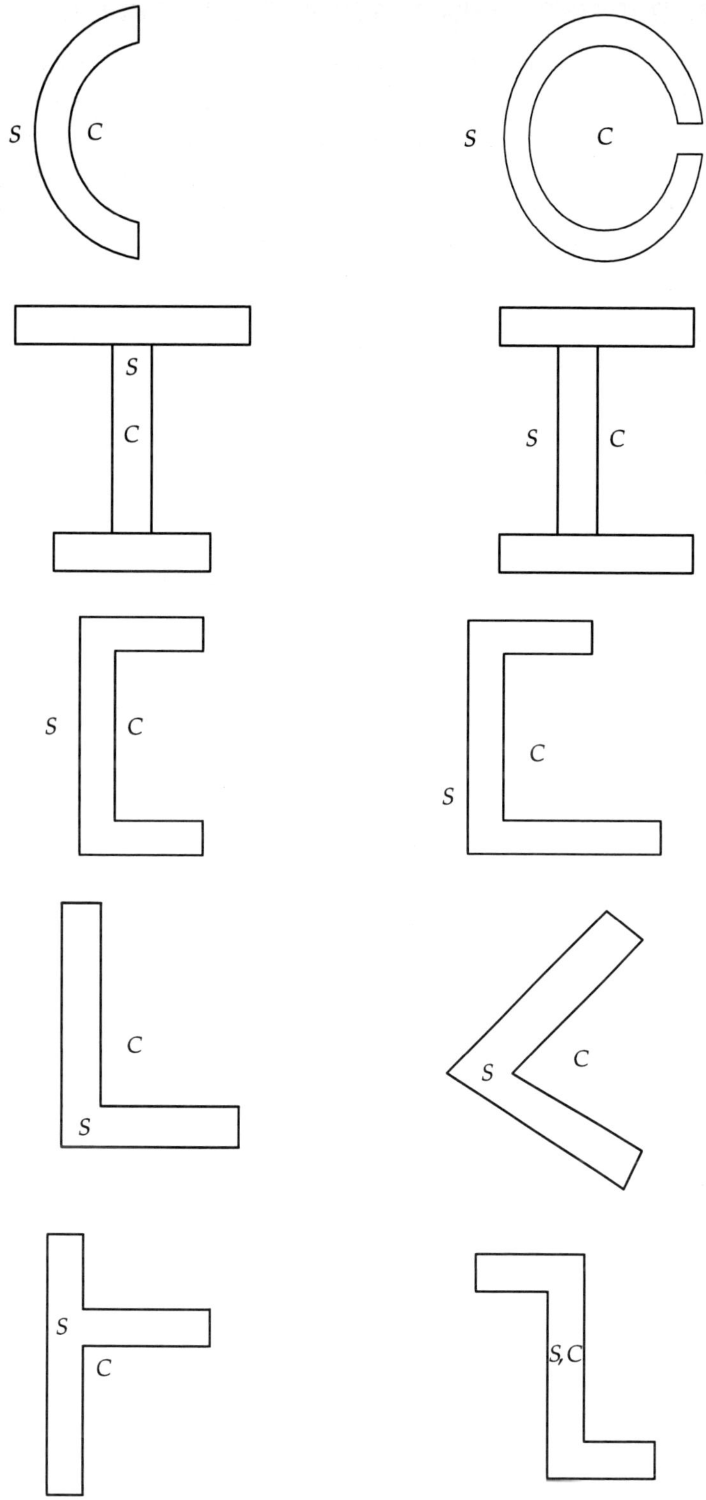
S
C
S
C
S
C
S
C
S
C
S
C
C
S
C
S
S
C
S
C
S,C

11.6 STRESSES DUE TO COMBINED BENDING AND TORSION

Let us consider a circular solid shaft of diameter (d) that is fixed at one end and free at the other end. It is subjected to bending moment (M) and twisting moment (T) at the free end as shown in Fig. 11.25. If we consider a small area on the surface of the shaft, two stresses act. One is maximum bending stress (σ_x)along the longitudinal direction due to bending moment (M), and another is maximum shear stress (τ_{xy}) due to twisting moment (T) acting perpendicular to the longitudinal axis.

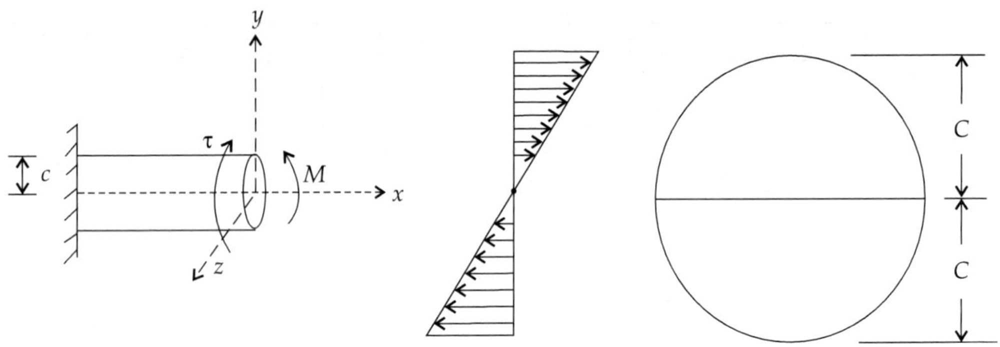

Fig. 11.25

As we already know that at the outermost surface of the shaft, bending normal stress will be maximum and bending shear stress is zero due to applied bending moment, M.

Maximum bending stress at the surface of the shaft,

$$\sigma_x = \frac{M}{I}C, \text{ where } C = \frac{d}{2}$$

$$\sigma_x = \frac{M}{I}\frac{d}{2}$$

$$I = \frac{\pi d^4}{64}$$

$$\sigma_x = \frac{32M}{\pi d^3}$$

Shear stress at any distance (r) from the center due to twisting moment, T

$$\tau = \frac{Tr}{J}$$

The maximum shear stress at $r = \frac{d}{2}$

$$\tau_{xy} = \frac{Td}{2J}$$

$$J = \frac{\pi d^4}{64}$$

$$\tau_{xy} = \frac{16T}{\pi d^3}$$

If we consider a small element on the surface, it is subjected to bi-axial stress as shown in Fig. 11.26. Then the maximum stress acting on the element is as

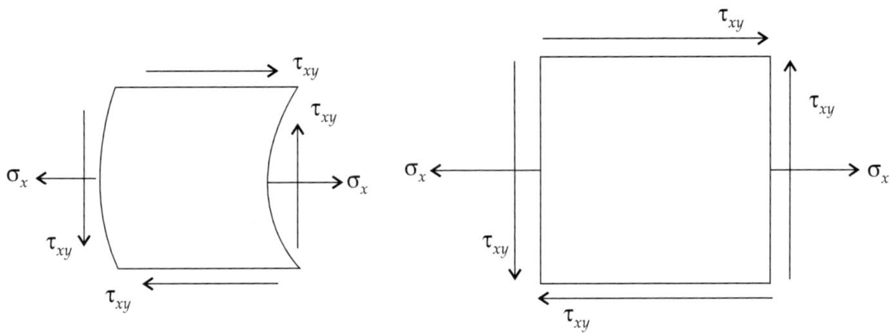

Fig. 11.26

Principal stresses at any point on the surface due to bi-axial stress,

$$\sigma_{1,2} = \frac{\sigma_x + \sigma_y}{2} \pm \sqrt{\left(\frac{\sigma_x - \sigma_y}{2}\right)^2 + \tau_{xy}^2}$$

Here, $\sigma_y = 0$

The principal stresses,

$$\sigma_{1,2} = \frac{\sigma_x}{2} \pm \sqrt{\left(\frac{\sigma_x}{2}\right)^2 + \tau_{xy}^2},$$

Putting, the value of σ_x and τ_{xy}

$$\sigma_{1,2} = \frac{16M}{\pi d^3} \pm \sqrt{\left(\frac{16M}{\pi d^3}\right)^2 + \left(\frac{16T}{\pi d^3}\right)^2} = \frac{16}{\pi d^3}\left(M \pm \sqrt{M^2 + T^2}\right)$$

Maximum shear stress,

$$\tau_{max} = \frac{\sigma_1 - \sigma_2}{2} = \pm\sqrt{\left(\frac{\sigma_x}{2}\right)^2 + \tau_{xy}^2}$$

Maximum shear stress of the shaft,

$$\tau_{max} = \frac{16}{\pi d^3}\sqrt{M^2 + T^2}$$

If we have to find the required diameter of the shaft knowing the allowable limit of the stresses for applied M and T, then we replace the principal stresses,

$$\sigma_1 = \sigma_{allowable} = \sigma_w \quad \text{and} \quad \tau_{max} = \tau_{allowable} = \tau_w$$

Then from normal stress condition required diameter of shaft,

$$d = \sqrt[3]{\frac{16}{\pi\sigma_w}\left(M \pm \sqrt{M^2 + T^2}\right)}$$

And from shear stress condition required diameter of shaft, $d = \sqrt[3]{\frac{16}{\pi\tau_w}\left(\sqrt{M^2 + T^2}\right)}$

Example 11.1: A solid shaft of diameter 150 mm is subjected to bending moment 20 kN-m and twisting moment 10 kN-m. Find out the stresses developed in the shaft.

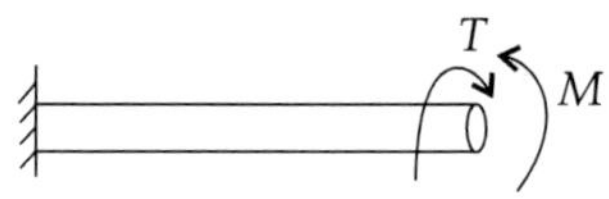

Fig. 11.27

Solution: $M = 20 \times 10^3$ N/m

$T = 10 \times 10^3$ N-m

$D = 0.150$ m

Maximum bending normal stress, $\sigma_x = \frac{32M}{\pi d^3}$

$$= \frac{32 \times 20 \times 10^3}{\pi \times 0.15^3}$$

$$= 60.36 \times 10^6 \text{ N/m}^2$$

$$= 60.36 \text{ MPa}$$

Shear stress, $\tau = \frac{16T}{\pi d^3}$

$$= \frac{16 \times 10 \times 10^3}{\pi \times 0.15^3}$$

$$= 15.09 \times 10^6 \text{ N/m}^2$$

$$= 15.09 \text{ MPa}$$

Principal stresses, $\sigma_{1,2} = \frac{16}{\pi d^3}\left(M \pm \sqrt{M^2 + T^2}\right)$

$$\sigma_{1,2} = \frac{16 \times 10^3}{\pi \times 0.150^3}\left(20 \pm \sqrt{20^2 + 10^2}\right)$$

$$\sigma_1 = 63.92 \times 10^6 \text{ N/m}^2$$

$$\sigma_2 = -3.56 \times 10^6 \text{ N/m}^2$$

$$\sigma_1 = 63.92 \text{ MPa}$$

$$\sigma_2 = -3.56 \text{ MPa}$$

Maximum shear stress, $\tau_{max} = \frac{16}{\pi d^3}\left(\sqrt{M^2 + T^2}\right)$

$$= \frac{16 \times 10^3}{\pi \times 0.15^3}\left(\sqrt{20^2 + 10^2}\right)$$

$$= 33.74 \times 10^6 \text{ N/m}^2$$

$$= 33.47 \text{ MPa}$$

or

Maximum normal bending stresses,

$$\sigma_x = \frac{M}{I}y = \frac{M}{I}\frac{d}{2} = \frac{20 \times 10^3 \times 0.15 \times 64}{2\pi \times (0.15)^4} = 60.36 \text{ MPa}$$

The shear stress on a section perpendicular to the axis due to torsion,

$$\tau = \frac{Tr}{J} = \frac{10 \times 10^3 \times 0.15 \times 32}{2 \times \frac{\pi}{4} \times (0.15)^4} = 15.09 \text{ MPa}$$

Principal stresses,

$$\sigma_{1,2} = \frac{\sigma_b}{2} \pm \frac{1}{2}\sqrt{(\sigma_b)^2 + 4\tau^2}$$

$$= \frac{60.36}{2} \pm \frac{1}{2}\sqrt{60.36^2 + 4 \times 15.09^2} = 63.92, -3.56$$

Maximum shear stress,

$$\tau_{max} = \frac{1}{2}\sqrt{(\sigma_b)^2 + 4\tau^2} = \frac{1}{2}\sqrt{60.36^2 + 4 \times 15.09^2} = 33.74 \text{ MPa}$$

Example 11.2: A hollow shaft of outer diameter 65 mm and inner diameter 60 mm is subjected to axial thrust of 2 kN, twisting moment of 600 N-m and bending moment of 200 N-m. Determine the maximum stresses developed in the shaft.

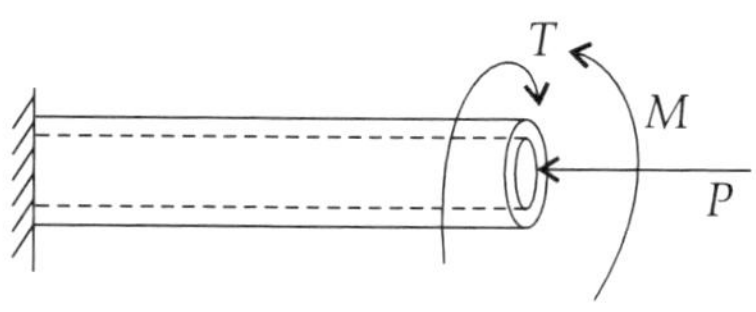

Fig. 11.28

Outside diameter

$d_0 = 0.065$ m

$d_i = 0.060$ m

$$A = \frac{\pi}{4}(d_0^4 - d_i^4) = 8.431 \times 10^{-7} \text{ m}^2$$

$$I = \frac{\pi}{64}(d_0^4 - d_i^4) = 0.5269 \times 10^{-7} \text{ m}^4$$

$$J = \frac{\pi}{32}(d_0^4 - d_i^4) = 1.054 \times 10^{-7} \text{ m}^4$$

$$P = 2 \times 10^3 \text{ N}$$

$$M = 200 \text{ N/m}$$

$$T = 600 \text{ N/m}$$

$$\sigma_c = \frac{P}{A} = \frac{2000}{8.43053 \times 10^{-7}} = 2372.197 \times 10^6 \text{ N/m}^2$$

$$\sigma_b = \frac{M}{I} \times \frac{d_0}{2} = \frac{200}{0.526908 \times 10^{-8}} \times \frac{0.065}{2} = 1233.612 \times 10^6 \text{ N/m}^2$$

$$\sigma_x = \sigma_b + \sigma_c = 3605.809 \times 10^6 \text{ N/m}^2$$

$$\tau = \frac{T}{J}\frac{d_0}{2} = \frac{600}{1.054 \times 10^{-7}} \times \frac{0.065}{2} = 185.009 \times 10^6 \text{ N/m}^2$$

$$\sigma_{1,2} = \frac{\sigma_x}{2} \pm \frac{1}{2}\left(\sqrt{\sigma_x^2 + 4\tau^2}\right)$$

$$= 1802.9045 \pm 1812.3712$$

$$\sigma_1 = 3615.27 \text{ MPa}$$

$$\sigma_2 = -9.4667 \text{ MPa}$$

$$\tau_{max} = \frac{1}{2}\left(\sqrt{\sigma_x^2 + 4\tau^2}\right) = 1812.3712 \text{ MPa}$$

Example 11.3: A thin wall steel tube of 3 m height and fixed at ground is used to carry a hoarding as shown in Fig. 11.29. The maximum wind pressure is 2.5 kN/m² acting perpendicular on the signboard. If the permissible stress is limited to 32 MPa and limited angle of twist at the top of the column is 6°, find the required diameter of the tube if thickness of the tube wall is limited to 4 mm. Take shear modulus, G = 80 GPa.

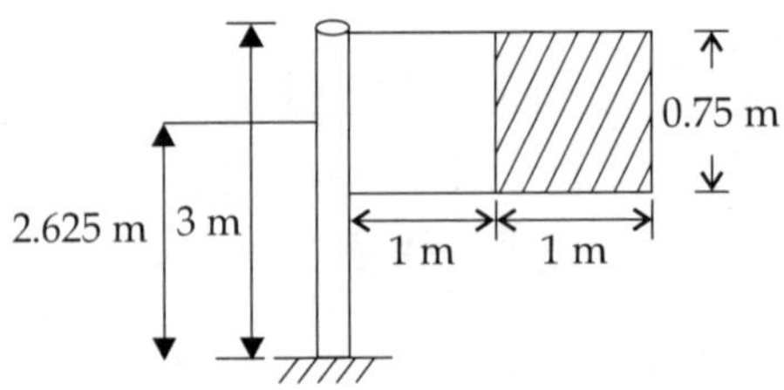

Fig. 11.29

Solution: The height of column, 3 m

The resultant torsion acting at the location, 2.625 m at the centroid of the signboard.

Area of the sign board, $A = (1 \times 0.75) = 0.75 \text{ m}^2$

Force on the sign board, $F = 0.75 \times 2.5 \times 10^3 = 1875$ N

Torsion acting at the column, $T = (1 + 0.5) \times 1875$ N-m $= 2812.5$ N-m

Neglecting the bending of the column, the shear stress developed in the outer fibre at location 2.625 m above the ground level

Let the thickness of the tube be t, and outside and inside diameter D and d, respectively.

We can write

$$t = \frac{D - d}{2}$$

$$d = D - 2t$$

Now polar moment of inertia,

$$J = \frac{\pi}{32}(D^4 - d^4)$$

$$J = \frac{\pi}{32}\left[(D^4 (D - 2t)^4\right]$$

$$= \frac{\pi D^4}{32}\left[1 - \left(1 - \frac{2t}{D}\right)^4\right]$$

$$= \frac{\pi D^4}{32}\left[1 - \left(1 - \frac{8t}{D} + \ldots\ldots\ldots\right)\right]$$

Higher term of $\frac{t}{D}$ neglected,

$$= \frac{\pi D^4}{32} \times \frac{8t}{D}$$

The polar moment of inertia,

$$J = \frac{\pi D^3 t}{4}$$

$$t = \frac{T}{J}\frac{D}{2}$$

$$t = \frac{4T}{\pi D^3 t}\frac{D}{2} = \frac{2T}{\pi D^2 t}$$

$T = 2812.5$ N-m, $t = 0.004$ m, allowable stress 32 MPa

$$D^2 = \frac{2T}{\pi t \tau_{al}}$$

$$D^2 = \frac{2 \times 2812.5}{\pi \times 0.004 \times 32 \times 10^6}$$

$$D = 118.33 \text{ mm}$$

Angle of twist limited to 6°

We know

$$0 = \frac{Tl}{GJ}$$

Putting the value of J

$$\theta = \frac{4Tl}{G\pi D^3 t}$$

Here,

$$\theta = \frac{6\pi}{180}\text{radian}$$

Hence, we get

$$D^3 = \frac{4Tl}{G\pi\theta t}$$

Here, $l = 2.625$ m

$$D^3 = \frac{4 \times 2812.5 \times 2.625 \times 180}{80 \times 10^6 \times \pi \times 0.004 \times 6\pi} = 0.280512 \times 10^{-3}$$

$$D = 0.065461 \text{ m or } 65.46 \text{ mm}$$

Required diameter of the column is 118.33 mm.

Outer diameter 118.33 mm and inner diameter 114.33 mm.

Practical purpose provide 120 mm outer diameter and inner diameter 116 mm.

11.7 STRESSES DEVELOPED IN CLOSED COILED HELICAL SPRING

A spring is generally made from thin metal rod or wire and acts as a device to store energy in the form of mechanical energy and restore the same slowly or rapidly depending upon its function. It also acts as support for vibrating body parts to absorb shock energy.

Springs are classified as

(i) Laminated or leaf spring

(ii) Helical spring

Laminated spring consists of a number of thin curved plates, each of same thickness and width but of different lengths all bent to same curvature (Fig. 11.30).

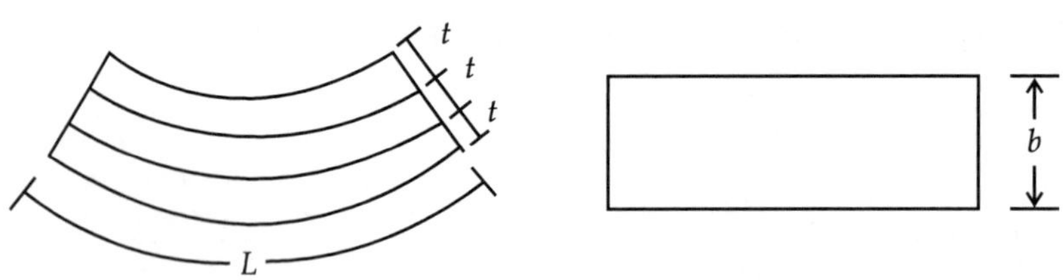

Fig. 11.30

Helical springs are formed by coiling a thin rod or a thick wier and mostly loaded under axial loading. Helical springs are classified as

(i) Close coiled helical spring

(ii) Open coiled helical spring

The coomonest and most used helical spring is close coiled helical spring. The slope of the helix of the coilis is very small and often less than 10° and in this cases the bending effects can be neglected and only torsional shear resistances are dominated.

On the other hand if slope of the helix of the coil is quite appreciable, then the bending as well as torsional shear resistance are predominant. This type of is helical spring is called an **open coiled spring.**

Helix angle: It is the angle which the axis of the spring wire makes with the horizontal line perpendicular to the axis of the spring.

Spring index: It is the ratio of the diameter of the spring (d) to the mean diameter of the coil, D.

Stiffness: Load per unit deflection of the spring and is denoted by K.

Sress developed in close coiled helical spring undera axial load, P.

The stresses developed in close coiled helical spring are discussed here. The following assumptions are made:

(i) An element of an axially loaded helical spring behaves like a straight bar under pure torsion.

(ii) The plane perpendicular to the bar axis does not wrap or distorts during deformation.

(iii) The effect of bending moment is neglected.

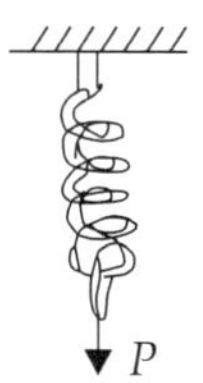

Fig. 11.31: Close coil spring

Let us consider a close coil spring as shown in Fig. 11.31.

Mean diameter of the coil = D

Diameter of the spring wire = d

Number of coils = n

Modulus of rigidity of the material = G

The twisting moment acting at any poiny of the wire T,

$$T = P.\frac{D}{2}$$

Due to this moment, the maximum shear stress developed at the surface of wire,

$$\tau = \frac{Tr}{J}$$

J = Polar moment of inetia of the spring wire,

$$J = \frac{\pi d^4}{32}$$

Maximum values of shear stress at the outer fibre,

$$\tau = \frac{Tr}{J} = \frac{\left(\frac{PD}{2}\right)\left(\frac{d}{2}\right)}{\left(\frac{\pi d^4}{32}\right)} = \frac{8PD}{\pi d^3}$$

Vertical shear stress, at any cross section,

$$\tau_v = \frac{V}{A} = \frac{P}{\left(\frac{\pi}{4}d^2\right)} = \frac{4P}{\pi d^2}$$

The maximum resultant stress shall occur in the outer fibre of the wire at a point, when the direction of two stresses, τ_v and are same.

$$\tau_{max} = \tau_v + \tau = \frac{8PD}{\pi d^3} + \frac{4P}{\pi d^2} = \frac{8PD}{\pi d^3}\left(1 + \frac{d}{2D}\right)$$

Now, if diameter d of the wire of the spring is very small compared to mean diameter of the coil D, then $d/2D$ term can be neglected.

Then maximum shear stress developed in the surface of the wire of the spring,

$$\tau_{max} = \frac{8PD}{\pi d^3}$$

Now, if there are n number coils then the total length of the spring under torsion, T,

$$L = \pi Dn$$

If θ is the angle of twist in spring wire then

$$\theta = \frac{TL}{GJ} = \frac{\left(\frac{PD}{2}\right)(\pi Dn)}{G\left(\frac{\pi d^4}{32}\right)} = 16\frac{PD^2 n}{Gd^4}$$

If the spring deflects δ amount due to the applied load (P), then the work done on the spring,

$$W = \frac{1}{2}P\delta$$

Therefore, strain energy (U) stored in the spring due to twisting moment (T),

$$U = \frac{1}{2}T\theta$$

Equating the work done on the spring and the energy stored in the spring

$$\frac{1}{2}P\delta = \frac{1}{2}T\theta$$

$$\delta = \frac{T\theta}{P}$$

Putting the value of T and θ

$$\delta = \frac{\left(\frac{PD}{2}\right)16\frac{PD^2n}{Gd^4}}{P} = \frac{8PD^3n}{Gd^4}$$

Load per unit deflection is defined as the stiffness K.

$$K = \frac{P}{\delta}$$

$$= \frac{Gd^4}{8D^3n}$$

Strain energy stored

$$U = \frac{1}{2}T\theta = \frac{4P^2D^3n}{Gd^4}$$

$$U = \frac{1}{G}\left(\frac{8PD}{\pi d^3}\right)^2 \pi Dn\left(\frac{\pi d^2}{4}\right)$$

Volume of spring = Length of spring × Cross section of the wire

$$V = (\pi Dn)\left(\frac{\pi d^2}{4}\right)$$

$$U = \frac{\tau^2}{4G}\text{(volume of spring)}$$

Stress developed in close coiled spring due to moment

Fig. 11.32

Consider a close coiled helical spring subjected to a twisting moment (M) about its axis as shown in Fig. 11.32. This torque will induce bending stress at the spring wire. This torque increases or decreases the curvature of each coil and thus this torque acts as bending moment in the coil. Due to this bending moment curvature of the coil will change.

Let, initial curvature of the spring coil before bending be R and after bending, R' Initial coil number n and after bending coil number n' respectively.

The change of curvature due to bending moment (M) is $= \frac{1}{R'} - \frac{1}{R}$

We know $\frac{M}{EI}$ = Change of curvature

Here, EI is the stiffness of the spring wire and M, the bending moment

$\sigma = \frac{M}{I} \times \frac{d}{2} \sigma = \frac{M}{I} \times \frac{d}{2}$ Then $M = EI\left(\frac{1}{R'} - \frac{1}{R}\right)$

The length of the spring (L) before and after bending will be same.

We can write $L = 2\pi nR = 2\pi n'R'$

$$\frac{1}{R} = \frac{2\pi n}{L} \text{ and } \frac{1}{R'} = \frac{2\pi n'}{L}$$

$$\frac{1}{R'} - \frac{1}{R} = \frac{2\pi(n' - n)}{L}$$

$$M = \frac{2\pi(n' - n)EI}{L}$$

Now wind up angle, $\phi = 2\pi(n' - n)$

$$M = \frac{\phi EI}{L}$$

$$\phi = \frac{ML}{EI}$$

Again $L = 2\pi nR$

Therefore, $\phi = \frac{2\pi nRM}{EI}$

For a circular cross section of the spring wire of diameter (d) moment of inertia (I),

$$I = \frac{\pi d^4}{64}$$

If D is mean diameter of the coil

Then $\phi = \frac{64nDM}{Ed^4}$

Bending stress developed at the surface of the spring wire, $\sigma = \frac{M}{I} \times \frac{d}{2}$

$$\sigma = \frac{32M}{\pi d^3}$$

Strain energy

The strain energy stored, $U = \frac{1}{2}M\phi$

$$U = \frac{1}{2} \times M \times \frac{64nDM}{Ed^4}$$

$$M = \frac{\pi d^3}{32}\sigma$$

Volume of spring, $V = \frac{\pi d^2}{4} \times L$

Length of spring, $L = \pi D n$

$$U = \frac{\sigma^2}{8E} \times \frac{\pi d^2}{4} \times L$$

Strain energy per unit volume, $u = \frac{\sigma^2}{8E}$

Example 11.4: A closely coiled spring is made of steel wire of diameter 10 mm. Mean diameter of the coil is 100 mm and the number coils is 12. An axial load of 200 N is applied at the end of the spring axially. Find the elongation of the spring, and the shearing stress developed in the spring. Take G = 82 GPa.

Fig. 11.33

Solution: $d = 10$ mm

$D = 100$ mm

$P = 200$ N

$$\theta = \frac{16PD^2N}{Gd^4}$$

$$\theta = \frac{16 \times 200 \times 0.1^2 \times 12}{82 \times 10^9 \times 0.01^4} = 0.4682 \text{ radian}$$

Deflection, $\delta = \frac{8PD^3n}{Gd^4}$

$$\sigma = \frac{8 \times 200 \times 0.1^3 \times 12}{82 \times 10^9 \times 0.01^4} = 0.0234 \text{ m}$$

Stiffness of the spring,

$$k = \frac{Gd^4}{8D^3n}$$

$$k = \frac{82 \times 10^9 \times 0.01^4}{8 \times 0.1^3 \times 12} = 8541.67 \text{ N/m}$$

$$\text{Torsional shear stress} = \frac{8PD}{\pi d^3}$$

$$\tau = \frac{8 \times 200 \times 0.1}{\pi \times 0.01^3} = 50929581.789 \text{ N/m}^2 = 50.93 \text{ MPa}$$

$$\text{Direct shear stress} = \frac{4P}{\pi d^2}$$

$$\tau_v = \frac{4 \times 200}{\pi \times 0.01^2} = 2546479.08 \text{ N/m}^2 = 2.55 \text{ MPa}$$

Maximum resultant shear tress at the surface of the coil due to torsional stress and direct stress,

$$\tau_{max} = \frac{8PD}{\pi d^3} + \frac{4P}{\pi d^2} = 50.93 + 2.55 = 53.48 \text{ MPa}$$

Example 11.5: A close coiled helical spring made of steel wire having diameter 12 mm and mean diameter 120 mm. The spring has 10 coils and is subjected to a torque of 20 N-m at the end. Determine the axial twist, maximum bending stress and strain energy stored in the spring. Take E = 200 GPa and G = 80 GPa

Solution: $d = 12$ mm

$D = 120$ mm

$n = 10$

$M = 20$ N-m

$$\phi = \frac{64nDM}{Ed^4}$$

$$\phi = \frac{64 \times 10 \times 0.120 \times 20}{200 \times 10^9 \times 0.012^4} = 0.3704 \text{ radian}$$

Shear stress,

$$\sigma = \frac{32M}{\pi d^3}$$

$$\sigma = \frac{32 \times 20}{\pi \times (0.012)^3} = 117.89 \times 10^6 \text{ N/m}^2$$

$$\text{Volume of spring, } V = \frac{\pi d^2}{4} \times L$$

$$L = \pi D n$$

$$V = \frac{\pi d^2}{4} \times \pi D n$$

$$= \frac{\pi^2 d^2 Dn}{4} = \frac{\pi^2 \times 0.012^2 \times 0.120 \times 10}{4} = 4.2637 \times 10^{-4}\ \text{m}^2$$

Strain energy stored,

$$U = \frac{\sigma^2}{8E} \times \text{Volume of spring} = \frac{117.89^2 \times 10^{12} \times 4.2637 \times 10^{-4}}{8 \times 200 \times 10^9} = 3.7035\ \text{J}$$

Example 11.5: Consider a simply supported rectangular beam uniformly loaded as shown in Fig. 11.34. Discuss bending stress (σ) and shear stress (τ) distribution at different points.

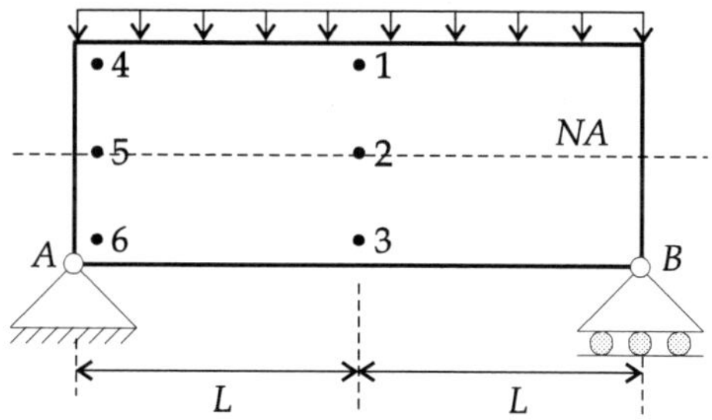

Fig. 11.34

Solution: Bending stress is directly proportional to bending moment, while shear stress is directly proportional to shear force.

At top and bottom of beams bending stress is maximum, while at NA it is zero.

Shear stress is maximum at NA, while it is zero at the top and bottom of the beam.

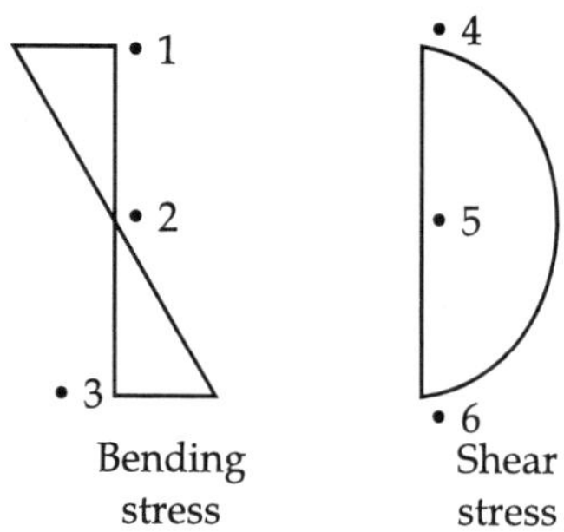

At 1 and 3 bending moment is maximum thus the compressive and tensile bending stresses are maximum.

At 4, 5 and 6 bending moment is zero, thus they are the points of zero bending stress.

Point 2 is on neutral axis, therefore bending stress at this point is zero.

Also at Point 2 shear force is zero, therefore shear stress at this point is zero.

Point 5 is on neutral axis and shear force is also maximum near the support, therefore shear stress at this point is maximum.

Example 11.6: Consider a simply supported T-section beam as shown in Fig. 11.35. (i) Calculate the maximum tensile and compressive bending stresses in the beam. (ii) Calculate the maximum tensile and compressive bending stresses, at the junction of the flange,-web joint, if the depth of flange is 2.5 cm. (iii) Calculate the total maximum force on the flange and moment of this force about neutral axis if the width of the flange is 10 cm.

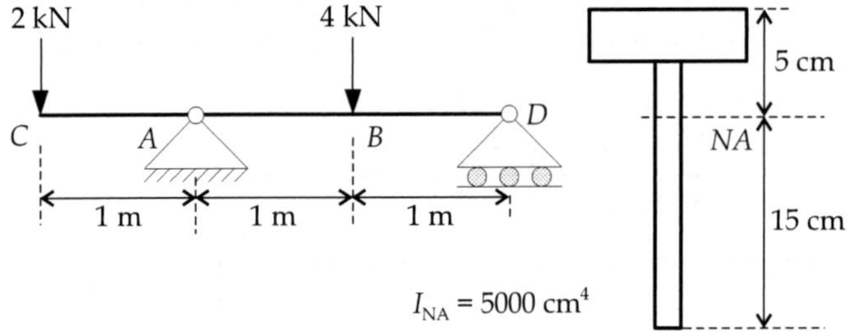

Fig. 11.35

Solution: (i) Maximum tensile and compressive bending stresses in the beam

Bending moment diagram of beam

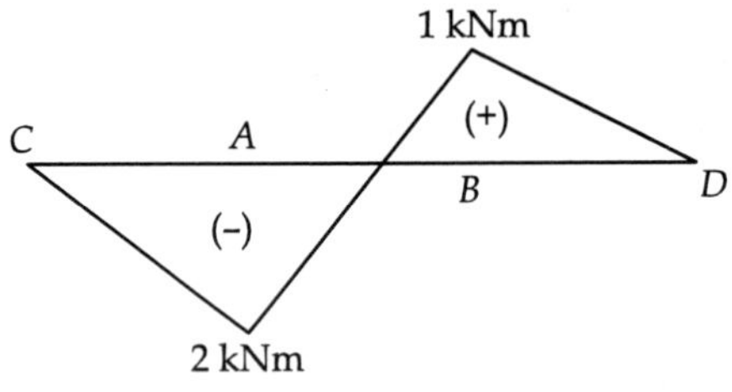

Since BM is negative at section A, the beam will be concave downward at section A and since the BM is positive at section B, the beam will be concave upward at section B. Therefore, for maximum tensile and compressive stresses we have to check the stresses at both the sections, since maximum BM is different and of opposite sign at both the sections.

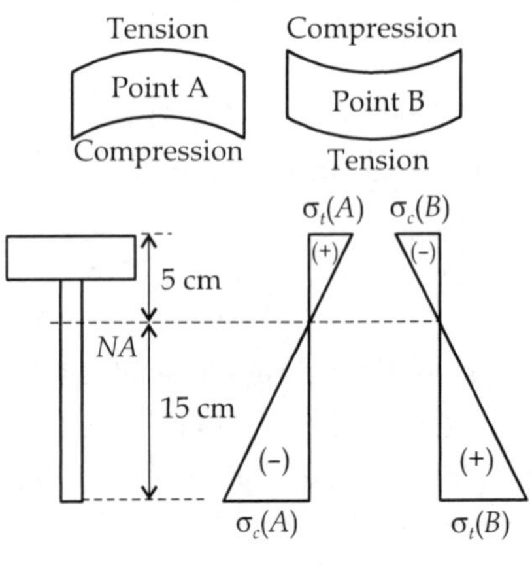

At section A

$$\sigma_t(A) = \frac{My}{I_{NA}} = \frac{2\times10^3\times5\times10^{-2}}{5000\times10^{-8}} = 2\times10^6\ \text{N/m}^2 = 2\ \text{MPa}$$

$$\sigma_c(A) = \frac{My}{I_{NA}} = \frac{2\times10^3\times15\times10^{-2}}{5000\times10^{-8}} = 6\times10^6\ \text{N/m}^2 = 6\ \text{MPa}$$

At section B

$$\sigma_t(B) = \frac{My}{I_{NA}} = \frac{1\times10^3\times15\times10^{-2}}{5000\times10^{-8}} = 3\times10^6\ \text{N/m}^2 = 3\ \text{MPa}$$

$$\sigma_c(B) = \frac{My}{I_{NA}} = \frac{1\times10^3\times5\times10^{-2}}{5000\times10^{-8}} = 1\times10^6\ \text{N/m}^2 = 1\ \text{MPa}$$

Thus, maximum tensile stress will occur at section B $\sigma_t(B) = 3$ MPa while maximum compressive stresses will occur at section A, $\sigma_C(A) = 6$ MPa.

(ii) Maximum tensile and compressive bending stresses, at the junction of the flange, web joint, when the depth of flange is 2.5 cm:

Bending stress diagram

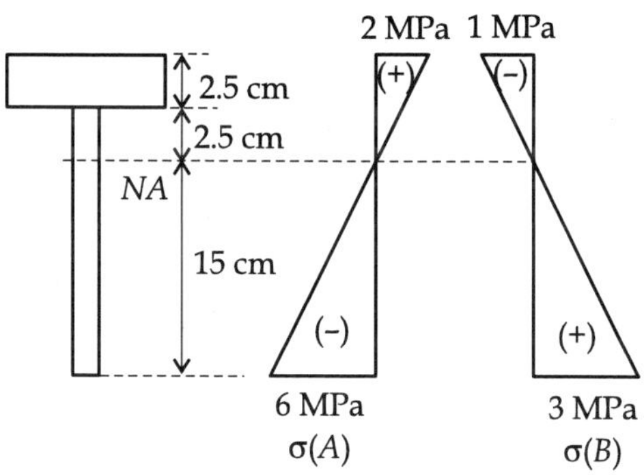

$$\sigma_t(\text{joint}) = \frac{2}{5}\times2.5 = 1\ \text{MPa}$$

$$\sigma_c(\text{joint}) = \frac{1}{5}\times2.5 = 0.5\ \text{MPa}$$

(iii) Total maximum force on the flange and moment of this force about neutral axis when the width of the flange is 10 cm.

Maximum force will be given by 2 MPa stress at section A.

$$F_t = A\times\sigma_{avg} = 25\times100\times\frac{2+1}{2} = 3750\ \text{N} = 3.75\ \text{kN}$$

$$M_{NA} = F_t\times\bar{y} = 3.75\times0.0375 = 0.14\ \text{kNm}$$

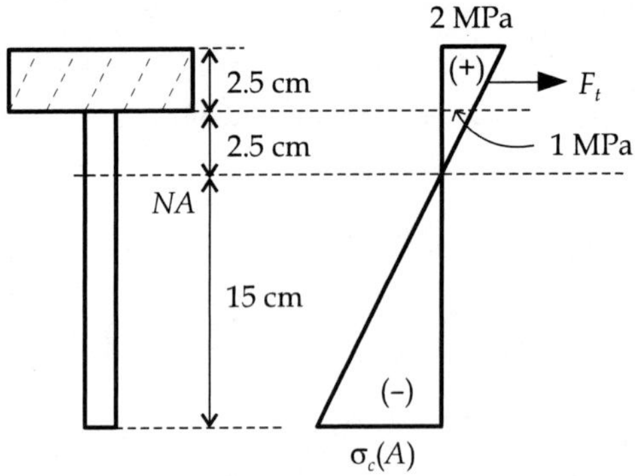

Example 11.7: A strip of copper 50 mm wide and 10 mm thick (Fig. 11.36) is bounded with another strip of aluminium of same size to form a bimetallic strip of 50 mm × 20 mm. If the composite strip is subjected to a pure bending moment of 0.5 kNm then calculate the radius of curvature of the strip and the maximum tensile and compressive stresses. (Given: $E_C = 1 \times 10^5$ MPa, $E_{al} = 0.5 \times 10^5$ MPa)

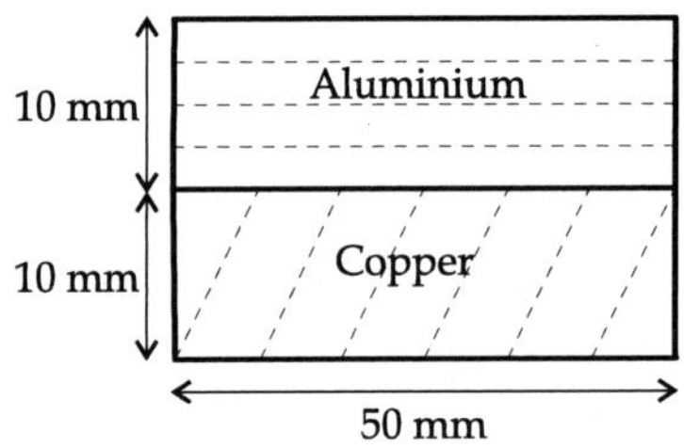

Fig. 11.36

Solution: Since sections of the beam are bounded together, therefore the whole section will bend about common neutral axis.

Let us convert the original section into an equivalent section in terms of copper as

$$\text{Modular ratio, } m = \frac{E_c}{E_{al}} = \frac{1 \times 10^5}{0.5 \times 10^5} = 2$$

$$\text{Equivalent width of copper strip} = \frac{50}{m} = \frac{50}{2} = 25 \text{ mm}$$

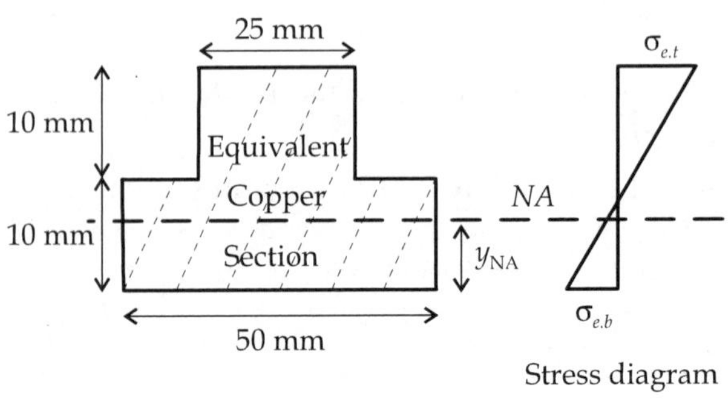

Let neutral axis be y_{NA} mm above the base of equivalent section

$$\bar{y} = \frac{50 \times 10 \times 5 + 25 \times 10 \times 15}{50 \times 10 + 25 \times 10} = 8.33 \text{ mm}$$

Moment of inertia of equivalent section about NA

$$I_{NA} = \frac{50 \times 10^3}{12} + 50 \times 10 \times (8.33 - 5)^2 + \frac{25 + 10^3}{12} + 25 \times 10 \times (15 - 8.33)^2 = 22917 \text{ mm}^4$$

Pure bending equation

$$\frac{M}{I} = \frac{E}{R} \Rightarrow R = \frac{E_c I_{NA}}{M} = \frac{1 \times 10^5 \times 22917}{0.5 \times 10^6}$$

$$\Rightarrow R = 4583.3 \text{ mm} = 4.583 \text{ m}$$

Since the beam is subjected to sagging bending moment, thus the stress developed at the top will be compressive and tensile at the bottom.

Stress in equivalent section,

$$\sigma_{e,b} = \frac{M}{I_{NA}} \times y_{\text{bottom}} = \frac{0.5 \times 10^6 \times 8.33}{22917} = 181.7 \text{ N/mm}^2$$

$$\sigma_{e,t} = \frac{M}{I_{NA}} \times y_{\text{top}} = \frac{0.5 \times 10^6 \times 11.67}{22917} = 254.6 \text{ N/mm}^2$$

Corresponding stress in aluminium,

$$\sigma_{al,t} = \frac{\sigma_{c,t}}{m} = \frac{254.67}{2} = 127.33 \text{ N/mm}^2$$

Example 11.8: A timber beam 10 cm × 20 cm is reinforced by two steel plate beams 10 cm × 1 cm at top and bottom as shown in Fig. 11.37. Calculate the resisting moment of flitched beam.

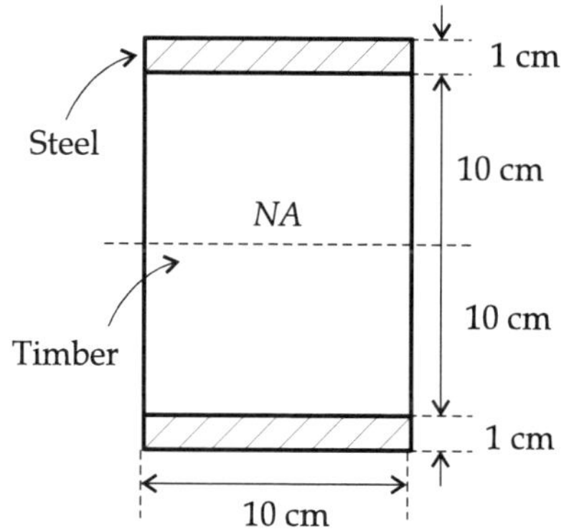

Fig. 11.37

(Given: Permissible bending stress in timber, σ_t = 10 MN/m^2; E_s = 250 GN/m^2; E_t = 25 GN/m^2)

Solution: Let $\sigma_{s,\,max}$ be the permissible bending stress in steel.

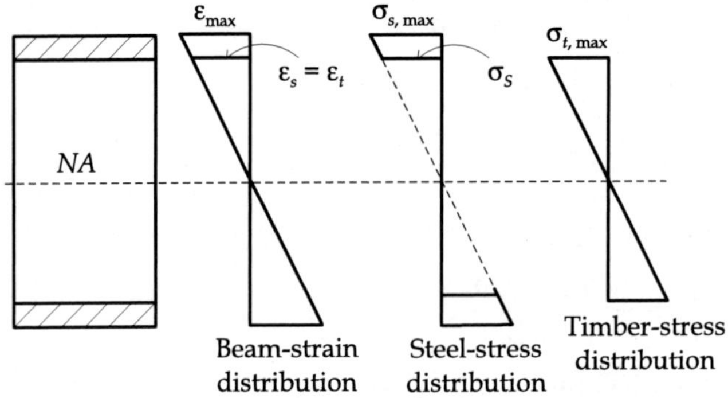

From similar triangle of steel stress diagram,

$$\frac{\sigma_s}{10} = \frac{\sigma_{s,max}}{11} \Rightarrow \sigma_s = \frac{10\sigma_{s,max}}{11}$$

Since steel and timber plates are joined, therefore strain at the junction point in both steel and timber must be same.

$$\varepsilon_s = \varepsilon_t \Rightarrow \frac{\sigma_s}{E_s} = \frac{\sigma_{t,max}}{E_t} \Rightarrow \frac{10\sigma_{s,max}}{11E_s} = \frac{\sigma_{t,max}}{E_t}$$

$$\Rightarrow \sigma_{s,max} = \frac{11 \times 250}{10 \times 25} \times 10 = 110 \text{ MN/m}^2$$

Moment of inertia,

$$I_{s,NA} = \frac{10 \times 22^3}{12} - \frac{10 \times 20^3}{12} = 2206.7 \text{ cm}^4 = 22.1 \times 10^{-6}\text{m}^4$$

$$I_{s,NA} = \frac{10 \times 20^3}{12} = 6666.7 \text{ cm}^4 = 66.7 \times 10^{-6}\text{m}^4$$

Bending moment resist by steel,

$$M_{s,max} = \frac{\sigma_{s,max}}{y_{max}} I_{s,NA} = \frac{110 \times 10^6}{0.11} \times 22.1 \times 10^{-6} = 22.1 \text{ kNm}$$

Bending moment resist by timber

$$M_{t,max} = \frac{\sigma_{t,max}}{y_{max}} I_{t,NA} = \frac{10 \times 10^6}{0.10} \times 66.7 \times 10^{-6} = 6.67 \text{ kNm}$$

Total moment resist by flitched beam

$$M_T = M_{s,max} + M_{t,max} = 22.1 + 6.67 = 28.8 \text{ kNm}$$

Example 11.9: A timber beam 10 cm × 20 cm is sandwitched between two steel plate beams of 1 cm × 20 cm as shown in Fig. 11.38. Calculate the resisting moment of the flitched beam.

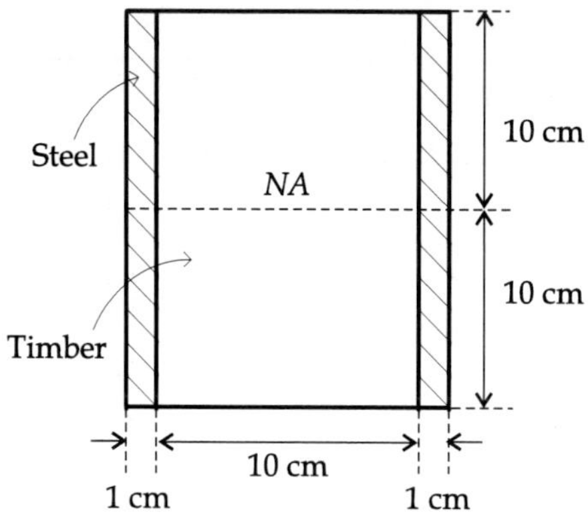

Fig. 11.38

(Given: Permissible bending stress in timber, σ_t = 10 MN/m^2; E_s = 250 GN/m^2; E_t = 25 GN/m^2))

Solution: $\sigma_{s,\,max}$ is the permissible bending stress in steel.

In this case maximum stress in both steel and timber will occur at the extreme ends (y = 11 cm).

Since steel and timber plates are joined, therefore strain at the extreme ends in both steel and timber must be same.

$$\varepsilon_s = \varepsilon_t \Rightarrow \frac{\sigma_{s,max}}{E_s} = \frac{\sigma_{t,max}}{E_t} \Rightarrow \sigma_{s,\,max} = \frac{250}{25} \times 10 = 100 \text{ MN/m}^2$$

Moment of inertia,

$$I_{s,NA} = \frac{12 \times 20^3}{12} - \frac{10 \times 20^3}{12} = 1333.3 \text{ cm}^4 = 13.3 \times 10^{-6} \text{ m}^4$$

$$I_{t,NA} = \frac{10 \times 20^3}{12} = 6666.7 \text{ cm}^4 = 66.7 \times 10^{-6} \text{ m}^4$$

Bending moment resist by steel,

$$M_{s,max} = \frac{\sigma_{s,max}}{y_{max}} I_{s,NA} = \frac{100 \times 10^6}{0.10} \times 13.3 \times 10^{-6} = 13.3 \text{ kNm}$$

Bending moment resist by timber,

$$M_{t,max} = \frac{\sigma_{t,max}}{y_{max}} I_{t,NA} = \frac{10 \times 10^6}{0.10} \times 66.7 \times 10^{-6} = 6.67 \text{ kNm}$$

Total moment resist by flitched beam,

$$M_T = M_{s,max} + M_{t,max} = 13.3 + 6.67 = 20.0 \text{ kNm}$$

Example 11.10: Consider a simply supported T-section beam as shown in Fig.11.39. Draw the shear stress distribution in beam.

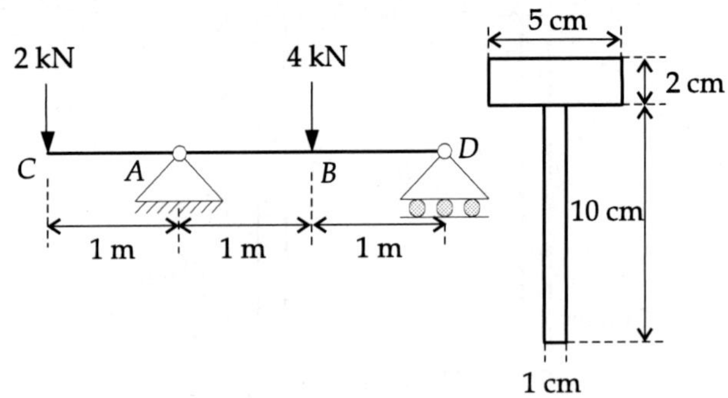

Fig. 11.39

Solution: Shear force diagram of beam

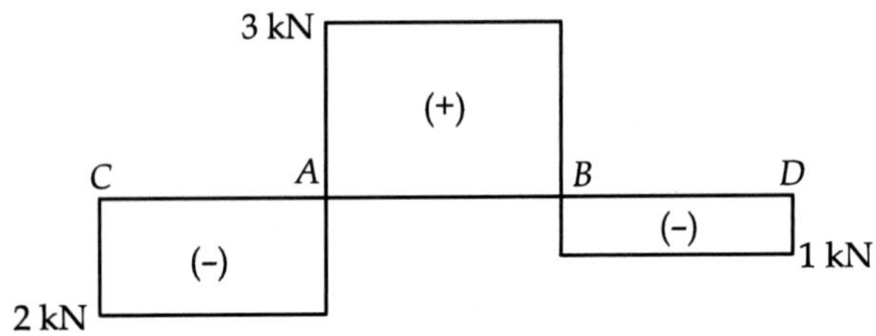

Since SF is maximum between AB, thus maximum shear stress will develop between AB under the shear force of 3 kN.

Neutral axis depth from top of flange,

$$y_{NA} = \frac{5\times2\times1+10\times1\times7}{5\times2+10\times1} = 4 \text{ cm}$$

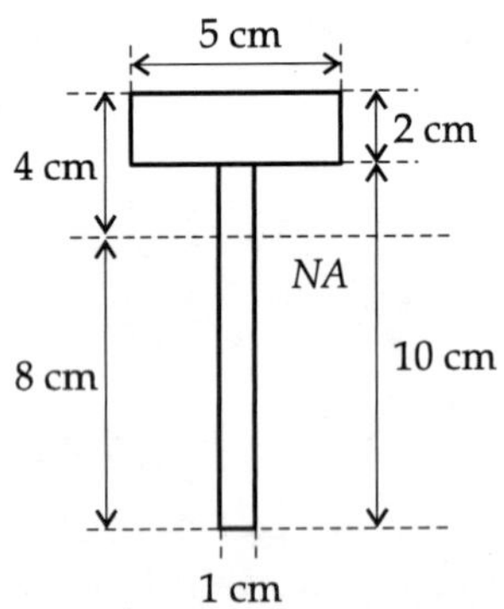

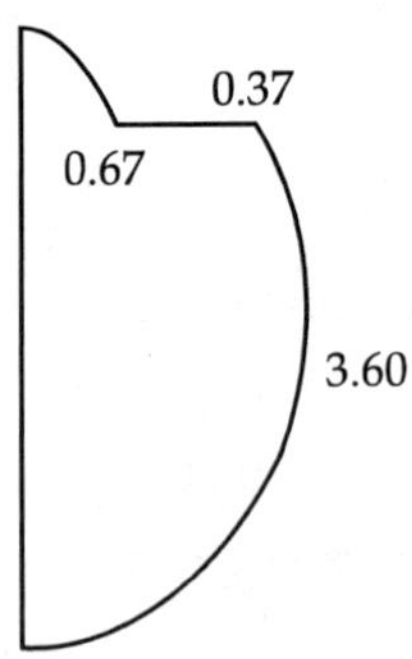

Moment of inertia about neutral axis,

$$I_{NA} = \frac{5\times2^3}{12} + (5\times2\times3^2) + \frac{1\times10^3}{12} + (1\times10\times3^2)$$

$$= 266.7 \text{ cm}^4 = 2.76\times10^{-6} \text{ m}^4$$

At the top and bottom of the beam shear tresses are zero.

Shear stress in flange near junction of flange-web,

$$\tau_f = \frac{VA\bar{y}}{Ib} = \frac{3\times10^3\times5\times2\times3\times10^{-6}}{2.67\times10^{-6}\times5\times10^{-2}} = 6.7\times10^5\ \text{N/m}^2$$

$= 0.67$ MPa

Shear stress in web near junction of flange-web,

$$\tau_w = \frac{VA\bar{y}}{Ib} = \frac{3\times10^3\times5\times2\times3\times10^{-6}}{2.67\times10^{-6}\times1\times10^{-2}} = 33.69\times10^5\ \text{N/m}^2$$

$= 3.37$ MPa

Maximum shear stress will occur at NA,

$$\tau_{max} = \frac{VA\bar{y}}{Ib} = \frac{3\times10^3\times(5\times2\times3+1\times2\times1)\times10^{-6}}{2.67\times10^{-6}\times1\times10^{-2}}$$

$= 35.95\times10^5\ \text{N/m}^2 = 3.60$ MPa

Example 11.11: A horizontal steel cantilever beam of length 0.5 m is fixed at one end and free at another supports a vertical concentrate load of 50 kN at the free end. The cross section of the cantilever beam is rectangular, 4 cm wide and 12 cm deep. Calculate the magnitude of principal stresses at fixed end (i) at the neutral axis, (ii) at 3 cm above the neutral axis, and (iii) at bottom of the beam (Fig. 11.40).

Solution:

(i) Magnitude of principal stresses at fixed end at the neutral axis:

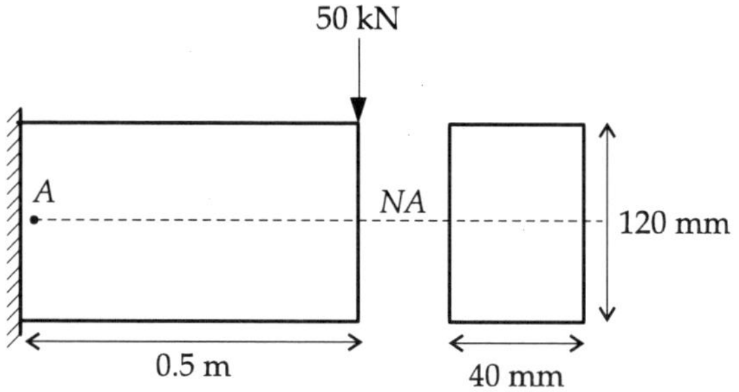

Fig. 11.40

Shear force is constant throughout the length, $S_A = 50$ kN

Bending moment at A,

$M_A = P\times L = 50\times0.5 = 25$ kNm

Bending stresses

Bending stress will be zero at neutral axis.

$$\sigma_A = \frac{M}{I}y = \frac{25\times10^6}{\left(\frac{40\times120^3}{12}\right)}\times0 = 0$$

Shear stress,

$$\tau_A = \frac{SA\bar{y}}{Ib} = \frac{50\times10^3\times40\times60\times30}{\left(\frac{40\times120^3}{12}\right)\times40} = 15.6\ \text{N/nm}^2$$

Positive shear will produce negative shear stress element, $\tau_{xy} = -\tau = -15.6\ \text{N/mm}^2$

Principal stresses,

$$\sigma_1, \sigma_2 = \frac{\sigma_x+\sigma_y}{2} \pm \sqrt{\left(\frac{\sigma_x-\sigma_y}{2}\right)^2 + \tau_{xy}^2}$$

$$\sigma_1, \sigma_2 = \frac{0+0}{2} \pm \sqrt{\left(\frac{0-0}{2}\right)^2 + (-15.6)^2}$$

$$\sigma_{1,2} = \pm\ 15.6\ \text{N/mm}^2$$

Direction of principal plane may be given as

$$\tan 2\theta_p = \frac{2\tau_{xy}}{\sigma_x-\sigma_y} = \frac{2\times(-15.6)}{0-0} = -\infty \qquad \theta_{P1} = \pm 45°\ \text{(clockwise)}$$

(ii) Magnitude of principal stresses at fixed end at 3 cm above the neutral axis:

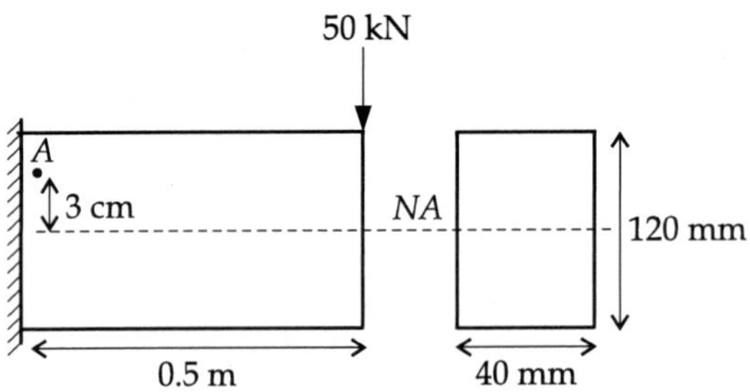

$S_A = 50$ kN and $M_A = 25$ kNm

Due to the vertical load of 50*kN*, the portion above NA will be in tension.

Normal bending stress and shear stress at a distance 3 cm above the NA, Bending stresses,

$$\sigma_A = \frac{M}{I}y = \frac{25\times10^6\times30}{\left(\frac{40\times120^3}{12}\right)} = 130.2\ \text{N/mm}^2$$

Shear stress,

$$\tau_A = \frac{SA\bar{y}}{Ib} = \frac{50\times10^3\times40\times30+45}{\left(\frac{40\times120^3}{12}\right)\times40} = 11.7\ \text{N/mm}^2$$

Positive shear will produce negative shear stress element,

$\tau_{xy} = -\tau = -11.7\ \text{N/mm}^2$

Principal stresses,

$$\sigma_1, \sigma_2 = \frac{\sigma_x + \sigma_y}{2} \pm \sqrt{\left(\frac{\sigma_x - \sigma_y}{2}\right)^2 + \tau_{xy}^2}$$

$$\sigma_1, \sigma_2 = \frac{130.2 \times 0}{2} \pm \sqrt{\left(\frac{130.2 - 0}{2}\right)^2 + (-11.7)^2}$$

$\sigma_1 = 131.25\ \text{N/mm}^2$ (tensile)

$\sigma_2 = -1.05\ \text{N/mm}^2$ (compressive)

Direction of principal plane may be given as

$$\tan 2\theta_p = \frac{2\tau_{xy}}{\sigma_x - \sigma_y} = \frac{2 \times (-11.7)}{130.2 - 0} = -0.18 \qquad \theta_{P1} = -5.1^\circ \text{ (clockwise)}$$

$\theta_{P2} = -90 + (-51^\circ) = -95.1$ (clockwise)

(iii) Magnitude of principal stresses at fixed end at bottom of the beam:

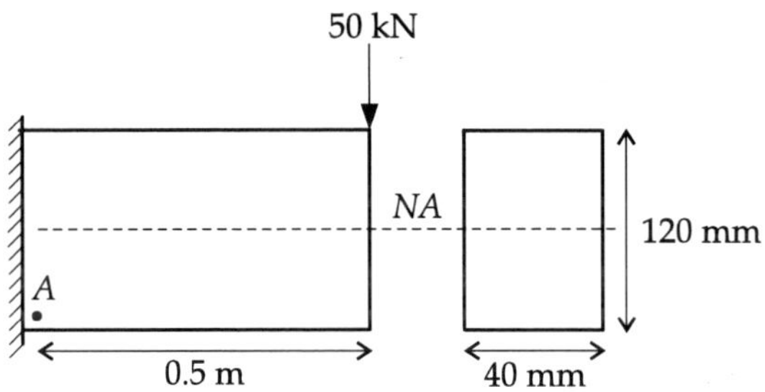

$S_A = 50$ kN and $M_A = 25$ kNm

Due to the vertical load of 50 kN, the portion below NA will be in compression.

Shear stress at top and bottom of the beam will be zero.

$\tau_A = 0$

Normal bending stress at bottom of the beam will be in compression (negative)

Bending stresses,

$$\sigma = \frac{M}{I} y = \frac{25 \times 10^6 \times 60}{\left(\frac{40 \times 120^3}{12}\right)} = 260.4\ \text{N/mm}^2\ \text{(C)}$$

Principal stresses,

$$\sigma_1, \sigma_2 = \frac{\sigma_x + \sigma_y}{2} \pm \sqrt{\left(\frac{\sigma_x - \sigma_y}{2}\right)^2 + \tau_{xy}^2}$$

$$\sigma_1, \sigma_2 = \frac{-260.4 \times 0}{2} \pm \sqrt{\left(\frac{-260.4 - 0}{2}\right)^2 + 0}$$

$\sigma_1 = 0$ and $\sigma_2 = 260.4\ \text{N/mm}^2$ (compressive)

Direction of the principal plane may be given as

$$\tan 2\theta_p = \frac{2\tau_{xy}}{\sigma_x - \sigma_y} = \frac{2 \times 0}{-260.4 - 0} = 0 \qquad \theta_{p1,\,p2} = 0°, 90°$$

Example 11.12: Consider a simply supported rectangular beam uniformly loaded as shown in Fig. 11.41. Calculate the principal stresses at a point *C*, 3 m from support *A* and 2 cm above NA.

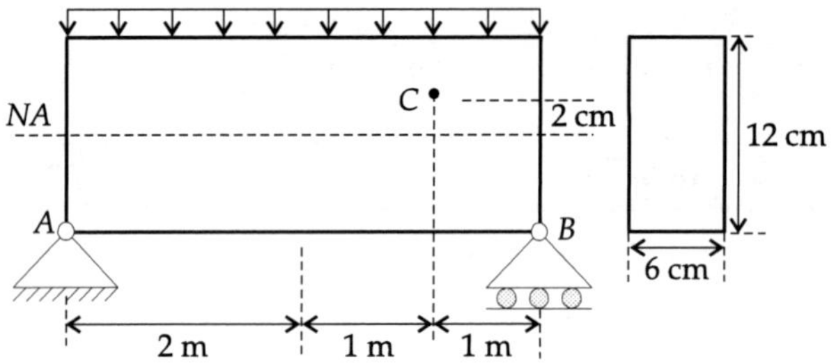

Fig. 11.41

Solution: Bending moment at *C*,

$$M_C = \frac{5 \times 4 \times 3}{2} - \frac{5 \times 3^2}{2} = 7.5 \text{ kNm}$$

Shear force at *C*,

$$V_C = \frac{5 \times 4}{2} - 5 \times 3 = -5 \text{ kNm}$$

Moment of inertia about neutral axis,

$$I_{NA} = \frac{6 \times 12^3}{12} = 864 \text{ cm}^4 = 8.64 \times 10^{-6} \text{ m}^4$$

Bending stress at *C*,

$$\sigma_C = \frac{My}{I_{NA}} = \frac{7.5 \times 10^3 \times 2 \times 10^{-2}}{8.64 \times 10^{-6}} = 1.74 \times 10^7 \text{ N/m}^2 = 17.4 \text{ MPa}$$

Shear stress at *C*,

$$\tau_C = \frac{VA\bar{y}}{Ib} = \frac{5 \times 10^3 \times (6 \times 4 \times 4 \times 10^{-6})}{8.64 \times 10^{-6} \times 6 \times 10^{-2}} = 9.26 \times 10^5 \text{ N/m}^2$$

$= 0.93$ MPa

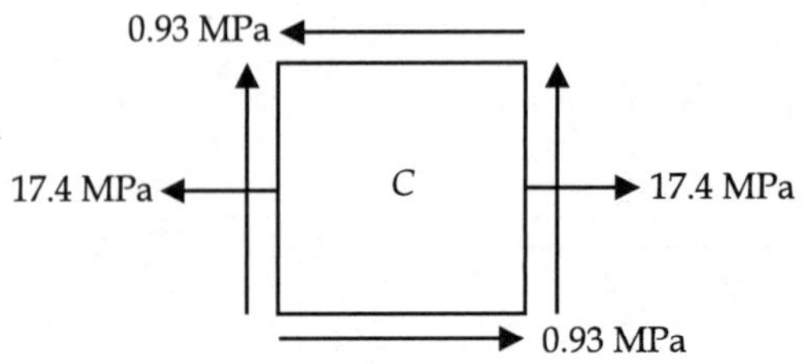

Principal stresses,

$$\sigma_{1,2} = \frac{\sigma_C}{2} \pm \sqrt{\frac{\sigma_C^2}{4} + \tau_C^2}$$

$$= \frac{17.4}{2} \pm \sqrt{\frac{17.4^2}{4} + 0.93^2} = -0.05 \text{ MPa and } 17.5 \text{ MPa}$$

Example 11.13: A rectangular steel beam 200 × 300 mm in crosssection is used as a simply supported beam over a span of 10 m and is subjected to a concentrated load of 10 kN at the centre inclined at 30° to the vertical axis as shown in Fig. 11.42.

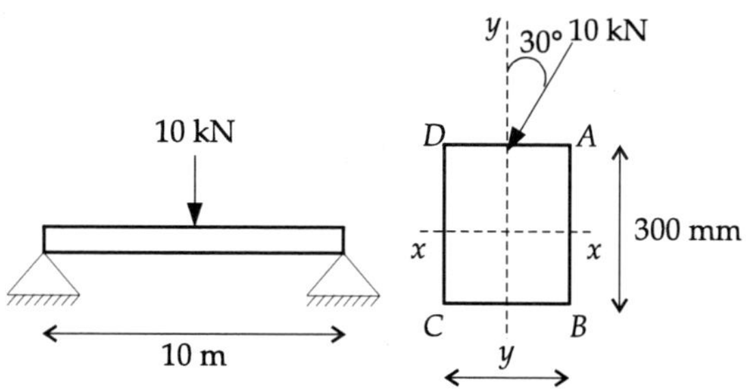

Fig. 11.42

Calculate the bending stresses (in N/mm^2) at four corners of the beam and resultant deflection of beam at the centre (Given: Modulus of elasticity of the beam is E = 210 GPa)

Solution:

Component of 10 kN in the direction of x-axis, $P_x = 10 \sin 30° = 5$ kN

Component of 10 kN in the direction of y-axis, $P_y = 10 \cos 30° = 5\sqrt{3}$ kN

Since loading is symmetrical and at centre, therefore,

$$H_A = H_B = \frac{5}{2} = 2.5 \text{ kN}$$

$$V_A = V_B = \frac{5\sqrt{3}}{2} = 2.5\sqrt{3} \text{ kN}$$

Maximum bending moment will occur at the centre, i.e., at $x = 5$ m from end, thus

Maximum bending moment due to horizontal load will be at the centre of the beam about y-axis,

$$M_y = H_A \times 5 = 2.5 \times 5 \text{ kN} = 12.5 \text{ kNm}$$

Maximum bending moment due to vertical load will be at the centre of the beam about x-axis,

$$M_x = V_A \times 5 = 2.5\sqrt{3} \times 5 \text{ kN} = 12.5\sqrt{3} \text{ kNm}$$

$$I_x = \frac{200 \times 300^3}{12} = 4.5 \times 10^8 \text{ mm}^4$$

$$I_y = \frac{300 \times 200^3}{12} = 2 \times 10^8 \text{ mm}^4$$

$$\sigma_{zx} = \frac{M_x}{I_x} \times y = \frac{12.5\sqrt{3} \times 10^6}{4.5 \times 10^8} \times 150 = 7.22 \text{ N/mm}^2$$

$$\sigma_{zy} = \frac{M_y}{I_y} \times x = \frac{12.5 \times 10^6}{2 \times 10^8} \times 100 = 7.22 \text{ N/mm}^2$$

Bending stress,

$$\sigma_z = \pm \frac{M_x}{I_x} \times y \pm \frac{M_y}{I_y} \times x = \pm\,\sigma_{zx} \pm \sigma_{zy}$$

Due to vertical component of load stresses at corners A and D will be compressive while stresses at corners B and C will be tensile.

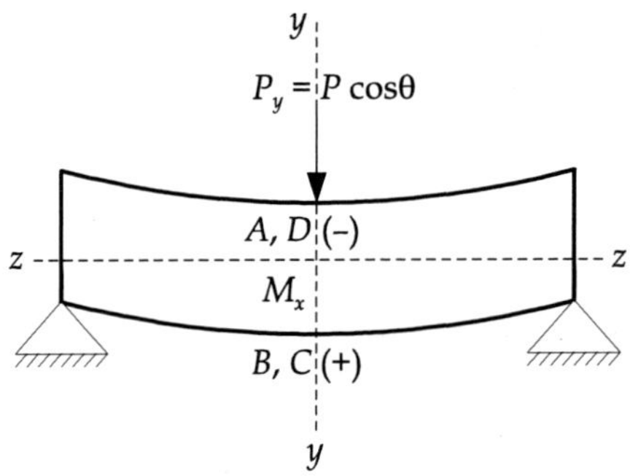

Due to the horizontal component of load stresses at corners A and B will be compressive while stresses at corners C and D will be tensile.

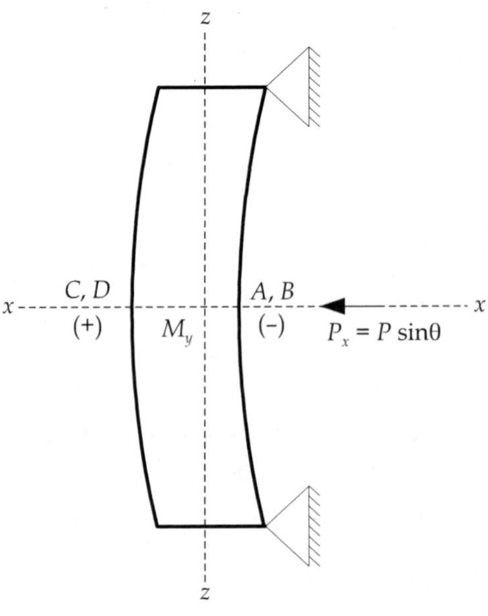

Stress at corner A,

$\sigma_A = -\sigma_{zx} - \sigma_{zy} = -7.22 - 6.25$

$= 13.5 \text{ N/mm}^2$ (compressive)

Stress at corner B,

$\sigma_B = +\sigma_{zx} - \sigma_{zy} = +7.22 - 6.25$

$= 1.0 \text{ N/mm}^2$ (tensile)

Stress at corner C,

$\sigma_A = \sigma_{zx} + \sigma_{zy} = 7.22 + 6.25$

$= 13.5 \text{ N/mm}^2$ (tensile)

Stress at corner D,

$\sigma_D = -\sigma_{zx} + \sigma_{zy} = -7.22 + 6.25$

$= 1.0 \text{ N/mm}^2$ (compressive)

Resultant deflection of beam at centre is given by

$$\delta_x = \frac{P_x L^3}{48EI_x} = \frac{5\times 10^3 \times (10\times 10^3)^3}{48\times 210\times 10^3 \times 4.5\times 10^8} = 1.1 \text{ mm}$$

$$\delta_y = \frac{P_y L^3}{48EI_y} = \frac{5\sqrt{3}\times 10^3 \times (10\times 10^3)^3}{48\times 210\times 10^3 \times 2\times 10^8} = 4.3 \text{ mm}$$

Resultant deflection, $\delta = \sqrt{\delta_x^2 + \delta_y^2} = \sqrt{1.1^2 + 4.3_y^2} = 4.44$ mm

Example 11.14: Consider the channel section as shown in Fig. 11.43. Calculate the value of shear centre, 'e'.

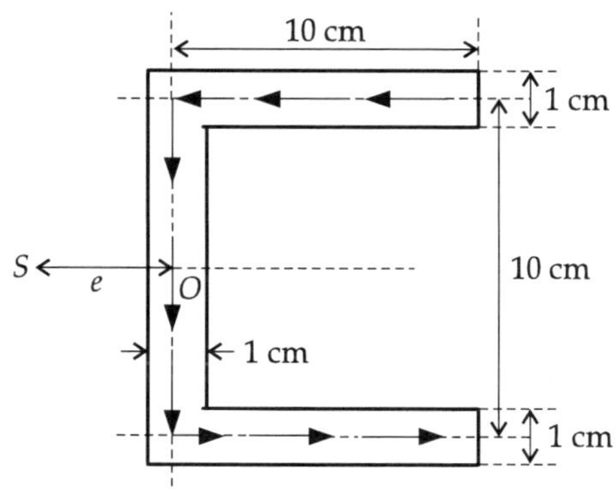

Fig. 11.43

Solution:

$t_f = t_w = 1$ cm; $b = 10$ cm; $h = 10$ cm;

For shear force, F_1

$A = t_f x$; $dA = t_f dx$; $\bar{y} = \frac{h}{2}$

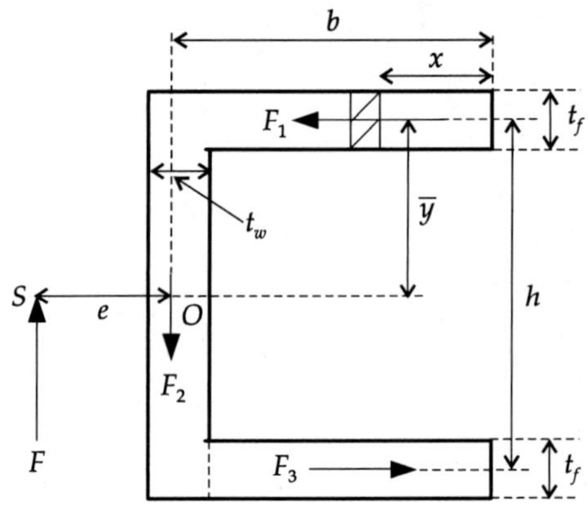

Note: Moment of inertia will be calculated by using centreline dimensions.

$$I = \frac{t_w h^3}{12} + 2\left[\frac{bt_f^3}{12} + bt_f\left(\frac{h}{2}\right)^2\right] \approx \frac{t_w h^3}{12} + \frac{bt_f h^2}{2}$$

$$\Rightarrow \qquad I = \frac{1\times 10^3}{12} + \frac{10\times 1\times 10^2}{2} = 583.3 \text{ cm}^4$$

$$F_1 = \int \frac{FA\overline{y}}{It} dA = \frac{F}{It_f}\int_0^b xt_f\left(\frac{h}{2}\right)t_f\,dx$$

$$\Rightarrow \qquad F_1 = \frac{Ft_f hb^2}{4I}$$

Due to symmetry of cross section along x-axis the shear force, $F_3 = F_1$

Taking moment about O,

$$F\times e = F_1 \times\left(\frac{h}{2}\right) + F_3\times\left(\frac{h}{2}\right) \quad e = \frac{t_f h^2 b^2}{4I} = \frac{1\times 10^2\times 10^2}{4\times 583.3} = 4.3 \text{ cm}$$

Example 11.15: Consider the channel section as shown in Fig. 11.44. Calculate the value of shear centre, 'e'.

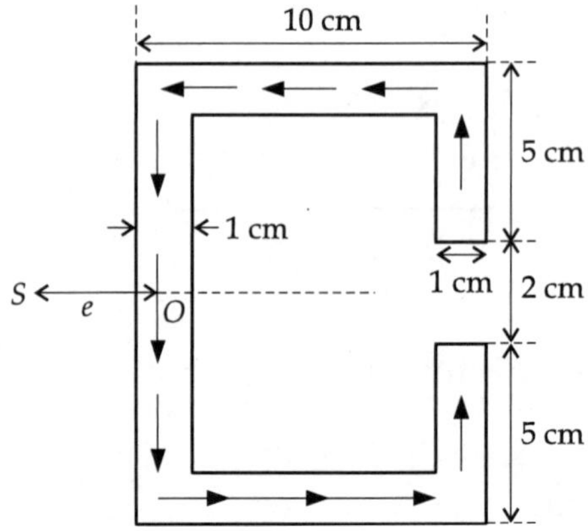

Fig. 11.44

Solution:

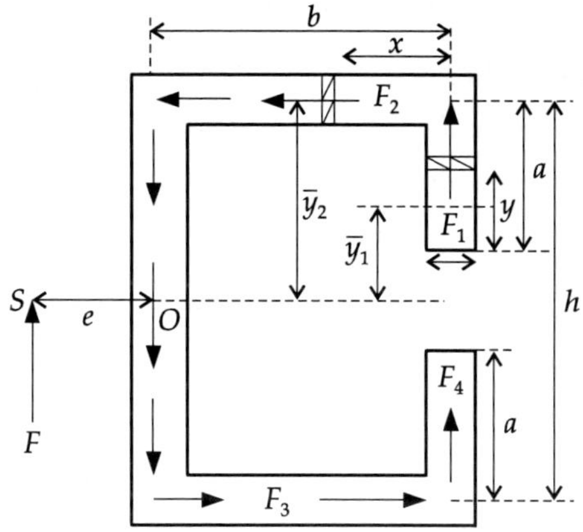

Using centreline dimensions:

$t_f = t_w = t = 1$ cm; $a = 5 - 0.5 = 4.5$ cm;

$b = 10 - 1 = 9$ cm; $h = 12 - 1 = 11$ cm;

$$I = 2\left[\frac{bt^3}{12} + bt\left(\frac{h}{2}\right)^2\right] + \frac{th^3}{12} + 2\left[\frac{ta^3}{12} + ta\left(\frac{h}{2} - \frac{a}{2}\right)^2\right]$$

$$I \approx \frac{bth^2}{2} + \frac{th^3}{12} + \frac{2ta^3}{3} + \frac{tah^2}{2} - ta^2h$$

For shear force, F_1

$$A = t\,y;\quad dA = t\,dy;\quad \bar{y}_1 = \frac{h}{2} - a + \frac{y}{2}$$

$$F_1 = \int \frac{FA\bar{y}}{It}dA = \frac{F}{It}\int_0^a t\,y\left(\frac{h}{2} - a + \frac{y}{2}\right)t\,dy$$

$$\Rightarrow F_1 = \frac{Fta^2}{I}\left(\frac{h}{4} - \frac{a}{3}\right)$$

For shear force, F_2

$$A\bar{y} = t\,x\left(\frac{h}{2}\right) + t\,a\left(\frac{h}{2} - \frac{a}{2}\right);\qquad dA = t\,dx;\qquad \bar{y} = \frac{h}{2}$$

$$F_2 = \int \frac{FA\bar{y}}{It}dA = \frac{F}{It}\int_0^b \left[t\,x\left(\frac{h}{2}\right) + t\,a\left(\frac{h}{2} - \frac{a}{2}\right)\right]t\,dx \qquad \Rightarrow F_2 = \frac{F}{I}\left[hb^2 + 2ab(h-a)\right]$$

Due to symmetry of cross section along x-axis the shear force, $F_3 = F_2$ and $F_4 = F_1$

Taking moment about O,

$$F \times e = 2 \times \left[F_1 \times (b) \times F_2 \times \left(\frac{h}{2}\right)\right]$$

$$\Rightarrow \quad e = \frac{2t}{I} \times \left[\frac{h^2b^2}{8} + \frac{abh^2}{4} - \frac{a^3b}{3} \right]$$

$$\Rightarrow \quad e = \frac{3h^2b^2 + 6abh^2 - 8a^3b}{6bh^2 + h^3 + 8a^3 + 6ah^2 - 12a^2h} \Rightarrow e = \frac{52245}{9188} \approx 5.7 \text{ cm}$$

Example 11.16: Consider the I-section section as shown in Fig. 11.45. Calculate the value of shear centre, '*e*'.

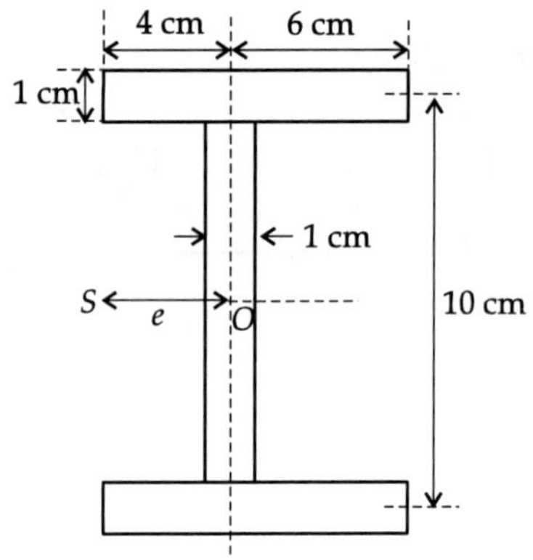

Fig. 11.45

Solution:

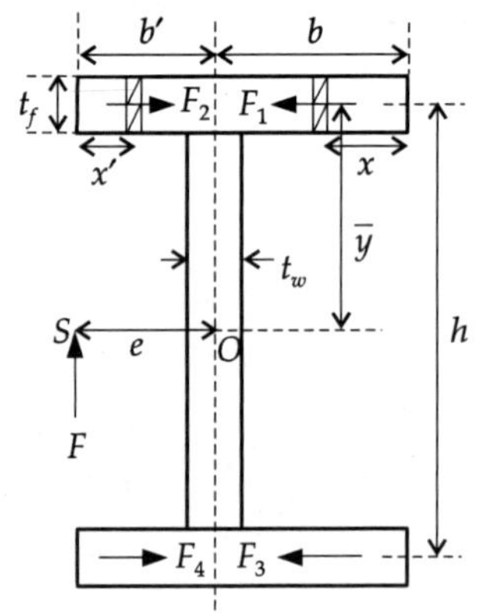

Using centre line dimensions:

$t_f = t_w = t = 1$ cm; $\quad b = 6$ cm;

$b' = 4$ cm; $\quad h = 10$ cm;

$$I = \frac{t_w h^3}{12} + 2\left[\frac{(b+b')t_f^3}{12} + (b+b')t_f \left(\frac{h}{2}\right)^2 \right]$$

$$\approx \frac{t_w h^3}{12} + \frac{(b+b')t_f h^2}{2} = \frac{1 \times 10^3}{12} + \frac{(4+6) \times 1 \times 10^2}{2}$$

$$= 583.3 \text{ cm}^4$$

For shear force, F_1

$$A = t_f x; \quad dA = t_f dx; \quad \bar{y} = \frac{h}{2}$$

$$F_1 = \int \frac{FA\bar{y}}{It} dA = \frac{F}{It_f} \int_0^b t_f x \left(\frac{h}{2}\right) t_f dy$$

$$\Rightarrow F_1 = \frac{F t_f b^2 h}{4I}$$

Similarly, shear force, F_2

$$F_2 = \int \frac{FA\bar{y}}{It} dA = \frac{F}{It_f} \int_0^{b'} t_f x' \left(\frac{h}{2}\right) t_f dy$$

$$\Rightarrow F_2 = \frac{F t_f b'^2 h}{4I}$$

Due to symmetry of cross section along x-axis the shear force, $F_3 = F_1$ and $F_4 = F_2$

Taking moment about O,

$$F \times e = 2 \times \left[F_1 \times \left(\frac{h}{2}\right) - F_2 \times \left(\frac{h}{2}\right) \right]$$

$$\Rightarrow e = \frac{t_f (b^2 - b'^2) h^2}{4I}$$

Note: At $b' = 0$, we get the eccentricity for channel section, and at $b = b'$, we get that the eccentricity for symmetrical I-section passes through centroid, i.e., $e = 0$.

For given un-symmetrical I-section,

$$e = \frac{t_f (b^2 - b'^2) h^2}{4I}$$

$$\Rightarrow e = \frac{1 \times (6^2 - 4^2) \times 10^2}{4 \times 583.3} \approx 0.86 \text{ cm}$$

Example 11.17: Consider the H-section as shown in Fig. 11.46. Calculate the value of shear centre, 'e'.

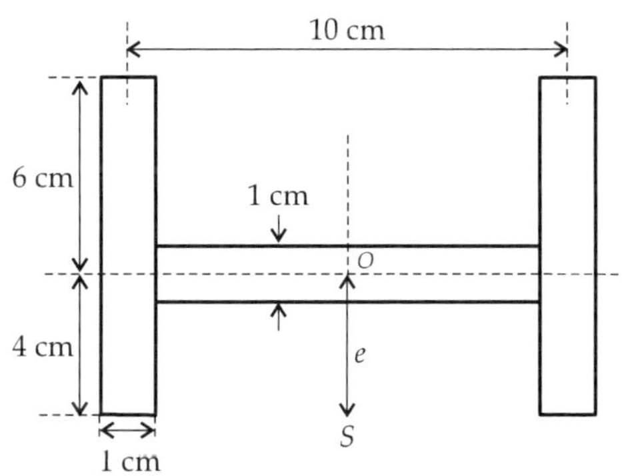

Fig. 11.46

Solution:

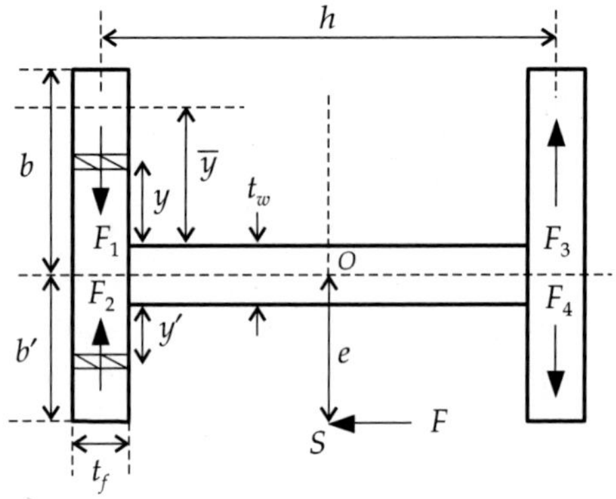

Using centre line dimensions:

$t_f = t_w = t = 1$ cm; $b = 6$ cm;

$b' = 4$ cm; $h = 10$ cm;

$$I = \frac{ht_w^3}{12} + 2\left[\frac{t_f b^3}{12} + t_f b\left(\frac{b}{2}\right)^2 + \frac{t_f b'^3}{12} + t_f b'\left(\frac{b'}{2}\right)^2\right]$$

$$\approx \frac{2t_f(b^3 + b'^3)}{3} = \frac{2\times 1\times(6^3 \times 4^3)}{3} = 186.7 \text{ cm}^4$$

For shear force, F_1

$$A = t_f(b-y); \quad dA = t_f dy; \quad \overline{y} = \frac{b+y}{2}$$

$$F_1 = \int \frac{FA\overline{y}}{It} dA = \frac{F}{It_f}\int_0^b t_f(b-y)\left(\frac{b+y}{2}\right)t_f dy$$

$$\Rightarrow F_1 = \frac{Ft_f b^3}{3I}$$

Similarly, shear force, F_2

$$F_2 = \frac{Ft_f b'^3}{3I}$$

Due to symmetry of cross section along x-axis the shear force, $F_3 = F_1$ and $F_4 = F_2$

Taking moment about O,

$$F \times e = 2\times\left[F_1 \times\left(\frac{h}{2}\right) - F_2 \times\left(\frac{h}{2}\right)\right]$$

$$\Rightarrow e = \frac{t_f(b^3 - b'^3)h}{3I}$$

Note: At $b' = 0$, we get the eccentricity for inverted channel section, and at $b = b'$, we get that the eccentricity for symmetrical H-section passes through centroid, i.e., $e = 0$

For given un-symmetrical H-section,

$$e = \frac{t_f(b^3 - b'^3)h}{3I}$$

$$\Rightarrow e = \frac{1 \times (6^3 - 4^3) \times 10}{3 \times 186.7} \approx 2.7\,\text{cm}$$

Example 11.18: Consider the H-section section as shown in Fig. 11.47. Calculate the ratio of shear centre distance, 'e_1/e_2'.

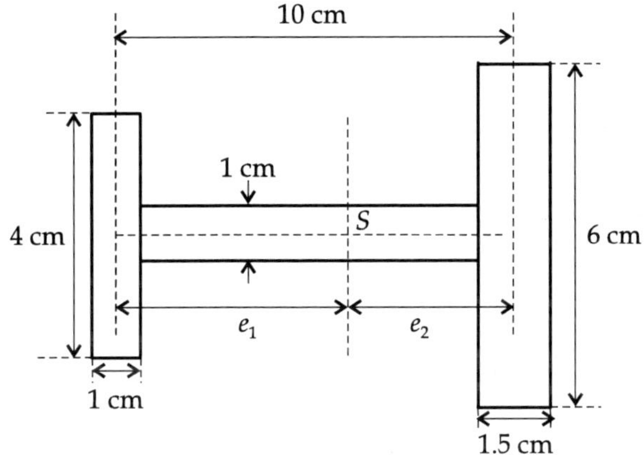

Fig. 11.47

Solution:

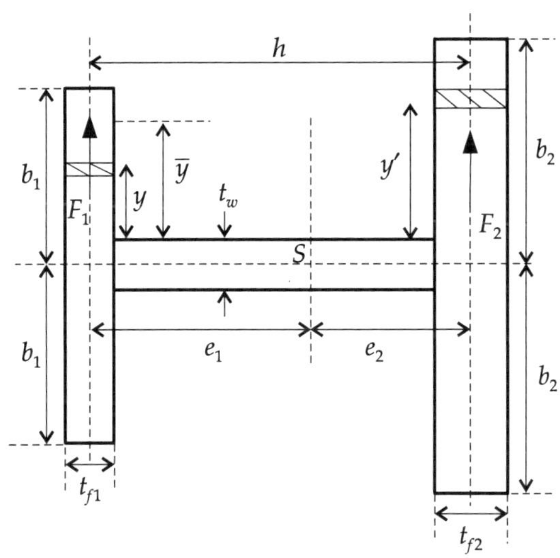

Using centreline dimensions:

$t_{f1} = 1$ cm; $t_{f2} = 1.5$ cm $b_1 = 2$ cm;

$b_2 = 3$ cm; $h = 10$ cm;

For shear force, F_1

$$A = t_{f1}(b_1 - y); \quad dA = t_{f1}dy; \quad \bar{y} = \frac{b_1 + y}{2}$$

$$F_1 = \int \frac{FA\bar{y}}{It} dA = \frac{F}{It_{f1}} \int_0^b t_{f1}(b_1 - y)\left(\frac{b_1 + y}{2}\right) t_{f1} dy$$

$$\Rightarrow F_1 = \frac{Ft_{f1}b_1^3}{3I}$$

Similarly, shear force, F_2

$$F_2 = \frac{Ft_{f2}b_2^3}{3I}$$

Taking moment about S,

$$F_1 \times e_1 = F_2 \times e_2$$

$$\Rightarrow \frac{e_1}{e_2} = \frac{t_2}{t_1} \times \left(\frac{b_2}{b_1}\right)^3 = \frac{1.5}{1} \times \left(\frac{3}{2}\right)^3 = 1.5^4$$

Example 11.19: Consider the semicircular section as shown in Fig. 11.48. Calculate the value of shear centre (e_s) of the semicircular section.

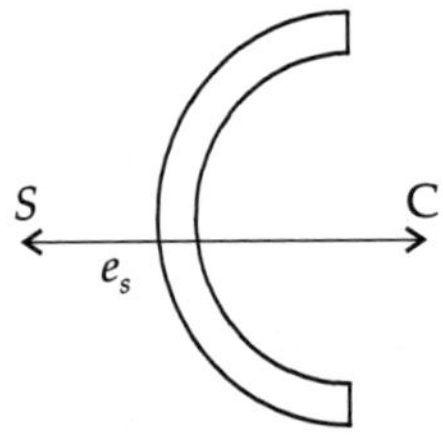

Fig. 11.48

Solution:

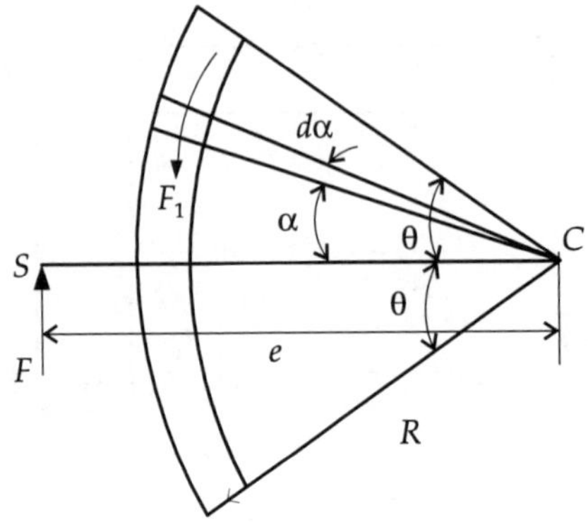

For shear force, F_1

$A = R(\theta - \alpha)t; \qquad dA = Rt\,d\alpha; \qquad \bar{y} = R\sin\alpha$

$I = tR^3\{\theta - \sin\theta\cos\theta\}$

$$F_1 = \int \frac{FA\bar{y}}{It} dA = \frac{F}{It}\int_{-\theta}^{\theta} R(\theta-\alpha)t.R\sin\alpha.Rtd\alpha$$

$$= \frac{FR^3t}{I}\int_{-\theta}^{\theta}(\theta-\alpha)\sin\alpha d\alpha$$

$$= \frac{2FR^3t}{I}(\theta\cos\theta - \sin\theta)$$

Taking moment about C,

$$F \times e = -F_1 \times R \Rightarrow e = \frac{2R^4t}{I}(\sin\theta - \theta\cos\theta)$$

$$\Rightarrow e = \frac{2R(\sin\theta - \theta\cos\theta)}{(\theta - \sin\theta\cos\theta)}$$

For semi-circular section $\theta = \pi/2 \Rightarrow e_s = \dfrac{4R}{\pi}$

Example 11.20: A tapered cantilever beam of solid circular section supports a load P at the free end (Fig. 11.49). Calculate how the diameter of the beam varies so that the maximum normal stress at every cross section will be equal to the allowable stress, σ_{allow}. (A beam meeting this condition is called a **fully stressed beam** or a **beam of constant strength.**)

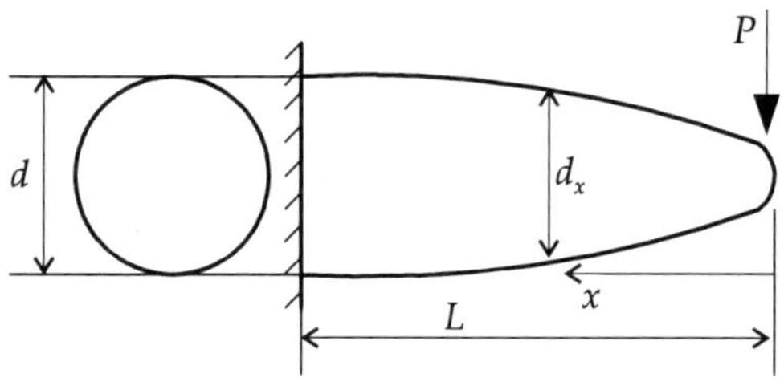

Fig. 11.49

Let d_x, M_x and Z_x be the diameter bending moment and section modulus of the beam at a distance x from the free end.

Then, $M_x = Px$, $Z_x = \dfrac{\pi d_x^3}{32}$

From flexure formula,

$$\sigma_x = \frac{M_x}{Z_x} = \frac{32Px}{\pi d_x^3} = \sigma_{\text{allow}}$$

Solving for the diameter of the beam,

$$d_x = \left(\frac{32Px}{\pi\sigma_{\text{allow}}}\right)^{1/3}$$

At the fixed end of the beam ($x = L$),

$$d = \left(\frac{32PL}{\pi\sigma_{\text{allow}}}\right)^{1/3}$$

Therefore, $d_x = d\left(\frac{x}{L}\right)^{1/3}$ i.e., diameter of fully stressed beam varies with the cube root of x.

Example 11.21: A tapered cantilever beam of a solid circular section supports a load P at the free end (Fig. 11.50). Determine the bending stress at the fixed support and the maximum bending stress, σ_{max}.

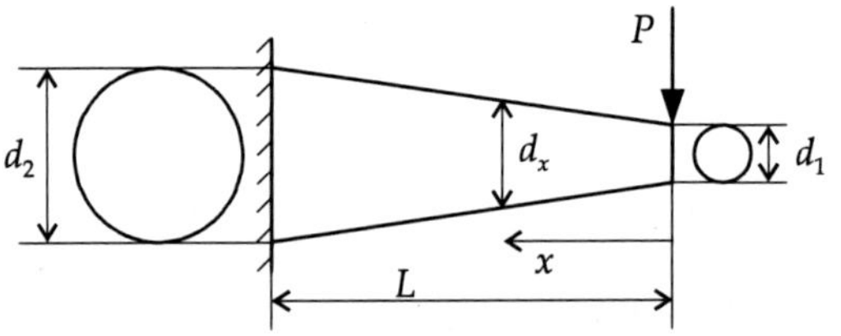

Fig. 11.50

Let d_x, M_x and Z_x be the diameter bending moment and section modulus of the beam at a distance x from the free end.

Then, $d_x = d_1 + (d_2 - d_1)\frac{x}{L}$

$M_x = Px$ and $Z_x = \frac{\pi d_x^3}{32}$

From flexure formula,

$$\sigma_x = \frac{M_x}{Z_x} = \frac{32Px}{\pi\left[d_1 + (d_2 - d_1)(x/L)\right]^3}$$

Putting $x = L$, we get stress at fixed end

$$\sigma_{\text{fixed}} = \frac{32PL}{\pi\left[d_1 + (d_2 - d_1)\right]^3}$$

The location of the cross section having the largest bending stress is obtained by taking derivative of stress (σ_x) with respect to x, i.e.,

$$\frac{d\sigma_x}{dx} = \frac{d}{dx}\left(\frac{32Px}{\pi\left[d_1 + (d_2 - d_1)(x/L)\right]^3}\right)$$

$$\Rightarrow x = \frac{d_1 L}{2(d_2 - d_1)}$$

The corresponding maximum stress, obtained by substituting, $x = \dfrac{d_1 L}{2(d_2 - d_1)}$ in σ_x,

$$\sigma_{max} = \frac{128PL}{27\pi(d_2 - d_1)d_1^2}$$

Example 11.22: Determine the ratio of the weights (or area) of the following sections (square of side a; circle of diameter b; and triangle of base = height = c), if all the sections are made of the same material and are of equal strength (or section modulus). Find which of them is more economical.

Section modulus:

Square

$$Z_S = \frac{I}{y_{max}} = \frac{a^3}{6}$$

Circle

$$Z_C = \frac{I}{y_{max}} = \frac{\pi d^3}{32} = \frac{\pi b^3}{32}$$

Triangle

$$Z_T = \frac{I}{y_{max}} = \frac{BH^2}{24} = \frac{c^3}{24}$$

Since all the sections are of equal strength (or section modulus), therefore,

$$Z_S = Z_C = Z_T \Rightarrow \frac{a^3}{6} = \frac{\pi b^3}{32} = \frac{c^3}{24}$$

$\Rightarrow a = 0.8383b = 0.63c$

$\Rightarrow b = 1.193a$ and $c = 1.5873a$

If W_S, W_C and W_T and A_S, A_C and A_T are the weights and areas of square, circle and triangle respectively, then

$A_S = a^2$; $A_C = \dfrac{\pi}{4}b^2 = 1.1178a^2$; $A_T = \dfrac{1}{2}c^2 = 1.26a^2$ Since, ratio of weight = ratio of area, i.e.,

$W_S : W_C : W_T = A_S : A_C : A_T \Rightarrow a^2 : 1.1178a^2 : 1.26a^2$

$\Rightarrow 1 : 1.12 : 1.26$

Thus, a square section is more economical than circle and triangle since for the same strength it has lowest weight.

Example 11.23: A thin steel wire of diameter 4 mm is bent over a circular drum of radius 400 mm as shown in Fig. 11.51. Calculate the maximum bending stress and bending moment in the wire. (Given: E = 200 GPA)

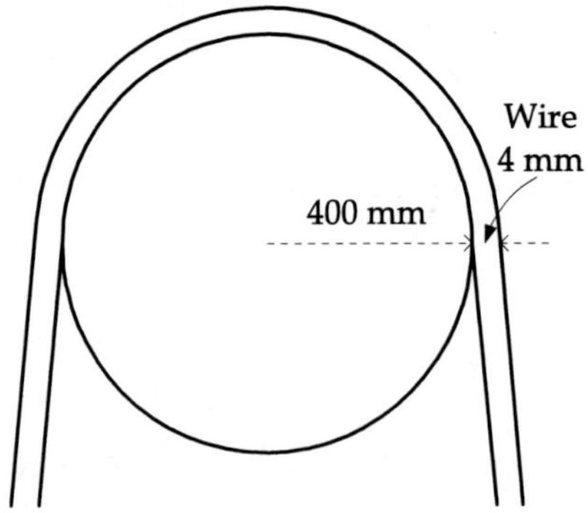

Fig. 11.51

Solution: Radius of drum, $r = 400$ mm

Diameter of steel wire, $d = 4$ mm

Radius of curvature, $R = r + \frac{d}{2} = 400 + 2 = 402$ mm

Maximum (tensile or compressive) stress will occur at the boundary of wire,

$$y_{max} = \frac{d}{2} = 2 \text{ mm}$$

$$\sigma_{max} = \frac{Ey_{max}}{R} = \frac{2 \times 10^5 \times 2}{402} = 995 \text{ MPA}$$

Section modulus, $Z = \frac{\pi d^3}{32} = \frac{\pi \times 4^3}{32} = 6.283 \text{ mm}^3$

Maximum bending moment in the wire,

$M_{max} = \sigma_{max} Z = 995 \times 6.283 = 6251.8 \text{ Nmm} = 6.25 \text{ Nm}$

Example 11.24: Calculate the area of the strongest rectangular section that can be cut from the circular log of timber of diameter, D.

Solution: The strength of any section depends on its flexural strength, i.e., larger be the section modulus (Z) more be the strength of section. Let the dimension of the strongest rectangular section cut from the circular log of diameter D be x and y then $D^2 = x^2 + y^2$ $\Rightarrow y^2 = D^2 - x^2$

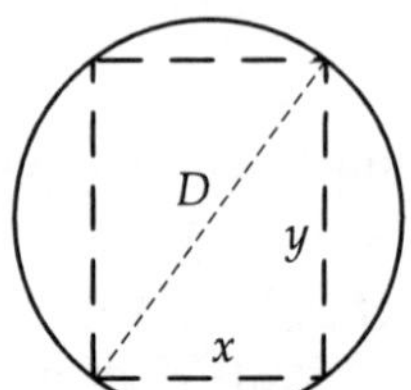

Fig. 11.52

Flexural strength of rectangular section,

$$Z = \frac{I}{y} = \frac{xy^3/12}{y/2} = \frac{xy^2}{6}$$

For Z to be maximum, $\frac{dZ}{dx} = 0$

$$\Rightarrow \frac{d}{dx}\left(\frac{xy^2}{6}\right) = 0 \Rightarrow \frac{d}{dx}\left\{\frac{1}{6}x(D^2 - x^2)\right\} = 0$$

$$\Rightarrow D^2 - 3x^2 = 0 \Rightarrow x = D/\sqrt{3}$$

$$y^2 = D^2 - x^2 \Rightarrow y = \sqrt{\frac{2}{3}}D$$

Therefore, the area of the strongest rectangular section from circular log of diameter D,

$$\text{Area} = x.y = \frac{D}{\sqrt{3}}.\sqrt{\frac{2}{3}}D = \frac{\sqrt{2}D^2}{3}$$

Example 11.25: Calculate the ratio of moment of resistance of a square beam section of the side 'a' is placed with one diagonal horizontal to the section placed with two sides are horizontal for same permissible bending stress.

Solution: Since $M = \sigma_{max}Z$, thus the MOR of any section will depend on its section modulus (Z) for same bending permissible stress.

When the beam is placed with one diagonal horizontal (Fig. 11.53)

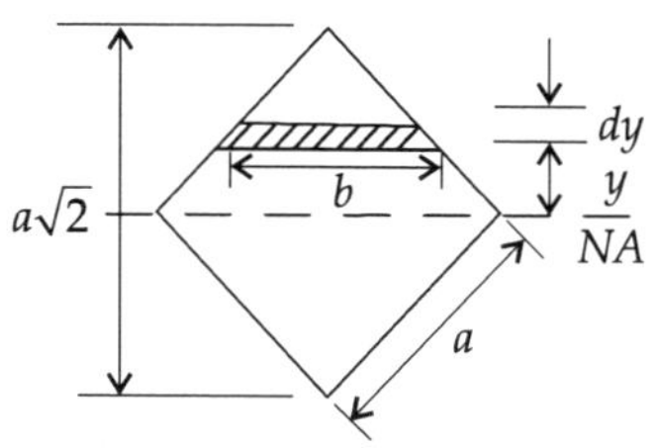

Fig. 11.53

Consider an elementary strip of thickness (dy) at an distance (y) from neutral axis. Let the width of the strip be b, then

$$\frac{b}{a\sqrt{2}} = \frac{a/\sqrt{2} - y}{a/\sqrt{2}} \Rightarrow b = a\sqrt{2} - 2y$$

Moment of inertia about the neutral axis,

$$I = 2\int_0^y (bdy)(y^2) = 2\int_0^{a/\sqrt{2}} (a\sqrt{2} - 2y)y^2dy \Rightarrow I = \frac{a^4}{12}$$

Distance to the extreme fibre, $y_{max} = \frac{d}{2} = \frac{a}{\sqrt{2}}$

Section modulus, $Z_1 = \frac{I}{y_{\max}} = \frac{a^4/12}{a\sqrt{2}/2} = \frac{a^3}{6\sqrt{2}}$

$$M_1 = \sigma_{\max} Z_1 = \sigma_{\max} \frac{a^3}{6\sqrt{2}}$$

When the beam is placed with two sides are horizontal

$$M_2 = \sigma_{\max} Z_2 = \sigma_{\max} \frac{a.a^3/12}{a/2} = \frac{a^3}{6}$$

$$\frac{M_1}{M_2} = \frac{\sigma_{\max} Z_1}{\sigma_{\max} Z_2} = \frac{a^3/6\sqrt{2}}{a^3/6} = 1/\sqrt{2}$$

PROBLEMS

1. Consider a simply supported rectangular section beam as shown in Fig. 11.54 (i) Calculate the maximum tensile and compressive bending stresses in the beam. (ii) Calculate the maximum shear stress in the beam.

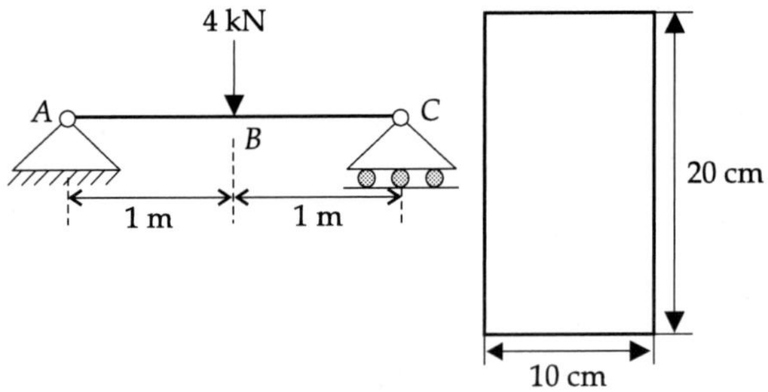

Fig. 11.54

Ans. $\sigma_{t,\max} = 3$ MPa $\tau_{\max} = 0.6$ MPa

2. A timber beam 10 cm × 20 cm is reinforced by two steel plate beams 10 cm × 1 cm at top and bottom as shown in Fig. 11.55. Calculate the resisting moment of flitched beam.

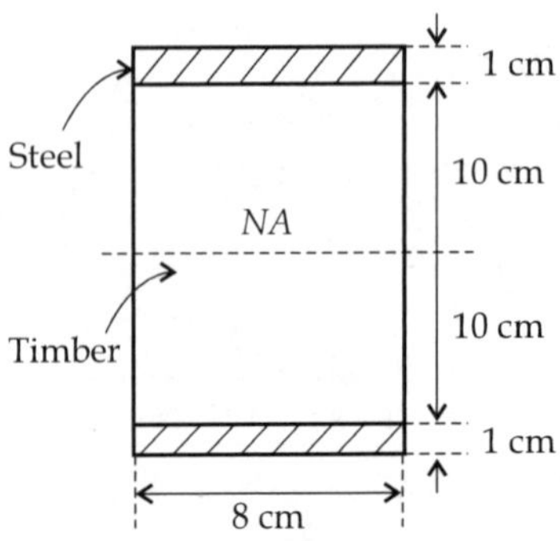

Fig. 11.55

(Given: Permissible bending stress in timber, $\sigma_t = 10$ MN/m^2; $E_s = 250$ GN/m^2; $E_t = 25$ GN/m^2))

Ans. $M = 23$ kNm

3. A timber beam 10 cm × 20 cm is sand-witched between two steel plate beams 1 cm × 20 cm as shown in Fig. 11.56. Calculate the resisting moment of flitched beam.

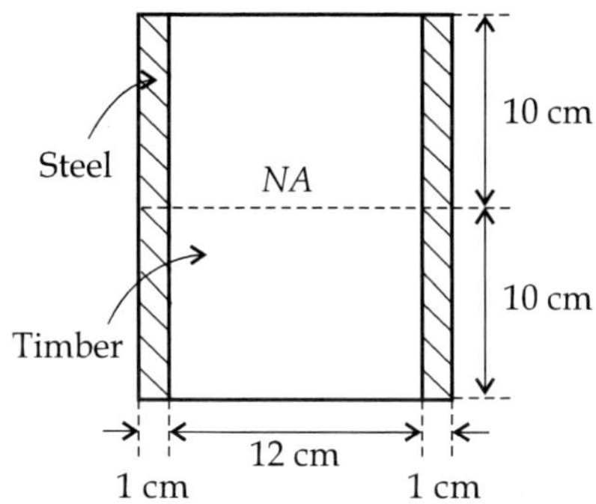

Fig. 11.56

(Given: Permissible bending stress in timber, $\sigma_t = 10$ MN/m^2; $E_s = 250$ GN/m^2; $E_t = 25$ GN/m^2))

Ans. $M = 21.3$ kNm

4. A horizontal steel cantilever beam of length 0.5 m is fixed at one end and free at another supports a vertical concentrate load of 50 kN at the free end. The cross section of the cantilever beam is rectangular, 4 cm wide and 12 cm deep. Calculate the magnitude of principal stresses at the fixed end (i) at the neutral axis, (ii) at 3 cm above the neutral axis, and (iii) at bottom of the beam (Fig. 11.57)

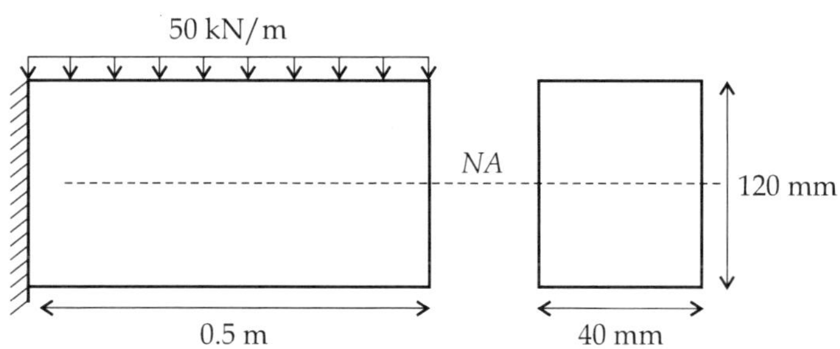

Fig. 11.57

Ans.

(i) Magnitude of principal stresses at fixed end at the neutral axis,

$\sigma_{1,2} = \pm 7.8$ N/mm^2

(ii) Magnitude of principal stresses at fixed end at 3 cm above the neutral axis,

$\sigma_1 = 65.6$ N/mm^2 (tensile)

$\sigma_2 = 0.52$ N/mm^2 (compressive)

(iii) Magnitude of principal stresses at fixed end at bottom of the beam,
$\sigma_1 = 0$ and $\sigma_2 = 130.4$ N/mm² (compressive)

5. Consider the simply supported rectangular beam uniformly loaded as shown in Fig. 11.58. Calculate the principal stresses at a point *C*, 3 m from support *A* and 2 cm above NA.

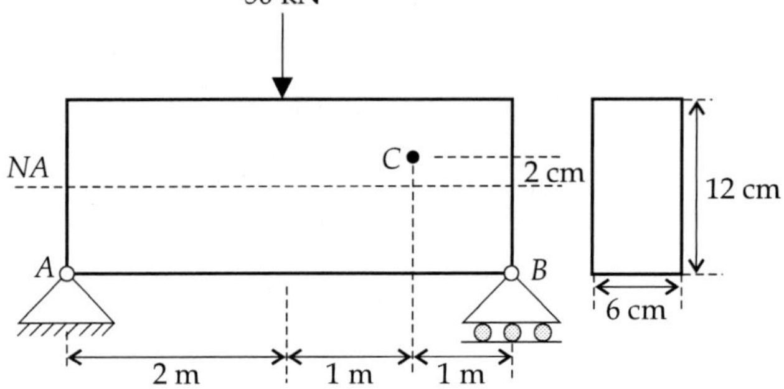

Fig. 11.58

Ans. $\sigma_{1,2}$ = –0.04 MPa and 58.4 MPa

6. Consider the circular tube as shown in Fig. 11.59. Calculate the value of shear centre 'e_t' of circular tube.

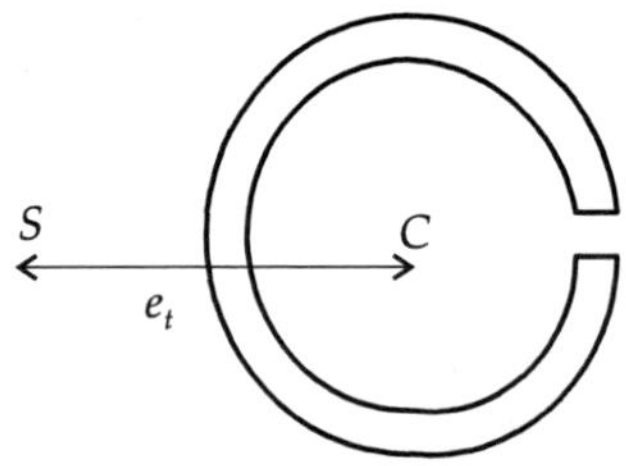

Fig. 11.59

Ans. For circular tube section, $\theta = \pi \Rightarrow e_t = 2R$

12
Buckling of Columns

12.1 INTRODUCTION

The vertical member of any structural system or frame or truss, which carries mainly axial compression loads are called struts and columns. The load carrying capacity of a compression member depends on its cross-sectional shape, length or height, end support conditions and initial geometrical condition of the member.

When an applied load is increased on a vertical member, it becomes large enough to cause the member to become unstable and failure may occur under buckling well before any crushing in the member. If the axial compressive load is gradually increased buckling occurs at some critical value of loading and the column deflects sideways suddenly. The sideway deflection occurs in any plane that is perpendicular to the line of action of axial loading. This critical value of loading is known as **buckling load** and the column begins to deflect in transverse direction. It mainly occurs in long columns even though the stresses that develop in the vertical member are below yield stress.

Since buckling occurs in vertical members due to axial loads, the main reasons of buckling are, (i) Member is not perfectly straight (Fig. 12.1 (a)), (ii) axial load is not perfectly along the centerline of the column (Fig. 12.1 (b)), and (iii) member is not uniform and homogeneous.

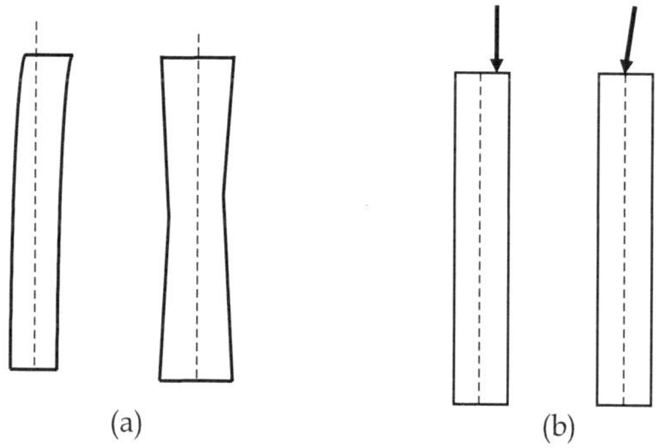

Fig. 12.1

12.2 TYPES OF COLUMN DEPENDING ON END CONDITIONS

There are four types of columns depending on end support condition:

(i) Both ends of the member are pinned or hinged.

(ii) One end is fixed and the other end is free.

(iii) Both ends are fixed or clamped.

(iv) One end is fixed and the other end is pinned.

In this section, we wish to find the critical load or buckling load for the above-mentioned columns with the following assumptions:

(i) Before application of axial load, the column was perfectly straight one.

(ii) The line of action of compressive loads exactly coincides with the unstrained axis of the column.

(iii) Flexural rigidity (EI) is uniform throughout the column and the cross section is uniform.

(iv) Column material is homogeneous, isotropic and obeys Hooke's law.

12.2.1 Both Ends of the Column are Pinned or Hinged

Let a column AB of length (l) and uniform cross section area (A) is hinged at both the ends, A and B. Let a compressive load (P) be applied at which the column begins to deflect in transverse direction. The deflected shape of the column is shown in Fig. 12.2 where the centerlines of the original straight column coincide with the x-axis. The origin point is A. Let y be the deflection at the section.

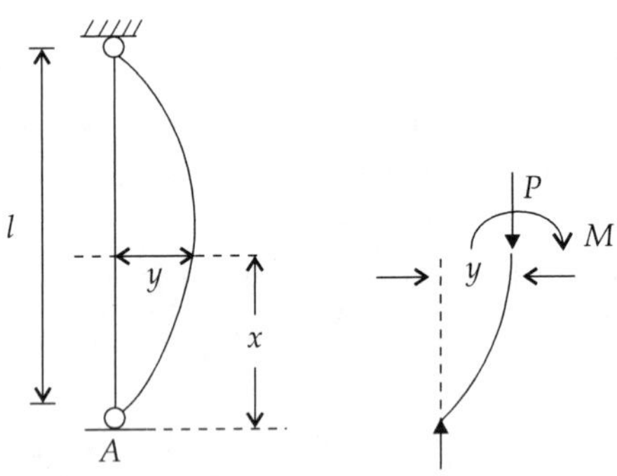

Fig. 12.2

Then bending moment (M) at the section, x-x

$$M = Py$$

Then we can write

$$EI\frac{d^2y}{dx^2} = -Py$$

$$EI\frac{d^2y}{dx^2} + Py = 0$$

$$\frac{d^2y}{dx^2}+\frac{P}{EI}y=0$$

Let $$k^2=\frac{P}{EI}$$

$$\frac{d^2y}{dx^2}+k^2y=0$$

Solution of the above differential equation,

$y = C_1 \cos kx + C_2 \sin kx$

where, C_1, C_2 are integration constants.

$$y=C_1\cos\left(x\sqrt{\frac{P}{EI}}\right)+C_2\sin\left(x\sqrt{\frac{P}{EI}}\right)$$

Putting the end boundary condition

at A, the deflection is zero, i.e.,

at $x = 0$, $y = 0$

$\therefore$ $$C_1=0$$

At B also, deflection is zero, i.e., at $x = l$, $y = 0$.

$$0=C_2\sin\left(l\sqrt{\frac{P}{EI}}\right)$$

$\because$ $C_1 = 0$, we conclude that C_2 cannot be zero, because if both C_1, C_2 are zero, the column will not bend at all.

Hence, $$\sin\left(l\sqrt{\frac{P}{EI}}\right)=0$$

$$C_2\sin kl=0$$

$$kl=n\pi$$

$$k=\frac{n\pi}{l}$$

$$n=0,1,2....$$

Taking the value of $n = 1$, the fundamental value,

$$\sqrt{\frac{P}{EI}}=\frac{\pi}{l}$$

$$P=\frac{\pi^2EI}{l^2}$$

The lowest value of P is known as Euler load or critical load or buckling load

The Euler buckling load, $P_{cr}=\dfrac{\pi^2EI}{l^2}$

12.2.2 One End of the Column is Fixed and Other End Free

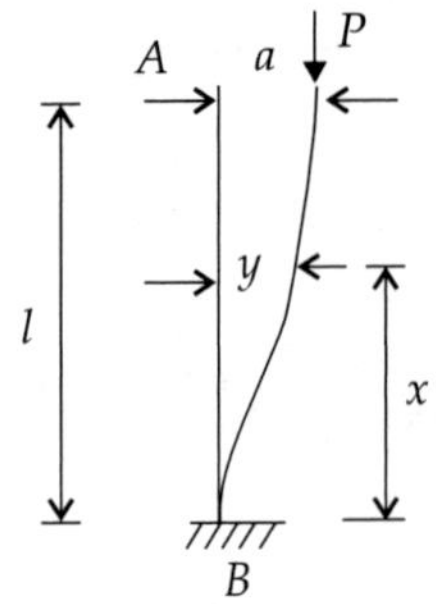

Fig. 12.3

A column (*A*)′ of length (*l*) whose lower end (*B*) is fixed, the upper end (*A*) bending free. Let *P* be the crippling load at which the column gets buckled (Fig. 12.3).

Let *a* be the deflection at the top end.

At any section distance (*x*) from the end *A*, the bending moment is given by $EI\frac{d\ y}{dx} = P(a - y)$ where, *y* is the deflection at *x*-*x*.

$$EI\frac{d^2y}{dx^2} + Py = Pa$$

$$\frac{d^2y}{dx^2} + \frac{P}{EI}y = \frac{Pa}{EI}$$

The solution to the above differential equation is: $y = C_1 \cos\left(x\sqrt{\frac{P}{EI}}\right) + C_2 \sin\left(x\sqrt{\frac{P}{EI}}\right) + a$

Where, C_1, C_2 are integration constant.

At *B*, deflection is zero, i.e., at $x = 0$, $y = 0$

$\therefore 0 = C_1 + a \Rightarrow C_1 = -a$

The slope at any section is given by $\frac{dy}{dx} = -C_1\sqrt{\frac{P}{EI}} \sin\left(x\sqrt{\frac{P}{EI}}\right) + C2\sqrt{\frac{P}{EI}} \cos\left(x\sqrt{\frac{P}{EI}}\right)$

At *B*, the slope is zero, i.e., at $x = 0$, $\frac{dy}{dx} = 0$

$$\therefore 0 = C_2\sqrt{\frac{P}{EI}} \Rightarrow C_2 = 0$$

At *A*, the deflection is *a*, i.e., at $x = b$, $y = a$

$$\therefore a = -a\cos\left(l\sqrt{\frac{P}{EI}}\right) + a \Rightarrow \cos\left(l\sqrt{\frac{P}{EI}}\right) = 0$$

$$\therefore\ l\sqrt{\frac{P}{EI}} = \frac{\pi}{2}, \frac{3\pi}{2}, \frac{5\pi}{2}, \ldots\ldots$$

For $n = 1$, $\quad l\sqrt{\frac{P}{EI}} = \frac{\pi}{2} \Rightarrow P = \frac{\pi^2 EI}{4l^2}$

The critical buckling load, $P_{cr} = \frac{\pi^2 EI}{4l^2}$

12.2.3 Both Ends of the Column are Fixed

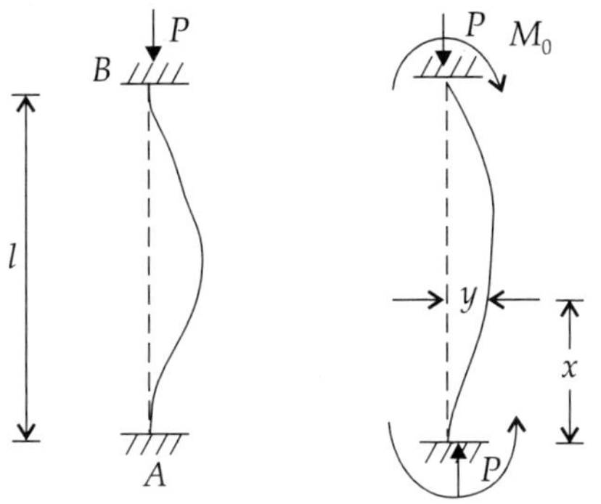

Fig. 12.4

A column (AB) of length (l) whose ends, A and B, are both fixed, let restraint moment (M_0) be at each end (Fig. 12.4).

Let P be the crippling load at which the column gets buckled. Consider a section (x-x) at a distance (x) from the lower end, A. The bending moment at the section (x-x) is given by

$$EI\frac{d^2y}{dx^2} = M_o - Py$$

$$EI\frac{d^2y}{dx^2} + Py = M_o$$

$$\frac{d^2y}{dx^2} + \frac{P}{EI}y = \frac{M_o}{EI}$$

Solution to the above differential equation is: $y = C_1 \cos\left(x\sqrt{\frac{P}{EI}}\right) + C_2 \sin\left(x\sqrt{\frac{P}{EI}}\right) + \frac{M_o}{P}$

where, C_1, C_2 are integration constants.

Slope at any section is given by $\frac{dy}{dx} = -C_1\sqrt{\frac{P}{EI}} \sin\left(x\sqrt{\frac{P}{EI}}\right) + C_2\sqrt{\frac{P}{EI}} \cos\left(x\sqrt{\frac{P}{EI}}\right)$

At A, the deflection is zero, i.e., at $x = 0$, $y = 0$

$$\therefore\ 0 = C_1 + \frac{M_o}{P} \Rightarrow C_1 = -\frac{M_o}{P}$$

At A, the slope is zero, i.e., at $x = 0, \frac{dy}{dx} = 0$

$$\therefore\ 0 = C_2\sqrt{\frac{P}{EI}} \Rightarrow C_2 = 0$$

At B, the deflection is zero, i.e., at $x = l, y = 0$

$$\Rightarrow 0 = \frac{M_o}{P}\,\text{ü}\,\left(l\sqrt{\frac{P}{EI}}\right) + \frac{M_o}{P} \Rightarrow \frac{M_o}{P}\left[1 - \cos\left(l\sqrt{\frac{P}{EI}}\right)\right] = 0$$

$$\Rightarrow \cos\left(l\sqrt{\frac{P}{EI}}\right) = 1$$

$$l\sqrt{\frac{P}{EI}} = 0, 2\pi, 4\pi, \ldots.$$

Taking lowest value, $l\sqrt{\frac{P}{EI}} = 2\pi \Rightarrow P = \frac{4\pi^2 EI}{l^2}$

$$P_{cr} = \frac{4\pi^2 EI}{l^2}$$

12.2.4 When One End of the Column is Fixed and the Other is Hinged

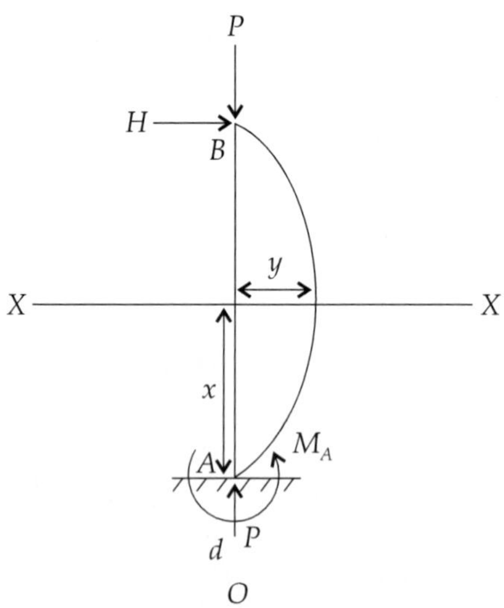

Fig. 12.5

A column (AB) of length (l), whose upper end (B) is hinged while its lower end (A) is fixed.

Let P be the crippling load at which the column gets buckled. Let there will be a restraint moment (M_a) at the lower fixed end, A. Due to restraint moment (M_b) at B, the hinge at B there must exist a horizontal force (H) at B.

Consider a section (x-x) at a distance (x) from the fixed end, A.

The bending moment at the section (x-x) is given by

$$EI\frac{d^2y}{dx^2} = -Py + H(l - x) \Rightarrow EI\frac{d^2y}{dx^2} + Py = H(l - x)$$

$$\Rightarrow \frac{d^2y}{dx^2} + \frac{P}{EI}y = \frac{H}{EI}(l - x)$$

Solution to the above differential equation is:

$$y = C_1 \cos\left(x\sqrt{\frac{P}{EI}}\right) + C_2 \sin\left(x\sqrt{\frac{P}{EI}}\right) + \frac{H}{P}(l - x)$$

where, C_1, C_2 are integration constants.

Slope of any section is given by $\frac{dy}{dx} = -C_1\sqrt{\frac{P}{EI}}\sin\left(x\sqrt{\frac{P}{EI}}\right) + C_2\sqrt{\frac{P}{EI}}\cos\left(x\sqrt{\frac{P}{EI}}\right) - \frac{H}{P}$

At A, the deflection is zero, i.e., at $x = 0$, $y = 0$

$$\therefore 0 = C_1 + 0 + \frac{Hl}{P} \Rightarrow C_1 = \frac{-Hl}{P}$$

At A, the slope is zero, i.e., at $x = 0$, $\frac{dy}{dx} = 0$

$$\therefore 0 = C_2\sqrt{\frac{P}{EI}} - \frac{H}{P} \Rightarrow C_2 = \frac{H}{P}\sqrt{\frac{EI}{P}}$$

At B, the deflection is zero, i.e., at $x = l$, $y = 0$

$$\therefore 0 = -\frac{H}{P}l\cos\left(l\sqrt{\frac{P}{EI}}\right) + \left(\frac{H}{P}\sqrt{\frac{EI}{P}}\right)\sin\left(l\sqrt{\frac{P}{EI}}\right)a$$

Simplifying, we get $\tan l\left(\sqrt{\frac{P}{EI}}\right) = l\sqrt{\frac{P}{EI}}$

Solution to the equation is $l\sqrt{\frac{P}{EI}} = 4.5$ radians

$$\frac{l^2P}{EI} = (4.5)^2 = 20.25 \Rightarrow P = \frac{20.25EI}{l^2} \Rightarrow P = \frac{2\pi^2EI}{l^2}$$

12.3 EFFECTIVE LENGTH OF A COLUMN

The effective length (L) of a column with given end condition is the length of an equivalent column of the same material and section with hinged ends having the value of the crippling load equal to that of the given column.

End conditions of column	Crippling load	Relation between effective length and actual length
Both ends hinged	$\frac{\pi^2 EI}{l^2} = \frac{\pi^2 EI}{L^2}$	$L = l$
One end fixed, one free	$\frac{\pi^2 EI}{4l^2} = \frac{\pi^2 EI}{L^2}$	$L = 2l$
Both ends fixed	$\frac{4\pi^2 EI}{l^2} = \frac{\pi^2 EI}{L^2}$	$L = \frac{l}{2}$
One end fixed, one end hinged	$\frac{2\pi^2 EI}{l^2} = \frac{\pi^2 EI}{L^2}$	$L = \frac{l}{\sqrt{2}}$

12.4 LIMITATION OF EULER'S FORMULA

According to Euler's formula, the crippling load,

$$P = \frac{\pi^2 EI}{L^2}$$

Putting $I = Ak^2$ where, k = Least radius of gyration of the column section.

$$\text{Crippling load, } P = \frac{\pi^2 EAk^2}{L^2}$$

$$\therefore \text{ Stress at failure } \frac{P}{A} = \pi^2 E \frac{k^2}{L^2} \text{ or } \frac{\pi}{A} = \frac{\pi^2 E}{(L/E)^2}$$

where, $\frac{L}{k} = \frac{\text{Effective length}}{\text{Least radius of gyration}}$ is called slenderness ratio of the column.

From the above formula it is clear that stress at failure will be high when slenderness ratio is small. But the stress at failure cannot be greater than the crushing stress for the column material. Hence, when the slenderness ratio is less than a certain limit, Euler's formula gives a value of the crippling load even greater than the crushing load.

Example 12.1: Consider a mild steel column, $E = 2 \times 10^5$ N/mm^2 The crushing stress for mild steel, f_c = 330 N/mm^2 Equating the crippling stress to the crushing stress corresponding to the minimum value of slenderness ratio, we have

$$\frac{\pi^2 E}{\left(\frac{L}{k}\right)^2} = f_e \Rightarrow \left(\frac{L}{k}\right)^2 = \frac{\pi^2 E}{f_e} = \frac{\pi^2 \times 2 \times 10^5}{330} \Rightarrow \frac{L}{k} \approx 80$$

Hence, for mild steel columns when the slenderness ratio is less than this limit, Euler's formula will not be valid.

12.5 RANKINE'S FORMULA

According to Rankine's experiment investigation the well defined relationship between the crippling load P_{cr}, crushing load $P_c = f_c A$ and Euler's bucking load $P_E = \frac{\pi^2 EI}{L^2}$ is given by

$$\frac{1}{P_{cr}} = \frac{1}{P_c} + \frac{1}{P_E}$$

$$\frac{1}{P_{cr}} = \frac{P_e P_E}{P_e + P_E}$$

$$= \frac{P_c}{1 + \frac{P_c}{P_E}}$$

Substituting $P_c = f_c A$ and $P_E = \frac{\pi^2 EI}{L^2}$

$$= \frac{f_c A}{1 + \left(\frac{f_c A L^2}{\pi^2 E A k^2}\right)}$$

$$= \frac{f_c A}{1 + \frac{f_c \left(\frac{L}{k}\right)^2}{\pi^2 E^2}} \qquad [\because I = 4k^2]$$

We get $P_{cr} = \dfrac{\pi^2 EI}{1 + f_c A \times \left(\frac{L^2}{\pi^2 EI}\right)}$

Thus, Rankine's Formula, m, where $a = \frac{f_c}{\pi^2 E}$ is constant for a given material and is called Rankine's constant.

12.6 AXIALLY LOADED COLUMN WITH INITIAL CURVATURE

A column AB of length (l) with both ends hinged. The column has an initial curvature having a central deflection, a.

Let at a distance (x) from the end (B), the initial deflection be (y'). Let us assume a sine curve for the initial profile of the centreline of the column, which satisfies the end conditions.

$$\therefore\ y' = a' \sin\frac{\pi x}{l} \Rightarrow \frac{dy'}{dx} = \frac{\pi a'}{l}\cos\frac{\pi x}{l} \Rightarrow \frac{d^2 y'}{dx^2} = -\frac{\pi^2 a'}{l^2}\sin\frac{\pi x}{l} \qquad \text{(i)}$$

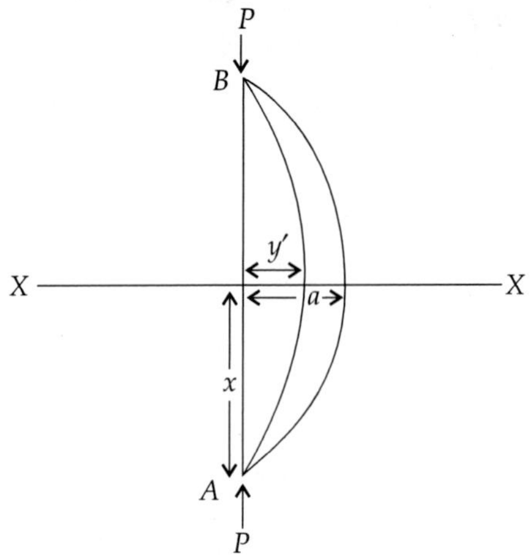

Fig. 12.6

When the loading on the column reaches the critical value (P), the column will deflect to the form ACB, so that the deflection of x changes from y' to y, due to bending moment, $M_0 = P\,y$.

$$\therefore EI\frac{d^2(y-y')}{dx^2} = -Py \Rightarrow \frac{d^2(y-y')}{dx^2} = \frac{-P}{EI}y$$

$$\Rightarrow \frac{d^2y}{dx^2} + \frac{\pi}{EI}y = \frac{d^2y'}{dx^2} = \frac{-\pi^2 a'}{l}\sin\frac{\pi x}{l} \qquad \text{(ii)}$$

Let the solution to the above differential equation be given by $y = C\,a'\sin\frac{\pi x}{l}$ where, C is integration constant.

$$\therefore \frac{dy}{dx} = C\,a'\frac{x}{l}\cos\frac{\pi x}{l} \Rightarrow \frac{d^2y}{dx^2} = -C\,a'\frac{x^2}{l^2}\sin\frac{\pi x}{l}$$

Substituting the expression for y and $\frac{d^2y}{dx^2}$ in equation (ii), we get

$$-C\,a'\frac{x^2}{l^2}\sin\frac{\pi x}{l} + \frac{P}{EI}C\,a'\sin\frac{\pi x}{l} = \frac{-\pi^2 a'}{l^2}\sin\frac{\pi x}{l} \Rightarrow C\left[\frac{\pi^2}{l^2} - \frac{P}{EI}\right] = \frac{\pi^2}{l^2}$$

$$\therefore C = \frac{\pi^2/l^2}{\frac{\pi^2}{l^2} - \frac{P}{EI}} = \frac{1}{1 - \frac{Pl^2}{\pi^2 EI}} = \frac{1}{1 - \frac{P}{P_o}}$$

Hence, the equation to the deflected form of the column is given by $y = \frac{P_c}{P_e - P}a'\sin\frac{\pi x}{l}$

(iii)

Since, the central deflection (a) will be maximum at the mid-section, x-x.

$\therefore$ At $x = l/2, y = a$

$$a = \frac{P_e}{P_e - P} a'$$

Maximum bending moment at the mid-section,

$$M_{max} = P\,a = \frac{PP_e}{P_e - P} a'$$

Maximum compressive stress $= P_{max} = P_{axial} + P_{bending} = \frac{P}{A} + \frac{M}{Z}$

$$\Rightarrow P_{max} = \frac{P}{A} + \frac{My_e}{Ak^2} = \frac{P}{A} + \frac{PP_e}{P_e - P}\frac{a'y_c}{AK^2}$$

$$\Rightarrow P_{max} = \frac{P}{A} + \left[1 + \frac{P_e}{P_e - P}.\frac{a'y_c}{k^2}\right] = P_o + \left[1 + \frac{P_e}{P_e - P}.\frac{a'y_c}{k^2}\right] = P_o + \left[1 + \frac{P_c}{P_e - P}\right].\frac{a'y_o}{k^2}$$

On rearranging, we get $\left[\frac{P_{max}}{P_o} - 1\right]\left[1 - \frac{P_o}{P_e}\right] = \frac{a'y_o}{k^2}$

12.7 LATERALLY LOADED STRUTS

Case (I): Strut hinged at both ends subjected to an axial thrust (P) and a transverse point load (W) at the centre:

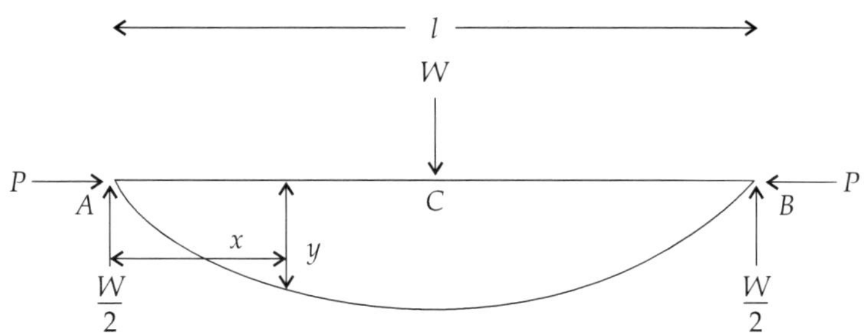

Fig. 12.7

Consider a section in AC at a distance x from the end A. The bending moment at the section is given by

$$EI\frac{d^2y}{dx^2} = -Py - \frac{W}{2}x \Rightarrow \frac{d^2y}{dx^2} + \frac{P}{EI}y = \frac{-W}{2EI}x$$

Solution to the above differential equation is $y = C_1 \cos x\sqrt{\frac{P}{EI}}\, C_2 \sin x\sqrt{\frac{P}{EI}} - \frac{Wx}{2P}$

Slope at this section is given by $\frac{dy}{dx} = -C_1\sqrt{\frac{P}{EI}} \sin x\sqrt{\frac{P}{EI}} + C_2\sqrt{\frac{P}{EI}} \cos x\sqrt{\frac{P}{EI}} - \frac{W}{2P}$

At $x = 0, y = 0 \Rightarrow C_1 = 0$

At $x = l/2, \frac{dy}{dx} = 0$

$$\Rightarrow 0 = C_2\sqrt{\frac{P}{EI}}\cos\frac{l}{2}\sqrt{\frac{P}{EI}} - \frac{W}{2P} \Rightarrow C_2 = \frac{W}{2P}\sqrt{\frac{EI}{P}}\sec\frac{l}{2}\sqrt{\frac{P}{EI}}$$

Hence, the deflection at any section in AC, is given by

$$y = \frac{W}{2P}\sqrt{\frac{EI}{P}}\sec\frac{l}{2}\sqrt{\frac{P}{EI}}\sin x\sqrt{\frac{P}{EI}} - \frac{Wx}{2P}$$

Maximum deflection at the centre is given by

At $x = \frac{l}{2}, y = y_{max}$

$$\therefore y_{max} = \frac{W}{2P}\sqrt{\frac{EI}{P}}\sec\frac{l}{2}\sqrt{\frac{P}{EI}}\sin\frac{l}{2}\sqrt{\frac{P}{EI}} - \frac{Wl}{4P} \Rightarrow \therefore y_{max} = \frac{W}{2P}\sqrt{\frac{EI}{P}}\tan\frac{l}{2}\sqrt{\frac{P}{EI}} - \frac{Wl}{4P}$$

$\therefore$ Maximum bending moment

$$M_{max} = -Py_{max} - \frac{W}{2}.\frac{l}{2} = \frac{-W}{2}\sqrt{\frac{EI}{P}}\tan\frac{l}{2}\sqrt{\frac{P}{EI}} = \frac{-W}{2}\sqrt{\frac{EI}{P}}\tan\theta, \qquad \text{where } \theta = \frac{l}{2}\sqrt{\frac{P}{EI}}$$

$$\therefore M_{max} = \frac{-W}{2}\sqrt{\frac{EI}{P}}\left[\frac{l}{2}\sqrt{\frac{P}{EI}} + \frac{1}{3}\frac{l^3}{8}\cdot\frac{P}{EI}\cdot\sqrt{\frac{P}{EI}}\right] \qquad \because \text{For small angle, } \tan\theta = \theta + \frac{\theta^3}{3}$$

$$M_{max} = -\left[\frac{Wl}{4} + \frac{Wl^3}{48EI}P\right]$$

Case (II): Strut hinged at both ends and subjected to an axial thrust (P) and a lateral uniformly distributed load of intensity (w) per unit run:

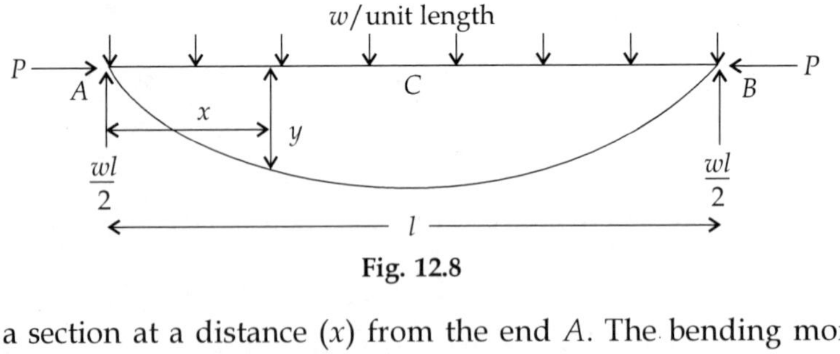

Fig. 12.8

Consider a section at a distance (x) from the end A. The bending moment at the section is given by

$$EI\frac{d^2y}{dx^2} = -Py + \frac{wx^2}{2} - \frac{wl}{2}x \Rightarrow \frac{d^2y}{dx^2} + \frac{P}{EI}y = \frac{-wx(l-x)}{2EI}$$

The solution to the above differential equation is

$$y = C_1\cos x\sqrt{\frac{P}{EI}} + C_2\sin x\sqrt{\frac{P}{EI}} - \frac{wx(l-x)}{2P} - \frac{wEI}{p^2}$$

$$\therefore \frac{dy}{dx} = -C_1\sqrt{\frac{P}{EI}}\sin x\sqrt{\frac{P}{EI}} + C_2\sqrt{\frac{P}{EI}}\cos x\sqrt{\frac{P}{EI}} + \frac{wx}{P} - \frac{wl}{2P}$$

At $x = 0, y = 0$

$$\therefore 0 = C_1 - \frac{wEI}{P^2} \Rightarrow C_1 = \frac{wEI}{P^2}$$

At $x = \frac{1}{2}, \frac{dy}{dx} = 0$

$$\therefore 0 = \frac{-wEI}{P^2}\sqrt{\frac{P}{EI}}\sin\frac{l}{2}\sqrt{\frac{P}{EI}} + C_2\sqrt{\frac{P}{EI}}.\cos\frac{l}{2}\sqrt{\frac{P}{EI}} \Rightarrow C_2 = C_1\tan\frac{l}{2}\sqrt{\frac{P}{EI}}$$

Substituting value of C_1, C_2 in the deflection equation, we get

$$y = \frac{wEI}{P^2}\left[\cos x\sqrt{\frac{P}{EI}} + \tan\frac{l}{2}\sqrt{\frac{P}{EI}}\sin x\sqrt{\frac{P}{EI}} 1\right] - \frac{wx(l-x)}{2P}$$

Maximum deflection will occur at C is given by

$$y = \frac{wEI}{P^2}\left[\cos\frac{l}{2}\sqrt{\frac{P}{EI}} + \tan\frac{l}{2}\sqrt{\frac{P}{EI}}.\sin\frac{l}{2}\sqrt{\frac{P}{EI}} - 1\right] - \frac{wl^2}{8P}$$

$$y_c = \frac{wEI}{P^2}\left[\sec\frac{l}{2}\sqrt{\frac{P}{EI}} - 1\right] - \frac{wl^2}{8P}$$

Maximum bending moment which will occur at 'C' is given by $M_c = -\frac{wl^2}{8} - Py_c$

$$= -\frac{wl^2}{8} - P\left[\frac{wEI}{P^2}\left(\sec\frac{l}{2}\sqrt{\frac{P}{EI}} - 1\right) - \frac{wl^2}{8P}\right] = -\frac{wEI}{P}\left[\sec\frac{l}{2}\sqrt{\frac{P}{EI}} - 1\right]$$

$$= -\frac{wEI}{P}[\sec\theta - 1], \quad \text{where } \theta = \frac{l}{2}\sqrt{\frac{P}{EI}}$$

$\because$ For small angle, $\sec\theta = 1 + \frac{\theta^2}{2} + \frac{5}{24}\theta^4$

$$\therefore M_c = \frac{wEI}{P}\left[\frac{1}{2}.\frac{l^2}{4}.\frac{P}{EI} + \frac{5}{24}.\frac{l^4}{16}.\frac{P^2}{E^2I^2}\right] \Rightarrow (M_c)_{\max} = -\left[\frac{wl^2}{8} + \frac{5}{384}\frac{wl^4}{EI}P\right]$$

Example 12.2: A 2 m long fixed ended column of square cross section is to be made of wood. Assuming E = 24 GPa and allowable stress being limited to 24 MPa, determine the size of the column to support 100 kN load safely.

Use factor of safety of '2' and Euler's crippling load for buckling.

Solution: Effective length = Actual length

= 2 m = 2000 mm

$E = 24 \text{ GPa} = 24000 \text{ N/mm}^2$

$f_c = 24 \text{ MPa} = 24 \text{ N/mm}^2$

From Euler's formula, $P_{cr} = \dfrac{\pi^2 EI}{L^2}$

Now, Working load $= \dfrac{\text{Critical load}}{\text{Factor of safety}}$

When load is 100 kN.

$$\Rightarrow 100 \times 1000 = \frac{\pi^2 EI}{2L^2}$$

Let a be the side of the square, then $I = \dfrac{1}{12} \times a \times (a)^3 = \dfrac{a^4}{23}$

$$\therefore\ 100000 = \frac{\pi^2 \times 24000}{2} \times \frac{a^4}{12} \times \frac{1}{(2000)^2}$$

$$\Rightarrow a^4 = \frac{100}{\pi^2} \times (2000)^2$$

$\therefore\ a \approx 80 \text{ mm}$

From the consideration of direct crushing, we have

$f_c \times A = 95000$

$\therefore\ 24 \times a^2 = 100000$

Hence, Euler's formula governs the design.

So use $80 \times 80 \text{ mm}^2$ square column.

Example 12.3: Compare the ratio of the strength of a solid steel column to that of a hollow of the same cross-sectional areas. The internal diameter of the hollow column is $\dfrac{1}{2}$ of the external diameter. The column having the same length are hinged at the ends.

Solution: Let suffix 'S' stand for solid column and suffix 'H' stand for hollow column.

$$P_s = \frac{\pi^2 E(Ak_s^2)}{l^2}$$

$$P_H = \frac{\pi^2 E(Ak_H^2)}{l^2}$$

$$\therefore \qquad \frac{P_H}{P_s} = \left(\frac{k_H}{k_s}\right)^2$$

Now for solid section, $k_s^2 = \dfrac{\pi}{64} d^4 \times \dfrac{4}{\pi d^4} = \dfrac{d^2}{16}$ and for hollow section, $k_H^2 = \dfrac{\frac{\pi}{64}(D^4 - D'^4)}{\frac{\pi}{4}(D^2 - D'^2)}$

$\dfrac{D^2 - D'^2}{16}$ = where, D and D' are external and internal diameters, respectively.

$$\therefore \quad \frac{P_H}{P_s} = \frac{k_H^2}{k_s^2}$$

$$= \frac{D^2 - D'^2}{d^2}$$

$$= \frac{D^2 - \frac{1}{4}D^2}{d^2}$$

$$= \frac{5}{4} \cdot \frac{D^2}{d^2}$$

Since cross-sectional area of the both is equal, we have

$$\text{or} \quad d^2 = \frac{3}{4}D^2$$

$$\text{or} \quad \frac{D^2}{d^2} = \frac{4}{3}$$

$$\therefore \quad \frac{P_H}{P_s} = \frac{5}{4} \cdot \frac{D^2}{d^2}$$

$$= \frac{5}{4} \times \frac{4}{3}$$

$$\Rightarrow \quad \frac{P_H}{P_s} = \frac{5}{3}$$

Example 12.4: A hollow cast iron column whose outside diameter is 200 mm with thickness of 20 mm is 5 m long and is fixed at both ends. Calculate the safe load by Rankine's formula using a factor of safety of 2. Find the ratio of Euler's-Rankine's loads. Take $E = 1 \times 10^5$ N/mm² and Rankine's constant $\dfrac{1}{1600}$ for both ends pinned case and $f_c = 550$ N/mm².

Solution: External diameter = 200 mm

Thickness = 20 mm

Internal diameter = 200 – 2 × 20 = 160 mm

$$I = \frac{\pi}{64}(200)^4 - \frac{\pi \times (160)^4}{64}$$

$$= 46369908 \text{ mm}^4$$

Since both ends are fixed, therefore,

$$\text{Effective length } L = \frac{1}{2} \times \text{Actual length}$$

$$= \frac{5}{2} = 2.5 \text{ m} = 2500 \text{ mm}$$

$$\text{Euler's critical load, } P_E = \frac{\pi^2 EI}{L^2}$$

$$= \frac{\pi^2 \times 1 \times 10^5 \times 463369908}{(2500)^2}$$

$$= 7322442 \text{ N}$$

$$\text{Now, } A = \frac{\pi}{4}(200^2 - 160^2)$$

$$= 11309.734 \text{ mm}^2$$

$$k^2 = \frac{I}{A}$$

$$= \frac{46369908}{11309.34}$$

$$\therefore \quad k = 64.031$$

$$\text{Rankine's critical load, } P_R = \frac{f_c A}{1 + a\left(\frac{L}{k}\right)^2}$$

$$= \frac{550 \times 11309.734}{1 + \frac{1}{1600} \times \left(\frac{2500}{64.0312}\right)^2}$$

$$= 3183916 \text{ N}$$

$$\therefore \quad \frac{P_E}{P_R} = \frac{7322442}{3183916} = 2.3$$

$$\text{Safe load using Rankine's formula} = \frac{P_R}{\text{Factor of safety}}$$

$$= \frac{3183916}{2}$$

$$v = 1591958 \text{ N}$$

$$= 1592 \text{ kN}$$

Example 12.5: A hollow cylindrical cast iron column is 4 m long, with both ends fixed. Design the column to carry an axial load of 250 kN. Use Rankine's formula and adopt a factor of safety of 5. The internal diameter may be taken as $0.8D$ Times the external diameter. Take $f_c = 550$ N/mm^2, $a = \frac{1}{1600}$.

Solution: Let the external diameter be D mm.

Internal diameter, $d = 0.8D$ mm

$$\text{Area of the section} = \frac{\pi}{4}(D^2 - d^2)$$

$$= \frac{\pi}{4}(D^2 - 0.64D^2)$$

$$= 0.09\,\pi D^2 \text{mm}^2$$

$$\text{Moment of inertia of the section, } I = \frac{\pi}{64}(D^2 - d^2)\text{mm}^4$$

$$\therefore \quad k^2 = \frac{I}{A}$$

$$= \frac{\frac{\pi}{64}(D^4 - d^4)}{\frac{\pi}{4}(D^2 - d^2)}\text{mm}^2$$

$$k^2 = \frac{D^2 + d^2}{16}$$

$$= \frac{D^2 + 0.64d^2}{16}$$

$$= \frac{164}{16}D^2$$

$$= 0.1025\,D^2\text{ mm}^2$$

Safe load on the column = 250 kN.

$\therefore$ Crippling load = Safe load $\times$ Factor of safety

$= 250 \times = 1250$ kN

$$\text{Effective length of the column, } L = \frac{l}{2}$$

$$= \frac{4000}{2} = 2000 \text{ mm}$$

$$\text{Applying Rankine's formula, } P = \frac{f_c A}{1 + a\frac{L^2}{k^2}}$$

$$\text{or, } 1250 \times 1000 = \frac{550 \times 0.09\pi D^2}{1 + \frac{1}{1600} \times \frac{2000 \times 2000}{0.1025D^2}}$$

$$= \frac{155.5D^4}{D^2 + 24390}$$

or $D^4 - 8038.58D^2 - 1.9606 \times 10^8 = 0$

Solving as a quadratic in D^2, we have $D^2 = \dfrac{8038.58 + \sqrt{8038.58^2 + 4 \times 1.9606 \times 10^8}}{2}$

$$= 18586.8$$

$\Rightarrow D = 136.3$ mm

$\therefore$ Internal diameter $= 0.8 \times 136.3$

$= 109.04$ mm

Example 12.6: A straight cylindrical bar is 15 mm diameter and 1.2 m long. It is freely supported at its two ends in a horizontal position and loaded at the centre with a concentrated load of 90 N. The central deflection is found to be 5 mm. If placed vertical and loaded along its axis, what load would cause it to buckle? What is the ratio of the maximum stress in the two cases?

Solution: $\delta = 5 = \dfrac{Wl^3}{48EI}$

$$= \frac{90 \times 1200}{48} \times \frac{l^2}{EI}$$

where, l, E and I are in mm units.

$$\therefore \quad \frac{EI}{l^2} = \frac{90 \times 1200}{48 \times 5} = 450$$

Now, $P_E = \dfrac{\pi^2 EI}{L^2}$

$$= \pi^2 \times 450 = 4441 \text{ N}$$

As a beam, $M_{max} = \dfrac{WL}{4}$

$$= \frac{90 \times 1200}{4}$$

$$= 27000 \text{ N/mm}$$

$$\therefore f_b = \frac{M}{Z} = \frac{27000}{\frac{\pi}{32}(15)^3}$$

$$= 81.5 \text{ N/mm}^2$$

As a strut, $f_E = \dfrac{P_E}{A}$

$$= \frac{4441}{\frac{\pi}{4}(15)^2}$$

$$= 25.13 \text{ N/mm}^2$$

$$\therefore \quad \frac{f_b}{f_E} = \frac{81.5}{25.13} = 3.24$$

Example 12.7: The section of a column consists of two channels (ISMC 350) and cover plates of size 500 mm × 10 mm. The distance between the two channels is 220 mm. Find its load carrying capacity by Rankine's formula, if it is used as a column of length 6 m with both ends hinged.

Given: f_c = 326 N/mm², $E = 2 \times 10^5$ N/mm² and factor of safety = 2

$$\text{Rankine's constant} = \frac{1}{7500}$$

For each ISMC 350

$A = 5366 \text{ mm}^2$

$I_{xx} = 100.08 \times 10^6 \text{ mm}^4$

$C_{yy} = 23.6 \text{ mm}$

$I_{yy} = 4.306 \times 10^6 \text{ mm}^4$

Solution: Symmetrical axes are the centroidal axes

$$I_{xx} = 2 \times 100.08 \times 10^6 + 2\left[\frac{1}{12} \times 500 \times 10^3 + 500 \times 10 \times (175+5)^2\right]$$

$$= 5.24243 \times 10^3 \text{ mm}^4$$

$$I_{yy} = 2[4.306 \times 10^6 + 6366 \times (110 + 23.6)^2] + 2 \times \frac{1}{12} \times 10 \times 500^3$$

$$= 4.085 \times 10^3 \text{ mm}^4$$

$\therefore \; I_{yy} < I_{xx}$

$\therefore \quad I = I_{yy} = 4.085 \times 10^3 \text{ mm}^4$

$= 2 \times 5366 + 2 \times 10 \times 500$

$= 20732 \text{ mm}^2$

$$k = \sqrt{\frac{I}{A}} = \sqrt{\frac{4.085 \times 10^3}{20732}}$$

$$= 14.37 \text{ mm}$$

Since both ends are hinged, therefore

Effective length = Actual length = 6 m

= 6000 mm

$$P_{cr} = \frac{326 \times 20732}{1 + \dfrac{1}{7500}\left(\dfrac{6000}{121.9687}\right)^2}$$

$= 54344699$ N

$= 5434.699$ kN

∴ Safe load carrying capacity is 5434.7 kN.

Example 12.8: Calculate the crippling stress, using Euler's formula for a pin-ended strut 2 m long consisting of a tube 7.5 cm outside diameter and wall thickness 2.5 mm. In compression test, a short length of this tube failed at a load of 31.5 tonnes and when tested as a strut with rounded ends, 2 m long, it failed at a load of 17.5 tonnes. Find from the data the value of constant (a) in the Rankine's formula.

Take Young's modulus = 2×10^6 kg/cm^2.

Solution: Effective length = 2 m = 200 cm

$E = 2 \times 10^6$ kg/cm^2

Outer diameter, $d_0 = 7.5$ cm

Inner diameter, $d_i = 7.5 - 2.025 = 7.0$ cm

Cross-sectional area of the tube

$$= \frac{\pi}{4}(7.5^2 - 7^2)$$

$$= 5.694 \text{ cm}^2$$

For the given section, $I = \frac{\pi}{64}(d_o^4 - d_i^4)$

$$= \frac{\pi}{64}(7.5^4 - 7^4)$$

$$= 37.457 \text{ cm}^4$$

Euler's crippling load , $P_{cr} = \frac{\pi^2 EI}{l^2}$

$$= \frac{\pi \times 2 \times 10^6 \times 37.457}{(200)^2}$$

$$= 18484.26 \text{ kg}$$

Euler's crippling stress, $P_c = \frac{P_{cr}}{A}$

$$= \frac{18484.26}{5.694}$$

$$= 3246.27 \text{ kg/cm}^2$$

Based on the specimen test, yield stress, $f_c = \frac{31.5 \times 1000}{5.694}$

$$= 5532.14 \text{ kg/cm}^2$$

Length of strut (effective), $l_c = 2$ m = 200 cm

Load at failure, $P = 17.4 \times 1000$

$= 17400$ kg

For the given tube, radius of gyration, $k = \sqrt{\frac{I}{A}} = \sqrt{\frac{37.457}{5.694}} = 2.565$ cm

Using Rankine's formula, we have $P = \dfrac{f_c A}{1 + a\left(\dfrac{l}{k}\right)^2}$

$$\text{or } 17400 = \frac{5532.14 \times 5.694}{1 + a\left(\dfrac{200}{2.564}\right)^2}$$

$$\text{or } 6079.74a = \frac{5532.14 \times 5.694}{17400} - 1 = 0.81$$

$$\therefore\ a = \frac{1}{7506}$$

Example 12.9: A strut of length (l) is encastered at lower end and its upper end is elastically supported against lateral deflection so that the resisting force is K times the end deflection. Show that the crippling load (P) is given by $\dfrac{\tan \alpha l}{\alpha l} = 1 - \dfrac{P}{Kl}$ when, $a^2 = \dfrac{P}{EI}$

Solution: Consider a section C of the strut AB at a height x from A. Let the deflection at C be y. If the deflection at end B is a, then the resisting force is Ka.

Bending moment at C is

$$M_C = Ka(l - x) - P(a - y)$$

$$\therefore\quad EI\frac{d^2y}{dx^2} = P(a - y) - Ka(l - x)$$

$$\text{or}\quad \frac{d^2y}{dx^2} + \frac{P}{EI}y = \frac{Pa}{EI} - \frac{Ka}{EI}(l - x)$$

Solution of the above differential equation is

$$y = C_1 \cos\sqrt{\frac{P}{EI}}x + C_2 \sin\sqrt{\frac{P}{EI}}x + a - \frac{Ka}{P}(l - x) \quad \text{... (i)}$$

To determine constants C_1, C_2 apply end conditions at A

When, $y = 0$, $x = 0$

$$\therefore\quad 0 = C_1 + a - \frac{Kal}{P}$$

$$\text{or}\quad C_1 = \frac{Kal}{P}$$

Substituting for C_1 and putting $\alpha^2 = \frac{P}{EI}$ in equation (i), we get

$$y = \left(\frac{Kal}{P} - a\right)\cos ax + C_2 \sin ax + a - \frac{Ka}{P}(l - x) \qquad \text{...(ii)}$$

$$\therefore \quad \frac{dy}{dx} = -\left(\frac{Kal}{P} - a\right)\alpha \sin ax + C_2 \alpha \cos \alpha + \frac{Ka}{P}$$

At A, when $x = 0$, $\frac{dy}{dx} = 0$

$$\therefore \qquad 0 = C_2\alpha + \frac{Ka}{P}$$

$$\Rightarrow \qquad C_2 = -\frac{Ka}{P\alpha}$$

Substituting for C_2 in equation (ii), we get

$$y = \left(\frac{Kal}{P} - a\right)\cos ax - \frac{Ka}{P\alpha}\sin \alpha x + a - \frac{Ka}{P}(l - x)$$

But at B when $x = l$ the deflection $y = a$

$$\therefore a = \left(\frac{Kal}{P} - a\right)wb\alpha l - \frac{Ka}{P\alpha}\sin \alpha l + a$$

$$\text{or } \tan \alpha l = \left(\frac{Kal}{P} - a\right)\frac{P\alpha}{K}$$

$$= al - \frac{P\alpha}{K}$$

$$= \alpha l\left[1 - \frac{P}{Kl}\right]$$

$$\frac{\tan \alpha l}{\alpha l} = 1 - \frac{P}{Kl}$$

Example 12.10: A strut of length (l) is so restrained at its ends, that the restraining moment of each end is K times the slope at each end. Show that the buckling load (P) is given by the equation $\frac{\alpha l}{2} = -\frac{P}{\alpha l}$ where, $a^2 = \frac{P}{EI}$

Solution: Let θ be the slope at each end, and M_0 be the restraining moment, so that $M_0 = K\theta$.

Let y be the deflection at a distance x from the lower end.

Let P be the buckling load.

The equation to the deflection shape of the column,

$$y = C_1 \cos\left(x\sqrt{\frac{P}{EI}}\right) + C_2 \sin\left(x\sqrt{\frac{P}{EI}}\right) + \frac{M_0}{P}$$

Putting $\frac{P}{EI} = \alpha^2$, we get

$$y = C_1 \cos \alpha x + C_2 \sin \alpha x + \frac{M_0}{P} \quad \text{...(i)}$$

$$\frac{dy}{dx} = -C_1\alpha \sin \alpha x + C_2 \alpha \cos \alpha x \quad \text{...(ii)}$$

At $x = 0, y = 0$

So from equation (i)

$$0 = C_1 + \frac{M_0}{P}$$

$$\Rightarrow \qquad C_1 = -\frac{M_0}{P}$$

At $x = \frac{l}{2}, \frac{dy}{dx} = 0$

From equation (ii), $0 = \frac{M_0}{P}\alpha \sin\frac{\alpha l}{2} + C_2\alpha.\cos\frac{\alpha l}{2}$

$$\Rightarrow C_2 = \frac{-M_0}{P}\tan\frac{\alpha L}{2}$$

At $x = 0, \frac{dy}{dx} = \theta$

$\therefore$ From equation (ii), we have $\theta = \frac{-M_0}{P}\tan\frac{\alpha L}{2}\cdot\alpha$

But $\quad M_0 = K\theta$

$$\therefore \qquad \theta = \frac{-K\theta}{P}.\tan\frac{\alpha l}{2}.\alpha$$

$$\therefore \tan\frac{\alpha l}{2} = \frac{-P}{K\alpha}$$

PROBLEMS

1. A 2 m long pin-ended column of square cross section is to be made of wood. Assuming E = 24 GPa and allowable stress being limited to 24 MPa, determine the size of the column to support 100 kN load safely. Use factor of safety of '2' and Euler's crippling load for buckling.

   Ans. 60 mm side

2. Compare the ratio of the strength of a solid steel column to that of a hollow of the same cross-sectional areas. The internal diameter of the hollow column is $\frac{1}{3}$

of the external diameter. The column having the same length are hinged at the ends.

Ans. 5:4

3. A hollow cast iron column whose outside diameter is 200 mm and has a thickness of 20 mm is 5 m long and is pinned ends. Calculated the safe load by Rankine's formula using a factor of safety of 2. Find the ratio of Euler's Rankine's loads. Take $E = 1 \times 10^5$ N/mm² and Rankine's constant $\frac{1}{1600}$ for both ends pinned case and $f_c = 550$ N/mm².

Ans. 1.42

4. A hollow cylindrical cast iron column is 4 m long, both ends being fixed. Find the axial load carry by the column. Use Rankine's formula and adopt a factor of safety of 5. The external diameter is 100 mm and the internal diameter may be taken as 0.8 times the external diameter. Take $f_c = 550$ N/mm², $a = \frac{1}{1600}$

Ans. 450 kN

5. A straight cylindrical bar is 15 mm diameter and 1.2 m long. It is freely supported at its two ends in a horizontal position and loaded at the centre with a concentrated load of 100 N. The central definition is found to be 5 mm. If placed vertical and loaded along its axis, what load would cause it to buckle? What is the ratio of the maximum stress in the two cases?

Ans. 3.25

6. The section of a column consists of two channels (ISMC 350) and cover plates of size 500 mm × 10 mm. The distance between the two channels is 220 mm. Find its load carrying capacity by Rankine's formula, if it is used as a column of length 6 m with both ends fixed.

   Given $f_c = 326$ N/mm², $E = 2 \times 10^5$ N/mm² and factor of safety = 2;

   Rankine's constant $= \frac{1}{7500}$

   For each ISMC 350:

   $A = 5366$ mm² $I_{xx} = 100.08 \times 10^6$ mm⁴ $C_{yy} = 23.6$ mm $I_{yy} = 4.306 \times 10^6$ mm⁴

Ans. 3127 kN

7. Calculate the crippling stress, using Euler's formula for a fixed ended strut 4 m long consisting of a tube 7.5 cm outside diameter and wall thickness of 2.5 mm. In compression test, the short length of this tube failed at a load of 31.5 tonnes and when tested as a strut with rounded ends, 2 m long, it failed at a load of 17.5 tonnes. Find from the data the value of constant a in the Rankine's formula. Take Young's modulus $= 2 \times 10^6$ kg/cm².

Ans. 1/7500

Appendix-1

Centre of Gravity, Centre of Mass, Centroids and Moment of Inertia

CENTRE OF GRAVITY

The forces exerted by gravity on each particle forming a body can be replaced by a single equivalent force equal to the weight of the body and applied at the *centre of gravity* (G) for the body, i.e., an object will remain at rest if it is balanced on any point along a vertical line passing through its centre of gravity.

Discrete Particles System

The coordinates of centre of gravity (G) are determined by the *principle of moments*, i.e., the moment of the sum of the weights about an axis equals the sum of moments of the discrete weights about the same axis.

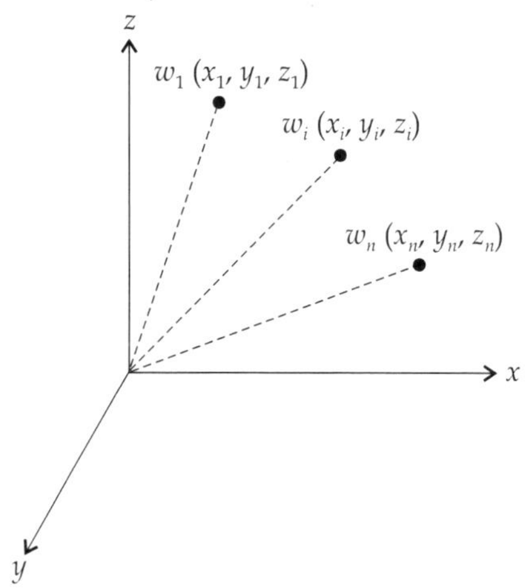

Fig. 1

Let w_i (Fig. 1) be the weight of n discrete particles with coordinates (x_i, y_i, z_i) and ($\overline{X}$, $\overline{Y}$, $\overline{Z}$) are the coordinates of the centre of gravity of the system of particles (Fig. 2), i.e., moments about z-axis gives $\overline{Z}\sum w_i = \sum(z_1 w_1 + ... z_i w_i + ... z_n w_n)$

$$\boxed{\overline{Z} = \frac{\sum(z_i\, w_i)}{\sum w_i}}$$

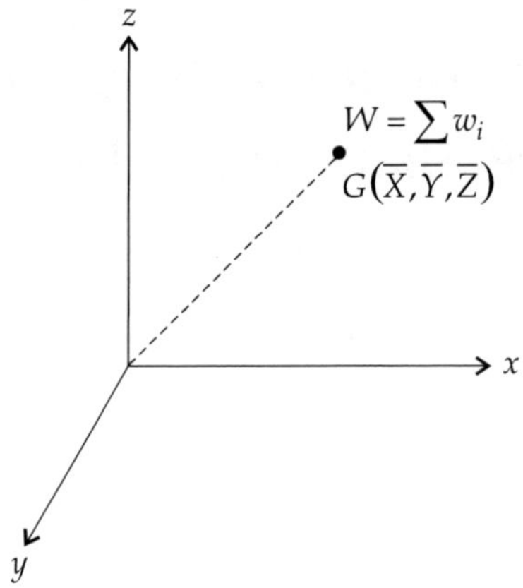

Fig. 2

Similarly, by taking moment about y- and x-axis, we get

$$\boxed{\overline{X} = \frac{\sum(x_i\, w_i)}{\sum w_i}} \qquad \boxed{\overline{Y} = \frac{\sum(y_i\, w_i)}{\sum w_i}}$$

Continuous System

The centre of gravity of continuous system (Fig. 3) is determined by treating a body as an infinite number of particles of differential weight (dw) with coordinates (x, y, z), the process of summation becomes one of integration. The total weight, $w = \int dw$.

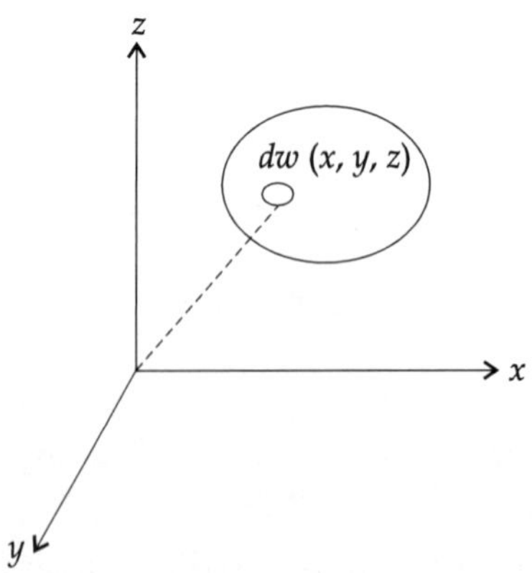

Fig. 3

Hence, the coordinates ($\overline{X}$, $\overline{Y}$, $\overline{Z}$) of the centre of gravity (Fig. 4) are

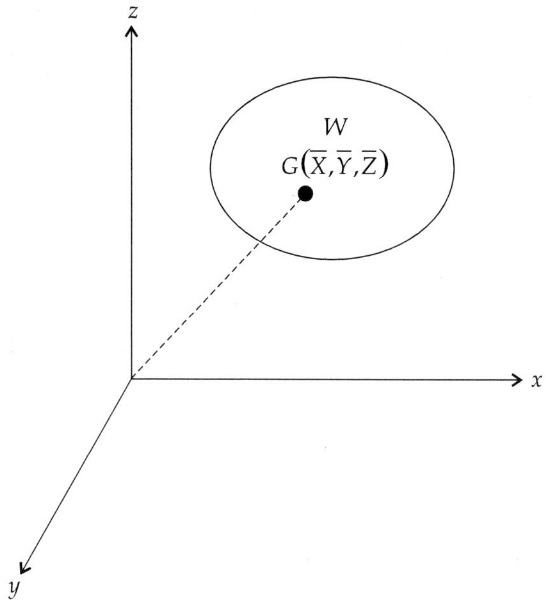

Fig. 4

$$\overline{X} = \frac{\int (xdw)}{\int dw}, \quad \overline{Y} = \frac{\int (ydw)}{\int dw}, \quad \overline{Z} = \frac{\int (zdw)}{\int dw}$$

CENTRE OF MASS (OR FIRST MOMENT OF MASS)

The location of the centre of mass depends on how the masses in the system are distributed. Centre of mass of a body is obtained by replacing the differential weight (dw) with differential mass (dm) into the relationship of centre of gravity. The total mass, $m = \sum m_i$ or $\int dm$.

The resulting equations for the centre of mass for continuous (or discrete) system are:

$$\overline{X} = \frac{\int (xdm)}{\int dm}, \quad \overline{Y} = \frac{\int (ydm)}{\int dm}, \quad \overline{Z} = \frac{\int (zdm)}{\int dm} \quad \text{or} \quad \overline{X} = \frac{\sum (x_i m_i)}{\sum m_i}, \quad \overline{Y} = \frac{\sum (y_i m_i)}{\sum m_i}, \quad \overline{Z} = \frac{\sum (z_i m_i)}{\sum m_i}$$

Note: If g, the acceleration of gravity, is constant then the centre of gravity and centre of mass coincide (since $w = mg$).

Example 1: Determine the location of centre of mass of the discrete system shown in Fig. 5.

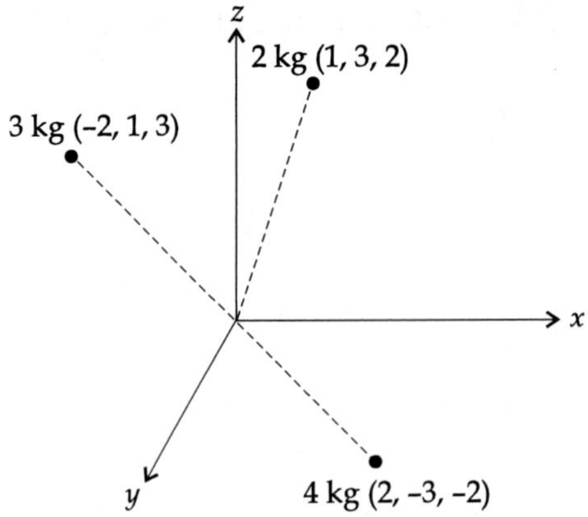

Fig. 5

$$m = \sum m_i = 2 + 3 + 4 = 9$$

$$\overline{X} = \frac{\sum (x_i m_i)}{\sum m_i} = \frac{2 \times 1 + 3 \times (-2) + 4 \times 2}{9} = \frac{4}{9}$$

$$\overline{Y} = \frac{\sum (y_i m_i)}{\sum m_i} = \frac{2 \times 3 + 3 \times 1 + 4 \times (-3)}{9} = \frac{-1}{3}$$

$$\overline{Z} = \frac{\sum (z_i m_i)}{\sum m_i} = \frac{2 \times 2 + 3 \times 3 + 4 \times (-2)}{9} = \frac{4}{9}$$

CENTROIDS

The centroid (C) of a continuous body is defined as the geometrical centre of the body. An axis through the centroid is called the centroidal axis, i.e., centroid of a body is obtained by replacing the differential weight (dw) with differential volume (dV) into the relationship of centre of gravity or by replacing the differential mass (dm) with differential volume (dV) into the relationship of centre of mass.

Centroid of a Volume

Coordinates for the centroid of a volume are

$$\boxed{\overline{X} = \frac{\int (x dV)}{\int dV}, \ \overline{Y} = \frac{\int (y dV)}{\int dV}, \ \overline{Z} = \frac{\int (z dV)}{\int dV}}$$

Note: If specific weight (γ) or density (ρ) of a body is uniform (homogeneous material) throughout the body then the centre of gravity (or centre of mass) and centroid coincide (since $w = \gamma V$ or $m = \rho V$).

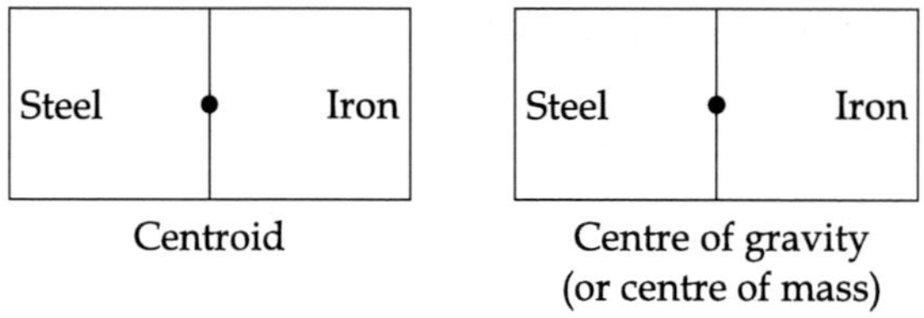

Fig. 6

Example 2: Determine the centre of gravity, centre of mass and centroid of the homogeneous right circular cone (Fig. 7).

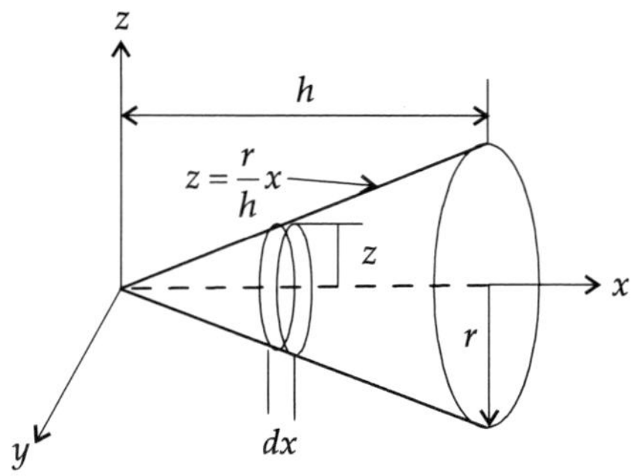

Fig. 7

Since γ or ρ is constant throughout the body because of homogeneous material, therefore centre of gravity, centre of mass and centroid are same.

By symmetry, y and z coordinates of centroid, are zero, i.e., $\overline{Y} = 0, \overline{Z} = 0$.

Since $\overline{X} = \dfrac{\int (x dV)}{\int dV}$, where, $dV = \pi z^2\, dx$

$$\overline{X} = \frac{\int x\pi z^2 dx}{\int \pi z^2 dx} = \frac{\int_0^h x\pi\left(\frac{rx}{h}\right)^2 dx}{\int_0^h \pi\left(\frac{rx}{h}\right)^2 dx} = \frac{3}{4}h$$

Centroid, $C\left(\overline{X} = \frac{3}{4}h, \overline{Y} = 0, \overline{Z} = 0\right)$

Centroid of an Area (First Moment of an Area) and Centroid of a Line

The concept of centroid can be applied to areas and lines also depending upon whether we can model the shape of the body involved as a line, or an area, or a volume. The difference is simply that differential elements of area (dA) and length (dL) replace the differential element of volume (dV).

Coordinates for the centroid of an area are

$$\overline{X} = \frac{\int (xdA)}{\int dA}, \ \overline{Y} = \frac{\int (ydA)}{\int dA}, \ \overline{Z} = \frac{\int (zdA)}{\int dA}$$

Example 3: Determine the centroid of the rectangular block (Fig. 8).

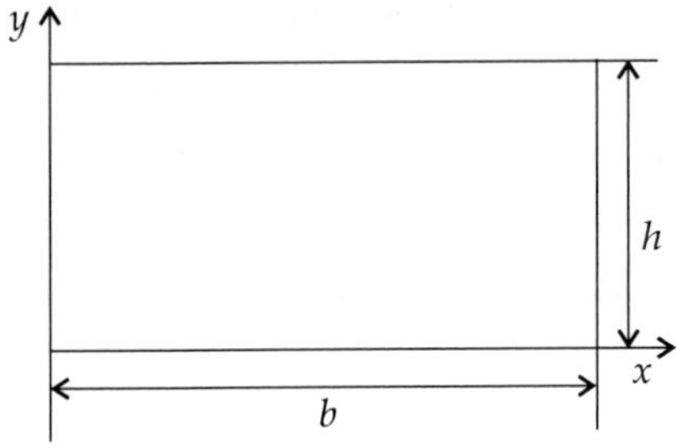

Fig. 8

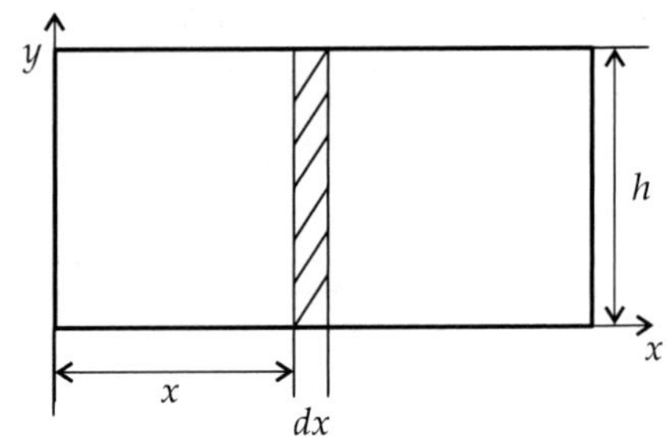

Fig. 9

Consider a strip dx along x-axis (Fig. 9) then $dA = hdx$.

X-coordinate of centroid is calculated as

$$\overline{X} = \frac{\int_0^b (xdA)}{\int_0^b dA} = \frac{\int_0^b (xhdx)}{\int_0^b hdx} = \frac{hb^2/2}{hb} = \frac{b}{2}$$

Similarly, y-coordinate of centroid by taking a strip (Fig. 10) along y-axis is

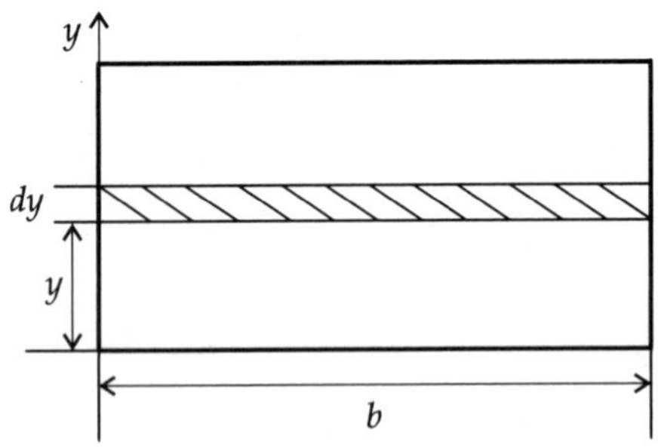

Fig. 10

$$\overline{Y} = \frac{\int_0^h (ydA)}{\int_0^h dA} = \frac{\int_0^h (ybdy)}{\int_0^h bdx} = \frac{bh^2/2}{bh} = \frac{h}{2}$$

Coordinates for the centroid of a line are

$$\overline{X} = \frac{\int (xdL)}{\int dL}, \ \overline{Y} = \frac{\int (ydL)}{\int dL}, \ \overline{Z} = \frac{\int (zdL)}{\int dL}$$

Example 4: Determine the centroid of the line segment (*ab*) of length (*L*) inclined at an angle θ with *x*-axis (Fig. 11).

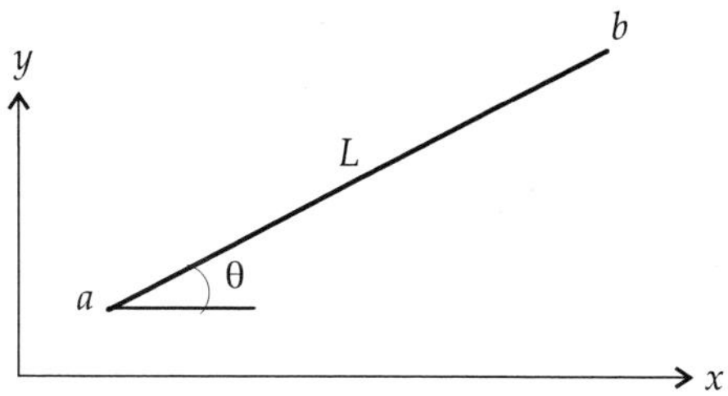

Fig. 11

Consider a small element (*dL*) (Fig. 12) along the segment.

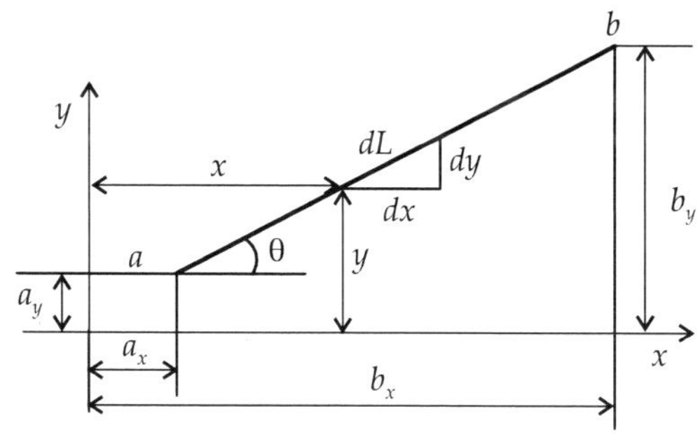

Fig. 12

Since, $dx = dL \cos\theta$ and $dy = dL \sin\theta$.
$b_x - a_x = L\cos\theta$ and $b_y - a_y = L\sin\theta$
Now *x* and *y*-coordinates of centroid are calculated as

$$\overline{X} = \frac{\int_{a_x}^{b_x} (xdL)}{\int_{a_x}^{b_x} dL} = \frac{\int_{a_x}^{b_x} (xdx)}{\int_{a_x}^{b_x} dx} = \frac{(b_x^2 - a_x^2)/2}{(b_x - a_x)}$$

$$= \frac{(b_x + a_x)}{2} = a_x + \frac{(b_x + a_x)}{2} = a_x + \frac{L\cos\theta}{2}$$

$$\overline{Y} = \frac{\int_{a_y}^{b_y} (ydL)}{\int_{a_y}^{b_y} dL} = \frac{\int_{a_y}^{b_y} (ydy)}{\int_{a_y}^{b_y} dy} = \frac{\left(b_y^2 - a_y^2\right)/2}{(b_y - a_y)}$$

$$= \frac{(b_y - a_y)}{2} = a_y + \frac{(b_y - a_y)}{2} = a_y + \frac{L\sin\theta}{2}$$

Note: If line is horizontal, i.e., $\theta = 0$; then

$\overline{X} = a_x + \frac{L}{2}$ and $\overline{Y} = a_y$

If the line starts from the origin also then

$\overline{X} = \frac{L}{2}$ and $\overline{Y} = 0$

Some Important Points

- An area (or section or body) with an axis of symmetry will find its first moment of area with respect to that axis is equal to zero, i.e., the centroid is located somewhere along that axis (Fig. 13. centroid of a semicircle is at its axis, passes through the centre and cut the semicircle in two parts.)

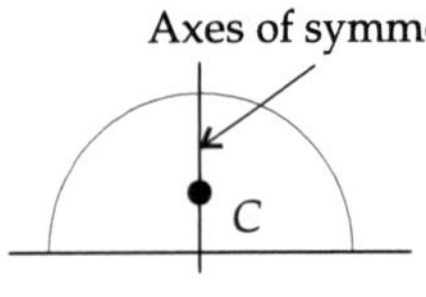

Fig. 13

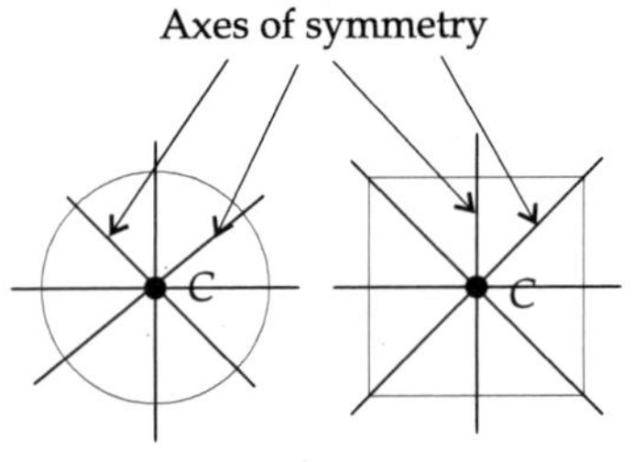

Fig. 14

- An area (or section or body) has two (or more) axes of symmetry; the centroid is located at the intersection of these axes. (Fig. 14. Centroid of a circle is at its centre and that of a square is at the intersection of two lines drawn connecting the midpoints of the parallel sides. The circle has an infinite number of lines of symmetry and the square has four.)
- The centroid of a section is not always within the area or material of the section (Fig. 15. *L*-shaped has its centroid located outside of the material of the section).

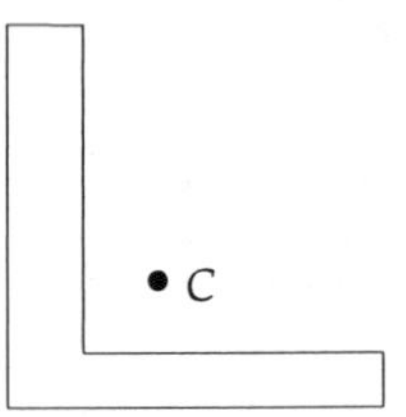

Fig. 15

COMPOSITE BODIES

A body consists of a combination of simple bodies. When a body is of more complex shape or unsymmetrical a simple method of determining the location of the centre of gravity (or centre of mass or centroid) may be used which divides the complex shape into smaller simple geometric shapes (treating it as a composite body) for which the centre of gravity (or centre of mass or centroid) locations may easily be determined.

First locate the centre of gravity (or centre of mass or centroid) for each sub-body. The resultant centre of gravity (or centre of mass or centroid) of the composite body is the same as the centre of gravity (or centre of mass or centroid) of a system of particles (sub-bodies).

Let B_1, B_2, B_3 be the three sub-bodies of a composite system (Fig. 16) and C_1, C_2, C_3 are the locations of their centre of gravity (or centre of mass or centroid), respectively (Fig. 17).

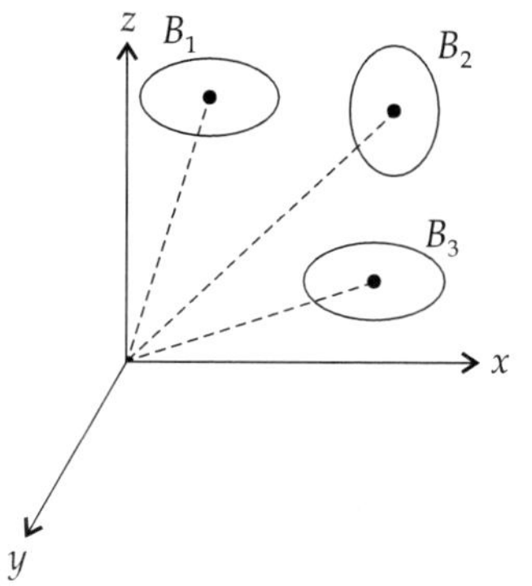

Fig. 16

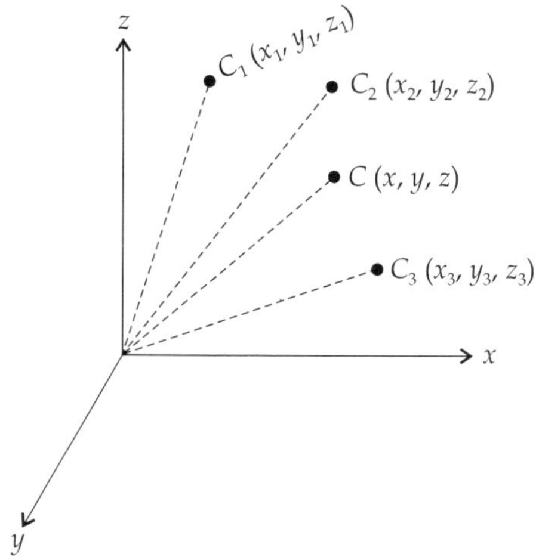

Fig. 17

Then the location of the centre of gravity (or centre of mass or centroid) of the composite body is given by

$$\overline{X} = \frac{x_1 \times B_1 + x_2 \times B_2 + x_3 \times B_3}{B_1 + B_2 + B_3}$$

$$\overline{Y} = \frac{y_1 \times B_1 + y_2 \times B_2 + y_3 \times B_3}{B_1 + B_2 + B_3}$$

$$\overline{Z} = \frac{z_1 \times B_1 + z_2 \times B_2 + z_3 \times B_3}{B_1 + B_2 + B_3}$$

where, B_1, B_2, B_3 can be considered as line, area, volume, mass, weight for calculating centroid of line, area, volume, centre of mass, centre of gravity, respectively.

Example 5: Determine the centroid of the composite area shown in Fig. 18.

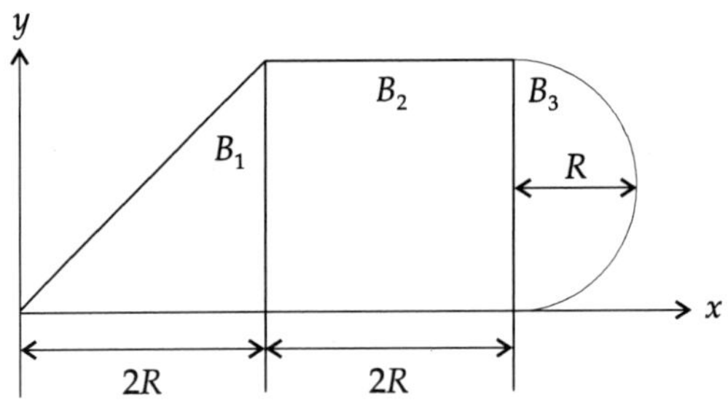

Fig. 18

Let C_1, C_2, C_3 be the centroids of the sub-areas B_1, B_2, B_3 (Fig. 19), whose centroids are well knowns.

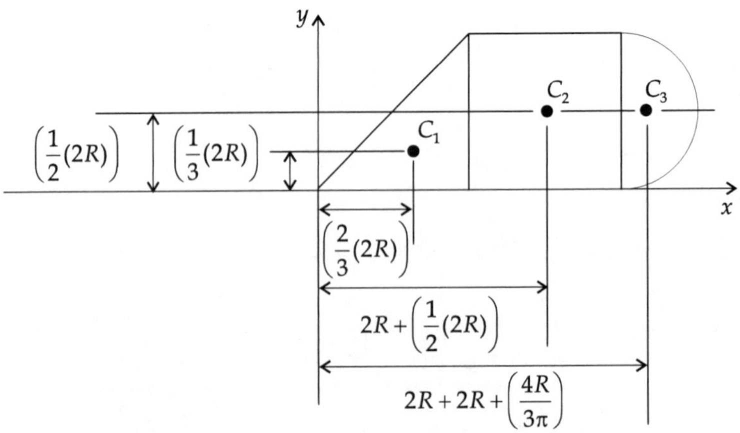

Fig. 19

Here, B_1, B_2, B_3 are the areas $2R^2$, $4R^2$, $\frac{\pi}{2}R^2$ respectively.

Then the location of the centroid of the composite body is given by

$$\overline{X} = \frac{\frac{4}{3}R \times 2R^2 + 3R \times 4R^2 + \left(4 + \frac{4}{3\pi}\right)R \times \frac{\pi}{2}R^2}{2R^2 + 4R^2 + \frac{\pi}{2}R^2}$$

$$= \left(\frac{92 + 12\pi}{36 + 3\pi}\right)R$$

$$\overline{Y} = \frac{\frac{2}{3}R \times 2R^2 + R \times 4R^2 + R \times \frac{\pi}{2}R^2}{2R^2 + 4R^2 + \frac{\pi}{2}R^2}$$

$$= \left(\frac{32 + 3\pi}{36 + 3\pi}\right)R$$

MOMENT OF INERTIA (SECOND MOMENT OF AREA)

The **moment of inertia (I)** is a term used to describe the capacity of a cross section to resist bending or how difficult it is to change its angular motion about that axis. It is always considered with respect to a reference axis such as *X-X* or *Y-Y*. It is a mathematical property of a section concerned with a surface area and how that area is distributed about the reference axis. The reference axis is usually a centroidal axis.

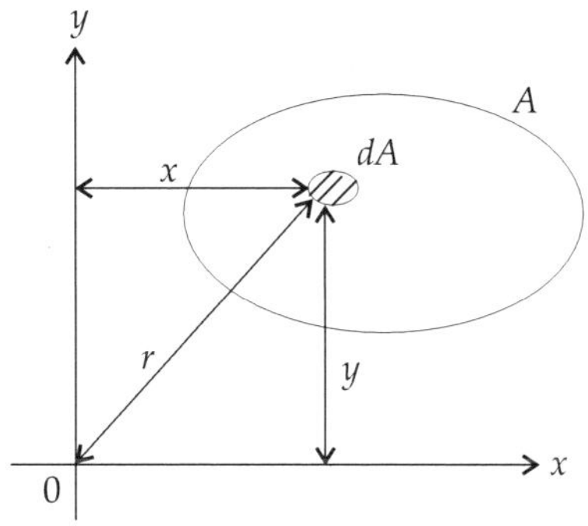

Fig. 20

The moment of inertia about an axis (Fig. 20) is obtained by multiplying the small area with its square of the perpendicular distance about that axis and integrating it over the whole area. The moment of inertia about the x-axis (I_x) and y-axis (I_y) are defined as

$$\boxed{I_x = \int_A y^2 dA \text{ and } I_y = \int_A x^2 dA}$$

Product of Inertia

The product of inertia is defined as

$$I_{xy} = I_{yx} = \int_A xy\,dA = \int_A yx\,dA$$

Polar Moment of Inertia

The polar moment of inertia (I_O) is the area moment of inertia about the origin (or z-axis) is given by

$$I_O = \int_A r^2 dA$$

where, r is the radial distance from the origin of the coordinate system to dA.

Since $r^2 = x^2 + y^2$, i.e., the polar moment of inertia is equal to the sum of moments of inertia about the x- and y-axes:

$$I_O = \int_A r^2 dA = \int_A x^2 dA + \int_A y^2 dA = I_y + I_x$$

Radius of gyration

The radius of gyration with respect to x-axis (r_x) and with respect to y-axis (r_y) are defined as

$$r_x = \sqrt{\frac{I_x}{A}} \text{ and } r_y = \sqrt{\frac{I_y}{A}}$$

The radius of gyration (r_O) with respect to origin (or z-axis) is defined as

$$r_O = \sqrt{\frac{I_O}{A}}$$

Since $I_O = I_y + I_x$,i.e., square of radius of gyration about z-axis is equal to the sum of the square of radius of gyration about the x- & y-axes: $r_O^2 = r_x^2 + r_y^2$

Note:

- The moments of inertia of an area has dimension length[4] and the radius of gyration has dimensions of length.
- Moments of inertia (I_x, I_y, I_O) and the radii of gyration (r_x, r_y, r_O)must have positive values for any area, they cannot be negative or zero.

- If an area is symmetric about either the x- or y-axis, its product of inertia is zero.
- Radius of gyration has no significant meaning. It is useful only in comparative purpose.

Example 6: Determine the moment of inertias, product of inertia, polar moment of inertia and radius of gyrations about the centroid of the rectangle as shown in Fig. 21.

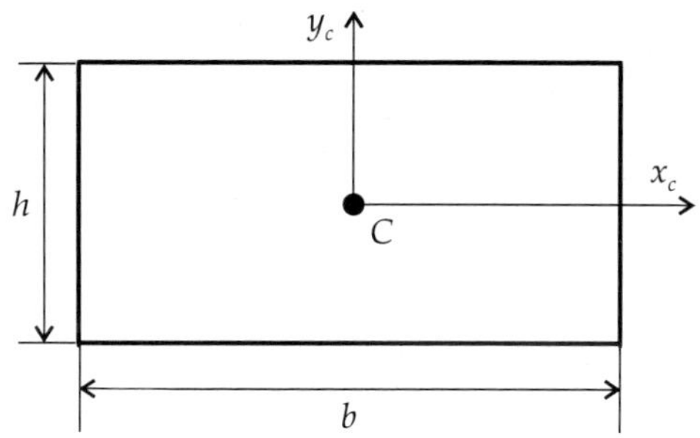

Fig. 21

Moment of Inertias

Consider a strip dx (Fig. 22) along x_c-axis at a distance x then moment of inertia about the y_c-axis

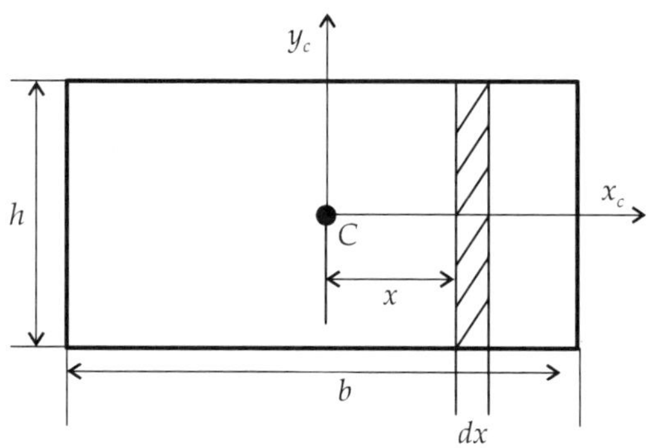

Fig. 22

$$I_{y_c} = \int_A x^2 dA = \int_{-b/2}^{b/2} x^2 h dx = \frac{bh^3}{12}$$

Similarly, we can calculate moment of inertia about the x_c-axis, i.e., $I_{x_c} = \frac{bh^3}{12}$.

Product of Inertia

Consider an element (Fig. 23) of small area $(dx.dy)$ then the product of inertia about the axes passing through centroid is

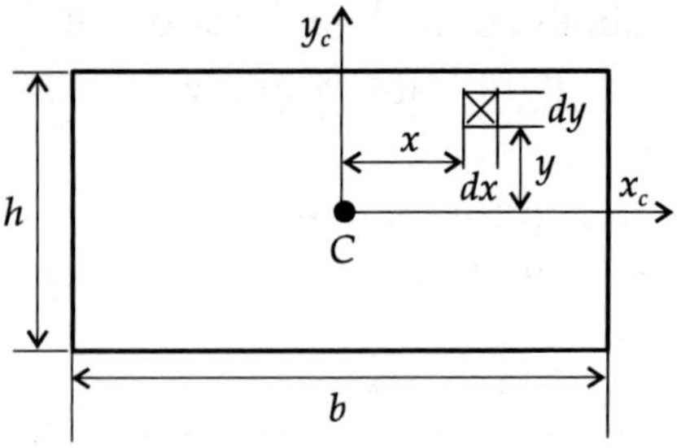

Fig. 23

$$I_{x_c y_c} = I_{y_c x_c} = \int_A yx dA = \int_{-h/2}^{h/2} \int_{-b/2}^{b/2} xy dx dy = 0$$

Since area is symmetrical about both the axes therefore, both the values of product of inertia is zero.

Polar Moment of Inertia

The polar moment of inertia (I_{Oc}) is the area moment of inertia about the origin (or z_c-axis) is

$$I_{Oc} = I_{x_c} + I_{y_c} = \frac{bh^3}{12} + \frac{b^3h}{12} = \frac{bh(h^2 + b^2)}{12}$$

Radius of Gyration

The radius of gyration with respect to x-axis (r_{x_c}) and with respect to y-axis (I_{y_c}) are

$$r_{x_c} = \sqrt{\frac{I_{x_c}}{A}} = \sqrt{\frac{(bh^3/12)}{bh}} = \frac{h}{2\sqrt{3}}$$

$$r_{y_c} = \sqrt{\frac{I_{y_c}}{A}} = \sqrt{\frac{(b^3h/12)}{bh}} = \frac{b}{2\sqrt{3}}$$

PARALLEL AXIS THEOREM

The parallel axis theorem is a relation between the moment of inertia about an axis passing through the centroid and the moment of inertia about any parallel axis.

Parallel axis theorem is useful:

- when moments of inertia of an area are known in terms of a particular coordinate system but we need their values in terms of a different coordinate system; and
- for determining the moments of inertia of a composite area when the moments of inertia of its parts are known.

According to parallel-axis theorem the moment of inertia of an area about an axis not through the centroid is equal to the moment of inertia about a parallel axis through the centroid of the area plus the product of the same area and the square of the distance between the two axes, i.e.,

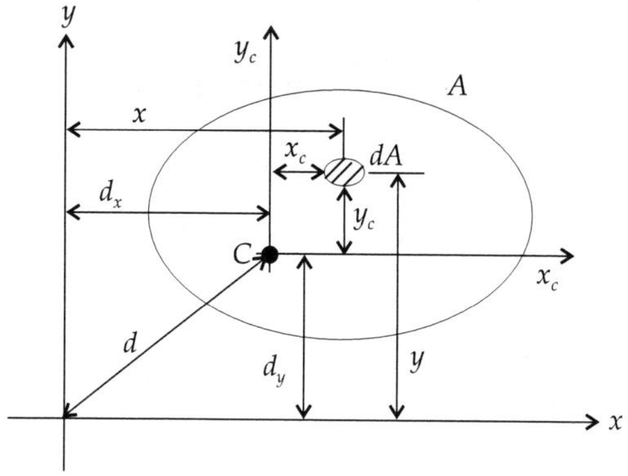

Fig. 24

Suppose that we know the moments of inertia of an area (A) (Fig. 24) in terms of a coordinate system x_c, y_c with its origin at the centroid (C) of the area and we want to determine the moments of inertia in terms of a parallel coordinate system x, y.

Moment of Inertia

Let in x_c, y_c coordinate system, the coordinates of the centroid of the area (A) are

$$\bar{x}_c = \frac{\int (x\, dA)}{\int dA}, \bar{y}_c = \frac{\int (y_c dA)}{\int dA}$$

But the origin of x_c, y_c coordinate system is located at the centroid of A, so

$$\bar{x}_c = \frac{\int (x\, dA)}{\int dA} = 0, \bar{y}_c = \frac{\int (y_c dA)}{\int dA} = 0$$

Now let in terms of the x, y coordinate system, the moment of inertia of A about the x-axis, $I_x = \int_A y^2 dA$

Since $y = y_c + d_y$, therefore

$$I_x = \int_A y^2 dA = \int_A (y_c + d_y)^2 dA$$

$$= \int_A y_c^2 dA + \int_A d_y^2 dA + 2d_y \int_A y_c dA$$

$$\boxed{I_x = I_{x_c} + d_y^2 A}$$

Since the first term (I_{xc}) is moment of inertia about x_c axis and the third term is equal to zero as explained above.

Similarly, the parallel axis theorem about y-axis gives

$$\boxed{I_y = I_{y_c} + d_x^2 A}$$

where, I_{yc} is the moment of inertia about y_c-axis.

Product of Inertia

The parallel-axis theorem for the product of inertia is:

$$\boxed{I_{xy} = I_{x_c y_c} + d_x d_y A}$$

where, $I_{x_c y_c}$ is the product of inertia about x_c, y_c axes.

Polar moment of linertia

The parallel-axis theorem for the polar moment of inertia is:

$$\boxed{I_O = I_{Oc} + \left(d_x^2 + d_y^2\right) A = I_{Oc} + d^2 A}$$

where, I_{Oc} is the polar moment of inertia about centroid (*C*).

PERPENDICULAR AXIS THEOREM

For a planar object, the moment of inertia about an axis perpendicular to the plane (Fig. 25) is the sum of the moments of inertia of two perpendicular axes through the same point in the plane of the object.

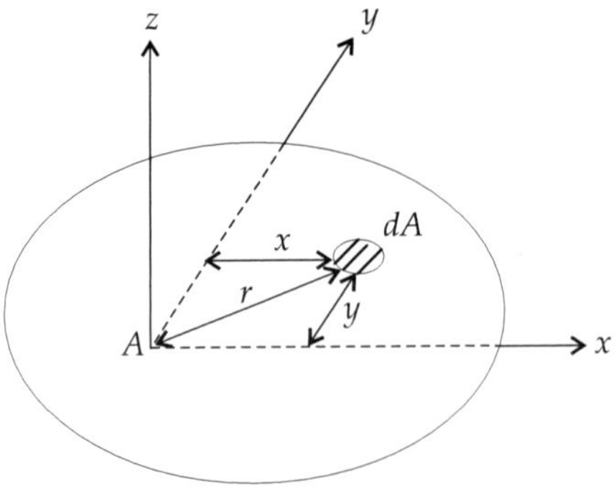

Fig. 25

Since $dI_x = y^2\, dA$, $dI_y = x^2\, dA$ and $dI_z = r^2 dA$

but $r^2 = x^2 + y^2$ therefore $dI_x + dI_y = (y^2 + x^2)dA = r^2\, dA = dI_z$

since this is true for each small area so by integrating it over the whole area, we get

$\boxed{I_z = I_x + I_y}$ (I_z is also known as polar moment of inertia).

The utility of this theorem goes beyond that of calculating moments of strictly planar objects. It is a very useful theorem for finding the moments of inertia of three-dimensional objects such as cylinders by breaking them up into planar disks and summing the moments of inertia of the composite disks.

Example 7: Determine the moment of inertias, product of inertia, polar moment of inertia and radius of gyrations about the axes passes through the base and left face of the rectangle as shown in Fig. 26.

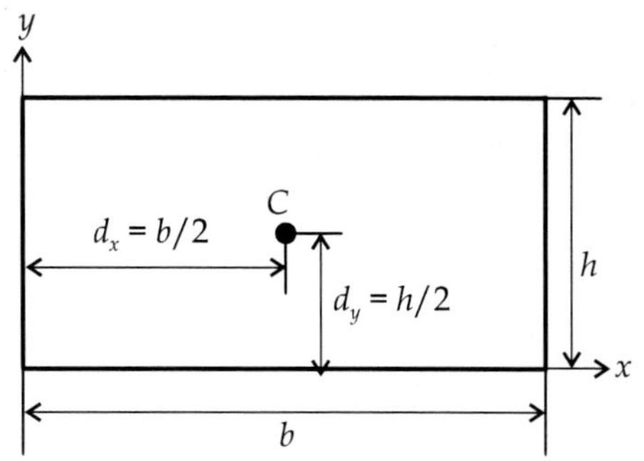

Fig. 26

Moment of Inertias

Method 1

Direct Integration

Consider a strip (dx) along x-axis at a distance x from y-axis and integrating it over the whole area, i.e.,

$$I_y = \int_A x^2 dA = \int_0^b x^2 h dx = \frac{hb^3}{3},$$

$$\text{similarly, } I_x = \frac{h^3 b}{3}$$

Method 2

By using the *parallel axis theorem*

Since we know the values of moment of inertias about the centroid, so by using parallel axis theorem moment of inertia about the left face and base of the rectangle are:

$$I_y = I_{y_c} + d_x^2 A = \frac{hb^3}{12} + (b/2)^2 bh = \frac{hb^3}{3}$$

$$I_x = I_{x_c} + d_y^2 A = \frac{h^3 b}{12} + (h/2)^2 bh = \frac{h^3 b}{3}$$

Product of Inertia

Method 1

Direct integration

Consider an element of small area ($dx \cdot dy$) then the product of inertia about axes passes through the left face and base of the rectangle is

$$I_{xy} = I_{yx} = \int_A yx dA = \int_0^h \int_0^b xy dx dy = \frac{b^2 h^2}{4}$$

Method 2

By using the *parallel axis theorem*

Since we know the value of the product of the inertia about the centroid, so by using parallel axis theorem product of inertia about the axes passes through the left face and base of the rectangle is:

$$I_{xy} = I_{x_c y_c} + d_x d_y A = 0 + \left(\frac{b}{2}\right)\left(\frac{h}{2}\right)bh = \frac{b^2h^2}{4}$$

Polar Moment of Inertia

Method 1

Direct integration

The polar moment of inertia (I_O) is the area moment of inertia about inertia about the axes passesthrough the left face and base of the rectangle is:

$$I_O = I_y + I_x = \frac{bh^3}{3} + \frac{b^3h}{3} = \frac{bh(b^2 + h^2)}{3}$$

Method 2

By using the *parallel axis theorem*

Since we know the value of polar moment of inertias about the centroid, so by using parallel axis theorem polar moment of inertia about the axes passes through left face and base of the rectangle is:

$$I_O = I_{Oc} + \left(d_x^2 + d_y^2\right)A$$

$$= \frac{bh(b^2 + h^2)}{12} + \left((h/2)^2 + (b/2)^2\right)bh$$

$$= \frac{bh(b^2 + h^2)}{3}$$

Radius of gyration

The radius of gyration with respect to x-axis (r_x) and with respect to y-axis (r_y) are:

$$r_x = \sqrt{\frac{I_x}{A}} = \sqrt{\frac{(bh^3/3)}{bh}} = \frac{h}{\sqrt{3}}$$

$$r_y = \sqrt{\frac{I_y}{A}} = \sqrt{\frac{(b^3/h)}{bh}} = \frac{b}{\sqrt{3}}$$

MOMENT OF INERTIA OF COMPOSITE BODIES

The moment of inertia of composite bodies about an axis can be calculated by adding together the moment of inertia of each of its sections about the same axis.

Consider the composite body (Fig. 27).

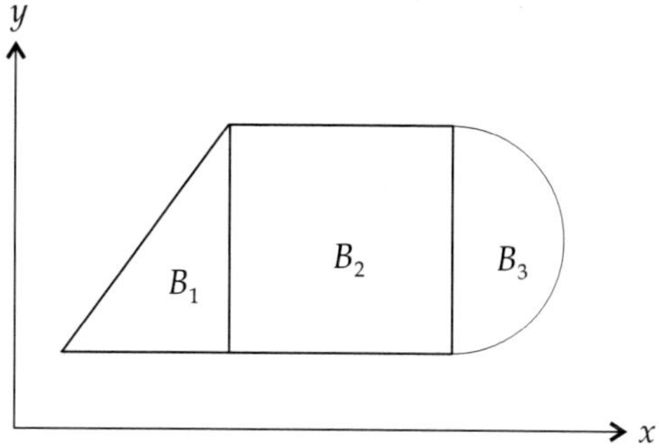

Fig. 27

Let B_1, B_2, B_3 be the sub-bodies and I_1, I_2, I_3 are its moment of inertia about an axis, respectively. Then moment of inertia (I_c) of a composite body about the same axis is:

$$\boxed{I_c = I_1 + I_2 + I_3}$$

Method for determining the moment of inertia of such a complex structure:

- Determine the centroids of the sub-bodies.
- Calculate the moment of inertia of the sub-body about its centroidal axes.
- Use the parallel axis theorem to move sub-body moment of inertia to axes where we have to find out the moment of inertia of the composite body.
- Sum the contributions of each sub-body to the overall moment of inertia of the composite body.

Note

- For determining the product of inertia of such a complex structure a similar approach to that used for calculating the moment of inertia of complex body is used but this approach is not valid for radius of gyration.
- The radius of gyration of a composite area is not equal to the sum of the radii of gyration of the component areas.
- The moments of inertia do not depend on how you divide a composite area into parts and you will often have a choice of convenient ways to divide a given area.
- Integration is an additive process, which is why the *moments of inertia* and *product of inertia* of composite areas can be determined by *adding* (or in the case of a cutout, *subtracting*) the moments of inertia of the parts, but the radii of gyration of composite areas cannot be determined by adding or subtracting the radii of gyration of the parts.

Example 8: Determine the moment of inertias about x-y axes of the composite area shown in Fig. 28.

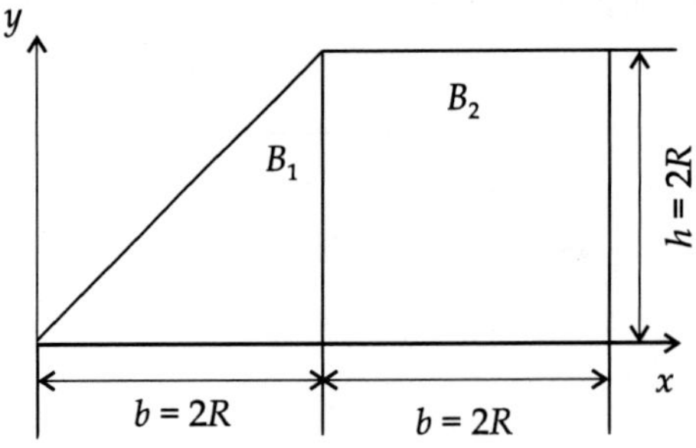

Fig. 28

Let C_1, C_2 and C be the values of centroids of sub-areas B_1, B_2 and composite area $B = B_1 + B_2$ respectively (Fig. 29).

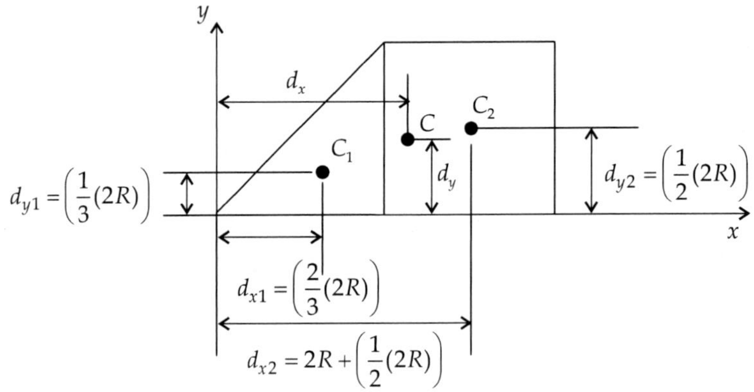

Fig. 29

As calculated above in Example 5 we know the values of:

$$d_{x1} = \frac{4R}{3} \quad \text{and} \quad d_{y1} = \frac{2R}{3}$$

$$d_{x2} = 3R \quad \text{and} \quad d_{y2} = R$$

$$d_x = \frac{22R}{9} \quad \text{and} \quad d_y = \frac{8R}{9}$$

Let I_{x_c1}, I_{y_c1}, I_{x_c2}, I_{y_c2}, I_{x1}, I_{y1} and I_{x2}, I_{y2} be the moment of inertias of sub-areas, B_1, B_2, about the centroid and x-y axes, respectively. I_{x_c}, I_{y_c} and I_x, I_y are the moment of inertias of the composite area B about their centroid and x-y axes, respectively.

Steps

- First find out the values of moment of inertia about the centroid of different sub-areas.
- Apply parallel axis theorem to transfer the moment of inertia of sub-areas to the point of interest.
- Add all the moment of inertias.

i.e., $I_{x_c1} = \dfrac{bh^3}{36} = \dfrac{4R^4}{9}, \quad I_{y_c1} = \dfrac{b^3h}{36} = \dfrac{4R^4}{9}$

$$I_{x_c2} = \frac{bh^3}{12} = \frac{4R^4}{3}, \quad I_{y_c2} = \frac{b^3h}{12} = \frac{4R^4}{3}$$

Now by parallel axis theorem

Therefore,

$$I_{x1} = I_{x_c1} + d_{x1}^2 A = \frac{4R^4}{9} + \left(\frac{4R}{3}\right)^2 2R^2 = 4R^4$$

$$I_{y1} = I_{y_c1} + d_{y1}^2 A = \frac{4R^4}{9} + \left(\frac{2R}{3}\right)^2 2R^2 = \frac{4R^4}{3}$$

Therefore,

$$I_{x2} = I_{x_c2} + d_{x2}^2 A = \frac{4R^4}{3} + (3R)^2 4R^2 = \frac{112R^4}{3}$$

$$I_{y2} = I_{y_c2} + d_{y2}^2 A = \frac{4R^4}{3} + (R)^2 4R^2 = \frac{16R^4}{3}$$

$$I_x = I_{x1} + I_{x2} = 4R^4 + \frac{112R^4}{3} = \frac{124R^4}{3}$$

$$I_y = I_{y1} + I_{y2} = \frac{4R^4}{3} + \frac{16R^4}{3} = \frac{20R^4}{3}$$

ROTATION OF AXES

Rotation of axes

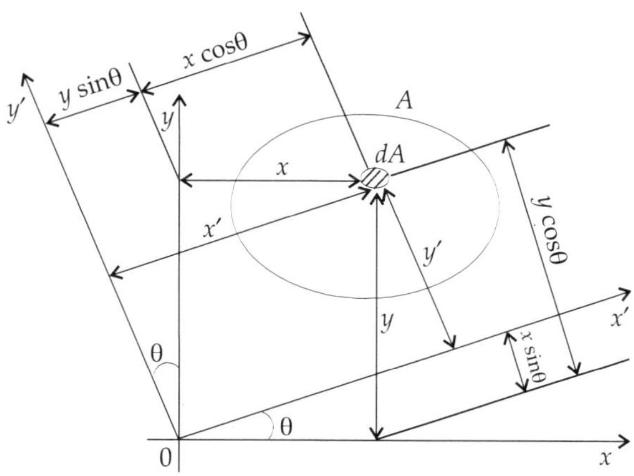

Fig. 30

Suppose that we know the moments of inertia of an area A (Fig. 30) in terms of the x, y coordinate system and we want to determine the moments of inertia in terms of the x', y' coordinate system at an angle of θ with x, y coordinate system then concept of rotation of axis is used as:

Moment of Inertia

The moment of inertia of the area with respect to the x' axis is:

$$I_{x'} = \int_A y'^2 dA = \int_A (y\cos\theta - x\sin\theta)^2 \, dA$$

$$= \cos^2\theta \int_A y^2 dA + \sin^2\theta \int_A x^2 dA - \sin 2\theta \int_A xy dA = I_x \cos^2\theta + I_y \sin^2\theta - I_{xy} \sin 2\theta$$

$$\boxed{I_{x'} = \left(\frac{I_x + I_y}{2}\right) + \left(\frac{I_x - I_y}{2}\right)\cos 2\theta - I_{xy} \sin 2\theta}$$

Similarly, the moment of inertia of the area with respect to the y'-axis is obtained by replacing $\theta \to \theta + \frac{\pi}{2}$.

$$\boxed{I_{y'} = \left(\frac{I_x + I_y}{2}\right) - \left(\frac{I_x - I_y}{2}\right)\cos 2\theta - I_{xy} \sin 2\theta}$$

Product of Inertia

The product of inertia of the area is obtained as

$$I_{x'y'} = \int_A x'y' dA$$

$$= \int_A (x\cos\theta + y\sin\theta)(y\cos\theta - x\sin\theta) dA$$

$$= \cos^2\theta \int_A xy dA - \sin^2\theta \int_A xy dA + \sin\theta\cos\theta \left(\int_A y^2 dA - \int_A x^2 dA\right)$$

$$= I_{xy} (\cos^2\theta - \sin^2\theta) + (I_x - I_y) \sin\theta \cos\theta$$

$$\boxed{I_{x'y'} = \frac{(I_x - I_y)}{2} \sin 2\theta + I_{xy} \cos 2\theta}$$

Polar Moment of Inertia

The polar moment of inertia in terms of the x', y' coordinate system is:

$$\boxed{I_{O'} = I_{x'} + I_{y'} = I_x + I_y = I_O}$$

- Thus, the value of the polar moment of inertia does not change by a rotation of the coordinate system.

Principal Moments of Inertia

Principal moments of inertia, I_{max}, I_{min}, are the maximum and minimum values of moment of inertia and the corresponding axes are defined as *principal axes*.

Let θ be the angle where the values of moment of inertia, $I_{x'}$ are maximum and minimum then

$$\frac{dI_{x'}}{d\theta} = (I_x - I_y)\sin 2\theta - 2\,I_{xy}\cos 2\theta = 0 \text{ which gives}$$

$$\boxed{\tan 2\theta = \frac{-2I_{xy}}{(I_x - I_y)}}$$

i.e., principal moments of inertia are obtained by putting the value of θ in moment of inertia, $I_{x'}$

$$\boxed{(I_{x'})_{\text{max,min}} = \left(\frac{I_x + I_y}{2}\right) \pm \sqrt{\left(\frac{I_x - I_y}{2}\right)^2 + I_{xy}^2}}$$

Note

- The angle θ for which $I_{x'}$ is a maximum $I_{y'}$ is a minimum; and at an angle θ for which $I_{x'}$ is a minimum, $I_{y'}$ is a maximum, i.e., $\boxed{(I_{x'})_{\text{max,min}} = (I_{y'})_{\text{min,max}}}$
- Since the tangent is a periodic function therefore the above equation does not yield a unique solution for the angle θ, i.e., it gives the orientation of the principal axes within an arbitrary multiple of 90° (θ + 90° n).
- The *product of inertia* corresponding to a set of principal axes is zero.

Example 9: Determine the moment of inertias, product of inertia, polar moment of inertia, radius of gyrations about the axes x'-y' of the square of side b and at an angle of 45° to x-y axes as shown in Fig. 31.

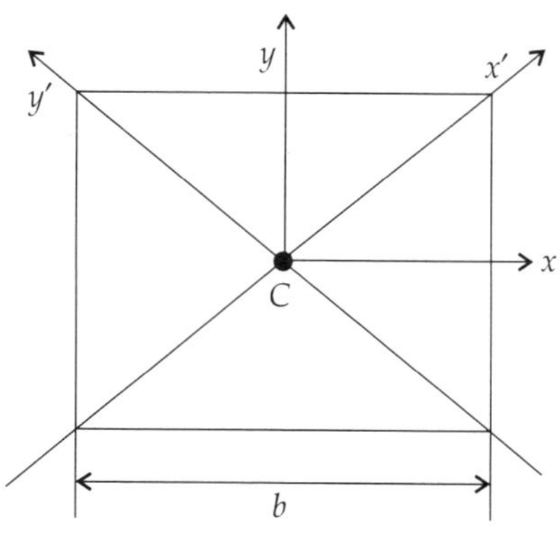

Fig. 31

Values of the moment of inertias, product of inertia, polar moment of inertia, radius of gyrations about the x-y (or x_c, y_c) axes through the centroid of the square of side b can be easily determined by putting $h = b$ in Example 6.

Now values about x'-y' axes are determined by rotation of axes as taking $\theta = 45°$:

Moment of Inertias

The moment of inertia of the area with respect to the x' and y' axes are:

$$I_{x'} = \left(\frac{I_x + I_y}{2}\right) + \left(\frac{I_x - I_y}{2}\right)\cos 2\theta - I_{xy}\sin 2\theta = \left(\frac{\frac{b^4}{12} + \frac{b^4}{12}}{2}\right) + \left(\frac{\frac{b^4}{12} - \frac{b^4}{12}}{2}\right)\cos 90 - 0 = \frac{b^4}{12}$$

$$I_{y'} = \left(\frac{I_x + I_y}{2}\right) + \left(\frac{I_x - I_y}{2}\right)\cos 2\theta - I_{xy}\sin 2\theta = \left(\frac{\frac{b^4}{12} + \frac{b^4}{12}}{2}\right) - \left(\frac{\frac{b^4}{12} - \frac{b^4}{12}}{2}\right)\cos 90 - 0 = \frac{b^4}{12}$$

Product of Inertia

$$I_{x'y'} = \frac{(I_x - I_y)}{2}\sin 2\theta + I_{xy}\cos 2\theta$$

$$= \left(\frac{\frac{b^4}{12} - \frac{b^4}{12}}{2}\right)\sin 90 + 0 = 0$$

Polar Moment of Inertia

$$I_{O'} = I_{x'} + I_{y'} = I_x + I_y = I_O = \frac{b^4}{6}$$

Radius of Gyration

The radii of gyration with respect to x-axis (r_{x_c}) and with respect to y-axis (r_{y_c}) are

$$r_{x'} = \sqrt{\frac{I_{x'}}{A}} = \sqrt{\frac{(b^4/12)}{b^2}} = \frac{b}{2\sqrt{3}}$$

$$r_{y'} = \sqrt{\frac{I_{y'}}{A}} = \sqrt{\frac{(b^4/12)}{b^2}} = \frac{b}{2\sqrt{3}}$$

MOHR'S CIRCLE OF INERTIA

Mohr's circle (Fig. 32) is a geometrical method to determine the values of moments of inertia and principal moment of inertia about axes.

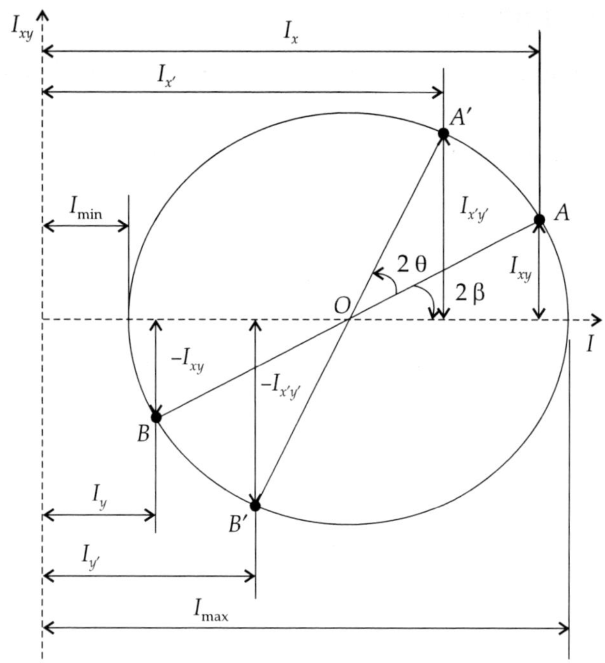

Fig. 32

Construction of the Mohr's Circle

- Draw the axes using I as the horizontal axis and I_{xy} as the vertical axis.
- Plot the points A and B whose coordinates are point A whose coordinates are (I_x, I_{xy}) and $(I_y, -I_{xy})$, respectively.
- Draw a straight line AB. Point O is located at the intersection between AB and the horizontal axis.
- Draw a circle whose origin is at point O and the length of its radius is the distance OA (or OB), i.e., points A and B are on the circumference of the circle.
- Suppose we want to determine the moments of inertia in terms of the x', y' coordinate system at an angle of θ with x, y coordinate system then the values of moments of inertia ($I_{x'}$, $I_{y'}$, $I_{x'y'}$) can be directly determined from the Mohr's circle by drawing a diameter $A'B'$ to the circle whose inclination is at an angle of θ (anticlockwise) from AB.
- Coordinates of points A' and B' on the circumference of the circle gives $(I_{x'}, I_{x'y'})$ and $(I_{y'}, I_{x'y'})$.
- The points where the circle circumference intersects the horizontal axis are the locations of the principal moment of inertia (I_{max}, I_{min}).
- 2β is the angle (clockwise) of principal moment of inertia with I-axis, i.e.,. tan 2β $= \dfrac{-2I_{xy}}{(I_x - I_y)}$.

MASS MOMENTS OF INERTIA

All the relationships defined above for moment of inertia of an area are also true for mass moment of inertia. Only the corresponding integrals for area (area distributions, dA) are replaced by mass integrations (mass distributions, dm).

i.e., $\boxed{dA \to dm}$

Properties of Plane Areas

A = Area

C = Centroid

(I_x, I_y) = Moment of inertia about x- and y-axes

I_{xy} = Product of inertia with respect to x- and y-axes

$I_p = I_x + I_y$ = Polar moment of inertia

I_{AA}, I_{BB} = Moment of inertia about axes A-A and B-B

Rectangle

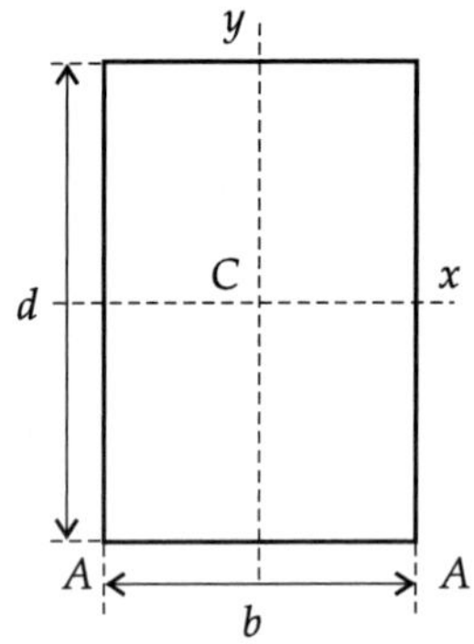

$$A = bd$$
$$C = (b/2, d/2)$$
$$I_x = \frac{bd^3}{12} \qquad I_y = \frac{bd^3}{12}$$
$$I_{xy} = 0 \qquad I_{AA} = \frac{bd^3}{3}$$
$$I_p = \frac{bd}{12}(d^2 + b^2)$$

Isosceles triangle

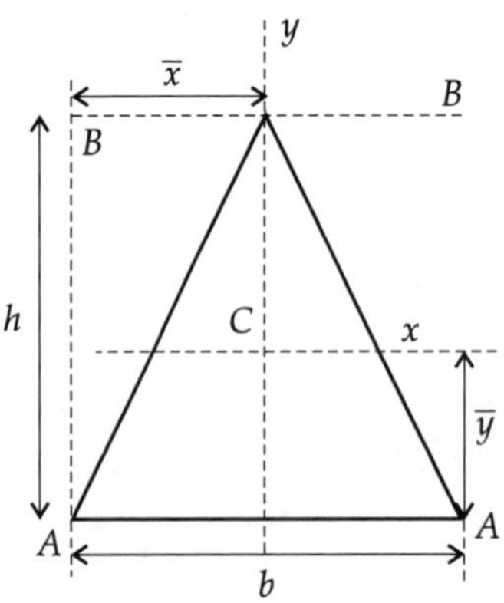

$$A = \frac{bh}{2} \quad I_{xy} = 0$$
$$C = (b/2, h/3)$$
$$I_x = \frac{bh^3}{36} \qquad I_y = \frac{hb^3}{48}$$
$$I_{AA} = \frac{bh^3}{12} \qquad I_{BB} = \frac{bh^3}{4}$$
$$I_p = \frac{bh}{144}(4h^2 + 3b^2)$$

Right triangle

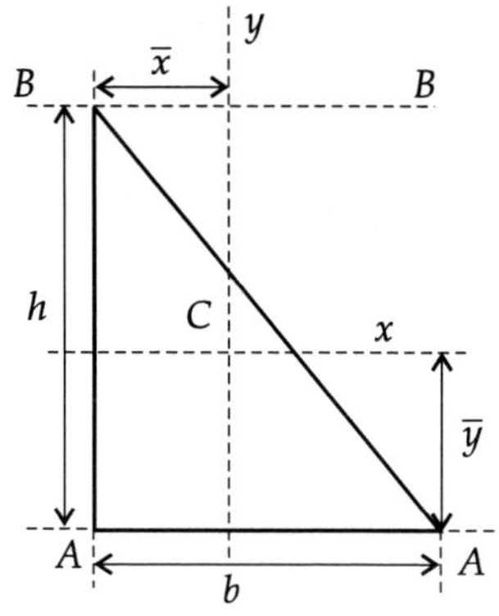

$$A = \frac{bh}{2} \quad C = (b/3, h/3)$$

$$I_x = \frac{bh^3}{36} \qquad I_y = \frac{hb^3}{36}$$

$$I_{AA} = \frac{bh^3}{12} \qquad I_{BB} = \frac{bh^3}{4}$$

$$I_p = \frac{bh}{36}(h^2 + b^2)$$

Circle

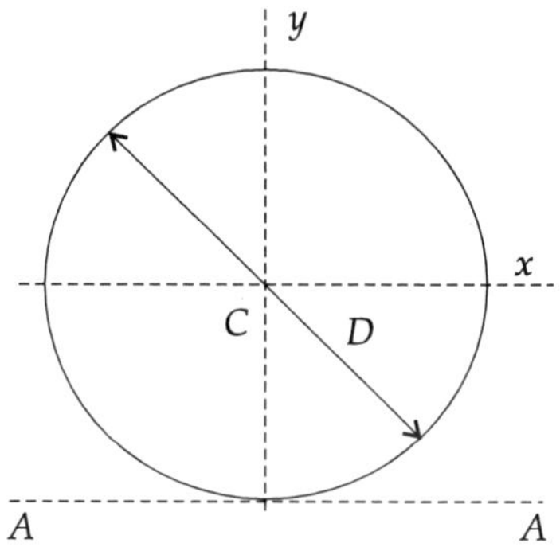

$$A = \frac{\pi d^2}{4} \quad C = (0, 0)$$

$$I_x = I_y = \frac{\pi d^4}{64} \quad I_{xy} = 0$$

$$I_p = \frac{\pi d^4}{32} \quad I_{AA} = \frac{5\pi d^4}{64}$$

Thin circular tube

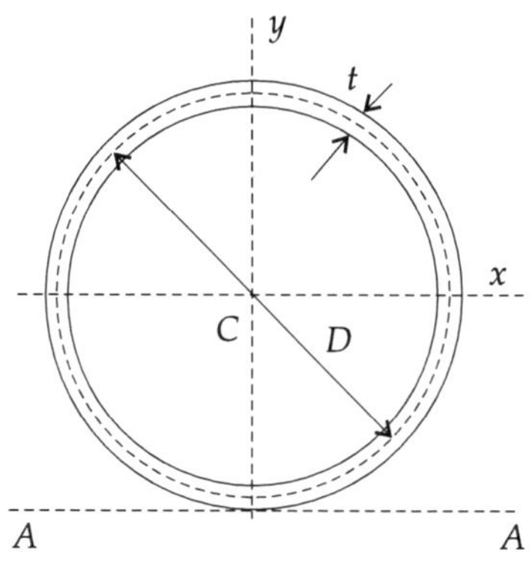

$$A = \pi dt \quad C = (0, 0)$$

$$I_x = I_y = \frac{\pi d^3 t}{8}$$

$$I_p = \frac{\pi d^3 t}{4} \quad I_{xy} = 0$$

Semicircle

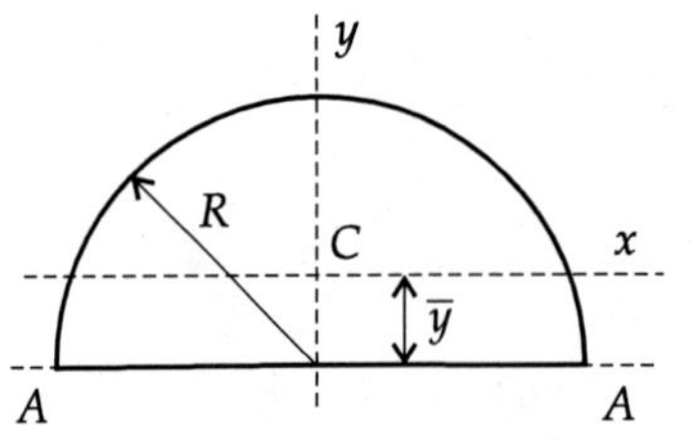

$$A = \frac{\pi R^2}{2} \quad C = (0, 4r/3\pi)$$

$$I_x \approx 0.11\, R^4 \quad I_y = \frac{\pi R^4}{8}$$

$$I_{AA} = \frac{\pi R^4}{8} \quad I_{xy} = 0$$

Parabolas

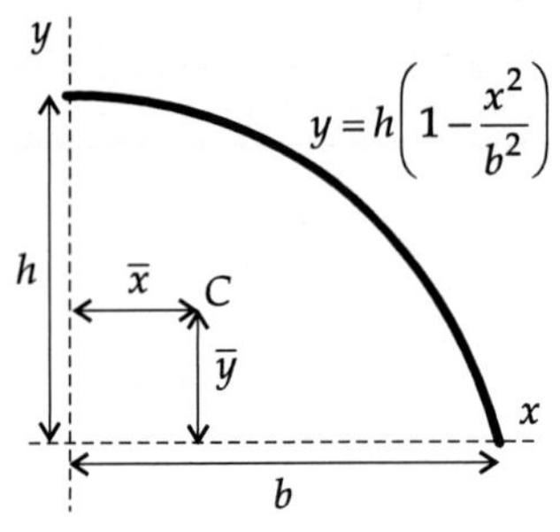

$$A = \frac{2bh}{3} \qquad I_{xy} = \frac{b^2h^2}{12}$$

$$C = (3b/8, 2h/5)$$

$$I_x = \frac{16bh^3}{155} \qquad I_y = \frac{2hb^3}{15}$$

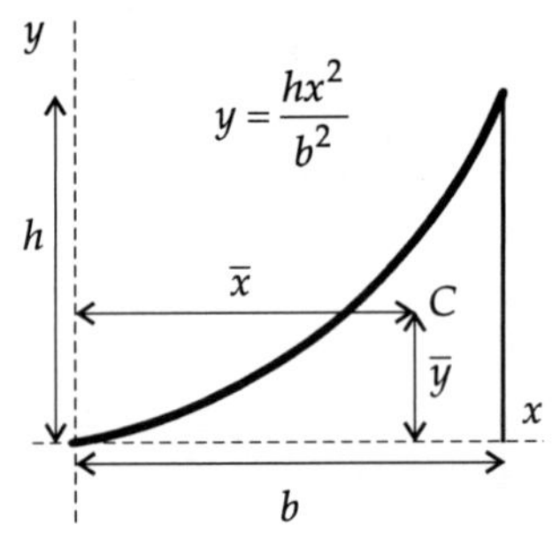

$$A = \frac{bh}{3} \qquad I_{xy} = \frac{b^2h^2}{12}$$

$$C = (3b/4, 3h/10)$$

$$I_x = \frac{bh^3}{21} \qquad I_y = \frac{hb^3}{5}$$

Appendix-2A

Competitive Objective Questions

1. Some structural members subjected sustained loads for long deform progressively with time, especially at elevated temperatures. What is such a phenomenon called?

   (a) Fatigue (b) Creep

   (c) Creep relaxation (d) Fracture [ESE-2008]

Ans. (b)

2. Consider the following statements:

   1. Failure occurs beyond elastic limit.
   2. Rupture takes place immediately after elastic limit.
   3. Permanent set occurs beyond elastic limit.

   Which of these are considered in the theories of failure?

   (a) 1, 2 and 3 (b) 1 and 3 only

   (c) 2 and 3 only (d) 1 and 2 only [ESE-2011]

Ans. (b)

3. Consider the following salient points in a stress-strain curve of a mild steel bar:

   1. Yield point
   2. Breaking point
   3. Yield plateau
   4. Proportionality limit
   5. Ultimate point

   The correct sequence in which they occur while testing the mild steel bar in tension from initial zero strain to failure is

   (a) 4, 1, 2, 3 and 5 (b) 1, 4, 3, 5 and 2

   (c) 4, 1, 3, 6 and 2 (d) 1, 4, 2, 3 and 5 [ESE-2011]

Ans. (c)

4. The material in which large deformation is possible before absolute failure by rupture takes place, is known as
   (a) Ductile (b) Plastic
   (c) Brittle (d) Elastic [ESE-2011]

Ans. (a)

5. If a material has identical elastic properties in all directions, it is said to be
   (a) elastic (b) isotropic
   (c) orthotropic (d) homogeneous [ESE-2012]

Ans. (b)

6. Which of the following statements are correct?
   1. In a ductile material elastic deformation is more predominant than plastic.
   2. Brittleness is not an absolute property of a metal.
   3. Rate of loading has no influence in changing the behavior of a metal from 'ductile' to brittle at low temperatures.
   4. Under hydrostatic compression, a brittle metal may change into a ductile one.

   (a) 1 and 2 (b) 2 and 3
   (c) 3 and 4 (d) 2 and 4 [ESE-2013]

Ans. (d)

7. Which one of the following represents constitutive relationship'?
   (a) Vertical displacements in a structure
   (b) Rotational displacements in a structure
   (c) System of force in equilibrium
   (d) Stress-strain behaviour of a material [ESE-2016]

Ans. (d)

8. In mild steel specimens subjected to tensile test cycle, the elastic limit in tension is raised and the elastic limit in compression is lowered. This is called
   (a) fatigue effect (b) Bauschinger effect
   (c) strain rate effect (d) None [ESE-2017]

Ans. (b)

9. Two identical circular rods of same diameter and same length are subjected to same magnitude of axial tensile force. One of the rods is of mild steel having the modulus of elasticity of 206 GPa. The other rod is of cast iron having the modulus of elasticity of 100 GPa. Assume both the materials to be homogeneous and isotropic and the axial force, cause the same amount stress in both the rods. The stresses developed are within the proportional limit of the respective materials. Which of the following observation is correct?

(a) Both rods elongate by the same amount
(b) Mild steel rod elongates more than cast iron rod
(c) Cast iron rod elongates more than the mild steel rod
(d) As the stresses are equal strains are also equal in both the rods

Ans. (c)

10. A steel bar of 40 mm × 40 mm square cross section is subjected to an axial compressive load of 200 kN. If the length of the bar is 2 m and E = 200 GPa, the elongation of the bar will be:

(a) 1.25 mm (b) 2.70 mm
(c) 4.05 mm (d) 5.40 mm [GATE-2006]

Ans. (a)

11. The number of independent elastic constant required to define the stress-strain relationship for an isotropic elastic solid is.....

(a) 1 (b) 2
(c) 3 (d) 4

Ans. (b)

12. A rod of length L and diameter D is subjected to a tensile load P. Which of the following is sufficient to calculate the resulting change in diameter?

(a) Young's modulus
(b) Shear modulus
(c) Poisson's ratio
(d) Both Young's modulus and shear modulus [GATE-2008]

Ans. (a)

13. If the Poisson's ratio of an elastic material is 0.4, the ratio of modulus of rigidity to Young's modulus is........

(a) 0.2 (b) 0.3
(c) 0.36 (d) 0.42 [GATE-2014]

Ans. (c)

14. For an isotropic material, the relationship between the Young's modulus (E), shear modulus (G) and Poisson's ratio (μ) is given by....

(a) $G = \frac{E}{2(1+\mu)}$ (b) $G = \frac{G}{2(1+\mu)}$

(c) $G = \frac{E}{(1+\mu)}$ (d) $G = \frac{E}{2(1-\mu)}$ [GATE-2007]

Ans. (a)

15. A rod is subjected to a uni-axial load within linear elastic limit. When the change in the stress is 200 MPa, the change in the strain is 0.001. If the Poisson's ratio of the rod is 0.3, the modulus of rigidity (in GPa) is .....

(a) 37 (b) 43

(c) 58 (d) 77 [GATE-2015]

Ans. (d)

16. What is total strain energy of a member subjected to an axial stress *f*(*E* = Young's modulus)?

(a) $\left(\frac{f^2}{2E}\right) \times$ volume of bar (b) $\left(\frac{f}{E}\right) \times$ volume of bar

(c) $\left(\frac{f^2}{E}\right) \times$ volume of bar (d) $\left(\frac{f}{2E}\right) \times$ volume of bar [ESE-2006]

Ans. (a)

17. What is the correct sequence of the following metals in the decreasing order of their Poisson's ratio?

1. Aluminium
2. Cast iron
3. Steel

Select the correct answer using the codes given below:

(a) 1-2-3 (b) 2-1-3

(c) 1-3-2 (d) 3-1-2 [ESE-2007]

Ans. (c)

18. In a body loaded under plane stress conditions, what is the number of independent stress components in order to completely specify the state of stress at a point?

(a) 3 (b) 4

(c) 6 (d) 9 [ESE-2008]

Ans. (a)

19. A composite system where the components are of equal length is subjected to temperature rise. Which one of the following stresses will develop in the component having highest coefficient of linear expansion?

(a) Compressive stress (b) Tensile stress

(c) Shear stress (d) Zero stress [ESE-2008]

Ans. (a)

20. What is the nature of stress in a ceiling fan rod?

(a) Bending (b) Tensile

(c) Compressive (d) Shear [ESE-2008]

Ans. (a)

21. Given E as the Young's modulus of elasticity of a material, what can be the minimum value of its bulk modulus of elasticity?

(a) $\frac{E}{2}$ (b) $\frac{E}{3}$

(c) $\frac{E}{4}$ (d) $\frac{E}{5}$ [ESE-2009]

Ans. (b)

22. Poisson's ratio is defined as the ratio of

(a) longitudinal stress and longitudinal strain

(b) lateral strain and longitudinal strain

(c) longitudinal stress and lateral stress

(d) lateral stress and longitudinal stress [ESE-2010]

Ans. (b)

23. The relationship between Young's modulus (E), modulus of rigidity (C) and bulk modulus K in an elastic material is given by the relation

(a) $E = \frac{9KC}{3K + C}$ (b) $E = \frac{3KC}{3K + C}$

(c) $E = \frac{9KC}{9K + C}$ (d) $E = \frac{3KC}{9K + C}$ [ESE-2013]

Ans. (a)

24. The symmetric of the stress tensor at a point in a body when at equilibrium is obtained from

(a) conservation of mass (b) force equilibrium equation

(c) moment equilibrium equations (d) conservation of energy [ESE-2014]

Ans. (c)

25. Which of the following statements are correct?

1. Strain in the direction of applied stress is known as longitudinal strain.
2. Tensile stress results in tensile strain in linear and lateral directions.
3. Strain in all directions perpendicular to the applied stress are known as lateral strains.
4. Ratio of change in volume to original volume is known as volumetric strain.

(a) 1, 2 and 3 only (b) 1, 3 and 3 only

(c) 3 and 4 only (d) 1, 2, 3 and 4 [ESE-2016]

Ans. (b)

26. Which of the following statements is/are correct?

1. In the infinitesimal strain theory, dilatation is taken as an invariant.
2. Dilatation is not proportional to the algebraic sum of all normal stresses.

3. The shearing modulus is always less than the elastic modulus.

(a) 1 only (b) 1 and 2 only

(c) 2 only (d) 1, 2 and 3 [ESE-2016]

Ans. (a)

27. A solid steel cube constrained on all six faces is heated so that the temperature rises uniformly by ΔT If the thermal coefficient of the material is a, Young's modulus is E and the Poisson's ratio is v, the thermal stress developed in the cube due to heating is

(a) $-\dfrac{\alpha(\Delta T)E}{(1-2u)}$ (b) $\dfrac{2\alpha(\Delta T)E}{(1-2u)}$

(c) $\dfrac{3\alpha(\Delta T)E}{(1-2u)}$ (d) $\dfrac{\alpha(\Delta T)E}{3(1-2u)}$ [GATE-2012]

Ans. (a)

28. A metal bar of length 100 mm inserted between two rigid supports and its temperature is increased by 10°C. If the coefficient of thermal expansion is 12×10 per °C and the Young's modulus is 2×10 MPa the stress in the bar is

(a) Zero (b) 12 MPa

(c) 24 MPa (d) 2400 MPa [GATE-2007]

Ans. (c)

29. A 200 mm long, stress free rod at room temperature is held between two immovable rigid walls. The temperature of the rod is uniformly raised by 250°C. If the Young's modulus and coefficient of thermal expansion are 200 GPa and 1×10^{-5}/°C, respectively, the magnitude of the longitudinal stress (in MPa) developed in the rod is...

(a) 200 (b) 300

(c) 500 (d) 600 [GATE-2014]

Ans. (c)

30. A circular rod of length L and area of cross section A has a modulus of elasticity E and coefficient of thermal α. One end of the rod is fixed and other end is free. If the temperature of the rod is increased by ΔT, then.... [GATE-2014]

(a) stress developed in the rod is $E\alpha\Delta T$ and strain developed in the rod is α, ΔT

(b) both stress and strain developed in the rod are zero

(c) stress developed in the rod is zero and strain developed in the rod is α, ΔT

(d) stress developed in the rod is $E\alpha\Delta T$ and strain developed in the rod is zero

Ans. (c)

31. A steel cube, with all faces free to deform, has Young's modulus, E, Poisson's ratio, v, and coefficient of thermal expansion, α. The pressure (hydrostatic

stress) developed within the cube, when it is subjected to a uniform increase in temperature, ΔT, is given by....

(a) 0 (b) $\dfrac{\alpha(\Delta T)E}{1-2u}$

(c) $-\dfrac{\alpha(\Delta T)E}{1-2u}$ (d) $\dfrac{\alpha(\Delta T)E}{3(1-2u)}$ [GATE-2014]

Ans. (a)

32. The creep strains are
    (a) caused due to dead loads only
    (b) caused due to live loads only
    (c) caused due to cyclic loads only
    (d) independent of loads [GATE-2013]

Ans. (a)

33. The flow stress (in MPa) of a material is given by $\sigma = 500\varepsilon^{0.1}$ where ε is true strain. The Young's modulus of elasticity of the material is 200 GPa. A block of thickness 100 mm made of this material is compressed to 95 mm thickness and then the load is removed. The final dimension of the block (in mm) is......
    (a) 95 (b) 102
    (c) 127 (d) 144 [GATE-2015]

Ans. (a)

34. The strain hardening exponent (n) of stainless steel SS304 with distinct yield and UTS values undergoing plastic deformation is
    (a) $n < 0$ (b) $n = 0$
    (c) $0 < n < 1$ (d) $n = 1$ [GATE-2015]

Ans. (c)

35. If permissible stress in plates of join through a pin as shown in the figure is 200 Mpa, then the width will be

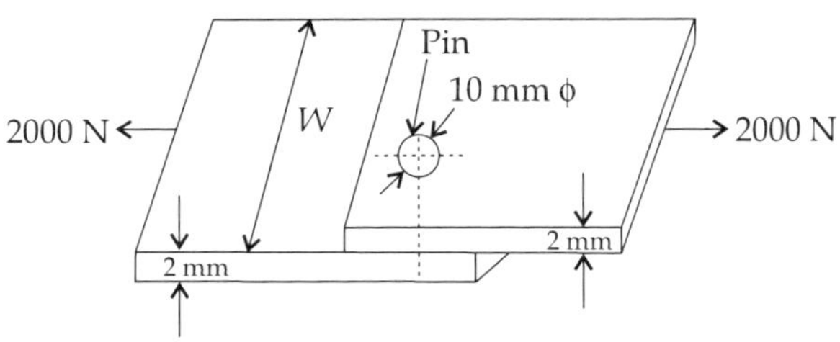

    (a) 15 mm (b) 20 mm
    (c) 18 mm (d) 25 mm

Ans. (a)

36. The state of plate stress in a plate of 100 mm thickness is: σ_{xx} = 100 N/mm^2, σ_{yy} = 200 N/mm^2, Young's modulus = 300 N/mm^2, Poisson's ratio = 0.3
The stress developed in the direction of thickness is

(a) Zero (b) 90 N/mm^2

(c) 100 N/mm^2 (d) 200 N/mm^2

Ans. (a)

37. In a tensile test, near the elastic limit zone, tensile stress

(a) increases at a faster rate

(b) decreases at a faster rate

(c) increases in linear proportional to the stress

(d) decreases in linear proportional to the stress [ESE-2006]

Ans. (b)

38. Which one of the following expresses the total elongation of a bar of length *L* with a constant cross section of *A* and modulus of elasticity *E* hanging vertically and subjected to its own weight, *W*?

(a) $\frac{WL}{AE}$ (b) $\frac{WL}{2AE}$

(c) $\frac{2WL}{AE}$ (d) $\frac{WL}{4AE}$ [ESE-2007]

Ans. (b)

39. A rod of length *l* tapers uniformly from a diameter *D* at one end to a diameter (*d*) at the other. The Young's modulus of the material is *E*. The extension caused by an axial load *P* is

(a) $\frac{4Pl}{\pi(D^2 - d^2)E}$ (b) $\frac{4Pl}{\pi(D^2 + d^2)E}$

(c) $\frac{4Pl}{\pi DdE}$ (d) $\frac{2Pl}{\pi DdE}$

Ans. (c)

40. Which of these statements are correct?

1. State of plane stress occurs at the surface
2. State of plane strain occurs in the surface
3. State of plane stress occurs in the interior part of the plate
4. State of plane strain occurs in the interior part of the plate

(a) 1 and 2 (b) 2 and 4

(c) 1 and 4 (d) 2 and 3 [ESE-2013]

Ans. (c)

41. Which of these statement is/are correct?
    1. On a principal plane, only normal stress acts
    2. On a principal plane, both normal and shear stresses act
    3. On a principal plane, only shear stress acts
    4. Isotropic state of stress is independent of frame of reference.

(a) 1 and 4 (b) 2 only
(c) 2 and 4 (d) 2 and 3 [GATE-2009]

Ans. (a)

42. In a plane stress condition the components of stress are σ_x = 20 MPa, σ_y = 80 MPa, τ_{xy} = 40 MPa The maximum shear stress (in MPa) at the point is

(a) 20 (b) 25
(c) 50 (d) 100 [GATE-2015]

Ans. (c)

43. A solid circular shaft of diameter 100 mm is subjected to an axial stress of 50 MPa. It is further subjected to a torque of 10 kNm. The maximum principal stress experienced on the shaft is close to

(a) 41 MPa (b) 82 MPa
(c) 164 MPa (d) 20 4MPa [GATE-2008]

Ans. (b)

44. The normal stresses at a point are σ_x = 10 MPa, σ_y = 2 MPa; the shear stress at this point is 4MPa. The maximum principal stress at this point is:

(a) 16 MPa (b) 14 MPa
(c) 11 MPa (d) 10 MPa [GATE-2014]

Ans. (c)

45. The state of stress at a point is given by σ_x = –6 MPa, σ_y = 4 MPa, τ_{xy} = –8 MPa The maximum tensile stress (in MPa) at the point is....

(a) 6.2 (b) 8.4
(c) 10.2 (d) 15.1 [GATE-2014]

Ans. (b)

46. A two-dimensional fluid element rotates like a rigid body. At a point within the element, the pressure is 1 unit. Radius of the Mohr's circle, characterized the state of stress at that point, is

(a) 0.5 unit (b) 0 unit
(c) 1 unit (d) 2 units [GATE-2008]

Ans. (b)

47. The state of stress at a point under plane stress condition is σ_{xx} = 40 MPa, σ_{yy} = 100 MPa, τ_{xy} = 40 MPa. The radius of the Mohr's circle representing the given state of stress in MPa is...

(a) 40 (b) 50
(c) 60 (d) 100 [GATE-2012]

Ans. (b)

48. Mohr's circle for the state of stress defined by $\begin{bmatrix} 30 & 0 \\ 0 & 30 \end{bmatrix}$ MPa is a circle with

(a) centre at (0, 0) and radius 30 MPa
(b) centre at (0, 0) and radius 60 MPa
(c) centre at (20, 0) and radius 30 MPa
(d) centre at (30, 0) and zero radius [GATE-2006]

Ans. (d)

49. Determine the maximum and minimum principal stresses respectively from the Mohr's circle

(a) +175 MPa, –175 MPa (b) +175 MPa, +175 MPa
(c) 0, -175 MPa (d) 0,0 [GATE-2003]

Ans. (b)

50. *E*, *G*, *K* and μ represent the elastic modulus, shear modulus, bulk modulus and Poisson's ratio, respectively of a linearly elastic, isotropic and homogeneous material. To express the stress-strain relations completely for this material, at least

(a) *E*, *G* and μ must be known
(b) *E*, *K* and μ must be known
(c) Any two of the four must be known
(d) All the four must be known [ESE-2006]

Ans. (c)

51. What is the relationship between the linear elastic properties; Young's modulus (*E*), rigidity modulus (*G*) and bulk modulus (*K*)?

(a) $\frac{1}{E} = \frac{9}{K} + \frac{3}{G}$ (b) $\frac{3}{E} = \frac{9}{K} + \frac{1}{G}$

(c) $\frac{9}{E} = \frac{3}{K} + \frac{1}{G}$ (d) $\frac{9}{E} = \frac{1}{K} + \frac{3}{G}$

Ans. (d)

52 What is the relationship between elastic constants *E*, *G* and *K*?

(a) $E = \frac{KG}{9K + G}$ (b) $E = \frac{9KG}{K + G}$

(c) $E = \frac{9KG}{K + 3G}$ (d) $E = \frac{9KG}{3K + G}$

Ans. (d)

53. If a piece of material neither expands nor contracts in volume when subjected to stress, then the Poisson's ratio must be

(a) Zero (b) 0.25

(c) 0.33 (d) 0.5

Ans. (d)

54. In a strained material, normal stresses on two mutually perpendicular planes are σ_x, σ_y (both alike) accompanied by a shear stress τ_{xy}. One of the principal stresses will be zero, only if

(a) $\tau_{xy} = \frac{\sigma_x \times \sigma_y}{2}$ (b) $\tau_{xy} = \sigma_x \times \sigma_y$

(c) $\tau_{xy} = \sqrt{\sigma_x \times \sigma_y}$ (d) $\tau_{xy} = \sqrt{\sigma_x^2 \times \sigma_y^2}$ [ESE-2006]

Ans. (c)

55. In a Mohr's circle, the radius of the circle is taken as

(a) $\sqrt{\left(\frac{\sigma_x + \sigma_y}{2}\right) + (\tau_{xy})^2}$ (b) $\sqrt{\left(\frac{\sigma_x - \sigma_y}{2}\right) + (\tau_{xy})^2}$

(c) $\sqrt{\left(\frac{\sigma_x - \sigma_y}{2}\right) - (\tau_{xy})^2}$ (d) $\sqrt{(\sigma_x - \sigma_y)^2 - (\tau_{xy})^2}$

where, σ_x, σ_y are normal stresses along x- and y-directions respectively, and τ_{xy} is the shear stress [ESE- 2006]

Ans. (b)

56. In a strained material one of the principal stresses is twice of the other. The maximum shear stress in the same case is τ_{max}. What is the value of the maximum principal stress?

(a) τ_{max} (b) $2\tau_{max}$

(c) $4\tau_{max}$ (d) $8\tau_{max}$ [ESE-2007]

Ans. (c)

57. For a general two-dimensional stress system, what are the coordinates of the centre of Mohr's circle?

(a) $\frac{\sigma_x - \sigma_y}{2}, 0$ (b) $0, \frac{\sigma_x + \sigma_y}{2}$

(c) $\frac{\sigma_x + \sigma_y}{2}, 0$ (d) $0, \frac{\sigma_x - \sigma_y}{2}$ [ESE-2007]

Ans. (c)

58. Maximum shear stress in a Mohr's circle

(a) is equal to the radius of Mohr's circle

(b) is greater than the radius of Mohr's circle

(c) is less than the radius of Mohr's circle

(d) any of the above [ESE-2008]

Ans. (a)

59. The principal stresses at a point in two-dimensional stress system are σ_1, σ_2 and corresponding principal strains are ε_1, ε_2. If E and μ denote Young's modulus and Poisson's ratio, respectively, then which one of the following is correct?

(a) $\sigma = E\varepsilon_1$ (b) $\sigma = \dfrac{E}{1-\mu^2}[\varepsilon_1 + \mu\varepsilon_2]$

(c) $\sigma = \dfrac{E}{1-\mu^2}[\varepsilon_1 - \mu\varepsilon_2]$ (d) $\sigma = E(\varepsilon_1 - \mu\varepsilon_2)$ [ESE-2008]

Ans. (b)

60. A point in a two-dimensional state of strain is subjected to pure shearing strain of magnitude γ_{xy} radians. Which one of the following is the maximum principal strain?

(a) γ_{xy} (b) $\dfrac{\gamma_{xy}}{\sqrt{2}}$

(c) $\dfrac{\gamma_{xy}}{2}$ (d) $2\gamma_{xy}$ [ESE-2008]

Ans. (c)

61. Assertion (*A*): Mohr's construction is possible for stresses, strains and area moment of inertia.

Reason (*R*): Mohr's circle represents the transformation of second-order tensor.

(a) both *A* and *B* are true and *R* is the correct explanation of *A*

(b) both *A* and *R* are true but *R* is not the correct explanation of *A*

(c) *A* is true but *R* is false

(d) *A* is false but *R* is true

Ans. (a)

62. Which of these statements is/are correct?

1. Two-dimensional stresses applied to a thin plate in its own plane represent the plane stress condition.
2. Under plane stress conditions, the strain in the directions perpendicular to the plane is zero.
3. Normal and shear stresses may occur simultaneously on a plane.

(a) 1 only (b) 1 and 2

(c) 2 and 3 (d) 1 and 3 [ESE-2009]

Ans. (d)

63. If the principal stresses and maximum shearing stresses are of equal numerical value at a point in a stressed body, the state of stress can be termed

(a) Isotropic (b) Uni-axial
(c) Pure shear (d) Generalized plane state of stress

[ESE-2010]

Ans. (c)

64. Statement (I): Mohr's circle of stress can be related to Mohr's circle of strain by some constant of proportionality.

Statement(II): The relationship is a function of yield strength of the material.

(a) Both statement (I) and statement (II) are individually true and statement (II) is the correct explanation of statement (I)
(b) Both statement (I) and statement (II) are individually true but statement (II) is NOT the correct explanation of statement (I)
(c) statement (I) is true but statement (II) is false
(d) Statement (I) is false but statement (II) is true [ESE-2012]

Ans. (c)

65. Mohr's circle construction is valid for both stress, as well as the area moment of inertia because:

(a) Both are tensors of the first order
(b) Both are tensors of the second order
(c) Both are axial vectors
(d) Both occur under plane stress condition [ESE-2013]

Ans. (b)

66. Consider the following statement related to Mohr's circle stresses in case of plane stress:

1. The construction is for variations of stress in a body.
2. The radius of the circle represents the magnitude of the maximum shearing stress.
3. The diameter represents the difference between the two principal stresses.

Which of the above statements are correct?

(a) 1, 2 and 3 (b) 2 and 3 only
(c) 1 and 3 only (d) 1 and 2 only [ESE-2015]

Ans. (b)

67. Polar moment of inertia (I_p) in cm^4, of a rectangular section having width, $b = 2$ cm and depth, $d = 6$ cm is..........

(a) 23 (b) 36
(c) 49 (d) 56 [GATE-2014]

Ans. (c)

68. According to maximum shear stress criterion, at what ratio of maximum shear stress to yield stress of material such that yielding of material does not take place?

(a) 2 (b) $2/\sqrt{3}$

(c) $1/\sqrt{3}$ (d) $1/2$ [ESE-2007]

Ans. (d)

69. If p_1 and p_2 are the principal stress at a point in a strained material with Young's modulus E and Poisson's ratio $1/m$, then what is the principal strain

(a) $\frac{p_1 + p_2}{mE}$ (b) $\frac{p_1 - p_2}{mE}$

(c) $\frac{p_1}{E} - \frac{p_2}{mE}$ (d) $\frac{p_1}{E} + \frac{p_2}{mE}$ [ESE-2006]

Ans. (c)

70. Mohr's stress circle helps in determining which of the following?

1. Normal stresses on one plane.
2. Normal and tangential stresses on two planes.
3. Principal stresses in all three directions.
4. Inclination of principal planes.

Select the correct answer using the codes given below:

(a) 1 and 2 only (b) 2 and 3 only

(c) 3 and 4 only (d) 2 and 4 only [ESE-2007]

Ans. (d)

71. Which of these statements is/are correct?

Mohr's strain circle can be drawn

1. for plane stress conditions
2. if strains in three directions are known
3. if strains on two mutually perpendicular planes are known

(a) 1, 2 and 3 (b) 2 and 3 only

(c) 1 and 2 only (d) 1 and 3 only [ESE-2007]

Ans. (a)

72. What is the angle between principal strain axis and maximum shear strain axis?

(a) 0° (b) 30°

(c) 45° (d) 90° [ESE-2007]

Ans. (c)

73. At what value of the ultimate shear strength, shall the material under the action of uniform axial tension fail due to shear?

(a) <0.5 times the ultimate tensile strength
(b) <0.7 times the ultimate tensile strength
(c) = ultimate tensile strength
(d) > the ultimate tensile strength [ESE-2007]

Ans. (a)

74. If a body carries two unlike principal stresses, what is the maximum shear stress?
(a) Half the difference of magnitude of the principal stresses
(b) Half the sum of the magnitude of principal stresses
(c) Difference of the magnitude of principal stresses
(d) Sum of the magnitude of principal stresses [ESE-2008]

Ans. (b)

75. In a biaxial strain system (ε_x, ε_y) what is the maximum engineering shearing strain?
(a) $\varepsilon_x + \varepsilon_y$ (b) $\varepsilon_x - \varepsilon_y$
(c) $\frac{\varepsilon_x + \varepsilon_y}{2}$ (d) $\frac{\varepsilon_x - \varepsilon_y}{2}$ [ESE-2008]

Ans. (b)

76. What would be the shape of the failure surface of a standard cast iron specimen subjected to torque?
(a) Cup and cone shape at the centre
(b) Plane surface perpendicular to the axis of the specimen
(c) Pyramid type wedge-shaped surface perpendicular to the axis of the specimen
(d) Helicoidally surface at 45° to the axis of the specimen. [ESE-2009]

Ans. (d)

77. What is the radius of Mohr's circle in cast of biaxial state of stress?
(a) Half the sum of the two principal stresses
(b) Half the difference of the two principal stresses
(c) Difference of the principal stresses
(d) Sum of the two principal stresses [ESE-2009]

Ans. (b)

78. Which of these statements is/are correct?
If there is a state of pure shear (τ) at a point then
1. The Mohr's circle is tangential to the y-axis.
2. The centre of the Mohr's circle coincides with the origin.

3. Unlike principal stresses are each numerically equal to τ.
4. Principal stresses are like.

(a) 1 only (b) 1 and 2
(c) 2 and 3 (d) 3 and 4 [ESE-2010]

Ans. (c)

79. Which of the following theories of failure is the most appropriate for a brittle material?
    (a) Maximum principal strain theory
    (b) Maximum principal stress thoery
    (c) Maximum shear stress theory
    (d) Maximum strain energy theory [ESE-2011]

Ans. (b)

80. Which one of the following statements is correct?
    (a) Principal stress is defined as the shear stresses on planes on which the normal stress is maximum or minimum.
    (b) The centre of Mohr's circle for a two-dimensional stress system always lies in the *y*-axis (adopting conventional axes notation).
    (c) The plane of maximum shear stress is inclined to the plane of principal stress at an angle of 45°.
    (d) In case of biaxial state of normal stresses, the normal stress on 45° plane is equal to the sum of normal stresses. [ESE-2011]

Ans. (c)

81. In a two-dimensional stress system, the radius of the Mohr's circle represents
    (a) maximum normal stress (b) minimum normal stress
    (c) minimum shear stress (d) maximum shear stress [ESE-2012]

Ans. (d)

82. In a plane stress problem, there are normal tensile stresses (σ_x, σ_y) with $\sigma_x > \sigma_y$, accompanied by shear stress (τ_{xy}) at a point in the *x*-*y* plane. If the minimum principal stress on a certain section is zero, then

    (a) $\tau_{xy} = \sqrt{\sigma_x . \sigma_y}$ (b) $\tau_{xy} = \sqrt{\dfrac{\sigma_x}{\sigma_y}}$

    (c) $\tau_{xy} = \sqrt{\sigma_x - \sigma_y}$ (d) $\tau_{xy} = \sqrt{\sigma_x + \sigma_y}$ [ESE-2012]

Ans. (a)

83. According to maximum shear stress failure theory, yielding occurs in the material when

(a) max. shear stress = yield stress
(b) max. shear stress = 2 times yield stress
(c) max. shear stress = 1/2 of yield stress
(d) max. shear stress = $\sqrt{2}$ times yield stress [ESE-2012]

Ans. (c)

84. For the design of a cast iron member, the most approprite theory of failure is
(a) Mohr's theory
(b) Rankine's theory
(c) Maximum stress theory
(d) Maximum shear energy theory [ESE-2012]

Ans. (b)

85. Which of the following statements are correct?
1. Experimental result on thin-walled tubes shows that maximum normal stress theory is valid.
2. Normal stress theory is valid in case of brittle materials.
3. In case of pure shear, St. Venant's theory is more valid than other theories.
4. Tresca's theory is very close to sub-staminate by experimental results.
5. In maximum stress theory, the yield focus is a hexagon.

(a) 1, 2 and 3 (b) 2, 4 and 5
(c) 1, 2 and 4 (d) 2, 3 and 4 [ESE-2013]

Ans. (b)

86. Consider the following statements:
Which of these statements are correct? If the planes at right angles carry only shear stress of magnitude (q) in a certain instance, then the
1. Diameter of Mohr's circle would be equal to $2q$.
2. Centre of Mohr's circle would lie at the origin
3. Principal stresses are unlike and at are of magnitude (q) each
4. Angle between the principal plane and maximum shear would be 45°

(a) 1, 2 and 3 only (b) 1, 2 and 4 only
(c) 3 and 4 only (d) 1, 2, 3, and 4 [ESE-2014]

Ans. (d)

87. Which of the following stresses is measured on inclined surface in Mohr's circle method?
(a) Principal stress (b) Normal stress
(c) Tangential stress (d) Maximum stress [ESE-2015]

Ans. (d)

88. Which of the following statements are correct for stresses acting on mutually perpendicular faces of a plane element?
    1. The sum of the normal stresses in mutually perpendicular planes is equal to the sum of the principal stresses.
    2. The shearing stresses in two mutually perpendicular planes are equal in magnitude and direction.
    3. Maximum shear stress is half of the difference between principal stresses.

    (a) 1, 2 and 3 (b) 1 and 2 only
    (c) 2 and 3 only (d) 1 and 3 only [ESE-2016]

Ans. (d)

89. Which of the statements are correct?
    1. The maximum shear stress is half of the normal stress in the case of unaxial stress field.
    2. In biaxial stress field, acted upon by normal stresses unaccompanied by shear stresses, the maximum shear stress is any one of the normal stresses.
    3. The Mohr's stress circle will be tangential to the vertical axis in the case of unaxial stress field.

    (a) 1, 2 and 3 (b) 1 and 2 only
    (c) 2 and 3 only (d) 1 and 3 only [ESE-2017]

Ans. (d)

90. The shear force in a beam subjected to pure positive bending is.......(positive/zero/negative) [GATE-1995]

Ans. (0)

91. A simply supported beam of length L is subjected to a varying distributed load, $\sin(3\pi x/L)$ Nm^{-1}, where the distance x is measured from the left support. The magnitude of the vertical reaction force in N at the left support is

    (a) Zero (b) $L/3\pi$
    (c) L/π (d) $2L/\pi$ [GATE-2013]

Ans. (a)

92. A simply supported beam AB of span L carries two concentrated loads W each at points $L/3$ from A and B. What is the SF in the middle one-third portion of the beam?

    (a) $W/2$ (b) $2W$
    (c) W (d) Zero [ESE-2008]

Ans. (d)

93. Shear span is defined as the zone where

    (a) Bending moment is zero (b) Shear force is zero
    (c) Shear force is constant (d) Bending moment is constant
    [ESE-2010]

Ans. (c)

94. Which of the following loads should be applied on a simply supported beam, so that the shear force is constant throughout its span?
   (a) UDL over the entire span
   (b) Two concentrated loads equally spaced in the span
   (c) A central concentrated load and a UDL over the entire span
   (d) A couple anywhere in the span [ESE-2013]

Ans. (d)

95. Which of the following statements are correct?
   1. All load carrying members transmit load through basic modes of tension compression, torsion and shear only.
   2. Beams carry transverse loads
   3. Combined loading is a complicated loading.
   4. Nature of internal forces decides the aspect of combined loading.

   (a) 1 and 2 (b) 2 and 3
   (c) 3 and 4 (d) 2 and 4 [ESE-2013]

Ans. (a)

96. Which of the following statements are correct? A simply supported beam is subjected to a couple at a section within its span. It will produce
   1. SF diagram of zero magnitude
   2. Uniformly varying triangular BM diagram.
   3. Sudden change in sign of BM at the point of application of the couple.
   4. Equal and opposite reactions at supports.

   (a) 1 and 2 (b) 2 and 3
   (c) 3 and 4 (d) 1 and 4 [ESE-2015]

Ans. (c)

97. In the analysis of beams subjected to loads, the point with nil bending moment can be a
   1. Point of contraflexure
   2. Point of maximum shear force
   3. Point of inflection

   Which of the above statements is/are correct?

   (a) 1 only (b) only
   (c) 3 only (d) 1, 2 and 3 [ESE-2016]

Ans. (a)

98. Consider the following statements in respect of arched construction made of voussoirs:

1. The superimposed load is transferred to the side walls only be the strength of cohesion of the mortar between the voussoirs.
2. The arch may fail under crushing when the compressive stress or thrust in it exceeds the safe crushing strength of the voussoirs material.
3. Every element in the arch is subjected to compression only.
4. Failure of the arch due to the sliding of any voussoirs past the adjacent one due to transverse shear can be avoided by reducing the height of the voussoirs.

Which of the above statements are correct?

(a) 1 and 4 only (b) 1 and 3 only

(c) 2 and 4 only (d) 2 and 3 only [ESE-2017]

Ans. (d)

99. The 'plane section remains plane' assumption in bending theory implies:

(a) strain profile is linear

(b) stress profile is linear

(c) both strain and stress profiles are linear

(d) shear deformations are neglected [GATE-2013]

Ans. (a)

100. A cantilever beam of length L, with uniform cross section and flexural rigidity. EI, is loaded uniformly by a vertical load, w per unit length. The maximum vertical deflection of the beam is given by

(a) $\frac{wL^4}{8EI}$ (b) $\frac{wL^4}{16EI}$

(c) $\frac{wL^4}{4EI}$ (d) $\frac{wL^4}{24EI}$ [GATE-2014]

Ans. (a)

101. The following statements are related to bending of beams:

I. The slope of the bending moment diagram is equal to the shear force.

II. The slope of the shear force diagram is equal to the load intensity.

III. The slope of the curvature is equal to the flexural rotation.

IV. The second derivative of the deflection is equal to the curvature.

The only FALSE statement is

(a) I (b) II

(c) III (d) IV [GATE-2012]

Ans. (c)

102. A cantilever beam of length L is subjected to a moment M at the free end. The moment of inertia of the beam's cross section about the neutral axis is I and the Young's modulus is E. The magnitude of the maximum deflection is....

(a) $\frac{ML^2}{2EI}$ (b) $\frac{ML^2}{EI}$

(c) $\frac{2ML^2}{EI}$ (d) $\frac{4ML^2}{EI}$ [GATE-2012]

Ans. (a)

103. The simply supported beam is subjected to a uniformly distributed load of intensity w per unit length, on half of the span from one end. The length of the span and the flexural stiffness are denoted as/and/El respectively. The deflection at mid-span of the beam is...

(a) $\frac{5}{6144}\frac{wl^4}{El}$ (b) $\frac{5}{768}\frac{wl^4}{El}$

(c) $\frac{5}{384}\frac{wl^4}{El}$ (d) $\frac{5}{192}\frac{wl^4}{El}$ [GATE-2012]

Ans. (b)

104. Match List-I (Loading condition) with List-II (Maximum slope) and select the correct answer using the codes given below the lists (E = Modulus of elasticity, I = Area moment of inertia):

List-I

A. Cantilever with concentrated load W at free end

B. Cantilever with *udl* (w/unit length) across the complete span ($W = wL$)

C. Simply supported beam with concentrated load W at the centre

D. Simply supported beam with *udl* (w/unit length) across complete span ($W = wL$)

List-II

1. $\frac{WL^2}{16EI}$ 2. $\frac{WL^2}{24EI}$

3. $\frac{WL^2}{2EI}$ 4. $\frac{WL^2}{6EI}$

Codes:

	A	B	C	D
(a)	1	2	3	4
(b)	3	4	1	2
(c)	1	4	3	2
(d)	3	2	1	4

[ESE-2006]

Ans. (b)

105. Match List-I (Load case) with List-II (Expression for slope/Deflection) and select the correct answer using the codes given below the lists (Flexural rigidity = EI)

List-I

A. Slope for tip load of W

B. Deflection for tip load of W

C. Slope for total UDL of W

D. Deflection for total UDL of W

List-II

1. $Wl^3/8EI$
2. $Wl^3/6EI$
3. $Wl^3/3EI$
4. $Wl^3/2EI$

Codes:

	A	B	C	D
(a)	4	2	3	1
(b)	1	3	2	4
(c)	4	3	2	1
(d)	1	2	3	4

[ESE-2007]

Ans. (c)

106. The compatibility conditions is terms of strain in a two-dimensional problem are associated with

(a) Stresses (b) Forces

(c) Properties of material (d) Deformations [ESE-2011]

Ans. (d)

107. If a cantilever beam of span L and flexural rigidity EI carries a moment M at the free end, the deflection at that end is

(a) $\dfrac{ML}{24EI}$ (b) $\dfrac{ML^2}{12EI}$

(c) $\dfrac{ML}{6EI}$ (d) $\dfrac{ML^2}{2EI}$ [ESE-2012]

Ans. (d)

108. Deflection limitations over beams are imposed because excessive deflection may cause

(a) undesirable twisting and distortion of end connections

(b) problems in drainage system

(c) psychological effect on users

(d) all of the above [ESE-2012]

Ans. (d)

109. The total deflections of a structure when subjected simultaneously to different sets of loading is equal to the sum of the deflection under each such set of loads as it is acting separately on the structure. Which one of the following justifies this statement?

    (a) Elastic limit is not exceeded
    (b) Elastic limit including bucking is not exceeded
    (c) Limit state is not exceeded
    (d) Proportional limit without bucking is not exceeded [ESE-2013]

Ans. (d)

110. A hinged support in a beam

    (a) becomes an internal hinge in a conjugate beam
    (b) changes to a free support in a conjugate beam
    (c) changes to a fixed support in a conjugate beam
    (d) remains as a hinged support in a conjugate beam [ESE-2014]

Ans. (d)

111. Maximum deflection of a cantilever beam of length L carrying uniformly distributed load (w) per unit length will be

    (a) wL^4/EI (b) $wL^4/4EI$
    (c) $wL^4/8EI$ (d) $wL^4/384EI$ [ESE-2008]

Ans. (c)

112. Match List-I (Formula/Theorem/Method) with List-II (Deals with topic) and select the correct answer using the codes given below the lists:

    List-I

    A. Clapeyron's theorem
    B. Macaulay's method
    C. Perry's formula

    List-II

    1. Deflection of beam
    2. Eccentrically loaded column
    3. Rivetted joins
    4. Continuous beam

    Codes:

	A	B	C
(a)	3	2	1
(b)	4	1	2
(c)	4	1	3
(d)	2	4	3

[ESE-2008]

Ans. (b)

113. A simply supported beam of length l is loaded by a uniformly distributed w over the entire span. It is propped at the mid-span so that the deflection at the centre is zero. The reaction at the prop is:

(a) $\frac{5}{16}w\ell$ (b) $\frac{1}{2}w\ell$

(c) $\frac{5}{8}w\ell$ (d) $\frac{1}{10}w\ell$ [ESE-2013]

Ans. (c)

114. A beam of length L and flexural rigidity EI is simply supported at the ends and carries a concentrated load W at the middle of the span. Another beam of identical length L and flexural rigidity EI is fixed horizontally at both ends and carries an identical concentrated load W at the mid-span. The ratio of the deflection of the first beam of that of the second beam is

(a) 1 (b) 2

(c) 0.25 (d) 4 [ESE-2014]

Ans. (d)

115. A solid circular shaft is subjected to a bending moment M and twisting moment T. What is the equivalent twisting moment T_e which will produce the same maximum shear stress as the above combination?

(a) $M^2 + T^2$ (b) $M + T$

(c) $\sqrt{M^2 + T^2}$ (d) $M - T$ [ESE-2007]

Ans. (c)

116. The bending moment (M) is constant over a length segment l of a beam the shearing force will also be constant over this length and is given by

(a) M/l (b) $M/2l$

(c) $M/4l$ (d) None of the above

Ans. (d)

117. A beam is said to be of uniform strength if

(a) The bending moment is the same throughout the beam

(b) The shear stress is the same throughout the beam

(c) The deflection is same throughout the beam

(d) The bending stress is the same at every section along its longitudinal axis

[ESE-2007]

Ans. (d)

118. Match List-I with List-II and select the correct answer using the codes given below the lists:

List-I

A. Subjected to bending moment at the end of a cantilever

B. Cantilever carrying uniformly distributed load over the whole length

C. Cantilever carrying linearly varying load from zero at the free end of maximum at the supports

D. A beam having a load at centre and supported as the ends

List-II

1. Triangle
2. Cubic parabola
3. parabola
4. Rectangle

Codes:

	A	B	C	D
(a)	1	2	3	4
(b)	4	3	2	1
(c)	1	3	2	4
(d)	4	2	3	1

[ESE-2011]

Ans. (b)

119. Statement (I): If the bending moment along the length of a beam is constant, then the beam's cross section will not experience any shear stress.

Statement (II): The shear force acting on the beam will be zero everywhere along its length.

(a) Both statement (I) and statement (II) are individually true and statement (II) is the correct explanation of statement (I)

(b) Both statement (I) and statement (II) are individually true but statement (II) is NOT the correct explanation of statement (I)

(c) Statement (I) is true but Statement (II) is false

(d) Statement (I) is false but Statement (II) is true [ESE-2012]

Ans. (a)

120. A beam simply supported at equal distance from the ends carries equal loads at each end. Which of the following statement is true?

(a) The bending moment is minimum at the mid-span

(b) The bending moment is minimum at the support

(c) The bending moment varies gradually between the supports

(d) The bending moment is uniform between the supports [ESE-2013]

Ans. (d)

121. Beams of uniform strength vary in section such that

(a) bending moment remains constant

(b) deflection remains constant

(c) maximum bending stress remains constant

(d) shear force remains constant [ESE-2006]

Ans. (c)

122. In the case of beams with circular cross section, what is the ratio of the maximum shear stress to average shear stress?

(a) 3:1 (b) 2:1

(c) 3:2 (d) 4:3 [ESE-2006]

Ans. (d)

123. When a rectangular section beam is loaded transversely along the length shear stress develops on

(a) Top fibre of rectangular beam

(b) Middle fibre of rectangular beam

(c) Bottom fibre of rectangular beam

(d) None of these [ESE-2007]

Ans. (b)

124. In I-section of a beam subjected to transverse shear force, the maximum shear stress is developed

(a) at the centre of the web

(b) at the top edge of the top flange

(c) at the bottom edge of the top flange

(d) None of the above [ESE-2008]

Ans. (a)

125. In case of a beam of circular cross section subjected to transverse loading, the maximum shear stress developed in the beam is greater than the average shear stress by

(a) 50% (b) 33%

(c) 25% (d) 10% [ESE-2008]

Ans. (b)

126. What is the shape of the shearing stress distribution across a rectangular cross section beam?

(a) Triangular

(b) Parabolic only

(c) Rectangular only

(d) A combination of rectangular and parabolic [ESE-2008]

Ans. (b)

127. A beam having rectangular cross section is subjected to an external loading. The average shear stress developed due to the external loading at a particular cross section is τ_{avg} What is the maximum shear stress developed at the same cross section due to the loading?

(a) $\frac{1}{2}\tau_{avg}$

(b) τ_{avg}

(c) $\frac{3}{2}\tau_{avg}$

(d) $2\tau_{avg}$

Ans. (c)

128. Match List-I with List-II and select the correct answer using the given below lists:

List-I

A. Point of inflection

B. Shearing strain

C. Section strain

D. Modulus of resilience

List-II

1. Strain energy
2. Equation of bending
3. Equation of torsion
4. Bending moment diagram

Codes:

	A	B	C	D
(a)	1	3	2	4
(b)	4	3	2	1
(c)	1	2	3	4
(d)	4	2	3	1

[ESE-2010]

Ans. (b)

129. A T-section beam is simply supported and subjected to a uniformly distributed load throughout its whole span. Maximum longitudinal stress in the beam occurs at

(a) Top fibre of the flange

(b) The junction of web and flange

(c) The midsection of the web

(d) The bottom fibre of the web [ESE-2011]

Ans. (d)

130. If E = elasticity modulus, I = moment of inertia about the neutral axis and M = bending moment in pure bending under the symmetric loading of a beam, the radius of curvature of the beam:

1. increases with E
2. increases with M

3. Decreases with I
4. Decreases with M

Which of these are correct?

(a) 1 and 3 (b) 2 and 3

(c) 3 and 4 (d) 1 and 4 [ESE-2013]

Ans. (d)

131. Statement (I): I-sections are best suited for carrying bending load in one lateral direction.

Statement (II): In the zone, in the vicinity of neutral axis of I-beams, I-section has the least material.

(a) Both statement (I) and statement (II) are correct and statement (II) is the correct explanation of statement (I)

(b) Both statement (I) and statement (II) are correct but statement (II) is NOT the correct explanation of statement (I)

(c) Statement (I) is true but statement (II) is false

(d) Statement (I) is false but statement (II) is true [ESE-2014]

Ans. (d)

132 Consider the following statement:

1. Cross section of a member of truss experience uniform stress.
2. Cross section of a beam experiences minimum stress.
3. Cross section of a beam experience linearly varying stress.
4. Cross section of truss members experience only compressive stress.

Which of the above statements are correct?

(a) 1 and 2 (b) 1 and 3

(c) 1 and 4 (d) 3 and 4 [ESE-2014]

Ans. (b)

133. The first moment of area about the axis of bending for a beam cross section is

(a) moment of inertia (b) section modulus

(c) shape factor (d) polar moment of inertia [GATE-2014]

Ans. (b)

134. Match items in columns I and II

Column I	Column II
P. Addendum	1. Cam
Q. Instantaneous centre of velocity	2. Beam
R. Section modulus	3. Linkage
S. Prime circle	4. Gear

(a) P-4, Q-2, R-3, S-1
(b) P-4, Q-3, R-2, S-1
(c) P-3, Q-2, R-1, S-4
(d) P-3, Q-4, R-1, S-2

Ans. (b)

135. If a beam of rectangular cross section is subjected to a vertical shear force, V, the shear force carried by the upper one-third of the cross section is....

(a) Zero
(b) $\frac{7V}{27}$
(c) $\frac{8V}{27}$
(d) $\frac{V}{3}$ [GATE-2006]

Ans. (b)

136. A symmetric I-section (with width of each flange = 50 mm, thickness of each flange = 10 mm, depth of web = 100 mm, and thickness of web = 10 mm) of steel is subjected to a shear force of 100 kN. Find the magnitude of the shear stress (in N/mm^2) in the web at its junction with the top flange.........

(a) 23
(b) 33
(c) 57
(d) 71 [GATE 2013]

Ans. (d)

137. The shear stress at the neutral axis in a beam of triangular section with a base of 40 mm and height 20 mm, subjected to a shear force of 3 kN is

(a) 3 MPa
(b) 6 MPa
(c) 10 MPa
(d) 20 MPa [GATE-2013]

Ans. (c)

138. The point within the cross sectional plane of a beam through which the resultant of the external loading on the beam has to pass through to ensure pure bending without twisting of the cross section of the beam is called...

(a) moment centre
(b) centroid
(c) shear centre
(d) elastic centre [GATE-2009]

Ans. (c)

139. Consider a simply supported beam of length, 50 h, with a rectangular cross section of depth, h, and width, $2h$. The beam carries a vertical point load, P, at its mid-point. The ratio of the maximum shear stress to the maximum bending stress in the beam is...

(a) 0.02
(b) 0.10
(c) 0.05
(d) 0.01 [GATE-2014]

Ans. (d)

140. Consider the following statements:

1. In a beam, the maximum shear stress occurs at the neutral axis of the beam cross section.

2. The maximum shear stress in a beam of circular cross section is 50% more than the average shear stress.
3. The maximum shear stress in a beam of triangular cross section, with its vertex upwards occurs at $b/6$ above the neutral axis.

Which of these statements are correct?

(a) 1, 2 and 3 (b) 2 and 3 only
(c) 1 and 2 only (d) 1 and 3 only [ESE-2007]

Ans. (d)

141. A cantilever beam of T cross section carries uniformly distributed load. Where does the maximum magnitude of the bending stress occur?

(a) At the top of cross section (b) At the junction of flange and web
(c) At the mid-depth point (d) At the bottom of the section

[ESE-2007]

Ans. (d)

142. A symmetrical I-section is subjected to shear force. At which location the shear stress induced across the section is maximum?

(a) Extreme fibres
(b) At the bottom of flanges in flanges
(c) At the bottom flanges in web portion
(d) At the neutral axis [ESE-2008]

Ans. (d)

143. A beam is made of two identical metal flats soldered together. What is the ratio of stiffness of this beam to the stiffness of a beam in which the two flats are not soldered and which acts independently?

(a) 2 (b) 4
(c) 6 (d) 8 [ESE-2008]

Ans. (b)

144. Consider the following statements for a beam of rectangular cross section and uniform flexural rigidity (EI) subjected to pure bending:

1. The bending stresses have the maximum magnitude at the top and bottom of the cross section.
2. The bending stresses vary linearly through the depth of the cross section.
3. The bending stresses vary parabolically through the depth of the cross section.

Which of these statements is/are correct?

(a) 1, 2 and 3 (b) 1 only
(c) 2 only (d) 1 and 2 only [ESE-2010]

Ans. (d)

145. Two beams carrying identical loads, simply supported, are having same depth but beam A has double the width as compared to that of beam B. The ratio strength of beam A to that of beam B is ...

(a) 1/2 (b) 1/4

(c) 2 (d) 4 [ESE-2010]

Ans. (c)

146. A steel beam is replaced by a corresponding aluminum beam of same cross-sectional shape and dimensions, and is subjected to same loading. The maximum bending stress will ...

(a) not alter

(b) increase

(c) decrease

(d) vary in proportion to their modulus of elasticity [ESE-2010]

Ans. (a)

147. A simply supported beam of span L carries a concentrated load W at its mid-span. If the width b of the beam is constant throughout the span, then with permissible bending stress as σ, the depth of the beam at mid-span will be ....

(a) $\frac{3WL}{2b\sigma}$ (b) $\sqrt{\frac{3WL}{2b\sigma}}$

(c) $\frac{6WL}{b\sigma}$ (d) $\sqrt{\frac{6WL}{b\sigma}}$ [ESE-2012]

Ans. (b)

148. Out of the two beams of the same material and same cross-sectional area, one is of circular cross section and the other is of square cross section. If each of these is subjected to bending moment of the same magnitude, then

(a) both sections would be equally strong

(b) both sections would be equally economical

(c) square section would be more economical than circular section

(d) square section would be less economical than circular section [ESE-2012]

Ans. (c)

149. A flitched beam composed of two different pieces, each having breadth b and depth d, supports an external load. This statement implies that

1. the two different material are rigidity connected
2. there will be relative movement between the two materials
3. for transforming into an equivalent signal material, section under 'strength' considerations, the depth is kept constant and only the breadth varies

Which of the above statements are correct?

(a) 1 and 2 only (b) 1 and 3 only

(c) 2 and 3 only (d) 1, 2 and 3 [ESE-2014]

Ans. (b)

150. A beam of square cross section is placed such that its neutral axis coincides with its diagonal, and it is subjected to a shear force *F*. What is the ratio of the maximum shear stress to the shear stress at the neutral axis?

(a) 9/8 (b) 8/9

(c) 7/8 (d) 8/7 [ESE-2015]

Ans. (a)

151. The bending stress in a T-beam section is maximum

1. At top fibre
2. At centroidal fibre
3. At bottom fibre

(a) 1 only

(b) 2 only

(c) 3 only

(d) At a level which is dependent on the loading condition [ESE-2016]

Ans. (c)

152. A long shaft of diameter *d* is subjected to twisting moment *T* at its ends. The maximum normal stress acting at its cross section is equal to

(a) Zero (b) $\frac{16T}{\pi d^3}$

(c) $\frac{32T}{\pi d^3}$ (d) $\frac{64T}{\pi d^3}$ [GATE-2006]

Ans. (a)

153. For a circular shaft of diameter to torque *T*, the maximum value of the shear stress is:

(a) $\frac{64T}{\pi d^3}$ (b) $\frac{32T}{\pi d^3}$

(c) $\frac{16T}{\pi d^3}$ (d) $\frac{8T}{\pi d^3}$ [GATE2006]

Ans. (c)

154. Two solid circular shafts of radii R_1 and R_2 are subjected to same torque. The maximum shear stresses developed in the two shafts are τ_1, τ_2. If $\frac{R_1}{R_2} = 2$, then $\frac{\tau_2}{\tau_1}$ is....

(a) 2 (b) 4
(c) 8 (d) 16 [GATE-2014]

Ans. (c)

155. A hollow circular shaft has outer diameter of 100 mm and a wall thickness of 25 mm. The allowed shear stress in the shaft is 125 MPa. The maximum torque the shaft can transmit is

(a) 46 kN-m (b) 24.5 kN-m
(c) 23 kN-m (d) 11.5 kN-m [GATE-2009]

Ans. (c)

156. A hollow shaft of 1 m length is designed to transmit a power of 30 kW at 700 rpm. The maximum permissible angle of twist in the shaft is 1°. The inner diameter of the shaft is 0.7 times the outer diameter. The modulus of rigidity is 80 GPa. The outside diameter (in mm) of the shaft is.......

(a) 44.5 (b) 56.2
(c) 66.1 (d) 75.5 [GATE 2015]

Ans. (a)

157. A hollow shaft $d_0 = 2d_i$ (where d_0 and d_i are the outer and inner diameters, respectively) needs to transmit 20 kW power at 3000 RPM. If the maximum permissible shear stress is 30 MPa, d_0 is

(a) 11.29 mm (b) 22.58 mm
(c) 33.87 mm (d) 45.16 mm [GATE-2015]

Ans. (b)

158. The maximum and minimum shear stresses in a hollow circular shaft of outer diameter 20 mm and thickness 2 mm, subjected to a torque of 92.7 N-m will be...

(a) 59 MPa and 47.2 MPa (b) 100 MPa and 80 MPa
(c) 118 MPa and 160 MPa (d) 200 MPa and 160 MPa [GATE-2007]

Ans. (b)

159. The maximum shear stress in a solid shaft of circular cross section having diameter d subjected to a torque T is τ. If the torque is increased by four times and the diameter of the shaft is increased by two times, the maximum shear stress in the shaft will be ...

(a) 2τ (b) τ
(c) $\frac{\tau}{2}$ (d) $\frac{\tau}{4}$ [GATE-2008]

Ans. (c)

160. What is the maximum torque transmitted by a hollow shaft of external radius R and internal radius r?

(a) $\frac{\pi}{16}(R^3 - r^3)f_s$ (b) $\frac{\pi}{2R}(R^4 - r^4)f_s$

(c) $\frac{\pi}{8R}(R^4 - r^4)f_s$ (d) $\frac{\pi}{32}(R^4 - r^4)f_s$ [ESE-2006]

Ans. (b)

161. Consider the following statements:

Maximum shear stress induced in a power transmitting shaft is

1. directly proportional to torque being transmitted .
2. inversely proportional to the cube of its diameter.
3. directly proportional to its polar moment of inertia.

Which of these statements are correct?

(a) 1, 2 and 3 (b) 1 and 3 only

(c) 2 and 3 only (d) 1 and 2 only [ESE-2008]

Ans. (d)

162. The ratio of torque carrying capacity of a solid shaft to that of a hollow shaft is given by

(a) $(1 - k^4)$ (b) $(1 - k^4)^{-1}$

(c) k^4 (d) $\frac{1}{K^4}$

where, $K = \frac{D_i}{D_0}$

D_i = Inside diameter of hollow shaft

D_0 = Outside diameter of hollow shaft

That material are the same. [ESE-2008]

Ans. (b)

163. A solid shaft transmits a torque T. The allowable sharing stress is τ What is the diameter of the shaft?

(a) $\sqrt[3]{\frac{16T}{\pi\tau}}$ (b) $\sqrt[3]{\frac{32T}{\pi\tau}}$

(c) $\sqrt[3]{\frac{16T}{\tau}}$ (d) $\sqrt[3]{\frac{T}{\tau}}$ [ESE-2008]

Ans. (a)

164. For the two shafts connected in parallel

(a) Torque in each shaft is the same

(b) Shear stress in shaft is the same

(c) Angle of first of each shaft is the same
(d) Torsional stiffness of each shaft is the same [ESE-2011]

Ans. (c)

165. The magnitude of shear induced in a shaft due to applied torque varies
(a) From maximum at the centre to zero at the circumference
(b) From zero at the centre to maximum at the circumference
(c) From maximum at the centre to minimum but not zero at the circumference
(d) From minimum but not zero at the centre, to maximum at the circumference [ESE-2012]

Ans. (b)

166. A power transmission solid shaft of diameter d length I and rigidity modulus G is subjected to a pure torque. The maximum strain energy/unit volume in the shaft is given by:

(a) $\frac{\tau_{max}^2}{4G}$ (b) $\frac{\tau_{max}^2}{2G}$

(c) $\frac{\tau_{max}^2}{5G}$ (d) $\frac{\tau_{max}^2}{5G}$ [ESE-2013]

Ans. (a)

167. Match List-I (Strain energy) with List-II (Gradually applied load) and select the correct answer using the codes given below the lists (σ = direct stress, τ = shear stress, M_x = Bending moment, E = Modulus of elasticity, G = Modulus of rigidity, I = Area moment of inertia, V = Volume).

List-I

A. $\frac{\sigma^2 V}{2E}$ B. $\frac{\tau^2 V}{2G}$

C. $\frac{\tau^2 V}{4G}$ D. $\int_0^L \frac{M_x^2 dx}{2EI}$

List-II

1. Axial load 2. Bending load
3. Shear load 4. Torsional load

Codes:

	A	B	C	D
(a)	2	3	4	1
(b)	1	4	3	2
(c)	2	4	3	1
(d)	1	3	4	2

[ESE-2006]

Ans. (d)

168. The failure surface of a standard cast iron torsion specimen, subjected to a torque is along
(a) the surface helicoidal at 45° to the axis of the specimen
(b) the curved surface at the grips
(c) the plane surface perpendicular to the axis of the specimen
(d) the curved surface perpendicular to the axis of the specimen [ESE-2006]

Ans. (a)

169. A close-coiled helical spring with n coils, mean radius R and diameter d is subjected to an axial load, W. What is the compression in spring?

(a) $\frac{64WR^3n}{Cd^3}$ (b) $\frac{64WR^3n}{Cd^4}$

(c) $\frac{32WR^3n}{Cd^3}$ (d) $\frac{32WR^3n}{Cd^4}$ [ESE-2007]

Ans. (b)

170. Two shafts, one of solid section and the other of hollow section, of same material and weight having same length are subjected to equal torsional force. What is the torsional stiffness of hollow shaft?
(a) Equal to that of the solid shaft
(b) Less than that of the solid shaft
(c) More than that of the solid shaft
(d) Exactly half of that of the solid shaft [ESE-2007]

Ans. (c)

171. A circular shaft is subjected to a bending moment M_b and a twisting moment M_t. What is the ratio of maximum shear stress and the maximum bending stress?

(a) $\frac{2M_t}{M_b}$ (b) $\frac{M_t}{M_b}$

(c) $\frac{M_t}{2M_b}$ (d) $\frac{1.5M_t}{M_b}$ [ESE: 2008]

Ans. (c)

172. A solid shaft has length and diameter L_s and D respectively. A hollow shaft of length L_h, external diameter D, and internal diameter d respectively. Both are of the same material. The ratio of torsional stiffness of hollow shaft to that of solid shaft is ....

(a) $\left[1+\left(\frac{d}{D}\right)^4\right].\frac{L_s}{L_h}$ (b) $\left[1-\left(\frac{d}{D}\right)^4\right].\frac{L_h}{L_s}$

(c) $\left[1-\left(\frac{d}{D}\right)^4\right].\frac{L_s}{L_h}$ (d) $\left[1-\left(\frac{D}{d}\right)^4\right].\frac{L_s}{L_h}$ [ESE: 2010]

Ans. (c)

173. A solid circular shaft subjected to a torque T produces maximum shear stress f_s which is the maximum principal value in the material. The corresponding diameter of the should be

(a) $\sqrt[3]{\frac{\pi \cdot f_s}{16 \cdot T}}$ (b) $\sqrt[3]{\frac{32 \cdot T}{\pi \cdot f_s}}$

(c) $\sqrt[3]{\frac{\pi}{16T \cdot f_s}}$ (d) $\sqrt[3]{\frac{16 \cdot T}{\pi . f_s}}$ [ESE-2010]

Ans. (d)

174. The polar moment of inertia of the cross section of the member is required to across the strength of the member in

(a) bending (b) torsion

(c) axial force (d) shear [ESE-2012]

Ans. (b)

175. In a circular shaft of diameter d, subjected to a torque T, then maximum shear stress inducted is ...

(a) proportional to d^3

(b) proportional to d^4

(c) inversely proportional to d^3

(d) inversely proportional to d^4 [ESE-2012]

Ans. (c)

176. Which of the following terms represents the torque corresponding to a twist of one radian in a shaft over its unit length?

(a) Torsional stress (b) Torsional rigidity

(c) Flexural rigidity (d) Moment of resistance [ESE-2012]

Ans. (b)

177. The polar modulus of a circular shaft diameter d is

(a) $\frac{\pi}{16}.d^3$ (b) $\frac{\pi}{32}.d^3$

(c) $\frac{\pi}{64}.d^3$ (d) $\frac{\pi}{128}.d^3$ [ESE-2012]

Ans. (a)

178. A solid shaft of circular cross section subjected to torque T which produces a maximum shear stress τ in the shaft. The diameter of the shaft will be

(a) $\sqrt[3]{\frac{16T}{\pi\tau}}$ (b) $\sqrt{\frac{\pi\tau}{16T}}$

(c) $\sqrt[3]{\frac{\pi\tau}{16T}}$ (d) $\sqrt{\frac{16T}{\pi\tau}}$ [ESE-2012]

Ans. (a)

179. The ratio of torsional moments of resistance of a solid circular shaft of diameter D to that of a hollow shaft with external diameter D and internal diameter d is

(a) $\frac{D^4}{D^4 - d^4}$ (b) $\frac{D^4 - d^4}{D^4}$

(c) $\frac{D^3}{D^3 - d^3}$ (d) $\frac{D^3 - d^3}{D^3}$ [ESE-2012]

Ans. (a)

180. A = Cross-sectional area

E = Young's modulus of elasticity

G = Modulus of inertia

I = Moment of inertia

J = Polar moment of inertia

Then torsional rigidity is given by

(a) AE (b) GE

(c) EI (d) GJ [ESE-2012]

Ans. (d)

181. A solid circular shaft has been subjected to a pure torsion moment. The ratio of maximum shear stress to maximum normal stress at any point would be ....

(a) 1:2 (b) 1:1

(c) 2:3` (d) 2:1 [ESE-2013]

Ans. (b)

182. A metal shaft of solid circular section rotates at 160 rpm and is subjected to a torque of 1500 Nm. What is the power, in kW, transmitted by the shaft?

(a) 32π (b) 16π

(c) 12π (d) 8π [ESE-2015]

Ans. (d)

183. A solid circular shaft has a diameter d. Its polar modulus will be

(a) $\frac{\pi}{16}d^2$ (b) $\frac{\pi}{64}d^3$

(c) $\frac{\pi}{16}d^3$ (d) $\frac{\pi}{32}d^2$ [ESE-2016]

Ans. (c)

184. Consider the following statements:

1. The shear stress distributed across the section of a circular shaft subjected to twisting varies parabolically.
2. The shear stress at the centre of a circular shaft under twisting moment is zero.

3. The shear stress at the external fibres of a circular shaft under twisting moment is maximum.

Which of the statements is/are correct?

(a) 1, 2 and 3 (b) 1 only
(c) 2 only (d) 3 only [ESE-2017]

Ans. (c, d)

185. A thin cylinder of inner radius 500 mm and thickness 10 mm is subjected to an internal pressure of 5 MPa. The average circumferential (hoop) stress in MPa is...

(a) 100 (b) 250
(c) 500 (d) 1000 [GATE-2011]

Ans. (b)

186. A thin-walled spherical shell is subjected to an internal pressure. If the radius of the shell is increased by 1% and the thickness is reduced by 1% with the internal pressure remaining the same, the percentage change in the circumferential (hoop) stress is

(a) 0 (b) 1
(c) 1.08 (d) 2.02 [GATE-2012]

Ans. (d)

187. A long thin-walled cylindrical, closed at both the ends, is subjected to an internal pressure. The ratio of the hoop stress (circumferential stress) to longitudinal stress developed in the shell is

(a) 0.5 (b) 1.0
(c) 2.0 (d) 4.0 [GATE-2013]

Ans. (c)

188. A thin gas cylinder with an internal radius of 100 mm is subject to an internal pressure of 10MPa. The maximum permissible working stress is restricted to 100 MPa. The minimum cylinder wall thickness (in mm) for safe design must be......

(a) 10 (b) 20
(c) 25 (d) 50 [GATE-2014]

Ans. (a)

189. A cylindrical tank with closed ends is filled with compressed air at a pressure of 500 kPa. The inner radius of the tank is 2 m, and it has wall thickness of 10 mm. The magnitude of maximum in-plane shear stress (in MPa) is...

(a) 5 (b) 25
(c) 50 (d) 100 [GATE-2015]

Ans. (b)

190. A gas is stored in a cylindrical tank of inner radius 7 m wall thickness 50 mm. The gage pressure of the gas is 2 MPa. The maximum shear stress (in MPa) in the wall is...

(a) 35 (b) 70

(c) 140 (d) 280 [GATE-2015]

Ans. (c)

191. A thin-walled cylindrical pressure vessel having a radius of 0.5 m and wall thickness of 25 mm is subjected internal pressure of 700 kPa. The hoop stress developed is...

(a) 14 MPa (b) 1.4 MPa

(c) 0.14 MPa (d) 0.014 MPa [GATE-2009]

Ans. (a)

192. A thin-walled long cylindrical tank of inside radius r is subjected simultaneously to internal gas pressure p and axial compressive force F at its ends. In order to produce 'pure shear' state of stress in the wall of the cylinder, F should be equal to...

(a) $p\pi r^2$ (b) $2p\pi r^2$

(c) $3p\pi r^2$ (d) $4p\pi r^2$ [GATE-2006]

Ans. (c)

193. A thin cylinder of thickness t, width b and internal radius r is subjected to a pressure p on the entire internal surface. What is the change in radius of the cylinder? (μ is the Poisson's ratio and E is the modulus of elasticity)?

(a) $\frac{p^2 r(2-\mu)}{Et}$ (b) $\frac{pr^2(2-\mu)}{Et}$

(c) $\frac{pr^2(2-\mu)}{2Et}$ (d) $\frac{p(1-\mu)}{Etr^2}$ [ESE-2006]

Ans. (c)

194. A thin cylinder of unit length thickness t and radius r is subjected to internal pressure p. What is the circumferential stress?

(a) $pr/2Et$ (b) $pr/2t$

(c) pr/t (d) $2pr/t$ [ESE-2008]

Ans. (c)

195. A thick cylinder of internal radius a and external radius b is subjected to internal pressure p as well as external pressure p.

The magnitude of circumferential stress developed is ....

(a) maximum at radius $r = a$ (b) maximum at radius $r = b$

(c) maximum at radius, $r = \sqrt{ab}$ (d) constant

Ans. (d)

196. Auto-frettage is a method of
   (a) improving the transmission of power from engine to the axial in automobiles
   (b) improving the strength of welded joints
   (c) Prestressing the cylinders before putting them in service
   (d) Improving the strength of riveted joints [ESE-2006]

Ans. (c)

197. Consider the following statements at the given point in the case of thick cylinder subjected to fluid pressure:
   1. Radial stress is compressive.
   2. Hoop stress is tensile.
   3. Hoop stress is compressive.
   4. Longitudinal stress is tensile and it varies along the length.
   5. Longitudinal stress is tensile and remains constant along the length of the cylinder.

   Which of these statements are correct?
   (a) 1, 2 and 4 only (b) 3 and 4 only
   (c) 1, 2 and 5 only (d) 1, 3 and 5 only [ESE-2006]

Ans. (c)

198. Consider the following statements

   In a thick walled cylindrical pressure vessel subjected to internal pressure, then tangential and radial stresses are ...
   1. Minimum at outer side
   2. Minimum at inner side
   3. Maximum at inner side and both reduce to zero at outer wall
   4. Maximum at inner wall but the radial stress reduced to zero at outer wall

   Which of these statements is/are correct?
   (a) 1 and 2 (b) 1 and 3
   (c) 1 and 4 (d) 4 only [ESE-2007]

Ans. (c)

199. Where does the maximum hoop stress in a thick cylinder under external pressure occur?
   (a) At the outer surface (b) At the inner surface
   (c) at the mid-thickness (d) At the $2/3^{rd}$ outer radius [ESE-2008]

Ans. (b)

200. Consider the following statements:
   1. The amount of interference needed to create a tight joint varies with diameter of the shaft.

2. An interference fit creates no stress state in the shaft.
3. The stress state in the hub is similar to a thick-walled cylinder with internal pressure.

Which of these statements are correct?

(a) 1, 2 and 3 (b) 1 and 2 only
(c) 2 and 3 only (d) 1 and 3 only [ESE-2008]

Ans. (d)

201. A seamless pipe of diameter d m is to carry fluid under a pressure of p kN/cm^2. The necessary thickness r of metal in cm. if the maximum stress is not to exceed σ kN/cm^2 is ...

(a) $t \geq \frac{pd}{2\sigma}$ cm (b) $t \geq \frac{100\,pd}{2\sigma}$ cm
(c) $t \geq \frac{pd}{2\sigma}$ cm (d) $t \leq \frac{100\,pd}{2\sigma}$ cm [ESE-2010]

Ans. (b)

202. A compression spring is made of music wire of 2 mm diameter having a shear strength and shear modulus of 800 MPa and 80 GPa, respectively. The mean coil diameter is 20 mm, free length is 40 mm and the number of active coils is 10. If the mean coil diameter is reduced to 10 mm, the stiffness of the spring is approximately...

(a) Decreases by 8 times (b) Decreases by 2 times
(c) Increases by 2 times (d) Increases by 8 times [GATE-2008]

Ans. (d)

203. A weighing machine consists of a 2 kg pan resting on spring. In this condition, with the pan resting on the spring, the length of the spring is 200 mm. When a mass of 20 kg is placed on the pan, the length of the spring becomes 100 mm. For the spring, the undeformed length I_o and the springs constant k (stiffness) are:

(a) $I_o = 220$ mm, $k = 1862$ N/m (b) $I_o = 210$ mm, $k = 1960$ N/m
(c) $I_o = 200$ mm, $k = 1960$ N/m (d) $I_o = 200$ mm, $k = 2156$ N/m

Ans. (b)

204. The ratio of Euler's bucking loads of columns with the same parameters having (i) both ends fixed, and (ii) both ends hinged is:

(a) 2 (b) 4
(c) 6 (d) 8 [GATE-2002, 1998, 2012]

Ans. (b)

205. A pin-ended column of length *L*, modulus of elasticity *E* and second moment of the cross-sectional area *I* is loaded centrically by a compressive load *P*. The critical bucking load (P_{cr}) is given by:

(a) $P_{cr} = \frac{EI}{\pi^2 L^2}$ (b) $P_{cr} = \frac{\pi^2 EI}{3L^2}$

(c) $P_{cr} = \frac{\pi EI}{L^2}$ (d) $P_{cr} = \frac{\pi^2 EI}{L^2}$ [GATE-2006]

Ans. (d)

206. Consider a steel (Young's modulus $E = 200$ GPa) column hinged on both sides. Its height is 1.0 m and cross-section is 10 mm × 20 mm. The lowest Euler's critical buckling load (in N) is............

(a) 1296 (b) 2100

(c) 2930 (d) 3290 [GATE-2015]

Ans. (d)

207. The piston rod of diameter 20 mm and length 700 mm in a hydraulic cylinder is subjected to a compressive force of 10 kN due to the internal pressure. The end conditions for the rod may be assumed as guided at the piston end and hinged at the other end. The Young's modulus is 200 GPa. The factor of safety for the piston rod is...

(a) 0.68 (b) 2.75

(c) 5.62 (d) 11.0 [GATE-2007]

Ans. (c)

208. A steel column, pinned at both ends, has a bucking load of 200 kN. If the column is restrained against lateral movement at its mid-height, its bucking load will be...

(a) 200 kN (b) 283 kN

(c) 400 kN (d) 800 kN [GATE-2007]

Ans. (c)

209. Two steel column *P* (length *L* and yield strength $f_y = 250$ MPa) and *Q* (length 2*L* and yield strength $f_y = 500$ MPa) have the same cross section and end conditions. The ratio of buckling load of column *P* to that of column *Q* is...

(a) 0.5 (b) 1.0

(c) 2.0 (d) 4.0 [GATE-2014]

Ans. (d)

210. With one fixed end and other free end, a column of length *L* buckles at load P_1. Another column of same length and same cross section fixed at both ends buckles at load P_2. The ratio of P_2/P_1 is:

(a) 1 (b) 2

(c) 4 (d) 16

Ans. (d)

211. Four vertical columns of same material, height and weight have the same end conditions. Which cross section will carry the maximum load?

(a) Solid circular section (b) Thin hollow circular section

(c) Solid square section (d) I-section [ESE-2009]

Ans. (b)

212. The end conditions of a column for which length of column is equal to the equivalent length are:

(a) Both the ends are hinged

(b) Both are ends are fixed

(c) One end fixed and other end free

(d) One end fixed and other end hinged [ESE-2013]

Ans. (a)

213. Statement (I): A circular cross-sectional column with diameter d is to be axially loaded in compression. For this the core of the section is considered to be concentrated circular area of diameter $\frac{d}{4}$

Statement (II): We can drill and take out a cylindrical volume of material with diameter $\frac{d}{4}$ in order to make the column lighter and still maintaining the same bucking (crippling) load carrying capacity.

(a) Both statement (I) and statement (II) are individually true and statement (II) is the correct explanation of statement (I)

(b) Both statement (I) and statement (II) are individually true but statement (II) is NOT the correct explanation of statement (I)

(c) Statement (I) is true statement (II) is false

(d) Statement (I) is false statement (II) is true [ESE-2013]

Ans. (c)

214. Which one of the following pairs is not correctly matched?

Boundary condition	Euler's bucking load
(a) Pin-Pin	$\frac{\pi^2 EI}{L^2}$
(b) Fixed-Fixed	$\frac{4\pi^2 EI}{L^2}$
(c) Fixed-Free	$\frac{0.25\pi^2 EI}{L^2}$
(d) Fixed-Pin	$\frac{\sqrt{2}\pi^2 EI}{L^2}$

Ans. (d)

215. For a masonry dam of base width b, at which location w.r.t. the central line, would the resultant loading intersect sections to avoid tension in any horizontal section?

(a) Outside of $b/6$ (b) Within $b/6$

(c) Within $b/8$ (d) At the central line [ESE-2007]

Ans. (b)

216. A circular column of external diameter D, and internal diameter d, carries an eccentric load such that tension is developed nowhere. What shall be the diameter of the core?

(a) $\frac{D^2 + d^2}{8D_2}$ (b) $\frac{D^2 - d^2}{8D_2}$

(c) $\frac{D^2 + d^2}{4D}$ (d) $\frac{D^2 - d^2}{4D}$ [ESE-2009]

Ans. (c)

217. A pedestal of rectangular cross section of size $B \times D$ is subjected to a vertical load P. For the stress to be zero at the farthest point in the first quadrant of X-Y axes of the section (with the origin at the symmetric midpoint) due to the load acting in the quadrant is bounded by the straight line,

(a) $\frac{6X}{B} - \frac{6Y}{D} = 1$ (b) $\frac{6X}{B} + \frac{6Y}{D} = -1$

(c) $\frac{6X}{B} + \frac{6Y}{D} = 1$ (d) $\frac{6X}{D} + \frac{6Y}{B} = 1$ [ESE-2011]

Ans. (c)

218. The base of a column is subjected to moment. If the intensity of bearing pressure due to axial load is equal to stress due to moment, then the bearing pressure between the base and the concrete is ....

(a) Uniform compression throughout

(b) Zero at one end and compression at the other end

(c) Tension at one end and compression at the other end

(d) Compression, varying as a parabolic profile [ESE-2011]

Ans. (b)

219. Consider a circular member of diameter D subjected to a compressive load P. For a condition of no tensile stress in the cross section, the maximum radial distance of the load from the central of the circle is

(a) $\frac{D}{6}$ (b) $\frac{D}{8}$

(c) $\frac{D}{12}$ (d) $\frac{D}{4}$ [ESE-2014]

Ans. (b)

220. U_1 and U_2 are the strain energies in a prismatic bar due to axial tensile force P_1 and P_2, respectively. The strain energy U stored in the same bar due to combined action of P_1 and P_2 will be....

(a) $U = U_1 + U_2$ (b) $U = U_1U_2$

(c) $U < U_1 + U_2$ (d) $U > U_1 + U_2$ [GATE-2007]

Ans. (d)

221. A mild steel specimen is under uniaxial tensile stress. Young's modulus and yield stress for mild steel are 2×10^5 MPa, 250 MPa, respectively. The maximum amount of strain energy per unit volume that can be stored in this specimen without permanent set is

(a) 156 Nmm/mm^3 (b) 15.6 Nmm/mm^2

(c) 1.56 Nmm/mm^3 (d) 0.156 Nmm/mm^2 [GATE-2008]

Ans. (d)

222. For a ductile material, toughness is a measure of....

(a) resistance to scratching

(b) ability to absorb energy up to fracture

(c) ability to absorb till elastic limit

(d) resistance to indentation [GATE-2013]

Ans. (b)

223. What is the strain energy stored in a body of volume V with stress s due to gradually apllied load?

(a) $\frac{\sigma E}{V}$ (b) $\frac{\sigma E^2}{V}$

(c) $\frac{\sigma V^2}{E}$ (d) $\frac{\sigma V^2}{2E}$ [ESE-2006]

Ans. (d)

224. If a material expands freely due to heating, it will develop

(a) thermal stress (b) tensile stress

(c) compressive stress (d) no stress [ESE-2007]

Ans. (d)

225. What is the expression for the strain energy due to bending of a cantilever beam (length l, modulus of elasticity E and moment of inertia I)?

(a) $\frac{P^2L^3}{3EI}$ (b) $\frac{P^2L^3}{6EI}$

(c) $\frac{P^2L^3}{4EI}$ (d) $\frac{P^2L^3}{48EI}$ [ESE-2009]

Ans. (b)

226. If a rod expands freely due to heating, it will develop....

(a) bending stress (b) thermal stress

(c) No stress (d) Compressive stress [ESE-2011]

Ans. (c)

227. A bar of copper and steel form a composite system which is heated through a temperature of 40°C. The stress induced in the copper bar is...

(a) Tensile (b) Compressive

(c) Both tensile and compressive (d) Shear [ESE-2012]

Ans. (b)

228. A circular bar L m long and d m in diameter is subjected to tensile force of F kN. Then the strain energy U will be, (where, E is the modulus of elasticity in kN/m^2)

(a) $\frac{4F^2}{\pi d^2}\cdot\frac{L}{E}$ (b) $\frac{F^2}{\pi d^2}\cdot\frac{L}{E}$

(c) $\frac{2F^2}{\pi d^2}\cdot\frac{L}{E}$ (d) $\frac{3F^2}{\pi d^2}\cdot\frac{L}{E}$ [ESE-2012]

Ans. (c)

229. An axially loaded bar is subjected to a normal stress of 173 MPa. The shear stress in the bar is....

(a) 75 MPa (b) 86.5 MPa

(c) 100 MPa (d) 122.3 MPa [GATE-2007]

Ans. (b)

230. A machine element is subjected to the following biaxial state of stress; σ_x = 80 MPa; σ_y = 20 MPa, τ_{xy} = 40 MPa. If the shear strength of the material is 100 MPa, the factor of safety as per Tresca's maximum shear stress theory is....

(a) 1.0 (b) 2.0

(c) 4.0 (d) 5.0 [GATE-2015]

Ans. (b)

231. According to von-Mises' distortion energy theory, the distortion energy under three- dimensional stress state is represented by...

(a) $\frac{1}{2E}[\sigma_1^2+\sigma_2^2+\sigma_3^2-2\nu(\sigma_1\sigma_2+\sigma_3\sigma_2+\sigma_1\sigma_3)]$

(b) $\frac{1-2\nu}{6E}[\sigma_1^2+\sigma_2^2+\sigma_3^2-2(\sigma_1\sigma_2+\sigma_3\sigma_2+\sigma_1\sigma_3)]$

(c) $\frac{1-\nu}{3E}[\sigma_1^2+\sigma_2^2+\sigma_3^2-(\sigma_1\sigma_2+\sigma_3\sigma_2+\sigma_1\sigma_3)]$

(d) $\frac{1}{3E}[\sigma_1^2+\sigma_2^2+\sigma_3^2-\nu(\sigma_1\sigma_2+\sigma_3\sigma_2+\sigma_1\sigma_3)]$ [GATE-2006]

Ans. (c)

232. A shaft is subjected to pure torsional moment. The maximum shear stress developed in the shaft is 100 MPa. The yield and ultimate strength of the shaft material in tension are 300 MPa and 450 MPa, respectively. The factor of safety using maximum distortion energy (von-Mise's) theory is......

(a) 1.25 (b) 1.73
(c) 2.11 (d) 3.14 [GATE-2014]

Ans. (b)

233. The homogeneous state of stress for a metal undergoing plastic deformation is

$$\begin{pmatrix} 10 & 5 & 0 \\ 5 & 20 & 0 \\ 0 & 0 & -10 \end{pmatrix}$$ where the stress component values are in MPa. Using von-Mises yield criterion, the value of estimated shear yield stress in MPa is...

(a) 9.50 (b) 16.07
(c) 28.52 (d) 49.41 [GATE-2012]

Ans. (b)

234. The uni-axial yield stress of a material is 300 MPa. According to von Mises criterion, the shear yield stress (in MPa) of the materials is...

(a) 173.3 (b) 201.3
(c) 405.4 (d) 501.8 [GATE-2015]

Ans. (a)

235. Which one of following NOT correct?

(a) Intermediate principal stress is ignored when applying the maximum principal stress theory
(b) The maximum shear stress theory gives the most accurate results amongst all the failure theories
(c) As per the maximum strain energy theory, failure occurs when the strain energy per unit volume exceeds a critical value
(d) As per the maximum distortion energy theory, failure occurs when the distortion energy per unit volume exceeds a critical value [GATE-2014]

Ans. (b)

236. Two thread bolts *A* and *B* of same material and length are subjected to identical tensile load. If the elastic strain energy stored in bolt *A* is 4 times that of bolt *B* and the mean diameter of bolt *A* is 12 mm, the mean diameter of bolt *B* in mm is....

(a) 16 (b) 24
(c) 36 (d) 48 [GATE-2013]

Ans. (b)

237. If the rivets are to be designed to avoid crushing failure, the maximum permissible load pin kN is...

(a) 7.50 (b) 15.00
(c) 22.50 (d) 30.00 [GATE-2013]

Ans. (c)

238. If the plates are to be designed to avoid tearing failure, the maximum permissible load pin kN is

(a) 83 (b) 125
(c) 167 (d) 501

Ans. (c)

239. A butt weld joints developed on steel plates having yield and ultimate tensile strength of 500 MPa and 700 MPa, respectively. The thickness of the plates is 8 mm and width is 20 mm. Improper selection of welding parameters caused an undercut of 3 mm depth along the weld. The maximum transverse tensile load (in kN) carrying capacity of the developed weld joint is.....

(a) 25 (b) 50
(c) 70 (d) 95 [GATE-2014]

Ans. (c)

Appendix-2B

ESE-GATE Objective Questions 2018-2020

1. A steel wire of 20 mm diameter is bent into a circular shape of 10 m radius. If E, the modulus of elasticity, is 2×10^6 kg/cm^2, then the maximum tensile stress induced in the wire is nearly

   (a) 2×10^3 kg/cm^2 (b) 4×10^3 kg/cm^2

   (c) 2×10^4 kg/cm^2 (d) 4×10^4 kg/cm^2 [ESE-2018]

Ans. (a)

Sol. Dia of steel wire = 20 mm

Radius of circular shape = 10 m

Modulus of elasicity, $E = 2 \times 10^6$ kg/cm^2

$\therefore$ From bending formula, $\dfrac{M}{I} = \dfrac{f}{y} = \dfrac{E}{R} \Rightarrow \dfrac{f}{y} = \dfrac{E}{R} \Rightarrow f = \dfrac{E}{R} \times y$

$$f = \frac{2 \times 10^6}{10} \times \frac{20}{2} \times 10^{-3} \text{ kg/cm}^2 \Rightarrow f = 2 \times 10^3 \text{ kg/cm}^2$$

2. A long rod of uniform rectangular section with thickness t, originally straight, is bent into the form of a circular arc with displacement d may be regarded as small as compared to the length l. The longitudinal surface strain is

   (a) $\dfrac{2td}{l^2}$ (b) $\dfrac{4td}{l^2}$

   (c) $\dfrac{8td}{l^2}$ (d) $\dfrac{16td}{l^2}$ [ESE-2018]

Ans. (b)

Sol. $\therefore$ From property of circle, $(2R - d) \times d = \left(\dfrac{1}{2}\right) \cdot \left(\dfrac{1}{2}\right)$

$\therefore$ for small 'θ' arch length and chord length are same: $d = \dfrac{l^2}{8R}$ or $R = \dfrac{l^2}{8d}$

From bending formula, $\frac{\sigma}{y}=\frac{E}{R}\Rightarrow\frac{\sigma}{t/2}=\frac{E}{l^2/8d}\Rightarrow\frac{\sigma}{E}=\varepsilon=\frac{y}{R}=\frac{4dt}{l^2}$

3. If strains on a piece of metal are ε_x = –120 μm/m, ε_y = –30 μm/m, γ = 120 μm/m, what is the maximum principal strain?
   (a) 0 (b) 50 μm/m
   (c) 75 μm/m (d) 150 μm/m [ESE-2018]

Ans. (d)

Sol. Given: ε_x = –120 μm/m, ε_y = –30 μm/m, γ = 120 μm/m

$$\therefore\ \varepsilon_{max}=\frac{\varepsilon_x-\varepsilon_y}{2}\pm\sqrt{\left(\frac{\varepsilon_x-\varepsilon_y}{2}\right)^2+\left(\frac{\gamma_{xy}}{2}\right)^2}=\frac{-120-30}{2}\pm\sqrt{\left(\frac{-120+30}{2}\right)^2+\left(\frac{120}{2}\right)^2}$$

$$=-75\pm\sqrt{42^2+60^2}\Rightarrow\varepsilon_{max}=-150\ \mu\text{m/m or }\varepsilon_{min}=0$$

Maximum magnitude of strain will be considered.

4. The state of stress at a point is given by: σ_x = 80 MPa, σ_y = 100 MPa, τ_{xy} = 60 MPa. If the yield strength for the material is 150 MPa, as determined in a uniaxial test, then the maximum shear stress is, nearly
   (a) 75.4 MPa (b) 127.4 MPa
   (c) 119.3 MPa (d) 104.0 MPa [ESE-2018]

Ans. (a)

Sol. Given σ_x = 80 MPa, σ_y = 100 MPa, τ_{xy} = 60 MPa, f_y = 150 MPa

$$\sigma_{1/2}=\frac{\sigma_x+\sigma_y}{2}\pm\sqrt{\left(\frac{\sigma_x-\sigma_y}{2}\right)^2+\tau_{xy}^2}\quad\sigma_1=\frac{80+100}{2}+\sqrt{\left(\frac{80-100}{2}\right)^2+60^2}=150.83\text{ MPa}$$

$$\sigma_2=\frac{80+100}{2}+\sqrt{\left(\frac{80-100}{2}\right)^2+60^2}=29.17\text{ MPa},$$

$$\tau_{max}=\max\left\{\left|\frac{\sigma_1-\sigma_2}{1}\right|,\left|\frac{\sigma_1}{2}\right|,\left|\frac{\sigma_2}{2}\right|\right\}=\tau_{max}=75.4\text{ MPa}$$

5. Principal stresses at a point in an elastic material are 1.5 σ (tensile), σ (tensile) and 0.5 σ (compressive). The elastic limit in tension is 210 MPa and μ = 0.3. The values of σ at failure when computed by maximum principal strain theory is nearly
   (a) 140.5 MPa (b) 145.5 MPa
   (c) 150.5 MPa (d) 155.5 MPa [ESE-2018]

Ans. (d)

Sol. σ_1 = 1.5σ (*T*), *E* = 210 MPa, σ_2 = σ (*T*), σ_3 = 0.5 σ (*c*), μ = 0.3

From maximum principal strain theory, $\frac{\sigma_1}{E} - \frac{\mu\sigma_2}{E} - \frac{\mu\sigma_3}{E} \le \frac{f_y}{E}$

$$\frac{1.5\sigma_2}{E} - \frac{0.3\times\sigma}{E} - 0.3\times\left(\frac{-0.5\sigma}{E}\right) \le \frac{f_y}{E} \Rightarrow \therefore f_y = 210 \text{ MPa}, \sigma = 155.5 \text{ MPa}$$

6. A slid shaft A of diameter D and length L is subjected to a torque T; another B of the same material and of the same length, but half the diameter, is also subjected to the same torque T. The ratio between the angles of twist of shaft B to that of shaft A is

   (a) 32 (b) 16

   (c) 8 (d) 4 [ESE-2018]

Ans. (b)

Sol. Shaft A, shaft B diameter, $d_1 = D$, $d_2 = \frac{D}{2}$

Length, $l = L$, Torque $= T$

$$\therefore \frac{T}{J} = \frac{\tau}{r} = \frac{C\theta}{L} \Rightarrow \theta = \frac{TL}{CJ} \Rightarrow \theta \propto \frac{1}{J} \Rightarrow \therefore J = \frac{\pi d^4}{32}$$

$$= \frac{\theta_1}{\theta_2} = \frac{d_2^4}{d_1^4} \Rightarrow \frac{\theta_A}{\theta_B} = \frac{1}{16} \Rightarrow \frac{\theta_A}{\theta_B} = \frac{1}{16}$$

7. The required diameter for a solid shaft to trasmit 400 kW at 150 rpm, with the working shear stress not to exceed 80 MN/m² is nearly

   (a) 125 mm (b) 121 mm

   (c) 117 mm (d) 113 mm [ESE-2018]

Ans. (c)

Sol. Power = 400 kW, N = 150 rpm

Shear stress = 80 MPa

$$P = \frac{2\pi NT}{60}, T = \frac{400\times10^3\times60}{2\pi\times150}, T = 25464.79 \text{ Nm}$$

$$\frac{T}{J} = \frac{\tau}{r} = \frac{25467.79\times10^3}{\frac{\pi}{32}\times d^4} = \frac{80}{d/2}\ d_3 = \left(\frac{25464.79\times16\times10^3}{\pi\times80}\right) \Rightarrow d = 117.4 \text{ mm}$$

8. The deformation of a vertically held bar of length L and cross section A is due to its self-weight only. If Young's modulus is E and the unit weight of the bar is γ, the elongation dL is

   (a) $\frac{\gamma L^3}{2E}$ (b) $\frac{EL^3}{2\gamma}$

   (c) $\frac{\gamma L^2}{2E}$ (d) $\frac{\gamma L^{32}}{2AE}$ [ESE-2018]

Ans. (c)

Sol. γ = unit wt of bar, E = Young's modulus of elasticity.

Total elongation, $dL = \dfrac{PL}{AE} \Rightarrow dL = \int_0^L \dfrac{\gamma A(L-y)dy}{AE}$

Integrating, we get $dL = \dfrac{\gamma L^2}{2E}$

9. For material, the modulus of rigidity is 100 GPa and the modulus of elasticity is 250 GPa. The value of the Poisson's ratio is

   (a) 0.20 (b) 0.25

   (c) 0.30 (d) 0.35 [ESE-2018]

Ans. (b)

Sol. $2G(1 + \mu)$, $E = 250$ GPa, $G = 100$ GPa

$$\mu = \frac{E}{2G} - 1 = \frac{250}{2 \times 100} - 1 \Rightarrow \mu = 0.25$$

10. In the case of a rectangular beam subjected to a transverse shearing force, the ratio of maximum shear stress to average shear stress is

    (a) 0.75 (b) 1.00

    (c) 1.25 (d) 1.50 [ESE-2018]

Ans. (d)

Sol. Average shear stress for a beam of cross section is $(d \times d) = \dfrac{V}{bd}$

Shear stress distribution for rectangular cross-section $= \dfrac{V}{2I}\left(\dfrac{d^2}{4} - y^2\right)$

Maximum will be at $y = 0$ (at N.A.) $\tau_{max} = \dfrac{V}{\dfrac{2bd^3}{12}} \times \dfrac{d^2}{4} = \dfrac{3}{2}\left(\dfrac{V}{bd}\right)$

So, $\dfrac{\tau_{max}}{\tau_{avg}} = 3/2 = 1.5$

11. A bar specimen of 36 mm diameter is subjected to a pull of 90 kN during a tension test. The extension on a gauge length of 200 mm is measured to be 0.089 mm and the change in diameter to be 0.0046 mm. The Poisson's ratio will be

    (a) 0.287 (b) 0.265

    (c) 0.253 (d) 0.241 [ESE-2019]

Ans. (a)

Sol. $d = 36$ mm, $P = 90$ kN, $l = 200$ mm, $\delta l = 0.089$, $\delta d = 0.0046$

Poisson's ratio $= \dfrac{0.0046/36}{0.089/200} = 0.287$

12. A steel rod 15 m long is at a temperature of 15°C. The values of $\alpha = 12 \times 10^{-6}/°C$, $E = 200\ GN/m^2$ are adapted. When the temperature is raised to 65°C, what is the free expansion of the length; and if this expansion of the rod is fully prevented, what is the temperature stress produced?

(a) 5 mm, 120 MN/m^2 (b) 9 mm, 120 MN/m^2

(c) 5 mm, 150 MN/m^2 (d) 9 mm, 120 MN/m^2 [ESE-2019]

Ans. (b)

Sol. $L = 15$ m $\Delta T = 65 - 15 = 50°C$, $\alpha = 12 \times 10^{6}/°C$, $E = 200$ GPa

$\delta l = \alpha \Delta T l = 12 \times 10^{-6} \times 50 \times 15000 = 9$, $\sigma = (E \propto \Delta T)$

$= 200 \times 10^3 \times 12 \times 10^{-6} \times 50 = 120$ MPa

13. A bar of uniform rectangular section of area A is subjected to an axial tensile load P; its Young's modulus is E and its Poisson's ratio is $1/m$. Its volumetric strain e_v is

(a) $\frac{P}{AM}\left(1+\frac{3}{m}\right)$ (b) $\frac{P}{AM}\left(1+\frac{2}{m}\right)$

(c) $\frac{P}{AM}\left(1-\frac{2}{m}\right)$ (d) $\frac{P}{AM}\left(1-\frac{1}{2m}\right)$ [ESE-2019]

Ans. (c)

Sol. $\varepsilon_x = \frac{\sigma_x}{E} - \gamma\frac{\sigma_1}{E} - \gamma\frac{\sigma_2}{E}\sigma_y = 0, \sigma_x = 0$

$$\varepsilon_x = \frac{P}{AE}, \varepsilon_x = \varepsilon_y = \frac{-1}{m}\frac{\sigma_x}{E}$$

$$\text{Volumetric strain} = \varepsilon_x + \varepsilon_y + \varepsilon_z = \frac{P}{AE} - \frac{2P}{mAE} = \frac{P}{AE}\left(1-\frac{2}{m}\right)$$

14. The normal stresses on two mutually perpendicular planes are 140 N/mm^2 (tensile) and 70 N/mm^2 (tensile). If the maximum shear stress is 45 N/mm^2, the shear stress on these planes will be nearly

(a) 20.9 N/mm^2 (b) 24.9 N/mm^2

(c) 28.3 N/mm^2 (d) 32.0 N/mm^2 [ESE-2019]

Ans. (c)

$$\textbf{Sol. } \tau_{max} = \sqrt{\left(\frac{\sigma_x - \sigma_y}{2}\right)^2 + \tau_{xy}^2} \Rightarrow 45^2 = \left(\frac{140-70}{2}\right)^2 + \tau_{xy}^2 \Rightarrow \sqrt{45^2 - 35^2} = 28.28 \text{ MPa}$$

15. The normal stresses on the two mutually perpendicular planes at a point are 120 MPa (tensile) and 60 MPa (tensile). If the shear stress across these planes is 30 MPa, the principal stresses will be nearly

(a) 124 MPa (tensile) and 24 MPa (compressive)

(b) 132 MPa (tensile) and 24 MPa (compressive)

(c) 124 MPa (tensile) and 48 MPa (tensile)

(d) 132 MPa (tensile) and 48 MPa (tensile) [ESE-2019]

Ans. (d)

Sol. $\sigma_x = 120$ MPa, $\sigma_y = 60$ MPa, $\tau_{xy} = 30$ MPa, $\sigma_{1,2} = C \pm R$

$$C = \frac{\sigma_x + \sigma_y}{2} = \frac{120 + 60}{2} = 90,\ R = \sqrt{\left(\frac{120 - 60}{2}\right)^2 + 30^2} = 42.46,$$

$\sigma_{1,2} = 90 \pm 42.46 = 132.46$ (or) 47.5 MPa

16. At a point in a material the stresses acting on two planes at right angles to each other are: $\sigma_x = 120$ MPa, $\sigma_y = -200$ MPa, $\tau_{xy} = -80$ MPa

The maximum shear stress on the element will be nearly

(a) 142 MPa (b) 155 MPa

(c) 167 MPa (d) 179 MPa [ESE-2019]

Ans. (d)

Sol. $\tau_{max} = \sqrt{\left(\frac{120 + 200}{2}\right)^2 + 80^2} = 178.9$ MPa

17. The principal stresses in the wall of a container are 40 MN/mm^2, 80 MN/mm^2 The normal makes an angle of 30° with a direction of maximum principal stress. The resultant stresses (in magnitude) in the plane will be nearly

(a) 84 MN/m^2 (b) 72 MN/m^2

(c) 64 MN/m^2 (d) 58 MN/m^2 [ESE-2019]

Ans. (b)

Sol. $\sigma_x = 80$ MPa, $\sigma_y = 40$ MPa, $\sigma_\theta = \sigma_x \cos^2\theta + \sigma_y \sin^2\theta + 2\tau_{xy} \cos\theta \sin\theta$

$\tau_\theta = (\sigma_y - \sigma_x) \cos\theta \sin\theta + \tau_{xy} (\cos^2\theta - \sin^2\theta)$, $\sigma_\theta = 80 \cos^2 30 + 40 \sin^2 30 = 70$ MPa

$\tau_\theta = (4 - 80)\cos 30 \sin 30 = 17.32$ MPa, $\sigma_R = \sqrt{70^2 + 17.32} = 72.11$ MPa

18. The change in shearing force between two points on the beam is equal to the area of points on the area of

(a) Loading diagram between the two points

(b) Shear force diagram between the two points

(c) Bending moment diagram between the two points

(d) M/EI diagram between the two points. [ESE-2019]

Ans. (a)

19. Which one of the following statements specifies shear flow?

(a) Flow of shear force along the beam

(b) It is the product of the shear stress at any level and the corresponding width b (of the section)

(c) Unbalanced force on any side of given section divided by area of section

(d) The deformation at any level due to sudden variation in shear stress

[ESE-2019]

Ans. (b)

20. The maximum shear stress induced in a solid circular shaft of diameter 15 cm, when the shaft transmits 150 kW power at 180 rpm will be

(a) 16 N/mm^2 (b) 14 N/mm^2

(c) 12 N/mm^2 (d) 10 N/mm^2 [ESE-2019]

Ans. (c)

Sol. $\tau = \frac{16T}{\pi d^3} = \frac{16}{\pi \times 0.15^3}\left(\frac{150\times 10^3}{\frac{180\times 2\pi}{60}}\right) = 12 \text{ MPa}$

21. A beam of triangular cross section is subjected to a shear force of 50 kN. The base width of the section is 250 mm and the height is 200 mm. The beam is placed with its base horizontal. The shear stress at neutral axis will be nearly

(a) 2.2 N/mm^2 (b) 2.7 N/mm^2

(c) 3.2 N/mm^2 (d) 3.7 N/mm^2 [ESE-2019]

Ans. (b)

Sol. $\tau_{NA} = \frac{4}{3}\frac{V}{\text{Area}} = \frac{4}{3}\frac{(50\times 10^3)}{\left(\frac{1}{2}\times 250\times 200\right)} \Rightarrow \tau_{NA} = 2.66 \text{ MPa} = 2.7 \text{ MPa}$

22. A timber beam, 100 mm wide and 150 mm deep, supports a UDL over a span of 2 m. If the safe stresses are not to exceed 28 MPa in bending and 2 MPa in shear, the maximum load that the beam can support is

(a) 16 kN/m (b) 20 kN/m

(c) 24 kN/m (d) 28 kN/m [ESE-2019]

Ans. (b)

Sol. $M_{max} = \frac{wl^2}{8}$, $V_{max} = \frac{wl}{2}$, $\tau_{max} = \frac{M_{max}}{bd^2/6} \pm 28 \text{ N/mm}^2 \; \frac{w(2000)^2}{8\times 100(150)^2/6} \le 28$

$w \le 21 \text{ N/mm}, \tau_{max} = \frac{3}{2}\times\frac{V_{max}}{bd} \le 2 \text{ N/mm}^2 = \frac{3}{2}\times\left(\frac{\frac{w\times 2000}{2}}{100\times 150}\right) \le 2 \Rightarrow w \le 20 \text{ N/mm}$

w should be min of the form the above two criteria.

$\Rightarrow w_{max} = 20 \text{ kN/m}$

23. A 1.5 m long column has a circular cross section of 50 mm diameter. Consider Rankine's formula with values of f_d = 560 N/mm², $\alpha = \frac{1}{1600}$ for pinned ends and factor of safety of 3. If one end of the column is fixed and the other end is free, the safe load will be

(a) 9948 N (b) 9906 N

(c) 9864 N (d) 9822N [ESE-2019]

Ans. (b)

Sol. $P = \frac{f_c A/F.O.S.}{1+a\lambda^2} = \frac{f_c A/F.O.S.}{1+\alpha\left(\frac{1}{r_{\min}}\right)^2}$, α for pinned end $= \frac{1}{1600} \Rightarrow$ α for one end fixed other free = 4α

$$P = \frac{\left[560 \times \frac{\pi}{4}(50)^2\right]/3}{1+\frac{4}{1600}\left(\frac{1.5\times 1000}{50/4}\right)^2},\ r_{\min} = \sqrt{\frac{I_{\min}}{A}} = \sqrt{\frac{\pi d^4}{64\times\frac{\pi d^2}{4}}} = d/4 = 9905.92 \text{ N}$$

24. The maximum shear stress across a circular section is

(a) (4/3) average shear stress (b) (3/2) average shear stress

(c) (5/4) average shear stress (d) (9/5) average shear stress [ESE-2019]

Ans. (a)

25. When a round bar material with diameter of 37.5 mm, length of 2.4 m, Young's modulus of 110 GN/m² and shear modulus of 42 GN/m² is stretched for 2.5 mm, its bulk modulus will be nearly

(a) 104 GN/m² (b) 96 GN/m²

(c) 84 GN/m² (d) 76 GN/m² [ESE-2020]

Ans. (b)

Sol. $E = \frac{9KG}{3K+G} \rightarrow 3KE + EG - 9KG \Rightarrow 9KG - 3KE = EG$

$$K = \frac{EG}{9G+3E} = \frac{110\times 42}{9\times 42 - 3\times 110} = 96.25 \text{ GN/m}^2$$

26. A punch of 20 mm diameter is used to punch a hole in 8 mm thick plate. If the force required to create a hole is 110 kN, the average shear stress in the plate will be nearly

(a) 410 MPa (b) 320 MPa

(c) 220 MPa (d) 140 MPa [ESE-2020]

Ans. (c)

Sol. Shear stress, $\tau = \frac{P}{A} = \frac{P}{\pi dt} = \frac{110 \times 10^3}{\pi \times 20 \times 8} = 219 \text{ MPa} \simeq 220 \text{ MPa}$

27. The strain energy U stored due to bending of the cantilever beam due to point load at the free end will be

(a) $\frac{W^2l^3}{6EI}$ (b) $\frac{W^2l^2}{6EI}$

(c) $\frac{W^3l^3}{36EI}$ (d) $\frac{W^2l^3}{36EI}$ [ESE-2020]

where, W = Concentrated load

L = Length of a cantilever

EI = Flexual rigidity

Ans. (a)

Sol. $U = \int_0^L \frac{(M_x)^2 dx}{2EI} = \int_0^L \frac{W^2x^2 dx}{2EI} = \frac{W^2l^3}{6EI}$

28. A steel bar 2 m long, 20 mm wide and 15 mm thick is subjected to a tensile load of 30 kN. If Poisson's ratio is 0.25 and Young's modulus is 200 GPa, an increase in volume will be

(a) 160 mm^3 (b) 150 mm^3

(c) 140 mm^3 (d) 130 mm^3 [ESE-2020]

Ans. (b)

Sol. Axial stress, $\sigma = \frac{P}{A} = \frac{30 \times 10^3}{20 \times 15} = 100 \text{ MPa}$

Volumetric strain, $\varepsilon_y = \frac{\sigma}{E}(1 - 2\mu)$

$\Rightarrow \therefore \Delta V = \frac{100}{200 \times 10^3}(1 - 0.5) \times 20 \times 15 \times 2000 = 150 \text{ mm}^3$

29. A bolt is under an axial thrust of 9.6 kN together with a transverse force of 4.8 kN. If factor of safety is 3, yield strength of both material is 270 N/mm² and Poisson's ratio is 0.3, its diameter as per maximum principal stress theory will be nearly

(a) 13 mm (b) 15 mm

(c) 17 mm (d) 19 mm

Ans. (a)

Sol. Axial stress, $\sigma = \frac{9.6 \times 10^3}{\frac{\pi}{4} d^2} = \frac{12.23 \times 10^3}{d^2} \text{ N/mm}^2$

Shear stress, $\tau = \dfrac{4.8\times10^3}{\dfrac{\pi}{4}d^2} = \dfrac{6.115\times10^3}{d^2}$ N/mm²

Maximum principal stress, $\sigma_1 = \dfrac{\sigma}{2} + \sqrt{\left(\dfrac{\sigma}{2}\right)^2 + \tau^2} = \dfrac{6115}{d^2} + \sqrt{\left(\dfrac{6115}{d^2}\right)^2 + \left(\dfrac{6115}{d^2}\right)^2}$

$$= \frac{1}{d^2}[6115 + 8647.916] = \frac{14762.9}{d^2} \text{ N/mm}^2$$

According to maximum principal stress theorem, $\sigma_1 \le \dfrac{\sigma_y}{FOS} \Rightarrow \dfrac{14762.9}{d^2} \le \dfrac{270}{3}$

$\Rightarrow \therefore\ d \ge 12.81$ m

30. In a material the principal stresses are 60 MN/m², 48 MN/m² and –36MN/m². When the values of E = 200 GN/m² and $\dfrac{1}{m} = 0.3$ the total strain energy per unit volume will be nearly

    (a) 43.5 kN/m³ (b) 35.5 kN/m³

    (c) 27.5 kN/m³ (d) 19.5 kN/m³ [ESE-2020]

Ans. (d)

Sol. Strain energy per unit volume $= \dfrac{1}{2E}\left[\sigma_1^2 + \sigma_2^2 + \sigma_3^2 - 2\mu(\sigma_1\sigma_2 + \sigma_2\sigma_3 + \sigma_3\sigma_1)\right]$

$$= \frac{1}{2\times200\times10^9}\left[60^2 + 48^2 + 36^2 - 0.6(60\times48 - 48\times36 - 60\times36)\right]\times10^{12}$$

$$= \frac{10^2}{4\times10^{11}}[7200 - 0.6(2880 - 1728 - 2160)] = \frac{10}{4}(7200 + 604.8) = 19.51 \text{ kNm/m}^3$$

31. At a point in a two-dimensional stress system, the stress on two mutually perpendicular planes are σ_{xx}, σ_{yy} and shear stress is τ_{xy}. One of the principal stresses will become zero when the value of shear stress, τ_{xy}, is

    (a) $\pm(\sigma_{xx}\sigma_{yy})$ (b) $\pm\sqrt{\sigma_{xx} - \sigma_{yy}}$

    (c) $\pm\sqrt{\sigma_{xx} + \sigma_{yy}}$ (d) $\pm\sqrt{\sigma_{xx}\sigma_{yy}}$ [ESE-2020]

Ans. (d)

Sol. $\sigma_r = \dfrac{\sigma_{xx} + \sigma_{yy}}{2}\sqrt{\left(\dfrac{\sigma_{xx} - \sigma_{yy}}{2}\right)^2 + \tau_{xy}^2} = +\text{ve} \ne 0$

$$\therefore\ \sigma_2 = 0 \Rightarrow \frac{\sigma_{xx} + \sigma_{yy}}{2}\sqrt{\left(\frac{\sigma_{xx} - \sigma_{yy}}{2}\right)^2 + \tau_{xy}^2} = 0 \Rightarrow \tau_{xy}^2 = \sigma_{xx}.\sigma_{yy} \Rightarrow \therefore\ \tau_{xy} = \pm\sqrt{\sigma_{xx}\sigma_{yy}}$$

32. The deflection δ of the closed coil helical spring is

(a) $\frac{WR^2n}{8Cd^3}$ (b) $\frac{64WR^3n}{Cd^3}$

(c) $\frac{128WR^3n}{Cd^3}$ (d) $\frac{64WR^2n}{Cd^3}$ [ESE-2020]

where, W is the axial load

R is the radius of the coil

n is the number of turns of coil

C is the modulus of rigidity

d is the diameter of the wire of the coil

Ans. (b)

Sol. Strain energy in spring, $\frac{T^2L}{2GJ} = \frac{P^2R^2(2\pi RN)}{2G\frac{\pi}{32}d^4} = \frac{32P^2R^3n}{Gd^4}$

by Castigliano's theorem, $\delta = \frac{\partial U}{\partial P} = \frac{\partial}{\partial P} = \frac{32P^2R^3n}{Gd^4} \Rightarrow \delta = \frac{64PR^3n}{Gd^4}$

$G = C =$ Modulus of rigidity

33. A closely-coiled helical spring of round steel wire 5 mm in diameter having 12 complete coils of 50 mm mean diameter is subjected to an axial load of 100 N. If modulus of rigidity is 80 GPa, the deflection of the spring will be

(a) 36 mm (b) 32 mm

(c) 28 mm (d) 24 mm [ESE-2020]

Ans. (d)

Sol. $\Delta = \frac{64PR^3n}{Gd^4} = \frac{64 \times 100 \times 25^3 \times 12}{80 \times 10^3 \times 5^4} \Rightarrow \Delta = 24$ mm

34. A hollow shaft of external and internal diameters as 100 mm and 40 mm respectively, is transmitting power at 120 rpm. If the shearing stress is not to exceed 50 MPa, the power the shaft can transmit will be

(a) 100 kW (b) 120 kW

(c) 140 kW (d) 160 kW [ESE-2020]

Ans. (b)

Sol. $\tau_{max} \frac{16T}{\pi d_0^3\left[1-\left(\frac{d_1}{d_0}\right)^4\right]} \Rightarrow 50 = \frac{16T}{\pi \times 1003^3\left[1-\left(\frac{40}{100}\right)^4\right]} \Rightarrow T = 9.56 \times 10^6$ Nmm

$\therefore P = \frac{2\pi NT}{60} = \frac{2\pi \times 120 \times 9.56 \times 10^3}{60} = 120$ kW

35. A circular beam of 100 mm diameter is subjected to a shear force of 30 kN. The maximum shear stress will be nearly

(a) 5.1 MPa (b) 6.3 MPa
(c) 7.5 MPa (d) 8.7 MPa [ESE-2020]

Ans. (a)

Sol. $\tau_{max} = \frac{4}{3} \cdot \frac{F}{A} = \frac{4}{3} \times \frac{30 \times 10^3}{\frac{\pi}{4}(100)^2} = 5.08 \text{ MPa}$

36. A cantilever beam *AB* is subjected to a point load of 12 kN at free end over a span of 6 m with $E = 2 \times 10^5 \text{ N/mm}^2$, $I_{xx} = 6 \times 10^7 \text{ mm}^4$. The deflection at the free end will be

(a) 80 mm (b) 72 mm
(c) 64 mm (d) 56 mm [ESE-2020]

Ans. (b)

Sol. $\Delta = \frac{PL^3}{3EI} = \frac{12 \times 10^3 \times 6000^3}{3 \times 2 \times 10^6 \times 6 \times 10^7} = 72 \text{ mm}$

37. A floor has to carry a load of 12 kN/m². The floor is supported on rectangular joists each 100 mm wide, 300 mm deep and 5 m long. If maximum stress in the joists should not exceed 8 MN/m², the centre to centre distance of joists will be

(a) 430 mm (b) 400 mm
(c) 360 mm (d) 320 mm [ESE-2020]

Ans. (d)

Sol. Let *h* be the distance between two joists.

$$\therefore \quad \frac{12 \times l \times h}{l} = w \Rightarrow w = 12 \text{ hkN/m}$$

$$\text{Now, } \sigma_{max} = \frac{M_{max}}{z} \Rightarrow 8 = \frac{wl^2}{8 \times \frac{bd^2}{6}} \Rightarrow 8 = \frac{6wl^2}{8bd^2} \Rightarrow \frac{64 \times 100 \times 300^2}{6 \times (5000)^2} = w$$

$\therefore w = 3.84 \text{ N/mm} \Rightarrow 12h = 3.84 \Rightarrow h = 0.32 \text{ m} = 320 \text{ mm}$

38. A simply supported wooden beam 100 mm wide, 250 mm deep and 3 long is carrying a uniformly distributed load of 40 kN/m. The maximum shear stress will be

(a) 2.4 MPa (b) 2.8 MPa
(c) 3.2 MPa (d) 3.6 MPa [ESE-2020]

Ans. (d)

Sol. Maximum S.F., $S = \frac{wL}{2} = \frac{40 \times 3}{2} = 60 \text{ kN} \Rightarrow \tau_{max} = 1.5\, \tau_{avg}$

$$= 1.5 \times \frac{60 \times 10^3}{100 \times 250} = 3.6 \text{ MPa}$$

39. A simply supported beam of span 8 m carries a uniformly distributed load of 24 kN/m run over the whole span. The beam is propped at the middle of the span. The values of $E = 200 \times 10^6$ kN/m², $I = 20 \times 10^{-5}$ m⁴. The amount by which the prop should yield in order to make all three reactions equal will be nearly

(a) 20 mm (b) 15 mm

(c) 10 mm (d) 5 mm [ESE-2020]

Ans. (b)

Sol. $3R = 24 \times 8 \Rightarrow R = 64$ kN

$$\Delta_g = \frac{5}{384}\frac{wL^4}{EI} - \frac{RL^3}{48EI} = \frac{1}{EI}\left[\frac{5}{384} \times 24 \times 8^4 - \frac{64 \times 8^3}{48}\right]$$

$$= \frac{1}{200 \times 10^9 \times 20 \times 10^{-5}}[1280 - 682.666] = 14.9 \text{ mm}$$

40. A solid circular beam with radius of 0.25 m and length of 2 m is subjected to a twisting moment of 20 kNm about the z-axis at the free end, which is the only load acting as shown in the figure. The shear stress component, τ_{xy}, at point M in the cross section of the beam at a distance of 1 m from the fixed end is

(a) 0.0 MPa (b) 0.51 MPa

(c) 0.815 MPa (d) 2.0 MPa [GATE-2018]

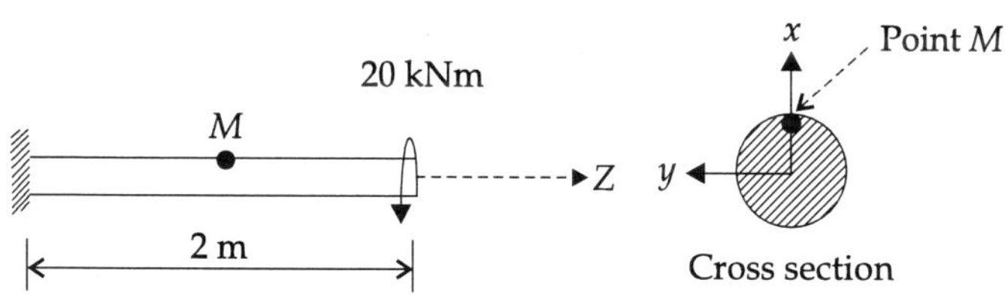

Ans. (a)

Sol. $\tau_{xy} = \tau_{yx} = 0 \because$ torsion applied is also in x–y plane

41. A cantilever beam of length of 2 m with a square section of side length 0.1 m is loaded vertically at the free end. The vertical displacement at the free end is 5 mm. The beam is made of steel with Young's modulus of 2.0×10^{11} N/m². The maximum bending stress at the fixed end of the cantilever is

(a) 20.0 MPa (b) 37.5 MPa

(c) 60.0 MPa (d) 75.0 MPa [GATE-2018]

Ans. (b)

Sol. $\sigma_{max} = \frac{M}{Z} = \frac{3125 \times 2}{(0.1)^3/6} = 37.5$ MPa

42. A cylinder of radius 250 mm and weight, $W = 10$ kN is rolled up an obstacle of height 50 mm by applying a horizontal force P at its centre as shown in the figure.

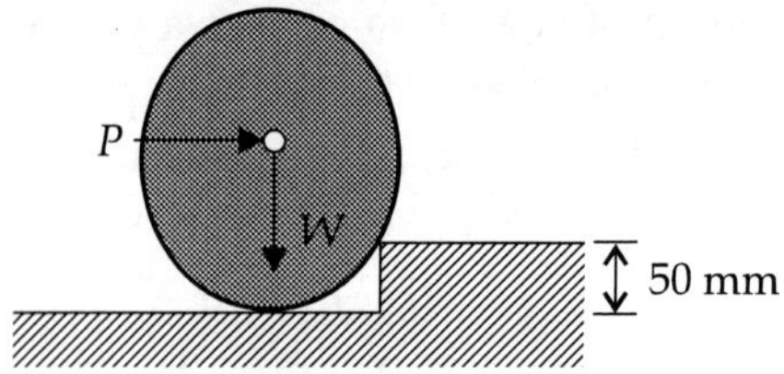

All interfaces are assumed frictionless. The minimum value of P is

(a) 4.5 kN (b) 5.0 kN

(c) 6.0 kN (d) 7.5 kN [GATE-2018]

Ans. (d)

Sol. $P \times 200 = W \times \sqrt{250^2 - 200^2} \Rightarrow P = 7.5$ kN

43. A plate in equilibrium is subjected to uniform stresses along its edges with magnitude σ_{xx} = 30 MPa, σ_{yy} = 50 MPa as shown in the figure.

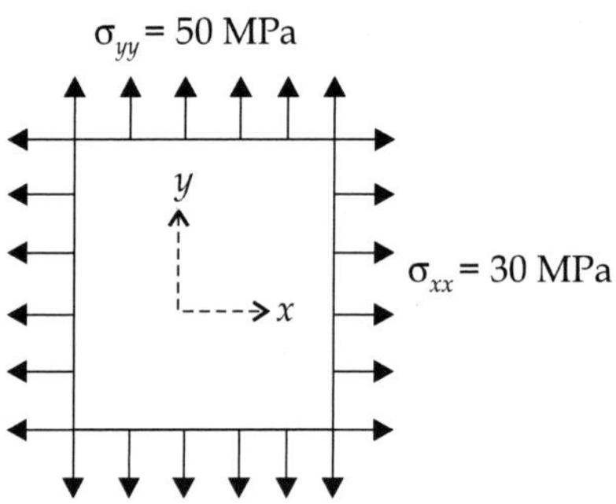

The Young's modulus of the material is 2×10^{11} N/m^2 and the Poisson's ratio is 0.3. If σ_{zz} is negligibly small and assumed to be zero, then the strain ε_{zz} is

(a) -120×10^{-6} (b) -60×10^{-6}

(c) 0.0 (d) 120×10^{-6} [GATE-2018]

Ans. (a)

Sol. $\varepsilon_{zz} = \frac{\sigma_{zz}}{E} - \mu\frac{\sigma_{xx}}{E} - \mu\frac{\sigma_{yy}}{E} = -\frac{0.3}{2 \times 10^5}(30 + 50) = -120 \times 10^{-6}$ MPa

44. The figure shows a simply supported beam PQ of uniform flexural rigidity, EI, carrying two moments, M and $2M$.

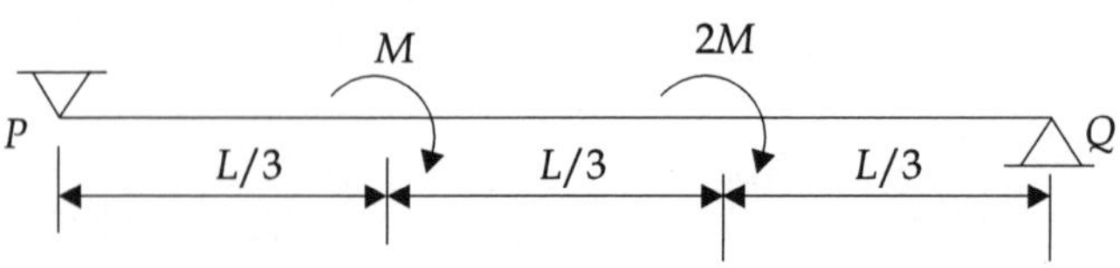

The slope at P will be

(a) 0 (b) $ML/(9EI)$

(c) $ML/(6EI)$ (d) $ML/(3EI)$ [GATE-2018]

Ans. (c)

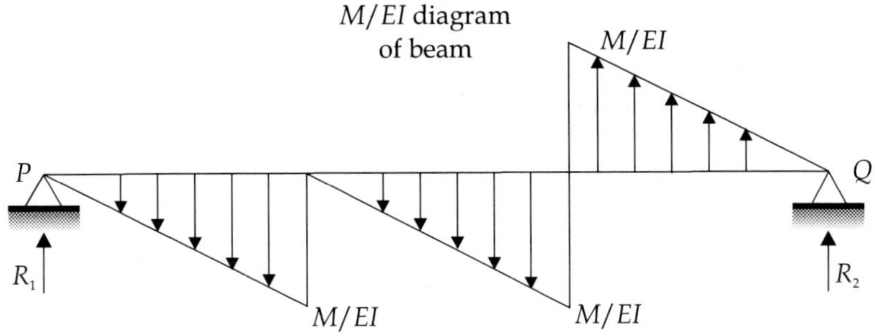

Sol. Slope at $P = R_1 = \dfrac{ML}{6EI}$

45. A vertical load of 10 kN acts on a hinge located at a distance of $L/4$ from the roller support Q of a beam of length L (see figure)

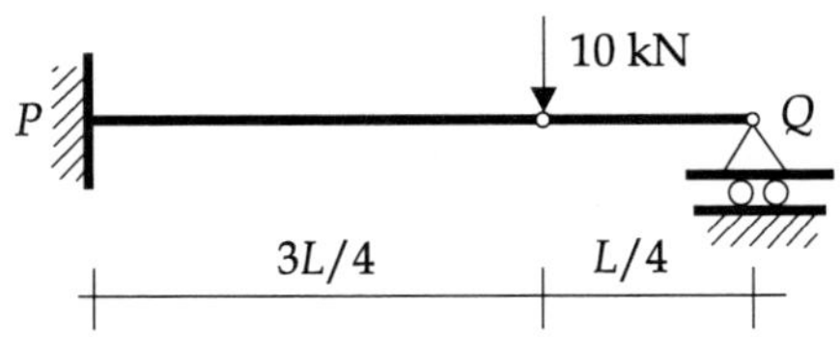

The vertical reaction at support Q is

(a) 0.0 kN (b) 2.5 kN

(c) 7.5 kN (d) 10.0 kN [GATE-2018]

Ans. (a)

Sol. $M_{\text{hinge}} = R_Q \times (L/4) = 0 \Rightarrow R_Q = 0$

46. Two rigid bodies of mass 5 kg and 4 kg are at rest on a frictionless surface until acted upon by a force of 36 N as shown in the figure.

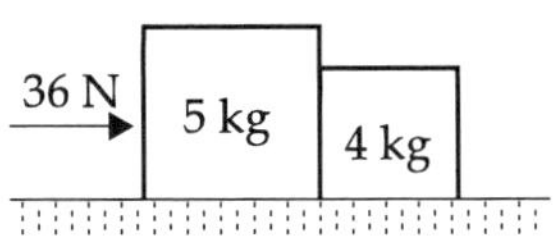

The contact force generated between the two bodies is

(a) 4.0 N (b) 7.2 N

(c) 9.0 N (d) 16.0 N [GATE-2018]

Ans. (d)

Sol. Acceleration of whole system $= \dfrac{36}{9} = 4$

Contact force

From 5 kg block: $36 - N = 5a \Rightarrow N = 36 - 20 = 16$ kN

or From 4 kg block: $N = ma = 4 \times 4 = 16$ kN

47. A prismatic beam *P-Q-R* of flexural rigidity, $EI = 1 \times 10^4$ kNm^2 is subjected to a moment of 180 kNm at *Q* as shown in the figure.

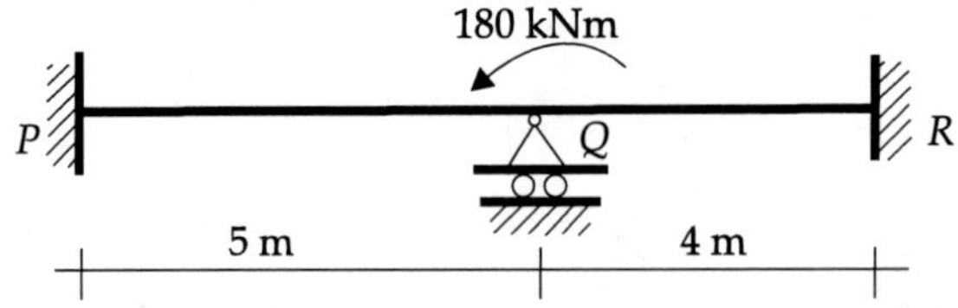

The rotation at *Q* (in rad, up to two decimal places) is............ [GATE-2018]

Ans. (0.01)

Sol. $K_Q = K_{QP} + K_{QR} = \frac{4EI}{5} + \frac{4EI}{4} = 1.8\ EI$

$$\theta_Q = \frac{M_Q}{K_Q} = \frac{180}{1.8 \times 10^4} = 0.01 \text{ rad}$$

48. An isolated concrete pavement slab of length *L* is resting on a frictionless base. The temperature of the top and bottom fibre of the slab are T_t, T_b, respectively. Given: the coefficient of thermal expansion = α and the elastic modulus = *E*. Assuming $T_t > T_b$ and the unit weight of concrete as zero, the maximum thermal stress is calculated as

(a) $\frac{E\alpha(T_t - T_b)}{2}$ (b) $La(T_t - T_b)$

(c) $E\alpha(T_t - T_b)$ (d) Zero [GATE-2019]

Ans. (d)

Sol. Since the surface is friction less therefore thermal stress will be zero.

49. An element is subjected to biaxial normal tensile strains of 0.0030 and 0.0020. The normal strain in the plane of maximum shear strain is

(a) Zero (b) 0.0050

(c) 0.0010 (d) 0.0025 [GATE-2019]

Ans. (d)

Sol. $(\in)_{\text{at max shear strain}} = \frac{\in_1 + \in_2}{2}$

$$= \frac{0.0030 + 0.002}{2} = 0.0025$$

50. A portal frame shown in figure (not drawn to scale) has a hinge support at joint *P* and a roller support at joint *R*. A point load of 50 kN is acting at joint *R* in the horizontal direction. The flexural rigidity, *EI*, of each member is 10^6 kNm^2. Under the applied load, the horizontal displacement (in mm, round off to-1 decimal place) of joint *R* would be.......... [GATE-2019]

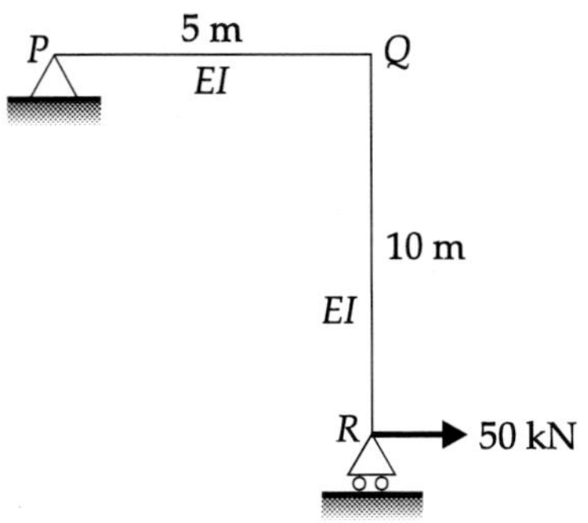

Ans. (25 mm)

Sol. $\sum M_P = 0$

$\Rightarrow -W \times 5 + 50 \times 10 = 0$

$\Rightarrow W = \dfrac{500}{5} = 100$ kN

$\because \delta = \int \dfrac{M.m.dx}{EI}$

Since, unit load at R is acting in the direction of 50 kN load, therefore reaction at $R = 2$ (downward)

Number	Limit	M	m	EI
PQ	0-5 m	$-100x + 500$	$-2x + 10$	10^6
QR	0-10 m	$50x$	X	10^6

$$\delta = \int_0^5 \frac{(-100x+500)(-2x+10)}{EI}dx + \int_0^{10} \frac{(50x)(x)}{EI}dx$$

$$= \int_0^5 \frac{(200x^2 - 1000x - 1000x + 5000)}{EI}dx + \int_0^{10} \frac{50x^2}{EI}dx$$

$$= \int_0^5 \frac{(200x^2 - 2000x + 5000)}{EI}dx + \int_0^{10} \frac{50x^2}{EI}dx$$

$$= \frac{1}{EI}\left[200 \times \frac{x^3}{3} - 2000\frac{x^2}{2} + 5000x\right]_0^5 + \frac{1}{EI} \times 50\left[\frac{x^3}{3}\right]_0^{10}$$

$$= \frac{1}{10^6}\left[\frac{200}{3} \times 125 - \frac{2000}{2} \times 25 + 5000 \times 5\right] + \frac{50}{10^6} \times \frac{1}{3} \times 1000$$

$= 2.5$ mm

51. A close thin-walled tube has thickness, t, mean enclosed area within the boundary of the centerline of tube's thickness, A_m and shear stress. τ. Torsional moment of resistance T, of the section would be

(a) $2\tau A_m t$ (b) $4\tau A_m t$

(c) $\tau A_m t$ (d) $0.5\tau A_m t$ [GATE-2019]

Ans. (a)

Sol. $\tau t = \dfrac{T}{2A_m} \Rightarrow T = 2\tau A_m t$

52. A long uniformly distributed load of 10 kN/m and a concentrated load of 60 kN are moving together on the beam *ABCD* as shown in the figure (not drawn to scale). The relative positions of the two loads are not fixed. The maximum shear force (in kN, round off to the nearest integer) caused at the internal hinge *B* due to the two loads is........ [GATE-2019]

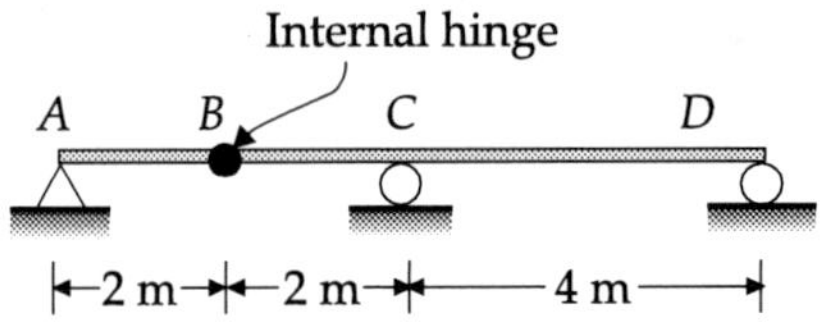

Ans. (70 kN)

Sol. $SF_{max} = 60 \times 1 + \left(\dfrac{1}{2} \times 1 \times 2\right) \times 10$

$= 60 + 10 = 70$ kN

53. A rigid, uniform, weightless, horizontal bar is connected to three vertical members *P*, *Q* and *R* as shown in the figure (not drawn to the scale). All three members have identical axial stiffness of 10 kN/mm. The lower ends of bars *P* and *R* rest on a rigid horizontal surface. When no load is applied, a gap of 2 mm exists between the lower end of the bar *Q* and the rigid horizontal surface. When a vertical load (*W*) is placed on the horizontal bar in the downward direction, the bar still remains horizontal and gets displaced by 5 mm in the vertically downward direction. [GATE-2020]

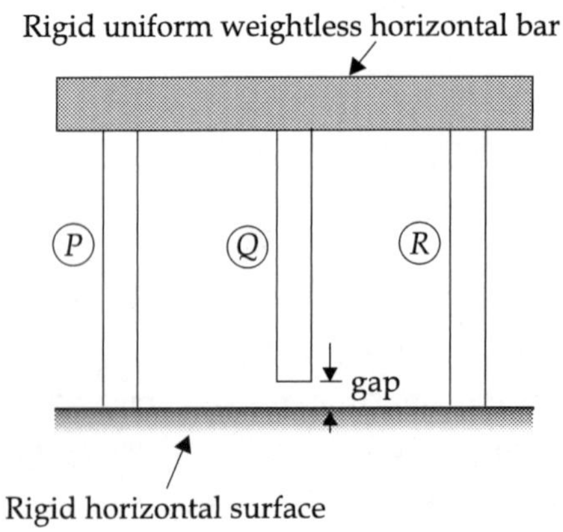

The magnitude of the load *W* (in kN, round off to the nearest integer), is ______.

Ans. (130)

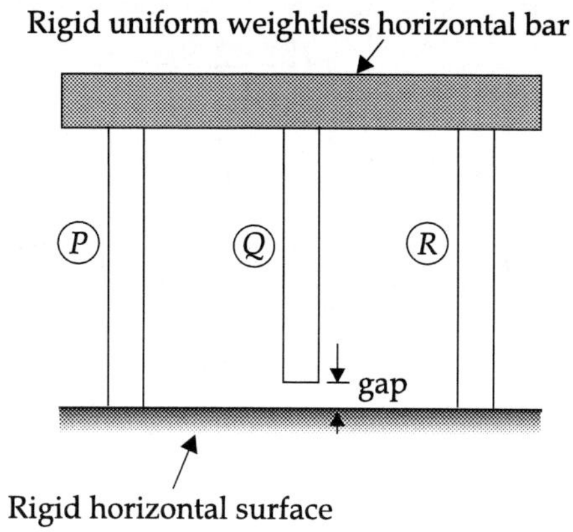

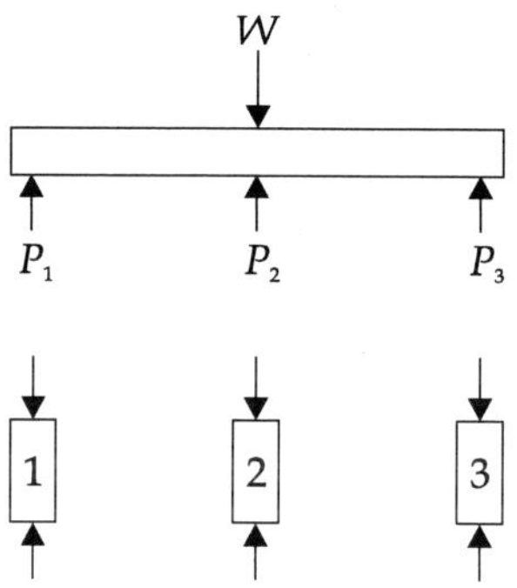

$$P_1 + P_1 + P_2 = W \quad \text{(i)}$$

$$P_1 = P_3$$

$$*\delta_1 = 5 \text{ mm} = \frac{P_1 L}{AE} \qquad \frac{AE}{L} = 10 \text{ kN/m}$$

$$\delta_2 = 3 \text{ mm} = \frac{P_2 L}{AE}$$

So,

$$P_1 = 10 \times 5 = 50 \text{ kN}$$

$$P_2 = 10 \times 3 = 30 \text{ kN}$$

$$W = 2(50) + 30 = 130 \text{ kN}$$

54. Distributed load(s) of 50 kN/m may occupy any position(s) (either continuously or in patches) on the girder PQRST as shown in the figure (not drawn to the scale)

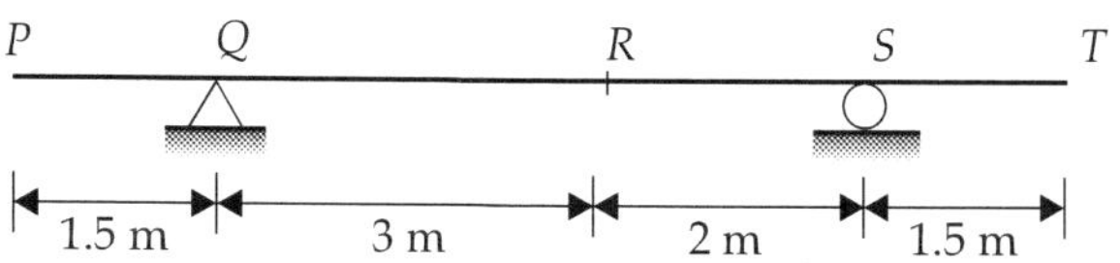

The maximum negative (hogging) bending moment (in kNm) that occurs at point R is

(a) 56.25 (b) 22.50

(c) 150.00 (d) 93.75 [GATE-2020]

Ans. (a)

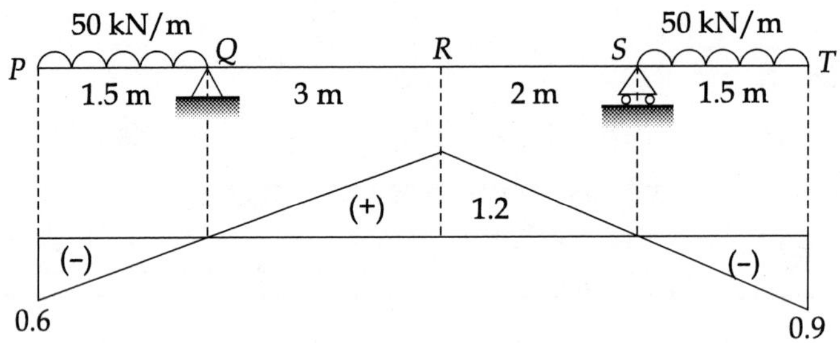

ILD for BM at R

To get maximum hogging BM at R, keep UDL over PQ and ST.

$$\text{Max. -ve BM at } R = 50\left[-\frac{1}{2}\times 1.5\times 0.6\right] + 50\left[-\frac{1}{2}\times 1.5\times 0.9\right]$$

$$= 56.25 \text{ kNm}$$

55. A cantilever beam PQ of uniform flexural rigidity (EI) is subjected to a concentrated moment M at R as shown in the figure.

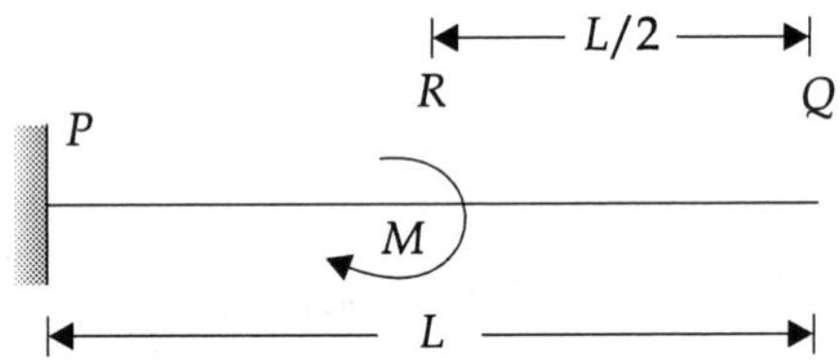

The deflection at the free end Q is

(a) $\frac{ML^2}{6EI}$ (b) $\frac{ML^2}{4EI}$

(c) $\frac{3ML^2}{4EI}$ (d) $\frac{3ML^2}{8EI}$ [GATE-2020]

Ans. (d)

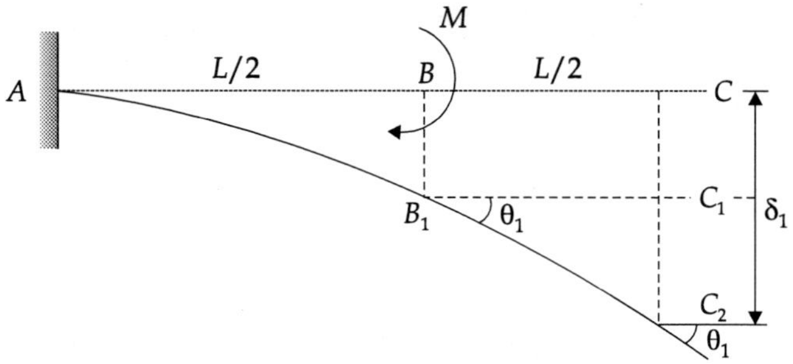

$$\delta_c = \delta_1 = CC_1 + C_1C_2$$
$$= BB_1 + C_1C_2$$
$$= \frac{M\left(\frac{L}{2}\right)^2}{2EI} + \frac{M\left(\frac{L}{2}\right)}{EI}\left(\frac{L}{2}\right)$$
$$\delta_c = \frac{3}{8}\frac{ML^2}{E_I}$$

56. A weightless cantilever beam of span L is loaded as shown in the figure. For the entire span of the beam, the material properties are identical and the cross section is rectangular with constant width.

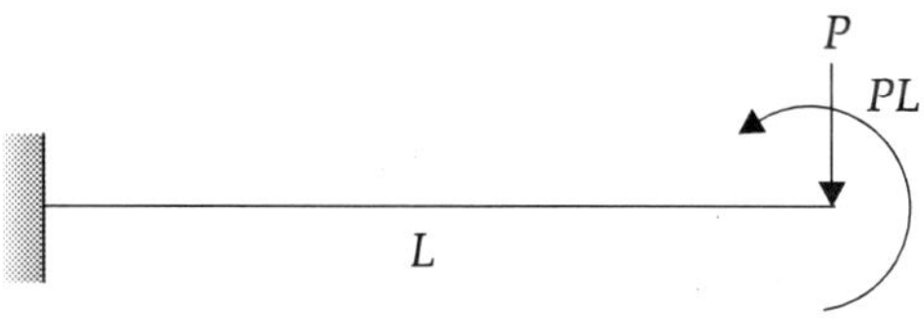

From the flexure-critical perspective, the most economical longitudinal profile of the beam to carry the given loads amongst the options given below is

(a) 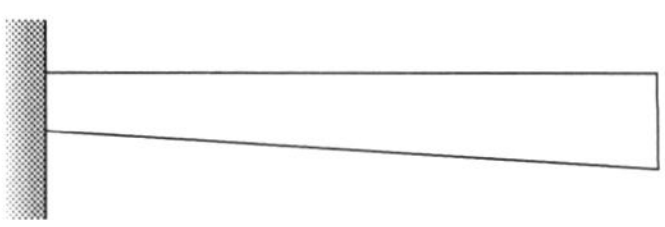

(b) 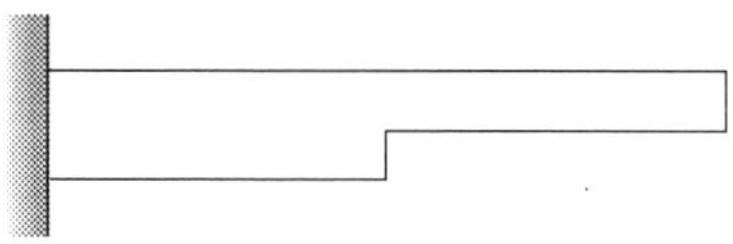

(c) 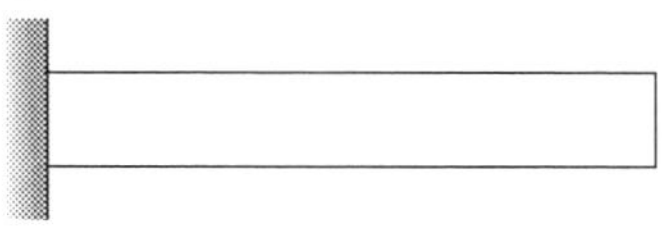

(d) 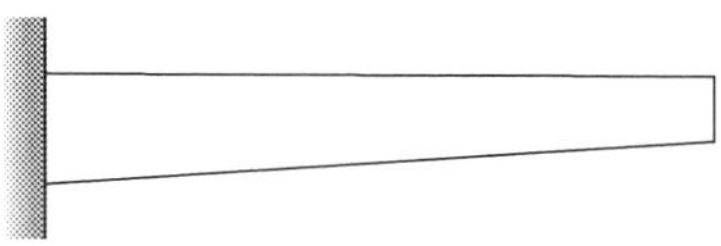

[GATE-2020]

Ans. (a)

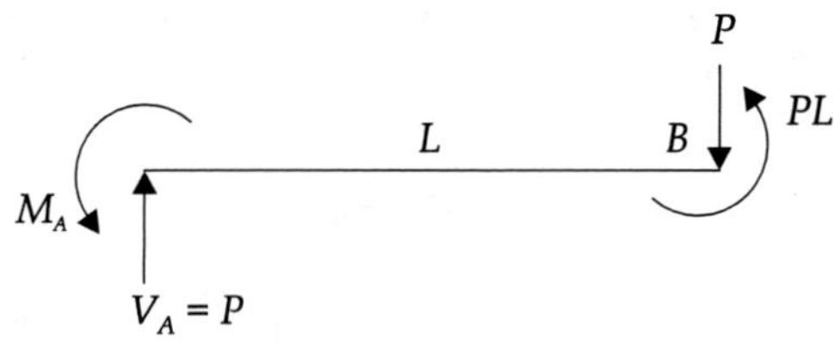

$$(-PL) + (PL) + (-M_A) = 0$$
$$M_A = 0$$

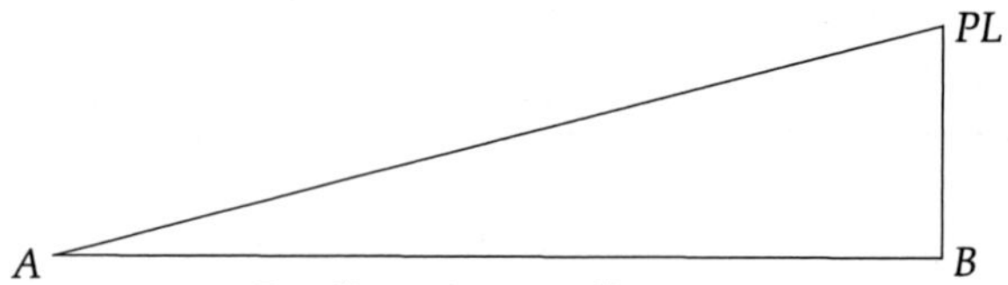

Bending moment diagram

The most economical maximum cross section is given where maximum bending moment occurs.

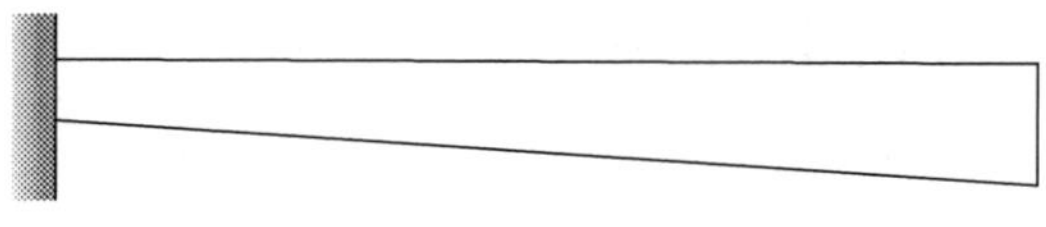

57. The maximum applied load on a cylindrical concrete specimen of diameter 150 mm and length 300 mm tested as per the split tensile strength test guidelines of IS 5816 : 1999 is 157 kN. The split tensile strength (in MPa, round off to one decimal place) of the specimen is _______. [GATE-2020]

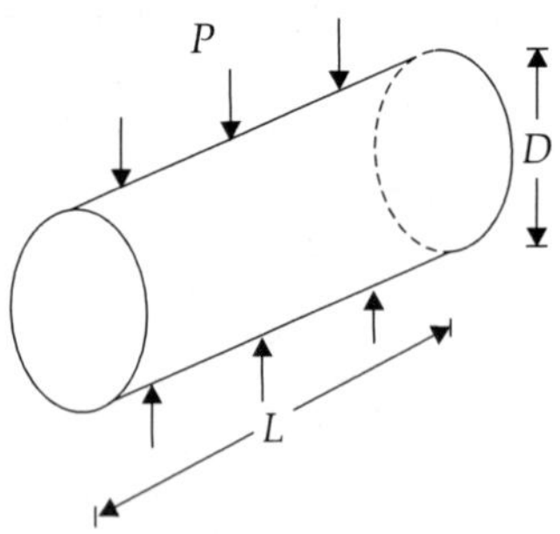

Ans. (2.2)

Sol. $P = 157$ kN

$D = 150$ mm

$L = 300$ mm

In split tensile strength test, split tensile strength of concrete,

$$f_{et} = \frac{2P}{\pi DL} = \frac{2 \times 157000}{\pi \times 150 \times 300}$$

$$= 2.22 \text{ N/mm}^2$$

58. The state of stress represented by Mohr's circle shown in the figure is

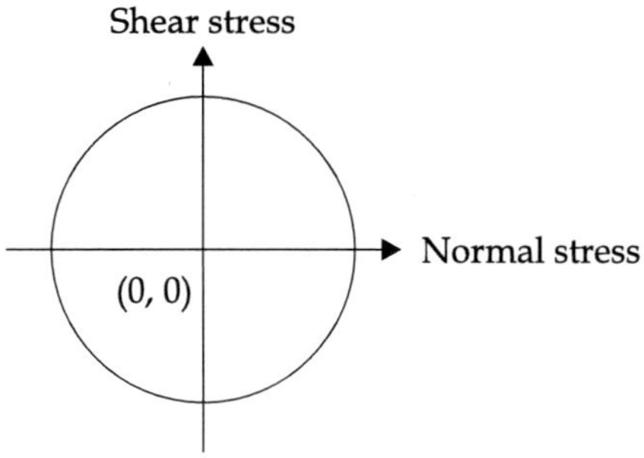

(a) hydrostatic stress
(b) uniaxial tension
(c) biaxial tension of equal magnitude
(d) pure shear [GATE-2020]

Ans. (d)

Sol.

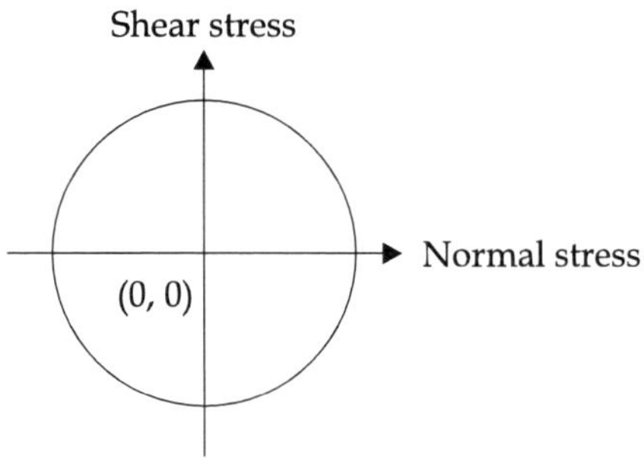

In pure shear condition, Mohr's circle has its centre at origin.

59. A prismatic linearly elastic bar of length, L, cross-sectional area A, and made up of a material with Young's modulus E, is subjected to axial tensile force as shown in the figures. When the bar is subjected to axial tensile forces P_1 and P_2, the strain energies stored in the bar are U_1 and U_2, respectively.

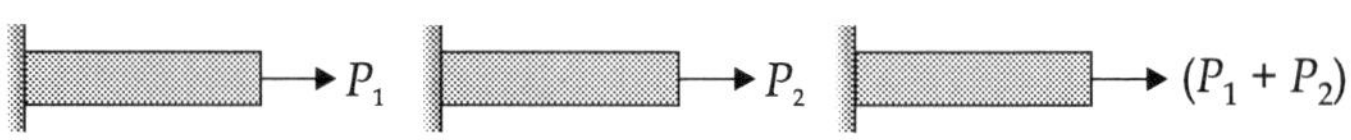

If U is the strain energy stored in the same bar when subjected to an axial tensile force $(P_1 + P_2)$, the correct relationship is

(a) $U = U_1 - U_2$ (b) $U = U_1 + U_2$

(c) $U < U_1 + U_2$ (d) $U > U_1 + U_2$ [GATE-2020]

Ans. (d)

Sol.

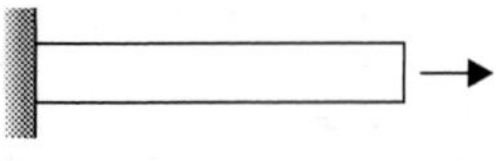 $\rightarrow P_1, U_1$	$U_1 = \dfrac{P_1^2 L}{2AE}$
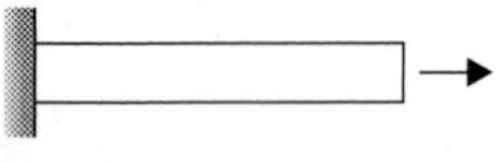 $\rightarrow P_2, U_2$	$U_2 = \dfrac{P_2^2 L}{2AE}$
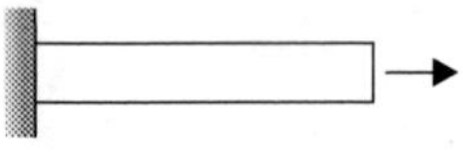 $\rightarrow (P_1 + P_2)$	$U = \dfrac{(P_1 + P_2)^2 L}{2AE}$

$$(P_1 + P_2)^2 > P_1^2 + P_2^2$$

$$U > U_1 + U_2$$

60. The planar structure RST shown in the figure is roller-supported at S and pin-supported at R. Members RS and ST have uniform flexural rigidity (EI) and S is a rigid joint. Consider only bending deformation and neglect effects of self-weight and axial stiffening.

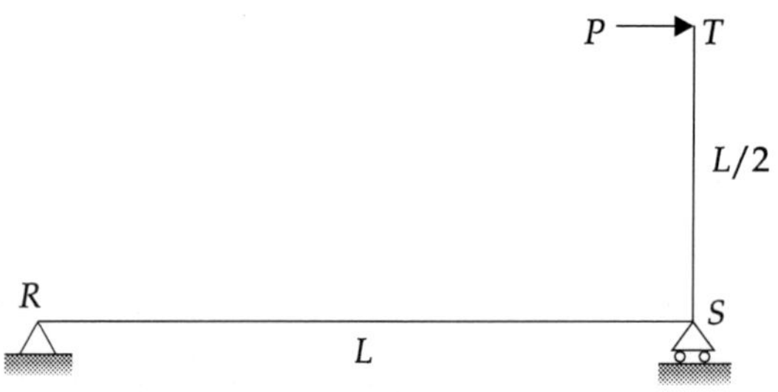

When the structure is subjected to a concentrated horizontal load P at the end T, the magnitude of rotation at the support R, is

(a) $\dfrac{PL}{6EI}$ (b) $\dfrac{PL^3}{12EI}$

(c) $\dfrac{PL^2}{6EI}$ (d) $\dfrac{PL^2}{12EI}$ [GATE-2020]

Ans. (d)

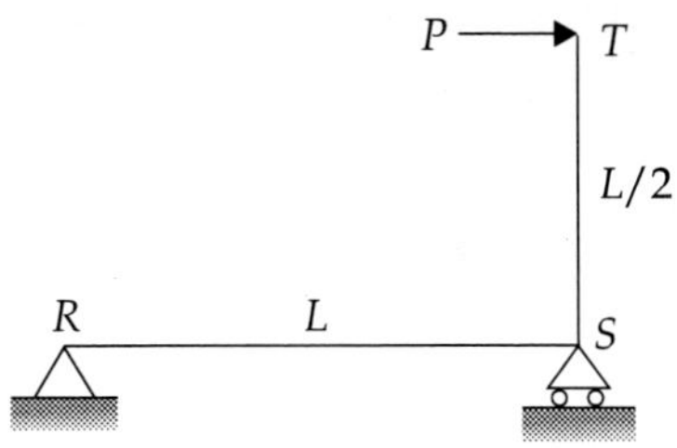

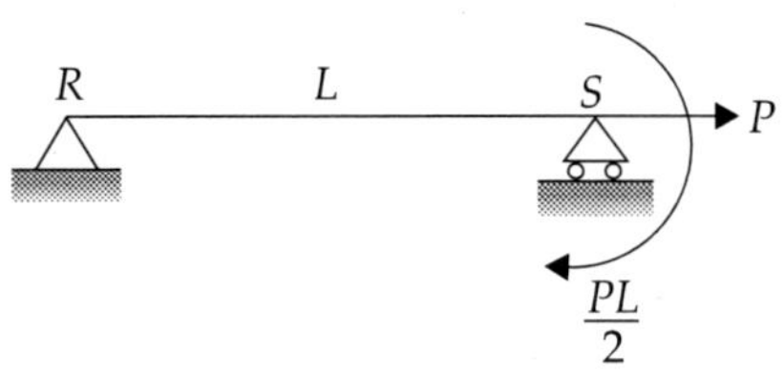

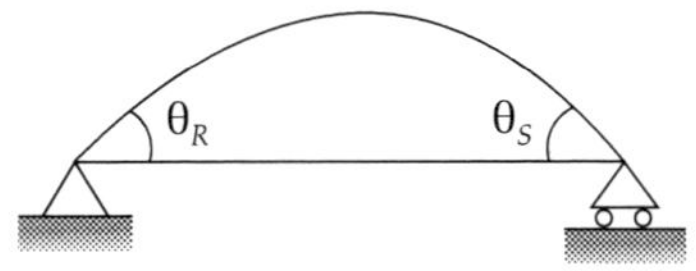

$$\theta_R = \frac{(PL/2)L}{6EI}$$

$$\theta_R = \frac{PL^2}{12EI}$$

Index